Teacher's Edition

# Life Skills Health

Shoreview, MN

## Publisher's Project Staff

Vice President of Curriculum and Publisher: Sari Follansbee, Ed. D.; Director of Curriculum Development: Teri Mathews; Managing Editor: Julie Maas; Senior Editor: Jody Peterson; Development Assistant: Bev Johnson; Director of Creative Services: Nancy Condon; Senior Designer: Diane McCarty; Project Coordinator/Designer: Jen Willman; Purchasing Agent: Mary Kaye Kuzma; Product Manager–Curriculum: Brian Holl

Copyright © 2007 by Pearson Education, Inc., publishing as Pearson AGS Globe, Shoreview, Minnesota 55126. All rights reserved. Printed in the United States of America. This publication is protected by copyright, and permission should be obtained from the publisher prior to any prohibited reproduction, storage in a retrieval system, or transmission in any form or by any means, electronic, mechanical, photocopying, recording, or likewise. For information regarding permission(s), visit www.agsglobe.com.

Pearson AGS Globe™ is a trademark of Pearson Education, Inc.
Pearson® is a registered trademark of Pearson plc.

ISBN 0-7854-4134-4

A 0 9 8 7 6 5 4 3 2 1

1-800-328-2560
www.agsglobe.com

# Contents

## Overview

Life Skills Health Overview ... T5
AGS Globe Science and Health Textbooks ... T6
Skill Track ... T7
Life Skills Health Students Text Highlights ... T8
Life Skills Health Teacher's Edition Highlights ... T10
Support for Students Learning English ... T12
Learning Styles ... T14
Teacher's Resource Library Highlights ... T15
Synopsis of the Scientific Research Base ... T16
Skills Chart ... T20
National Health Education Standards Correlation ... T22

## Lesson Plans

How to Use This Book: A Study Guide ... x
Unit 1 Mental and Emotional Health ... xviii
    Chapter 1 Maintaining Health and Wellness ... 2
    Chapter 2 Managing Emotions ... 26
    Chapter 3 Maintaining Mental Health ... 48
    Chapter 4 Mental Health Problems ... 74
Unit 2 Personal Health and Family Life ... 96
    Chapter 5 Identifying Human Body Systems ... 98
    Chapter 6 Maintaining Personal Hygiene and Fitness ... 126
    Chapter 7 The Life Cycle and Human Development ... 148
    Chapter 8 The Family ... 174
Unit 3 Nutrition ... 200
    Chapter 9 The Role of Diet in Health ... 202
    Chapter 10 Choosing Healthy Foods ... 228
Unit 4 Preventing and Controlling Diseases and Disorders ... 250
    Chapter 11 Disease—Causes and Protection ... 252
    Chapter 12 Preventing AIDS and Sexually Transmitted Diseases ... 268
    Chapter 13 Recognizing Common Diseases ... 284
Unit 5 Use and Misuse of Substances ... 314
    Chapter 14 Recognizing Medicines and Drugs ... 316
    Chapter 15 Dealing with Drug Dependence ... 344
Unit 6 Injury Prevention and Safety Promotion ... 366
    Chapter 16 Reducing Risks of Injury ... 368
    Chapter 17 Applying First Aid to Injuries ... 394
    Chapter 18 Preventing Violence and Resolving Conflicts ... 416

# Contents, continued

| | | |
|---|---|---|
| Unit 7 Health and Society | | 442 |
| Chapter 19 | Consumer Health | 444 |
| Chapter 20 | Public Health | 466 |
| Chapter 21 | Environmental Health | 484 |

Appendix A: Body Systems..........516
Appendix B: Nutrition Tables..........520
Appendix C: Fact Bank..........522
Glossary..........528
Index..........541
Photo and Illustration Credits..........555

## Teacher's Resources

Midterm and Final Mastery Tests..........556
Teacher's Resource Library Answer Key..........559
Activities..........559
Alternative Activities..........561
Workbook Activities..........563
Community Connections..........566
Home Connections..........566
Self-Assessment Activities..........566
Self-Study Guides..........566
Chapter Outlines..........567
Chapter Mastery Tests..........570
Unit Mastery Tests..........576
Scoring Rubric for Mastery Tests..........576

*Life Skills Health*

# Life Skills Health

**Life Skills Health** is designed to help students and young adults learn about health. The textbook addresses the issues and decisions encountered by teenagers. Topics discussed include mental health and emotions; exercise; personal hygiene; the human body systems; family health; nutrition; disease prevention; the use and misuse of drugs and medicines; injury prevention and conflict resolution; and consumer, public, and environmental health. Written to meet national standards, the textbook offers students who read below grade level the opportunity to sharpen their abilities to evaluate health information, make decisions, set goals, and solve problems. The text uses simple sentence structure and provides assistance with difficult vocabulary to enhance comprehension.

Short, concise lessons hold students' interest. Clearly stated objectives presented at the beginning of each lesson focus on what students will learn in the lessons. Illustrations and photos enhance students' understanding of the content. Lesson, Chapter, and Unit Reviews include open-ended questions to encourage students to use critical-thinking skills. Self-assessment activities at the beginning of each chapter gauge students' knowledge of chapter content and get students thinking about what they will learn. Health in Your Life and Decide for Yourself features provide everyday applications of health topics and give students the opportunity to set health-related goals and make healthy decisions.

# AGS Globe Science and Health Textbooks

*Enhance your science program with AGS Globe textbooks*—an easy, effective way to teach students the practical skills they need. Each AGS Globe textbook meets your science or health curriculum needs. These exciting, full-color books use student-friendly text and real-world examples to show students the relevance of science and health in their daily lives. Each book provides comprehensive coverage of skills and concepts. The short, concise lessons will motivate even your most reluctant students. With readabilities of all the texts at or below fourth-grade reading level, your students can concentrate on learning the content. AGS Globe is committed to making learning accessible to all students.

**For more information on AGS Globe textbooks and worktexts:**
call 800-328-2560, visit our Web site at www.agsglobe.com,
or e-mail AGS Globe at mail@agsglobe.com

# Skill Track Software

*Skill Track* monitors student progress and helps schools meet the demands of adequate yearly progress (AYP). Students using AGS Globe curriculum access multiple-choice assessments to see how well they understand the content of each textbook lesson and chapter. With timely and ongoing feedback from individual student and class reports, teachers can make informed instructional decisions. Administrators can use the reports to support teacher effectiveness and parents can keep up to date on what their students are learning.

Simple to use, Skill Track is secure and confidential. Students enter through two paths—lesson by lesson, or at the end of a chapter or unit. Either way, mastery is assessed by a variety of multiple-choice items; parallel forms are available for chapter assessments. Hundreds of items cover the content and skills in each textbook. Students may retake any assessment as often as necessary and scores are reported for each attempt. Accordingly, teachers can identify areas in need of reinforcement and practice for individual learners as well as the class.

**For more information about Skill Track:**
call **800-328-2560** or visit our Web site at **www.agsglobe.com**

# Student Text Highlights

- Each lesson is clearly labeled to help students focus on the skill or concept to be learned.
- Vocabulary terms are boldfaced and then defined in the margin at the top of the page and in the glossary.

**Alcohol**
A drug that slows down the central nervous system

**Drug**
A chemical substance other than food that changes the way the mind and body work

Body language plays an important role in classrooms and other settings. You show interest in a teacher or speaker by pointing your body toward that person.

- Goals for Learning at the beginning of each chapter identify learner outcomes.
- Self-Assessment activities at the beginning of each chapter gauge students' prior knowledge of health content.
- Many features reinforce and extend student learning beyond the lesson content.
- Technology and Society helps students make connections between health and technology and society.

### Goals for Learning
- To describe physical, social, and emotional health
- To recognize principles of good health
- To identify causes of health risks
- To evaluate and recognize the power to change personal health

### Technology and Society
About 75 percent of teens use the Internet regularly. In spite of fears that teens online might avoid communicating with people, the opposite is true. Teens use e-mail and instant messaging to stay in touch with friends and family. Most also spend about the same amount of time with people as before.

### Health Myth
**Myth:** It is possible to avoid all stress.
**Fact:** Stress is a part of every person's life. You can manage stress but not eliminate it.

### Link to >>>
**Biology**
Exercise causes many positive changes in your body. It makes your heart beat at a healthy higher rate. This higher rate strengthens your heart. Then your circulation, the flow of blood through your body, is better. The blood more efficiently delivers substances your body needs throughout your body. Then you have more energy, and you are less likely to become ill.

- Decide for Yourself offers real-world applications of health topics.
- Health in the World features a look at world-wide health topics and issues.

### Decide for Yourself
Tanya's best friend came to her with a problem. The friend often felt faint and shaky. She wanted to eat strange foods sometimes. Other times, she couldn't eat at all. Her toes had started feeling numb. "I haven't told anyone," she said. "I feel so awful today. I just can't make it to soccer practice. Please cover for me. Tell them I couldn't come. Don't tell them why. I don't want to get kicked off the team."

Tanya knew her friend's mom had diabetes, which meant that her body did not have the right amount of insulin. Tanya thought her friend might have diabetes, too. Her friend needed to see a doctor.

Her friend had begged Tanya not to say anything to anybody. Tanya didn't want her friend to get kicked off the team.
On a sheet of paper, write the answers to the following questions. Use complete sentences.
1. What are Tanya's possible choices in this situation?
2. Tanya's friend is afraid she will be kicked off the soccer team. Why else might she be scared to admit she is sick?
3. Which choice do you think Tanya should make?

### HEALTH IN THE WORLD
**The World Health Organization**
As part of the United Nations, the World Health Organization (WHO) identifies health issues around the world. The WHO represents 192 countries. All of these countries work together to fight disease and to work on other health issues. Some programs that the WHO supports address the health needs of specific countries. Other programs address worldwide health issues. Each program responds to an immediate health crisis. The WHO works to stop the spread of AIDS. It also brings health care workers to regions in need and provides medicine to the poorest regions of the world. The following is an example of one of the WHO's many programs.

**Niger**
Niger is a country between Algeria, Chad, and Nigeria in Africa. It has a population of about 11.6 million people. More than 60 percent of its people live on less than one dollar a day. Most people do not live beyond the age of 44. More than 80 percent of the people cannot read. In 2004, a huge number of grasshoppers and a lack of rain totally destroyed the year's crop in many regions. About 3.6 million people, including nearly 200,000 children, were left without enough food.

People did not have safe drinking water. These poor living conditions resulted in widespread diseases.

A team of experts from the WHO arrived in Niger to study the problem. They worked with Niger's leaders. The WHO asked members of the United Nations for $1.3 million to help the people of Niger get medicine and food. The project in Niger was part of the WHO's efforts in many African countries to improve health.

On a sheet of paper, write the answers to the following questions. Use complete sentences.
1. Why is the WHO important?
2. How does the WHO bring countries together?
3. What do you think is the biggest challenge for the WHO?

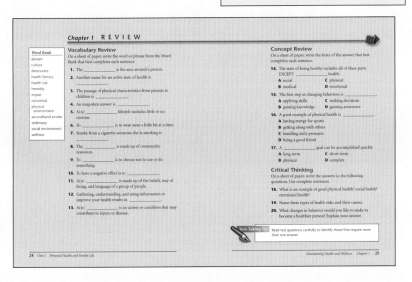

- ◆ Health in Your Life helps students relate chapter content to everyday life.

- ◆ Health at Work provides examples of some health careers.

- ◆ Lesson Review questions allow students to check their understanding of key concepts presented in the text.

- ◆ Summaries at the end of each chapter and unit highlight main ideas for students.

- ◆ Chapter and Unit Reviews allow students and teachers to check for skill mastery. They cover the objectives in the Goals for Learning at the beginning of each chapter.

- ◆ Test-Taking Tips at the end of each Chapter Review help reduce test anxiety and improve test scores.

Life Skills Health   T9

# Teacher's Edition Highlights

The comprehensive, wraparound Teacher's Edition provides instructional strategies at point of use. Everything from preparation guidelines to teaching tips and strategies are included in an easy-to-use format. Activities are featured at point of use for teacher convenience.

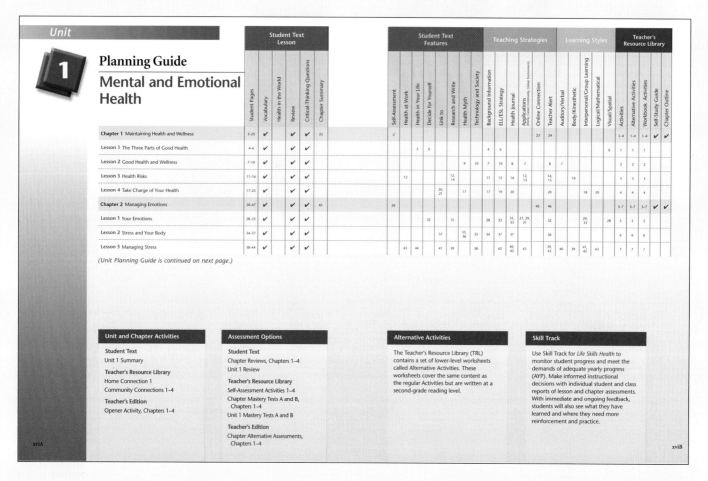

## Unit Planning Guides

- The Planning Guide saves valuable preparation time by organizing all materials for each unit.
- A complete listing of chapters and lessons allows you to preview each unit quickly.
- Assessment options are highlighted for easy reference. Options include:
    - Chapter Reviews
    - Unit Reviews
    - Self-Assessment Activities
    - Chapter and Unit Mastery Tests, Forms A and B
    - Chapter Alternative Assessments
    - Midterm and Final Tests
- Page numbers of Student Text and Teacher's Edition features help customize lesson plans to your students.
- Many teaching strategies and learning styles are listed to support students with diverse needs.
- Activities for the Teacher's Resource Library are listed.

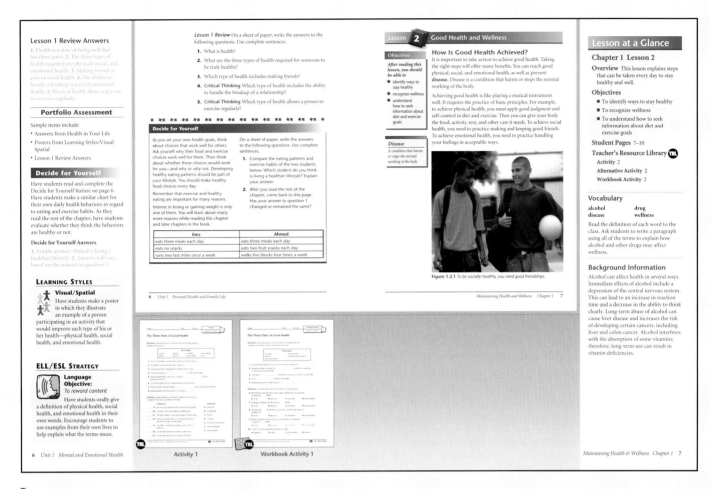

## Lessons

◆ Quick overviews of chapters and lessons save planning time.

◆ Lesson objectives are listed for easy reference.

◆ Page references are provided for convenience.

◆ Easy-to-follow lesson plans in three steps save time: Warm-Up Activity, Teaching the Lesson, and Reinforce and Extend.

◆ Teacher Alerts highlight content that may need further explanation.

◆ Health Journal activities give students an opportunity to write about health.

◆ Applications: Five areas of application—At Home, Career Connection, Global Connection, In the Community, and In the Environment—help students relate science to the world outside the classroom. Applications motivate students and make learning relevant.

◆ A Portfolio Assessment, which appears at the end of each lesson, lists items the student has completed for that lesson.

◆ Online Connections list relevant Web sites.

◆ Learning Styles provide teaching strategies to help meet the needs of students with diverse ways of learning. Modalities include Auditory/Verbal, Visual/Spatial, Body/Kinesthetic, Logical/Mathematical, and Interpersonal/Group Learning. Additional teaching activities are provided for ELL students.

◆ Answers are provided in the Teacher's Edition for all reviews in the Student Text. Answers to the Teacher's Resource Library and Student Workbook are provided at the back of this Teacher's Edition and on the TRL CD-ROM.

◆ Worksheet, Workbook Activity, and Test pages from the Teacher's Resource Library are shown at point of use in reduced form.

*Life Skills Health* **T11**

# Support for Students Learning English

Increasing numbers of students learning English are among the students in most schools and classrooms. The purpose of the ELL/ESL Strategy feature in this Teacher's Edition is to incorporate the language and content needs of English Language Learners in a regular and explicit manner.

ELL/ESL Strategy activities promote English language acquisition in the context of content area learning. Students should not be separated or isolated for these activities and interaction with English-speaking peers is always encouraged.

The ELL/ESL Strategy helps the teacher scaffold the content and presentation in relation to students' language and skill proficiency. Each activity suggests to the teacher some ideas about how to adjust the presentation of content to meet the varying needs of diverse learners, including students learning English. *Scaffolding* refers to structuring the introduction of vocabulary, concepts, and skills by providing additional supports or modifications based on students' needs. Ideally, these supports become less necessary as students' language proficiency increases and their knowledge and skill level becomes more developed.

### ELL/ESL Strategy

**Language Objective:**
*To reword content*
Have students orally give a definition of physical health, social health, and emotional health in their own words. Encourage students to use examples from their own lives to help explain what the terms mean.

Each activity includes a language objective and strategy related to *listening, speaking, reading,* or *writing*. The language objective and activity relate to one or more content objectives listed in the Teacher's Edition under Lesson at a Glance. Some examples of language objectives include: reading for meaning, understanding different styles or purposes of writing, identifying and practicing common grammar structures, learning vocabulary specific to the content area, preparing and giving a group presentation, speaking in front of a group, or discussing an assigned topic as a small group.

## Strategies That Support English Learners

- Identify and build on prior knowledge or experience; start with what's familiar and elaborate to include new content and new connections, personal associations, cultural context
- Use visuals and graphic organizers—illustrations, photos, charts, posters, graphs, maps, tables, webs, flow charts, timelines, diagrams
- Use hands-on artifacts (realia) or manipulatives
- Provide *comprehensible input*—paraphrase content, give additional examples, elaborate on student background knowledge and responses; be aware of rate of speech, syntax, and language structure and adjust accordingly
- Begin with lower-level, fact recall questions and move to questions that require higher-order critical-thinking skills (application, hypothesis, prediction, analysis, synthesis, evaluation)
- Teach vocabulary—pronunciations, key words or phrases, multiple meanings, idioms/expressions, academic or content language
- Have students create word banks or word walls for content (academic) vocabulary
- Teach and model specific reading and writing strategies—advance organizers, main idea, meaning from context, preview, predict, make inferences, summarize, guided reading
- Support communication with gestures and body language
- Teach and practice functional language skills—negotiate meaning, ask for clarification, confirm information, argue persuasively
- Teach and practice study skills—structured note-taking, outlining, use of reference materials
- Use cooperative learning, peer tutoring, or other small group learning strategies
- Plan opportunities for student interaction—create a skit and act it out, drama, role play, storytelling
- Practice self-monitoring and self-evaluation—students reflect on their own comprehension or activity with self-checks

## How Do AGS Globe Textbooks Support Students Learning English?

AGS Globe is committed to helping all students succeed. For this reason, AGS Globe textbooks and teaching materials incorporate research-based design elements and instructional methodologies configured to allow diverse students greater access to subject area content. Content access is facilitated by controlled reading level, coherent text, and vocabulary development. Effective instructional design is accomplished by applying research to lesson construction, learning activities, and assessments.

AGS Globe materials feature key elements that support the needs of students learning English in sheltered and immersion settings.

| Key Elements | AGS Globe Features |
| --- | --- |
| Lesson Preparation | ◆ Content- and language-specific objectives |
| Building Background | ◆ Warm-Up Activity<br>◆ Explicit vocabulary instruction and practice with multiple exposures to new words<br>◆ Background information; building on prior knowledge and experience |
| Comprehensible Input | ◆ Controlled reading level in student text (Grades 3–4)<br>◆ Highlighted vocabulary terms with definitions<br>◆ Student glossary with pronunciations<br>◆ Clean graphic and visual support<br>◆ Content links to examples<br>◆ Sidebar notes to highlight and clarify content<br>◆ Audio text recordings (selected titles)<br>◆ Alternative Activity pages (Grade 2 reading level) |
| Lesson Delivery | ◆ Teaching the Lesson/3-Step Teaching Plan<br>◆ Short, skill- or content-specific lessons<br>◆ Orderly presentation of content with structural cues |
| Strategies | ◆ ELL/ESL Strategy activities<br>◆ Learning Styles activities<br>◆ Writing prompts in student text<br>◆ Teaching Strategies Transparencies provide additional graphic organizers<br>◆ Study skills: Self-Study Guides, Chapter Outlines |
| Interaction | ◆ Vocabulary-building activities<br>◆ Language-based ELL/ESL Strategy activities<br>◆ Learning Styles activities<br>◆ Reinforce and Extend activities |
| Practice/Application | ◆ Skill practice or concept application in student text<br>◆ Reinforce and Extend activities<br>◆ Career, home, and community applications<br>◆ Student Workbook<br>◆ Multiple TRL activity pages |
| Review and Assessment | ◆ Lesson reviews, chapter reviews, unit reviews<br>◆ Skill Track monitors student progress<br>◆ Chapter, Unit, Midterm, and Final Mastery Tests |

For more information on these key elements, see Echevarria, J., Vogt, M., & Short, D. (2004). *Making content comprehensible for English language learners: The SIOP model* (2nd ed.). Boston, MA: Allyn & Bacon.

# Learning Styles

Differentiated instruction allows teachers to address the needs of diverse learners and the variety of ways students process and learn information. The Learning Styles activities in this Teacher's Edition provide additional teaching strategies to help students understand lesson content by teaching or expanding upon the content in a different way. The activities are designed to help teachers capitalize on students' individual strengths and learning styles.

The Learning Styles activities highlight individual learning styles and are classified based on Howard Gardner's theory of multiple intelligences: Auditory/Verbal, Body/Kinesthetic, Interpersonal/Group Learning, Logical/Mathematical, and Visual/Spatial. In addition, the various writing activities suggested in the Student Text are appropriate for students who fit Gardner's description of Verbal/Linguistic intelligence.

Following are examples of activities featured in the *Life Skills Health* Teacher's Edition:

## *Body/Kinesthetic*

Students learn from activities that include physical movement, manipulatives, or other tactile experiences.

> ### LEARNING STYLES
>
> **Body/Kinesthetic**
> Have students work in groups to create a skit that depicts examples of positive and negative health behaviors. After each group performs its skit, have the class identify the health behaviors shown in the skit and classify them as positive or negative.

## *Auditory/Verbal*

Students learn by listening to text read aloud or from an audiorecording, and from other listening or speaking activities. Musical activities related to the content may help auditory learners.

> ### LEARNING STYLES
>
> **Auditory/Verbal**
> Have students write a song, a jingle, or a poem that incorporates all of the steps to staying healthy. Invite students to share their work with the class.

## *Interpersonal/Group Learning*

Students learn from working with at least one other person or in a cooperative learning group on activities that involve a process and an end product.

> ### LEARNING STYLES
>
> **Interpersonal/Group Learning**
> Have students work in groups to come up with a health-related problem a teenager might be experiencing and then use the steps in the decision-making process to solve it. Students should present the results of their work to the class. Have classmates discuss if they would have tried to solve the problem with a different solution.

## *Logical/Mathematical*

Students learn by using logical/mathematical thinking and problem solving in relation to the lesson content.

> ### LEARNING STYLES
>
> **Logical/Mathematical**
> Have students read and complete the Health Assessment Chart on page 21. They will have to keep track of their points, total them, and score themselves in each area of health. Then they should add the three scores together to find out their Overall Health Score. Encourage students to answer honestly. They can use this assessment as a starting point to change behaviors that do not positively affect their health.

## *Visual/Spatial*

Students learn by viewing or creating illustrations, graphics, patterns, or additional visual demonstrations beyond what is in the text.

> ### LEARNING STYLES
>
> **Visual/Spatial**
> Have students make a poster in which they illustrate an example of a person participating in an activity that would improve each type of his or her health—physical health, social health, and emotional health.

# Teacher's Resource Library Highlights

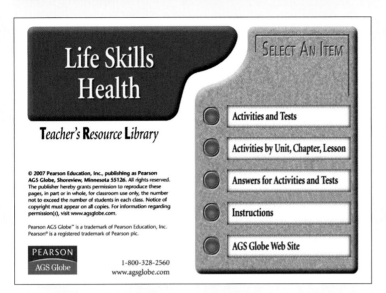

**TRL** All of the activities you'll need to reinforce and extend the text are conveniently located on the AGS Globe Teacher's Resource Library (TRL) CD-ROM. All of the reproducible activities pictured in the Teacher's Edition are ready to select, view, and print. You can also preview other materials by linking directly to the AGS Globe Web site.

## Activities
Lesson activities are available to reinforce lesson content and vocabulary.

## Alternative Activities
These activities cover the same content as the Activities but are written at a second-grade reading level.

## Community Connections
Relevant activities help students extend their knowledge to the real world and reinforce concepts covered in class.

## Home Connections
These activities include unit content for students and families to share.

## Self-Study Guides
These assignment guides provide teachers with the flexibility to use individualized instruction or independent study.

## Chapter Outlines
Outlines for each chapter help students focus on and review main ideas.

## Self-Assessment Activities
Assessments and activities for students to assess their knowledge, set chapter learning goals, and describe new learning.

## Mastery Tests
Chapter, Unit, Midterm, and Final Mastery Tests are convenient assessment options. Critical-thinking items are included

## Answer Key
All answers to reproducible activities are included in the TRL and in the Teacher's Edition.

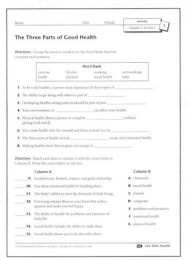

Activities | Workbook Activities

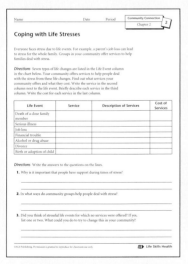

Community Connections | Mastery Tests

# Synopsis of the Scientific Research Base

| Research-Based Principles | AGS Globe Textbooks | References |
|---|---|---|
| **Standards Alignment** | | |
| Subject area instruction needs to be based on skills, concepts, and processes represented by common standards for that subject area. | ◆ Textbook content and skills aligned with national standards and state grade-level or course-specific content standards, where available | Matlock, L., Fielder, K., & Walsh, D. (2001). Building the foundation for standards-based instruction for all students. *Teaching Exceptional Children, 33*(5), 68–72.<br>Miller, S. P., & Mercer, C. D. (1997). Educational aspects of mathematics disabilities. *Journal of Learning Disabilities, 30*(1), 47–56.<br>Reys, R., Reys, B., Lapan, R., Holliday, G. & Wasman, D. (2003). Assessing the impact of standards-based middle grades mathematics curriculum materials on student achievement. *Journal of Research in Mathematics Education, 34*(1), 74–95. |
| **Readability** | | |
| Many students struggle to learn from core content-area textbooks that are written too high above their reading level. Students need access to textbooks written at a level they can read and understand, where the reading level is within the students' range of comprehension. | ◆ Grade 4.0 or lower readability using the Spache formula<br>◆ Controlled vocabulary matched to student reading ability and use of synonyms to replace non-essential difficult words above grade 4<br>◆ Simple sentence structures<br>◆ Limited sentence length | Allington, R. L. (2002). You can't learn much from books you can't read. *Educational Leadership, 60*(3), 16–19.<br>Chall, J. S., & Conard, S. S. (1991). *Should textbooks challenge students? The case for easier or harder textbooks.* New York: Teachers College Press.<br>*Readability calculations.* (2000). Dallas: Micro Power & Light Company. |
| **Language Complexity and Sequence** | | |
| Students struggling with vocabulary and text comprehension need textbooks with accessible language. | ◆ Simple, direct language using an active voice<br>◆ Clear organization to facilitate understanding<br>◆ Explicit language signals to show sequence of and links between concepts and ideas | Anderson, T. H., & Armbruster, B. B. (1984). Readable texts, or selecting a textbook is not like buying a pair of shoes. In R. C. Anderson, J. Osborne, & R. J. Tierney (Eds.), *Learning to read in American schools* (pp. 151–162). Hillsdale, NJ: Lawrence Erlbaum Associates, Inc.<br>Curtis, M. E. (2002, May 20). *Adolescent reading: A synthesis of research.* Paper presented at the Practice Models for Adolescent Literacy Success Conference, U.S. Department of Education. Washington, DC: National Institute of Child Health and Human Development. Retrieved September 15, 2003, from http://216.26.160.105/conf/nichd/synthesis.asp<br>McAlpine, L., & Weston, C. (1994). The attributes of instructional materials. *Performance Improvement Quarterly, 7*(1), 19–30.<br>Seidenberg, P. L. (1989). Relating text-processing research to reading and writing instruction for learning disabled students. *Learning Disabilities Focus, 5*(1), 4–12. |
| **Vocabulary Use and Development** | | |
| Students need content-related vocabulary instruction in the context of readable and meaningful text. | ◆ New vocabulary boldfaced on first occurrence, used in context, and defined in a sidebar<br>◆ Glossary with pronunciation, definition, and relevant graphic illustrations for all vocabulary words<br>◆ Direct vocabulary instruction introduced in the Teacher's Edition and reinforced in context throughout<br>◆ Multiple exposures to new vocabulary in text and practice exercises | Ciborowski, J. (1992). *Textbooks and the students who can't read them: A guide to teaching content.* Cambridge, MA: Brookline.<br>Kameenui, E. J., & Simmons, D. C. (1990). *Designing instructional strategies.* Columbus, OH: Merrill Publishing Company.<br>Marzano, R. J. (1998). *A theory-based meta-analysis of research on instruction.* Aurora, CO: Mid-Continent Research for Education and Learning. Retrieved October 1, 2003, from http://www.mcrel.org/topics/productDetail/asp?productID=83<br>McAlpine, L., & Weston, C. (1994). The attributes of instructional materials. *Performance Improvement Quarterly, 7*(1), 19–30.<br>National Reading Panel. (2000). *Teaching children to read: An evidence-based assessment of the scientific research literature on reading and its implications for reading instruction.* Reports of the subgroups. Washington, DC: National Institute of Child Health and Human Development.<br>Taylor, S. E., Frackenpohl, H., White, C. E., Nieroroda, B. W., Browning, C. L., & Birsner, E. P. (1989). *EDL core vocabularies in reading, mathematics, science, and social studies.* Austin, TX: Steck-Vaughn. |

| Research-Based Principles | AGS Globe Textbooks | References |
|---|---|---|

## Text Organization: Presentation and Structure

Students need an uncluttered page layout, with easy-to-read print, that clearly directs the reader to main ideas, important information, examples, and comprehensive practice and review.

Reading comprehension is improved by structural features in the text that make it easier for learners to access the content.

*Print characteristics and page layout:*
- Serif font for body copy; sans serif font for boxed features, examples
- Maximum line length of 5" for ease of reading
- Unjustified (ragged) right margins
- Major/minor column page design presents primary instructional information in the major column and support content in the sidebar or in a box

*Presentation characteristics:*
- Lesson introductions, summaries
- Explicit lesson titles, headings, and subheadings label and organize main ideas
- Signals alert readers to important information, content connections, illustrations, graphics
- Cues (e.g., boldface type) highlight important information

*Text structure:*
- Lesson heads in question or statement format guide comprehension
- Text written to explicitly link facts and concepts within and across lessons; text cohesiveness
- Each skill or concept linked to direct practice and review

Armbruster, B. B., & Anderson, T. H. (1988). On selecting "considerate" content area textbooks. *Remedial and Special Education, 9*(1), 47–52.

Beck, I. L., McKeown, M. G., & Grommoll, E. W. (1989). Learning from social studies texts. *Cognition and Instruction, 6*(2), 99–158.

Chambliss, M. J. (1994). Evaluating the quality of textbooks for diverse learners. *Remedial and Special Education, 15*(5), 348–362.

Ciborowski, J. (1992). *Textbooks and the students who can't read them: A guide to teaching content.* Cambridge, MA: Brookline.

Dickson, S. V., Simmons, D. C., & Kameenui, E. J. (1995). *Text organization and its relation to reading comprehension: A synthesis of the research* (Technical Report No. 17) and *Text organization: Curricular and instructional implications for diverse learners* (Technical Report No. 18). National Center to Improve the Tools of Educators. Eugene, OR: University of Oregon. Retrieved January 26, 2000, from http://idea.uoregon.edu/~ncite/documents/techrep/tech17.html and http://idea.uoregon.edu/~ncite/documents/techrep/tech18.html

Dickson, S. V., Simmons, D. C., & Kameenui, E. J. (1998). Text organization: Research bases *and* Text organization: Instructional and curricular basics and implications. In D. C. Simmons & E. J. Kameenui (Eds.), *What reading research tells us about children with diverse learning needs: Bases and basics* (pp. 239–278; 279–294). Mahwah, NJ: Lawrence Erlbaum Associates, Inc.

Mansfield, J. S., Legge, G. E., & Bane, M. C. (1996). Psychophysics of reading. XV: Font effects in normal and low vision. *Investigative Ophthalmology and Vision Science, 37,* 1492–1501.

McAlpine, L., & Weston, C. (1994). The attributes of instructional materials. *Performance Improvement Quarterly, 7*(1), 19–30.

McNamara, D. S., Kintsche, E., Songer, N. B., & Kintsche, W. (1996). Are good texts always better? Interactions of text coherence, background knowledge, and levels of understanding in learning from text. *Cognition and Instruction, 14*(1), 1–43.

Tyree, R. B., Fiore, T. A., & Cook, R. A. (1994). Instructional materials for diverse learners: Features and considerations for textbook design. *Remedial and Special Education, 15*(6), 363–377.

## Differentiated Instruction and Learning Styles

Student learning is more successful when tasks are aligned with academic skill levels and developmental stage, and adjustments are made to allow students multiple means to engage and express their learning strengths and styles at appropriate levels of challenge and support.

Differentiated instruction allows teachers to organize instruction to adjust for diverse learning needs within a classroom.

Learning activities that capitalize on students' learning styles can structure planning for individual differences based on multiple intelligences theory.

- Multiple features, including Learning Styles activities, help teachers match assignments to students' abilities and interests
- Variety of media to select from—print, audio, visual, software
- Step-by-step, part-by-part basic content and skill-level lessons in the Student and Teacher's Editions
- Alternative Activities written at a Grade 2 (Spache) readability in the Teacher's Resource Library
- Variety of review materials, activities, sidebars, and alternative readings
- Multiple assessments—lesson or chapter reviews, end-of-chapter tests, cumulative midterm/final mastery tests, alternative assessment items

*Learning Styles activities include:*
- Auditory/Verbal
- Body/Kinesthetic
- Interpersonal/Group Learning
- Logical/Mathematical
- Visual/Spatial

*ELL/ESL Strategies provide support for students who are learning English and lesson content concurrently.*

Allington, R. L. (2002). You can't learn much from books you can't read. *Educational Leadership, 60*(3), 16–19.

Carnine, D. (1994). Introduction to the mini-series: Diverse learners and prevailing, emerging, and research-based educational approaches and their tools. *School Psychology Review, 23*(3), 341–350.

Forsten, C., Grant, J., & Hollas, B. (2003). *Differentiating textbooks: Strategies to improve student comprehension and motivation.* Peterborough, NH: Crystal Springs Books.

Gardner, H. (1983). *Frames of mind: The theory of multiple intelligences.* New York: Harper and Row.

Gersten, R., & Baker, S. (2000). The professional knowledge base on instructional practices that support cognitive growth for English-language learners. In R. Gersten, E. P. Schiller, & S. Vaughn (Eds.), *Contemporary special education research: Syntheses of the knowledge base on critical instructional issues* (pp. 31–80). Mahwah, NJ: Lawrence Erlbaum Associates, Inc.

Hall, T. (2002, June). *Effective classroom practices report: Differentiated instruction.* Wakefield, NJ: National Center on Accessing the General Curriculum. Retrieved September 29, 2003, from http://www.cast.org/cac/index.cfm?i=2876

Lazear, D. (1999). *Eight ways of knowing: Teaching for multiple intelligences* (3rd ed.). Arlington Heights, IL: Skylight Training and Publishing.

Orlich, D. C., Harder, R. J., Callahan, R. C., & Gibson, H. W. (2001). *Teaching strategies: A guide to better instruction* (6th ed.). Boston: Houghton Mifflin Company.

Roderick, M. & Camburn, E. (1999). Risk and recovery from course failure in the early years of high school. *American Educational Research Journal, 36*(2), 303–343.

Tomlinson, C. A. (1999). *The differentiated classroom: Responding to the needs of all learners.* Alexandria, VA: Association for Supervision and Curriculum Development.

Life Skills Health **T17**

# Synopsis of the Scientific Research Base, continued

| Research-Based Principles | AGS Globe Textbooks | References |
|---|---|---|

## Instructional Design: Lesson Structure and Learner Support Strategies

**Research-Based Principles**

Instruction that includes the components of effective instruction, utilizes effective strategies and interventions to facilitate student learning, and aligns with standards improves learning for all students, especially diverse learners and students who are struggling.

Elements of effective instruction:

*Step 1: Introduce the lesson and prepare students to learn*
*Step 2: Provide instruction and guided practice*
*Step 3: Provide opportunities for applied practice and generalization*

Organizational tools:
*Advance organizers*
*Graphic organizers*

Instructional process techniques:
*Cooperative learning*
*Student self-monitoring and questioning*
*Real-life examples*
*Mnemonics*

**AGS Globe Textbooks**

**Step 1: Introduce the lesson and prepare students to learn**
*In the Student Edition:*
- "How to Use This Book" feature explicitly teaches text organization
- Chapter and lesson previews with graphic and visual organizers
- Goals for Learning
- Sidebar notes review skills and important facts and information

*In the Teacher's Edition:*
- Lesson objectives
- Explicit 3-Step Teaching Plan begins with "Warm-Up Activity" to inform students of objectives, connect to previous learning and background knowledge, review skills, and motivate students to engage in learning

**Step 2: Provide instruction and guided practice**
*In the Student Edition:*
- Short, manageable lessons break content and skills into smaller, step-by-step, part-by-part pieces
- Systematic presentation of lesson concepts and skills
- Chapter and lesson headings presented as questions or statements
- Graphic organizers arrange content visually—charts, graphs, tables, diagrams, bulleted lists, arrows, graphics, mnemonics, illustrations, and captions
- Models or examples link directly to the explanation of the concept
- Multiple opportunities for direct practice throughout

*In the Teacher's Edition:*
- 3-Step Teaching Plan for each lesson includes "Teaching the Lesson" with direct instruction, and helps teachers present and clarify lesson skills and concepts through guided practice and modeling of important ideas
- Supplemental strategies and activities, including hands-on modeling, transparencies, graphic organizers, visual aids, learning styles

**Step 3: Provide opportunities for applied practice and generalization**
*In the Student Edition:*
- Each skill or concept lesson is followed by direct practice or review questions
- Multiple exercises throughout
- Generalization and application activities in sidebars and lessons link content to real-life applications
- Chapter reviews and summaries highlight major points

**References**

Allsopp, D. H. (1990). Using modeling, manipulatives, and mnemonics with eighth-grade math students. *Teaching Exceptional Children, 31*(2), 74–81.

Chambliss, M. J. (1994). Evaluating the quality of textbooks for diverse learners. *Remedial and Special Education, 15*(5), 348–362.

Ciborowski, J. (1992). *Textbooks and the students who can't read them: A guide to teaching content.* Cambridge, MA: Brookline.

Cole, R. W. (Ed.). (1995). *Educating everybody's children: Diverse teaching strategies for diverse learners.* Alexandria, VA: Association for Supervision and Curriculum Development.

Curtis, M. E. (2002, May 20). *Adolescent reading: A synthesis of research.* Paper presented at the Practice Models for Adolescent Literacy Success Conference, U.S. Department of Education. Washington, DC: National Institute of Child Health and Human Development. Retrieved September 15, 2003, from http://216.26.160.105/conf/nichd/synthesis.asp

Dickson, S. V., Simmons, D. C., & Kameenui, E. J. (1995). *Text organization: Curricular and instructional implications for diverse learners* (Technical Report No. 18). National Center to Improve the Tools of Educators. Eugene, OR: University of Oregon. Retrieved January 26, 2000, from http://idea.uoregon.edu/~ncite/documents/techrep/tech18.html

Dixon, R. C., Carnine, D. W., Lee, D., Wallin, J., & Chard, D. (1998). *Review of high quality experimental mathematics research: Report to the California State Board of Education.* Sacramento, CA: California State Board of Education.

Jarrett, D. (1999). *The inclusive classroom: Mathematics and science instruction for students with learning disabilities—It's just good teaching.* Portland, OR: Northwest Regional Educational Laboratory.

Johnson, D. W., Johnson, R. T., & Stanne, M. B. (2000, May). *Cooperative learning methods: A meta-analysis.* Minneapolis: The Cooperative Learning Center, University of Minnesota. Retrieved October 29, 2003, from http://www.cooplearn.org/pages/cl-methods.html

Kameenui, E. J., & Simmons, D. C. (1990). *Designing instructional strategies.* Columbus, OH: Merrill Publishing Company.

Lovitt, T. C., & Horton, S. V. (1994). Strategies for adapting science textbooks for youth with learning disabilities. *Remedial and Special Education, 15*(2), 105–116.

Marzano, R. J. (1998). *A theory-based meta-analysis of research on instruction.* Aurora, CO: Mid-Continent Research for Education and Learning. Retrieved October 1, 2003, from http://www.mcrel.org/topics/productDetail/asp?productID=83

Marzano, R. J., Pickering, D. J., & Pollock, J. E. (2001). *Classroom instruction that works: Research-based strategies for increasing student achievement.* Alexandria, VA: Association for Supervision and Curriculum Development.

Miller, S. P., & Mercer, C. D. (1993). Mnemonics: Enhancing the math performance of students with learning difficulties. *Intervention in School and Clinic, 29*(2), 78–82.

Montague, M. (1997). Cognitive strategy instruction in mathematics for students with learning disabilities. *Journal of Learning Disabilities, 30*(2), 164–177.

Reiser, R. A., & Dick, W. (1996). *Instructional planning: A guide for teachers* (2nd ed.). Boston: Allyn and Bacon.

Roderick, M., & Camburn, E. (1999). Risk and recovery from course failure in the early years of high school. *American Educational Research Journal, 36*(2), 303–343.

Steele, M. (2002). Strategies for helping students who have learning disabilities in mathematics. *Mathematics Teaching in the Middle School, 8*(3), 140–143.

Swanson, H. L. (2000). What instruction works for students with learning disabilities? Summarizing the results from a meta-analysis of intervention studies. In R. Gersten, E. P. Schiller, & S. Vaughn (Eds.), *Contemporary special education research: Syntheses of the knowledge base on critical instructional issues* (pp. 1–30). Mahwah, NJ: Lawrence Erlbaum Associates, Inc.

Tyree, R. B., Fiore, T. A., & Cook, R. A. (1994). Instructional materials for diverse learners: Features and considerations for textbook design. *Remedial and Special Education, 15*(6), 363–377.

Vaughn, S., Gersten, R., & Chard, D. J. (2000). The underlying message in LD intervention research: Findings from research syntheses. *Exceptional Children, 67*(1), 99–114.

| Research-Based Principles | AGS Globe Textbooks | References |
|---|---|---|

**Instructional Design: Lesson Structure and Learner Support Strategies,** *continued from previous page*

| | | |
|---|---|---|
| | *In the Teacher's Edition:*<br>◆ *3-Step Teaching Lesson Plan* concludes with "Reinforce and Extend" to reinforce, reteach, and extend lesson skills and concepts<br>◆ Unit or chapter projects link and apply unit or chapter concepts<br>◆ Multiple supplemental/alternative activities for individual and group learning and problem solving<br>◆ Career, home, and community application exercises<br><br>*In the Teacher's Resource Library:*<br>◆ Multiple exercises in Student Workbook and reproducibles offer applications, content extensions, additional practice, and alternative activities at a lower (Grade 2 Spache) readability<br><br>*Skill Track Online:*<br>◆ Monitors student learning and guides teacher feedback to student | |

### Ongoing Assessment and Tracking Student Progress

| Research-Based Principles | AGS Globe Textbooks | References |
|---|---|---|
| Textbooks can incorporate features to facilitate and support assessment of learning, allowing teachers to monitor student progress and provide information on mastery level and the need for instructional changes.<br><br>Assessment should measure student progress on learning goals over the course of a lesson, chapter, or content-area textbook.<br><br>Students and teachers need timely and ongoing feedback so instruction can focus on specific skill development. | ◆ Test-taking tips and strategies for students who benefit from explicit strategy instruction<br>◆ Lesson and chapter reviews check student understanding of content<br>◆ Workbook and reproducible lesson activities (Teacher's Resource Library) offer additional monitoring of student progress<br>◆ Discussion questions allow teachers to monitor student progress toward lesson objectives<br>◆ Self-Study Guides (Teacher's Resource Library) allow teacher and student to track individual assignments and progress<br>◆ Chapter assessment activities and curriculum-based assessment items correlate to chapter Goals for Learning:<br>• Chapter reviews<br>• End-of-chapter tests<br>• Cumulative Midterm and final mastery tests<br>• Alternative chapter assessments<br>• Skill Track Online assesses and tracks individual student performance by lesson and chapter | Deshler, D. D., Ellis, E. S., & Lenz, B. K. (1996). *Teaching adolescents with learning disabilities: Strategies and methods* (2nd ed.). Denver, CO: Love Publishing Company.<br><br>Jarrett, D. (1999). *The inclusive classroom: Mathematics and science instruction for students with learning disabilities—It's just good teaching.* Portland, OR: Northwest Regional Educational Laboratory.<br><br>Reiser, R. A., & Dick, W. (1996). *Instructional planning: A guide for teachers* (2nd ed.). Boston: Allyn and Bacon.<br><br>Tyree, R. B., Fiore, T. A., & Cook, R. A. (1994). Instructional materials for diverse learners: Features and considerations for textbook design. *Remedial and Special Education, 15*(6), 363–377. |

**For more information on the scientific research base for AGS Globe Textbooks, please go to www.agsglobe.com or call Customer Service at 800-328-2560 to request a research report.**

*Life Skills Health*

# Life Skills Health Skills Chart

## CHAPTER

| Health Content | 1 | 2 | 3 | 4 | 5 | 6 | 7 | 8 | 9 | 10 | 11 | 12 | 13 | 14 | 15 | 16 | 17 | 18 | 19 | 20 | 21 |
|---|---|---|---|---|---|---|---|---|---|---|---|---|---|---|---|---|---|---|---|---|---|
| Causes of Disease | | | | | | | | | | | ✓ | ✓ | ✓ | | | | | | | | |
| Consumer Health | | | | | | | | | | | | | | | | | | | ✓ | | |
| Diet and Food Choices | ✓ | | | | | | | | ✓ | ✓ | | | | | | | | | | | |
| Disease Prevention | | | | | | | | | | | ✓ | ✓ | ✓ | | | | | | | | |
| Environmental Health | | | | | | | | | | | | | | | | | | | | | ✓ |
| Exercise and Fitness | ✓ | | | | | ✓ | | | | | | | | | | | | | | | |
| Family Health | | | | | | | | ✓ | | | | | | | | | | | | | |
| First Aid and Safety | | | | | | | | | | | | | | | | ✓ | ✓ | ✓ | | | |
| The Human Body | | | | | ✓ | | | | | | | | | | | | | | | | |
| The Life Cycle and Human Development | | | | | | | | ✓ | | | | | | | | | | | | | |
| Medicines and Drugs | | | | | | | | | | | | | | ✓ | ✓ | | | | | | |
| Mental and Emotional Health | ✓ | ✓ | ✓ | ✓ | | | | | | | | | | | | | | | | | |
| Nutrition | | | | | | | | | ✓ | ✓ | | | | | | | | | | | |
| Personal Hygiene | | | | | | ✓ | | | | | | | | | | | | | | | |
| Physical Health | ✓ | | | | | ✓ | | | ✓ | ✓ | ✓ | ✓ | ✓ | | | | | | | | |
| Public Health and Health Care | | | | | | | | | | | | | | | | | | | | ✓ | |
| Social Health | ✓ | | | | | | | ✓ | | | | | | | | | | | | ✓ | |
| Violence and Conflict Resolution | | | | | | | | | | | | | | | | | | ✓ | | | |

| Process Skills | 1 | 2 | 3 | 4 | 5 | 6 | 7 | 8 | 9 | 10 | 11 | 12 | 13 | 14 | 15 | 16 | 17 | 18 | 19 | 20 | 21 |
|---|---|---|---|---|---|---|---|---|---|---|---|---|---|---|---|---|---|---|---|---|---|
| Accessing Information | ✓ | ✓ | ✓ | ✓ | ✓ | ✓ | ✓ | ✓ | ✓ | ✓ | ✓ | ✓ | ✓ | ✓ | ✓ | ✓ | ✓ | ✓ | ✓ | ✓ | ✓ |
| Analyzing Information | ✓ | ✓ | ✓ | ✓ | ✓ | ✓ | ✓ | ✓ | ✓ | ✓ | ✓ | ✓ | ✓ | ✓ | ✓ | ✓ | ✓ | ✓ | ✓ | ✓ | ✓ |
| Communicating | ✓ | ✓ | ✓ | ✓ | ✓ | ✓ | ✓ | ✓ | ✓ | ✓ | ✓ | ✓ | ✓ | ✓ | ✓ | ✓ | ✓ | ✓ | ✓ | ✓ | ✓ |
| Demonstrating Skills Learned | ✓ | ✓ | ✓ | ✓ | ✓ | ✓ | ✓ | ✓ | ✓ | ✓ | ✓ | ✓ | ✓ | ✓ | ✓ | ✓ | ✓ | ✓ | ✓ | ✓ | ✓ |
| Describing | ✓ | ✓ | ✓ | ✓ | ✓ | ✓ | ✓ | ✓ | ✓ | ✓ | ✓ | ✓ | ✓ | ✓ | ✓ | ✓ | ✓ | ✓ | ✓ | ✓ | ✓ |
| Evaluating Causes and Effects | | | | ✓ | | ✓ | | ✓ | | ✓ | ✓ | ✓ | ✓ | | | ✓ | ✓ | | ✓ | | ✓ |
| Following Written Directions | ✓ | ✓ | ✓ | ✓ | ✓ | ✓ | ✓ | ✓ | ✓ | ✓ | ✓ | ✓ | ✓ | ✓ | ✓ | ✓ | ✓ | ✓ | ✓ | ✓ | ✓ |
| Researching and Writing About Health | ✓ | ✓ | ✓ | ✓ | ✓ | ✓ | ✓ | ✓ | ✓ | ✓ | ✓ | ✓ | ✓ | ✓ | ✓ | ✓ | ✓ | ✓ | ✓ | ✓ | ✓ |

| Thinking Skills | \multicolumn{21}{c}{CHAPTER} |
| --- | --- | --- | --- | --- | --- | --- | --- | --- | --- | --- | --- | --- | --- | --- | --- | --- | --- | --- | --- | --- | --- |
| | 1 | 2 | 3 | 4 | 5 | 6 | 7 | 8 | 9 | 10 | 11 | 12 | 13 | 14 | 15 | 16 | 17 | 18 | 19 | 20 | 21 |
| Applying Information | ✓ | ✓ | ✓ | ✓ | ✓ | ✓ | ✓ | ✓ | ✓ | ✓ | ✓ | ✓ | ✓ | ✓ | ✓ | ✓ | ✓ | ✓ | ✓ | ✓ | ✓ |
| Comparing and Contrasting | ✓ | ✓ | ✓ | ✓ | ✓ | ✓ | ✓ | ✓ | ✓ | ✓ | ✓ | ✓ | ✓ | ✓ | ✓ | ✓ | ✓ | ✓ | ✓ | ✓ | ✓ |
| Drawing Conclusions | ✓ | ✓ | ✓ | ✓ | ✓ | ✓ | ✓ | ✓ | ✓ | ✓ | ✓ | ✓ | ✓ | ✓ | ✓ | ✓ | ✓ | ✓ | ✓ | ✓ | ✓ |
| Explaining Ideas and Concepts | ✓ | ✓ | ✓ | ✓ | ✓ | ✓ | ✓ | ✓ | ✓ | ✓ | ✓ | ✓ | ✓ | ✓ | ✓ | ✓ | ✓ | ✓ | ✓ | ✓ | ✓ |
| Formulating Questions | ✓ | ✓ | ✓ | ✓ | ✓ | ✓ | ✓ | ✓ | ✓ | ✓ | ✓ | ✓ | ✓ | ✓ | ✓ | ✓ | ✓ | ✓ | ✓ | ✓ | ✓ |
| Identifying and Solving Problems | ✓ | ✓ | ✓ | ✓ | ✓ | ✓ | ✓ | ✓ | ✓ | ✓ | ✓ | ✓ | ✓ | ✓ | ✓ | ✓ | ✓ | ✓ | ✓ | ✓ | ✓ |
| Interpreting Visuals | ✓ | ✓ | ✓ | ✓ | ✓ | ✓ | ✓ | ✓ | ✓ | ✓ | ✓ | ✓ | ✓ | ✓ | ✓ | ✓ | ✓ | ✓ | ✓ | ✓ | ✓ |
| Learning Health Vocabulary | ✓ | ✓ | ✓ | ✓ | ✓ | ✓ | ✓ | ✓ | ✓ | ✓ | ✓ | ✓ | ✓ | ✓ | ✓ | ✓ | ✓ | ✓ | ✓ | ✓ | ✓ |
| Making Decisions | ✓ | ✓ | ✓ | ✓ | ✓ | ✓ | ✓ | ✓ | ✓ | ✓ | ✓ | ✓ | ✓ | ✓ | ✓ | ✓ | ✓ | ✓ | ✓ | ✓ | ✓ |
| Organizing Information | ✓ | ✓ | ✓ | ✓ | ✓ | ✓ | ✓ | ✓ | ✓ | ✓ | ✓ | ✓ | ✓ | ✓ | ✓ | ✓ | ✓ | ✓ | ✓ | ✓ | ✓ |
| Recalling Facts | ✓ | ✓ | ✓ | ✓ | ✓ | ✓ | ✓ | ✓ | ✓ | ✓ | ✓ | ✓ | ✓ | ✓ | ✓ | ✓ | ✓ | ✓ | ✓ | ✓ | ✓ |
| Recognizing Cause and Effect | ✓ | ✓ | ✓ | ✓ | ✓ | ✓ | ✓ | ✓ | ✓ | ✓ | ✓ | ✓ | ✓ | ✓ | ✓ | ✓ | ✓ | ✓ | ✓ | ✓ | ✓ |
| Recognizing Main Ideas | ✓ | ✓ | ✓ | ✓ | ✓ | ✓ | ✓ | ✓ | ✓ | ✓ | ✓ | ✓ | ✓ | ✓ | ✓ | ✓ | ✓ | ✓ | ✓ | ✓ | ✓ |
| Recognizing Relationships | ✓ | ✓ | ✓ | ✓ | ✓ | ✓ | ✓ | ✓ | ✓ | ✓ | ✓ | ✓ | ✓ | ✓ | ✓ | ✓ | ✓ | ✓ | ✓ | ✓ | ✓ |
| Understanding Concepts | ✓ | ✓ | ✓ | ✓ | ✓ | ✓ | ✓ | ✓ | ✓ | ✓ | ✓ | ✓ | ✓ | ✓ | ✓ | ✓ | ✓ | ✓ | ✓ | ✓ | ✓ |

*Life Skills Health*

# National Health Education Standards Correlation

## STANDARD A  Health Promotion and Disease Prevention

Students will comprehend concepts related to health promotion and disease prevention:

**Life Skills Health**

- Analyze how behavior can impact health maintenance and disease prevention.

    Pages 3, 5–12, 14–25, 27, 30–32, 34–43, 53–64, 66–73, 76–91, 93–95, 113, 122, 126–147, 153, 161–162, 166, 169–173, 177–181, 183, 186, 189–195, 197–199, 202–227, 228–245, 247–249, 251–254, 257, 258, 260–261, 263, 265–267, 268–269, 271–283, 284–291, 294–297, 299–301, 303–309, 311–313, 315, 320–343, 344–346, 350–361, 363–394, 398–400, 416–417, 419, 424–437, 439–441, 443–453, 455–456, 463–465, 472, 479, 484–485, 487–492, 495–515

- Describe the interrelationships of mental, emotional, social, and physical health throughout adulthood.

    Pages 1–12, 14–25, 26–47, 48–73, 74–95, 97–99, 110–111, 134–135, 138–139, 142, 145, 166, 176–178, 180–181, 186–195, 222, 232, 279, 324, 328–331, 333–343, 345–351, 356–361, 381, 383–384, 416–437

- Explain the impact of personal health behaviors on the functioning of body systems.

    Pages 7, 14, 16, 34, 36, 38, 80, 83, 100–125, 128–135, 143, 161–162, 211–218, 222–224, 254–258, 270–271, 274, 276–281, 286–291, 299–300, 321, 324–343, 346, 488, 490–492, 501–503

- Analyze how the family, peers, and community influence the health of individuals.

    Pages 5, 7, 12, 13, 29–31, 41–43, 52–55, 60–65, 67–71, 77–80, 84, 86–89, 140, 143, 150–154, 174–195, 231–238, 254, 275–281, 322, 326, 328, 348–351, 357–359, 376, 385, 389–391, 398–400, 422–423, 424–435, 446–465, 466–483

- Analyze how the environment influences the health of the community.

    Pages 5, 12, 161, 256, 297, 472, 489–513

- Describe how to delay onset and reduce risks of potential health problems during adulthood.

    Pages 100–103, 122, 127–147, 169, 170, 203, 222–225, 286–291, 293–307, 323–341, 376, 391, 444–456

- Analyze how public health policies and government regulations influence health promotion and disease prevention.

    Pages 12, 92, 153, 196, 204–205, 239–243, 246, 310, 317, 321, 438, 459, 460, 466–483, 496, 503–510, 512

- Analyze how the prevention and control of health problems are influenced by research and medical advances.

    Pages 154, 260–261, 263, 310, 319, 320, 472, 473

    *Technology and Society*: pages 10, 35, 64, 84, 115, 131, 163, 182, 242, 263, 276, 288, 322, 352, 386, 405, 433, 461, 473, 503 *Health Myths*: pages 9, 17, 35, 36, 55, 86, 87, 101, 107, 131, 143, 159, 178, 181, 218, 222, 234, 236, 258, 260, 271, 273, 289, 294, 304, 326, 330, 348, 353, 374, 410, 411, 424, 426, 449, 460, 470, 474, 494, 505 *Research and Write*: pages 12, 31, 39, 65, 81, 82, 113, 132, 166, 186, 213, 218, 230, 235, 258, 262, 273, 274, 290, 296, 325, 340, 350, 357, 380, 384, 396, 412, 421, 430, 448, 458, 471, 473, 489, 500

## STANDARD B  Health Information, Products and Services

Students will demonstrate the ability to access valid health information and health-promoting products and services.

**Life Skills Health**

- Evaluate the validity of health information, products, and services.

    Pages 5, 12, 161, 256, 297, 472, 489–513

- Demonstrate the ability to evaluate resources from home, school, and community that provide valid health information.

    The opportunity to explore this concept can be found on pages 452–456, and 466–483.

- Evaluate factors that influence personal selection of health products and services.

    Pages 234–243, 447, 450, 471–473

- Demonstrate the ability to access school and community health services for self and others.

    The opportunity to explore this concept can be found on pages 452–456, and 471–483.

- Analyze the cost and accessibility of health care services.

    Pages 457–461, 471–483

- Analyze situations requiring professional health services.

    Pages 43, 85–89, 161, 190–192, 261, 278, 286–295, 299, 330–331, 356–357, 396–397, 402–412, 427, 432, 452–453, 471–472

*Life Skills Health*

## STANDARD C  Reducing Health Risks

Students will demonstrate the ability to practice health-enhancing behaviors and reduce health risks.

### Life Skills Health

- **Analyze the role of individual responsibility for enhancing health.**

  Pages 3, 5–12, 14–25, 27, 30–32, 34–43, 53–64, 66–73, 76–91, 93–95, 113, 122, 126–147, 153, 161–162, 166, 169–173, 177–181, 183, 186, 189–195, 197–199, 202–227, 228–245, 247–249, 251–254, 257, 258, 260–261, 263, 265–267, 268–269, 271–283, 284–291, 294–297, 299–301, 303–309, 311–313, 315, 320–343, 344–346, 350–361, 363–394, 398–400, 416–417, 419, 424–437, 439–441, 443–453, 455–456, 463–465, 472, 479, 484–485, 487–492, 495–515

- **Evaluate a personal health assessment to determine strategies for health enhancement.**

  Pages 20-21

- **Analyze the short-term and long-term consequences of safe, risky and harmful behaviors.**

  Pages 8, 14–16, 18–20, 32, 38–43, 57–73, 76–79, 82–91, 113, 119, 122, 128–135, 138–139, 142–245, 152, 153, 161–162, 176–183, 189–193, 204–205, 210–227, 229–232

- **Develop strategies to improve or maintain personal, family and community health.**

  Pages 55–71, 78–79, 85–89, 100–103, 122, 127–147, 161, 169, 170, 177–183, 187–195, 203, 222–225, 230, 232, 236–247, 254, 261, 278–279, 286–291, 293–307, 323–341, 348–359, 367–391, 395–411, 429–435, 444–456, 466–473, 478–483

- **Develop injury prevention and management strategies for personal, family, and community health.**

  Pages 136–139, 366–393, 394–415

- **Demonstrate ways to avoid and reduce threatening situations.**

  Pages 189–192, 422–423, 429–435

- **Evaluate strategies to manage stress.**

  Pages 38–45, 135, 232

## STANDARD D  Influences on Health

Students will analyze the influence of culture, media, technology, and other factors on health.

### Life Skills Health

- **Analyze how cultural diversity enriches and challenges health behaviors.**

  Pages 13, 65, 231, 351

- **Evaluate the effect of media and other factors on personal, family, and community health.**

  Specifically media: pages 51, 86, 88, 233–242, 447, 476 Other factors: pages 5, 7, 12, 13, 29–31, 41–43, 52–55, 60–65, 67–71, 77–80, 84, 86–89, 140, 143, 150–154, 161, 174–195, 231–238, 254, 256, 275–281, 297, 322, 326, 328, 348–351, 357–359, 376, 385, 389–391, 398–400, 422–423, 424–435, 446–405, 466–483, 489–513

- **Evaluate the impact of technology on personal, family, and community health.**

  Pages 155, 160, 167–171, 260–263, 310, 318–323, 326, 371–372, 381–382, 454, 572–477, 505 and *Technology and Society*: pages 10, 35, 64, 84, 115, 131, 163, 182, 242, 263, 276, 288, 322, 352, 386, 405, 433, 461, 473, 503

- **Analyze how information from the community influences health.**

  Pages 12, 191, 353–358, 386, 389–390, 433–434, 447, 452–456, 467–472, 474, 476–481, 487–503, 507

## STANDARD E  Using Communication Skills to Promote Health

Students will demonstrate the ability to use interpersonal communication skills to enhance health.

### Life Skills Health

- **Demonstrate skills for communicating effectively with family, peers, and others.**

  The opportunity to explore this concept can be found on pages 39, 43, 52–54, 57, 59–64, 69, 78, 79, 84, 177–178, 180, 186, 279, 331, 351, 356, 358, 419, and 430–433.

- **Analyze how interpersonal communication affects relationships.**

  The opportunity to explore this concept can be found on pages 39, 43, 52–54, 57, 59–64, 69, 78, 79, 84, 177–178, 180, 186, 279, 331, 351, 356, 358, 419, and 430–433.

- **Demonstrate healthy ways to express needs, wants, and feelings.**

  The opportunity to explore this concept can be found on pages 32–33, 35, 38–43, 50–55, 57–71, 79, 177–178, 180, 279, 331, 351, 356–359, 419, and 429–435

- **Demonstrate ways to communicate care, consideration, and respect of self and others.**

  The opportunity to explore this concept can be found on pages 29–33, 38–43, 48–73, 79, 127–147, 153, 177–178, 180, 279, 331, 351, 356, 357–359, 419, and 429–435.

- **Demonstrate strategies for solving interpersonal conflicts without harming self or others.**

  The opportunity to explore this concept can be found on pages 57, 61–72, 79, 279, 331, 351, 356–359, 419, 424, and 429–435.

*Life Skills Health*

# National Health Education Standards Correlation, continued

## STANDARD E  Using Communication Skills to Promote Health (continued)

- Demonstrate refusal, negotiation, and collaboration skills to avoid potentially harmful situations.

  The opportunity to explore this concept can be found on pages 279, 331, 351, 356, 357–359, 419, and 430–432.

- Analyze the possible causes of conflict in schools, families, and communities.

  Pages 57, 60, 419, 424–428

- Demonstrate strategies used to prevent conflict.

  The opportunity to explore this concept can be found on pages 57–64, 419, 424–426, and 429–435.

## STANDARD F  Setting Goals for Good Health

Students will demonstrate the ability to use goal-setting and decision-making skills to enhance health.

**Life Skills Health**

- Demonstrate the ability to utilize various strategies when making decisions related to health needs and risks of young adults.

  Pages 17–20, 22, 23, 53, 54, 182, 269–283, 331, 351, 356–359, 419, 430–432 and specifically Decide For Yourself on pages 6, 32, 56, 88, 119, 142, 154, 184, 214, 237, 257, 279, 300, 331, 384, 412, 432, 450, 472, and 497.

- Analyze health concerns that require collaborative decision making.

  Pages 182, 269–283

- Predict immediate and long-term impact of health decisions on the individual, family, and community.

  The opportunity to explore this concept can be found on pages 17–23, 38–45, 48–71, 85–89, 140–145, 200–209, 228–232, 268–283, 284–301, 334–343, 344–361, 366–393, 429–437, 453–456 and specifically *Decide For Yourself* on pages 6, 32, 56, 88, 119, 142, 154, 184, 214, 237, 257, 279, 300, 331, 384, 412, 432, 450, 472, and 497.

- Implement a plan for attaining a personal health goal.

  The opportunity to explore this concept can be found on pages 6–8, 17–20, 56, 58, 142, 154, and 352.

- Evaluate progress toward achieving personal health goals.

  The opportunity to explore this concept can be found on pages 6–8, 17–20, 56, 58, 142, 154, and 352.

- Formulate an effective plan for lifelong health.

  Pages 6–8, 17–20, 56, 58, 140–145, 154, 200–209, 228–232, 352, 356–359, 381–391, 442–445, 504–509; examples: *Health In Your Life*: pages 5, 44, 52, 86, 122, 133, 153, 191, 219, 235, 261, 279, 297, 351, 375, 399, 422, 456, 476, 507

## STANDARD G  Health Advocacy

Students will demonstrate the ability to advocate for personal, family, and community health.

**Life Skills Health**

- Evaluate the effectiveness of communication methods for accurately expressing health information and ideas.

  Pages 208, 218, 234–243, 445–449, 455–456, 466–483.

- Express information and opinions about health issues.

  *Decide For Yourself*: pages 6, 32, 56, 88, 119, 142, 154, 184, 214, 237, 257, 279, 300, 331, 384, 412, 432, 450, 472, 497; Health In Your Life: pages 5, 44, 52, 86, 122, 133, 153, 191, 219, 235, 261, 279, 297, 351, 375, 399, 422, 456, 476, 507 Research and Write: pages 12, 31, 39, 65, 81, 82, 113, 132, 166, 186, 213, 218, 230, 235, 258, 262, 273, 274, 290, 296, 325, 340, 350, 357, 380, 384, 396, 412, 421, 430, 448, 458, 471, 473, 489, 500; also please view *Critical Thinking* exercises at the end of each chapter and unit.

- Utilize strategies to overcome barriers when communicating information, ideas, feelings, and opinions about health issues.

  The opportunity to explore this concept can be found on pages 7, 32, 52, 61–71, 140, 191–192, and 353.

- Demonstrate the ability to influence and support others in making positive health choicesl.

  The opportunity to explore this concept can be found on pages 29–32, 39, 43, 50–54, 57, 59–70, 78, 79, 84, 177–178, 180, 186, 279, 331, 351, 356, 358, 419, and 430–433.

- Demonstrate the ability to work cooperatively when advocating for healthy communities.

  The opportunity to explore this concept can be found on pages 7, 12, 59, 350, 389, 433–434, and 478–480.

- Demonstrate the ability to adapt health messages and comunication techniques to the characteristics of a particular audience. health.

  The opportunity to explore this concept can be found on pages 12, 31, 39, 65, 81, 82, 113, 132, 166, 186, 213, 218, 230, 235, 258, 262, 273, 274, 290, 296, 325, 340, 350, 357, 380, 384, 396, 412, 421, 430, 448, 458, 471, 473, 489, and 500.

# Life Skills Health

Shoreview, MN

**Photo credits** for this textbook can be found on page 555.

The publisher wishes to thank the following educators for their helpful comments during the review process for *Life Skills Health*. Their assistance has been invaluable.

**Heidi Ann Coe,** SELPA Program Specialist, Morongo Unified School District, 29 Palms, CA; **Susan G. Helman,** Special Education Teacher, Alma High School, Alma, MI; **Alva Jones, Ed.S.,** Educational Consultant, Evans, GA; **Russell Laya,** Special Education Teacher, Roy Miller High School, Corpus Christi, TX; **Deik Maxwell, M.S., M.A.,** Department Chair of Special Education, R. Rex Parris High School, Palmdale, CA; **Brenda P. Seigler,** ESE Teacher, Mandarin High School, Jacksonville, FL

## Publisher's Project Staff

Vice President of Curriculum and Publisher: Sari Follansbee, Ed. D.; Director of Curriculum Development: Teri Mathews; Managing Editor: Julie Maas; Senior Editor: Jody Peterson; Development Assistant: Bev Johnson; Director of Creative Services: Nancy Condon; Senior Designer: Diane McCarty; Project Coordinator/Designer: Jen Willman; Purchasing Agent: Mary Kaye Kuzma; Product Manager–Curriculum: Brian Holl

Copyright © 2007 by Pearson Education, Inc., publishing as Pearson AGS Globe, Shoreview, Minnesota 55126. All rights reserved. Printed in the United States of America. This publication is protected by copyright, and permission should be obtained from the publisher prior to any prohibited reproduction, storage in a retrieval system, or transmission in any form or by any means, electronic, mechanical, photocopying, recording, or likewise. For information regarding permission(s), visit www.agsglobe.com.

Pearson AGS Globe™ is a trademark of Pearson Education, Inc.
Pearson® is a registered trademark of Pearson plc.

ISBN 0-7854-4133-6

A 0 9 8 7 6 5 4 3 2 1

1-800-328-2560
www.agsglobe.com

# Contents

**How to Use This Book: A Study Guide . . . . . . . . x**

### Unit 1 — Mental and Emotional Health . . . . . . . . . . . xviii

**Chapter 1    Maintaining Health and Wellness . . . . 2**
Lesson 1    The Three Parts of Good Health . . . . . . . . . . 4
Lesson 2    Good Health and Wellness . . . . . . . . . . . . . . 7
Lesson 3    Health Risks . . . . . . . . . . . . . . . . . . . . . . . . 11
Lesson 4    Take Charge of Your Health . . . . . . . . . . . . . 17
◆ Chapter 1 Summary . . . . . . . . . . . . . . . . . . . . . . . . . . 23
◆ Chapter 1 Review . . . . . . . . . . . . . . . . . . . . . . . . . . . . 24
◆ Test-Taking Tip . . . . . . . . . . . . . . . . . . . . . . . . . . . . . 25

**Chapter 2    Managing Emotions . . . . . . . . . . . . . . 26**
Lesson 1    Your Emotions . . . . . . . . . . . . . . . . . . . . . . . 28
Lesson 2    Stress and Your Body . . . . . . . . . . . . . . . . . . 34
Lesson 3    Managing Stress . . . . . . . . . . . . . . . . . . . . . . 38
◆ Chapter 2 Summary . . . . . . . . . . . . . . . . . . . . . . . . . . 45
◆ Chapter 2 Review . . . . . . . . . . . . . . . . . . . . . . . . . . . . 46
◆ Test-Taking Tip . . . . . . . . . . . . . . . . . . . . . . . . . . . . . 47

**Chapter 3    Maintaining Mental Health . . . . . . . . 48**
Lesson 1    Influences on Mental Health . . . . . . . . . . . . 50
Lesson 2    Emotional Health . . . . . . . . . . . . . . . . . . . . 55
Lesson 3    Healthy Relationships . . . . . . . . . . . . . . . . . . 61
Lesson 4    Becoming More Emotionally Healthy . . . . . . 66
◆ Chapter 3 Summary . . . . . . . . . . . . . . . . . . . . . . . . . . 71
◆ Chapter 3 Review . . . . . . . . . . . . . . . . . . . . . . . . . . . . 72
◆ Test-Taking Tip . . . . . . . . . . . . . . . . . . . . . . . . . . . . . 73

**Chapter 4    Recognizing Mental Health Problems 74**
Lesson 1    Characteristics of Poor Mental Health . . . . . 76
Lesson 2    Mental Disorders . . . . . . . . . . . . . . . . . . . . . 80
Lesson 3    Treating Mental Disorders . . . . . . . . . . . . . . 85
◆ Chapter 4 Summary . . . . . . . . . . . . . . . . . . . . . . . . . . 89
◆ Chapter 4 Review . . . . . . . . . . . . . . . . . . . . . . . . . . . . 90
◆ Test-Taking Tip . . . . . . . . . . . . . . . . . . . . . . . . . . . . . 91
Health in the World: The World Health Organization . . . . 92
Unit 1 Summary . . . . . . . . . . . . . . . . . . . . . . . . . . . . . . 93
Unit 1 Review . . . . . . . . . . . . . . . . . . . . . . . . . . . . . . . . 94

## Unit 2   Personal Health and Family Life . . . . . . . . . . 96

### Chapter 5   Identifying Human Body Systems . . . 98
Lesson 1   The Skeletal and Muscular Systems . . . . . . . 100
Lesson 2   The Nervous System and Sense Organs . . . 104
Lesson 3   The Endocrine System . . . . . . . . . . . . . . . . . 110
Lesson 4   The Circulatory and Respiratory Systems . . 112
Lesson 5   The Digestive and Excretory Systems . . . . . 116
Lesson 6   The Body's Protective Covering . . . . . . . . . . 120
◆ Chapter 5 Summary . . . . . . . . . . . . . . . . . . . . . . . . . . . 123
◆ Chapter 5 Review . . . . . . . . . . . . . . . . . . . . . . . . . . . . . 124
◆ Test-Taking Tip . . . . . . . . . . . . . . . . . . . . . . . . . . . . . . . 125

### Chapter 6   Maintaining Personal Hygiene and Fitness . . . . . . . . . . . . 126
Lesson 1   Hygiene for Good Health . . . . . . . . . . . . . . 128
Lesson 2   Exercise and Physical Fitness . . . . . . . . . . . 134
Lesson 3   Personal Fitness Plan . . . . . . . . . . . . . . . . . . 140
◆ Chapter 6 Summary . . . . . . . . . . . . . . . . . . . . . . . . . . . 145
◆ Chapter 6 Review . . . . . . . . . . . . . . . . . . . . . . . . . . . . . 146
◆ Test-Taking Tip . . . . . . . . . . . . . . . . . . . . . . . . . . . . . . .147

### Chapter 7   The Life Cycle and Human Development . . . . . . . . . . . . . . . . . . . 148
Lesson 1   The Life Cycle and Adolescence . . . . . . . . . 150
Lesson 2   Reproduction . . . . . . . . . . . . . . . . . . . . . . . . 156
Lesson 3   Pregnancy and Childbirth . . . . . . . . . . . . . . 160
Lesson 4   Heredity and Genetics . . . . . . . . . . . . . . . . . 167
◆ Chapter 7 Summary . . . . . . . . . . . . . . . . . . . . . . . . . . . 171
◆ Chapter 7 Review . . . . . . . . . . . . . . . . . . . . . . . . . . . . . 172
◆ Test-Taking Tip . . . . . . . . . . . . . . . . . . . . . . . . . . . . . . . 173

**Chapter 8   The Family................ 174**
Lesson 1   The Family Life Cycle, Dating,
           and Marriage....................... 176
Lesson 2   Parenting and Family Systems........... 182
Lesson 3   Problems in Families ................. 187
◆ Chapter 8 Summary ....................... 193
◆ Chapter 8 Review......................... 194
◆ Test-Taking Tip ........................... 195
Health in the World: Overpopulation and Health....... 196
Unit 2 Summary................................ 197
Unit 2 Review ................................. 198

## Unit 3

**Nutrition ................... 200**

**Chapter 9   The Role of Diet in Health ....... 202**
Lesson 1   A Healthy Diet ..................... 204
Lesson 2   Carbohydrates, Fats, and Protein......... 210
Lesson 3   Vitamins, Minerals, and Water ......... 216
Lesson 4   Special Dietary Needs................. 221
◆ Chapter 9 Summary ....................... 225
◆ Chapter 9 Review......................... 226
◆ Test-Taking Tip ........................... 227

**Chapter 10   Choosing Healthy Foods.......... 228**
Lesson 1   Healthy Eating Patterns
           and Food Choices ................... 230
Lesson 2   How the Media Influences
           Eating Patterns ..................... 233
Lesson 3   Food Labels and Food Additives ......... 239
◆ Chapter 10 Summary ...................... 243
◆ Chapter 10 Review ........................ 244
◆ Test-Taking Tip ........................... 245
Health in the World: International Red Cross ......... 246
Unit 3 Summary................................ 247
Unit 3 Review ................................. 248

**Unit 4 Preventing and Controlling Diseases and Disorders . . . . . . . . . . . . . . . . . 250**

**Chapter 11 Disease—Causes and Protection . . . 252**
Lesson 1   Causes of Disease . . . . . . . . . . . . . . . . . . . . . 254
Lesson 2   How the Body Protects Itself from Disease . . . . . . . . . . . . . . . . . . . . . . . . 259
◆ Chapter 11 Summary . . . . . . . . . . . . . . . . . . . . . . . . . . . 265
◆ Chapter 11 Review . . . . . . . . . . . . . . . . . . . . . . . . . . . . . 266
◆ Test-Taking Tip . . . . . . . . . . . . . . . . . . . . . . . . . . . . . . . . 267

**Chapter 12 Preventing AIDS and Sexually Transmitted Diseases . . . . . . . . . . . . 268**
Lesson 1   AIDS. . . . . . . . . . . . . . . . . . . . . . . . . . . . . . . 270
Lesson 2   Sexually Transmitted Diseases . . . . . . . . . . . . 275
◆ Chapter 12 Summary . . . . . . . . . . . . . . . . . . . . . . . . . . . 281
◆ Chapter 12 Review . . . . . . . . . . . . . . . . . . . . . . . . . . . . . 282
◆ Test-Taking Tip . . . . . . . . . . . . . . . . . . . . . . . . . . . . . . . . 283

**Chapter 13 Recognizing Common Diseases . . . . 284**
Lesson 1   Cardiovascular Diseases and Problems. . . . 286
Lesson 2   Cancer . . . . . . . . . . . . . . . . . . . . . . . . . . . . . 292
Lesson 3   Diabetes . . . . . . . . . . . . . . . . . . . . . . . . . . . . 298
Lesson 4   Arthritis, Epilepsy, and Asthma . . . . . . . . . . 302
◆ Chapter 13 Summary . . . . . . . . . . . . . . . . . . . . . . . . . . . 307
◆ Chapter 13 Review . . . . . . . . . . . . . . . . . . . . . . . . . . . . . 308
◆ Test-Taking Tip . . . . . . . . . . . . . . . . . . . . . . . . . . . . . . . . 309

Health in the World: The Need for Vaccinations . . . . . . . . 310
Unit 4 Summary . . . . . . . . . . . . . . . . . . . . . . . . . . . . . . . . . 311
Unit 4 Review . . . . . . . . . . . . . . . . . . . . . . . . . . . . . . . . . . . 312

## Unit 5 — Use and Misuse of Substances .......... 314

### Chapter 14 Recognizing Medicines and Drugs .. 316
- Lesson 1 Medicines ............................. 318
- Lesson 2 Tobacco ............................... 324
- Lesson 3 Alcohol ................................ 328
- Lesson 4 Stimulants, Depressants, Narcotics, and Hallucinogens ..................... 332
- Lesson 5 Other Dangerous Drugs ................ 338
- ◆ Chapter 14 Summary ........................... 341
- ◆ Chapter 14 Review ............................. 342
- ◆ Test-Taking Tip ................................ 343

### Chapter 15 Dealing with Drug Dependence .... 344
- Lesson 1 Drug Dependence—The Problems ....... 346
- Lesson 2 Drug Dependence—The Solutions ....... 352
- Lesson 3 Avoiding Drug Use .................... 356
- ◆ Chapter 15 Summary ........................... 359
- ◆ Chapter 15 Review ............................. 360
- ◆ Test-Taking Tip ................................ 361

Health in the World: Drug Trafficking .............. 362
Unit 5 Summary .................................. 363
Unit 5 Review ................................... 364

## Unit 6 — Injury Prevention and Safety Promotion ... 366

### Chapter 16 Reducing Risks of Injury .......... 368
- Lesson 1 Reducing Risks at Home ............... 370
- Lesson 2 Reducing Risks Away from Home ....... 376
- Lesson 3 Reducing Risks on the Road ........... 381
- Lesson 4 Safety During Natural Disasters ........ 385
- ◆ Chapter 16 Summary ........................... 391
- ◆ Chapter 16 Review ............................. 392
- ◆ Test-Taking Tip ................................ 393

**Chapter 17 Applying First Aid to Injuries** ...... 394
Lesson 1    First Aid Basics ....................... 396
Lesson 2    First Aid for Life-Threatening Emergencies. ........................ 401
Lesson 3    First Aid for Poisoning and Other Problems. ..................... 407
◆ Chapter 17 Summary ........................... 413
◆ Chapter 17 Review ............................. 414
◆ Test-Taking Tip ................................ 415

**Chapter 18 Preventing Violence and Resolving Conflicts** .............. 416
Lesson 1    Defining Violence .................... 418
Lesson 2    Causes of Violence ................... 424
Lesson 3    Preventing Violence .................. 429
◆ Chapter 18 Summary ........................... 435
◆ Chapter 18 Review ............................. 436
◆ Test-Taking Tip ................................ 437
Health in the World: Doctors Without Borders......... 438
Unit 6 Summary ..................................... 439
Unit 6 Review ....................................... 440

## Unit 7    Health and Society ................... 442

**Chapter 19 Consumer Health** ................ 444
Lesson 1    Being a Wise Consumer ............... 446
Lesson 2    Seeking Health Care .................. 452
Lesson 3    Paying for Health Care ................ 457
◆ Chapter 19 Summary ........................... 463
◆ Chapter 19 Review ............................. 464
◆ Test-Taking Tip ................................ 465

**Chapter 20 Public Health**. ................... 466
Lesson 1    Public Health Problems ............... 468
Lesson 2    U.S. Public Health Solutions ........... 474
Lesson 3    Health Promotion and Volunteer Organizations ...................... 478
◆ Chapter 20 Summary ........................... 481
◆ Chapter 20 Review ............................. 482
◆ Test-Taking Tip ................................ 483

| Chapter 21 | Environmental Health | 484 |
|---|---|---|
| Lesson 1 | Health and the Environment | 486 |
| Lesson 2 | Air Pollution and Health | 490 |
| Lesson 3 | Water Pollution and Health | 495 |
| Lesson 4 | Other Environmental Problems and Health | 499 |
| Lesson 5 | Protecting the Environment | 504 |

◆ Chapter 21 Summary ............................. 509
◆ Chapter 21 Review .............................. 510
◆ Test-Taking Tip ................................ 511
Health in the World: Health Care and Aging .......... 512
Unit 7 Summary .................................. 513
Unit 7 Review ................................... 514

## Appendix A: Body Systems ............. 516

## Appendix B: Nutrition Tables ........... 520

## Appendix C: Fact Bank ................ 522

## Glossary ........................... 528

## Index .............................. 541

## Photo and Illustration Credits ........... 555

## Using This Section

### How to Use This Book: A Study Guide

**Overview** This section may be used to introduce the study of health, to preview the book's features, and to review effective study skills.

### Objectives
- To introduce the study of health science
- To preview the student textbook
- To review study skills

**Student Pages** x-xvii

**Teacher's Resource Library** TRL
How to Use This Book 1–6

## Introduction to the Book

Have volunteers read aloud the two paragraphs of the introduction. Discuss with students why studying health is important and what kinds of information people can learn from studying health.

## How to Study

Read aloud each bulleted statement, pausing to discuss with students why the suggestion is a part of good study habits. Distribute copies of the "Study Habits Survey" (How to Use This Book 1) to students. Read the directions together and then have students complete the survey. After they have scored their surveys, ask them to make a list of the study habits they plan to improve. After three or four weeks, have students complete the survey again to see if they have improved their study habits. Encourage them to keep and review the survey every month or so.

Give students an opportunity to become familiar with the textbook features and the chapter and lesson organization and structure of *Life Skills Health*.

List the following text features on the board: Table of Contents, Unit Opener, Chapter Opener, Lesson, Lesson Review, Chapter Summary, Chapter Review, Health in the World, Unit Summary, Unit Review, Appendix A: Body Systems, Appendix B: Nutrition Tables, Appendix C: Fact Bank, Glossary, and Index.

Remind the students that they can use the Table of Contents to help identify and locate major features in the text. They also can use the Index to identify specific topics and the text pages on which they are discussed.

x    *How to Use This Book: A Study Guide*

## How to Use This Book: A Study Guide

Welcome to the study of health. Everyone wants to have good health and wellness. Studying health helps us learn ways to promote wellness. It helps us identify causes of health problems and ways to prevent them.

As you read this book, you will learn about promoting emotional, physical, and social health.

### *How to Study*
- Plan a regular time to study.
- Study in a quiet place with good lighting where you will not be distracted.
- Gather all the books, pencils, and paper you need to complete your assignments.
- Decide on a goal. For example: "I will finish reading and taking notes on Chapter 1, Lesson 1, by 8:00."
- Take a 5- to 10-minute break every hour to keep alert.

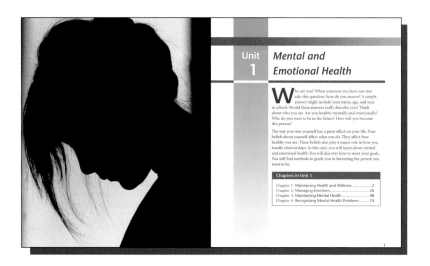

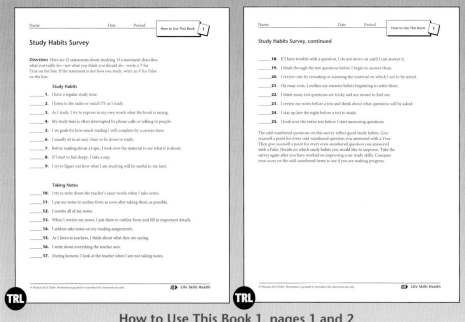

How to Use This Book 1, pages 1 and 2

## Before Beginning Each Unit

- Read the title and the opening text.
- Study the photograph. What does it say to you about health?
- Read the titles of the chapters in the unit.
- Look at the headings of the lessons and paragraphs to help you locate main ideas.
- Read the chapter and unit summaries to help you identify key issues.

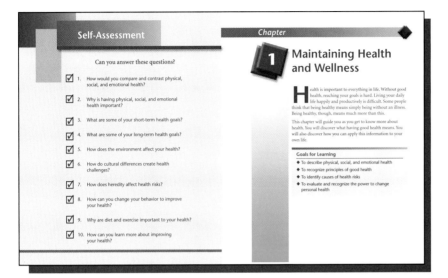

## Before Beginning Each Chapter

- Read the chapter title.
- Take the Self-Assessment to rate your knowledge of that health topic.
- Study the goals for learning. The chapter review and tests will ask questions related to these goals.

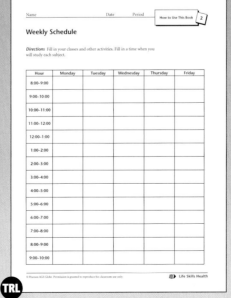

## Before Beginning Each Unit

When students begin their study of Unit 1, you might wish to have them read aloud and follow each of the bulleted suggestions on page xi. Actually trying the suggestions will help them understand what they are supposed to do and will help them recognize how useful the suggestions are when previewing a unit. At the beginning of subsequent units, refer students to page xi and encourage them to follow the suggestions. You might wish to continue to do this as a class each time you begin a unit or to allow students to work independently.

In addition to the suggestions on page xi, the text in the Teacher's Edition that accompanies each Unit Opener offers teaching suggestions and questions for unit introduction. The text also includes a list of the chapters in the unit and Teacher's Resource Library (TRL) materials for the unit, as well as a list of outside resource materials for teachers and students including books, videos/DVDs, CD-ROMS/software, and Web sites.

A Unit Opener organizes information in an easy-to-read format. To help students organize their time and work in an easy-to-follow format, have them fill out the "Weekly Schedule" (How to Use This Book 2). Encourage them to keep the schedule in a notebook or folder where they can refer to it easily. Suggest that they review the schedule periodically and update it as necessary.

## Before Beginning Each Chapter

Chapter Openers organize information in easy-to-read formats. When students begin their study of Chapter 1, have them turn to page 2. Read aloud the first bulleted statement on page xi. Have a volunteer find and read aloud the Chapter 1 title. Read aloud the second bulleted statement and have volunteers take turns reading aloud the Chapter 1 Self-Assessment items on page 2. Finally, read aloud the third bulleted statement and have students take turns reading aloud the Chapter 1 Goals for Learning on page 3. Discuss with students why knowing these goals can help them when they are studying the chapter.

# Before Beginning Each Lesson

With students, read through the information in "Before Beginning Each Lesson." Then assign each of the four lessons in Chapter 1 to a small group of students. Have them restate the lesson title in the form of a statement or a question. Then have them make a list of the bold words in their lesson. Explain to students that these words are important to the content in their lesson. Have groups note any subheads in their lesson.

Encourage them to pay attention to any illustrations in their lesson. Explain that these visuals will help them understand the lesson content.

After their survey of the lesson, have each group report to the class on their findings. Then have each group turn to the Lesson Review for their lesson. Explain that the review provides an opportunity to determine how well they have understood the lesson content.

## Note These Features

Use the information on pages xii–xiii to identify features included in each chapter. As a class, locate examples of these features in Chapter 1. Read the examples and discuss their purpose.

Tell the students that each chapter has features that relate health to their life and to the world around them.

Notes in the margins extend the information presented in the text. For example, the note on page 8 of Chapter 1 gives useful information about guidelines for drinking the right amount of water.

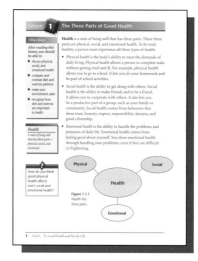

## Before Beginning Each Lesson

Read the lesson title and restate it in the form of a question. For example:

**Read:** Lesson 1 The Three Parts of Good Health
**Write:** What are the three parts of good health?

Look over the entire lesson, noting:
- objectives
- figures
- tables
- bold words
- text organization
- questions in the margins
- lesson review

### Also note these features:

- **Health in Your Life** — A description of how a health topic applies to you
- **Health at Work** — A health career and its requirements
- **Decide for Yourself** — A topic that allows you to practice health skills you have learned
- **Health Myth** — A correction of a false statement that many people think is true
- **Link to** — A subject such as math or literature related to the chapter topic
- **Technology and Society** — A technological advance related to the chapter topic

An exercise on researching and writing about a health topic

 **Note:** A short fact or tip about health, fitness, or nutrition

 **Application Question:** A question linking the text and your life

 A widespread real-life health problem and its possible solutions

## As You Read the Lesson

- Read the lesson headings. The subheads are questions.
- Read the paragraphs that follow to answer the question.
- Before moving on to the next heading, see if you can answer the question. If you cannot, reread the section to look for the answers.

## Using the Bold Words

**Wellness**
An active state of health

Knowing the meaning of all the boxed words in the left column will help you understand what you read. These words appear in **bold** type the first time they appear in the text and are defined in the paragraph.

**Wellness** is an active state of health.

All of the words in the left column are also defined in the Glossary.

**Wellness** (wel´ nis) An active state of health (p. 9)

Have a volunteer read the Application Question on page 4 of Chapter 1. Let students know that Application questions will help them relate what they learn in a lesson to their everyday life.

Have a volunteer read the title of the Health in the World feature on page 92. Ask students: How does this topic relate to the world around you? Emphasize how health is an important part of life. Point out to students that they will find a Health in the World feature at the end of each unit.

Note that all of these features have teaching suggestions in the Teacher's Edition text.

Now that students are familiar with the unit, chapter, and lesson structure of the book, distribute copies of "Finding Information" (How to Use This Book 4). Direct students to use their textbook to fill in the information on the worksheet. Point out that there are many possible answers for most of the features in the list, but for several features, there is only one correct answer. You might wish to remind students of a book feature that is particularly useful for completing this activity—the Table of Contents. When students have finished filling in the chart, ask volunteers to read some of their answers and have the other students check to see that the page numbers and titles are correct.

## As You Read the Lesson

Read aloud the statements in the section "As You Read the Lesson." Have students preview lessons in Chapter 1 and note lesson titles and subheads. Remind students as they study each lesson to follow this study approach.

## Using the Bold Words

Read aloud the information about how to use bold words. Make sure students understand what the term *bold* means. Explain to students that the words in bold are important vocabulary terms. Then ask them to look at the boxed words in Chapter 1. Have a volunteer read the boxed term *health* and then find and read the sentence in the text in which that word appears in bold type. Have another volunteer read the definition of the word in the box.

Point out that boxed words may appear on other pages in a lesson besides the first page. Explain that vocabulary terms appear in a box on the same page that they are used in the text.

Distribute copies of "Word Study" (How to Use This Book 5) to students. Suggest that as they read, students write unfamiliar words, their page numbers, and their definitions in the chart. Point out that having such a list will be very useful for reviewing vocabulary before taking a test.

## Taking Notes

Before reading the information in this section on page xiv, ask students why note-taking is an important study skill. Encourage them to share what method they use to take notes during class discussions or when reading. Ask them to explain why they use the method they do. Then have volunteers read the information page xiv. Suggest that students who do not have a method for taking notes try one of the methods mentioned and see how it works for them. Students can use How to Use This Book 7 or Graphic Organizer 5 to help them.

## Preparing for Tests

Encourage students to offer their opinions about tests and their ideas on test-taking strategies. Ask students: What do you do to study for a test? List their comments on the board. Then ask a volunteer to read the set of bulleted statements under "Preparing for Tests." Add these suggestions to the list on the board if they are not already there.

Discuss why each suggestion can help students when they are taking a test. Lead students to recognize that these suggestions, along with the Test-Taking Tips in their textbooks, can help them improve their test-taking skills.

## *Taking Notes*

Taking notes during class and as you read this book is helpful.

- Use headings to label the main sections of your notes.
- Summarize the important information, such as main ideas and supporting details.
- Use short phrases.
- Use your own words to describe, explain, or define things.
- Try taking notes using a three-column format. Use the first column to write headings or vocabulary words. Use the middle column to write the main information. Use the last column to draw diagrams, write questions, record homework assignments, or for other purposes.
- Right after taking notes, review them to fill in possible gaps.

| Vocabulary | Definition | Additional information |
|---|---|---|
| Health | A state of being that has three parts | The three parts of health are physical, social, and emotional |

## *Preparing for Tests*

The Summaries and Reviews for lessons, chapters, and units can help you prepare to take tests.

- Read the summaries from your text to make sure you understand the chapter's and unit's main ideas.
- Make up a sample test of items you think may be on the test. You may want to do this with a classmate and share your questions.
- Review your notes and test yourself on words and key ideas.
- Practice writing about some of the main ideas from the chapter or unit.
- Answer the questions under Vocabulary Review.
- Answer the questions under Concept Review.
- Write what you think about the questions under Critical Thinking.

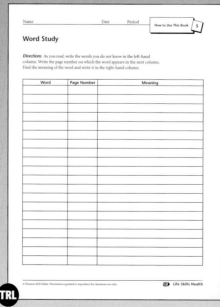

How to Use This Book 5

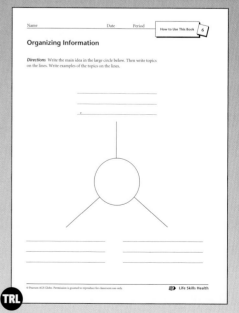

How to Use This Book 6

## Using the Test-Taking Tip

Read the Test-Taking Tip with each Chapter Review of the text.

**Test-Taking Tip:** After you have completed a test, reread each question and answer. Ask yourself: Have I answered the question completely?

## Using Graphic Organizers

A graphic organizer is a way to show information visually. It can help you see how ideas are related to each other. A graphic organizer can help you study for a test or organize information before you write. Here are some examples.

**Concept Map**

A concept map includes a main idea and related concepts. Each concept is written in a circle or box. The organization of concepts in the map shows how they are related.

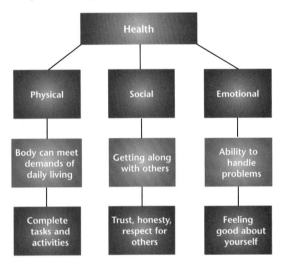

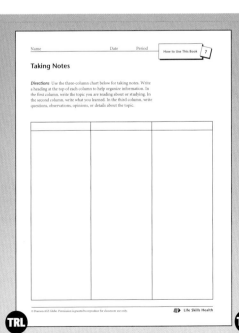
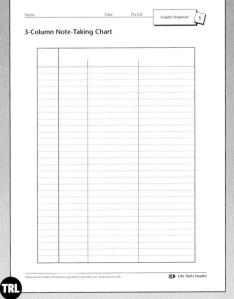

## Using the Test-Taking Tip

Have students find a Test-Taking Tip in the Chapter Reviews. Remind them that Chapter Reviews can be found at the end of each chapter. Ask several volunteers to read aloud the tips they find in the Chapter Reviews. Discuss how using the tips can help students study and take tests more effectively.

## Using Graphic Organizers

Explain to students that graphic organizers provide ways of visually organizing information to make it easier to understand and remember. Tell students that they can use a variety of organizers to record information for a variety of purposes. Encourage them to create graphic organizers to help them understand a concept in the text, compare several things, visualize how information is related, summarize information, and study for a test.

### Concept Maps

Concept maps help organize information such as main ideas and related details. Encourage students to study the concept map on page xv and write sentences based on it. Write an example on the board. *(One component of health is emotional health.)* Pass out "Organizing Information" (How to Use This Book 6). Explain to students that the graphic organizer shown is a type of concept map that represents the same information shown in the concept map on page xv of the student book. Have students turn to Chapter 1 and fill in details about physical, social, and emotional health for the example concept map. Students can use the blank concept map at the bottom of "Organizing Information" to help them organize the concepts in any lesson or chapter.

## Flowcharts

Tell students that flowcharts can help them visualize the order and number of steps in a particular process or procedure. Have students look at the example of the flowchart on page xvi. Ask volunteers to describe how this flowchart helps them understand the process of infection by a pathogen.

## Column Charts

Explain to students that charts can contain lists of information that include every detail that is known about the category. The information might show similarities and/or differences between each of the headings. Explain to students that it is best to include the same type of details in each column. For example, the description column of the genetic disorders column chart provides the same general type of description of each disorder. This type of organization provides an easy reference tool.

### Flowchart

A flowchart can be used to show the steps in a process. Flowcharts can be vertical or horizontal. Each step is shown in order in a box. Arrows connect the boxes to show how to get from one step to the next.

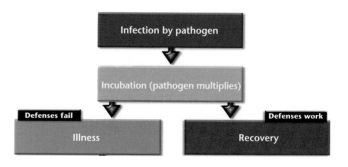

### Column Chart

A column chart or a table is a good way to record information and organize it into groups or categories. Each column has a heading. Under each heading, you can write details, questions, and other information. You can also use a column chart to take notes.

| Genetic Disorders and Their Causes | | |
|---|---|---|
| Disorder | Description | Cause |
| Hemophilia | Blood does not clot, or clump, normally | Recessive abnormal gene |
| Dwarfism | Long bones do not develop properly | Dominant abnormal gene |
| Cystic fibrosis | Abnormally thick mucus, constant respiratory infections | Two recessive genes |
| Sickle cell disease | Abnormally shaped red blood cells, weakness, irregular heart action | Two recessive genes |

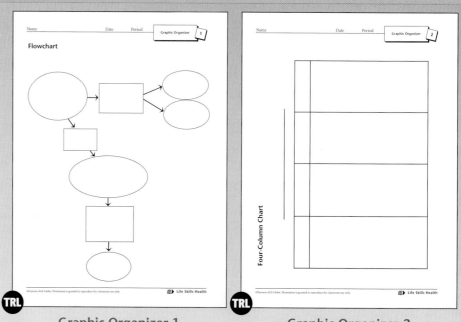

Graphic Organizer 1          Graphic Organizer 2

**Venn Diagram**

A Venn diagram is used to compare and contrast two ideas or objects. A Venn diagram has two circles of equal size that partially overlap. The circles represent the two ideas or objects being compared. Information about each idea or object is written in the circles. Things that the ideas or objects have in common are found where the circles intersect.

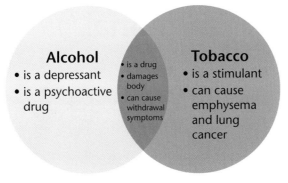

**Graph**

A graph is used to make comparisons and identify patterns among data. Graphs come in many forms such as line graphs, bar graphs, and circle graphs.

*Venn Diagrams*

Show students how a Venn diagram is useful for comparing and contrasting information. Draw a Venn diagram on the board. Explain how to use the diagram to compare and contrast two items, such as a ball and a globe. Discuss how the diagram on page xvii clearly shows the similarities and differences between alcohol and tobacco.

*Graphs*

Draw a sample circle graph, line graph, and bar graph depicting the same data on the board. Discuss with students how each type of graph provides a different visual representation of the same data. As you draw the graphs, remind students that the horizontal line is called the $x$-axis and the vertical line is called the $y$-axis.

Encourage students to refer back to the pages in this section, "How to Use This Book," as often as they wish while using this textbook.

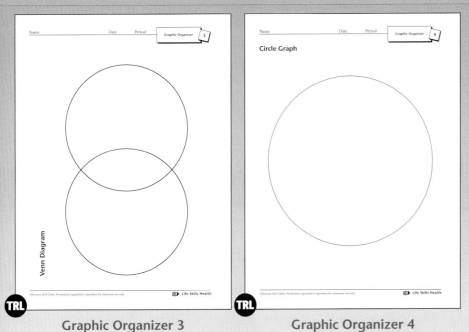

**Graphic Organizer 3**     **Graphic Organizer 4**

# Unit 1

## Planning Guide
## Mental and Emotional Health

| | Student Pages | Vocabulary | Health in the World | Review | Critical-Thinking Questions | Chapter Summary |
|---|---|---|---|---|---|---|
| **Chapter 1** Maintaining Health and Wellness | 2–25 | ✔ | | ✔ | ✔ | 23 |
| Lesson 1 The Three Parts of Good Health | 4–6 | ✔ | | ✔ | ✔ | |
| Lesson 2 Good Health and Wellness | 7–10 | ✔ | | ✔ | ✔ | |
| Lesson 3 Health Risks | 11–16 | ✔ | | ✔ | ✔ | |
| Lesson 4 Take Charge of Your Health | 17–22 | ✔ | | ✔ | ✔ | |
| **Chapter 2** Managing Emotions | 26–47 | ✔ | | ✔ | ✔ | 45 |
| Lesson 1 Your Emotions | 28–33 | ✔ | | ✔ | ✔ | |
| Lesson 2 Stress and Your Body | 34–37 | ✔ | | ✔ | ✔ | |
| Lesson 3 Managing Stress | 38–44 | ✔ | | ✔ | ✔ | |

*(Unit Planning Guide is continued on next page.)*

### Unit and Chapter Activities

**Student Text**
Unit 1 Summary

**Teacher's Resource Library**
Home Connection 1
Community Connections 1–4

**Teacher's Edition**
Opener Activity, Chapters 1–4

### Assessment Options

**Student Text**
Chapter Reviews, Chapters 1–4
Unit 1 Review

**Teacher's Resource Library**
Self-Assessment Activities 1–4
Chapter Mastery Tests A and B, Chapters 1–4
Unit 1 Mastery Tests A and B

**Teacher's Edition**
Chapter Alternative Assessments, Chapters 1–4

| | Student Text Features | | | | | | | | Teaching Strategies | | | | | | Learning Styles | | | | | Teacher's Resource Library | | | | |
|---|---|---|---|---|---|---|---|---|---|---|---|---|---|---|---|---|---|---|---|---|---|---|---|---|
| | Self-Assessment | Health at Work | Health in Your Life | Decide for Yourself | Link to | Research and Write | Health Myth | Technology and Society | Background Information | ELL/ESL Strategy | Health Journal | Applications (Home, Career, Community, Global, Environment) | Online Connection | Teacher Alert | Auditory/Verbal | Body/Kinesthetic | Interpersonal/Group Learning | Logical/Mathematical | Visual/Spatial | Activities | Alternative Activities | Workbook Activities | Self-Study Guide | Chapter Outline |
| | 2 | | | | | | | | | | | | 23 | 24 | | | | | | 1–4 | 1–4 | 1–4 | ✓ | ✓ |
| | | | | | 5 | 6 | | | 4 | | 6 | | | | | | | | 6 | 1 | 1 | 1 | | |
| | | | | | | 9 | 10 | 7 | 10 | 8 | 7 | | 8 | 7 | | | | | 2 | 2 | 2 | | |
| | | 12 | | | 12, 14 | | | | 11 | 15 | 16 | 12, 13 | | 14, 15 | | 16 | | | | 3 | 3 | 3 | | |
| | | | | | 20, 22 | | 17 | | 17 | 19 | 20 | | | 20 | | | 18 | 20 | | 4 | 4 | 4 | | |
| | 26 | | | | | | | | | | | | 45 | 46 | | | | | | 5–7 | 5–7 | 5–7 | ✓ | ✓ |
| | | | 32 | | | 31 | | | 28 | 32 | 31, 33 | 27, 29, 31 | | 32 | | | 29, 33 | | 28 | 5 | 5 | 5 | | |
| | | | | | 37 | | 35, 36 | 35 | 34 | 37 | 37 | | | 36 | | | | | | 6 | 6 | 6 | | |
| | | 43 | 44 | | 41 | 39 | | 38 | | 42 | 40, 42 | 43 | | 39, 43 | 40 | 39 | 41, 42 | 43 | | 7 | 7 | 7 | | |

## Alternative Activities

The Teacher's Resource Library (TRL) contains a set of lower-level worksheets called Alternative Activities. These worksheets cover the same content as the regular Activities but are written at a second-grade reading level.

## Skill Track

Use Skill Track for *Life Skills Health* to monitor student progress and meet the demands of adequate yearly progress (AYP). Make informed instructional decisions with individual student and class reports of lesson and chapter assessments. With immediate and ongoing feedback, students will also see what they have learned and where they need more reinforcement and practice.

# Unit 1

## Planning Guide
## Mental and Emotional Health *(continued)*

| | Student Pages | Vocabulary | Health in the World | Review | Critical-Thinking Questions | Chapter Summary |
|---|---|---|---|---|---|---|
| **Chapter 3** Maintaining Mental Health | 48–73 | ✔ | | ✔ | ✔ | 71 |
| Lesson 1 Influences on Mental Health | 50–54 | ✔ | | ✔ | ✔ | |
| Lesson 2 Emotional Health | 55–60 | ✔ | | ✔ | ✔ | |
| Lesson 3 Healthy Relationships | 61–65 | ✔ | | ✔ | ✔ | |
| Lesson 4 Becoming More Emotionally Healthy | 66–70 | ✔ | | ✔ | ✔ | |
| **Chapter 4** Recognizing Mental Health Problems | 74–91 | ✔ | | ✔ | ✔ | 89 |
| Lesson 1 Characteristics of Poor Mental Health | 76–79 | ✔ | | ✔ | ✔ | |
| Lesson 2 Mental Disorders | 80–84 | ✔ | | ✔ | ✔ | |
| Lesson 3 Treating Mental Disorders | 85–88 | ✔ | 92 | ✔ | ✔ | |

| Student Text Features | | | | | | | | Teaching Strategies | | | | | | Learning Styles | | | | | Teacher's Resource Library | | | | |
|---|---|---|---|---|---|---|---|---|---|---|---|---|---|---|---|---|---|---|---|---|---|---|---|
| Self-Assessment | Health at Work | Health in Your Life | Decide for Yourself | Link to | Research and Write | Health Myth | Technology and Society | Background Information | ELL/ESL Strategy | Health Journal | Applications (Home, Career, Community, Global, Environment) | Online Connection | Teacher Alert | Auditory/Verbal | Body/Kinesthetic | Interpersonal/Group Learning | Logical/Mathematical | Visual/Spatial | Activities | Alternative Activities | Workbook Activities | Self-Study Guide | Chapter Outline |
| 48 | | | | | | | | | | | | 71 | 72 | | | | | | 8–11 | 8–11 | 8–11 | ✔ | ✔ |
| | | 52 | | 51 | 53 | | | 50 | 52 | | 54 | | 53 | | | | | 53 | 8 | 8 | 8 | | |
| | | | 56 | | 55 | | | 55 | 57 | 57, 58 | 58 | | | | | 59 | 58 | 59 | 9 | 9 | 9 | | |
| | | | | 65 | 65 | | 64 | 61 | 62 | | 63, 64 | | | | 63 | | | | 10 | 10 | 10 | | |
| | 69 | | | 67 | | 70 | | 66 | 67 | | | | 66, 69 | 67 | | | | | 11 | 11 | 11 | | |
| 74 | | | | | | | | | | | | 89 | 90 | | | | | | 12–14 | 12–14 | 12–14 | ✔ | ✔ |
| | 78 | | | 79 | | | | 76 | 77 | 76 | 77, 78 | | 76 | 76 | 78 | 77 | | | 12 | 12 | 12 | | |
| | | | | 82 | 81, 82 | | 84 | 80 | 82 | 81 | 83, 84 | | 81 | | | | 83 | 83 | 13 | 13 | 13 | | |
| | | 86 | 88 | | | 86, 87 | | 85 | 87 | 86 | | | 88 | | | | | | 14 | 14 | 14 | | |

## Unit at a Glance

### Unit 1: Mental and Emotional Health
pages xviii–95

### Chapters

1. **Maintaining Health and Wellness**
   pages 2–25
2. **Managing Emotions**
   pages 26–47
3. **Maintaining Mental Health**
   pages 48–73
4. **Recognizing Mental Health Problems**
   pages 74–91

**Unit 1 Summary** page 93

**Unit 1 Review** pages 94–95

**Audio CD** 🎧

**Skill Track for Life Skills Health**

**Teacher's Resource Library** TRL

  Home Connection 1

  Unit 1 Mastery Tests A and B

  (Answer Keys for the Teacher's Resource Library begin on page 559 of this Teacher's Edition.)

## Other Resources

### Books for Teachers

American Medical Association. *American Medical Association Family Medical Guide.* Hoboken, NJ: John Wiley & Sons, 2004.

Kahn, Ada P. and Jan Fawcett. *The Encyclopedia of Mental Health.* New York: Facts on File, 2001.

### Books for Students

Bellenir, Karen. *Mental Health Information for Teens: Health Tips About Mental Health.* Detroit: Omnigraphics, 2001. (self-esteem, peer pressure, and common mental illnesses)

### CD-ROM/Software

*Health.* Bethesda, MD: Discovery Communications, 2000 (1-800-627-9399). (teen brains, healthy skin, hormones, nutrition)

### Videos and DVDs

*Eating Disorders* (22 minutes). Venice, CA: TMW Media Group, 2002 (1-800-262-8862).

*Teen Depression* (22 minutes). Venice, CA: TMW Media Group, 2002 (1-800-262-8862). (warning signs, getting help)

### Web Sites

http://win.niddk.nih.gov/publications/take_charge.htm (taking charge of one's health, teen health issues)

www.iub.edu/%7Ecafs/adol/adol.html (health and mental health issues, health risk factors, conflicts and violence,)

# Unit 1: Mental and Emotional Health

Who are you? When someone you have just met asks this question, how do you answer? A simple answer might include your name, age, and year in school. Would these answers really describe you? Think about who you are. Are you healthy mentally and emotionally? Who do you want to be in the future? How will you become this person?

The way you view yourself has a great effect on your life. Your beliefs about yourself affect what you do. They affect how healthy you are. These beliefs also play a major role in how you handle relationships. In this unit, you will learn about mental and emotional health. You will discover how to meet your goals. You will find methods to guide you in becoming the person you want to be.

### Chapters in Unit 1

Chapter 1: Maintaining Health and Wellness .................. 2
Chapter 2: Managing Emotions ..................................... 26
Chapter 3: Maintaining Mental Health ........................ 48
Chapter 4: Recognizing Mental Health Problems ......... 74

## Introducing the Unit

Have students read the text on page 1. Ask each student to make a list of phrases that describes the kind of person he or she wants to be. Ask students to classify the phrases into physical and mental/emotional characteristics. Discuss how one category affects the other. *(Sample answer: When you have good mental and emotional health, you feel good about yourself and want to exercise to stay fit. Exercising helps you feel good physically. It also makes you feel good about how you look, which improves your mental and emotional health. Exercise helps relieve stress, which improves mental and emotional health.)*

### HOME CONNECTION

The Home Connection unit activity gives students practical experience with concepts taught in the *Life Skills Health* student text. Students complete the Home Connection activity outside the classroom with the help of family members. These worksheets appear on the Life Skills Health Teacher's Resource Library (TRL) CD-ROM.

### CAREER INTEREST INVENTORY

The AGS Publishing Harrington-O'Shea Career Decision-Making System-Revised (CDM) may be used with the chapters in this unit. Students can use the CDM to explore their interests and identify careers. The CDM defines career areas that are indicated by students' responses on the inventory.

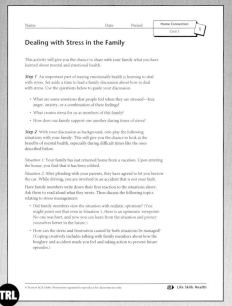

Home Connection 1

# Chapter at a Glance

## Chapter 1: Maintaining Health and Wellness
pages 2–25

## Lessons

1. The Three Parts of Good Health
   pages 4–6
2. Good Health and Wellness
   pages 7–10
3. Health Risks
   pages 11–16
4. Take Charge of Your Health
   pages 17–22

**Chapter 1 Summary** page 23

**Chapter 1 Review** pages 24–25

**Audio CD**

**Skill Track for Life Skills Health**

**Teacher's Resource Library**
  Activities 1–4
  Alternative Activities 1–4
  Workbook Activities 1–4
  Self-Assessment 1
  Community Connection 1
  Chapter 1 Self-Study Guide
  Chapter 1 Outline
  Chapter 1 Mastery Tests A and B
  (Answer Keys for the Teacher's Resource Library begin on page 559 of this Teacher's Edition.)

## Opener Activity

Ask students to think about what the word *health* means to them. Then have students define the word *health* in their own words. Ask volunteers to share their definitions and as a class come up with a single definition for the word *health*. After students read the chapter, have them review the definition. Students can revise the definition as needed.

2  Unit 1  Mental and Emotional Health

# Self-Assessment

## Can you answer these questions?

☑ 1. How would you compare and contrast physical, social, and emotional health?

☑ 2. Why is having physical, social, and emotional health important?

☑ 3. What are some of your short-term health goals?

☑ 4. What are some of your long-term health goals?

☑ 5. How does the environment affect your health?

☑ 6. How do cultural differences create health challenges?

☑ 7. How does heredity affect health risks?

☑ 8. How can you change your behavior to improve your health?

☑ 9. Why are diet and exercise important to your health?

☑ 10. How can you learn more about improving your health?

Self-Assessment 1       Community Connection 1

# Chapter 1
# Maintaining Health and Wellness

Health is important to everything in life. Without good health, reaching your goals is hard. Living your daily life happily and productively is difficult. Some people think that being healthy means simply being without an illness. Being healthy, though, means much more than this.

This chapter will guide you as you get to know more about health. You will discover what having good health means. You will also discover how you can apply this information to your own life.

## Goals for Learning

- ◆ To describe physical, social, and emotional health
- ◆ To recognize principles of good health
- ◆ To identify causes of health risks
- ◆ To evaluate and recognize the power to change personal health

3

Chapter 1 Self-Study Guide, pages 1–2

Chapter 1 Outline, pages 1–2

## Introducing the Chapter

Make a large three-column table on the board with column titles "Type of Health," "Definition," and "How to Stay Healthy." Under "Type of Health" make three rows titled "Physical Health," "Social Health," and "Emotional Health." Explain to students that in Chapter 1 they will learn about the three types or parts of health. Have them brainstorm to provide a definition for each type and then fill in health behaviors they can follow to stay healthy for each one. After students have completed the chapter, have them review the table. Have them make revisions and additions to the table where needed based on their new knowledge.

## Notes and Questions

Ask volunteers to read the notes and questions that appear in the margins throughout the chapter. Then discuss them with the class.

## Self-Assessment

Have students complete the Self-Assessment worksheet before and after reading the chapter. Before reading the chapter, have students fill in the "Before" column. Ask students to identify their goals for learning. To get ideas for setting goals, students might use the chapter introductory material on page 3, the checklist on page 2, or the questions on the Self-Assessment worksheet. Students can use the back of the worksheet if they need more space to write.

Collect the Self-Assessment worksheets and pass them out again at the end of the chapter. Have students fill in the "After" column. Ask them to identify at least four major points they have learned. Again, suggest they use the back of the worksheet if they need more space to write. You may want to collect and review the worksheets, but return them to students so they have a record of their goals and accomplishments.

*Maintaining Health & Wellness* Chapter 1   3

# Lesson at a Glance

## Chapter 1 Lesson 1

**Overview** This lesson describes the three parts of health—physical health, social health, and emotional health.

### Objectives
- To discuss physical, social, and emotional health
- To compare and contrast diet and exercise patterns
- To make your environment safer
- To recognize how diet and exercise are important to health

**Student Pages** 4–6

**Teacher's Resource Library**
- Activity 1
- Alternative Activity 1
- Workbook Activity 1

## Vocabulary

**health**

Read and discuss the definition of the word *health*. Ask a volunteer to look up the word in a dictionary. Have students write a sentence using the word in context.

## Background Information

Lead poisoning is a serious condition that can affect both physical and social health. Symptoms of lead poisoning include reduced energy and appetite, headaches, sleeping problems, and anemia. Children can be exposed to lead in paint on walls, toys, and furniture, in soil, and in drinking water.

 **Warm-Up Activity**

Ask for two volunteers to act out a skit. Away from the rest of the class, explain that the two students should act out a situation in which the first student wants the second student to give up plans to clean the garage so they can go on a bike ride together. After the skit, ask the class to identify which type of health—physical, social, or emotional—each student was working on.

---

## Lesson 1 — The Three Parts of Good Health

### Objectives
After reading this lesson, you should be able to
- discuss physical, social, and emotional health
- compare and contrast diet and exercise patterns
- make your environment safer
- recognize how diet and exercise are important to health

**Health**
*A state of being well that has three parts—physical, social, and emotional*

How do you think good physical health affects one's social and emotional health?

**Health** is a state of being well that has three parts. These three parts are physical, social, and emotional health. To be truly healthy, a person must experience all three types of health.

- *Physical health* is the body's ability to meet the demands of daily living. Physical health allows a person to complete tasks without getting tired and ill. For example, physical health allows you to go to school. It lets you do your homework and be part of school activities.

- *Social health* is the ability to get along with others. Social health is the ability to make friends and to be a friend. It allows you to cooperate with others. It also lets you be a productive part of a group, such as your family or community. Social health comes from behaviors that show trust, honesty, respect, responsibility, fairness, and good citizenship.

- *Emotional health* is the ability to handle the problems and pressures of daily life. Emotional health comes from feeling good about yourself. You show emotional health through handling your problems, even if they are difficult or frightening.

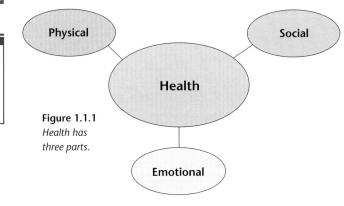

**Figure 1.1.1** Health has three parts.

Unit 1 *Personal Health and Family Life*

Exercise and healthy eating go hand in hand. Making healthy food choices gives you energy to exercise. Exercising releases chemicals in your brain that reduce appetite and make you feel happy. Then you are less likely to overeat. Overeating can make you feel tired. Because you're not overeating, you have more energy to exercise. What a positive circle!

The physical, social, and emotional parts of health all work together. If one part gets weak, the other parts may weaken, too. For example, you might have trouble getting along with others if you are sick. In a similar way, problems you may have can affect how you feel physically. For example, you may feel extra tired when you have stress at home or at school. You may notice you are more likely to get upset when you are worried about something.

### Health in Your Life

**Your Environment**

Your environment, or surroundings, can affect your health. You can have a big effect on the things around you, like air, land, water, plants, and animals. Copy the list on the right. Use it to improve your community's environment. Add to it as you learn more about ways to make your environment safer.

On a sheet of paper, write the answers to the following questions. Use complete sentences.

1. How can your family members help you apply these ideas?
2. How can you influence your community to make environmental changes?
3. Why should you add to your list?

- Do not leave water in containers outside your home. Standing water creates a breeding ground for mosquitoes. Mosquitoes can give people diseases.
- Do not use lead-based paint. Young children can eat paint chips and be poisoned.
- Store gasoline for your lawn mower in safe, approved containers. Gasoline in unsafe containers can cause fire dangers.
- Do not smoke. Smoking creates health risks for you and those around you.
- Do not dump used motor oil into sewers, lakes, or streams. It can kill wildlife and leave behind harmful chemicals.

*Maintaining Health and Wellness* Chapter 1

 **Teaching the Lesson**

To help students understand that a balance must be maintained between the three parts of health, draw a triangle with three equal sides on the chalkboard. Label each corner with one type of health (physical, social, and emotional). After you read the scenarios below, have volunteers draw a new triangle that represents how health may be out of balance in each scenario.

Scenario 1: A person spends all of her free time training for a marathon.

Scenario 2: A friend begins dating someone and spends all of his free time with that person.

Scenario 3: A classmate has just lost a parent. He spends days in bed with the shades drawn.

### Reinforce and Extend

### Health in Your Life

Have students read and complete the Health in Your Life feature on page 5. Ask students to describe which things on the list they currently practice. Have students brainstorm changes they can make in their lives to start these practices.

**Health in Your Life Answers**

1. Answers will vary but should include content from the checklist and references to the actions of the student and family members. 2. Answers will vary but should clearly reference specific environmental issues in the community, as well as actions the student can take to influence positive change. 3. Answers will vary but should show an awareness of the importance of specific actions to improve the environment of the home and community.

# Lesson 1 Review Answers

1. Health is a state of being well that has three parts. 2. The three types of health required are physical, social, and emotional health. 3. Making friends is part of social health. 4. The ability to handle a breakup is part of emotional health. 5. Physical health allows a person to exercise regularly.

## Portfolio Assessment

Sample items include:
- Answers from Health in Your Life
- Posters from Learning Styles/Visual Spatial
- Lesson 1 Review Answers

## Decide for Yourself

Have students read and complete the Decide for Yourself feature on page 6. Have students make a similar chart for their own daily health behaviors in regard to eating and exercise habits. As they read the rest of the chapter, have students evaluate whether they think the behaviors are healthy or not.

### Decide for Yourself Answers

1. Sample answer: Ahmed is living a healthier lifestyle. 2. Answers will vary, based on the answers to question 1.

## LEARNING STYLES

### Visual/Spatial

Have students make a poster in which they illustrate an example of a person participating in an activity that would improve each type of his or her health—physical health, social health, and emotional health.

## ELL/ESL STRATEGY

### Language Objective:
To reword content

Have students orally give a definition of physical health, social health, and emotional health in their own words. Encourage students to use examples from their own lives to help explain what the terms mean.

---

*Lesson 1 Review* On a sheet of paper, write the answers to the following questions. Use complete sentences.

1. What is health?
2. What are the three types of health required for someone to be truly healthy?
3. Which type of health includes making friends?
4. **Critical Thinking** Which type of health includes the ability to handle the breakup of a relationship?
5. **Critical Thinking** Which type of health allows a person to exercise regularly?

### Decide for Yourself

As you set your own health goals, think about choices that work well for others. Ask yourself why their food and exercise choices work well for them. Then think about whether these choices would work for you—and why or why not. Developing healthy eating patterns should be part of your lifestyle. You should make healthy food choices every day.

Remember that exercise and healthy eating are important for many reasons. Interest in losing or gaining weight is only one of them. You will learn about many more reasons while reading this chapter and later chapters in the book.

On a sheet of paper, write the answers to the following questions. Use complete sentences.

1. Compare the eating patterns and exercise habits of the two students below. Which student do you think is living a healthier lifestyle? Explain your answer.
2. After you read the rest of the chapter, come back to this page. Has your answer to question 1 changed or remained the same?

| Inez | Ahmed |
|---|---|
| eats three meals each day | eats three meals each day |
| eats no snacks | eats two fruit snacks each day |
| runs two fast miles once a week | walks five blocks four times a week |

# Lesson 2: Good Health and Wellness

### Objectives
After reading this lesson, you should be able to
- identify ways to stay healthy
- recognize wellness
- understand how to seek information about diet and exercise goals

**Disease**
A condition that harms or stops the normal working of the body

## How Is Good Health Achieved?

It is important to take action to achieve good health. Taking the right steps will offer many benefits. You can reach good physical, social, and emotional health, as well as prevent **disease.** Disease is a condition that harms or stops the normal working of the body.

Achieving good health is like playing a musical instrument well. It requires the practice of basic principles. For example, to achieve physical health, you must apply good judgment and self-control in diet and exercise. Then you can give your body the food, activity, rest, and other care it needs. To achieve social health, you need to practice making and keeping good friends. To achieve emotional health, you need to practice handling your feelings in acceptable ways.

**Figure 1.2.1** *To be socially healthy, you need good friendships.*

---

## Lesson at a Glance

### Chapter 1 Lesson 2

**Overview** This lesson explains steps that can be taken every day to stay healthy and well.

### Objectives
- To identify ways to stay healthy
- To recognize wellness
- To understand how to seek information about diet and exercise goals

**Student Pages** 7–10

**Teacher's Resource Library**
  Activity 2
  Alternative Activity 2
  Workbook Activity 2

### Vocabulary

| | |
|---|---|
| alcohol | drug |
| disease | wellness |

Read the definition of each word to the class. Ask students to write a paragraph using all of the terms to explain how alcohol and other drugs may affect wellness.

### Background Information

Alcohol can affect health in several ways. Immediate effects of alcohol include a depression of the central nervous system. This can lead to an increase in reaction time and a decrease in the ability to think clearly. Long-term abuse of alcohol can cause liver disease and increases the risk of developing certain cancers, including liver and colon cancer. Alcohol interferes with the absorption of some vitamins; therefore, long-term use can result in vitamin deficiencies.

**Warm-Up Activity**

Ask students to give examples of things people can do to achieve good health. Encourage students to list things that can be done every day, every week, or every year (e.g., doctor/dental check up). Be sure to have students give examples for social and emotional health as well as physical health.

 **Teaching the Lesson**

Help students learn the eight steps to staying healthy by making a two-column table. The first column should be titled "Steps to Stay Healthy." The second column should be titled "How to Practice." Have students fill in the table with examples of how they can carry out each step in their daily lives.

Help students become more familiar with the Wellness Scale by having them make their own chart showing the Wellness Scale. Encourage them to be creative by making their own color code and drawing a scenario that represents each level on the scale.

 **Reinforce and Extend**

### AT HOME

Encourage students to discuss the eight steps to staying healthy with their family members. They should come up with a plan with the help of their family to follow the steps. All family members should be able to participate in the plan.

### LEARNING STYLES

 **Auditory/Verbal**
Have students write a song, a jingle, or a poem that incorporates all of the steps to staying healthy. Invite students to share their work with the class.

*Alcohol*
A drug that slows down the central nervous system

*Drug*
A chemical substance other than food that changes the way the mind and body work

There is no "right" amount of water everyone needs daily. Needs can differ for reasons such as illness, environment, and exercise level. A good rule is to drink eight 8-ounce glasses daily. Drink more in hot weather or if you are exercising hard. What if you don't like to drink water? You might be able to substitute other fluids. Ask your health care professional for suggestions.

## How Can You Stay Healthy?

You can control many ways of achieving and maintaining good health. By practicing healthy behaviors, you can help increase the quality and length of your life. What are some of these healthy behaviors?

**Eight Basic Steps to Staying Healthy**

1. Exercise three to five times during the week.
2. Eat three balanced, healthy meals each day, including breakfast.
3. Choose to eat healthy snacks, such as fruits, vegetables, and low-fat yogurt.
4. Drink enough water every day.
5. Maintain a healthy weight for your height and age.
6. Sleep at least eight hours each night.
7. Do not smoke, drink **alcohol,** or use other **drugs** your health care professional has not recommended.
8. Seek medical and dental care and advice from qualified health care professionals.

All of these behaviors can help you prevent health problems and maintain good health.

**Figure 1.2.2** *Exercise three to five times during the week to stay healthy.*

| Wellness |
| --- |
| *An active state of health* |

## What Is Wellness?

**Wellness** is an active state of health. To achieve wellness, a person moves toward balancing physical, social, and emotional health. The Wellness Scale shows how to chart a person's level of wellness. At the far right of the scale, the chart shows an excellent state of physical, social, and emotional health. The middle indicates average health. At the far left, the chart indicates very poor health and even death.

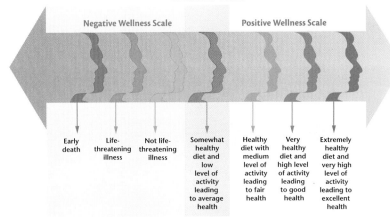

**Figure 1.2.3** *The Wellness Scale*

As you read on, you will learn even more about achieving good health and wellness. As you read, consider your own health. Think about how it relates to the three parts of health. Where do you stand on the Wellness Scale? Then identify ways you can achieve and maintain good health. Develop ideas to help others do the same.

### Health Myth

**Myth:** People should not eat carbohydrates.

**Fact:** People need carbohydrates. Healthy carbohydrates are an important part of a balanced diet. They give you energy. Some foods that contain carbohydrates are healthier than others. For example, sugar is not as healthy as oatmeal or whole-grain pasta.

## TEACHER ALERT

Remind students that the Wellness Scale is a continuum. People may move from one level to another during different times of their lives. Have students study the scale. Ask them to give examples of some serious illnesses that may place someone in the negative part of the scale. Ask them to describe how much time they think people spend exercising each day at a low level, high level, and very high level of activity.

## HEALTH JOURNAL

Have students make a list of things they do every day to maintain good health and wellness. Students should use the eight steps to staying healthy as a guide. Students should also include changes they could make in their health behaviors that would improve their health.

### Health Myth

Have students read the Health Myth feature on page 9. Discuss the issue that some popular diets call for people to cut out carbohydrates from their diets.

Carbohydrates provide the brain and nervous system with energy. Carbohydrates, such as whole grains, are also a source of fiber for the body. The recommendation of the National Institutes of Health is that people get between 40–60 percent of their daily calories from carbohydrates.

## Lesson 2 Review Answers

1. wellness 2. illness 3. behaviors
4. The basketball player is getting enough exercise. Reasoning will vary but should reflect the idea that staying healthy means exercising three to five times each week. 5. Juanita will need energy. Answers will vary as to the foods she should and should not eat; however, responses should show knowledge that foods high in sugar and fat would not be good choices, and foods such as fruits, vegetables, and low-fat yogurt would be good choices.

### Portfolio Assessment

Sample items include:
- Songs or poems from Learning Styles Auditory/Verbal
- List from Health Journal
- Lesson 2 Review answers

### Technology and Society

Have students read the Technology and Society feature on page 10. Have students carry out a search on the Internet in class using the search words suggested in the feature. Have them identify a reputable Web site and summarize the information they found.

### ELL/ESL Strategy

**Language Objective:** *To practice note-taking skills*

Have students practice their note-taking skills by making an outline of the lesson using the bold heads as outline entries. For each entry, students should fill in at least two details that help describe or define the entry.

---

**Word Bank**
behaviors
illness
wellness

**Lesson 2 Review** On a sheet of paper, write the word from the Word Bank that best completes each sentence.

1. Improving and balancing physical, social, and emotional health helps a person achieve _____.
2. Taking steps to achieve good physical, social, and emotional health helps you prevent _____.
3. Practicing healthy _____ helps you live a longer and better life.

On a sheet of paper, write the answers to the following questions. Use complete sentences.

4. **Critical Thinking** Suppose a student plays basketball six days a week. Is the student getting enough exercise to stay healthy? Explain your answer.
5. **Critical Thinking** Juanita is getting ready to go to soccer practice. She is choosing her snack. Name two foods that would be good choices. Name two foods that would not be good choices. Explain your answers.

### Technology and Society

What is a healthy weight for you? How much should you exercise? Web sites are waiting with answers. Use an Internet search engine. Sites ending in *.gov*, *.edu*, or *.org* are usually the most trustworthy. Evaluate information carefully. Enter terms such as *weight range*, *exercise*, or *physical exercise*. Be sure to discuss personal needs with your health care professional before making major changes. Remember, everyone's health needs are different.

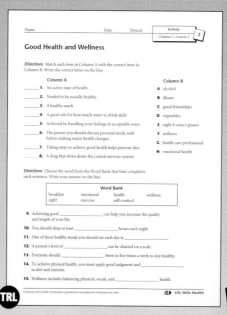

Activity 2

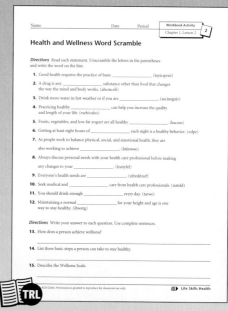

Workbook Activity 2

# Lesson 3: Health Risks

### Objectives
*After reading this lesson, you should be able to*
- understand what a health risk is
- identify causes of health risks related to heredity and environment
- recognize causes of health risks related to culture and behavior
- develop the power to make changes that will improve health

**Health risk**
*An action or a condition that may contribute to injury or disease*

**Heredity**
*Passage of physical characteristics from parents to children*

**Physical environment**
*The area around a person*

## What Is a Health Risk?
A **health risk** is an action or a condition that may contribute to injury or disease. Some causes of these risks include:

- heredity
- environment, both physical and social
- culture
- behavior, including drug use and sexual activity

## How Does Heredity Cause Risk?
**Heredity** is the passage of physical characteristics from parents to children. Heredity can be a cause of health risks. It can make a person more likely to get certain diseases. It can make a person unable to fight a disease. For example, some people are born with a likelihood they will get heart disease. They cannot change their heredity. They can, though, change their lifestyle to make becoming ill less likely.

## How Does Environment Cause Risk?
Environment can also create health risks. The **physical environment,** or the area around a person, can hold a number of risks. The physical environment includes air, water, land, plants, animals, and buildings. How might the physical environment cause a health risk? A young child might live in a home that has been painted with lead-based paint. The child might eat small bits of paint that have chipped off the walls. The lead could poison the child and cause serious illness or death.

Other environmental health risks also exist. For example, a nonsmoker might get lung disease from breathing in another person's cigarette smoke. Still another environmental health risk can come from poor air quality caused by harmful substances that factories release into the air.

*Maintaining Health and Wellness* Chapter 1  11

---

## Lesson at a Glance

### Chapter 1 Lesson 3

**Overview** This lesson describes different health risks and their causes.

### Objectives
- To understand what a health risk is
- To identify causes of health risks related to heredity and environment
- To recognize causes of health risks related to culture and behavior
- To develop the power to make changes that will improve health

**Student Pages** 11–16

**Teacher's Resource Library**
Activity 3
Alternative Activity 3
Workbook Activity 3

| | |
|---|---|
| abstain | nonverbal |
| culture | physical environment |
| deteriorate | secondhand smoke |
| frustrate | sedentary |
| health risk | social environment |
| heredity | tobacco |
| impair | |

Have students work in pairs to create a word-find puzzle using all of the vocabulary terms. Have groups exchange the word finds and solve them.

### Background Information
Smoking tobacco can increase a person's risk of developing heart disease including high blood pressure and angina; cancers including lung, mouth, esophagus, bladder, and kidney; and lung diseases including emphysema and bronchitis. People who are regularly exposed to secondhand smoke have an increased risk of developing lung cancer. Children who inhale secondhand smoke have higher rates of respiratory infections and asthma.

*Maintaining Health & Wellness* Chapter 1  11

### Warm-Up Activity

Review the definition of a health risk. Ask students to make a list of things they feel are health risks. As students read the chapter, have them add to the list.

### Teaching the Lesson

To help students organize the material in this lesson, have them make a web diagram for each of the three main types of health risks—heredity, environment, and behavior. For each web diagram, students should make branches that include examples of each type of risk.

### Reinforce and Extend

#### Research and Write

Have students read and complete the Research and Write feature on page 12. Have students work in pairs to complete the assignment. Provide students with a sample business letter so they can model their own letters after the format. Have students share the results of their work with the rest of the class.

#### CAREER CONNECTION

Have students pretend that they are environmental protection specialists. Have them make a list of things they would inspect in their communities such as local lake, pond, or river water, or air quality. Have them give reasons why they chose these things to inspect.

#### Health at Work

Have students read the Health at Work feature on page 12. Environmental protection specialists may work at the county, state, or federal level of government. The job may include going into the field to take samples of water or air.

---

**Social environment**
*Community resources such as access to doctors, hospitals, family counseling, and after-school programs*

The **social environment** is made up of community resources. A poor social environment can cause health risks. For example, people might not have safe neighborhoods or access to healthy food. Finding jobs that pay a fair wage might be difficult. Some communities do not offer good health care or other important social services. People who are ill may not have enough money to pay a doctor. If the community does not offer health care to people who cannot pay for it, this is a health risk.

Risks in the social environment can be lowered. Community education and attention to the needs of all people can help. Community leaders can work together with police officers to make a neighborhood safer. Elected officials and doctors can help communities set up neighborhood clinics that provide health care.

#### Research and Write

Research an environmental challenge in your community. Think about the quality of air, water, and land as you research. E-mail city council members or other elected local officials for additional information. Then write a letter to your local newspaper. Suggest ways to help your community deal with the challenge.

#### Health at Work

**Environmental Protection Specialist**

The Environmental Protection Agency (EPA) is a government agency that works to protect human health and the environment. Environmental protection specialists work for the EPA. They check the quality and safety of water, air, land, and other environmental features. The specialists work to make sure that environmental laws and procedures are being followed.

The EPA requires at least a four-year college degree of most of its environmental protection specialist jobs. A postgraduate degree, or education after college, is necessary for some positions.

| Culture |
|---|
| Beliefs, way of living, and language of a group of people |
| **Frustrate** |
| To block from something wanted |
| **Nonverbal** |
| Unspoken |

## How Does Culture Cause Risk?

A **culture** is made up of the beliefs, way of living, and language of a group of people. People's culture can affect how they receive, understand, and use health information. Imagine you have recently moved to another country and are ill. You go to the doctor, but the doctor does not speak your language. You try to tell the doctor what is wrong. She might not understand what you are trying to explain. She might give you information written in a language you do not understand.

Additional problems may arise when sick people cannot understand where to go to get health care. Even if they know where to go, they might have problems inside the health care office. Perhaps they cannot speak or read the language on the signs in the office. They might become scared or **frustrated,** or blocked from getting what they want. They might leave the office before getting the care they need. This causes a health risk to them. It can also cause a risk to others who might catch their untreated illness.

Suppose a health care provider's belief system and way of doing things differ from the patient's. What if a health care provider does not understand what is often done and what is never done in a specific culture? These situations can cause risks. Even the differences in facial expressions and other **nonverbal,** or unspoken, communication can cause risks. For example, in Bulgaria, nodding the head up and down means "no." Shaking the head from side to side means "yes."

Here are some ways cultural risks can be decreased:

- Communities can offer health care providers who speak the languages and understand the cultures of the people they serve.

- Directions and information can be printed in more than one language.

- Information can be made available in many ways. Possible ways to communicate information include TV, radio, newspapers, Web sites, and bulletin boards at community centers, sports arenas, schools, and churches.

*Maintaining Health and Wellness* Chapter 1  13

## IN THE ENVIRONMENT

Have students evaluate the safety of their own environment—at school, at home, and in the community. Students should focus on the physical environment as well as the social environment. Tell students they do not need to share their evaluations if they do not feel comfortable doing so.

## IN THE COMMUNITY

Have students make a flyer for a community bulletin board that advertises a medical clinic that serves the needs of a multicultural community. Students should assume that there are people in the community who speak a language other than English as their first language. Students should also assume that people in the community came to the United States from other countries and may have a variety of cultural beliefs.

## GLOBAL CONNECTION

Have students research how medical care systems works in a country other than the United States. Students should answer questions such as "How do people get to see doctors?" "How is medication given out?" and "How do people make payments?" You may choose to assign countries so there is no overlap. Have students present the results of their research to the class.

### Research and Write

Have students read the Research and Write feature on page 14. Ask them to make a list of questions they need to have answered before they can make their brochure. Have them make the brochure as if their community were a model for other communities interested in solving their own challenges to health care.

#### TEACHER ALERT

Spend time discussing the meaning of *sedentary* with students. Ask students to describe a sedentary lifestyle. Have students provide examples of activities that are sedentary. (*sitting in front of the television, playing video or computer games, working at the computer*) Ask students to identify which activities they do that are sedentary. Have them rate their lifestyles as sedentary or active and explain why.

---

**Tobacco**
*Dried leaves from certain plants that are used for smoking or chewing*

**Sedentary**
*Including little or no exercise*

### Research and Write

Work with a partner to prepare a brochure. List cultural challenges to health care in your own community or one nearby. For each challenge, list possible solutions. Use a variety of research sources, including local newspapers and the Internet. If possible, conduct interviews of local health care professionals, neighbors, and city officials.

## What Are the Causes of Behavioral Risks?

You know you cannot change your heredity. You may not be able to change every risk in your environment. Still, you can make changes to improve your health and the health of others. You can do this through understanding the causes of behavioral risks. Then you can change the action that causes the risks. You can greatly decrease your risks by living a healthy lifestyle.

### Examples of Behavioral Risks
- making poor food choices
- using poisonous substances carelessly
- riding without a seat belt
- failing to observe safety rules in the home
- failing to observe safety rules during sports activities
- engaging in unsafe sexual activity
- using **tobacco** (smoking or chewing), alcohol, and other drugs not advised by your health care professional
- having a **sedentary** lifestyle, a lifestyle including little or no exercise, which can cause sadness, weight gain, and an inability to fight disease

You have learned a great deal about different types of health risks. Read on to learn more about the causes of behavioral risks.

### Alcohol

Alcohol is a drug which affects every major system in the body. Drinking alcohol can cause dizziness and poor driving ability. Alcohol use also can cause problems with the senses and memory. Some dangers of long-term alcohol use include disease and damage to body parts such as the heart, liver, stomach, colon, and mouth.

**Secondhand smoke**
Smoke from a cigarette or cigar someone else is smoking

**Deteriorate**
Wear away a little bit at a time

**Abstain**
Choose not to use or do something

**Impair**
Have a negative effect on

### Tobacco

Tobacco use also has harmful effects. It can increase heart rate and blood pressure to very dangerous levels. Tobacco use causes a buildup of material in the blood vessels or lungs. This buildup can result in life-threatening diseases. Smoking also can cause serious coughing and diseases that interfere with breathing.

Smokers do not cause damage only to themselves. Nonsmokers can be injured through **secondhand smoke,** smoke from a cigarette or cigar someone else is smoking. Secondhand smoke can give a nonsmoker the same kinds of health problems the smoker has.

### Other Drugs

Taking drugs that have not been advised by your health care professional can cause many serious health problems. Using the wrong drugs can even lead to death. Drugs can cause heartbeat and breathing to speed up or slow down. They can cause body temperature to rise or fall. They can cause body organs to **deteriorate,** or wear away a little bit at a time. Drugs can cause people to feel confusion and panic. This can lead people to make unhealthy decisions. Some drugs also can cause dependence. A person using the drugs may not be able to stop.

Never take a drug that has been prescribed for someone else. Even legal, over-the-counter drugs can cause serious health effects if taken in large amounts or used incorrectly.

You will learn more about the effects of alcohol, tobacco, and other drugs in Chapters 14 and 15.

### Failure to Abstain

Failure to **abstain** from, or choose not to have, a sexual relationship can result in unexpected pregnancy. This can have serious effects on the mother, father, their families, and the child. Failure to abstain can also result in diseases.

Using alcohol or other drugs can **impair,** or have a negative effect on, a person's judgment. Alcohol or other drugs might cause a person to do something dangerous. The person's judgment is impaired.

Why is it dangerous to take a drug that has been prescribed for someone else?

## TEACHER ALERT

Remind students that many prescribed drugs have side effects. If drugs are taken that are not prescribed for them, they risk having a reaction to the drug or having the drug interact with any other medications they may be taking.

## ELL/ESL STRATEGY

**Language Objective:** *To practice pronunciation*

Pair students learning English with those who are already proficient in English. Have them use the dictionary to look up the pronunciation of the vocabulary words in this lesson. They can make index cards for each word with the word on the front and the phonetic spelling on the back. Students can take turns practicing how to pronounce the words correctly.

## Lesson 3 Review Answers

1. D 2. C 3. A 4. Nonverbal communication is unspoken communication. Examples will vary but might include smiling, laughing, or frowning. 5. Answers will vary but should include a health risk in the home and a viable solution.

### Portfolio Assessment

Sample items include:
- Flyer from In the Community
- Letter from Research and Write on page 12
- Brochure from Research and Write on page 14
- Lesson 3 Review answers

### HEALTH JOURNAL

Have students evaluate their own health behaviors against the health risks covered in this lesson. Have them explain what changes they can make to reduce risks to their health.

### LEARNING STYLES

#### Body/Kinesthetic

Have students work in groups to create a skit that depicts examples of positive and negative health behaviors. After each group performs its skit, have the class identify the health behaviors shown in the skit and classify them as positive or negative.

---

### Failure to Observe Safety Rules in the Home

Specific actions you take or do not take in the home also can cause health risks. Keep your home in good repair. For example, fix loose carpet on stairs. Clean up spills right away. Do not use something inappropriate, such as boxes, in place of a ladder. Make sure appliances are correctly wired and electrical outlets are not overloaded.

Clean chimneys and place screens in front of fireplaces when fires are burning. Keep fire extinguishers handy and install smoke detectors and carbon monoxide detectors. These examples represent only a small number of ways to observe safety rules and ways to be safer.

**Lesson 3 Review** On a sheet of paper, write the letter of the answer that best completes each sentence.

1. Alcohol is a drug that affects the brain and _____.
   A speeds up heart rate
   B increases blood pressure
   C solves emotional problems
   D slows down the central nervous system

2. The physical environment includes all of the following EXCEPT _____.
   A animals    C language
   B air        D land

3. A sedentary lifestyle is a lifestyle with little _____.
   A exercise   C risk
   B food       D water

On a sheet of paper, write the answers to the following questions. Use complete sentences.

4. **Critical Thinking** What is nonverbal communication? Provide two examples.

5. **Critical Thinking** What is a health risk you have seen in your home or someone else's home? What solutions can you offer?

16  Unit 1  Personal Health and Family Life

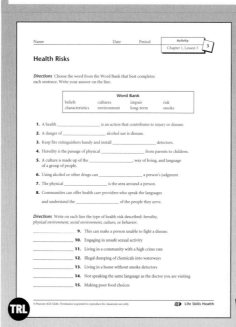

Activity 3

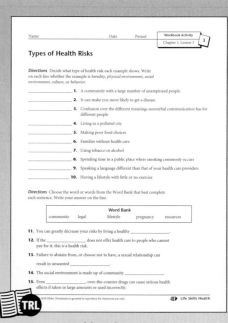

Workbook Activity 3

## Lesson 4: Take Charge of Your Health

**Objectives**

*After reading this lesson, you should be able to*

- distinguish between long-term and short-term goals
- develop health literacy
- solve health problems and change behaviors
- evaluate your own health

You have learned there are risks to health. You know you can control some, but not all, of the causes of these risks. What if you cannot control a cause? You still have the power to make choices that will lessen the risks. You can set goals and change your behaviors by using many "tools." These tools include awareness, knowledge, decision making, and application of skills. Making positive decisions and setting goals promote good health throughout your life.

### How Do You Set Goals?

As you think about promoting good health, think about your short-term and long-term goals. A short-term goal is one you can reasonably accomplish quickly. For example, you can choose to eat healthy foods and to exercise regularly during upcoming weeks. A long-term goal is one you accomplish over a long period of time. For example, you can plan to arrive at your desired weight within the upcoming year.

Setting goals is important. It gives you a positive direction. When you set goals, you figure out what you want to accomplish. Then you can figure out how to move forward to reach the goals. When you set goals, you can develop daily steps to take toward a healthy lifestyle. You can control your life and work to make yourself healthy and happy.

**Health Myth**

**Myth:** You should choose just one kind of exercise and stick with it.

**Fact:** You are likely to get bored with only one kind of exercise. Then you might stop exercising completely. Also, a variety of exercises will benefit a variety of muscles.

---

## Lesson at a Glance

### Chapter 1 Lesson 4

**Overview** This lesson describes how to set goals and solve health problems

**Objectives**

- To distinguish between long-term and short-term goals
- To develop health literacy
- To solve health problems and change behaviors
- To evaluate your own health

**Student Pages** 17–22

**Teacher's Resource Library**
  Activity 4
  Alternative Activity 4
  Workbook Activity 4

### Vocabulary

**health literacy**

Have students read the definition of the vocabulary term and then use it in a sentence.

### Background Information

Health literacy is a national issue. The U.S. Department of Health and Human Services (HHS) defines health literacy as "the degree to which individuals have the capacity to obtain, process, and understand basic health information and services needed to make appropriate health decisions." This is part of the Healthy People 2010 initiative being undertaken by the Health Resources and Services Administration, a division of HHS. Having inadequate health literacy can interfere with a patient's ability to take medication properly or see a doctor regularly.

### Health Myth

Have students read the Health Myth feature on page 17. Have individual students make lists of their favorite types of exercise. Have them make up a weekly schedule of exercise in which they are not repeating any exercise more than twice in one week.

## Warm-Up Activity

Have students brainstorm to describe the process by which they make a decision. Have them explain what things they consider when making a decision. Then have students compare their process with the decision-making process described later in the lesson. Have them revise their process after they have read the lesson.

## Teaching the Lesson

To help students organize the material about goal setting, have them make a Venn diagram to compare and contrast short-term goals and long-term goals.

To help students organize the material about making decisions and solving problems, have them make a flow chart that sequences and describes the steps involved in making a decision.

## Reinforce and Extend

### LEARNING STYLES

#### Interpersonal/Group Learning

Have students work in groups to come up with a health-related problem a teenager might be experiencing and then use the steps in the decision-making process to solve it. Students should present the results of their work to the class. Have classmates discuss if they would have tried to solve the problem with a different solution.

---

*Health literacy*
*Gathering, understanding, and using information to improve health results*

How do you decide which goals to set? How do you decide which decisions to make to help you reach those goals? You learn all you can about a healthy lifestyle. You work actively toward **health literacy.** You work to gather and understand information. Then you use the information to improve your health. You can use the information to develop skills for making good decisions and solving problems. This helps you begin to work toward your goals.

### How Do You Change Behaviors?

**Gaining Awareness**

Being aware means paying attention to the signals your mind and body send you. Do you feel tired all the time? Does your stomach often hurt? Are you constantly sad? These are all signals that deserve your attention. You can work toward solving the problems these signals point out. You have the power to make decisions that will help you feel healthier.

**Gaining Knowledge**

You can use your power to learn more about health. Health literacy will guide you in this direction. You can read information in books, in magazines, and online. You can watch DVDs and use CDs that provide health information. You can also ask teachers, family members, and health care professionals for information.

### How Do You Make Decisions and Solve Problems?

After you gain awareness and knowledge, you are ready to make decisions. You will consider all you have learned as you move forward. Through thinking about decisions, you will gain skills that can help you in the future.

**Applying Skills**

You will learn how best to deal with your symptoms based on what you have been told and how you have made your decision. You can apply these same skills in the future to help you gain knowledge and make decisions about your health. Read on to see how one student made decisions to solve a problem.

Lou felt sad all the time. He had no energy. His clothes were starting to fit tightly. He became aware he had a health problem when he listened to the signals his mind and body were sending him. Lou followed a five-step process. He solved his problem and made a decision about what to do to feel better.

1. **Define the problem.** Lou thought about his tired feelings, sadness, and tight clothes. He asked, "Why is this a problem? How did the problem start? How long have I had the problem?" He did not blame himself. He did not blame anybody else. He just looked for answers to his questions.

    Lou collected as much information as he could. He gained knowledge through reading information online and in his health book at school. He learned that his diet was not healthy. He also learned that he was not getting enough exercise.

2. **Come up with possible solutions.** Lou knew from information he had gathered that he needed to eat healthier foods. He needed more exercise. He asked a doctor at his community center about the best way to move forward. The doctor suggested several different meal plans. The doctor also suggested many kinds of exercise for Lou to try.

3. **Consider the consequences of each solution.** Lou thought about each meal plan the doctor had suggested. One would include a great deal of preparation time at home, time Lou did not have. He knew he probably would not follow that meal plan. He decided to try a meal plan that would require less preparation time. One kind of exercise the doctor had suggested was swimming. Lou knew the weather would become cold soon. The community pool would close, so he probably would not continue that exercise. He chose to do an exercise program at home instead.

## ELL/ESL Strategy

**Language Objective:** *To practice writing skills*

Have students write a paragraph in which they describe a time when they had to make a decision. Have them share the paragraph with another student who is proficient in English. The reader should verbally summarize what is written in the paragraph to the writer. The writer should use the summary to make sure his or her writing was clear and understandable. If not, have the two students work together to revise the paragraph.

### Link to

**Biology.** Have students read the Link to Biology feature on page 20. The U.S. Department of Health and Human Services recommends that teenagers get one hour of moderate-intensity exercise most days of the week, if not daily. Have students make an entry in their Health Journal about how much exercise they currently get. If students feel they need more exercise, have them write a plan describing how they could reach that goal.

#### TEACHER ALERT

Spend time discussing with students the idea that they have the power to change their behavior and make positive decisions and choices. As they read through the textbook this year, they will learn certain skills that will enable them to do this. These are skills that they can use and practice every day, even with the smallest decision or choice they make. Have student volunteers share an example of a time when they made a positive choice and how they felt about it afterward.

4. **Put your plan into action.** Lou thought about how to make his solution work. He knew he would have to plan his meals and cut out unhealthy snacks. He knew he would have to set aside exercise time at least three days a week. He made his plans and put them into action.

5. **Evaluate the outcome.** After eating healthier foods and exercising, Lou found he was not feeling sad anymore. He had more energy. His clothes were fitting better. He found, though, that he was growing bored with his exercise routine. He was afraid he would not continue it, so he explored other types of exercise.

Consider your own health as you use the Health Assessment Chart on the next page to uncover valuable information. The chart will help you assess your own health and wellness. It will help you rate behaviors and choices that affect your health. The chart also will guide you in making choices to maintain or change behaviors.

### Link to >>>

**Biology**

Exercise causes many positive changes in your body. It makes your heart beat at a healthy higher rate. This higher rate strengthens your heart. Then your circulation, the flow of blood through your body, is better. The blood more efficiently delivers substances your body needs throughout your body. Then you have more energy, and you are less likely to become ill.

## HEALTH ASSESSMENT CHART

**Directions:** Create three columns on a sheet of paper. Copy the titles from the columns below. Write the numbers as they appear in the columns. Then read each statement. If it is mostly true for you, write *yes* next to its number. If it is mostly false for you, write *no*.

Answer honestly to learn more about yourself. The results are your own, and you do not need to share them.

Follow instructions at the bottom of each column to find your score for the type of health. Then add all three scores to check your overall health.

**Overall Health Score**
30–36 = excellent health
25–29 = above average health
20–24 = average health
19 or below = below average health

### PHYSICAL HEALTH

1. I eat a balanced diet made up of different foods every day.
2. I avoid unhealthy snacks.
3. I do not use tobacco or alcohol. I do not take drugs unless my health care professional has advised me to take them.
4. I get at least 8 hours of sleep a night.
5. I rarely feel tired or run down.
6. I exercise energetically, getting my heart to beat quickly, at least three times a week.
7. I work to develop muscle tone at least three times a week, using resistance exercises and weight lifting.
8. I do stretches at least three times a week to make me more flexible.
9. I try different kinds of exercise.
10. I relax at least 10 minutes a day.
11. I get medical and dental checkups at least twice a year.

Score one point for each *yes* answer.
10–11 = excellent physical health
8–9 = good, but physical health could use improvement
6–7 = physical health could use quite a bit of improvement
5 or lower = physical health needs a great deal of improvement

### SOCIAL HEALTH

1. I meet people and make friends often.
2. I have at least one close friend.
3. I can say no to my friends if they want me to do something I do not want to do.
4. I balance having my way with allowing others to have theirs.
5. I respect the right of others not to be like me.
6. I work cooperatively with others.
7. If I have a problem with others, I face it. I try to work through the problem with them.
8. I am comfortable communicating with most adults.
9. I am comfortable talking with most young men and women my own age.
10. I practice good citizenship.
11. I am fair and trustworthy when dealing with others.

Score one point for each *yes* answer.
10–11 = excellent social health
8–9 = good, but social health could use improvement
6–7 = social health could use quite a bit of improvement
5 or lower = social health needs a great deal of improvement

### EMOTIONAL HEALTH

1. I try to accept my feelings of love, fear, anger, and sadness.
2. I can tell when I feel as though I am under pressure.
3. I try to find ways to deal with pressure and control my reaction to it.
4. I try to have a positive outlook.
5. I ask for help when I need it.
6. I can discuss problems with my friends and family members.
7. I can accept compliments.
8. I give compliments.
9. I can accept and use constructive comments.
10. I take responsibility for my actions.
11. I am honest with myself and others.

Score one point for each *yes* answer.
10–11 = excellent emotional health
8–9 = good, but emotional health could use some improvement
6–7 = emotional health could use quite a bit of improvement
5 or lower = emotional health needs a great deal of improvement

---

## LEARNING STYLES

**Logical/Mathematical**
Have students read and complete the Health Assessment Chart on page 21. They will have to keep track of their points, total them, and score themselves in each area of health. Then they should add the three scores together to find out their Overall Health Score. Encourage students to answer honestly. They can use this assessment as a starting point to change behaviors that do not positively affect their health.

## HEALTH JOURNAL

Have students identify and describe which behaviors they think are most important to their own physical, social, and emotional health.

# Lesson 4 Review Answers

1. The five steps for solving a problem and making a decision are defining the problem, coming up with possible solutions, considering the consequences of each solution, putting your plan into action, and evaluating the outcome. 2. Health literacy is gathering, understanding, and using information to improve your health. 3. Answers should include at least three of the following: awareness, knowledge, decision making, and applying skills. 4. Answers will vary. Answers should indicate an understanding of the distinction between short-term and long-term goals. 5. It is important to decide whether or not your decision solved the problem so you can use this experience when trying to solve future problems.

## Portfolio Assessment

Sample items include:
- Health Assessment from Learning Styles Logical/Mathematical
- Paragraph from Health Journal
- Lesson 4 Review answers

### Link to

**Language Arts.** Have students read the Link to Language Arts feature on page 22. Suggest that they keep a food journal for one week making daily entries in their Health Journals. After the week is over, have them review their entries to determine any patterns. Have them identify any changes they may make based on what they have learned.

---

**Lesson 4 Review** On a sheet of paper, write the answers to the following questions. Use complete sentences.

1. What are the five steps for solving a problem and making a decision?
2. What is health literacy?
3. What are three tools for changing behaviors?
4. **Critical Thinking** Name one of your short-term goals and one of your long-term goals.
5. **Critical Thinking** Why is it important to evaluate the outcome of a decision?

### Link to ▶▶▶

**Language Arts**

Keeping a daily journal of the food you eat and your exercise can help you achieve health goals. (You may be surprised by the end of a week to see what you have actually eaten!) When you list the foods you are eating and the exercise you have done, be sure to include your feelings while you eat—happy, sad, nervous, or scared. Your journal can help you recognize patterns and choose behaviors to continue or discontinue.

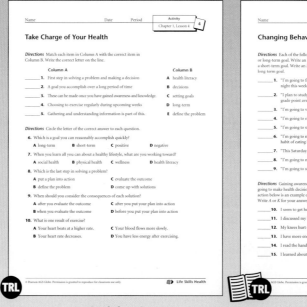

# Chapter 1 SUMMARY

- People need good health to accomplish the things they need and like to do.
- Being healthy means much more than just not being ill.
- A person must have physical health, social health, and emotional health to be truly healthy.
- People should take basic steps to prevent disease and stay healthy.
- A health risk can cause disease. Many health risks relate to heredity, the environment, and behavior.
- The physical environment includes air, water, land, plants, animals, and buildings. Cigarette smoke is an example of an environmental health risk.
- The social environment is made up of community resources. Lack of health care for those who cannot pay creates a risk in the social environment.
- Cultural risks occur when health information is not received, understood, or used properly. Risks can be decreased when health care providers understand languages and cultures of community members.
- Behavioral risks relate to action or inaction. You have the power to control behavioral risks.
- You can set short-term and long-term goals as you make health decisions.

| Vocabulary | | |
|---|---|---|
| abstain, 15 | health, 4 | physical environment, 11 |
| alcohol, 8 | health literacy, 18 | secondhand smoke, 15 |
| culture, 13 | health risk, 11 | sedentary, 14 |
| deteriorate, 15 | heredity, 11 | social environment, 12 |
| disease, 7 | impair, 15 | tobacco, 14 |
| drug, 8 | nonverbal, 13 | wellness, 9 |
| frustrate, 13 | | |

*Maintaining Health and Wellness* Chapter 1

## Chapter 1 Summary

Have volunteers read aloud each Summary Item on page 23. Ask volunteers to explain the meaning of each item. Direct students' attention to the Vocabulary box on the bottom of page 23. Have them read and review each term and its definition.

### ONLINE CONNECTION

For more information about environmental risks to young people, have students go to the Environmental Protection Agency Web site at: http://yosemite.epa.gov/ochp/ochpweb.nsf/frmChemicals

If students want to learn more about how much and what type of exercise they should be getting every day, have them go to the Centers for Disease Control (CDC) Web site at: www.cdc.gov/nccdphp/dnpa/physical/recommendations/young.htm

For students who are interested in learning more about a healthy diet, direct them to the U. S. Food and Drug Administration's Web site for information about the food pyramid: www.mypyramid.gov

## Chapter 1 Review

Use the Chapter Review to prepare students for tests and to reteach content from the chapter.

## Chapter 1 Mastery Test

The Teacher's Resource Library includes two forms of the Chapter 1 Mastery Test. Each test addresses the chapter Goals for Learning. An optional third page of additional critical-thinking items is included for each test. The difficulty level of the two forms is equivalent.

## Review Answers

**Vocabulary Review**

1. physical environment 2. wellness
3. heredity 4. nonverbal 5. sedentary
6. deteriorate 7. secondhand smoke
8. social environment 9. abstain 10. impair 11. culture 12. health literacy
13. health risk

### TEACHER ALERT

In the Chapter Review, the Vocabulary Review includes a sample of the chapter's vocabulary terms. The activity will help determine students' understanding of key vocabulary terms and concepts presented in the chapter. Other vocabulary terms used in the chapter are listed below.

| | |
|---|---|
| alcohol | frustrate |
| disease | health |
| drug | tobacco |

24    Unit 1    Mental and Emotional Health

---

## Chapter 1 REVIEW

**Word Bank**
abstain
culture
deteriorate
health literacy
health risk
heredity
impair
nonverbal
physical environment
secondhand smoke
sedentary
social environment
wellness

### Vocabulary Review

On a sheet of paper, write the word or phrase from the Word Bank that best completes each sentence.

1. The _____ is the area around a person.

2. Another name for an active state of health is _____.

3. The passage of physical characteristics from parents to children is _____.

4. An unspoken answer is _____.

5. A(n) _____ lifestyle includes little or no exercise.

6. To _____ is to wear away a little bit at a time.

7. Smoke from a cigarette someone else is smoking is _____.

8. The _____ is made up of community resources.

9. To _____ is to choose not to use or do something.

10. To have a negative effect is to _____.

11. A(n) _____ is made up of the beliefs, way of living, and language of a group of people.

12. Gathering, understanding, and using information to improve your health results in _____.

13. A(n) _____ is an action or condition that may contribute to injury or disease.

24    Unit 1    Personal Health and Family Life

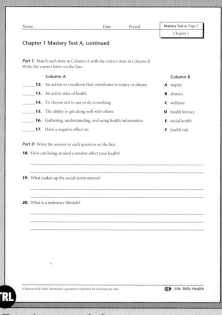

Chapter 1 Mastery Test A, pages 1–3

## Concept Review

On a sheet of paper, write the letter of the answer that best completes each sentence.

14. The state of being healthy includes all of these parts EXCEPT _____ health.
    - A social
    - B medical
    - C physical
    - D emotional

15. The first step in changing behaviors is _____.
    - A applying skills
    - B gaining knowledge
    - C making decisions
    - D gaining awareness

16. A good example of physical health is _____.
    - A having energy for sports
    - B getting along with others
    - C handling daily pressures
    - D being a good friend

17. A _____ goal can be accomplished quickly.
    - A long-term
    - B physical
    - C short-term
    - D complex

## Critical Thinking

On a sheet of paper, write the answers to the following questions. Use complete sentences.

18. What is an example of good physical health? social health? emotional health?

19. Name three types of health risks and their causes.

20. What changes in behavior would you like to make to become a healthier person? Explain your answer.

**Test-Taking Tip** Read test questions carefully to identify those that require more than one answer.

### Concept Review
14. B  15. D  16. A  17. C

### Critical Thinking
18. Sample answer: An example of good physical health is being able to wake up in the morning feeling refreshed and able to easily get to school on time. An example of good social health is feeling close to family members and enjoying their company. An example of emotional health is being able to get a bad grade on a test and use it as motivation for working harder and improving on the next test. 19. Answers will vary but should reflect knowledge of causes of risk as they relate to heredity, environment, culture, and behavior. 20. Answers will vary but should reflect knowledge of healthy diet, exercise, and social and emotional health.

## ALTERNATIVE ASSESSMENT

Alternative Assessment items correlate to the student Goals for Learning at the beginning of this chapter.

- Alternative Assessment items correlate to the student Goals for Learning at the beginning of this chapter.
- Have students orally define physical, social, and emotional health. Have them explain why each is important to overall health.
- Have students make a list of steps they can take to stay healthy.
- Have students orally explain different types of health risks. For each risk, have students give an example.
- Have students write a paragraph outlining the steps involved in making decisions and solving problems.

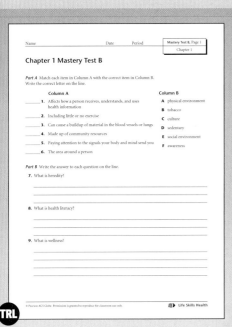

Chapter 1 Mastery Test B, pages 1–3

# Chapter at a Glance

## Chapter 2: Managing Emotions
pages 26–47

### Lessons
1. Your Emotions
   pages 28–33
2. Stress and Your Body
   pages 34–37
3. Managing Stress
   pages 38–44

**Chapter 2 Summary** page 45

**Chapter 2 Review** pages 46–47

Audio CD 🎧

Skill Track for
Life Skills Health

Teacher's Resource Library **TRL**
   Activities 5–7
   Alternative Activities 5–7
   Workbook Activities 5–7
   Self-Assessment 2
   Community Connection 2
   Chapter 2 Self-Study Guide
   Chapter 2 Outline
   Chapter 2 Mastery Tests A and B
   (Answer Keys for the Teacher's Resource Library begin on page 559 of this Teacher's Edition.)

## Opener Activity

Have students brainstorm a list of emotions such as joy, sadness, anger, fear, and surprise. Ask volunteers to role-play a person who is feeling a particular emotion. Then have classmates guess the emotion. Discuss situations in which the emotion is likely to occur.

26  Unit 1  Mental and Emotional Health

# Self-Assessment

## Can you answer these questions?

☑ 1. What is emotional health?

☑ 2. What kinds of needs do people have?

☑ 3. In what way do people show their needs?

☑ 4. How does your body respond to stress?

☑ 5. What life events cause stress?

☑ 6. How is fear linked to stress?

☑ 7. What are some common responses to anger?

☑ 8. What can you do to handle stress?

☑ 9. Why is paying attention to emotional signals important?

☑ 10. What suggestions for coping with stress would you give a friend?

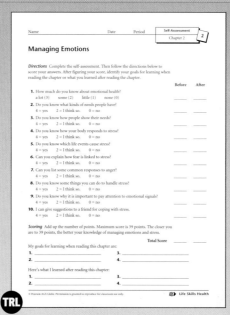

Self-Assessment 2

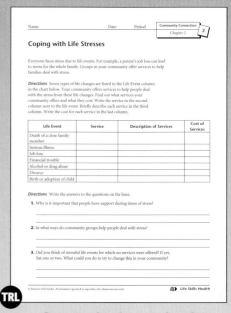

Community Connection 2

# Chapter 2: Managing Emotions

Being healthy involves both the body and the mind. If you are excited about something, that feeling can affect your body in a positive way. If you are worried about something, that too can affect your body. Your mind affects your body.

We also know that the opposite is true—the body affects the mind. For example, if you are well rested, you are usually able to think clearly. However, if you are tired or feel ill, you may not be able to think as clearly.

In this chapter, you will learn about keeping your mind and body healthy by understanding emotions and stress. You will also learn about common needs that people have and how these needs affect your emotions.

### Goals for Learning

- ◆ To explain the purpose of emotions and how they happen
- ◆ To describe five different levels of basic human needs
- ◆ To list three ways mental health affects physical health
- ◆ To create a stress management plan

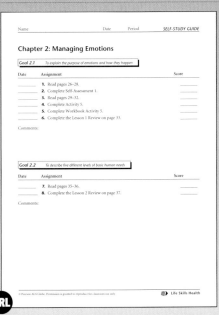

Chapter 2 Self-Study Guide, pages 1–2    Chapter 2 Outline, pages 1–2

## Introducing the Chapter

Have students write down what physical changes they notice in their bodies when they feel different emotions including fear, anxiety, and anger. Discuss the idea that different emotions can cause physical changes in the body such as rapid heartbeat, sweaty palms, loss of appetite, or blushing. Continue the discussion by adding that the physical state of the body can affect emotions, reactions, and the ability to think. For example, being tired or feeling ill may affect a student's ability to do well on a test.

Have students read the Goals for Learning. Explain that in this chapter students will learn about emotions and stress and healthy ways to cope with stress.

## *Notes and Questions*

Ask volunteers to read the notes and questions that appear in the margins throughout the chapter. Then discuss them with the class.

## *Self-Assessment*

Have students complete the Self-Assessment worksheet before and after reading the chapter. Before reading the chapter, have students fill in the "Before" column. Ask students to identify their goals for learning. To get ideas for setting goals, students might use the chapter introductory material on page 27, the checklist on page 26, or the questions on the Self-Assessment worksheet. Students can use the back of the worksheet if they need more space to write.

Collect the Self-Assessment worksheets and pass them out again at the end of the chapter. Have students fill in the "After" column. Ask them to identify at least four major points they have learned. Again, suggest they use the back of the worksheet if they need more space to write. You may want to collect and review the worksheets, but return them to students so they have a record of their goals and accomplishments.

# Lesson at a Glance

## Chapter 2 Lesson 1

**Overview** This lesson explains how a person's basic needs are related to emotions.

### Objectives

- To explain three ways people show emotion
- To list Maslow's five human needs
- To describe what self-actualization means

**Student Pages** 28–33

**Teacher's Resource Library**

Activity 5
Alternative Activity 5
Workbook Activity 5

## Vocabulary

| | |
|---|---|
| attachment | hierarchy |
| body language | reaction |
| emotion | romantic love |
| esteem | self-actualization |
| family love | self-concept |
| friendship | self-esteem |
| guilt | |

Read the vocabulary words and discuss their meanings with the class. Ask students to write sentences, leaving blanks for the missing vocabulary words. Then have students exchange papers and write in the missing words.

## Background Information

A person can help strengthen self-esteem in the following ways: behaving in ways that are consistent with values; concentrating on realistic, positive thoughts about oneself; focusing on one's own strong points; learning to express negative thoughts in positive ways; and looking at mistakes as opportunities for learning.

---

## Lesson 1 Your Emotions

### Objectives

*After reading this lesson, you should be able to*

- explain three ways people show emotion
- list Maslow's five human needs
- describe what self-actualization means

**Emotion**
Feeling

**Reaction**
Response

**Body Language**
*The way people move and hold their heads, arms, and legs that gives clues about what they think and feel*

Body language plays an important role in classrooms and other settings. You show interest in a teacher or speaker by pointing your body toward that person.

The same experience can cause different people to react, or respond, in different ways. This happens because every human being has **emotions,** or feelings. Emotions are a person's individual **reactions** to an experience. A reaction is a response. Your emotions are a special part of who you are.

Emotions can be difficult to understand. Emotions start with thoughts. When you think about something, you may feel a certain way about it. Your thinking triggers an emotion. This emotion may then cause your body to react. It may cause a physical change. For example, if you think about going to a party, you may feel excited when it is time to leave. Then your heart may beat faster. Your thoughts about people and events trigger the emotions that cause physical changes in your body.

### What Do Emotions Do?

Emotions are natural. They serve a purpose. Emotions signal reactions to events and experiences. They let you know when something is right, as well as when things are wrong. Learning about your feelings can help you understand yourself better.

People's emotions show up in the way the people look, act, and sound. You often can tell how people feel by how loudly and forcefully they speak. **Body language,** or the way people move and hold their heads, arms, and legs, also gives clues about how people feel.

All people have basic needs, including emotional needs. Understanding these needs can help you understand yourself and your emotions.

**Esteem**
*Value or worth; how one sees oneself or others*

**Hierarchy**
*Order from most to least important*

## What Basic Needs Do People Have?

Abraham Maslow studied what humans need. He observed that people have five basic needs. They are physical, safety, belonging, **esteem,** and self-actualization needs. Maslow put the needs in a **hierarchy,** or order, beginning with the most basic needs. Physical needs are at the bottom of the hierarchy. Safety needs come next, and so on. Maslow's hierarchy of needs appears in Figure 2.1.1.

**Figure 2.1.1** *Maslow's hierarchy of needs*

Maslow said that most people spend their lives working on the first three needs. Each day, you work to meet the needs on the first three levels. Then you can work on meeting the next higher level of needs. For example, a person must first satisfy the physical need to eat—that need is basic to staying alive. Once that need is met, a person can satisfy the need to find a safe place to live. Later, that person can satisfy the need to be with others and to belong to groups. As you read these explanations of needs, think about where you are in the hierarchy.

 **Warm-Up Activity**

Ask students to think about the difference between something they need and something they want. As students share their ideas, write a list of "needs" and "wants" on the board. After reading and discussing the lesson, ask students how they might change the list of "needs" and "wants."

 **Teaching the Lesson**

Help students understand a hierarchy by having them make their own hierarchy using fruits as an example. Explain that in a hierarchy, the broadest category is the base and each level becomes more defined and smaller. Tell them the hierarchy levels for the fruits should be All Fruits, Citrus Fruits, and Orange Citrus Fruits, and have them fill in examples for each level.

Help students learn the needs at each level of Maslow's hierarchy by having them create a mnemonic using the first letter of the need for each level. For example: **M**aslow's **P**lan **S**eems **B**est for **E**veryone's **S**ake.

 **Reinforce and Extend**

### GLOBAL CONNECTION

 Researchers have learned that people in all parts of the world express their emotions with the same facial expressions. Across all cultures and socioeconomic backgrounds, people smile when they are happy, frown when they are sad, and raise their eyebrows and open their mouths wide when they are surprised.

### LEARNING STYLES

 **Visual/Spatial**
Have students make a poster about Maslow's hierarchy of needs. Have students divide a triangle into five horizontal sections. Students can use magazine pictures or drawings to illustrate each of the needs.

## LEARNING STYLES

**Interpersonal/Group Learning**

Have students work in small groups to brainstorm a list of qualities that make a person a good friend to another person. Ask a volunteer from each group to read their list to the class.

## IN THE ENVIRONMENT

As a class, choose a controversial environmental issue involving a conflict between needs. For example, the logging industry can provide jobs for people and can have a positive effect on economic growth. On the other hand, logging can lead to habitat destruction, which can threaten the survival of some species. Discuss the emotional as well as the scientific and pragmatic viewpoints of the issue and then ask students to share their own viewpoints.

## CAREER CONNECTION

Discuss Maslow and his career as a noted psychologist. Have students research careers in the field of psychology and write a report about one such career. Ask volunteers to present their reports to the class.

---

*Friendship*
Love based on choices

*Attachment*
Emotional bond or tie to others

*Family love*
Love based on attachment and support

*Romantic love*
Strong physical and emotional attraction between two people

### Physical Needs
Physical needs are everything necessary to stay alive—water, food, and oxygen. Regularly eating the right foods meets many of your physical needs.

### Safety Needs
Safety needs are protection needs. That means protecting yourself from danger and from the conditions around you. Being sheltered from harm meets your basic safety needs.

### Belonging Needs
There are two kinds of belonging—belonging in a place and belonging with people. Belonging in a place helps you feel secure and meet other needs such as safety. This may be in your home, school, town, or country.

Needing to belong with people leads you to form **friendships** or join groups. Friendships, or love based on choices, help you feel healthy and secure. A relationship based on need is called an **attachment,** which is an emotional bond or tie to others.

People experience different kinds of love in their need to belong with people. **Family love** is based on attachment and support. Family members often feel safe when they are in contact with one another and support one another.

You are born into your family, but you can choose your friends. You usually choose friends because you have common interests and goals. The more similar you are to someone, the easier it is to be friends. You can have many friends. Usually, however, you have only a few friends to whom you feel especially close.

**Romantic love** is a strong physical and emotional attraction between two people. When it is part of belonging, romantic love is based on need. Two people may feel they are in love because they want to be together and are unhappy when they are apart.

**Self-esteem**
*Self-respect; how one feels about oneself*

**Self-concept**
*Ideas one has about oneself*

### Research and Write

Use the Internet to learn more about Maslow's hierarchy of needs. Work with a partner to list the many ways people satisfy each need. Use pictures from magazines to create a collage that shows what you learned. Share your collage with your class.

### Esteem Needs

If you are able to meet your needs for survival, safety, and belonging, you then seek esteem. Esteem is the value or worth you place on yourself or others. People need two kinds of esteem—the esteem of others and **self-esteem,** or how you feel about yourself. You feel the esteem of others when family members express their love for you. You feel it when friends tell you that you have done a good job. Another word for self-esteem is self-confidence.

Self-esteem, or self-respect, is your sense of being valuable and worthwhile. Feeling happy and content is usually part of high self-esteem. Feeling unimportant and unsatisfied is usually part of low self-esteem. Your **self-concept,** or your ideas about who you are, determines your self-esteem. Good self-concept usually leads to high self-esteem. Poor self-concept usually leads to low self-esteem. Anything that affects your self-concept can influence your self-esteem. Your self-concept and self-esteem continue to develop throughout life.

**Figure 2.1.2**
*Volunteering to help others is a way to meet belonging needs of both the volunteer and the person being helped.*

### HEALTH JOURNAL

Ask students to write a paragraph about their self-concept and their self-esteem. Do they feel they have high self-esteem or low self-esteem? Why? What actions can they take to improve their self-esteem?

### AT HOME

Encourage students to write a list of ways they can help friends or family members strengthen their self-esteem. Allow students to keep their lists private if they prefer to do so.

### Research and Write

Have students read and complete the Research and Write feature on page 31. Encourage students to think realistically about how they might meet some of these needs and to include these as examples in the collage.

## TEACHER ALERT

Point out to students that not everyone satisfies their needs to the same degree. Some people may spend unusually large amounts of time satisfying their physical needs, safety needs, and esteem needs and spend little effort satisfying other needs. Other people choose to live simply and satisfy fewer physical needs. They may spend more time fulfilling self-actualization needs.

## Decide for Yourself

Have students read and complete the Decide for Yourself feature on page 32. Discuss examples of bullying and how upsetting being the victim of bullying can be. Have students role-play a situation in which someone is being bullied. They should be sure to include examples of how to react to a bully.

### Decide for Yourself Answers

1. Answers will vary but should include the idea that bullying makes one feel bad about oneself. 2. They have low self-esteem and try to feel better by picking on others. 3. Answers will vary. Students should present a concrete plan for working with other students and school officials to reduce the prevalence of bullying behavior at their school.

## ELL/ESL STRATEGY

**Language Objective:** *To construct sentences*

Pair students learning English with those who are proficient in English. Have pairs write sentences that identify a need and how it is met. Students should use their own needs as examples.

---

**Self-actualization**
*Achieving one's possibilities*

### Self-Actualization Needs

If you are able to satisfy all the other needs, you can begin to work on **self-actualization.** Self-actualization means achieving your possibilities. People who have satisfied their physical, safety, belonging, and esteem needs are able to become self-actualized. They can go beyond themselves to give to others. For example, you might feel good about yourself when you spend time volunteering at a hospital.

### Decide for Yourself

A bully is someone who tries to hurt someone else to gain power. Bullies often have low self-esteem and try to make themselves feel better by picking on others. Bullying can take many forms. Physical bullying includes physically harming someone. Emotional bullying includes spreading rumors or ignoring someone on purpose. Verbal bullying is name-calling or making fun of someone. Racist and sexual bullying involve offensive words and actions. Cyber bullying includes the use of technology such as cell phones or computers to threaten or send hurtful messages. Here are some ways you can deal with bullying behavior:

- Remember that the bully is at fault, not you. No one deserves to be bullied.
- Do not use physical force to stop a bully. This only encourages a physical response from the bully.
- Ignore the bully and walk away. A bully looks for a reaction. Do not provide one.
- In times when walking away is not possible, compliment the bully. This will buy you some time to walk away.
- Talk about what happened with a trusted adult who can help.
- Join positive groups. Stay with friends who treat others with respect. Be sure you treat others with respect.
- Work with your student council and school leaders to stop bullying at your school.

On a sheet of paper, write the answers to the following questions. Use complete sentences.

1. How might bullying harm a person's self-esteem?
2. Why do most bullies pick on others?
3. Make a plan that shows how you can work with others to stop bullying at your school.

**Guilt**
*A feeling of having done something wrong*

## How Do Needs Lead to Different Emotions?

Emotions help you realize what you need. For example, if you often feel frightened, your need for safety is not being met. Your feelings or emotions are closely connected to your needs. Sometimes you may feel unpleasant emotions such as **guilt** or shame. Guilt is the feeling of having done something wrong. Shame is an emotion coming from a strong sense of guilt, embarrassment, or unworthiness. If you make a mistake in front of friends, you may feel shame or embarrassment. Your need to belong is what leads to your feeling of embarrassment. While people have the same needs, they may have different ways of showing those needs. For example, you may wear a certain style of clothing to feel you belong. Another person may behave in a certain way to feel secure or socially accepted. You both have a need to belong, but you each show that need in a different way.

*Lesson 1 Review* On a sheet of paper, write the letter of the answer that best completes each sentence.

1. Emotions start with _____.
   A actions          C body language
   B thoughts         D guilt

2. Self-actualization means _____.
   A meeting physical needs   C achieving possibilities
   B setting goals            D feeling unimportant

3. According to Maslow, the type of need people try for when their physical and safety needs are met is _____.
   A self-actualization   C belonging
   B esteem               D love

On a sheet of paper, write the answers to the following questions. Use complete sentences.

4. **Critical Thinking** How can your words and body language affect your relationships with others?

5. **Critical Thinking** What are two ways a person can satisfy his or her need for self-actualization?

*Managing Emotions* Chapter 2  33

## Lesson 1 Review Answers

**1.** B **2.** C **3.** C **4.** Your words and body language can show your emotions. If you show positive emotions, people will want to be around you. If you show negative emotions, people may want to avoid you. **5.** You can satisfy self-actualization needs by helping others or doing volunteer work. Self-actualization means achieving your possibilities.

## Portfolio Assessment

Sample items include:
- Poster from Learning Styles/Visual Spatial
- Paragraph from Health Journal
- Lesson 1 Review answers

### LEARNING STYLES

**Interpersonal/Group Learning**
Have individual students make nine index cards: three with *self-esteem* written on them, three with *self-concept* written on them, and three with *self-actualization*. Then ask them to write three descriptions or examples of each of the terms on separate cards (for a total of 18 cards). Ask students to turn the cards facedown. Working in pairs or groups of three, students can take turns picking up two cards from each other's set until they match each of the words with the descriptions.

### HEALTH JOURNAL

Ask students to write about a time when they or someone they know was bullied. Have them describe how the situation was handled. What did they do in the situation? What things would they do differently now that they have learned more about bullying?

*Managing Emotions* Chapter 2  33

# Lesson at a Glance

## Chapter 2 Lesson 2

**Overview** This lesson explains how stress affects the body and gives examples of common stressful events for teenagers.

### Objectives
- To identify three physical reactions to stress
- To explain the fight-or-flight response
- To list five life events that cause stress

**Student Pages** 34–37

**Teacher's Resource Library**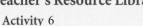
 Activity 6
 Alternative Activity 6
 Workbook Activity 6

## Vocabulary

stress   stress response

Write the words *stress* and *stress response* on the board. Read the definitions aloud. Ask students to describe events in school that cause them stress and the stress responses they experience as a result. Write their responses on the board.

## Background Information

The stress response can help a person achieve peak-level performance. For example, it might help a runner win a race or a football player make a touchdown. The stress response might also help a person perform extraordinary feats—such as moving objects of great weight or running extraordinarily fast—in emergency situations. In one case, a woman lifted a car high enough to free a child pinned beneath it.

In the stress response, the brain alerts the endocrine glands to release adrenaline, a hormone, into the bloodstream. The immediate release of adrenaline can provide the extra strength and speed a person needs for emergency action.

34   Unit 1 *Mental and Emotional Health*

---

## Lesson 2  Stress and Your Body

### Objectives
After reading this lesson, you should be able to
- identify three physical reactions to stress
- explain the fight-or-flight response
- list five life events that cause stress

**Stress**
*A state of physical or emotional pressure*

**Stress response**
*Automatic physical reactions to stress*

Some people overeat during times of stress. Others eat too little. Eating healthy foods like whole grains, fruits, and vegetables regularly will help your body handle stress better.

Emotions begin with thoughts. Emotions may cause physical changes or reactions. When basic needs are threatened, people experience **stress.** Stress is a state of physical or emotional pressure. It is a reaction to anything that places demands on the body or mind to which people must adapt. Understanding how stress affects your mind and your body is important.

Stress is normal, and it can be either good or bad. For example, if you are excited about a project and do a good job on it, you benefit from good stress. Happy events such as graduations, dates, and weddings can all be sources of good stress. Good stress helps people to accomplish goals and to change. Usually we hear about bad stress because its effects are harmful. Examples of bad stress are family problems or the death of a loved one. Bad stress can interfere with healthy living.

### How Do People React to Stress?

When you face stress, you try to find out what is wrong and whether it is dangerous. For example, if your teacher seems unhappy and says she wants to talk with you, you may feel stressful. You think you may hear bad news. First, ask yourself what might be wrong and if it could be serious. Then decide what your reactions will be.

Most of your reactions may be automatic and physical. You may feel worried and notice that your heart is beating faster. This is called the **stress response.** The stress response affects your body in ways you may not be able to control. Your heart beats faster, you sweat, and your breathing quickens. These are automatic physical reactions to stress. Other examples are blushing, gasping in surprise, and crying out. Stronger reactions to stress are headaches, stomach pain, sleeping problems, and nervous feelings.

34   Unit 1   *Personal Health and Family Life*

### Health Myth

**Myth:** It is possible to avoid all stress.

**Fact:** Stress is a part of every person's life. You can manage stress but not eliminate it.

Exercise is a good way to relieve stress. Take a walk, dance to your favorite music, or work out at a gym or health club. Not only will you feel better physically, you will have a sense of emotional well-being.

A person's response to stress often takes one of two forms: fight or flight. When something threatens or angers you, you may feel an impulse to fight it. If you fight, you attack what threatens you. Fighting may simply be using words to stand up for what you believe. The other response is to flee. If you flee, you leave the situation, but nothing changes. The fight-or-flight impulse is natural and happens without your thinking about it.

You can also react to stress by thinking about what to do. For example, if you enter a dark room and hear a noise, you may react by jumping back. Your body becomes calmer after your first reaction. You may then decide to turn on a light to see what is making the noise. You find the noise is nothing to fear. What prompted the stress was imagined, and you were able to overcome it. Overcoming the stress helps you feel more at ease.

Relating to peers and adults in positive ways helps you avoid stress. Avoid negative comments when talking with others. Carefully choose your words during a disagreement. Make good risk decisions. For example, avoid high-risk behaviors such as drug use. Taking charge of your mental and physical health will help you manage stress and promote your emotional maturity. By doing so, you become a good role model and help promote positive health behaviors among your peers.

### Technology and Society

In biofeedback, electronic sensors are placed on a person's head or muscles. Computers then track and measure muscle activity. By learning to use positive thoughts, people can learn to manage pain and stress.

Biofeedback is used to treat migraine headaches, sleeplessness, hyperactivity, attention deficit disorders, and chemical dependency. Athletes use biofeedback to control anxiety before games and sporting events.

---

 **Warm-Up Activity**

Invite students to give examples of situations that may be enjoyable for some people but cause unpleasant pressure for others. One example might be performing in a school play.

 **Teaching the Lesson**

To help students process the material in the lesson, have them write a summary paragraph, including the definitions of vocabulary terms in their own words. Then have them illustrate the material presented under each head with personal examples.

To help students understand what happens in the body as the brain reacts to a perceived threat, have them create a flowchart of the steps described in the text under the head *What Physical Reactions Does Stress Cause?*

 **Reinforce and Extend**

### Health Myth

Have students read the Health Myth feature on page 35. Remind students that it is essential to have coping skills to deal with stress in life rather than allow stress to stop them from doing things they might enjoy. For example, a person who wants to go to a party may feel that the stress of interacting with people and making conversation is too overwhelming. Have students discuss positive actions they could take to deal with stress in this situation.

### Technology and Society

Ask a volunteer to read the Technology and Society feature on page 35 aloud. Then invite students to relax their muscles and take slow, deep breaths. After several minutes, ask students to share how they feel.

## TEACHER ALERT

Point out to students that stress is generated by both positive and negative events in life. People are often unaware of this and may be puzzled as to why they feel stressed, upset, or unsettled as they experience positive events or life changes. Have students analyze the examples given in Table 2.2.1. Which of the examples are generally considered to be negative events? *(death, divorce, involvement with alcohol and other drugs, deformity)* Which of the examples are generally considered to be positive events? *(outstanding achievement, college acceptance, start of dating)* Ask students to explain why a person might experience stress as a result of each of the positive events listed in the table. *(Changes occur, people may have different expectations, may be moving to unfamiliar surroundings, routines change)* Have students give some more examples of positive events that can cause stress.

## Health Myth

Have students read the Health Myth feature on page 36. The bacterium *Helicobacter pylori* is responsible for the development of stomach ulcers. Ulcers can be successfully treated with antibiotics and other drugs. But stress can irritate an ulcer and cause other gastrointestinal problems. Stress can lead to stomach cramps and bloating, diarrhea, constipation, or changes in eating habits, such as eating too much or not at all.

---

**Health Myth**

**Myth:** Stress causes ulcers.

**Fact:** An ulcer is a sore found on the lining inside the stomach. Today, doctors think that bacteria cause ulcers. The bacteria harm the protective lining in the stomach. Doctors believe the bacteria are spread through person-to-person contact.

## What Physical Reactions Does Stress Cause?

When you interpret something as a threat, your thought processes trigger your brain into action. Your brain triggers your body to fight or flight. Your brain also sends a chemical message to the body system that controls bodily processes. The chemicals go directly into your bloodstream. Some chemical messages cause faster breathing, sweating, and a rush of blood to your arms, legs, and head. Other chemical messages from your brain strengthen your muscles for endurance in case the stress lasts a long time.

Normally, physical reactions to stress help you resist it. The chemicals that your body produces help you react quickly. They help you remain alert while you choose the best way to respond to the threat. While good stress can have a positive effect on your body, bad stress can have a harmful effect.

## What Life Events Cause Stress?

A number of events can cause stress. Table 2.1.1 lists a few events that commonly cause stress for teens.

| Table 2.1.1 Common Stressful Life Events for Teens ||
|---|---|
| Death of a parent | An outstanding personal achievement |
| A visible deformity | Acceptance to college |
| Parents' divorce or separation | Being a senior in high school |
| Involvement with alcohol or other drugs | A change in acceptance by peers |
| Death of a brother or sister | A change in parents' financial status |
| Pressure of getting good grades | Start of dating |

*Lesson 2 Review* On a sheet of paper, write the letter of the answer that best completes each sentence.

1. The three automatic physical reactions to stress are _____.
   A blood rushing, muscle strengthening, and faster breathing
   B blood rushing, muscle weakening, and faster breathing
   C blood slowing, muscle strengthening, and slowed breathing
   D blood slowing, muscle strengthening, and fast breathing

2. An example of good stress is _____.
   A a fight with a friend
   B getting a bad grade
   C getting a promotion
   D changing schools

3. Bad stress affects your body _____.
   A positively
   B negatively
   C very little
   D the same way good stress does

On a sheet of paper, write the answers to the following questions. Use complete sentences.

4. **Critical Thinking** How can you avoid stress when relating to peers and adults?

5. **Critical Thinking** What is the fight-or-flight response?

### Link to >>> Biology

When embarrassed or nervous, do you blush? Blushing is a reddening of the face caused by nerves that widen blood vessels. More blood flows and causes the skin to look red. Some people also blush in their ears, neck, and chest. Some people blush easier than others do. Blushing can cause a feeling of heat as the blood flows through the vessels under the skin.

Managing Emotions   Chapter 2   37

## Lesson 2 Review Answers

**1.** A **2.** C **3.** B **4.** Answers will vary but should include avoiding negative comments, choosing words carefully, and making good risk decisions. **5.** When threatened, a person might stay and fight or flee. The fight-or-flight response occurs without a person thinking about it.

### Portfolio Assessment

Sample items include:
- Paragraph and flowchart from Teaching the Lesson
- List from Health Journal
- Lesson 2 Review answers

### Link to

**Biology.** Have students read the Link to Biology feature on page 37. Ask volunteers to describe under what circumstances they blush. How do they feel physically when they blush? *(skin may feel hot, palms may sweat)* Remind students that blushing is a normal response to feeling a strong emotion such as anger, embarrassment, or guilt.

### HEALTH JOURNAL

Have students make a list of different events in their lives that cause them to feel stress. Ask them to include a description of how they feel physically when they are stressed and how stress changes their reactions to other events in their daily life.

### ELL/ESL STRATEGY

**Language Objective:** *To summarize reading using one's own words*

Have students use their own words to provide a summary of the information presented in Table 2.2.1 on page 36. They may want to include examples of events listed on the table such as "I won my first ribbon at last week's track meet."

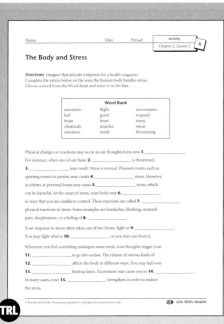

Activity 6

Workbook Activity 6

# Lesson at a Glance

## Chapter 2 Lesson 3

**Overview** This lesson describes some emotions that are linked to stress and explains ways to handle stressful emotions and relieve stress.

## Objectives

- To understand which emotions are related to stress
- To create a plan to manage stress
- To identify three ways to relieve stress

**Student Pages** 38–44

**Teacher's Resource Library**

Activity 7

Alternative Activity 7

Workbook Activity 7

## Vocabulary

| | |
|---|---|
| aggression | fear |
| anger | frustration |
| anxiety | grief |
| cope | relief |
| depression | well-being |

Have one student read a definition of a word and a partner name the word that matches the definition. Have the pairs continue this process until all the words are reviewed. The students can then switch roles and repeat the activity.

## Background Information

In 1969, Dr. Elisabeth Kübler-Ross published her now-famous book *On Death and Dying* in which she mapped out the five stages of grief—denial, anger, bargaining, depression, and acceptance—that people usually experience. She based her work on her studies of the experiences of dying people and their families. These stages can differ for different people, but they do offer general guidelines for understanding the grieving process. After her death, in 2004, her final book, *On Grief and Grieving*, was published.

38  Unit 1  Mental and Emotional Health

---

## Lesson 3 Managing Stress

**Objectives**

*After reading this lesson, you should be able to*

- understand which emotions are related to stress
- create a plan to manage stress
- identify three ways to relieve stress

**Fear**
*A strong feeling of fright; awareness of danger*

**Well-being**
*State of being physically and emotionally healthy*

**Anxiety**
*A feeling of uneasiness or fearful concern*

Several emotions are related to stress. Understanding these emotions and how the body reacts to some of them is helpful.

### Which Emotions Are Linked to Stress?

The emotions most commonly linked to stress are fear, anxiety, anger, grief, and relief.

### Fear

You feel **fear** when something important to your survival and **well-being** is threatened. Well-being is the state of being physically and emotionally healthy. Fear is a strong feeling of fright or awareness of danger. It causes you to want to escape. For example, you would feel fear if a car turned in front of you just as you crossed a street. Anything that might cause you physical or emotional pain can cause fear.

Your body reacts to fear. You may notice muscle tightness, especially in your legs. Your hands may shake nervously. You may bite your fingernails or feel like crying. To deal with fear, you can try to change the situation causing the fear or you can talk about the fear. Talking with a trusted person is a healthy way to deal with fear.

### Anxiety

An emotion related to fear is **anxiety,** a feeling of uneasiness or fearful concern. For example, when you are nervous about a test, you may feel anxious. You do not really feel fear, because the test cannot hurt you, but you may worry about getting a low grade. Anxiety is very unpleasant. The best way to deal with anxiety is to identify its source and take effective action. For example, if you are anxious about giving a speech, you can practice it. If you feel well prepared, you will be more relaxed and self-confident when giving your speech.

38  Unit 1  Personal Health and Family Life

**Anger**
*Strong feeling of displeasure*

**Aggression**
*Any action that intends harm to someone*

## Anger

**Anger** is a strong feeling of displeasure. Everyone feels angry at times. Strong anger, however, can lead to the desire to attack or be aggressive. **Aggression** is any action that intends harm to someone. Frustration, or a feeling of being blocked from something you want, often causes aggression. You might push someone who blocks your way. You might hit the vending machine that keeps your money. Aggression is a common impulse, but it is not an effective response to frustration. Aggression is likely to cause more problems than it solves.

When you are angry, you have a complaint about someone or something. The best way to express your anger is to say why you are angry and what you want to happen. You can do that without accusing or blaming another person by using "I" messages. For example, you could try saying, "I feel hurt when you tease me," instead of, "You hurt me when you tease me." Sometimes this alone solves the problem. Talking about your complaint is a more successful way to handle anger than aggression is.

Uncontrolled anger keeps you from thinking through your decisions clearly. Working out a solution and talking about feelings helps manage anger. Staying calm is another way to control anger. Self-talk also helps. For example, you can calm your body by telling yourself, "Stay calm. Relax. Stay in control." Taking deep breaths and breathing in and out slowly relaxes your muscles. You can redirect the anger by doing something physical such as taking a walk, exercising, or dancing to music. All of these actions can help you think more clearly and consider what to do when you feel angry.

### Research and Write

All children experience anger. For example, they may fight over a toy or yell at each other. Talk to preschool and elementary school teachers about positive ways to help children cope with their anger. Develop a creative booklet to show children what you learned. Share your booklet with an area preschool or elementary class, or share it with a child care center.

*Managing Emotions* Chapter 2 **39**

 **Warm-Up Activity**

Ask students to identify the emotions they feel when they experience stress. List the emotions on the chalkboard. At the end of the lesson, refer back to the list and ask students to suggest ways to deal with the emotions listed.

 **Teaching the Lesson**

To help students organize the material in this lesson, have them make a table that contains information about the five emotions linked to stress. Column titles could include "Emotion," "Definition," and "Ways to Cope."

**3 Reinforce and Extend**

### TEACHER ALERT

Be sure students understand the difference between fear and anxiety, as they are closely related and often may be felt at the same time. The ways of coping with fear and anxiety are also similar.

### LEARNING STYLES

 **Body/Kinesthetic**

Have students work in pairs to write and act out a skit in which they use I-messages to express their anger and frustration about something. The skit should include the problem or cause of the anger, a dialogue between two or more people using I-messages, and a positive and realistic resolution.

### Research and Write

Have students read and complete the Research and Write feature on page 39. Encourage students to write out a list of questions they would like answered before they interview the teachers. Encourage students to be creative and think about their audience as they design their booklets. A comic-strip design or cartoon-like characters may be appealing to preschoolers.

## LEARNING STYLES

**Auditory/Verbal**
Have students work in small groups. Ask a volunteer in each group to describe denial and explain how a person experiencing this step of the grieving process might feel. Encourage others in the group to listen carefully and write a summary sentence in their own words. Ask group members to take turns acting as the volunteer spokesperson while the others listen. Continue the activity for each step of the grieving process.

## HEALTH JOURNAL

Have students write a journal entry about a loss they have experienced, whether it was the death of a family member, the loss of a pet, or a friend moving away. Encourage students to describe how they felt and if they can recognize the five stages of grief in their feelings. If they would prefer, students may choose not to share this entry.

---

*Grief*
A feeling of sorrow; feelings after a loss

*Depression*
A state of deep sadness

*Relief*
A light, pleasant feeling after something painful or distressing is gone

### Grief
Loss is a common source of stress that you cannot fight or escape. You experience loss when a relationship breaks up, a good friend moves away, or someone dies. The feelings you experience after a loss are part of **grief,** or sorrow. Grief can be a mixture of several emotions, including fear, anger, anxiety, and guilt.

Every culture expresses grief. However, each culture deals with grief in its own way. Traditions direct how long people mourn the death of an individual. Different cultures have different rules for what they wear or how they act following someone's death. Respecting the way people from different cultures deal with grief is important.

Grief is a natural healing process. Usually a person goes through these reactions to grief:

- Denial—pretending a loss has not occurred
- Anger—feeling angry about the loss
- Bargaining—wishing for a second chance
- **Depression**—feeling deep sadness
- Acceptance—accepting something over which you had no control

Each reaction to grief can take a long time. The length of time is different for each person. Each person may go through the reactions in any order or may go back and forth among them. Talking about the painful emotions felt from the loss helps. Depression is the most common of these reactions. A depressed person sometimes needs professional help.

### Relief
**Relief** is the light, pleasant feeling you experience when stress is gone. Sometimes, stress returns. Then you only feel relief for a little while. You may be nervous, because you know the stress can return. For example, you feel relief because your math test was delayed until tomorrow. But the relief does not last long, because you still have to take the test. Learning to deal with stress brings relief that lasts. Relief is an emotion you can work to achieve.

**Cope**
*To deal with or overcome problems and difficulties*

### How Can Stressful Emotions Be Handled?

All emotions are normal. Emotions are neither good nor bad. Your mental health depends on how you handle an emotion. Pay attention to emotional signals and take positive action. Recognize what you do best and be willing to ask for help when you need it. If you need help with your homework, for example, ask a friend, parent, or teacher.

Being direct works best. Dropping hints or hoping others will guess what you need or how you feel usually does not work. Expressing your thoughts and feelings in an open and honest way helps improve communication between you and others. It also helps reduce problems that might arise from misunderstandings.

You can do a great deal to help yourself and others stay emotionally healthy, especially at school. Understand and follow school rules. Participate in your school's efforts to promote healthy living through your actions and words. Encourage others to live healthy lives.

### What Are Some Ways to Relieve Stress?

To **cope** is to find a way to overcome stress or other problems. Coping with a situation relieves the stress for a while and brings relief. You will always have to deal with problems and emergencies. Coping with one threat will not solve all your problems. However, it can reduce the number of your daily problems. When an emotion signals a need, the healthiest response is to cope.

#### Link to ➤➤➤
**Social Studies**

Different cultures and religions deal with death in different ways. In the Jewish faith, a person's body is buried in a wooden coffin within 24 hours of death. Mourning takes place for seven days afterward. Among Buddhists, bodies are burned until they become ash on the day of the funeral. The urn holding a person's ashes is honored for 35 days before burial. Among Muslims, the dead are wrapped in a plain cloth and buried in a raised grave soon after death.

## LEARNING STYLES

**Interpersonal/Group Learning**

Have students work in small groups to present several examples of situations when a person feels stressful emotions. Ask them what actions they might take to handle the emotion being described. Have each person in the group present one situation and a positive way to handle it.

### Link to

**Social Studies.** Have students read the Link to Social Studies feature on page 41. Ask students why they think cultures and religions would have such specific ways of dealing with death. *(Possible answers: It allows people to grieve. It helps people to feel closure.)* Ask student volunteers to share any traditions their culture or religion practices when someone dies.

## HEALTH JOURNAL

Have students identify a source of stress in their lives. Then have them write and carry out a plan to solve the problem and help relieve the stress. After they have carried out the plan have them record how they feel. Do they feel less stressed? Was the plan a success? What did they learn from the experience?

## ELL/ESL STRATEGY

**Language Objective:** *To summarize reading material*

Help students become more proficient in English by using sentence frames. Provide sentence frames such as the following to help students review lesson content:

When you identify your _____, you can determine how best to meet them. _____ is finding a way to deal with stress.

Decide on a way to solve a _____.

## LEARNING STYLES

**Interpersonal/Group Learning**

Ask each group of students to discuss common problems for teenagers. Examples might include difficulties budgeting time or saying no to negative peer pressure. Ask one volunteer in each group to list the problems discussed and then read the list. Instruct each group to choose one problem to enact. Have students work together to write a skit that shows a positive way to cope with the problem. Then ask students to perform their skit.

**Figure 2.3.1** *Studying is a way to handle school-related stress and improve your outlook for the future.*

You can do three things to cope with a problem:

1. Identify what causes the problem.
2. Decide on a way to solve the problem.
3. Carry out your plan as a way to cope.

> What is something positive you have learned from a success? What is something positive you have learned from a failure?

Coping helps you feel better and relieves anxiety. For example, you might feel anxious about finding a job. To cope with the problem, plan ways to get help—check the newspaper or online job ads and ask trusted adults for suggestions. Set a plan and act on it. Your chance for success in finding a job is better because you are able to cope with the problem.

When you solve a problem successfully, you relieve stress. Learning to cope with one problem may help you deal with other problems. You can learn from your successes as well as your mistakes.

Experience and time also help relieve stress. Often when people try something new, they find it stressful. The more you do things and gain experience, the less stress you will associate with those things. For example, a first date can be stressful. The more you date an individual or a variety of people, the more experience you have in dating. While the first date with a new person may still be stressful, the stress may be less powerful with experience.

**C** Calm down
**O** Outline options
**P** Pick a plan
**E** Evaluate your plan

How do you think planning ahead helps you relieve the stress of homework assignments?

Here are some other actions that can relieve stress:
- Pay attention to your emotional signals.
- Identify what you need—make sure your needs are reasonable.
- Take responsibility for meeting your needs.
- Ask respectfully for what you need.
- Get the rest of what you need honestly and fairly.
- Respect others, even when they may do or say things that are difficult for you to understand or accept.
- Seek help through talking with trusted individuals.
- Use positive self-talk—keep an "I can" attitude.
- Engage in physical activity.
- Set personal goals for maintaining a healthy body.
- Plan ahead.

### Health at Work

**Mental Health Specialist**

Do you think of yourself as caring and trustworthy? Then why not become a mental health specialist? Mental health specialists give emotional support and assistance to individuals and families. They help people work through marriage, family, or job-related problems. Some mental health specialists work just with infants and their families. They help new parents work through the changes that come with having a baby. They also assist families in getting help for any mental health or physical problems the new baby may have. Mental health specialists must have a college degree in social work, psychology, or a related field. Infant mental health specialists often have training in speech and language therapy, physical therapy, nursing, or child welfare. Being a mental health specialist can be a very rewarding career.

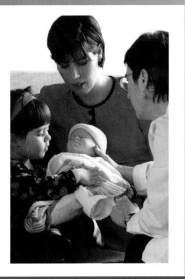

## LEARNING STYLES

### Logical/Mathematical

Have students make a graphic organizer, such as a web diagram, that features the actions listed that can relieve stress and promote mental health.

## LEARNING STYLES

### Auditory/Verbal

Have students work in pairs. One student should describe a scenario of a teenager who feels overwhelmed with stress. Have the student name some causes of the stress (for example, stress related to school, changes in life situation, social pressures, etc.). Ask the student's partner to listen carefully, summarize the causes of stress presented, and suggest steps the teenager can take to cope with the stress. Then have the first student summarize the coping steps suggested by the partner.

### Health at Work

Have volunteers read the Health at Work feature on page 43 aloud to the class. Discuss the skills a good listener needs such as paying close attention to what another person is saying, not interrupting, listening to a person's tone of voice, watching body language, and staying calm. Ask students why good listening skills would be important for a mental health specialist.

### IN THE COMMUNITY

Invite a representative from a community mental health agency to speak to the class about programs and services available to teenagers. Point out that such services are confidential.

## Lesson 3 Review Answers

1. grief 2. coping 3. relief 4. Answers will vary but should include the concepts of acting as a positive role model for classmates and following school rules. 5. Coping provides relief because it helps an individual overcome stress. Coping involves identifying a problem, deciding on the best way to solve the problem, and carrying out the plan.

### Portfolio Assessment

Sample items include:
- Table from Teaching the Lesson
- Stress management plan from Health in Your Life
- Lesson 3 Review answers

### Health in Your Life

Have students read and complete the Health in Your Life feature on page 44. After answering the questions, have students write a complete stress management plan in their Health Journals.

**Health in Your Life Answers**

1. A stress management plan will help you deal with day-to-day problems.
2. Identifying the cause of stress is important so you can avoid it.
3. Relaxation allows a person to make the decisions needed to better handle stress.

---

**Word Bank**
coping
grief
relief

**Lesson 3 Review** On a sheet of paper, write the word from the Word Bank that best matches each description.

1. Feeling you might have after a loss
2. Dealing with or overcoming problems
3. A feeling after stress is gone

On a sheet of paper, write the answers to the following questions. Use complete sentences.

4. **Critical Thinking** What can you do at school to help yourself and others stay emotionally healthy?
5. **Critical Thinking** How does coping help relieve stress?

### Health in Your Life

**Stress Management Plan**

Everyone experiences stress. Finding a way to manage stress will help you lead a healthy life. One effective way to manage stress is to have a stress management plan. Making a stress management plan will help you recognize the things in your life that cause stress. It will help you plan how you will manage that stress. Here is how you can have an effective stress management plan:

- List five things that create the most stress for you.
- Write one way you can avoid stress.
- List five things that happen to your body when you feel stress.
- Write one idea for what you will do when your body shows signs of stress.
- List five things that you can do to make you feel more relaxed.
- Put your plan into action!

On a sheet of paper, write the answers to the following questions. Use complete sentences.

1. How will a stress management plan help you?
2. Why is identifying what causes you stress important?
3. Why is relaxation important?

44   Unit 1   Personal Health and Family Life

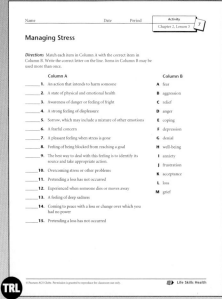

Activity 7

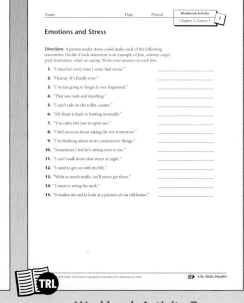

Workbook Activity 7

44   Unit 1   Mental and Emotional Health

# Chapter 2 SUMMARY

- Emotions begin with thoughts and may cause physical changes or reactions.
- Five basic human needs are physical, safety, belonging, esteem, and self-actualization needs.
- Physical and safety needs are the most basic needs.
- The two kinds of belonging needs are belonging in a place and belonging with people.
- Esteem needs involve the esteem of others and self-esteem, which your self-concept determines.
- Self-actualization needs involve achieving your possibilities and going beyond yourself to give to others.
- While people have the same needs, they may have different ways of showing those needs.
- Stress is a natural mental and physical reaction to anything that requires you to adapt.
- Your brain triggers physical reactions to stress. It sends chemical messages into your bloodstream that may cause faster breathing, blushing, sweating and other physical reactions.
- Anger can lead to aggression. The best way to handle anger is to talk about it.
- Emotions that are often linked to stress are fear, anxiety, anger, grief, and relief.
- All emotions are normal. How you handle emotions is important to your mental health.
- Coping with stress involves thinking through a problem, deciding how to solve the problem, and carrying out your decision.

| Vocabulary | | |
|---|---|---|
| aggression, 39 | esteem, 29 | relief, 40 |
| anger, 39 | family love, 30 | romantic love, 30 |
| anxiety, 38 | fear, 38 | self-actualization, 32 |
| attachment, 30 | friendship, 30 | self-concept, 31 |
| body language, 28 | grief, 40 | self-esteem, 31 |
| cope, 41 | guilt, 33 | stress, 34 |
| depression, 40 | hierarchy, 29 | stress response, 34 |
| emotion, 28 | reaction, 28 | well-being, 38 |

## Chapter 2 Summary

Have volunteers read aloud each Summary item on page 45. Ask volunteers to explain the meaning of each item. Direct students' attention to the Vocabulary box on the bottom of page 45. Have them read and review each term and its definition.

### ONLINE CONNECTION

For more information about stress management, have students go to the National Library of Medicine and National Institutes of Health Medical Encyclopedia Web site at: www.nlm.nih.gov/medlineplus/ency/article/001942.htm

This site, offered by GirlsHealth.gov and specifically meant for girls, will help students learn how to handle stress in everyday and unusual situations: http://www.4girls.gov/mind/stress.htm

Another Web site that is geared toward the stresses that teens may face is from the American Academy of Child and Adolescent Psychiatry. Note that this site is designed for parents and teachers, not students. The site contains a list of things teens can do to manage stress: www.aacap.org/publications/factsfam/66.htm

## Chapter 2 Review

Use the Chapter Review to prepare students for tests and to reteach content from the chapter.

## Chapter 2 Mastery Test

The Teacher's Resource Library includes two forms of the Chapter 2 Mastery Test. Each test addresses the chapter Goals for Learning. An optional third page of additional critical-thinking items is included for each test. The difficulty level of the two forms is equivalent.

## Review Answers

**Vocabulary Review**

1. emotions 2. attachment 3. hierarchy
4. self-esteem 5. stress 6. stress response
7. friendship 8. depression 9. anger
10. aggression 11. grief 12. cope 13. relief

### Teacher Alert

In the Chapter Review, the Vocabulary Review activity includes a sample of the chapter's vocabulary terms. The activity will help determine students' understanding of key vocabulary terms and concepts presented in the chapter. Other vocabulary terms used in the chapter are listed below.

| | |
|---|---|
| anxiety | guilt |
| body language | reaction |
| esteem | romantic love |
| family love | self-actualization |
| fear | self-concept |
| friendship | well-being |

46 Unit 1 Mental and Emotional Health

---

# Chapter 2 REVIEW

**Word Bank**
aggression
anger
attachment
cope
depression
emotions
friendship
grief
hierarchy
relief
self-esteem
stress
stress response

## Vocabulary Review

On a sheet of paper, write the word or phrase from the Word Bank that best completes each sentence.

1. Individual reactions to experiences are _____.
2. A relationship based on need is called a(n) _____.
3. Maslow identified a(n) _____ of needs.
4. Self-concept determines _____.
5. A state of physical or emotional pressure is _____.
6. Blushing is an example of a physical _____.
7. Love based on choices is _____.
8. A state of deep sadness is _____.
9. Feelings most often linked to stress are fear, anxiety, _____, grief, and relief.
10. Anger can lead to _____, which is any action intended to harm someone.
11. Another word for deep sorrow is _____.
12. Finding positive ways to _____ with stress helps relieve anxiety.
13. A light, pleasant feeling that comes after stress is gone is _____.

## Concept Review

On a sheet of paper, write the letter of the answer that best completes each sentence.

14. When you are sheltered from harm, your _____ need is being satisfied.
   A belonging
   B safety
   C self-actualization
   D physical

46 Unit 1 Personal Health and Family Life

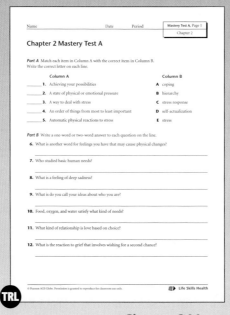
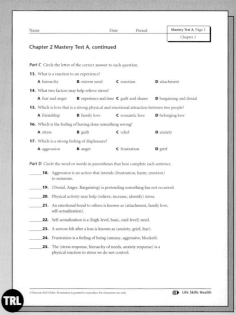

Chapter 2 Mastery Test A, pages 1–3

**15.** People experience _____ when their basic needs are threatened.
   A relief
   B aggression
   C grief
   D stress

**16.** Fear is a common response when a person's _____ is threatened.
   A well-being
   B anxiety
   C anger
   D grief

**17.** Coping with stress includes identifying the problem, deciding how to solve the problem, and _____.
   A gaining relief
   B carrying out the plan
   C thinking through the issue
   D solving another problem

## Critical Thinking

On a sheet of paper, write the answers to the following questions. Use complete sentences.

**18.** How can stress be both good and bad?

**19.** Why do you think coping with stress and other emotions is important to your physical and mental health?

**20.** What could you do to help a friend who feels sad and upset?

**Test-Taking Tip** If you know you will have to define certain terms on a test, write each term on one side of a card. Write its definition on the other side. Use the cards to test yourself, or work with a partner and test each other.

*Managing Emotions* Chapter 2 47

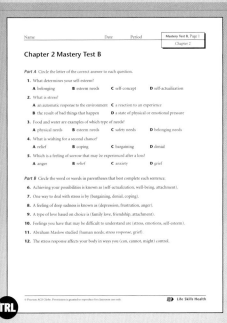
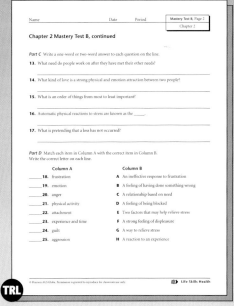

Chapter 2 Mastery Test B, pages 1–3

### Concept Review
14. B  15. D  16. A  17. B

### Critical Thinking
**18.** Answers will vary but should include the idea that stress can be good if it serves to better a situation, such as improve performance in a sports event, an audition, or an important test. Stress can be bad when the effects harm an individual. **19.** Answers will vary but should include the ideas that coping can prevent serious harm to the body, foster good mental health, and maintain good relationships. It also can bring relief and the ability to deal with other problems as they arise. **20.** Answers will vary but should include positive examples of ways to help a friend cope with a stress-related event. A friend can help another person simply by offering to listen when someone needs to talk about his or her feelings.

### ALTERNATIVE ASSESSMENT

Alternative Assessment items correlate to the student Goals for Learning at the beginning of this chapter.

- Have students orally define emotions and explain how they happen.
- Have students draw Maslow's hierarchy of needs and provide a brief explanation of each need.
- Have students write a paragraph describing three ways mental health can affect physical health.
- Have students choose one thing in their lives that causes them stress and complete a stress management plan for it.

*Managing Emotions* Chapter 2 47

## Chapter at a Glance

### Chapter 3: Maintaining Mental Health
pages 48–73

### Lessons

1. Influences on Mental Health
   pages 50–54
2. Emotional Health
   pages 55–60
3. Healthy Relationships
   pages 61–65
4. Becoming More Emotionally Healthy
   pages 66–70

**Chapter 1 Summary** page 71

**Chapter 1 Review** pages 72–73

**Audio CD**

**Skill Track for Life Skills Health**

**Teacher's Resource Library**
- Activities 8–11
- Alternative Activities 8–11
- Workbook Activities 8–11
- Self-Assessment 3
- Community Connection 3
- Chapter 3 Self-Study Guide
- Chapter 3 Outline
- Chapter 3 Mastery Tests A and B

(Answer Keys for the Teacher's Resource Library begin on page 559 of this Teacher's Edition.)

### Opener Activity

Ask students to suggest words that describe a mentally healthy person. Write the words on the chalkboard and discuss them. Help students develop a definition for good mental health based on the words they chose. Save the definition for later reference.

48  Unit 1  Mental and Emotional Health

## Self-Assessment

### Can you answer these questions?

 1. How would you describe yourself?

 2. What early experiences have influenced your personality?

 3. What is social esteem?

 4. What actions can you take to maintain and improve your social health?

 5. What are some of your values and beliefs?

 6. How do you manage your impulses?

 7. What are healthy ways to cope with emotional challenges?

 8. How can you solve problems through compromise?

 9. How are communication skills important to your well-being?

 10. How are citizenship skills important to your well-being?

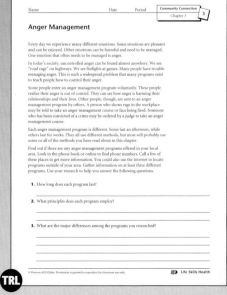

Self-Assessment 3     Community Connection 3

# Chapter 3
# Maintaining Mental Health

Many things can affect your mental and emotional health. Your personal well-being has many sources. Learning about yourself can make a difference in your emotional well-being. It also can make a difference in how you look at things and in the quality of your relationships.

In this chapter, you will learn what affects mental health. You will learn about the characteristics, or qualities, of mentally healthy people and healthy relationships. You will also learn some ways that you can improve and maintain good mental and emotional health.

### Goals for Learning

◆ To describe well-being and ways to promote it
◆ To identify ways to reach and maintain good emotional health
◆ To explain the characteristics of healthy relationships
◆ To describe ways to accept yourself and others

## Introducing the Chapter

Have students leaf through the chapter, noting the lesson titles and boldfaced subheads. Discuss what sort of information students think might be presented in this chapter. Ask students why this information might be important to everyone and not just to people who are currently having problems. *(Anyone might have problems at some time in the future; some of the information can help people keep problems from developing.)*

Have students read the Goals for Learning. Explain that in this chapter students will learn how to recognize and promote good mental health and healthy relationships.

### Notes and Questions

Ask volunteers to read the notes and questions that appear in the margins throughout the chapter. Then discuss them with the class.

### Self-Assessment

Have students complete the Self-Assessment worksheet before and after reading the chapter. Before reading the chapter, have students fill in the "Before" column. Ask students to identify their goals for learning. To get ideas for setting goals, students might use the chapter introductory material on page 49, the checklist on page 48, or the questions on the Self-Assessment worksheet. Students can use the back of the worksheet if they need more space to write.

Collect the Self-Assessment worksheets and pass them out again at the end of the chapter. Have students fill in the "After" column. Ask them to identify at least four major points they have learned. Again, suggest they use the back of the worksheet if they need more space to write. You may want to collect and review the worksheets, but return them to students so they have a record of their goals and accomplishments.

Chapter 3 Self-Study Guide, pages 1–2     Chapter 3 Outline, pages 1–2

## Lesson at a Glance

### Chapter 3 Lesson 1

**Overview** This lesson explains what influences mental health and describes actions that promote well-being.

### Objectives

- To explain how personality develops
- To identify influences on self-concept
- To describe how the different areas of well-being work together
- To promote physical, emotional, social, and personal well-being

**Student Pages** 50–54

**Teacher's Resource Library**

Activity 8
Alternative Activity 8
Workbook Activity 8

### Vocabulary

| | |
|---|---|
| personality | temperament |
| social esteem | well-being |

Write definitions for the vocabulary words on the chalkboard. Ask a student to read a definition and challenge the class to name the word that matches the definition. Continue until all the words are reviewed.

### Background Information

Many theories of personality have been developed to explain how and why people think, feel, and behave as they do. They fall under three broad orientations: psychoanalytical, behavioristic, and humanistic. Sigmund Freud used psychoanalysis to explore the unconscious. He felt the answers laid buried inside the person and had to be brought to the surface with the help of a skilled therapist. B.F. Skinner observed a person's behavior. Humanists, like Abraham Maslow and Carl Rogers, studied both consciousness and experience. The study of personality remains in the area of theory rather than science.

---

## Lesson 1  Influences on Mental Health

**Objectives**

*After reading this lesson, you should be able to*

- explain how personality develops
- identify influences on self-concept
- describe how the different areas of well-being work together
- promote physical, emotional, social, and personal well-being

**Personality**
*All of one's behavioral, mental, and emotional characteristics*

**Temperament**
*A person's emotional makeup*

How can a person who gets teased a lot develop a good self-concept?

What kind of person are you? Friendly, helpful, cooperative? Grouchy, fearful, selfish? You may at different times be all of these things. Your thoughts, feelings, and behaviors form your **personality.** As you grow older, your personality will change. This is because your thoughts and behaviors will change as a result of your experiences and opportunities.

### How Does Personality Develop?

When you were a baby, your personality had not yet developed. However, you did have a certain temperament. **Temperament** is your emotional makeup. A baby's temperament appears soon after birth. As children grow, they develop new thoughts and feelings. They learn new behaviors. Their temperament remains the same, but they begin to develop their personality.

Early experiences may influence a child's personality. For example, a baby needs close contact with a caregiver to feel secure. If that need is met, the baby is content. Whether this need is met as a baby may influence how that person relates to others later.

### What Influences Your Self-Concept?

Several factors influence self-concept. One is your own natural positive esteem. Another is what other people tell you about yourself. Suppose your music teacher says you have a good sense of rhythm. Part of your self-concept is that you have musical talent. The actions of others also help shape self-concept. Suppose team captains want to choose you for every game. You might think of yourself as a good athlete.

The messages you get from others are called social messages. These messages may not always match your self-concept. For example, you may think of yourself as shy and not interesting. Your friends, however, may think you are quiet and sincere. They say you have a good sense of humor. In this case, you may decide to change your belief about yourself.

50  *Unit 1  Mental and Emotional Health*

**Social esteem**
*How others value a person*

Sometimes, you will not want to change a belief about yourself. Suppose you believe you are good at baseball, but you drop the ball during a game. Your classmate tells you that you are not a good player. You made four other good plays, so you decide not to change your belief.

Social messages play a part in forming your self-concept. They reflect your **social esteem,** or how others value you. If others have a good opinion of you it is more likely you will have high self-esteem. Unfortunately, this also works the other way. If others do not seem to value you, it is more likely you will have low self-esteem.

**Figure 3.1.1** *Images in the media are an important influence on self-concept.*

One important influence on self-concept is media such as TV and magazines. Many advertisements show people who are beautiful, talented, muscular, or just the right size. If you compare yourself to the people in these ads, you might feel that you do not measure up. This can lead to a negative self-concept. You can promote, or support, a positive self-concept by not comparing yourself to the people you see in the ads. Focus on your positive qualities. Remember that you are special and unique, or one of a kind.

### What Is Well-Being?

Well-being is feeling happy, healthy, and content. Well-being involves balancing many different needs and behaviors. If you feel healthy and content, then many areas of your well-being are working together. These include your physical, emotional, social, and personal well-being.

### Link to ➤➤➤
**Biology**

The brain contains chemicals that carry messages between it and other parts of your body. These chemicals are present at birth and influence temperament. Brain chemicals can affect whether you tend to be irritable or calm.

 **Warm-Up Activity**

Have students suggest the names of three or four fictional characters that most or all of the class is familiar with. Good sources are a popular television series or a book students are currently reading in school. Then have students suggest words that describe each character's personality. Ask students the following questions about each character: Is the character physically healthy? How well does the character handle stress and problems that arise? How well does the character get along with the other characters? How does the character see himself or herself?

 **Teaching the Lesson**

Help students understand that one should not base one's self-concept on comparisons to ideal examples. A student playing basketball in gym class should not be upset because she doesn't play as well as a professional. A student singing along with his MP3 player should not be depressed because he doesn't sing as well as his favorite pop star. Help students see that it would be just as foolish to compare their appearance to pictures of models and movie stars.

 **Reinforce and Extend**

### Link to

**Biology.** Have a volunteer read aloud the Link to Biology feature on page 51. Students who wish to learn more about these brain chemicals can research some of these: serotonin, neuropeptides, endogenous opioids, oxytocin, and vasopressin.

## Health in Your Life

Have students read the Health in Your Life feature on page 52. Point out to students that people often use support systems for help. If your wrist were sore, you might ask your coach or the school nurse to look at it. If a friend were on crutches for a couple of weeks, you might help by carrying the friend's books to class.

**Health in Your Life Answers**

1. Talking to people helps you have good mental health by helping you understand things more clearly. 2. It is a good idea to have a list of names ready when you need support so that you will not have to spend time looking them up and you will be more likely to call. 3. Sample answer: I felt down about a bad grade in math. If I had talked it over with someone, I might not have been upset for as long as I was.

## ELL/ESL STRATEGY

**Language Objective:** *To describe by example*

Have students form groups composed of a student learning English and several students with strong English skills. Provide each group with a scenario that illustrates a social message being given to someone. Examples include someone being put down or someone being cheered on in a race. Ask students to perform a skit of the scenario and have their classmates identify the social message being depicted and how a person might react to it. Then have students discuss both healthy and unhealthy ways of dealing with the situation. Groups can repeat the scenario and add an example of a healthy way of dealing with the message.

---

Physical well-being is your body's ability to meet the demands of daily living. You are physically healthy and free from pain and illness.

Emotional well-being is your ability to handle problems and stress in daily life. You look at life in a positive way and are in control of your emotions.

Social well-being is your ability to get along well with others. You can successfully work, play, and talk with others many times every day. Social well-being promotes healthy relationships.

Your sense of personal well-being depends on whether you are satisfied with your values and beliefs. Your values and beliefs increase your well-being if they are right for you and respectful of others. Your system of beliefs can guide you when you feel confused. It can help you make good decisions.

### Health in Your Life

**Create Your Own Support System**

An important part of mental health is asking others for support. You might have a problem you cannot solve on your own. You might feel down. Talking to someone helps you see things more clearly. Make a list of people you can contact for these times. Write down the names, phone numbers, and e-mail addresses of your closest friends. Include other trusted adult family members and maybe your school counselor or the school nurse.

In turn, be a health advocate for your friends by supporting their health. You can support your friends' health by offering to listen when they need to talk. Make sure your friends have your contact information. Let them know you are there for them.

On a sheet of paper, write the answers to the following questions. Use complete sentences.

1. How can talking to people help you have good mental health?
2. Why is it a good idea to have a list of names ready when you need support?
3. Think about a time you needed support but did not ask for it. How do you think talking might have helped?

### How Are the Forms of Well-Being Related?

Your overall well-being depends on how all these areas of well-being work together. For example, when you are physically healthy, you can handle emotional challenges better. A good mental attitude can promote physical health. Problems in one area of well-being can affect another area. For example, you may feel depressed if you do not feel well physically. You may not want to be with others if you are worried about something.

### What Actions Promote Your Well-Being?

You can promote your well-being when you pay attention to how its different parts work together. Problems in one area may cause problems in another. The key to mental health is to support all four areas of your well-being. You do this by practicing healthy behaviors in all four areas.

Physical well-being means eating right, resting, and exercising. If you do become ill, knowing what to do can help you get well faster. For example, when you know that rest helps you get over the flu, you may sleep more.

For your emotional well-being, you need to learn how to handle stress. Many times, physical activity helps to relieve stress. This can be as simple as taking a walk. Understanding and dealing with your different emotions will also help relieve your stress.

To promote your social well-being, you can improve your understanding of others. You can learn about what others need. Some social skills that you can learn are:

- communication skills—listen, ask questions, be honest in what you say, and pay attention when others speak
- friendship skills—be a friend, ask others for advice, show respect and loyalty, and keep promises and confidences
- citizenship skills—follow the rules of society, respect others, do your part, and volunteer to help

> **Research and Write**
>
> What you choose as entertainment can help you have good mental health. It also can have negative health effects. Work in small groups to research healthy forms of entertainment. Then prepare posters with messages about healthy forms of entertainment. Aim the messages at people who are your age. Display the posters in the cafeteria, gym, or hallway.

---

### Research and Write

Have students read and conduct the Research and Write feature on page 53. The best source for finding the necessary information is the Internet. Show students how to use a search engine and keywords to find helpful Web sites. Some newspapers list "family" shows in the weekly television guide. Students can also look for reviews in magazines.

### LEARNING STYLES

**Visual/Spatial**

Draw a concept map on the chalkboard that illustrates how physical, emotional, social, and personal well-being work together to promote overall well-being. Discuss how a change in one kind of well-being might affect other kinds of well-being.

### TEACHER ALERT

Help students think of things people can do to improve their personal well-being, physical well-being, emotional well-being, and social well-being. Point out that what works for each individual may be different. Help students see that the first three involve a change in behavior or attitude of the person, but improving social well-being requires changing the behavior and attitude of others.

# Lesson 1 Review Answers

1. One important early influence on personality development is whether a baby has close contact with a parent or caregiver. 2. The four areas of well-being are physical, emotional, social, and personal. 3. Some skills that promote social well-being are communication skills, friendship skills, and citizenship skills. 4. Sample answer: My history teacher told me I had written a good paper. That helped me feel smart. 5. Sample answer: The student might keep to herself instead of making new friends. She might decide not to get close to anyone else for a while.

## Portfolio Assessment

Sample items include:
- Health in Your Life answers
- Lesson 1 Review answers

## IN THE ENVIRONMENT

Many different elements of the environment can affect a person's well-being, both physical and emotional. These elements are not limited to the environment outdoors. For example, "sick building syndrome" is caused by some substance, often unknown, in a building that affects the physical health of the people inside. Toxic mold in homes can also damage a person's physical and mental health. Have students research an environmental factor that affects people's well-being. Invite them to give oral presentations describing the effects of the factor and what, if anything, can be done to prevent or control it.

---

For your personal well-being, you need to decide what your values and beliefs are. You need to figure out what is important to you and act in a way that is true to your beliefs. Is it right to help a friend? Is it right to copy answers on a test? Is it right to stay up too late and oversleep the next day? You face many similar questions and choices every day. Your sense of personal well-being can help you find the answers.

Working on all these areas of well-being helps promote your mental health. Maintaining your mental health can be difficult. Sometimes, you will be more successful than other times. However, maintaining your mental health is important to your overall well-being.

***Lesson 1 Review*** On a sheet of paper, write the answers to the following questions. Use complete sentences.

1. What is one important early influence on personality development?
2. What are the four areas of well-being?
3. What are some skills that promote social well-being?
4. **Critical Thinking** Think of a positive social message that someone has given you recently. The person could be a classmate, teacher, or family member. How did that message affect your self-concept?
5. **Critical Thinking** A student is depressed because her best friend moved away. How might her depression affect her social health?

54   Unit 1   Mental and Emotional Health

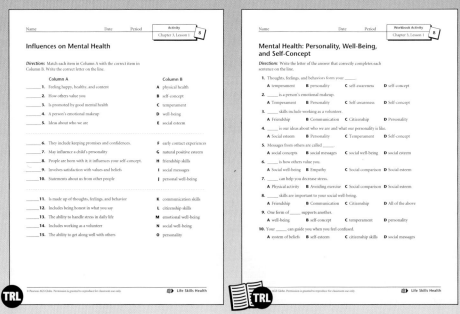

## Lesson 2: Emotional Health

### Objectives

After reading this lesson, you should be able to
- describe how to maintain healthy relationships
- explain how to set and reach goals
- use problem-solving skills to settle conflicts
- identify healthy ways of thinking

Emotional well-being is an ongoing process rather than an unchanging state of mind. Everyone has occasional setbacks and problems. Emotionally healthy people learn to solve problems and adjust to their world. People can learn abilities and behaviors to make themselves emotionally healthy.

### How Can a Person Learn and Change?

People learn most things through experience. They change their behavior when they discover a better way to act.

When you make a mistake, you may feel upset or embarrassed. You can overcome this feeling by reminding yourself that everyone makes mistakes. You can also analyze, or figure out, what you learned from your mistake. Then you can decide to act differently next time.

### How Can Relationships Be Maintained?

Problems in relationships are common. After all, people do not always want the same things or see things the same way. Emotionally healthy people learn how to communicate well with others. They usually can balance their own wishes with those of others. They learn to get along. When they have problems with others, they work through the problems. They pay attention to the needs of friends and family members.

Relationships change as you mature, or grow older. When you were young, you depended completely on others for love, care, and survival. As you mature, you form ties to other people. Emotionally healthy people value these ties. At the same time, they think for themselves. For example, you may ask your friends for advice about an important decision. You do that because you value their opinions.

In the end, however, you must decide for yourself what is right for you. Emotionally healthy people can form bonds with others and still follow their own belief system.

### Health Myth

**Myth:** Conflicts in relationships are not normal.

**Fact:** Conflicts are common and natural because people have different wants and needs. They see things in different ways. You cannot avoid conflict. What you can do is deal with conflict in a peaceful way. The skills you learn now to solve conflicts will help you the rest of your life.

---

## Lesson at a Glance

### Chapter 3 Lesson 2

**Overview** This lesson explains how people can develop and maintain emotional health

### Objectives
- To describe how to maintain healthy relationships
- To explain how to set and reach goals
- To use problem-solving skills to settle conflicts
- To identify healthy ways of thinking

**Student Pages** 55–60

**Teacher's Resource Library**
Activity 9
Alternative Activity 9
Workbook Activity 9

### Vocabulary

compromise          rational
disadvantage        realistic
impulse             resilient
optimism            symptom
pessimism

Ask students to review the definitions, use each word in a sentence, and then describe a situation that illustrates each vocabulary word.

### Background Information

People who do not know how to set realistic goals and achieve them are likely to feel that they have no control over their lives. They might find that the decisions they make do not get them what they want. People who learn to set and achieve realistic goals are more likely to make the best decisions for themselves, succeed in school, and plan well for the future.

 **Warm-Up Activity**

Have students read page 55. Ask them to identify events or experiences that might cause people to change. For example, making a mistake, having a health problem, or having an argument with a friend might cause changes. For each event or experience, ask what a person might learn from it.

### Health Myth

Have students read the Health Myth feature on page 55. Ask one or two volunteers to describe conflicts they have had recently with friends or family members. Then have the class suggest ways the conflict might be resolved peacefully.

**Teaching the Lesson**

Help students understand that some goals are realistic and others are not. Ask volunteers for examples of self-improvement goals that are too pessimistic. Then ask for examples of self-improvement goals that are unrealistically optimistic. Discuss how each example could be turned into a realistic goal.

**3 Reinforce and Extend**

### Decide for Yourself

Prepare several photos from a magazine or newspaper to bring to class. Each photo should depict someone with an obvious problem. Advertisements and stills from movies often show this.

Bring in the photos on the day students will read Decide for Yourself. Have them read the text on page 56. Then ask them to look at each picture and identify the problem each person has. Challenge students to prepare a goal-setting plan for each person. Students can work in groups and each group work with one picture.

**Decide for Yourself Answers**

1. Sample answer: My goal is to learn how to play the guitar and then perform in a public setting. 2. Sample answer: Here are the steps I need to follow: (a) buy or rent a guitar; (b) find a music teacher; (c) learn and practice playing the guitar; and (d) arrange a public performance. Here is my timeline: After I find a guitar and a teacher, I will practice for 15 months before I plan on playing publicly. Then I will arrange a public performance in 2 months. Here are potential problems: I need to save money for a guitar and lessons; I need to manage my time so that I can do everything I need to do; and I need to overcome my shyness in order to give a public concert. I will keep a journal to check my progress.

---

**Realistic**
*Practical or reasonable*

**Pessimism**
*Tending to expect the worst possible outcome*

**Optimism**
*Tending to expect the best possible outcome*

## The Importance of Goals

For good emotional health, you need to set **realistic,** or reasonable, goals. Realistic goals are balanced between **pessimism** and unrealistic **optimism.** Pessimism is tending to expect the worst possible outcome. If you feel pessimistic about a goal, you may give up without trying. Optimism is tending to expect the best possible outcome. If your goal is too optimistic, you may not be able to reach it.

### Decide for Yourself

Following the steps in this sequence chart can help you set and reach goals.

**Setting Goals**

Write down your goal.
↓
List the steps you need to reach your goal.
↓
Set up a time line for completing each step.
↓
Identify any problems that might get in your way.
↓
Check your progress as you work toward your goal.

On a sheet of paper, write the answers to the following questions. Use complete sentences.

1. List a goal you would like to achieve.
2. Go through the "Setting Goals" steps, using your specific goal.

*Compromise*
Agree by giving in a little

*Rational*
Realistic or reasonable

*Impulse*
The urge to react before taking time to think

## How Can Conflicts Be Solved?

Conflicts happen for many reasons. Suppose you argue with a friend about how to raise money for your school. You think your plan is best. Your friend disagrees. You both become frustrated, or upset. Frustration is a common feeling when you are kept from reaching a goal. Emotionally healthy people find a way to solve conflicts. One way is to **compromise.** Both sides agree to give in a little. In that way, no one "loses." Another way to solve a conflict is to look at other choices. You and your friend could find out about other ways to raise money. Then you can work together to decide which way is best.

**Rational** thinking is important in solving problems. Something rational is realistic or reasonable. For example, imagine you want to buy a new music system. Your parents think it costs more than you can afford. You have to think rationally. Do you have enough money, or can you earn the amount you need? If you borrow the money, will you be able to pay it back? After careful thought, you may decide that your plan can work. Or you may decide that the item is too expensive to buy.

## How Can Emotions Be Managed?

Remember that emotions signal needs. However, these signals are not always adjusted to the correct "volume." For example, when you are tired, you may react too strongly, or overreact, to a situation. If you were rested, you might react with more control.

Everyone has **impulses.** An impulse is the urge to react before taking time to think. For example, if someone bumps into you, your impulse may be to bump that person back. Emotional health depends on being able to think about the consequences and learning to control your impulses. Consider what might happen if you react a certain way. In this example, you can predict that if you followed through with your impulse and bumped the person back, you might get into an argument or even a fight. You think about it and realize that the person did not bump you on purpose, and you let it go. You have thought about the consequence and controlled your impulse.

## ELL/ESL Strategy

**Language Objective:**
*To describe by example*

Have students form groups composed of a student learning English and several students with strong English skills. Have each group devise a scenario in which conflict arises over a lack of verbal communication. Then have students act out the scenario. If they have time, encourage them to act it out several times with each one taking a turn pretending to be the one who has no English at all. Have each group devise a way to resolve the conflict nonverbally. Ask them to discuss how body language and nonverbal communication helped them.

## Health Journal

Have students describe in their journals several realistic personal goals. Suggest that for each goal they list steps they could take to achieve that goal.

## HEALTH JOURNAL

Have students describe in their journals different methods they have used in the past year to cope with stress and anxiety, as well as with disappointment, hardship, or failure.

## GLOBAL CONNECTION

Ask students to research ways in which members of the Peace Corps help in other parts of the world. Encourage them to explore how Peace Corps volunteers work in areas with different values and how coping skills are important for volunteers. Interested students may want to find out more about the education and skills required for members of the Peace Corps.

## LEARNING STYLES

### Logical/Mathematical
Draw a web diagram on the chalkboard and have students copy it into their notebooks. Put "Healthy Ways of Thinking" at the center of the web. Draw four branches off the center and label each one with one of the subheads on pages 58–59. Draw enough branches off each subhead to include each of the bulleted items listed underneath it. (The bulleted items for the subhead "Healthy Explanations" appear on page 60.)

---

*Disadvantage*
A condition or situation that makes reaching a goal harder

## What Are Some Healthy Ways of Thinking?

Most actions and emotions begin with thoughts. For example, if you think someone has insulted you, you may feel angry. If you believe the person did not mean to insult you, you may not feel angry. Emotionally healthy people think in ways that help them adjust and reach their goals. They have four helpful thinking styles—realistic optimism, meaningful values, coping, and healthy explanations.

### Realistic Optimism
Realistic means practical. When emotionally healthy people think with realistic optimism, they:

- set and work toward goals that are challenging but reachable
- recognize what can and cannot be controlled in their life
- have faith in themselves and hope that events will work out for the best

### Meaningful Values
Emotionally healthy people make sure their actions are based on their values. They look on **disadvantages** as challenges to be overcome. A disadvantage is a condition or situation that makes reaching a goal harder. Emotionally healthy people:

- develop values that become part of them and guide their thinking and behavior
- reflect on their own feelings and thoughts instead of worrying about how others might judge them
- face and overcome disadvantages
- find meaning in their work or actions

### Coping
People respond, or react, in two ways to stress—avoidance or coping. Sometimes, avoidance is the best response. For example, you should get away from someone who tries to harm you. Most often, however, avoidance does not remove the stress.

**Symptom**
*A bodily reaction that signals an illness or a physical problem*

**Resilient**
*Able to bounce back from misfortune or hardship*

If you avoid a problem, you may develop **symptoms** of illness. A symptom is a bodily reaction that signals an illness or a physical problem. For example, you may feel anxious about a long assignment. You could avoid the feeling by playing video games and trying to forget about it. You will probably still continue to feel anxious. You may even develop other symptoms, such as a headache or an upset stomach. Coping deals with and removes stress and anxiety.

Emotionally healthy people cope by:

- learning from experiences, mistakes, and successes
- knowing when to change behaviors that are not working
- asking for help from family members, school counselors, or other trusted adults
- trying again after a failure

Figure 3.2.1 *People can overcome disadvantages.*

**Healthy Explanations**
Sometimes a loss or a disappointment can result in stress. **Resilient** people tend to bounce back or recover easily from misfortune or hardship. How do they do it? Resilient people have a positive way of looking at things. They cope by talking to friends and family. They often find that volunteering in their communities and giving back to others less fortunate is helpful. Resilient people believe they have some control over their lives. They accept that they cannot control everything.

## LEARNING STYLES

**Visual/Spatial**
Direct students' attention to the photograph on page 59. Point out to students that many people face and overcome physical disadvantages to achieve success. Encourage interested students to write a story based on one of the players in the picture. Have them include the person's achievements and how he maintained a positive attitude to overcome his disadvantage.

## LEARNING STYLES

**Interpersonal/Group Learning**
Encourage groups of students to act out a situation in which a person chooses behavior based on his or her own strong values in spite of negative pressure from some peers to act otherwise. Examples include saying no to drugs and refusing to play a cruel practical joke on someone.

## Lesson 2 Review Answers

1. optimism  2. meaningful values  3. compromise  4. Sample answer: The student could ask for extra help from the English teacher or a family member. The student could see passing English as a challenging goal she will work hard to reach.  5. Sample answer: I would think about the consequences. If I say something mean or hit the person, a fight might happen. I could consider that maybe it was an accident and let it go.

## Portfolio Assessment

Sample items include:
- Decide for Yourself answers
- Web diagram from Learning Styles Logical/Mathematical
- Lesson 2 Review answers

---

Emotionally healthy people use a positive thinking style to explain events. You may wonder why unpleasant events occur. The answers you come up with can influence the quality of your life. For example, a student receives a poor grade on a test. He could react by thinking "This isn't fair! Why is it happening to me?" A more positive reaction would be "That was a hard test! I think I can do better next time."

Emotionally healthy people who have positive attitudes cope by:

- looking on the positive side
- talking to friends, family, or trusted adults
- helping someone else or volunteering in the community
- letting go of the worries about things they cannot change

**Lesson 2 Review** On a sheet of paper, write the word or phrase in parentheses that best completes each sentence.

1. A person who usually expects the best possible outcome is showing (optimism, pessimism, compromise).

2. People who base their behavior on their values are showing the thinking style called (meaningful values, healthy explanations, coping).

3. One way to solve a conflict is through (avoidance, compromise, pessimism).

On a sheet of paper, write the answers to the following questions. Use complete sentences.

4. **Critical Thinking** A student gets kicked off the soccer team because she did not pass her English class. How might she use coping skills to deal with what happened?

5. **Critical Thinking** Suppose you trip over someone's foot in the hall as you're going to your next class. Your impulse is to react angrily because you think the person tripped you on purpose. What might you do to manage your anger?

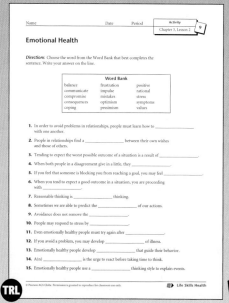

Activity 9

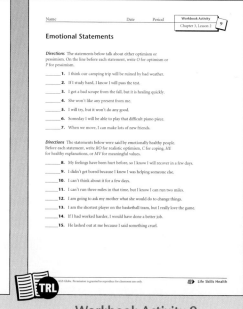

Workbook Activity 9

# Lesson 3  Healthy Relationships

**Objectives**

*After reading this lesson, you should be able to*

- identify the characteristics of healthy relationships
- explain how to maintain relationships
- describe successful speaking, listening, and nonverbal communication skills
- describe how self-esteem affects relationships

**Intimate**
*Very personal or private*

**Nurture**
*Help someone or something grow or develop*

Everyone has a need to belong to a group. The most common way people meet that need is to form and maintain close relationships. Some relationships, such as those with family members, are chosen for you. You choose other relationships, such as friendships. Healthy relationships can help make your life happier.

## What Makes a Relationship Healthy?

A healthy, close relationship has these characteristics:

- emotional attachment
- mutual, or common, dependence between partners
- satisfaction of both partners' needs

The needs that a close relationship satisfies include **intimate** communication and the need to **nurture**. Something intimate is very personal or private. Intimate communication is being able to confide in someone you trust. To nurture means to help someone or something grow or develop.

## What Is the Basis of Friendships?

The most common close relationship is friendship. Most friendships are based on something people have in common. You make friends because you share interests and activities. You can more easily plan time doing things with someone who enjoys what you do. You might also share similar backgrounds with your friends. You might share the same values and goals. Sharing common values and goals can strengthen your self-concept and lift your self-esteem.

## How Can Relationships Be Maintained?

As close as two people may be, they are still individuals with separate lives and experiences. Staying friends can be difficult when one or both of you grow and change.

---

# Lesson at a Glance

## Chapter 3 Lesson 3

**Overview** This lesson describes healthy relationships and explains how communication skills and self-esteem contribute to relationships.

### Objectives

- To identify the characteristics of healthy relationships
- To explain how to maintain relationships
- To describe successful speaking, listening, and nonverbal communications skills
- To describe how self-esteem affects relationships

**Student Pages** 61–65

**Teacher's Resource Library**
  Activity 10
  Alternative Activity 10
  Workbook Activity 10

## Vocabulary

intimate
nonverbal communication
nurture

Write the vocabulary words on the chalkboard and have students read the definitions aloud. Then ask students to name relationships to which the words would apply (*nurture*: parent and child; *intimate*: husband and wife; *nonverbal communication*: coach and player).

## Background Information

Communication can be nonverbal. For example, sympathy can be communicated by touch. A friendly greeting can be communicated by looking at a person in the eye, smiling, or shaking hands. Facial expressions, body movements, and posture are all forms of body language that communicate feelings to others.

**Warm-Up Activity**

Have students work in pairs. Designate one person in each pair the sender and the other the receiver. Have each sender think of a piece of information to communicate to the receiver—for example, "Oh, that's a cool shirt you're wearing." Then have each sender deliver the information verbally, but with facial expression, posture and stance, or tone of voice that conveys the opposite information. For example, the sender might say "Oh, *that's* a cool shirt you're wearing," sarcastically while rolling his eyes. Then ask the senders to describe what message they actually received.

 **Teaching the Lesson**

As students read the lesson, have them write down the main idea of each paragraph. Also have them write down any questions they may have about the material.

Be sure students understand what it means to accept yourself—to know what you believe and to feel good about yourself. When they accept themselves, they aren't easily influenced by what others think about them.

 **Reinforce and Extend**

## ELL/ESL Strategy

**Language Objective:** *To summarize*

As you work through the lesson, have students learning English take notes. Tell them to look at each heading, read the section, and write a sentence or two to summarize the section. Then ask them to answer the question in their own words.

Help any student who cannot answer one or more questions by discussing the specific sections. Help students to understand the question and the section and to find the vocabulary to answer the question as needed.

---

Take time for yourself—it helps you accept yourself. Continue to keep up with your personal interests and goals. Favorite subjects or hobbies can satisfy you even when you are alone.

Since everyone changes throughout life, you must work at maintaining relationships. One action you can take is to keep communicating with each other. Writing or talking can help you avoid misunderstandings. Both will help you keep learning about each other. Another action you can take is to show an interest in your friend's new hobbies or activities. In turn, you can share your own new hobbies or activities with your friend.

Self-acceptance helps maintain a strong, healthy relationship. When you accept yourself, accepting others is easier. You do not feel threatened by others' differences. Healthy relationships allow people to be themselves. Suppose you are interested in photography, but your friends are not. If your relationships are healthy, your friends will respect your need to spend time on your hobby by yourself. Relationships that limit people are unhealthy. Relationships in which people are asked to go against their belief systems also are unhealthy. Suppose a friend dislikes you because you will not go along with something. That is an unhealthy relationship.

**Figure 3.3.1** *Sharing your interests with friends helps maintain relationships.*

### What Makes Communication Successful?

Communication is a basic part of relationships. You must be able to clearly state your beliefs, opinions, and feelings. Good communication skills strengthen relationships and reduce conflicts. They bring you closer to friends and family. Communication can mean talking, writing, touching, listening, smiling, and even knowing when to be silent.

Every communication has a sender and a receiver. The sender sends a message to the receiver with a specific purpose or intent.

**Nonverbal communication**
*Using one's body to send messages*

Communication is effective, or successful, when the sender's intent matches the effect on the receiver. Suppose you want to compliment your friend. If she feels good because of the words you say, your communication was effective.

Successful communication includes three basic skills: speaking, listening, and body language. Body language, or **nonverbal communication,** is using your body to send messages. Examples of body language are arm or head movements, looking away from the speaker, nodding your head, and frowning.

To communicate effectively:

- Think about what you want to say before you say it.
- Speak as clearly as possible.
- Avoid statements that put down the other person.
- Use "I" messages. "I" messages clearly state how you feel. "I" messages help other people understand how their behavior affects you. They are different from "you" messages, which tend to make other people feel angry, hurt, or embarrassed. Suppose a friend is often late for the movies. A "you" message would be "You're always late! Can't you get anywhere on time?" This message might make your friend angry. An "I" message would be "I like seeing the beginning of a movie and when you're late, I miss it." This message would tell your friend that it is important to you to be on time.

To listen effectively:

- Avoid interrupting the speaker.
- Give the speaker your attention. Do not fidget or look away.
- Focus on what the speaker is saying. Do not think about what you are going to say next. You might miss something important.

## In the Community

Have students list various activities that people can share. Items might range from hobbies like photography, to sports, to board or video games, to supposedly solitary activities such as watching movies. When students have completed their lists, they should note what opportunities for each activity, if any, are available in your area. Some things, like a nearby movie theater, students will be able to note easily. Others, such as a camera club in a neighboring town, will require research. Students may want to suggest how new opportunities might be made available.

## Learning Styles

**Body/Kinesthetic**
Have volunteers create and perform a pantomime. Direct them to convey emotions only through body language. Then ask classmates to identify the emotions being pantomimed.

## Career Connection

Point out to students that although American Sign Language (ASL) is a complete language with a large vocabulary of words, it also fits the definition of "nonverbal communication" presented here. Students will undoubtedly realize that ASL is "spoken" with the hands. Explain that body language and facial expressions are used to modify the words presented by the hands. Almost all urban and suburban areas have organizations that provide ASL translators where needed. You may want to contact one to set up an interview, either in person or over the phone, for interested students. Students can also research careers involving ASL (sometimes referred to as "Amslan") on the Internet.

### Technology and Society

Have students read the Technology and Society feature on page 64. Lead a discussion with students as to whether frequent access to computers and the Internet separates them from others or helps to bring them together. Mention time spent surfing the Web instead of interacting with other people, as well as friendships maintained over a distance.

### AT HOME

Ask students to take notes on three different conversations they observe—but do not participate in—at home over the next several days. They may choose conversations between family members, family members and friends, or characters on a television show or movie. Students should note participants' body language and other forms of nonverbal communication. Students should also note things each participant does that promote effective communication, as well as things each one does that hinder it. Have students bring their findings to class to compare with the findings of

More than 60 percent of communication is nonverbal. If body language does not match the words spoken, people tend to believe the nonverbal communication.

### Technology and Society

About 75 percent of teens use the Internet regularly. In spite of fears that teens online might avoid communicating with people, the opposite is true. Teens use e-mail and instant messaging to stay in touch with friends and family. Most also spend at least the same amount of time with people as they did before.

- Repeat in your own words what the speaker said. By doing this, you make sure you understand the speaker's intended message. You might start by saying "What you mean is" or "What I'm hearing is."

To use nonverbal communication effectively:

- Make sure your body language matches the message you intend. Suppose you tell a friend that you are happy to see him. You frown and your arms are crossed. Your body language shows the opposite of what you are saying.
- Pay attention to your tone of voice. If you use positive words with a negative or insincere tone, your friend may feel insulted instead of complimented.

### What Makes Communication Unsuccessful?

Ineffective, or unsuccessful, communication has three possible sources. They are twisted sending, twisted receiving, and poor communication conditions. Twisted sending can be poor word choices or not knowing what to say. It can mean your body language is sending the opposite message from your words. Twisted receiving usually results from poor listening. Suppose you think you know what the teacher is going to say. You may pretend to listen while your attention is elsewhere. Later, you realize that you have no idea what the teacher really said.

Poor communication conditions may include noise, interruptions, and lack of time. For example, suppose your friend's cell phone rings while you are talking with her. She answers the call, and you lose your train of thought. You forget what you were going to tell her.

**Research and Write**

In small groups, research other communication skills. You might survey people to gather their favorite tips. Then write and perform a skit in which you show effective communication skills. Make sure you use nonverbal communication to help get your message across.

## How Does Self-Esteem Affect Relationships?

Relationships meet some, but not all, needs. Expecting too much from a friend or partner is easy. Romantic fairy tales and movies make people hope another person can make them happy. The truth is that happiness comes from your own thoughts, feelings, and actions. Likewise, other people should not depend on you for their happiness.

Relationships are healthy when both people have high self-esteem. People who accept themselves and have meaningful goals are better able to give affection to others. You promote healthy relationships when you learn about and appreciate yourself. Continue to develop your interests and improve your skills. The more you like yourself, the healthier your relationships will be.

*Lesson 3 Review* On a sheet of paper, write the answers to the following questions. Use complete sentences.

1. What is the basis of most friendships?
2. What are two things you can do to maintain a relationship?
3. What three things can make communication unsuccessful?
4. **Critical Thinking** A friend said she would call you. You are upset that she did not. What "I" message would express your feelings?
5. **Critical Thinking** A friend says, "My girlfriend makes me very happy." Can that be true? Explain your answer.

### Link to ➤➤➤
**Culture**

Nonverbal communication does not always mean the same things in all cultures. For example, in some cultures, looking someone in the eye is disrespectful. Smiling may mean the person is really angry. Touching a stranger may be considered rude. Learning how people from other cultures communicate can help you avoid conflicts.

*Maintaining Mental Health* Chapter 3  65

## Lesson 3 Review Answers

1. The basis of most friendships is sharing interests and activities. 2. Two things you can do to maintain a relationship are to keep communicating with others and to take an interest in the other person's hobbies or activities. 3. Three things that can make communication unsuccessful are twisted sending, twisted receiving, and poor communication conditions. 4. Sample answer: I was upset that you did not call me. I was frustrated because I changed my schedule based on the expectation that you would call me. 5. The statement is not correct because your friend's happiness depends on his own thoughts, feelings, and actions, not on someone else. A healthy romantic relationship can be an important part of a person's life, but true happiness comes from within a person.

### Portfolio Assessment

Sample items include:
- Written skit from Research and Write
- Lesson 3 Review answers

### Research and Write

Have students read and conduct the Research and Write feature on page 65. You may want to arrange for students to interview other students or willing teachers during the school day, or have them interview friends, family members, and other adults as homework. Schedule class time for students to present their skits.

### Link to

**Culture.** Have a volunteer read aloud the Link to Culture feature on page 65. Students who want to find out more about this extensive subject can start at the public library. A number of books, most of them aimed at international businesspeople, describe cultural differences in nonverbal communication. Another good source of information is the Internet. Students may want to enter "nonverbal communication" + "cultural differences" into their favorite search engine.

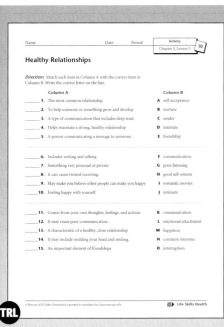

Activity 10

Workbook Activity 10

*Maintaining Mental Health* Chapter 3  65

# Lesson at a Glance

## Chapter 3 Lesson 4

**Overview** This lesson describes self-awareness, social comparison, and self-acceptance and explains how to help improve relationships.

### Objectives

- To explain what self-awareness and social comparison are
- To describe how you can improve your self-acceptance
- To describe how self-acceptance improves relationships
- To identify ways to accept yourself

**Student Pages** 66–70

**Teacher's Resource Library**

  Activity 11
  Alternative Activity 11
  Workbook Activity 11

## Vocabulary

| | |
|---|---|
| discriminate | self-awareness |
| empathy | social comparison |
| prejudice | social support |

Have students write clues for the vocabulary words. For example: She felt sad. I felt _____ with her. Allow students to show their clues to classmates who name the word for each clue.

## Background Information

This lesson discusses how students can better accept themselves and others. Whether your particular community is homogeneous or diverse, it will be helpful for students to be able to accept others who are different from them.

### TEACHER ALERT

Students should show good judgment when comparing themselves to others and accepting individual differences. For example, accepting someone who is of a different religion can be a good thing. However, trying to fit in with someone who steals or does drugs is not a good thing.

66  Unit 1  Mental and Emotional Health

---

## Lesson 4  Becoming More Emotionally Healthy

**Objectives**

*After reading this lesson, you should be able to*

- explain what self-awareness and social comparison are
- describe how you can improve your self-acceptance
- describe how self-acceptance improves relationships
- identify ways to accept yourself

**Self-awareness**
*Understanding oneself as an individual*

**Social comparison**
*Observing other people to determine how to behave*

You can perform better when you accept and feel good about yourself. Your mind will work better, too. Your relationships will be healthier. Improving your self-acceptance and relationships with others can support your mental health. Two things that can help you improve your self-acceptance are **self-awareness** and **social comparison.**

### What Is Self-Awareness?

Self-awareness is understanding yourself as an individual. You understand your feelings and goals from moment to moment. For example, what are you feeling right now as you read this book? You may be comfortable, or you may be tired and unable to concentrate. If you are not comfortable, you may decide to find a better place to study. Being self-aware will help you make a good choice and will increase your chances for success.

Self-awareness also means understanding how you feel about things. How do you feel about your abilities, appearance, and behavior? What are your goals? The more you understand yourself, the better you can understand others.

### What Is Social Comparison?

Social comparison is observing other people to help you decide how to behave. You can use social comparison to improve your self-acceptance. Suppose you are taking a new computer class and do not know what to do. You notice other students ask permission before using the computers. You copy their behavior and in that way learn for yourself.

### How Does Self-Acceptance Improve Relationships?

As you accept yourself more, others will be more comfortable with you. Spending time with people who have poor self-acceptance can be uncomfortable. They may constantly put themselves down. You may try to help them see themselves differently, but they may not accept your help.

66  Unit 1  Mental and Emotional Health

**Prejudice**
Negative, or unfavorable, opinions formed about something or someone without enough experience or knowledge

**Discriminate**
Treat differently on the basis of something other than individual worth

What is one thing teenagers can do to stop prejudice?

People who feel good about themselves do not expect others to build them up or entertain them. People who accept themselves are more likely to accept individual differences, get to know a variety of people, and give and accept help.

## Accept Individual Differences

By nature, people are different from one another. Some differences are easy to see, such as skin color, gender, and height. Some differences are not visible, such as thoughts and feelings. For example, some people are born with a more sensitive temperament than others. Some people have different personal beliefs than do others. Everyone has different talents and abilities.

Unfortunately, people may form negative, or unfavorable, opinions about someone who is different from themselves. This is called **prejudice.** They may not even know this person but may still dislike him or her. Prejudiced people tend to **discriminate** against others who are different from themselves. They treat others differently on the basis of something other than individual worth. Prejudice creates unhealthy situations. Suppose you learn that people you have never met dislike you. You are different from them in some way. You, in turn, may find it hard to like these people because you believe they are being unfair. This cycle becomes difficult to break. Prejudice can destroy the fairness, honesty, and goodwill of a school and community.

### Link to ➤➤➤
**Language Arts**
Tony Morrison's novel *The Bluest Eye* focuses on two African American girls from different families. One girl is abused and feels ugly for being black. She believes that she can be happy only if she becomes white. She prays for blue eyes. Her sadness makes her ill. The other girl likes the way she looks. Her self-acceptance helps her stay healthy.

## Warm-Up Activity

Ask students to give examples of how people are different. Keep the discussion abstract and don't let students discuss individuals. Bring in examples from the community of how different groups contribute to a richer community experience (e.g., different ways different cultures celebrate holidays).

## Teaching the Lesson

Discuss how an understanding of oneself can lead to a better understanding of others. Have students read pages 66–67 to learn about improving self-acceptance and accepting individual differences.

## Reinforce and Extend

### Link to

**Language Arts.** Have a volunteer read aloud the Link to Language Arts feature on page 67. Suggest that students check out and read *The Bluest Eye*.

You may want to ask the school librarian to compile a list of young adult books available in the school library that deal with self-acceptance and acceptance of others among young people.

## ELL/ESL Strategy

**Language Objective:**
*To practice common grammar structures*

Have English learners write dialogue for a short scene in which someone is subjected to discrimination. The scene may be completely fictional, or it may be based on something that actually happened to the student, a friend, or a family member. Invite them to present their scenes to the class. Discuss emotionally healthy ways of dealing with discrimination.

## Learning Styles

**Auditory/Verbal**

Prepare several context-clue sentences for each vocabulary word, but leave the vocabulary word out. Divide the class into teams. Read one of your sentences to each team in turn. Let team members confer, then suggest the vocabulary word that belongs in that sentence.

---

*Empathy*
*Recognition and understanding of other people's feelings*

You can learn to value differences in people. You can do this by truly getting to know a person. Differences in people add richness to your relationships. Suppose you get to know someone from another country. You may find you are developing new tastes in food and music. You may come to appreciate and respect different points of view.

### Get to Know a Variety of People

You can overcome prejudices when you get to know members of different groups. Working together toward a common goal is a way to learn cooperation. Through this, you discover how you are alike. You begin to have **empathy** with one another. You recognize and understand other people's feelings. Close relationships are healthiest when they are based on empathy.

Here are some other ways you can get to know a variety of people:

- Get to know a person of the opposite sex as a friend rather than as a possible date.
- Meet people from different cultural backgrounds or religions through common interests such as music or sports.
- Establish relationships with older adults—grandparents, family friends, or others. They can offer practical advice for problem solving and decision making based on their experiences.

**Figure 3.4.1** *One way to overcome prejudices is by learning about people from different cultural backgrounds.*

## Give and Accept Help

Friends help you do things, solve problems, and deal with hard times. Help for those you care about is called **social support.** To improve your relationships with others, you can learn how to give and accept social support. This is also another way to improve your emotional health.

When a friend is in need, you may empathize and want to help. Suppose your best friend has had a painful loss. However, your friend may not be willing to ask you for help. You can still do many things to support your friend. You can take notes in a class your friend had to miss. You can offer to take care of a pet or run an errand. It is best to offer something particular. For example, you might ask "Can I help by taking care of your younger brother for you today?"

Sometimes, you are the one to accept social support when you need help. Needing help is not a sign of weakness or failure. Everyone needs help at times. When you let others give you social support, your friendships grow stronger.

> **Social support**
> Help for those you care about

### Health at Work

#### High School Counselor

High school counselors help students handle problems. They also help students develop realistic school and career goals. They advise students about college majors and technical schools. They help provide students with life skills that will improve their personal and social health. Counselors are trained to help students deal with personal problems. For example, they help students avoid drug use and learn how to solve conflicts. Counselors are dedicated to their jobs and deal with problems every day. A master's degree in counseling is usually required. All states require school counselors to be certified in counseling.

# Lesson 4 Review Answers

1. C 2. B 3. B 4. Sample answer: I would bring her school assignments and help her do them. I would offer to do her chores. 5. Sample answer: Some students at my school will not sit with students who have recently come to this country. I would encourage the prejudiced people to talk with the other students and get to know them. They might discover that the new students are not so different.

## Portfolio Assessment

Sample items include:
- Job descriptions from Health at Work
- Lesson 4 Review answers

## Health Myth

Have students read the Health Myth feature on page 70. Ask students what they might do if they had a sore throat, the flu, or a broken arm. Point out how many of their responses involved seeing a doctor—a health care professional. Discuss briefly why it would be advisable to see a doctor under these circumstances. Then point out that a counselor or a psychiatrist is simply another type of health professional—a mental health professional. To avoid seeing a counselor when it might be helpful, or even necessary, makes no more sense than avoiding a doctor when you have the flu or avoiding seeing a dentist when you have chipped a tooth.

---

**Health Myth**

**Myth:** A person who seeks help from a mental health counselor is weak or mentally ill.

**Fact:** At one time or another, everyone has problems and needs help. Seeking help from a counselor does not mean a person is mentally ill. People who seek help are strong because they have the courage to face their problems and try to solve them.

---

**Lesson 4 Review** On a sheet of paper, write the letter of the answer that best completes each sentence.

1. When you observe other people to decide how to behave, you are practicing _____.
   A empathy        C social comparison
   B prejudice      D social support

2. A person who discriminates against others is likely to _____.
   A feel empathy for others
   B have poor self-acceptance
   C have different abilities from others
   D have a high degree of self-acceptance

3. Self-accepting people _____.
   A make friends only with people who are like them
   B get to know older adults, such as grandparents
   C refuse social support to show how tough they are
   D look only to themselves to decide how to behave

On a sheet of paper, write the answers to the following questions. Use complete sentences.

4. **Critical Thinking** A friend has missed several days of school because of illness. What are two things you could do to show your friend social support?

5. **Critical Thinking** Describe one form of prejudice you have observed at your school. Suppose you could talk with the people who are prejudiced. What would you say to them?

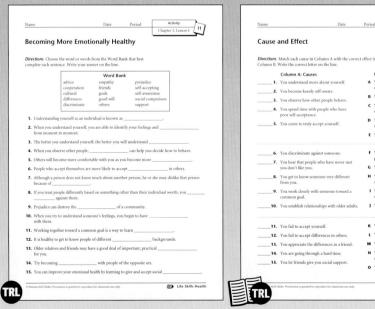

# Chapter 3 SUMMARY

- Personality development is affected by temperament, meaningful relationships formed early in life, and self-concept.

- What other people tell you about yourself and how they act toward you help shape your self-concept. The media also help form your self-concept.

- The areas of well-being are physical, social, emotional, and personal well-being. The four areas of well-being balance and influence one another.

- You can promote your well-being by staying healthy and learning how to handle stress. You can also promote your well-being by improving your understanding of others and forming beliefs and values.

- Emotional well-being is an ongoing process that you work on throughout your life.

- Characteristics of emotional health include the ability to learn and change through experiences, balancing personal needs with the needs of others, setting and reaching goals, solving conflicts rationally, controlling impulses, and thinking in healthy ways.

- Healthy relationships involve emotional attachment, mutual dependence, and satisfaction of both partners.

- Maintaining healthy relationships involves successful communication and good self-esteem.

- You can become more emotionally healthy by improving your self-acceptance through self-awareness and social comparison.

- Self-acceptance helps you improve relationships with others. You can learn about and value differences among people.

### Vocabulary

| | | |
|---|---|---|
| compromise, 57 | nurture, 61 | resilient, 59 |
| disadvantage, 58 | optimism, 56 | self-awareness, 66 |
| discriminate, 67 | personality, 50 | social comparison, 66 |
| empathy, 68 | pessimism, 56 | social support, 69 |
| impulse, 57 | prejudice, 67 | social esteem, 51 |
| intimate, 61 | rational, 57 | symptom, 59 |
| nonverbal communication, 63 | realistic, 56 | temperament, 50 |

*Maintaining Mental Health*  Chapter 3

## Chapter 3 Summary

Have volunteers read aloud each Summary item on page 71. Ask volunteers to explain the meaning of each item. Direct students' attention to the Vocabulary box on the bottom of page 71. Have them read and review each term and its definition.

### ONLINE CONNECTION

For more information about mental health, have students go to the section of the National Institute of Mental Health's Web site that is devoted specifically to the mental health of children and adolescents: www.nimh.nih.gov/healthinformation/childmenu.cfm

The Web site of New York Online Access to Health (NOAH) is a readable collection of information on all health topics. Topics on mental health of interest to teens may be found at: www.noah-health.org/en/mental/teens.html

## Chapter 3 Review

Use the Chapter Review to prepare students for tests and to reteach content from the chapter.

## Chapter 3 Mastery Test

The Teacher's Resource Library includes two forms of the Chapter 3 Mastery Text. Each test addresses the chapter Goals for Learning. An optional third page of additional critical-thinking items is included for each test. The difficulty level of the two forms is equivalent.

## Review Answers

**Vocabulary Review**

1. temperament  2. well-being  3. optimism  4. compromise  5. rational  6. impulses  7. symptom  8. resilient  9. intimate  10. nonverbal communication  11. social comparison  12. prejudice  13. discriminate  14. empathy

### TEACHER ALERT

In the Chapter Review, the Vocabulary Review activity includes a sample of the chapter's vocabulary terms. The activity will help determine students' understanding of key vocabulary terms and concepts presented in the chapter. Other vocabulary terms used in the chapter are listed below.

| | |
|---|---|
| disadvantage | realistic |
| nurture | self-awareness |
| pessimism | social esteem |
| personality | social support |

72  Unit 1  *Mental and Emotional Health*

---

# Chapter 3  REVIEW

**Word Bank**
- compromise
- discriminate
- empathy
- impulses
- intimate
- nonverbal communication
- optimism
- prejudice
- rational
- resilient
- social comparison
- social esteem
- symptom
- temperament

## Vocabulary Review

On a sheet of paper, write the word or phrase from the Word Bank that best completes each sentence.

1. A person's emotional makeup is the person's _____.
2. The way others value you is your _____.
3. Realistic goals are balanced between pessimism and unrealistic _____.
4. When people work out a solution to a problem by giving in a little, they _____.
5. Solving problems requires _____ thinking.
6. Emotionally healthy people learn to control their _____.
7. A(n) _____ may develop if a person avoids dealing with a difficult situation.
8. People who are _____ bounce back from disappointments.
9. Being able to confide in someone you trust is the basis of _____ communication.
10. Shaking your head and smiling are examples of _____.
11. Observing others to determine how to behave is called _____.
12. A negative opinion formed without enough experience or knowledge is _____.
13. Prejudiced people tend to _____ against people who are different from themselves.
14. Trying to identify and understand other people's feelings is _____.

72  Unit 1  *Mental and Emotional Health*

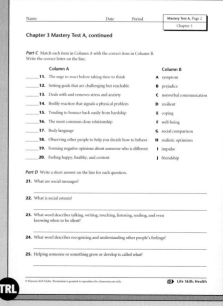

Chapter 3 Mastery Test A, pages 1–3

## Concept Review

On a sheet of paper, write the letter of the answer that best completes each sentence.

15. The factor that has the most influence on a child's personality development is _____.
    A beliefs
    B temperament
    C values
    D well-being

16. An example of how two different forms of well-being affect each other is _____.
    A eating healthful foods gives a person energy
    B good communication skills help a person be a friend
    C having many friends helps a person handle stress
    D strong values guide a person in making decisions

17. The difference between coping and avoidance as a way to deal with stress is that _____.
    A coping deals with stress, while avoidance does not
    B coping is more likely to lead to symptoms of illness
    C avoidance is more likely to get rid of the source of stress
    D avoidance means asking for help from friends or family

## Critical Thinking

On a sheet of paper, write the answers to the following questions. Use complete sentences.

18. What can you do to teach someone to be less prejudiced?

19. Are you better at sending or receiving communications? How can you become better at either sending or receiving?

20. Is it healthy to feel that you must control everything in a relationship? Explain your answer.

 When studying for a test, use a marker to highlight important facts and terms in your notes.

*Maintaining Mental Health* Chapter 3 73

### Concept Review
15. B 16. C 17. A

### Critical Thinking
18. You can encourage people who are prejudiced to get to know the people toward whom they feel prejudice. You can suggest that they work toward a common goal. Getting to know a variety of types of people can help someone to overcome his or her prejudice.
19. Sample answer: I am better at sending. I can become better at receiving by not interrupting the speaker and by making sure I understand what the person is saying. (Encourage students to cite specific communication skills in their responses.) 20. It is not healthy because healthy relationships allow people to be themselves. If you try to control your friends, they will feel you do not accept and respect them. In healthy relationships, people balance their own wishes with those of others.

## ALTERNATIVE ASSESSMENT

Alternative Assessment items correlate to the student Goals for Learning at the beginning of this chapter.

- Have students orally describe well-being and suggest several ways to promote it.
- Have students make a chart describing how to reach good mental health and three ways to maintain good emotional health.
- Have students design a poster that presents the characteristics of a healthy relationship.
- Have students orally describe ways to accept themselves and others.

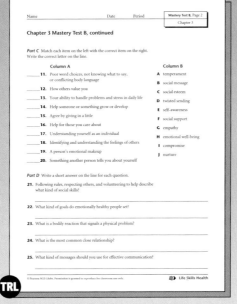

Chapter 3 Mastery Test B, pages 1–3

## Chapter at a Glance

### Chapter 4: Recognizing Mental Health Problems
pages 74–91

### Lessons
1. Characteristics of Poor Mental Health
   pages 76–79
2. Mental Disorders
   pages 80–84
3. Treating Mental Disorders
   pages 85–88

**Chapter 4 Summary** page 89

**Chapter 4 Review** pages 90–91

**Audio CD**

**Skill Track for Life Skills Health**

**Teacher's Resource Library**
- Activities 12–14
- Alternative Activities 12–14
- Workbook Activities 12–14
- Self-Assessment 4
- Community Connection 4
- Chapter 4 Self-Study Guide
- Chapter 4 Outline
- Chapter 4 Mastery Tests A and B

(Answer Keys for the Teacher's Resource Library begin on page 559 of this Teacher's Edition.)

## Opener Activity

Invite students to discuss reasons why mental illness has historically carried a stigma. Ask students what could be done today to help people better understand mental illness. *(better education to disprove myths and stereotypes, changing the way mental illnesses are portrayed in movies and on television)*

74   Unit 1 Mental and Emotional Health

## Self-Assessment

### Can you answer these questions?

1. What does having good mental health mean?
2. What are self-defeating behaviors?
3. What defense mechanisms do you use?
4. Why are dysfunctional relationships unhealthy?
5. What is a substance abuse disorder?
6. How can anxiety disorders be treated?
7. What is a phobia?
8. How many people suffer from clinical depression?
9. Are teens at risk for suicide?
10. How serious is an eating disorder?

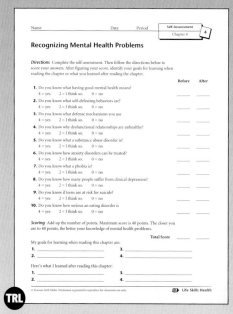

Self-Assessment 4

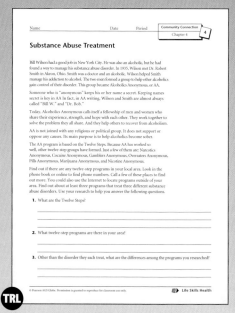

Community Connection 4

# Chapter 4
## Recognizing Mental Health Problems

Have you ever heard someone called "crazy"? What did you think that meant? Have you wondered about the difference between mentally healthy and unhealthy behavior? Mental health covers a wide range of behaviors. The most helpful sign of mental health is how well a person gets along in his or her world. Everyone has difficulties at times. While calling someone "crazy" is never OK, a person whose behavior is often unhealthy may have mental health problems. The person may even have a mental disorder. Most mental health problems and disorders can be treated successfully.

In this chapter, you will learn about some of the symptoms of poor mental health and some self-defeating behaviors. You will learn about mental disorders and what can be done for them. You will learn some ways to improve your emotional well-being.

### Goals for Learning

♦ To identify some characteristics of poor mental health
♦ To explain self-defeating behaviors
♦ To identify substance abuse disorders, anxiety disorders, affective disorders, and thought disorders, and how these disorders are treated
♦ To describe three kinds of eating disorders

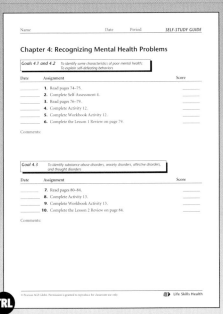

Chapter 4 Self-Study Guide, pages 1–2    Chapter 4 Outline, pages 1–2

## Introducing the Chapter

Ask students to define what mental health means to them. Then ask them about a book or movie they are familiar with that portrays a person with a mental disorder. Provide an example such as *Sling Blade, Forrest Gump,* or *Rain Man.* Discuss the symptoms of the disorder, the self-defeating behavior the person may have exhibited, the problems the person had getting along in the world, if and how the person coped with the problems, and the final outcome. Then discuss how dealing with mental disorders and getting help can lead to a better life.

Have students read the Goals for Learning. Explain that in this chapter students will learn how to recognize mental health problems and the different ways they are treated.

### Notes and Questions

Ask volunteers to read the notes and questions that appear in the margins throughout the chapter. Then discuss them with the class.

### Self-Assessment

Have students complete the Self-Assessment worksheet before and after reading the chapter. Before reading the chapter, have students fill in the "Before" column. Ask students to identify their goals for learning. To get ideas for setting goals, students might use the chapter introductory material on page 75, the checklist on page 74, or the questions on the Self-Assessment worksheet. Students can use the back of the worksheet if they need more space to write.

Collect the Self-Assessment worksheets and pass them out again at the end of the chapter. Have students fill in the "After" column. Ask them to identify at least four major points they have learned. Again, suggest they use the back of the worksheet if they need more space to write. You may want to collect and review the worksheets, but return them to students so they have a record of their goals and accomplishments.

# Lesson at a Glance

## Chapter 4 Lesson 1

**Overview** This lesson describes characteristics of poor mental health, discusses self-defeating behaviors and defense mechanisms, and describes dysfunctional relationships.

## Objectives

- To explain the difference between mental health and mental illness
- To describe self-defeating behaviors
- To identify your own defense mechanisms
- To describe dysfunctional relationships

**Student Pages** 76–79

**Teacher's Resource Library**

Activity 12
Alternative Activity 12
Workbook Activity 12

## Vocabulary

| | |
|---|---|
| abnormal | projection |
| defense mechanism | psychologist |
| denial | repression |
| displacement | self-defeating |
| dysfunctional | behavior |

Have students write a sentence using each vocabulary term. Then tell them to write the sentence again, substituting the definition of the term for the term itself. Ask volunteers to share one sentence with the class and then have the class identify the vocabulary term or definition.

## Background Information

A phobia is a mental disorder triggered by specific experiences and objects. Some phobias include acrophobia (fear of heights), agoraphobia (fear of public places), ailurophobia (fear of cats), astraphobia (fear of lightning), nyctophobia (fear of darkness), and xenophobia (fear of strangers).

76  Unit 1  Mental and Emotional Health

---

## Lesson 1: Characteristics of Poor Mental Health

**Objectives**

*After reading this lesson, you should be able to*

- explain the difference between mental health and mental illness
- describe self-defeating behaviors
- identify your own defense mechanisms
- describe dysfunctional relationships

**Abnormal**
*Unusual or different from normal*

**Psychologist**
*A person who studies mental and behavioral characteristics*

**Self-defeating behavior**
*An action that blocks a person's efforts to reach goals*

What is an adjustment you have made recently?

A person is usually neither completely mentally healthy nor completely mentally ill. Mentally healthy people may have problems, make mistakes, and behave in self-defeating ways. People with mental health problems still do many things well. The difference has to do with patterns of behavior that are either normal or **abnormal.** Abnormal means not usual.

### How Do Behaviors Indicate Mental Health?

A **psychologist** is someone who studies mental and behavioral characteristics. Most psychologists evaluate mental health by determining the level of normal behavior. If a person has many abnormal behaviors, that person may not be mentally healthy. The more abnormal a behavior or thinking pattern is, the greater the possible risk of mental illness.

Psychologists define normal and abnormal behavior in many ways. One way is to judge how well a person adjusts to life's changes. People who meet their basic needs and personal goals show a normal pattern of behavior. They get along independently and maintain personal relationships. People who have difficulty getting along with others or who depend too much on others show abnormal behavior patterns. People whom failure easily defeats also show behavior that is considered abnormal.

### What Are Self-Defeating Behaviors?

**Self-defeating behaviors** block a person from reaching his or her own goals. For example, suppose you want to try out for the swim team. You know you must practice and increase your speed. The day before tryouts, you practice for a long time. Then you decide to practice even more that evening. You end up being tired and do not perform well at tryouts the next day. The decision to practice too much was self-defeating. It actually blocked you from reaching your goal.

76  Unit 1  Mental and Emotional Health

**Defense mechanism**
*A mental device one uses to protect oneself*

**Dysfunctional**
*Harmed or unhealthy*

**Repression**
*A refusal to think about something that upsets you*

**Denial**
*A refusal to believe or accept something negative or threatening*

**Projection**
*Assuming that another person has one's own attitudes, feelings, or purposes*

**Displacement**
*Shifting an emotion from its real object to a safer or more immediate one*

Sometimes self-defeating behavior is a sign of poor mental health. Such behavior can take many different forms. Two common forms of self-defeating behavior are **defense mechanisms** and **dysfunctional** relationships. A defense mechanism is a mental device that people use to protect themselves. A dysfunctional relationship is harmed in some way, or unhealthy.

### Defense Mechanisms

People use various defense mechanisms to hide from their problems instead of solving them. The most common defense mechanisms are **repression, denial, projection,** and **displacement.**

- Repression is a refusal to think about something that upsets you. For example, you may "forget" about a dental appointment that you don't want to go to.

- Denial is a refusal to believe or accept something negative or threatening. For example, an older person you know is very ill, but you do not believe that the person might die.

- Projection is assuming that someone has the same attitudes, feelings, or purposes as you. For example, if you are angry with your brother, you may project your anger on him. You might say "What are you so mad about?"

- Displacement is shifting an emotion from its real object to a safer one. For example, your supervisor made you stay late to fix a problem at work. You are angry but afraid to say anything to the supervisor. When your mother calls, you yell at her for asking a simple question. You displaced your anger from its real object to a safer one.

You can probably identify many examples of these defense mechanisms in your own and others' behavior. Using defense mechanisms is often normal, but using them too much leads to self-defeating behavior.

*Recognizing Mental Health Problems Chapter 4* **77**

### Warm-Up Activity

Challenge students to name some characteristics and behaviors that are associated with mentally healthy people. Write the list on the chalkboard. Keep a copy of the list for use with the next lesson.

### Teaching the Lesson

Have a volunteer read the first paragraph of the lesson aloud. Ask students to raise a hand if they feel completely physically healthy. Ask those students with their hands raised if any of them have a slight headache, have a runny nose, or feel a little tired. Continue with questions such as these to see how many students actually do feel completely physically healthy.

As you work through the lesson, help students see that mentally healthy people might still engage in self-defeating behavior or be involved in dysfunctional relationships, just as a physically healthy person might have a headache or a sore wrist.

#### TEACHER ALERT

Point out that using defense mechanisms to some degree is normal and helpful in everyday situations. For example, suppose you are angry with your teacher. It may be unwise to express that anger to your teacher and safer to direct it toward someone or something else.

#### HEALTH JOURNAL

Ask students to recall a time when they may have used one of the defense mechanisms described on page 77. Have students describe in their journals what problem they were trying to avoid, what they did, and what the outcome was. Encourage students to add a few sentences to describe what they might have done to solve the problem instead of hiding from it.

#### LEARNING STYLES

**Auditory/Verbal**

Have a volunteer read aloud the descriptions for the four defense mechanisms one at a time. After each description, ask students to provide an example of that mechanism in addition to the one given in the text. Students may make up hypothetical examples, or describe actions taken by a character in a book, movie, or television show.

*Recog. Mental Health Prob. Chapter 4* **77**

# 3 Reinforce and Extend

## ELL/ESL Strategy

**Language Objective:** *To write sentences describing qualities of trusted friends and family members*

Direct students to identify trusted people to whom teens can talk. These people might include parents, siblings, other relatives, teachers, counselors, school nurses, members of the clergy, and so on. With the help of a student proficient in English, ask students to write sentences describing qualities of a trusted friend or family member.

## In the Environment

Challenge students to list environmental stress factors that might cause feelings of anxiety for mentally healthy people. Examples might include prolonged exposure to noise pollution, experiences with natural disasters such as tornadoes or floods, and so on.

## Learning Styles

**Interpersonal/Group Learning**

Have students work in groups. Assign each group a defense mechanism. Have students role-play a situation that clearly demonstrates the mechanism. Examples include "forgetting" about a school assignment, denying the possibility of the death of a loved one, and yelling at someone for being late because you're angry that you yourself were late. Then have others in the class guess what defense mechanism was being enacted.

---

Writing down your thoughts and feelings can help you deal with them. You can write letters that you do not mail. You can also write stories about something that has happened.

You can watch your use of defense mechanisms by reviewing your behavior. For example, if you often "forget" important or unpleasant events, you may be using repression too much. If you often say "This isn't serious" when talking about difficulties, you may be overusing denial. Realizing that you use these defense mechanisms is the first step toward changing your behavior.

### Dysfunctional Relationships

In many dysfunctional relationships, people meet each other's needs in what turn out to be destructive ways. A common form of dysfunction in a relationship is poor communication. For example, you are worried about something but refuse to confide in a close friend about it. Your friend cannot offer help or understand what is wrong. If you continue to refuse help, others may feel pushed away and may withdraw from you. Unfortunately, you may be cutting yourself off from others' support when you need it most.

### Health at Work

**Resident Assistant**

Resident assistants work with residents in mental health settings, hospitals, and nursing homes. When residents need help, assistants answer the call. They may check a resident's temperature, blood pressure, pulse, and breathing rate. Sometimes they bathe residents or give massages. Resident assistants help people who must stay in bed or a wheelchair. They deliver and serve meals and help residents eat when necessary. They socialize with residents, make them comfortable, and lift their spirits. Sometimes resident assistants are called nurses' aides or nurse assistants. Training varies from on-the-job training to formal training programs. The training programs may take several weeks or more to complete.

---

### Health at Work

Have students read the Health at Work feature on page 78. Point out that a resident assistant is an assistant to the residents of a hospital, nursing home, or mental health facility. Have students brainstorm a short list of things residents of each facility might require assistance with that residents of the other two might not.

Give "I" messages respectfully and kindly to maintain communication in relationships.

You can avoid this self-defeating pattern of shutting out other people. You do this by identifying the people you trust. When you are troubled, you can talk with these people about your feelings. A helpful hint is to start your sentences with "I" messages. For example, "I feel nervous" or "I worry that you'll get upset with me." Starting a statement with "I" makes it easier for others to understand and respond to you.

Figure 4.1.1 *When you are angry, you may project your anger on the person you are angry with.*

When you have problems, you may make heavy demands on a friend or family member. You may want the person to reassure you or spend time with you. If you do this too often, the other person may become worn out and refuse to help you. Expecting another person to make you happy is unreasonable. People who love you will give you help, but only you can be responsible for your happiness.

If you feel you make too many demands on friends or family members, you can work on changing your behavior. You can try to help yourself. By learning to take care of yourself, you can become more independent. You can also become less demanding and can improve your mental health.

### Link to >>>
**Language Arts**
Journaling is one way to deal with emotional problems. By writing your thoughts and feelings down, and connecting them with the events in your life, you can gain valuable insights. Some artists and famous people have published their journals. These expressions about their lives are part of their craft.

*Lesson 1 Review* On a sheet of paper, write the answers to the following questions. Use complete sentences.

1. What is one way to decide if a person is mentally healthy?
2. What are self-defeating behaviors?
3. What are four defense mechanisms?
4. **Critical Thinking** Give an example of a dysfunctional relationship.
5. **Critical Thinking** What is an example of an "I" message?

## LEARNING STYLES

**Body/Kinesthetic**
Have students work in pairs. Have students write a short skit in two scenes. The first scene shows two characters in a dysfunctional relationship, and the second shows the same characters in a functional one. Have students perform their skits for the class.

### Link to
**Language Arts.** Have a volunteer read the Link to Language Arts feature on page 79. Have students research and write a short list of journals that were written with the intent of publication. Ask how these would differ from private journals, such as their Health Journals.

## AT HOME

Have students write a letter of encouragement to a family member who has a problem to solve. Suggest that students discuss with their family how such a letter might help a person's self-esteem and lead to good communication within the family.

## Lesson 1 Review Answers

1. A person is mentally healthy if he or she adjusts well to life's changes.
2. Self-defeating behaviors are actions that block you from reaching your goal.
3. Repression, denial, projection, and displacement are defense mechanisms.
4. An example of a dysfunctional relationship would be one in which a person who cannot take care of himself pairs up with someone with low self-esteem. These people meet each other's needs in destructive ways.
5. An example of an "I" message is "I feel nervous about giving my speech in front of the class."

## Portfolio Assessment

Sample items include:
- List from Link to Language Arts
- Letter from Application: At Home
- Lesson 1 Review answers

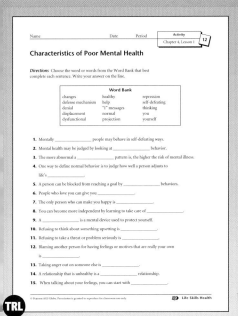

Activity 12

Workbook Activity 12

# Lesson at a Glance

## Chapter 4 Lesson 2

**Overview** This lesson describes a number of different mental disorders.

## Objectives

- To explain what substance abuse and anxiety disorders are
- To describe the symptoms of clinical depression
- To explain thought disorders
- To describe three eating disorders

**Student Pages** 80–84

**Teacher's Resource Library**

Activity 13
Alternative Activity 13
Workbook Activity 13

## Vocabulary

| | |
|---|---|
| addicted | hallucination |
| affective disorder | panic attack |
| anxiety disorder | phobia |
| bipolar disorder | substance abuse |
| clinical depression | disorder |
| delusion | suicide |
| eating disorder | thought disorder |

Ask students which of the terms they have heard before. Encourage them to explain what they think each term means. Then read each definition with students.

## Background Information

Anorexia is an eating disorder. Most people with anorexia are females who are high achievers. They perceive themselves as being too heavy, even though they are extremely thin. People with anorexia may think about and talk about food, but they do not eat an adequate amount of food. They deprive their bodies of essential nutrients. Many people with anorexia also exercise to an extreme. Anorexia can lead to severe malnutrition and can even cause death by starvation.

---

# Lesson 2 — Mental Disorders

### Objectives

*After reading this lesson, you should be able to*

- explain what substance abuse and anxiety disorders are
- describe the symptoms of clinical depression
- explain thought disorders
- describe three eating disorders

**Substance abuse disorder**
*An unhealthy dependence on alcohol or other drugs*

**Addicted**
*Unable to stop using alcohol or another drug*

**Anxiety disorder**
*A mental problem that makes normal life difficult because of intense anxiety*

**Panic attack**
*A feeling of terror that comes without warning and includes symptoms of chest pain, rapid heartbeat, sweating, shaking, or shortness of breath*

While many mental health problems are not considered serious, some are quite serious. The more you know about the most common mental disorders, the less frightening they may seem. A disorder is a sickness or a disease. More knowledge can also help you promote your own and others' mental health.

## What Are Substance Abuse Disorders?

A **substance abuse disorder** is an unhealthy dependence on alcohol or other drugs. Most substance abuse begins as a defensive behavior pattern. For example, a person may drink alcohol because he thinks he will feel less shy and self-conscious. Someone else may use another drug because it helps her feel happy and confident. Alcohol and other drugs never solve the real problem. Since the problem still exists, the user must rely on the drug more and more. Many substances are habit forming. You can become **addicted** to alcohol and certain other drugs. That means you cannot stop using the drug.

## What Are Anxiety Disorders?

Anxiety is a feeling of uneasiness or fearful concern. All people have anxiety at some time. An **anxiety disorder** is a mental problem that makes normal life difficult because of intense anxiety. An extreme symptom of an anxiety disorder is a **panic attack**. A panic attack is a feeling of terror that comes without warning. Its symptoms include chest pain, rapid heartbeat, sweating, shaking, or shortness of breath. It can be triggered by a particular place or situation or may result from increasing stress.

There are different types of anxiety disorders. One is social anxiety disorder. This disorder makes people intensely afraid of situations where they might be embarrassed. These are usually social situations. People with social anxiety disorder may sweat heavily, blush, tremble, and have heart palpitations. They also may become confused.

80   Unit 1   Mental and Emotional Health

**Phobia**
*Excessive fear of an object or a specific situation*

**Affective disorder**
*A mental problem in which a person has disturbed or uncontrolled emotions*

**Clinical depression**
*An affective disorder involving long-lasting, intense sadness*

**Suicide**
*The act of killing oneself*

**Bipolar disorder**
*An affective disorder involving an uncontrolled shift from feeling too much energetic emotion to feeling very depressed*

**Research and Write**

There are other types of anxiety disorders than the ones discussed in this lesson. Research these other types and write an article that explains them.

Another type of anxiety disorder is a **phobia.** A phobia is an excessive fear of an object or a specific situation. An example is claustrophobia. This is a fear of small spaces. People with claustrophobia will have difficulty doing ordinary things such as riding in elevators.

## What Are Affective Disorders?

An **affective disorder** is a mental disorder in which a person has disturbed or uncontrolled emotions. The most common affective disorder is **clinical depression.**

Many people have mild depression. For example, you may feel depressed if you break up with someone. For most people, the level of depression is related to the loss. If it is not a serious loss, you will feel better and go back to normal activities quickly.

Clinical depression is a long-lasting, intense sadness. It comes from a change in brain chemistry. That means it has a physical cause.

People with clinical depression have trouble handling normal activities. They may not eat and might lose weight. They often have trouble sleeping, or they may want to sleep all day. They find it difficult to concentrate and have poor performance in school. Clinically depressed people cannot take pleasure in activities they used to enjoy. They feel worthless and guilty. They have low energy levels and withdraw from other people.

In any given 1-year period, almost 10 percent of Americans have clinical depression. In its most dangerous extremes, clinical depression may lead an individual to attempt **suicide,** or the act of killing oneself. Depression affects many teens. Suicide is the third leading cause of death among 15- to 24-year-olds.

Another example of an affective disorder is **bipolar disorder.** This disorder causes people to experience a wide shift in feeling. Their behavior changes from one extreme of overly energetic emotion to the other extreme of serious depression.

*Recognizing Mental Health Problems    Chapter 4    81*

### Teacher Alert

Tell students that the depression mode of bipolar disorder is very similar to clinical depression. The manic mode is much more than just "high spirits." A person in the manic phase may be overly euphoric, extremely irritable, or easily distracted. A person may speak very quickly, trying to keep up with racing thoughts that jump from one subject to the next. The manic phase may be marked by poor judgment, a lack of sleep, spending sprees, or overly aggressive behavior.

### 1 Warm-Up Activity

Ask students to brainstorm a list of phobias about which they have heard *(Though students may not know the names, they may have heard of fear of heights, fear of water, or fear of certain animals such as snakes or flying insects.)* Tell students that irrational means "not based on logic or reasoning." Ask how these phobias might be considered irrational. *(People might be afraid of things that actually pose no danger to them. A person looking out the window of a tall building would not be in any danger, but might be terrified nonetheless.)* Then ask how a fear or phobia might seem to be rational. *(Phobias may seem rational to those who have them because of early childhood experiences that are difficult to overcome. For example, a person with a phobia of water may have nearly drowned as a child. To this person, a fear of water makes sense.)*

### 2 Teaching the Lesson

Have students make an outline of the material presented in the lesson. The headings should relate to the headings in the text with room for important points to be added.

As you finish each section, have students fill in their outlines. Then ask volunteers to share their outlines with the class. Discuss the difference between the main points of each section and supporting details.

### Health Journal

Ask students to recall a time when they were sad or depressed. Have them write a journal entry describing the cause and how they felt. Ask them to consider whether they may have suffered from clinical depression at that time. Have them to write reasons why they do or do not think so.

### Research and Write

Have students read the Research and Write feature on page 81. You may want to have students perform this research in pairs. Collect students' articles in a class-published "magazine" about mental health.

## 3 Reinforce and Extend

### ELL/ESL Strategy

**Language Objective:** *To participate in a class discussion of an assigned topic*

Discuss with students how cultural attitudes might affect the way people treat those who have depression or other forms of mental illness. Include in your discussion ways that cultural attitudes might affect if, when, or how people seek help for emotional problems.

### Research and Write

Collect pamphlets, brochures, or other materials from mental health facilities or organizations that ask the reader questions or present information that might easily be reworded as a question (for example, "If you ever feel as if you. . . ."). Have students read the Research and Write feature on page 82, then distribute the information you have collected.

### Link to

**Geography.** Have students read the Link to Geography feature on page 82. Suggest students do research to see if there is a connection between SAD rates and suicide rates. Students should research rates for areas with short daylight hours in winter (for example, Scandinavian countries or Alaska) as well as areas that are often overcast (Seattle, Washington; Rochester or Ithaca, New York; etc.). Have students compare these rates with those from high-sunshine areas, such as Florida or southern California.

---

*Thought disorder*
A mental problem in which a person has twisted or false ideas and beliefs

*Hallucination*
A twisted idea about an unreal object or event

*Delusion*
A false belief

## What Are Thought Disorders?

A **thought disorder** is the most serious kind of mental disorder. Symptoms of thought disorders are twisted or false ideas and beliefs. An example of a twisted idea is a **hallucination,** which is an idea about an unreal object or event. For example, a person who has hallucinations might see things that do not exist. An example of a twisted or false belief is a **delusion.** A person who has delusions might falsely believe that he or she is an entirely different person.

The combination of twisted ideas and false beliefs keeps a person with thought disorders out of touch with reality. This loss of touch with reality may range from mild to severe. A person with thought disorders may have only a few unrealistic ideas or may seem to live in a dream world. The person may say things that make no sense and may do things that seem very odd. The behavior can be frightening and dangerous. This is the type of mental illness most people think of when they say someone is "crazy."

### Research and Write

A part of diagnosing depression is asking questions. Research the questions that doctors ask patients who might be depressed. Create an informational brochure that includes these questions.

### Link to ▶▶▶

**Geography**

Scientists have found that seasonal affective disorder (SAD) affects about five million Americans. This condition causes depression, difficulty sleeping, and energy loss. Less daylight during winter months causes SAD. Northern areas of the United States have more cases of SAD than southern areas. Alaska has the highest rate, at 20 percent. Florida's rate of 1.4 percent is the lowest. Decreased daylight seems to reduce the amount of an important chemical in some people's brains. Recent tests show that treatment with light can restore the chemical. People sit in front of a light box for 30 minutes a day. When they do, they find their SAD symptoms are greatly reduced.

**Eating disorder**
*An attempt to cope with psychological problems through eating habits*

## What Are Eating Disorders?

**Eating disorders** are attempts to cope with psychological problems through eating habits. People falsely believe that they do not look good and have to change their appearance. This focus on appearance may lead to an eating disorder. An eating disorder can be a way of avoiding the pain of regular life. Every feeling and struggle can be like a war between the individual and food. Eating disorders are serious. They can affect females and males. They severely challenge physical health. People can die from eating disorders. Look at Table 4.2.1 below to learn about three eating disorders.

| Disorder | Description | Characteristics | Consequences |
|---|---|---|---|
| Anorexia | Emotional problem characterized by severe weight loss | • Extreme dieting, food rituals, not eating<br>• Compulsive exercising<br>• Frequent weighing<br>• Intense fear of becoming fat | • Malnutrition<br>• Menstruation stops<br>• Lowered metabolic rate<br>• Poor temperature regulation<br>• Heart problems<br>• Death |
| Bulimia | Emotional problem involving bingeing (eating large amounts of food in a short time) followed by purging (ridding one's body of the food) or severe dieting | • Fear or inability to stop eating<br>• Vomiting or use of laxatives<br>• Secretly storing up food<br>• Extreme concentration on appearance<br>• Feeling of being out of control<br>• Depression | • Enlarged or ruptured stomach<br>• Eroded tooth enamel<br>• Pneumonia from inhaling vomit<br>• Behavioral problems<br>• Psychological problems<br>• Danger of substance abuse |
| Binge Eating | Emotional problem involving bingeing but no purging | • Continually snacking<br>• Large meals and frequent snacks while bingeing<br>• Great feelings of guilt and shame for feeling out of control<br>• Inability to stop eating during binges | • Obesity (condition of being extremely overweight)<br>• Heart disease<br>• Diabetes<br>• Some kinds of cancer<br>• Reduced life span |

**Table 4.2.1** *Eating disorders*

## LEARNING STYLES

**Logical/Mathematical**

Have students make a table summarizing the information in this lesson about mental disorders and their treatment. You may want to have students work in groups to complete the table.

## LEARNING STYLES

**Visual/Spatial**

Allow students to skim through fashion and teen magazines, taking note of the ads. Discuss whether the ads send a subtle message that thinner is better. Suggest students work in pairs to redesign a particular ad—images and language—to send a healthier message.

## IN THE COMMUNITY

Inform students that people who have substance abuse problems can get help from the National Council on Alcoholism and Drug Dependence by calling the following number: 1-800-622-2255. The agency also provides information specifically for teens about alcohol and drugs at the following web site: www.ncadd.org. Visit this site and take a test to find out how alcohol or drugs may be affecting your life.

### Technology and Society

Have students read the Technology and Society feature on page 84. Suggest students find one or two sites from each of the three categories and write a brief report describing what each site has to offer.

### GLOBAL CONNECTION

Have students research the rate of teen suicides in other countries. Suggest that students use the library and the Internet as sources of information. Have volunteers share what they learn by giving an oral presentation to the class.

### Lesson 2 Review Answers

**1.** social **2.** phobia **3.** addiction **4.** No, the person will not be able to reach a challenging goal. The person will have low energy, low motivation, and an inability to take pleasure in the activities he or she used to enjoy. **5.** Anorexia, bulimia, and compulsive overeating are eating disorders. These eating disorders are alike because they are all attempts to cope with psychological problems through eating habits. Treatment for each of these eating disorders requires professional help. Anorexia is characterized by severe weight loss. Bulimia involves bingeing and purging. Compulsive overeating involves bingeing but not purging.

### Portfolio Assessment

Sample items include:
- Outline from Teaching the Lesson
- Article from Research and Write on page 81
- Table from Learning Styles Logical/Mathematical
- Lesson 2 Review answers

---

**Word Bank**
addiction
phobia
social

**Lesson 2 Review** On a sheet of paper, write the word from the Word Bank that best completes each sentence.

1. A person with _____ anxiety disorder may tremble, sweat, or have heart palpitations.
2. An extreme fear of an object is a(n) _____.
3. An inability to stop drinking alcohol is a(n) _____.

On a sheet of paper, write the answers to the following questions. Use complete sentences.

4. **Critical Thinking** Suppose a person has clinical depression. Will he or she be able to reach a challenging goal? Explain your answer.
5. **Critical Thinking** Name three eating disorders and describe how they are alike and how they are different.

### Technology and Society

The Internet has opened new avenues for people with mental disorders. Here are some things available online:

- **Information.** The Internet lists many sites for people with mental disorders. Some offer information. Other describe support groups, their work, and how to get in touch with them.
- **Communication.** Online magazines and newsletters are written by, for, and about people with mental disorders.
- **Community.** People with mental disorders can use online chat rooms and bulletin boards. They can share experiences and concerns. In online forums, experts speak on related topics and answer questions from online audiences.

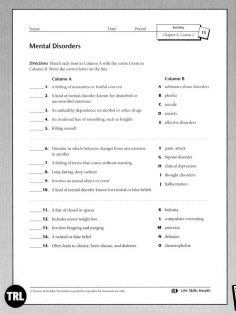

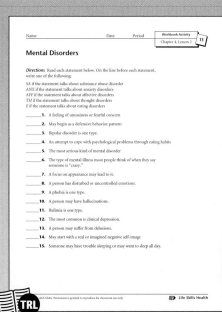

## Lesson 3  Treating Mental Disorders

### Objectives

After reading this lesson, you should be able to
- explain how mental disorders are treated
- describe psychotherapy
- describe behavior modification

**Psychotherapy**
Psychological treatment for mental or emotional disorders

**Behavior modification**
Form of psychotherapy that teaches a person to replace a less effective behavior pattern with a more effective one

**Cognitive therapy**
Form of psychotherapy that teaches a person to replace destructive thoughts with positive ones

Trained mental health professionals are the best people to treat mental disorders. Symptoms of mental health problems can be relieved so the person is better able to live well. Many disorders can be cured, or people can learn to cope with them.

### How Are Substance Abuse Disorders Treated?

People who abuse substances can get help. How well they respond to treatment depends on their determination to get well and regain control. They must decide never to use the drug again. The most important step is admitting a problem exists. For this reason, most treatment programs begin with self-awareness and admitting the need for help. Treatment may involve individual counseling, behavior change programs, or support groups.

### How Are Anxiety Disorders Treated?

Anxiety disorders are really many different kinds of problems. Some are physical and some are psychological. Physical problems can be treated with medicine that a physician orders. Medicine can reduce the symptoms long enough for the person to learn how to prevent further problems. Psychological problems can be treated with **psychotherapy,** including behavior modification. Psychotherapy is psychological treatment for mental or emotional disorders. Counseling is a type of psychotherapy. People talk over their problems with a therapist or counselor. The counselor or therapist listens more than he or she talks. **Behavior modification** is another form of psychotherapy. It teaches a person to replace less effective behavior patterns with more effective ones. For example, through behavior modification, a person can learn to relax and avoid panic by using deep breathing. **Cognitive therapy** is also a way of treating anxiety disorders. In this type of therapy, people are taught to replace destructive thoughts with positive ones.

*Recognizing Mental Health Problems* Chapter 4  85

---

## Lesson at a Glance

### Chapter 4  Lesson 3

**Overview** This lesson describes how various mental disorders are treated

### Objectives
- To explain how mental disorders are treated
- To describe psychotherapy
- To describe behavior modification

**Student Pages** 85–88

**Teacher's Resource Library**
Activity 14
Alternative Activity 14
Workbook Activity 14

### Vocabulary
behavior modification
cognitive therapy
psychotherapy

Have students look up each term in the Glossary. As a class, discuss each word. Then have students write sentences using each word.

### Background Information

Throughout human history, people with mental illness have been maltreated, shunned, or—at best—ignored. Treatments for mental illness ranged from restraint and incarceration to such harmful procedures as bleeding and submersion in ice baths. Starting in the 1930s, many patients were subjected to prefrontal lobotomy--a procedure that ranged from delicate surgery to scrambling the brain with an ice pick. Electroshock therapy ("EST"), also known as electroconvulsive therapy, was used in 1930s to treat schizophrenia, and later to alleviate depression and as a method of discipline. It is only recently that psychopharmaceuticals, or drugs, have produced humane treatments and positive results. These drugs are usually prescribed in connection with psychotherapy. EST is still occasionally used, but only after the patient has been given anesthesia and muscle relaxants.

Recog. Mental Health Prob.  Chapter 4  85

### 1. Warm-Up Activity

Have students brainstorm ways to overcome short bouts of sadness in oneself or others. *(tell jokes, send letters, talk it over, listen, and so on)* Then ask whether these methods might help someone suffering from clinical depression. *(probably not)* Have students suggest things that might help alleviate clinical depression.

### 2. Teaching the Lesson

Have students make an outline similar to the one they made for the previous lesson. The headings should again relate to the headings in the text.

#### Health Myth

Have students read the Health Myth feature on page 86. Point out that clinical depression is an illness that prevents its own treatment. People with the flu can take cold medicine. People with appendicitis can have surgery. But people with clinical depression sometimes are not able to do much that might be beneficial to them and need to be helped into treatment.

#### Health in Your Life

Collect samples of magazine ads for antidepressant medications and bring them to class. Have students read the Health in Your Life feature on page 86. Distribute the magazine ads. Ask students to describe television or radio ads for antidepressants they may have seen or heard. Discuss which ads might be most helpful and least helpful to a person who suffers from clinical depression but does not know it. Have volunteers read the disclaimers (usually in small print) on the magazine ads. What side effects are possible? Discuss whether advertising such products to the general public is a good idea.

**Health in Your Life Answers**

1. Sample answer: People become aware of disorders before being diagnosed.
2. Sample answer: People might only imagine they have disorders that they hadn't even thought about before.

---

**Health Myth**

**Myth:** People with clinical depression should go out and do something fun.

**Fact:** People with clinical depression simply cannot get up and do things. This is a part of their illness. They have low energy and no motivation. They cannot take pleasure in things they used to enjoy.

## How Are Affective Disorders Treated?

Depression is an affective disorder. One way to deal with or prevent mild depression is to become involved in a goal-directed activity. Working to reach a reasonable goal helps set a new direction. The depressed person gains a sense of involvement in life and of being busy. Some examples are caring for a pet or taking up a hobby such as painting, photography, or writing stories. Physical activity also helps to relieve mild depression.

Clinical depression can be treated with help from professionals. Medical help includes medicines that work to bring the brain chemistry back to normal. Treatment also includes psychotherapy to help people see their problems and change their behavior patterns.

As with clinical depression, bipolar disorder requires professional help for treatment.

### Health in Your Life

**Advertising of Medications**

Many people have suffered from clinical depression or an anxiety disorder without being diagnosed. They have not gotten help. In recent years, advertising for antidepressant and anti-anxiety medicines has helped with this. Ads for these medicines have appeared on TV, in magazines and newspapers, on billboards, and on the Internet.

By telling what a medicine is for, the ads explain to people what clinical depression or anxiety disorders are. Possible negative side effects of a medicine are quickly described at the end of TV commercials or in small print in ads. The person watching can then go to a doctor for diagnosis and treatment.

On a sheet of paper, write the answers to the following questions. Use complete sentences.

1. What is an advantage of advertising medicines?
2. What is a disadvantage of advertising medicines?

---

### CAREER CONNECTION

Invite students to discuss how a school counselor might help a student who feels depressed. Emphasize that counselors are trained to watch for warning signs of suicide. Encourage interested students to find out more about a career as a school counselor by writing to the American School Counselor Association, 801 N. Fairfax St., Alexandria, VA 22314.

**Health Myth**

**Myth:** People with eating disorders should just use some willpower.

**Fact:** Willpower cannot be applied to an eating disorder. Many people with eating disorders may have great willpower in other areas. People with eating disorders need professional help to treat their problems.

Should mentally ill people be allowed to have jobs and live by themselves?

### How Are Thought Disorders Treated?

Only mental health professionals can treat thought disorders. Some medicines can reduce the symptoms, but there is no cure yet for thought disorders. Most individuals with thought disorders respond to treatment. Their symptoms may become fewer or may completely disappear. Some people do not respond to treatment and continue to have symptoms throughout their lives.

### How Are Eating Disorders Treated?

People with eating disorders require professional help for treatment. People who have eating disorders often resist professional help. If you know someone who may have an eating disorder, encourage the person to get help. Let the person know you care about his or her well-being. Encourage the person to talk about his or her feelings. You can be a good role model by not criticizing your own body or appearance.

### How Does Society Deal with Mental Illness?

Society has gradually changed the way it treats people who are mentally ill. Few people today believe that a person can catch mental illness from someone else. Yet, society still has great difficulty accepting mental illness without fear. People tend to fear what they do not understand. Other people's prejudices against mental illness can influence attitudes. For example, a person might find it hard to befriend someone whom others think is "sick" or "crazy."

Fortunately, developments in psychology and medicine have helped change attitudes toward people who are mentally ill. Almost everyone knows or is related to someone who has had psychological problems. For most of these problems, people can get help that will lessen or solve these problems.

---

## 3 Reinforce and Extend

### ELL/ESL Strategy

**Language Objective:** *To research an assigned topic*

Ask groups of students to work together to find out about places in their community that offer help for mental problems. The list could include suicide prevention centers, mental health clinics, mental health counselors, clergy, telephone help-lines, and so on. Have students use the information to make a directory.

### Teacher Alert

Discuss attitudes toward mental health treatment with the class. Relate seeking treatment for a mental health problem to seeking treatment for a physical health problem. A person would not hesitate to seek treatment for a physical health problem such as pneumonia or a broken arm. Similarly, no one should hesitate to seek treatment for a possible mental health problem.

### Health Myth

Have students read the Health Myth feature on page 87. Briefly discuss addictions such as drug, alcohol, or nicotine addiction. Ask whether most addicts are able to stop their destructive behavior on their own. Then expand the discussion to include eating disorders. Help students see that people suffering from eating disorders are just as unlikely to be able to simply stop.

# Lesson 3 Review Answers

1. help 2. cognitive therapy 3. physical activity 4. Sample answer: This is not a good plan for treating the disorder. People who abuse a substance must decide never to use the drug again. By drinking alcohol only on the weekends, this person is avoiding the problem. 5. Sample answer: People with thought disorders need treatment from mental health professionals because these conditions are very serious. Most people with thought disorders respond to treatment. If they don't seek help, people with thought disorders may be dangerous to themselves or to others.

## Decide for Yourself

Collect samples of magazine ads featuring models of both sexes. Have students read the Decide for Yourself feature on page 88. Distribute the magazine ads. Ask students if they know anyone who is as thin as the female models or as well-muscled as the male models. Point out to students that fashion designers prefer to hire models who are very thin because their clothes can best be seen when they hang straight down. Selection of models has nothing to do with the health or beauty of the model.

**Decide for Yourself Answers**

1. Sample answer: Popular culture and the media have a big influence on people and the mental disorders they develop, especially eating disorders. There is a lot of pressure on me to try to be thin and look good all the time. 2. Answers will vary based on students' answers to question 1.

## Portfolio Assessment

Sample items include:
- Outline from Teaching the Lesson
- Health in Your Life answers
- Decide for Yourself answers
- Lesson 3 Review answers

---

**Lesson 3 Review** On a sheet of paper, write the word of phrase in parentheses that best completes each sentence.

1. Many treatment programs for substance abuse disorders begin with the person admitting the need for (confidence, help, willpower).

2. People who replace destructive thoughts with positive ones are using (cognitive therapy, defense mechanisms, hallucinations).

3. Mild depression can be relieved with (clinical depression, guilt, physical activity).

On a sheet of paper, write the answers to the following questions. Use complete sentences.

4. **Critical Thinking** An alcohol addict decides to drink alcohol only on weekends. Is this a good plan for treating this substance abuse disorder? Explain your answer.

5. **Critical Thinking** Why do people with thought disorders need treatment from mental health professionals?

### Decide for Yourself

What kinds of messages about your body have you received from popular culture and the media? Popular culture glorifies thinness. All kinds of media—newspapers, magazines, billboards, movies, and TV—show only thin models and actors.

Every person's body is different. Suppose we all ate exactly the same thing and exercised the same way for a year. We still would not all look alike.

Here are some things to remember:
- Media images and messages rarely reflect reality.
- Ads have one purpose: to convince you to buy something.
- Ads often create an emotional experience. They show you what the advertiser thinks you want to see.

On a sheet of paper, write the answers to the following questions. Use complete sentences.

1. How much influence do you think popular culture and the media have on mental disorders? Explain your answer.

2. We would not all look alike even if we ate exactly the same thing and exercised exactly the same way. Why do you think this is true?

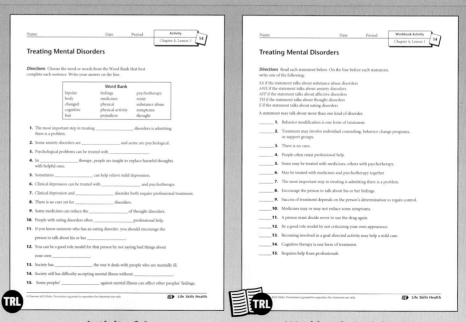

# Chapter 4 SUMMARY

- Mental health includes a broad range of behaviors.
- Normal or abnormal behavior is best judged by how well a person adjusts to meeting basic needs and goals.
- Patterns of self-defeating behaviors may indicate abnormal behavior.
- Two common self-defeating behavior patterns are defense mechanisms and dysfunctional relationships.
- Defense mechanisms are mental devices people use to protect themselves.
- People with severe emotional problems may need professional help.
- Substance abuse disorders involve a dependence on alcohol and other drugs.
- A phobia, or irrational fear of an object or situation, is one kind of anxiety disorder.
- One kind of affective disorder is clinical depression.
- Thought disorders, the most serious kind of mental disorder, involve twisted thoughts and false beliefs.
- Eating disorders are a type of mental problem involving overemphasis on appearance to avoid the pain of regular life.
- Society's attitudes toward mental illness have changed, but some people still fear it.
- Substance abuse disorders can be treated if an individual with the disorder is determined to get well.
- Anxiety disorders can be treated with medication and psychotherapy.
- People who are clinically depressed require help from professionals to get better.

## Vocabulary

| | | |
|---|---|---|
| abnormal, 76 | defense mechanism, 77 | projection, 77 |
| addicted, 80 | delusion, 82 | psychologist, 76 |
| affective disorder, 81 | denial, 77 | psychotherapy, 85 |
| anxiety disorder, 80 | displacement, 77 | repression, 77 |
| behavior modification, 85 | dysfunctional, 77 | self-defeating behavior, 76 |
| bipolar disorder, 81 | eating disorder, 83 | substance abuse disorder, 80 |
| clinical depression, 81 | hallucination, 82 | suicide, 81 |
| cognitive therapy, 85 | panic attack, 80 | thought disorder, 82 |
| | phobia, 81 | |

## Chapter 4 Summary

Have volunteers read aloud each Summary item on page 89. Ask volunteers to explain the meaning of each item. Direct students' attention to the Vocabulary box on the bottom of page 89. Have them read and review each term and its definition.

## ONLINE CONNECTION

 The Web site of the National Institute of Mental Health provides information on all the disorders covered in this chapter and many more. Information includes description of the disorder and of treatment, as well as how to find treatment in your area.

www.nimh.nih.gov

A section of the Web site of the Centers for Disease Control and Prevention is devoted to mental health. This page contains links to information about treatments, a state-by-state list of mental health organizations, statistics, and publications:

www.cdc.gov/mentalhealth/index.htm

## Chapter 4 Review

Use the Chapter Review to prepare students for tests and to reteach content from the chapter.

## Chapter 4 Mastery Test

The Teacher's Resource Library includes two forms of the Chapter 4 Mastery Test. Each test addresses the chapter Goals for Learning. An optional third page of additional critical-thinking items is included for each test. The difficulty level of the two forms is equivalent.

## Review Answers

**Vocabulary Review**

1. psychologist 2. phobia 3. self-defeating behavior 4. defense mechanisms 5. dysfunctional 6. substance abuse disorder 7. anxiety disorder 8. affective disorder 9. thought disorder 10. suicide 11. hallucination 12. repression 13. eating disorder

### TEACHER ALERT

In the Chapter Review, the Vocabulary Review activity includes a sample of the chapter's vocabulary terms. The activity will help determine students' understanding of key vocabulary terms and concepts presented in the chapter. Other vocabulary terms used in the chapter are listed below.

| | |
|---|---|
| abnormal | therapy |
| addicted | delusion |
| behavior modification | denial |
| bipolar disorder | displacement |
| clinical depression | panic attack |
| cognitive | projection |
| | psychotherapy |

---

## Chapter 4 REVIEW

**Word Bank**
- affective disorder
- anxiety disorder
- defense mechanisms
- dysfunctional
- eating disorder
- hallucination
- phobia
- psychologist
- repression
- self-defeating behavior
- substance abuse disorder
- suicide
- thought disorder

### Vocabulary Review

On a sheet of paper, write the word or phrase from the Word Bank that best completes each sentence.

1. A(n) _____ is someone who studies mental and behavioral characteristics.
2. An irrational fear of an object is a(n) _____.
3. A(n) _____ blocks a person's efforts to reach goals.
4. Mental devices people use to protect themselves are _____.
5. A(n) _____ relationship is one that works in an unhealthy way.
6. A(n) _____ is an unhealthy dependence on alcohol and other drugs.
7. A panic attack is a symptom of a(n) _____.
8. Clinical depression is one kind of _____.
9. Symptoms of a(n) _____ are twisted ideas or false beliefs.
10. Extreme clinical depression may lead to _____.
11. A twisted idea about an unreal object or event is a(n) _____.
12. The refusal to think about something upsetting is _____.
13. A(n) _____ is an attempt to cope with psychological problems through eating habits.

90 Unit 1 Mental and Emotional Health

Chapter 4 Mastery Test A, pages 1–3

## Concept Review

On a sheet of paper, write the answers to the following questions. Use complete sentences.

14. How do most psychologists evaluate mental health?
15. What are two forms of self-defeating behavior?
16. Why are some relationships dysfunctional?
17. What is clinical depression and how is it treated?
18. What are two consequences of eating disorders?

## Critical Thinking

On a sheet of paper, write the answers to the following questions. Use complete sentences.

19. Do you feel anxiety when you have to give a speech? How could you handle the anxiety?
20. If you believe a friend is feeling depressed, what could you do to help?

**Test-Taking Tip** Avoid waiting until the night before a test to study. Plan your study time so you can get a good night's sleep the night before a test.

*Recognizing Mental Health Problems* Chapter 4 91

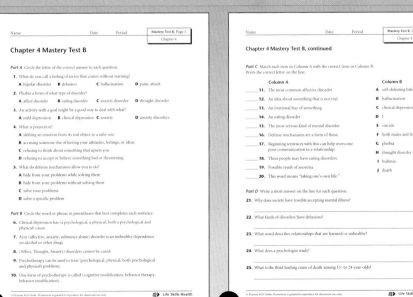

Chapter 4 Mastery Test B, pages 1–3

### Concept Review

**14.** Psychologists look for the balance between normal and abnormal behaviors. The more abnormal behaviors a person engages in, the more likely the person is to be mentally ill. **15.** Defense mechanisms and dysfunctional relationships are two forms of self-defeating behavior. **16.** Some relationships are dysfunctional because they involve people who meet each other's needs in what turn out to be destructive ways. **17.** Clinical depression is an affective disorder that comes from a change in brain chemistry, so it has a physical cause. It is characterized by intense, long-term sadness, loss of energy, sleep problems, and inability to feel pleasure. It can be treated with medicine and psychotherapy. **18.** Eating disorders can severely challenge physical health. Eating disorders can even cause death. Anorexia can cause malnutrition. Bulimia can cause stomach problems, such as an enlarged or ruptured stomach. Compulsive overeating can lead to heart disease.

### Critical Thinking

**19.** Answers will vary. Most students feel at least a little anxiety when having to give a speech. They might handle it by breathing deeply, preparing well, or talking it over with a friend. **20.** You could listen to the friend, be nonjudgmental, and if necessary, encourage the friend to get professional help. You could also help your friend find information and resources to learn about mild depression and the warning signs of clinical depression.

## ALTERNATIVE ASSESSMENT

Alternative Assessment items correlate to the student Goals for Learning at the beginning of this chapter.

- Have students create a poster listing some characteristics of poor mental health.
- Have students orally define and describe self-defeating behaviors.
- Have students write and perform a script for a radio health spot that describes substance abuse disorders, anxiety disorders, and thought disorders. The spot should also describe how each of these disorders is treated.
- Have students write an encyclopedia article that describes three kinds of eating disorders.

# HEALTH IN THE WORLD

## The World Health Organization

As part of the United Nations, the World Health Organization (WHO) identifies health issues around the world. The WHO represents 192 countries. All of these countries work together to fight disease and to work on other health issues. Some programs that the WHO supports address the health needs of specific countries. Other programs address worldwide heath issues. Each program responds to an immediate health crisis. The WHO works to stop the spread of AIDS. It also brings health care workers to regions in need and provides medicine to the poorest regions of the world. The following is an example of one of the WHO's many programs.

## Niger

Niger is a country between Algeria, Chad, and Nigeria in Africa. It has a population of about 11.6 million people. More than 60 percent of its people live on less than one dollar a day. Most people do not live beyond the age of 44. More than 80 percent of the people cannot read. In 2004, a huge number of grasshoppers and a lack of rain totally destroyed the year's crop in many regions. About 3.6 million people, including nearly 200,000 children, were left without enough food.

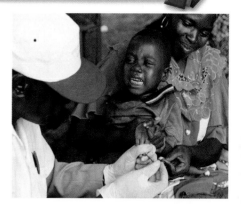

People did not have safe drinking water. These poor living conditions resulted in widespread diseases.

A team of experts from the WHO arrived in Niger to study the problem. They worked with Niger's leaders. The WHO asked members of the United Nations for $1.3 million to help the people of Niger get medicine and food. The project in Niger was part of the WHO's efforts in many African countries to improve health.

On a sheet of paper, write the answers to the following questions. Use complete sentences.

1. Why is the WHO important?
2. How does the WHO bring countries together?
3. What do you think is the biggest challenge for the WHO?

# Unit 1 SUMMARY

- People need good health to accomplish the things they need and like to do.

- A person must have physical health, social health, and emotional health to be truly healthy.

- A health risk can cause disease. Many health risks relate to heredity, the environment, and behavior.

- You can set short-term and long-term goals as you make health decisions.

- Emotions begin with thoughts and may cause physical changes or reactions.

- Five basic human needs are physical, safety, belonging, esteem, and self-actualization needs.

- Stress is a mental and physical reaction to anything that requires you to change.

- Coping with stress involves thinking through a problem, deciding how to solve the problem, and carrying out your decision.

- All emotions are normal. How you handle emotions is important to your mental health.

- Personality development is affected by temperament, meaningful relationships formed early in life, and self-concept.

- Maintaining your emotional well-being is an ongoing process that you work on throughout your life.

- Characteristics of emotional health include the ability to learn and change through experiences and balancing personal needs with the needs of others. They also include setting and reaching goals, solving conflicts rationally, controlling impulses, and thinking in healthy ways.

- Healthy relationships involve emotional attachment, mutual dependence, and the satisfaction of both partners.

- Normal or abnormal behavior is best judged by how well a person adjusts to meeting basic needs and goals.

- Substance abuse disorders involve a dependence on alcohol and other drugs. This type of disorder can be treated if the individual is determined to get well.

- One kind of affective disorder is clinical depression. People who are clinically depressed require help from professionals to get better.

- Society's attitudes toward mental illness have changed, but some people still fear it.

## Unit 1 Summary

Have volunteers read aloud each Summary item on page 93. Ask volunteers to explain the meaning of each item.

# Unit 1 Review

Use the Unit Review to prepare students for tests and to reteach content from the unit.

## Unit 1 Mastery Test

The Teacher's Resource Library includes two forms of the Unit 1 Mastery Test. An optional third page of additional critical-thinking items is included for each test. The difficulty level of the two forms is equivalent.

## Review Answers

**Vocabulary Review**

1. sedentary 2. self-esteem 3. symptom 4. phobia 5. physical environment 6. social comparison 7. wellness 8. bipolar disorder 9. prejudice 10. defense mechanism 11. depression 12. stress

**Concept Review**

13. D 14. A 15. B 16. C

---

# Unit 1 REVIEW

**Word Bank**
bipolar disorder
defense mechanism
depression
phobia
physical environment
prejudice
sedentary
self-esteem
social comparison
stress
symptom
wellness

## Vocabulary Review

On a sheet of paper, write the word or phrase from the Word Bank that best completes each sentence.

1. People who do not get much physical activity have a(n) _____ lifestyle.

2. The value that you place on yourself is your _____.

3. A bodily reaction that signals an illness or a physical problem is a(n) _____.

4. The extreme fear of an object or a specific situation is a(n) _____.

5. The area around you, including the air, land, and buildings, is your _____.

6. Learning about proper behavior by watching other people is _____.

7. Balancing physical, social, and emotional health is a way to achieve _____.

8. If one's feelings shift uncontrollably from energetic emotions to serious depression, the person may have _____.

9. Having a negative opinion about a group of people before even meeting them is a form of _____.

10. A mental device that people use to hide from their problems is a(n) _____.

11. A common reaction to grief is a state of deep sadness called _____.

12. People must change in response to physical or emotional pressure, and this reaction is _____.

94  Unit 1 Mental and Emotional Health

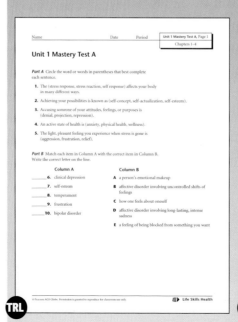
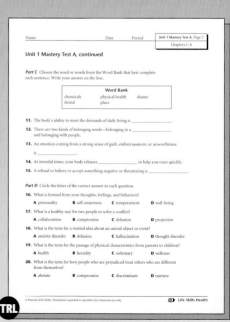

Unit 1 Mastery Test A, pages 1–3

## Concept Review

On a sheet of paper, write the letter of the answer that best completes each sentence.

13. Heredity, environment, culture, and behavior all can be causes for _____.
    A biofeedback
    B community resources
    C secondhand smoke
    D health risks

14. Abraham Maslow's list of human needs includes physical, safety, _____, esteem, and self-actualization needs.
    A belonging
    B fight-or-flight
    C hierarchy
    D aggression

15. One way that people can solve a conflict is by agreeing to _____.
    A discriminate
    B compromise
    C have delusions
    D use body language

16. Anorexia and bulimia are types of _____.
    A hallucinations
    B clinical depression
    C eating disorders
    D panic attacks

## Critical Thinking

On a sheet of paper, write the answers to the following questions. Use complete sentences.

17. What are some negative effects of choosing to smoke cigarettes?
18. What are the characteristics of a person who meets his or her self-actualization needs?
19. Why is empathy an important part of a close relationship?
20. What is a self-defeating behavior? How can it be avoided?

## Critical Thinking

**17.** Sample answer: Smoking can increase a person's heart rate and blood pressure to dangerous levels. Smoking can interfere with breathing, and it can contribute to diseases such as lung cancer. Smoking can lead to an addiction to nicotine. Secondhand smoke is a health hazard for nonsmokers who are in an environment with smokers. Smoking can alienate friends and family members. **18.** Sample answer: After people have met other basic needs—physical, safety, belonging, and esteem needs—they can work on meeting their self-actualization needs. Self-actualized individuals are able to look beyond themselves and help other people. A self-actualized person might volunteer at a food kitchen, teach a literacy class, or organize a fundraiser to benefit a charity. **19.** Sample answer: Having empathy for someone means that you understand that person's feelings. For example, suppose that you auditioned for a play but were not given an important role. You might have a friend who tried out for the soccer team but was not selected. Your experience with disappointment can help you to empathize with your friend, who is also feeling disappointed. When you recognize and acknowledge a person's feelings, your relationship with that person can grow stronger. Having empathy helps people communicate well with one another. **20.** Sample answer: A self-defeating behavior blocks a person from reaching his or her goals. The first step in avoiding such behavior is to recognize it. Writing in a diary or journal can help people understand their own behavior. Openly discussing problems can help people avoid self-defeating behavior. Suppose that a poet had a goal of getting her writing published in a famous poetry magazine. This poet also had a fear of rejection. Instead of submitting her poetry for consideration, she said to herself, "I'm not going to submit my work because the editors probably won't like it." The poet is using a defense mechanism to protect herself from getting her feelings hurt. This is self-defeating behavior because it prevents her from reaching her goal. A dysfunctional relationship is also a form of self-defeating behavior.

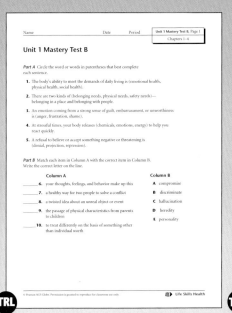
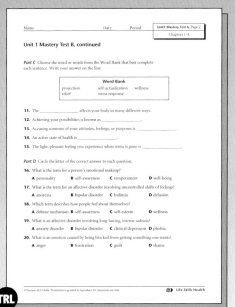

Unit 1 Mastery Test B, pages 1–3

# Unit 2

## Planning Guide
## Personal Health and Family Life

| | Student Pages | Vocabulary | Health in the World | Review | Critical-Thinking Questions | Chapter Summary |
|---|---|---|---|---|---|---|
| **Chapter 5** Identifying Human Body Systems | 98–125 | ✔ | | ✔ | ✔ | 123 |
| Lesson 1 The Skeletal and Muscular Systems | 100–103 | ✔ | | ✔ | ✔ | |
| Lesson 2 The Nervous System and Sense Organs | 104–109 | ✔ | | ✔ | ✔ | |
| Lesson 3 The Endocrine System | 110–111 | ✔ | | ✔ | ✔ | |
| Lesson 4 The Circulatory and Respiratory Systems | 112–115 | ✔ | | ✔ | ✔ | |
| Lesson 5 The Digestive and Excretory Systems | 116–119 | ✔ | | ✔ | ✔ | |
| Lesson 6 The Body's Protective Covering | 120–124 | ✔ | | ✔ | ✔ | |
| **Chapter 6** Maintaining Personal Hygiene and Fitness | 126–147 | ✔ | | ✔ | ✔ | 145 |
| Lesson 1 Hygiene for Good Health | 128–133 | ✔ | | ✔ | ✔ | |
| Lesson 2 Exercise and Physical Fitness | 134–139 | ✔ | | ✔ | ✔ | |
| Lesson 3 Personal Fitness Plan | 140–144 | | | ✔ | ✔ | |

*(Unit Planning Guide is continued on next page.)*

### Unit and Chapter Activities

**Student Text**
Unit 2 Summary

**Teacher's Resource Library**
Home Connection 2
Community Connection 5–8

**Teacher's Edition**
Opener Activity, Chapters 5–8

### Assessment Options

**Student Text**
Chapter Reviews, Chapters 5–8
Unit 2 Review

**Teacher's Resource Library**
Self-Assessment Activities 5–8
Chapter Mastery Tests A and B, Chapters 5–8
Unit 2 Mastery Tests A and B

**Teacher's Edition**
Chapter Alternative Assessments, Chapters 5–8

| | Student Text Features | | | | | | | | Teaching Strategies | | | | | | Learning Styles | | | | | Teacher's Resource Library | | | | |
|---|---|---|---|---|---|---|---|---|---|---|---|---|---|---|---|---|---|---|---|---|---|---|---|---|
| Self-Assessment | Health at Work | Health in Your Life | Decide for Yourself | Link to | Research and Write | Health Myth | Technology and Society | Background Information | ELL/ESL Strategy | Health Journal | Applications (Home, Career, Community, Global, Environment) | Online Connection | Teacher Alert | Auditory/Verbal | Body/Kinesthetic | Interpersonal/Group Learning | Logical/Mathematical | Visual/Spatial | Activities | Alternative Activities | Workbook Activities | Self-Study Guide | Chapter Outline |
| 98 | | | | | | | | | | | | 123 | 124 | | | | | | 15–20 | 15–20 | 15–20 | ✓ | ✓ |
| | | | | 101 | | 101 | | 100 | 102 | | 102, 103 | 100 | | | 103 | | | | 15 | 15 | 15 | | |
| | 107 | | | 108 | | 107 | | 104 | 106 | | 107, 108 | 106 | | | | | 106 | | 16 | 16 | 16 | | |
| | | | | | | | | 110 | 111 | 110 | | | | 111 | | | | | 17 | 17 | 17 | | |
| | | | | | 113 | | 115 | 112 | 114 | | 114 | | | | | | 115 | | 18 | 18 | 18 | | |
| | | | 119 | | 118 | | | 116 | 118 | 117 | | | | | | 118 | | | 19 | 19 | 19 | | |
| | | 122 | | | | | | 120 | 121 | | | | 121 | | | | | | 20 | 20 | 20 | | |
| 126 | | | | | | | | | | | | 145 | 146 | | | | | | 21–23 | 21–23 | 21–23 | ✓ | ✓ |
| | 133 | | 129 | 132 | 131 | 131 | | 128 | 130 | 130 | 130, 132 | 129 | | | | 132 | | 131 | 21 | 21 | 21 | | |
| | | | 139 | | 138 | | | 134 | 136 | | 135, 137, 138 | 136 | | | 136 | | 138 | | 22 | 22 | 22 | | |
| | 143 | | 142 | | | | | 140 | 143 | 142 | | 141 | | 141 | | | | | 23 | 23 | 23 | | |

## Alternative Activities

The Teacher's Resource Library (TRL) contains a set of lower-level worksheets called Alternative Activities. These worksheets cover the same content as the regular Activities but are written at a second-grade reading level.

## Skill Track

Use Skill Track for *Life Skills Health* to monitor student progress and meet the demands of adequate yearly progress (AYP). Make informed instructional decisions with individual student and class reports of lesson and chapter assessments. With immediate and ongoing feedback, students will also see what they have learned and where they need more reinforcement and practice.

# Unit 2

## Planning Guide
## Personal Health and Family Life (continued)

| | Student Pages | Vocabulary | Health in the World | Review | Critical-Thinking Questions | Chapter Summary |
|---|---|---|---|---|---|---|
| **Chapter 7** The Life Cycle and Human Development | 148–175 | ✔ | | ✔ | ✔ | 171 |
| Lesson 1 The Life Cycle and Adolescence | 150–155 | ✔ | | ✔ | ✔ | |
| Lesson 2 Reproduction | 156–159 | ✔ | | ✔ | ✔ | |
| Lesson 3 Pregnancy and Childbirth | 160–166 | ✔ | | ✔ | ✔ | |
| Lesson 4 Heredity and Genetics | 167–170 | ✔ | | ✔ | ✔ | |
| **Chapter 8** The Family | 174–195 | ✔ | | ✔ | ✔ | 193 |
| Lesson 1 The Family Life Cycle, Dating, and Marriage | 176–181 | ✔ | | ✔ | ✔ | |
| Lesson 2 Parenting and Family Systems | 182–186 | ✔ | | ✔ | ✔ | |
| Lesson 3 Problems in Families | 187–192 | ✔ | 196 | ✔ | ✔ | |

96C

| Student Text Features | | | | | | | | Teaching Strategies | | | | | | Learning Styles | | | | | Teacher's Resource Library | | | | |
|---|---|---|---|---|---|---|---|---|---|---|---|---|---|---|---|---|---|---|---|---|---|---|---|
| Self-Assessment | Health at Work | Health in Your Life | Decide for Yourself | Link to | Research and Write | Health Myth | Technology and Society | Background Information | ELL/ESL Strategy | Health Journal | Applications (Home, Career, Community, Global, Environment) | Online Connection | Teacher Alert | Auditory/Verbal | Body/Kinesthetic | Interpersonal/Group Learning | Logical/Mathematical | Visual/Spatial | Activities | Alternative Activities | Workbook Activities | Self-Study Guide | Chapter Outline |
| 148 | | | | | | | | | | | | 171 | 172 | | | | | | 24–27 | 24–27 | 24–27 | ✓ | ✓ |
|  | 155 | 153 | 154 | | | | | 150 | 152 | 151, 152 | 151 | | | 152 | 153 | | | | 24 | 24 | 24 | | |
|  | | | | | 157 | | | 156 | 157 | | 158 | | 158, 159 | | | | | | 25 | 25 | 25 | | |
|  | | | | 161, 163 | 166 | | 163 | 161 | 162 | | 164, 165 | | 163, 165 | | | 164 | 164 | | 26 | 26 | 26 | | |
|  | | | | | 170 | 169 | | 167 | 169 | | | | | | | | | 168 | 27 | 27 | 27 | | |
| 174 | | | | | | | | | | | | 193 | 194 | | | | | | 28–30 | 28–30 | 28–30 | ✓ | ✓ |
|  | | | | | | 178, 181 | | 176 | 177 | 179, 180 | 178 | | 178, 179 | | 180 | 177 | | | 28 | 28 | 28 | | |
|  | | | 184 | 183 | 185, 186 | | 182 | 182 | 185 | 185 | 185 | | | | 184 | | 186 | | 29 | 29 | 29 | | |
|  | 192 | 191 | | 190 | | | | 187 | 189 | 191 | 188, 189 | | 189 | 190 | | | | 191 | 30 | 30 | 30 | | |

# Unit at a Glance

## Unit 2: Personal Health and Family Life
pages 96–199

### Chapters

5. Identifying Human Body Systems
   pages 98–125

6. Maintaining Personal Hygiene and Fitness
   pages 126–147

7. The Life Cycle and Human Development
   pages 148–173

8. The Family
   pages 174–195

Unit 2 Summary  page 197

Unit 2 Review  pages 198–199

Audio CD

Skill Track for Life Skills Health

Teacher's Resource Library  TRL

Home Connection 2

Unit 2 Mastery Tests A and B

(Answer Keys for the Teacher's Resource Library begin on page 559 of this Teacher's Edition.)

## Other Resources

### Books for Teachers

Mc Tavish, Sandra. *Life Skills: 225 Ready-to-Use Health Activities for Success and Well-Being.* New York: John Wiley & Sons, 2003. (ready-to-use worksheets on health topics)

Tortora, Gerard J. *Introduction to the Human Body.* 7th ed. New York: John Wiley & Sons, 2006. (study of the human body, system by system)

### Books for Students

Abrahams, George, and Sheila Ahlbrand. *Boy v. Girl? How Gender Shapes Who We Are, What We Want, and How We Get Along.* Minneapolis, MN : Free Spirit Publishing, 2002. (explore gender, accept self)

Leifer, Gloria, and Heidi Hartston. *Growth and Development Across the Lifespan: A Health Promotion.* Philadelphia: W.B. Saunders, 2004. (growth and development information with health promotion theme)

### CD-ROM/Software

*Becoming an Adult.* Family Health Series, vol. 4. Toronto: Core Learning, 2003 (1-800-270-4643). (multimedia content, worksheets, and tests to support life cycle studies)

### Videos and DVDs

*Family and Friends* (24 minutes). Silver Spring, MD: Discovery Communications, 2004 (1-800-627-9399). (VHS and DVD) (social aspects of family and friend relationships)

### Web Sites

http://healthlibrary.stanford.edu/ (resources from Stanford Hospital health library)

www.APAHelpCenter.org (help center information from the American Psychological Association)

# Unit 2: Personal Health and Family Life

What do you think of when you hear the word *family*? Research shows that today's families are often made up differently than families of the past. Still, many things about the family have remained the same. It is within the family group that you learn to relate to others. You learn to care for other people.

What does it mean to be healthy? Decisions you make every day determine your physical, social, and mental well-being. This unit provides information about how your body systems are related to one another. What else will you learn through this unit? You will discover how the parts of the body affect the whole. You will find information about the life cycle—from reproduction and birth through old age. You will learn about different family systems and family relationships.

### Chapters in Unit 2

Chapter 5: Identifying Human Body Systems ............... 98
Chapter 6: Maintaining Personal Hygiene and Fitness .................. 126
Chapter 7: The Life Cycle and Human Development .............................. 148
Chapter 8: The Family ................................ 174

## Introducing the Unit

Have students read the text on page 97. Ask them to respond to the question in the first paragraph. Make a list of their answers on the board. Ask students if they think families today differ from families in the past and, if so, how they differ. *(Sample answer: Single-parent families and blended families are more common today than in the past.)* Have students explain how their family's health is related to their own health. *(Sample answer: Since you are part of a family, the state of health of your family might be reflected in your own health. If your family is not healthy, this could cause you to feel stress, which could affect your health.)*

### HOME CONNECTION

The Home Connection unit activity gives students practical experience with concepts taught in the *Life Skills Health* student text. Students complete the Home Connection activity outside the classroom with the help of family members. These worksheets appear on the Life Skills Health Teacher's Resource Library (TRL) CD-ROM.

### CAREER INTEREST INVENTORY

The AGS Publishing Harrington-O'Shea Career Decision-Making System-Revised (CDM) may be used with the chapters in this unit. Students can use the CDM to explore their interests and identify careers. The CDM defines career areas that are indicated by students' responses on the inventory.

Home Connection 2

# Chapter at a Glance

## Chapter 5: Identifying Human Body Systems
pages 98–125

### Lessons

1. The Skeletal and Muscular Systems
   pages 100–103
2. The Nervous System and Sense Organs
   pages 104–109
3. The Endocrine System
   pages 110–111
4. The Circulatory and Respiratory Systems
   pages 112–115
5. The Digestive and Excretory Systems
   pages 116–119
6. The Body's Protective Covering
   pages 120–122

**Chapter 5 Summary** page 123

**Chapter 5 Review** pages 124–125

**Audio CD**

**Skill Track for Life Skills Health**

**Teacher's Resource Library** TRL

   Activities 15–20

   Alternative Activities 15–20

   Workbook Activities 15–20

   Self-Assessment 5

   Community Connection 5

   Chapter 5 Self-Study Guide

   Chapter 5 Outline

   Chapter 5 Mastery Tests A and B

   (Answer Keys for the Teacher's Resource Library begin on page 559 of this Teacher's Edition.)

## Opener Activity

Have students leaf through the chapter and describe the diagrams and illustrations (not photographs) they see. Ask them what information they think will be covered in this chapter. Have them write down any questions they might have. As they read the chapter, students should record the answers to their questions.

98    Unit 2    Personal Health & Family Life

# Self-Assessment

## Can you answer these questions?

☑ 1. Why is exercising important?

☑ 2. How do muscles and bones work together?

☑ 3. How does the nervous system work with the other body systems?

☑ 4. How do sense organs give information to the brain?

☑ 5. How does the endocrine system work with the other body systems?

☑ 6. What does the respiratory system do for the rest of the body?

☑ 7. How do the respiratory and circulatory systems work together?

☑ 8. What happens to food after you eat it?

☑ 9. How does the body rid itself of liquid waste?

☑ 10. How does the skin protect the body?

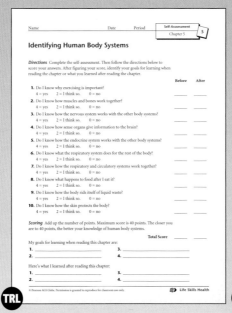

Self-Assessment 5

Community Connection 5

# Chapter 5
# Identifying Human Body Systems

Your body is amazing. All your body systems work together as one unit to keep you in good health. When your body systems stay within limits, your body is in a state of balance. For example, your body temperature should not to be too high or too low. Your body systems always work together to stay in balance.

This chapter will tell you about your body's systems and how they work together. You will learn about the skeletal and muscular systems. Then you will learn about the nervous system, the sense organs, and the endocrine system. Next, you will learn about the respiratory and circulatory systems and the digestive and excretory systems. Finally, you will learn about the body's protective covering—the skin, hair, and nails.

### Goals for Learning

- ◆ To describe the purpose of the muscular and skeletal systems
- ◆ To explain the parts of the nervous system and how information comes through the sense organs
- ◆ To explain how the endocrine glands work with the nervous system
- ◆ To describe respiration and how the body circulates blood
- ◆ To explain the digestive and excretory systems
- ◆ To describe the purpose and structure of the skin, hair, and nails

## Introducing the Chapter

Ask a volunteer to read the introductory information. Discuss what students already know about the body systems. Then discuss the Goals for Learning. Ask students to give examples of organs in each of the systems mentioned in the text.

### Notes and Questions

Ask volunteers to read the notes and questions that appear in the margins throughout the chapter. Then discuss them with the class.

### TEACHER'S RESOURCE

The AGS Publishing Human Body Systems Transparencies may be used with this chapter. The transparencies add an interactive dimension to expand and enhance the *Life Skills Health* chapter content.

### Self-Assessment

Have students complete the Self-Assessment worksheet before and after reading the chapter. Before reading the chapter, have students fill in the "Before" column. Ask students to identify their goals for learning. To get ideas for setting goals, students might use the chapter introductory material on page 99, the checklist on page 98, or the questions on the Self-Assessment worksheet. Students can use the back of the worksheet if they need more space to write.

Collect the Self-Assessment worksheets and pass them out again at the end of the chapter. Have students fill in the "After" column. Ask them to identify at least four major points they have learned. Again, suggest they use the back of the worksheet if they need more space to write. You may want to collect and review the worksheets, but return them to students so they have a record of their goals and accomplishments.

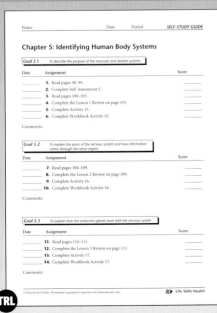

Chapter 5 Self-Study Guide, pages 1–2       Chapter 5 Outline, pages 1–2

# Lesson at a Glance

## Chapter 5 Lesson 1

**Overview** This lesson describes the functions of the skeletal and muscular systems and how these systems work together.

### Objectives

- To describe the parts of bones
- To identify the purpose of bones
- To analyze how the muscular and skeletal systems work together
- To distinguish between voluntary and involuntary muscles

**Student Pages** 100–103

**Teacher's Resource Library**

Activity 15

Alternative Activity 15

## Vocabulary

| contract | nutrient |
|---|---|
| ligament | tendon |
| marrow | |

Write each vocabulary word on the board. Then read each definition randomly and have students choose the matching term. If students answer incorrectly, let them try again. Provide the correct response if necessary. After each response, have students use the term in an original sentence.

## Background Information

Our bones form the skeletons that support us. Our muscles move those bones. Muscles move by contracting and relaxing. Muscles are connected to bones by tough, fibrous tissues called tendons. When a muscle contracts, it pulls on the tendon it is attached to, which in turn pulls on the bone. In order for the bone to be moved back to its original position, it must be pulled by a muscle and tendon on its other side.

Like tendons, ligaments are composed of tough, fibrous tissue. Ligaments connect bones to cartilage or bones to bones. Cartilage is tough connective tissue. It covers the ends of bones and cushions joints as they move. The tip of your nose and your ear lobes are made of cartilage.

---

## Lesson 1 — The Skeletal and Muscular Systems

### Objectives

*After reading this lesson, you should be able to*

- describe the parts of bones
- identify the purpose of bones
- analyze how the muscular and skeletal systems work together
- distinguish between voluntary and involuntary muscles

**Nutrient**
*A part of food that helps the body function and grow*

Two body systems that work together closely are the skeletal and muscular systems. The skeletal system is your body's system of connected bones, or your skeleton. It provides the frame for your body and protects your organs. The muscular system allows your body to move.

Figure A.1 in Appendix A shows the skeletal system. Figure A.2 in Appendix A shows the muscular system.

### What Is the Purpose of Bones?

Your body has more than 200 bones. Your bones provide your body's structure. They also protect your organs, store important minerals, and produce certain blood cells.

Bones are living material. They are made of cells, the basic structure of life. Each cell has its own job to do. Cells that do the same job form tissues. Bone cells make up bone tissue. Because bones are living tissue, they need **nutrients,** just as other parts of your body do. A nutrient is a part of food that helps the body function and grow properly. You get nutrients from the food you eat. Your blood carries the nutrients to your bones.

The size of the bones in your body ranges from large to very small. There are four basic kinds of bones: long, short, flat, and irregular. Long bones are in arms and legs. Short bones are in such places as wrists and ankles. Flat bones are in such places as the ribs. Irregular bones are in such places as fingers and toes.

### How Are Bones Joined?

Long bones have larger ends that form a joint with another bone. Joints allow for several kinds of movement. For example, the joint that has the greatest range of motion is the ball-and-socket joint. You have a ball-and-socket joint in each of your hips and shoulders. This joint allows you to move your arms and legs forward and backward. It also allows your legs and arms to move from side to side and in a circle.

100   Unit 2   *Personal Health and Family Life*

---

### TEACHER ALERT

Tell students that the longest bone in their body is the femur, or thighbone. In the next section, they will read about three tiny bones inside their ears. Taken together, these three bones—the incus, malleus, and stapes (sometimes called the anvil, hammer, and stirrup)—are smaller than an orange seed.

**Ligament**
*Tough band of stretchy tissue that holds joints together or keeps organs in place*

**Marrow**
*The matter inside bones, which forms blood*

**Contract**
*Shorten*

### Health Myth

**Myth:** Cracking your knuckles will make them larger.

**Fact:** Cracking knuckles pushes the joint out of and then back into its normal position. Thick fluid fills the space between bones. The sound comes from gas bubbles popping in the fluid. This habit will not make your knuckles larger. You might make your grip weaker if you crack your knuckles too often.

Your knee joint is a hinge joint. It is similar to a hinge on a door. You can bend your leg back at the knee, but you can't bend it forward after you straighten it again. Pivot joints are in your elbows and between your head and spinal column. They move in the same way hinge joints do but also can rotate.

Tough bands of tissue called **ligaments** hold joints together. Ligaments are stretchy. They move easily.

Long bones are somewhat hard on the outside. Inside, however, is a soft substance called **marrow.** Bone marrow forms blood. Bone marrow makes both red blood cells and white blood cells. As these cells become worn-out or damaged, the marrow replaces them. You will learn about blood cells in Lesson 4.

### How Do the Muscular and Skeletal Systems Work Together?

Joints allow your body's skeletal frame to move. However, all body movements also depend on muscles. Your body has about 600 muscles, which do many things. They move your bones, pump blood, and carry nutrients. They also move air in and out of your lungs.

A muscle is made up of fibers grouped together. When muscles **contract,** or shorten, they produce movement. Muscles hold your skeleton in place, and they also produce body heat. They act on messages they receive from your body's nervous system. You will learn about the nervous system in Lesson 2.

#### Link to >>>

**Social Studies**

Very old bones can show us how some people died long ago. Here is one example. Herculaneum was a Roman city destroyed when Mount Vesuvius erupted in 79 B.C. Modern scientists discovered bodies in beachside storerooms there. First, scientists thought the people had died because they could not breathe. Later tests on the bones proved the people had died immediately from the heat. The bones showed that people had not curled up or moved in pain. They did not have time to react.

---

 **Warm-Up Activity**

Read the lesson title and ask students to name the bones and muscular systems with which they are familiar. Ask students which bones of their skeleton they can feel and which ones they can't feel. Ask why some bones can be felt through the skin and others cannot. *(Some are covered by muscles and some are not.)*

 **Teaching the Lesson**

Have students make a three-column KWL chart to show what they already know about muscles and bones and what they want to find out. Have students complete the third column, showing what they have learned, as you work through the lesson with them.

### Health Myth

Have students read the Health Myth feature on page 101. Tell students the fluid in their knuckles—called "synovial fluid"—lubricates the joints. The fluid has oxygen, carbon dioxide, and nitrogen dissolved in it. When you tug on a finger, you increase the size of the gap between bones. This reduces the pressure on the fluid, which draws the oxygen, carbon dioxide, and nitrogen out of solution, forming a bubble. The bubble pops again almost immediately, causing the "cracking" sound. However, it takes about 25 to 30 minutes for the gas to be completely redissolved. Until it is, you cannot crack your knuckles again.

### Link to

**Social Studies.** Ask a volunteer to read aloud the Link to Social Studies feature on page 101. Students can find photos of Herculaneum at http://wings.buffalo.edu/AandL/Maecenas/italy_except_rome_and_sicily/herculaneum/thumbnails_contents.html. More photographs and a brief description of the event can be found at http://volcano.und.nodak.edu/vwdocs/volc_images/img_vesuvius.html

# 3 Reinforce and Extend

## ELL/ESL Strategy

**Language Objective:** *To identify and build on prior knowledge*

Have students form pairs composed of a student learning English and a student with strong English skills. Have each student flex his or her forearm up and down while using the other hand to feel what is happening on both the front and the back of the upper arm. Have each pair of students produce a drawing that shows how they think the muscles that control the forearm are arranged and how they work. Students then can search the Internet or the school library for information on these three muscles: brachialis, biceps brachii, and triceps brachii. (Note: *biceps* and *triceps* are not plurals. There is no such thing as a "bicep" or a "tricep.") After completing their research, students can discuss how accurate their original drawings were.

## Global Connection

Any type of physical activity requires the coordination of bone and muscle movement. One of the most popular forms of physical activity is dance. Have students learn about the dances of a particular culture by examining their origin and meaning. They may want to investigate a particular dance, such as the Irish jig, the Spanish flamenco, the Hopi rain dance, the Yoruba apala dance, or the Japanese bugaku. As students investigate the dance, they might learn a few basic steps and demonstrate these for the class.

---

**Tendon**
*A strong set of fibers joining muscle to bone or muscle to muscle*

How can you keep your cardiac muscles healthy?

It is important to warm up muscles before you exercise. Warming up helps oxygen and blood flow better through your body. It prepares your heart for exercise, and it keeps you from hurting yourself. Cooling down after exercise is also important. Cooling down slows your heart rate and stretches your muscles while they are warm.

Your body has three basic types of muscles—smooth, skeletal, and cardiac. Smooth muscles are involuntary. They work even though you don't think about making them work. Some smooth muscles are in the walls of your stomach and of your blood vessels. They move food, waste, and blood through your body.

Skeletal muscles are voluntary muscles. You control them. You decide what they will do. For example, if you decide to stand, walk, jump, or run, your voluntary muscles move. They react to your decision. Skeletal muscles are connected to the skeletal system. Tough tissues called **tendons** usually attach skeletal muscles to bones. A tendon is a strong set of fibers joining muscle to bone or muscle to muscle.

Cardiac muscles are in the walls of your heart. They contract regularly to pump blood throughout your body. Cardiac muscles are similar to smooth muscles because they are involuntary.

**Figure 5.1.1** *Muscles make it possible for you to play basketball.*

## How Do Muscles Work?

Muscles work by contracting and relaxing. When a muscle contracts, it pulls on a tendon. The tendon acts on the bone to produce a movement. Some muscles work in pairs. When one contracts, the other one relaxes. You can feel this happening when you bend your arm at the elbow. The muscle on the top of the upper arm contracts. At the same time, the muscle on the bottom of the upper arm relaxes. When you extend your arm, the opposite happens.

## What Is Muscle Tone?

Some muscles never relax completely. They are somewhat contracted all the time. This is because of muscle tone. When you are in good health, a constant flow of messages runs from your nerves to your muscles. This helps you keep good muscle tone. Exercise and healthy eating are important for good muscle tone.

**Word Bank**
cardiac
joints
marrow

*Lesson 1 Review* On a sheet of paper, write the word from the Word Bank that best completes each sentence.

1. The soft inside part of a bone is called _____.
2. Your body's skeletal frame can move because of your _____.
3. The three basic types of muscles in your body are smooth, skeletal, and _____.

On a sheet of paper, write the answers to the following questions. Use complete sentences.

4. **Critical Thinking** What would happen if you had no bones? Explain your answer.
5. **Critical Thinking** How are voluntary and involuntary muscles different?

*Identifying Human Body Systems* Chapter 5  103

## Lesson 1 Review Answers

1. marrow 2. joints 3. cardiac 4. The skin, muscles, and organs would fall into a heap because the body would have no structure. 5. Involuntary muscles move even though you don't think about making them move. Voluntary muscles are under your control.

## Portfolio Assessment

Sample items include:
- KWL chart from Teaching the Lesson
- Drawing from ELL/ESL Strategy
- Lesson 1 Review answers

### LEARNING STYLES

**Body/Kinesthetic**
Have students hold a snap-type clothespin between their thumb and forefinger and count the number of times they can quickly pinch the clothespin open and closed before becoming tired. Then have them switch hands and repeat the process. Ask if they were able to repeat the process more times when using the hand with which they write. Explain that the small muscles in their writing hand get more exercise than the muscles in their other hand and, therefore, tire more slowly.

### IN THE COMMUNITY

Have an athletic trainer, sports medicine physician, or sports physical therapist visit your class to discuss athletic injuries common to teenagers and adults.

Activity 15

Workbook Activity 15

*I.D.ing Human Body Systems* Chapter 5  103

# Lesson at a Glance

## Chapter 5 Lesson 2

**Overview** This lesson describes the differences between the central and peripheral nervous systems. The sense organs are also discussed.

### Objectives

- To list the parts of the nervous system
- To explain how the brain controls the body
- To describe the peripheral nervous system
- To understand how the sense organs work

**Student Pages** 104–109

**Teacher's Resource Library**

Activity 16

Alternative Activity 16

Workbook Activity 16

## Vocabulary

| | |
|---|---|
| auditory nerve | peripheral nervous system |
| brain stem | pupil |
| cerebellum | receptor cell |
| cerebrum | reflex |
| cornea | retina |
| eardrum | spinal column |
| iris | spinal cord |
| lens | taste bud |
| medulla | |
| optic nerve | |

Read the words and discuss their meanings with the class. Ask students to write sentences, leaving blanks for the missing vocabulary words. Have students exchange papers and write in the missing words.

## Background Information

Helen Keller is an outstanding example of a person who overcame physical disabilities. She lost her sight and hearing as a result of a serious illness when she was young. With the help of her devoted teacher, Keller learned to communicate through touch. She listened to people by placing her hand on their nose, lips, and throat. At first, she communicated through sign language, but later she learned to speak. She became a lecturer and worked on behalf of visually impaired people around the world.

104  Unit 2  Personal Health & Family Life

---

## Lesson 2 — The Nervous System and Sense Organs

### Objectives

After reading this lesson, you should be able to

- list the parts of the nervous system
- explain how the brain controls the body
- describe the peripheral nervous system
- understand how the sense organs work

**Cerebrum**
*The part of the brain that lets a person read, think, and remember*

The nervous system is the body's communication network. It sends messages throughout your body. Figure A.3 in Appendix A shows the nervous system. The nervous system has two parts—the central nervous system and the peripheral nervous system. The central nervous system is made up of the brain and the spinal cord.

### How Does the Brain Control the Body?

Your brain receives messages from your nerves and sends messages through the nerves to all parts of your body. Your brain is like a computer and a chemical factory combined. It can process and store information. It produces and uses chemicals to send signals. The brain has three main parts that work together to control your body. Figure 5.2.1 shows the parts of the brain.

The **cerebrum** is the part of the brain that lets a person read, think, and remember. It is the largest part of the brain. The cerebrum is divided into two halves. The right half controls the movement on the left side of your body. It also is the site for artistic skills and instinctive thinking. Instinctive thinking relates to actions that happen instantly without your thinking about them. The left half controls the movement on the right side of your body. It is the site for math and language skills and logical, or sensible, thinking.

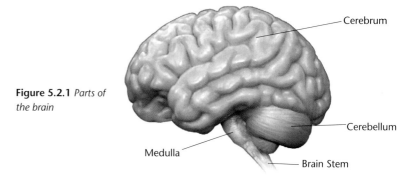

**Figure 5.2.1** Parts of the brain

104  Unit 2  Personal Health and Family Life

**Cerebellum**
*The part of the brain that controls balance and helps coordinate muscular activities*

**Brain stem**
*The part of the brain that connects the cerebrum to the spinal cord*

**Medulla**
*The part of the brain that controls the body's automatic activities*

**Spinal cord**
*Major pathway the brain uses to send messages to the body*

**Spinal column**
*Series of small bones that surround and protect the spinal cord*

**Peripheral nervous system**
*Network of peripheral nerves carrying messages between the brain and spinal cord and the rest of the body*

Why is wearing a helmet important when riding a bike?

The other two parts of your brain are much smaller than your cerebrum. They are the **cerebellum** and **brain stem**. The cerebellum lies between the cerebrum and brain stem. Your cerebellum controls balance and helps coordinate muscular activities such as walking. The brain stem connects the cerebrum to the spinal cord.

One part of the brain stem—the **medulla**—controls the body's automatic activities. These include breathing, digesting food, circulating blood, swallowing, coughing, and sneezing.

### How Does the Spinal Cord Assist the Brain?

The **spinal cord** is a major pathway your brain uses to send messages to your body. Your spinal cord is a large batch of long nerve cells wound together. It connects to your brain stem and extends to the lower part of your back. The **spinal column** is made up of a series of small bones. These bones surround and protect your spinal cord.

The nerves that make up the spinal cord send and receive messages. These nerves receive messages from the brain and send them to other sets of nerves.

### What Does the Peripheral Nervous System Do?

The **peripheral nervous system** is a network of peripheral nerves. They carry messages between your brain and spinal cord and the rest of your body. *Peripheral* means "located away from the center."

One part of the peripheral nervous system helps control your body's automatic activities. It helps your body do what it must to remain stable, or under control. It also helps your body act in an emergency. For example, suppose you are in a dangerous situation and need to run fast. The peripheral nervous system signals your body to speed up your breathing and heartbeat so you can act on the signal to run. This happens in the body's stress response. Afterward, the peripheral nervous system slows the workings of your body and returns your body to normal.

 **Warm-Up Activity**

Have students imagine that they have to do without one of their sense organs for the day. Students should list five everyday tasks that they would ordinarily do using that particular sense organ. They then should describe how the absence of the sense organ would affect these tasks.

 **Teaching the Lesson**

As you complete each section, discuss with students what would happen if the functioning of a person's nervous system was impaired. Discuss brain and spinal cord injuries as well as the loss of a sense. (The term for loss of the sense of smell is anosmia, and the term for loss of the sense of taste is ageusia.) Then discuss what students can do to protect themselves against such injuries.

# 3  Reinforce and Extend

## TEACHER ALERT

Help students understand that reflex actions such as jerking your hand away from something hot do not involve the brain. When your fingers sense heat, they send a message along sensory neurons to the spinal column. The spinal column relays this message to the brain. It also sends another message telling your fingers to move through motor neurons back down to your fingers. This is why you usually don't realize something is hot until after your hand has jerked away.

## ELL/ESL STRATEGY

**Language Objective:**
*To teach and practice functional language skills*

Have students form pairs composed of a student learning English and a student with strong English skills. Using the diagrams on pages 106 and 108, have one student explain to the other how the eye and the ear work. Prompt the student receiving the explanation to ask for clarification of anything he or she feels is not completely explained.

## LEARNING STYLES

**Visual/Spatial**
Have a volunteer stand in front of the class to illustrate a reflex. Hold a book, ball, or other object and make a movement as if to throw the object to the student, but do not actually throw the object. Ask the class to observe what happens. One observation will likely be that the volunteer blinked. Point out that blinking is a reflex. Ask why blinking is an important reflex. *(It helps protect the eyes from injury.)*

---

**Reflex**
Automatic response

**Cornea**
Part of the eye that light passes through

**Pupil**
Dark center part of the eye that adjusts to let in the correct amount of light

---

A special part of your peripheral nervous system controls your **reflexes**. Reflexes are automatic responses to something such as heat or pain. For example, when you touch something hot, you jerk your hand away without thinking. Your nerves have sent a message to the muscles in your hand. A person cannot stop reflex actions from happening.

### How Do the Sense Organs Work?
Nerves throughout your body carry messages to your brain. Your brain receives messages and then sends signals to other parts of your body. When messages come from outside your body, your sense organs receive them. The sense organs are your eyes, ears, tongue, nose, and skin.

### The Sense of Sight
Your eye is your organ of sight. Look at Figure 5.2.2 to see the parts of the eye. How does the eye work? First, light enters the eye through the **cornea**. The cornea sends the light to the **pupil**, the dark center of the eye. The pupil can adjust its size. It gets smaller in bright light and larger in dim light. This causes the pupil to let the right amount of light into the eye.

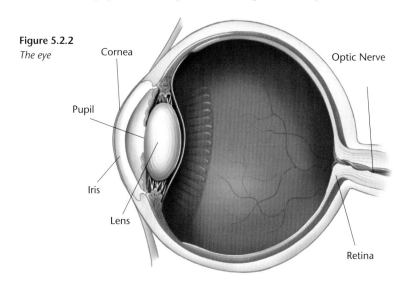

**Figure 5.2.2** The eye

106   Unit 2   *Personal Health and Family Life*

**Lens**
*The soft, clear tissue behind the pupil*

**Retina**
*Part of the eye that sends information about light*

**Optic nerve**
*Part of the eye that sends information to the brain*

**Iris**
*Colored part of the eye around the pupil*

Behind your pupil is the soft, clear tissue called the **lens**. The lens helps direct the light energy onto the **retina**. The retina contains special cells that send the light information to the **optic nerve**. This is the nerve that sends the information to the brain. Then the brain changes the light information into understandable pictures. This entire process happens faster than you can blink.

Another part of your eye is the **iris**. The iris is the colored part surrounding the pupil that gives your eye its color.

### Health Myth

**Myth:** Reading in dim light is bad for your eyes.

**Fact:** Reading without enough light may be hard for you. Even so, reading in dim light will not cause any part of the eye to wear out. You will not harm your eyes by reading in dim light.

▼◄▲▼◄▲▼◄▲▼◄▲▼◄▲▼◄▲▼◄▲▼◄▲▼◄▲▼◄▲▼◄▲▼

### Health at Work

**Optician**

Many workers help people care for their eyes. Opticians help people who need eyeglasses or contact lenses. What does an optician do? First, an ophthalmologist or optometrist writes a prescription for eyeglasses.

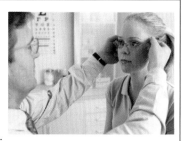

The optician helps the patient choose a pair of frames. Then the optician uses the prescription to make lenses that fit into the frames. Afterward, the optician makes sure the eyeglasses fit the patient. The optician can also make contact lenses for a patient. Patients can learn from an optician how to take care of their glasses and contact lenses.

This career requires a high school diploma and a great deal of on-the-job training. Many community colleges offer formal training for students interested in becoming opticians. Some states require opticians to have special licenses.

*Identifying Human Body Systems* Chapter 5 **107**

### Health Myth

Ask a volunteer to read aloud the Health Myth feature on page 107. Explain to students that two things happen when you move from a lit area to a dim area. First, the muscles of the iris relax, which causes the pupil to become larger. This happens almost immediately. Second, the rod and cone cells and the nerve cells in the retina must adapt to the lower light levels. This takes about ten minutes. Once these adjustments occur, the eye operates in dim light exactly the way it operates in bright light. However, reading in dim light might cause temporary eye strain, the symptoms of which include headaches and dry eyes.

### Health at Work

Have students read the Health at Work feature on page 107. Then discuss the job description with students. Ask for reasons why someone might want to become an optician. Also ask what personality traits would be useful to a person who is considering becoming an optician. *(It would be helpful to aspiring opticians if they enjoyed working with and helping people.)*

### CAREER CONNECTION

Ask students if they know what an optometric assistant is. An optometric assistant is most likely the person who greets people when they have their eyes examined. The optometric assistant records the person's history and prepares the person for the eye exam. During the exam, the optometric assistant may perform certain eye tests, including color vision, depth perception, nearsightedness, and farsightedness. Ask students to discuss the advantages and disadvantages of choosing this career.

### Link to

**Arts.** Have students read the Link to Arts feature on page 108. Have students make a list of songs or other pieces of music to which they have emotional reactions. Ask them to discuss which songs or pieces of music make them feel happy, sad, silly, or frightened?. You might want to ask students to bring in samples of these pieces to play for the class.

## IN THE ENVIRONMENT

There is evidence that exposure to loud rock music is damaging to hearing. Workers in noisy factories and rock musicians are two groups of people who are especially susceptible to hearing impairment. Discuss what other jobs might have a loud work environment that could cause hearing damage to employees. Ask students to brainstorm steps that employers and employees can take to prevent hearing loss. *(Affected workers include airport ground crews and construction workers. Some of these workers wear earplugs to prevent hearing loss.)*

**Eardrum**
*Thin piece of tissue stretched across the ear canal*

**Auditory nerve**
*Part of the ear that sends sound information to the brain*

### The Sense of Hearing

You may not notice that the air vibrates, or shakes, when a sound is made. Your ears, however, do notice. Figure 5.2.3 shows the parts of the ear. Your outer ear picks up air vibrations and sends them through the ear canal to your **eardrum.** Your eardrum is a thin piece of tissue stretched across the ear canal. When your eardrum receives the vibrations, it also vibrates.

Eardrum vibrations cause three small bones in the middle ear to vibrate. These bones—the hammer, anvil, and stirrup—pass the vibrations to a snail-like organ in the inner ear. There, tiny cells transfer the vibrations to your **auditory nerve.** The auditory nerve is the part of the ear that sends information to your brain. Your brain interprets the message and tells you what kind of sound you have heard. All of this happens in the time it takes for the sound to be made.

**Figure 5.2.3**
*The ear*

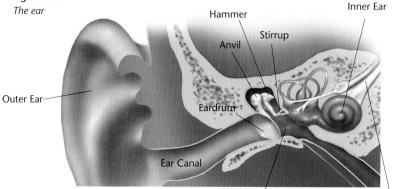

### Link to

**Arts**

Scientists have learned that music vibrations act as triggers through the ears to the brain. Vibrations trigger different groups of brain cells. Different feelings result from setting off different cell groups. This is why some music can make you feel sad and some can make you feel happy.

**Receptor cell**
Cell that receives information

**Taste bud**
Receptor on the tongue that recognizes tastes

### The Senses of Smell and Taste

Both your tongue and your nose contain **receptor cells,** or cells that receive information. The receptor cells in your nose send messages through nerves. These nerves include the olfactory nerve. Your olfactory nerve is connected to your brain. Your nose can pick up thousands of different odors.

Your tongue, however, can recognize only four kinds of taste. The four tastes are sweet, salty, sour, and bitter. Different **taste buds** are receptors for each kind of taste. The taste buds send messages to your brain through your nerves. The taste buds are located on different parts of your tongue. They tell you how something tastes. Different parts of your tongue can recognize each of the four tastes.

### The Sense of Touch

Your skin, your sense organ for touch, is your body's largest organ. Sense receptors all over your skin receive different sensations. You have receptors for touch, pressure, pain, heat, and cold. Your sense receptors send messages through the nerves to your spinal cord and brain. This is how you determine whether something is hot, cold, rough, or smooth. Your fingertips and lips are the most sensitive parts of your body because they have the greatest number of sense receptors.

*Lesson 2 Review* On a sheet of paper, write the answers to the following questions. Use complete sentences.

1. Which half of the cerebrum controls artistic skills?
2. What is the main job of the peripheral nervous system?
3. What is a reflex?
4. **Critical Thinking** What might happen if the pupil of your eye could not change size?
5. **Critical Thinking** Why is being able to feel heat and pain important?

*Identifying Human Body Systems* Chapter 5 **109**

### Lesson 2 Review Answers

1. The right half of the cerebrum controls artistic skills. 2. The main job of the peripheral nervous system is to carry messages between the brain and spinal cord and the rest of the body. 3. A reflex is an automatic response to something such as heat or pain. 4. You would not be able to see clearly because the pupil would not let in the right amount of light. 5. If people could not feel heat and pain, they might keep doing something (such as touching a hot pan) that would cause them more harm. They would not know when they needed medical help. Burns and cuts could get worse, especially if they were on a part of the body that people do not often see, such as the bottom of the foot or the back of the neck.

### Portfolio Assessment

Sample items include:
- List from Warm-up Activity
- List or samples of music from Link to Arts
- Lesson 2 Review answers

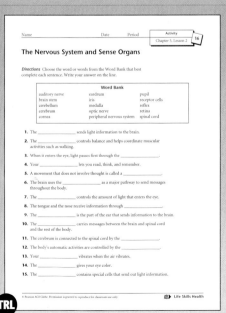

Activity 16          Workbook Activity 16

## Lesson at a Glance

### Chapter 5 Lesson 3

**Overview** This lesson describes the functions of the endocrine system and how it controls activities in the human body.

### Objectives

- To understand what the endocrine system does
- To recognize how the pituitary gland affects the body's growth
- To explain the effect of the thyroid gland
- To describe the job of the adrenal glands

**Student Pages** 110–111

**Teacher's Resource Library**

Activity 17

Alternative Activity 17

Workbook Activity 17

### Vocabulary

| | |
|---|---|
| adrenal gland | hormone |
| adrenaline | metabolism |
| gland | |

Have a student choose a vocabulary word and read aloud the sentence in the text in which the word appears in bold type. Then have the student read the definition in the vocabulary box. Repeat this procedure for each vocabulary word.

### Background Information

The word hormone is derived from a Greek word meaning "to arouse to activity" or "to excite." The term was first used by British physiologist E. H. Starling in 1905 to describe substances secreted by the stomach in the digestive process.

### 1 Warm-Up Activity

Explain that glands in the endocrine system produce chemical messengers called hormones that work together to control body activities. Adrenaline, a hormone secreted by the adrenal glands, is released into the bloodstream when a person is feeling stressed or afraid. Encourage students to share experiences of their own reactions to fear, excitement, happiness, and other emotions.

---

## Lesson 3 The Endocrine System

**Objectives**

*After reading this lesson, you should be able to*

- understand what the endocrine system does
- recognize how the pituitary gland affects the body's growth
- explain the effect of the thyroid gland
- describe the job of the adrenal glands

**Gland**
*A group of cells or an organ that produces a substance and sends it into the body*

**Hormone**
*A chemical messenger that helps control how body parts do their jobs*

**Metabolism**
*Rate at which cells produce energy*

### What Does the Endocrine System Do?

The endocrine system is made up of **glands.** A gland is a group of cells or an organ that produces a special substance and sends it into the body. Figure 5.3.1 shows these glands. Endocrine glands produce **hormones** and send them directly into the bloodstream. Hormones are chemical messengers that help control the working of different body systems.

### How Do Glands Affect the Body's Growth?

One of the endocrine glands is the pituitary gland. It is a small gland the base of the brain. The pituitary gland produces several hormones. One of these is a growth hormone. Too much or too little of this hormone can affect a person's body size. For example, too much grow hormone can make bones unusually long. The person becomes much taller than average.

### What Does the Thyroid Gland Do?

Another endocrine gland is the thyroid gland. It is located in the neck. The hormone it produces affects all tissues. Each cell in your body turns food into energy. The rate at which cells produce energy is called **metabolism.** The cells depend on the thyroid to change food into energy. Too much of the hormone can cause a person to have too much energy. This signals high metabolism. Too little of the hormone may result in low metabolism.

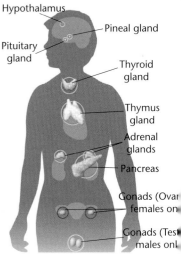

**Figure 5.3.1** *The endocrine system*

---

### 2 Teaching the Lesson

Have volunteers read each paragraph in the lesson aloud. After each paragraph is read, have another volunteer paraphrase it, presenting the same information in his or her own words.

Have students write a paragraph about each vocabulary word, describing what the term means and how it relates to the human body. (Students may deal with *adrenaline* and *adrenal gland* in the same paragraph.)

### HEALTH JOURNAL

Building on the Warm-Up Activity, ask students to recall a time when they were very scared and/or faced an emergency. Have students describe in their journals how they felt physically at the time. Ask them consider their energy level, whether their hands or knees trembled, and how their voice and breathing were affected.

**Adrenal gland**
Part of the endocrine system that releases several hormones

**Adrenaline**
Hormone released into the bloodstream that causes the heart to beat faster and blood pressure to rise in an emergency

---

Sleep affects all the functions of your body. Lack of sleep can cause you to feel sad and make concentrating hard. When you do not get enough sleep, smaller amounts of growth hormone are released, affecting growth. These are only a few of the many possible effects of lack of sleep.

---

This causes a person to feel tired. The pituitary gland produces a hormone that causes the thyroid to act.

## What Do the Adrenal Glands Do?

The system also contains two **adrenal glands,** located on top of each kidney. These glands produce a mixture of hormones that affect the working of many body systems. These hormones help the body keep the proper water balance. One adrenal hormone helps a person deal with stress.

The adrenal gland also produces a hormone called **adrenaline.** Adrenaline causes your heart to beat faster and blood pressure to rise. It is released into the blood in greater amounts when a person feels fear or faces an emergency.

**Lesson 3 Review** On a sheet of paper, write the letter of the answer that best completes each sentence.

1. The pituitary gland mainly affects _____.
   - A taste
   - B thought
   - C vision
   - D growth

2. The adrenal glands help a person deal with _____.
   - A hearing
   - B stress
   - C smell
   - D balance

3. Cells depend on the thyroid to change food into _____.
   - A energy
   - B cell
   - C hormones
   - D liquids

On a sheet of paper, write the answers to the following questions. Use complete sentences.

4. **Critical Thinking** Suppose that a person's adrenal glands are not producing enough adrenaline. What might happen to the person in an emergency?

5. **Critical Thinking** Suppose that a person's thyroid gland is not producing enough hormone. What are two possible effects?

*Identifying Human Body Systems* Chapter 5 111

---

## 3 Reinforce and Extend

### ELL/ESL Strategy

**Language Objective:** Learning vocabulary specific to the content area

Have students create a word bank that includes all the vocabulary words in the lesson, their pronunciations, and their definitions. (Note that pronunciation keys may vary according to the spelling rules of each student's native language.) If necessary, point out that the e in "adrenaline" is short, but the one in "adrenal" is long. Help students find relationships among the words—tell them, for example, that the adrenal gland is a gland and that adrenaline is a hormone produced by the adrenal gland.

### Learning Styles

**Auditory/Verbal**
Choose one or more students to summarize the lesson, aloud and in their own words. Students should speak extemporaneously and address their summaries to the entire class.

### Lesson 3 Review Answers

1. D 2. B 3. A 4. Sample answer: The person might not feel the need to move quickly to safety. 5. Sample answer: The person would probably feel tired and gain weight.

### Portfolio Assessment

Sample items include:
- Paragraphs from Teaching the Lesson
- Health Journal entry
- Lesson 3 Review answers

---

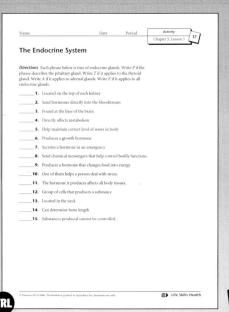

Activity 17

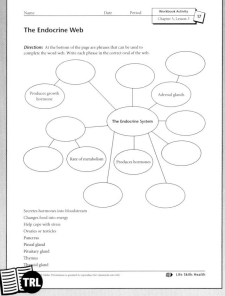

Workbook Activity 17

*I.D.ing Human Body Systems* Chapter 5 111

# Lesson at a Glance

## Chapter 5 Lesson 4

**Overview** This lesson describes the circulatory and respiratory systems and how these systems work together.

### Objectives

- To explain why the body needs blood
- To recognize how the respiratory and circulatory systems work together
- To understand how the body takes in and processes oxygen

**Student Pages** 112–115

**Teacher's Resource Library**

Activity 18

Alternative Activity 18

Workbook Activity 18

## Vocabulary

| | |
|---|---|
| alveoli | diaphragm |
| artery | exhale |
| blood pressure | inhale |
| bronchi | platelet |
| bronchioles | pulse |
| capillary | respiration |
| carbon dioxide | trachea |
| chest cavity | vein |
| clotting | |

Read the words and discuss their meanings with the class. Have students draw simple diagrams of the respiratory system and of blood vessels. Then challenge them to use the vocabulary words to label the parts of the diagrams.

## Background Information

In 1900, it was found that during blood transfusions, elements in one person's blood could cause the red blood cells in another person's blood to clump together or to break down. This led to the classification of blood types into three groups: A, B, and O. A fourth type, AB, was found in 1902. In 1939 a transfusion of type O blood caused the red cells in another patient's type O blood to clump together. This was caused by the Rh factor. All blood types are now classified as Rh-positive or Rh-negative. Since then, over 250 additional factors that affect blood compatibility have been discovered.

112  Unit 2  Personal Health & Family Life

---

## Lesson 4 The Circulatory and Respiratory Systems

### Objectives

*After reading this lesson, you should be able to*

- explain why the body needs blood
- recognize how the respiratory and circulatory systems work together
- understand how the body takes in and processes oxygen

**Platelet**
*Small part in the blood that helps with clotting*

**Carbon dioxide**
*A gas that is breathed out of the lungs*

**Clotting**
*Sticking together to form a plug and stop a wound from bleeding*

**Pulse**
*Regular beat of the heart*

All systems in your body work together. You know that all systems depend on your nervous system. They also depend on your circulatory and respiratory systems. The circulatory system includes the heart, and three kinds of blood vessels. It is your body's transportation system. Figure A.4 in Appendix A shows the circulatory system.

### Why Does the Body Need Blood?

The circulatory system moves blood through the body. Blood carries food and oxygen to every cell. Cells use the food and oxygen to do their work. The blood also carries waste products away from the cells.

Blood is made up of red blood cells, white blood cells, and **platelets.** Red blood cells carry oxygen to all parts of your body. They also remove **carbon dioxide,** a gas, from these parts. White blood cells work to keep your body healthy by fighting disease and germs. Usually, your body has fewer white blood cells than red blood cells. However, white blood cells increase when your body is fighting germs. Platelets are the smallest parts in blood. They prevent the body from losing blood through a wound by **clotting,** or sticking together to form a plug. Clotting stops bleeding.

### How Does Blood Move Through the Body?

The heart pumps blood to all parts of your body. Your heart has been beating every minute since you were born, and even before that. In fact, your heart beats about 72 times a minute. This regular beat is your **pulse.**

Your heart is in the left side of your chest. It has four chambers—two on the left side and two on the right. Blood from all parts of your body flows into the right side of your heart. This blood contains carbon dioxide. The blood must get rid of the carbon dioxide through the lungs. The right side of your heart pumps the blood to the lungs, where you breathe out the carbon dioxide.

112  Unit 2  Personal Health and Family Life

**Artery**
*Vessel that carries blood away from the heart*

**Capillary**
*Tiny blood vessel that connects arteries and veins*

**Vein**
*Vessel that carries blood back to the heart*

**Blood pressure**
*Force of blood on the walls of blood vessels*

**Research and Write**

Research the ways that smoking is dangerous to your circulatory and respiratory systems. Check the American Heart Association and American Lung Association Web sites. Then talk to someone in your community who provides health care. Make a fact sheet to share with your classmates. Provide warnings about the dangers of smoking to all body systems.

In your lungs, blood gets rid of carbon dioxide and picks up oxygen. Then the blood travels to the left side of your heart. From there, your heart pumps blood to all parts of your body to get oxygen to the cells.

Blood vessels distribute blood, or send it, throughout your body. You have three kinds of blood vessels—**arteries, capillaries,** and **veins.** Arteries are the largest blood vessels and carry blood away from your heart. They have thick, three-layered walls. Arteries branch into smaller vessels. These small branches regulate, or control, the flow of blood into the capillaries.

Capillaries are the smallest blood vessels. Capillaries connect arteries and veins. They have thin walls. These walls allow nutrients and oxygen to pass from the blood to the body cells. Capillaries also pick up waste products from the cells.

When blood leaves the capillaries, it travels to veins. Veins carry the blood back to your heart. Veins that receive blood from capillaries are small. They become larger, though, as they come closer to your heart. This passage of the blood throughout your body happens quickly. It takes about one minute for blood to travel through your whole body.

### What Is Blood Pressure?

**Blood pressure** is the force of blood on the walls of blood vessels. Pressure is created as your heart pumps blood to all parts of your body. Your body needs to keep a certain level of blood pressure. Many things affect blood pressure, including age, tobacco use, diet, exercise, and heredity.

### How Do the Respiratory and Circulatory Systems Work Together?

The circulatory system carries blood to all areas of your body. That blood carries oxygen to the cells. The cells need oxygen to break up nutrients in the cells. Energy is released from the nutrients. Then the respiratory system gets the oxygen into your body. It also gets rid of carbon dioxide, a waste product. Together, the respiratory and circulatory systems give all your cells the oxygen they need to survive.

### Warm-Up Activity

Discuss the meaning of heart rate. Show students how to find their pulse and have them count their heartbeats for ten seconds. Tell them to multiply this number by 6 to find their heart rate for one minute. Next, have students run in place for ten seconds and find their heart rate again. Ask them to explain the relationship between exercise and heart rate. *(During exercise, the heart beats faster to supply the muscles with oxygen and remove the waste products produced by the activity of muscles.)*

### Teaching the Lesson

Help students understand that the circulatory system and the respiratory system are two separate systems. The lungs are not part of the circulatory system. The circulatory system consists of the heart, arteries, capillaries, veins, and blood. This system sends blood through the lungs, but it also sends blood through every other organ in all body systems.

Explain that the circulatory system can be thought of as two big loops. One loop runs from the heart to the lungs and back again. The other loop runs from the heart to every other part of the body and back again. Blood starts the first loop depleted of oxygen and ends it containing oxygen. Blood then starts the second loop containing oxygen and ends it depleted of oxygen.

**Research and Write**

Have students read and conduct the Research and Write feature on page 113. The URL for the American Heart Association's home page is www.americanheart.org. The URL for the American Lung Association's home page is www.lungusa.org. You might want to have students work with partners or in small groups to complete this activity.

# 3  Reinforce and Extend

## ELL/ESL Strategy

**Language Objective:**
*To teach specific reading strategies*

Point out to students learning English that each section of this lesson is headed by a question. As you finish each section, have English language learners summarize it. Be sure they include the answer to each question.

## At Home

Ask students to take the heart rates of their family members before and after exercise. (Students should be certain that family members who participate in this exercise do not have heart conditions.) Discuss how to analyze the heart rates to try to discover a pattern concerning the differing ages of family members. Have students bring their findings to class to compare with the findings of other students.

---

**Respiration**
Breathing in oxygen and breathing out carbon dioxide

**Trachea**
A long tube running from the nose to the chest; windpipe

**Bronchus**
Breathing tube (plural is bronchi)

**Bronchiole**
A tube that branches off the bronchus

**Respiration** is breathing in oxygen and breathing out carbon dioxide. The respiratory system is a system of tubes and organs that allows you to breathe. You need to breathe about 20 times every minute. Figure 5.4.1 shows the respiratory system.

### How Does the Body Take in Oxygen?

When you breathe, you take in air through your nose or mouth. The air flows down through a long tube called the **trachea,** or windpipe. The trachea divides into two branches called **bronchi,** or breathing tubes. Each one leads into one of the lungs—the major breathing organs. In the lungs, each of the bronchi divides. Then it divides again to form a network of tubes called **bronchioles.**

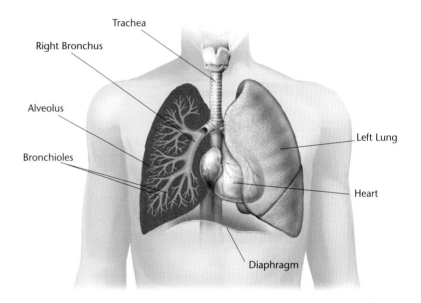

**Figure 5.4.1** *The respiratory system*

**Alveolus**
Air sac with thin walls (plural is alveoli)

**Diaphragm**
Band of muscle tissue beneath the respiratory organs

**Chest cavity**
Part of the body that includes ribs and muscles that surround the heart and lungs

**Inhale**
Breathe in

**Exhale**
Breathe out

At the end of each bronchiole is a cluster of air sacs with thin walls. These are called **alveoli.** You have about 300 million alveoli in your lungs. The alveoli are covered with a network of capillaries. The thin walls of the alveoli and capillaries allow the gases to change places. That is, the oxygen goes in, and the carbon dioxide comes out.

Your lungs need help from your **diaphragm** and **chest cavity.** The diaphragm is a band of muscle tissue beneath your respiratory organs. Your chest cavity includes ribs and muscles that surround your heart and lungs. When you **inhale,** or breathe in, your rib muscles and diaphragm contract. This enlarges your chest cavity and allows air to rush in. When you **exhale,** or breathe out, your rib muscles and diaphragm expand. This forces the air out.

*Lesson 4 Review* On a sheet of paper, write the answers to the following questions. Use complete sentences.

1. What do white blood cells do?
2. What are the three kinds of blood vessels?
3. What happens to rib muscles and the diaphragm when you inhale and hold your breath?
4. **Critical Thinking** Why is clotting important?
5. **Critical Thinking** Why is blood pressure important?

---

**Technology and Society**

A treadmill can be good for more than just exercise. An electrocardiograph (EKG) treadmill can tell how well the heart is working. Special sensors hooked up to a patient's body send information to a computer. Treadmill speed and level are increased and decreased. A doctor reads the information to determine whether the patient is at risk of heart problems.

---

Identifying Human Body Systems  Chapter 5  **115**

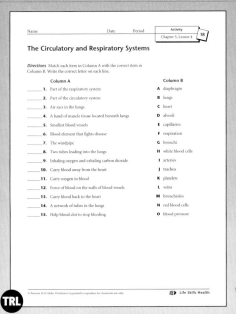
Activity 18

Workbook Activity 18

## LEARNING STYLES

**Logical/Mathematical**
Have students list the sequence of events as a drop of blood moves through the body, heart, and lungs. You may want to suggest a common starting point, such as the left ventricle of the heart. At each step of the sequence, students should describe what happens to or in the blood.

### Technology and Society

Have students read the Technology and Society feature on page 115. Tell students that it describes a medical device used by doctors to test patients' health. But a simpler sort of device is often found in gyms. Ask students if they have ever used a stationary cycle that measured heart rate. If so, ask them to describe what sort of data it provided. Invite interested students to do an Internet search on exercise-equipment technology and share their results with the class.

### Lesson 4 Review Answers

**1.** White blood cells work to keep the body healthy by fighting disease and germs. **2.** The three kinds of blood vessels are arteries, capillaries, and veins. **3.** Clotting is important because without it, people would die because they could not stop bleeding. **4.** Blood pressure is important because if blood pressure is too low, blood is not properly distributed throughout the body. **5.** Rib muscles and the diaphragm contract and do not relax.

### Portfolio Assessment

Sample items include:
- Labeled diagrams from Vocabulary
- Fact sheet from Research and Write
- Data from Application At Home
- List and descriptions from Learning Styles Logical/Mathematical
- Results from Technology and Society
- Lesson 4 Review answers

# Lesson at a Glance

## Chapter 5 Lesson 5

**Overview** This lesson describes the functions of the digestive and excretory systems.

### Objectives
- To describe how the digestive system breaks down food
- To explain how the body gets rid of solid and liquid waste

**Student Pages** 116–119

**Teacher's Resource Library**
- Activity 19
- Alternative Activity 19
- Workbook Activity 19

## Vocabulary

| | |
|---|---|
| anus | liver |
| bile | pancreas |
| digestion | rectum |
| enzyme | saliva |
| esophagus | small intestine |
| excretory system | ureter |
| feces | urethra |
| gall bladder | urinary bladder |
| insulin | urine |
| kidney | villi |
| large intestine | |

Write the vocabulary words on the board. Read the definition of each word and ask a volunteer to identify the word defined. When students identify a word correctly, erase it from the board and continue.

## Background Information

A person can have just one kidney and still function normally. However, when a person loses both kidneys due to an accident or disease, the person needs either kidney dialysis—in which a machine removes wastes are removed from the blood—or a kidney transplant. Kidneys from close relatives are preferred for transplants. The relative might have a genetically similar immune system, which makes a person's body less likely to reject the new kidney.

---

## Lesson 5: The Digestive and Excretory Systems

### Objectives

*After reading this lesson, you should be able to*
- describe how the digestive system breaks down food
- explain how the body gets rid of solid and liquid waste

**Digestion**
*Breaking down food into smaller parts and changing it to a form that cells can use*

**Saliva**
*A liquid in the mouth*

**Enzyme**
*Special chemical that breaks down food*

Your body needs food for energy, growth, and repair. Your body must break down food into substances that cells can use. Breaking down food into smaller parts and changing it to a form that cells can use is called **digestion.** The digestive system is the system that breaks down food for your body's use. Figure 5.5.1 shows the digestive organs.

### How Does the Digestive System Break Down Food?

When your mouth waters at the sight, smell, or thought of food, it produces **saliva.** Saliva is a liquid in your mouth. It contains an **enzyme,** a special chemical that breaks down food. Enzymes are in all your digestive organs. When you take a bite of an apple, your tongue pushes the food around. Your teeth help you chew the food into small pieces. The food mixes with saliva. The enzyme in saliva begins to break down the food.

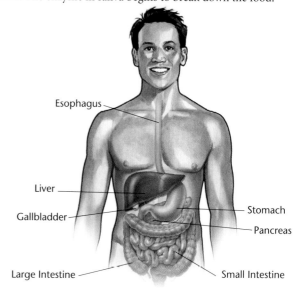

**Figure 5.5.1** *The digestive system*

116 Unit 2 Personal Health and Family Life

*Esophagus*
Long tube that connects the mouth and the stomach

*Small intestine*
A curled-up tube just below the stomach where most of the breakdown of food takes place

*Villus*
Fingerlike bulge that absorbs food after it has been broken down into chemicals (plural is villi)

*Liver*
A large organ that produces bile

*Gallbladder*
Digestive organ attached to the liver that stores bile

*Bile*
Liquid in the liver that breaks down fats

*Pancreas*
A gland that produces insulin and enzymes

*Insulin*
Hormone produced in the pancreas that helps cells use sugar

*Large intestine*
Tube connected to the small intestine

Food next enters your **esophagus,** a long tube that connects your mouth and your stomach. Muscles in the walls of your esophagus push the food along to your stomach. Your stomach walls give off juices containing enzymes that break down food more. Your stomach also breaks down food through twisting and churning it. Food stays in your stomach for 3 to 4 hours. During that time, the enzymes change solid food into partly liquid form.

From your stomach, food enters the **small intestine.** The small intestine is a curled-up tube just below your stomach. Most of the breakdown of food into chemicals takes place in your small intestine. The small intestine is lined with millions of tiny, fingerlike bulges called **villi.** Villi absorb the food that has been broken down into chemicals. Villi contain tiny blood vessels that are connected to the rest of your bloodstream. The chemicals from the broken-down food enter your blood vessels in the villi. Your bloodstream then sends the chemicals to all parts of your body.

Your **liver** and **gallbladder** are also digestive organs. Your liver is a large organ that produces **bile.** This liquid breaks down fats, such as butter. Your gallbladder is a small pouch attached to the liver. The gallbladder stores bile. When bile is needed for digestion, it is pushed into the small intestine.

Your **pancreas** also helps in digestion. This gland produces a hormone called **insulin,** which helps cells use sugar. Your pancreas also gives off enzymes that break down foods. Most food moves through your small intestine in 1 to 4 hours. Some foods are digested very quickly.

## How Does the Body Get Rid of Solid Waste?

Anything that the villi do not absorb in the small intestine moves on to the **large intestine.** The large intestine is a tube connected to the small intestine. It helps your body by gathering and removing waste materials that are left over after digestion. Through its walls, the large intestine absorbs water and nutrients from the waste material.

## Warm-Up Activity

Pass a tray of unsalted crackers around the room. Discuss the nutrients in a cracker. *(Carbohydrates, which are made up of starches, sugars, and fiber, are one of the main nutrients in crackers.)* Ask students to take a cracker and chew it for a much longer length of time than they usually would before swallowing. Did they notice anything happening to the taste of the cracker as they chewed it for a long time? *(Students should notice that the cracker begins to taste sweeter.)* Explain to students that the main function of the digestive system is to break down food into simpler substances that can be transported by the blood to all parts of the body. Point out that enzymes in the saliva began to break down the starch in the cracker into maltose, a type of sugar.

## Teaching the Lesson

As students read the lesson, have them write down the main idea of each paragraph. Encourage them to ask questions about any concept they do not understand. Have them write down any additional information that helps them understand the concept.

Go through the lesson as a class. Have volunteers explain the concepts in each paragraph. Clear up any misconceptions students may have.

## HEALTH JOURNAL

Have students describe in their journals what happens in their bodies just before they eat. They might want to make an observation just before their next meal and then write their entry. You might suggest that they pay special attention to the small intestine ("rumbling" sound) and the inside of the mouth (salivation).

# 3 Reinforce and Extend

## ELL/ESL Strategy

**Language Objective:**
*To use visuals*

Have English language learners create a flowchart tracing the digestive process. They should include the following terms in order:

mouth

esophagus

stomach

small intestine

large intestine

rectum

anus

Students should also include information about what occurs at each step of the digestive process. They can use this flowchart later to help study for a test.

## Research and Write

Have students read and conduct the Research and Write feature on page 118. If there is a hospital nearby, you might want to ask a surgeon to speak to the class, either in person or over the phone, about organ transplants.

## Learning Styles

**Interpersonal/Group Learning**

Display a 1-liter bottle filled with water. Ask small groups of students to predict how many 1-liter bottles would be filled with the blood that the kidneys filter in one day. Have the group share their predictions. Inform students that the kidneys of an average person filter approximately 1,600 liters of blood every day. Have students calculate how many liters of blood would be filtered each week. *(11,200 liters)*

---

*Feces*
Solid waste material remaining in the large intestine after digestion

*Rectum*
Lower part of the large intestine

*Anus*
Opening through which solid waste leaves the body

*Excretory system*
System that allows the body to eliminate liquid and solid waste

*Kidney*
Organ in the excretory system where urine forms

### Research and Write

Doctors sometimes perform organ transplants. That is, they take one person's organ and put it in another person's body. With a partner, research the challenges these doctors face. Then write a short article to explain the challenges. Look for ideas to help make organ transplants work better.

---

The solid material left in the large intestine is called **feces.** Feces are stored in the **rectum,** or lower part of the large intestine. Feces leave the body through an opening called the **anus.**

## How Does the Body Get Rid of Liquid Waste?

Your body is good at using nutrients it needs. It can also eliminate, or get rid of, the rest. The **excretory system** allows the body to eliminate liquid and solid waste. Figure 5.5.2 shows the excretory system. This system removes water and salts through your sweat glands.

The excretory system also takes waste products out of your blood. The main excretory organs that do this are your two **kidneys.** The kidneys are on either side of your spine in your lower back. The kidneys take waste products out of your blood. Then they return water and minerals to your blood.

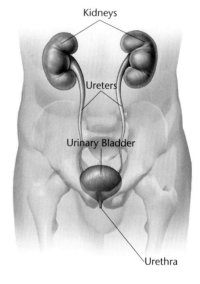

**Figure 5.5.2** *The excretory system*

**Ureter** Tube through which urine passes from a kidney to the urinary bladder

**Urinary bladder** Bag that stores urine

**Urine** Liquid waste product formed in the kidneys

**Urethra** Tube through which urine passes out of the body

As your blood supplies nutrients to your cells, the cells form waste products. These wastes flow through cell walls into your bloodstream. When blood circulates through your kidneys, wastes are strained out. The blood then travels back to your heart through your veins.

Waste products move out of your kidneys through tubes called **ureters** to your **urinary bladder.** This bag can stretch to hold this liquid waste, which is called **urine.** Urine passes out of your body through the **urethra.**

**Lesson 5 Review** On a sheet of paper, write the answers to the following questions. Use complete sentences.

1. What happens when your mouth waters?
2. What is the esophagus?
3. What do enzymes do to solid food?
4. **Critical Thinking** Why are villi important?
5. **Critical Thinking** What do you think would happen if the kidneys stopped working properly?

### Decide for Yourself

Tanya's best friend came to her with a problem. The friend often felt faint and shaky. She wanted to eat strange foods sometimes. Other times, she couldn't eat at all. Her toes had started feeling numb. "I haven't told anyone," she said. "I feel so awful today. I just can't make it to soccer practice. Please cover for me. Tell them I couldn't come. Don't tell them why. I don't want to get kicked off the team."

Tanya knew her friend's mom had diabetes, which meant that her body did not have the right amount of insulin. Tanya thought her friend might have diabetes, too. Her friend needed to see a doctor.

Her friend had begged Tanya not to say anything to anybody. Tanya didn't want her friend to get kicked off the team.

On a sheet of paper, write the answers to the following questions. Use complete sentences.

1. What are Tanya's possible choices in this situation?
2. Tanya's friend is afraid she will be kicked off the soccer team. Why else might she be scared to admit she is sick?
3. Which choice do you think Tanya should make?

*Identifying Human Body Systems* Chapter 5 119

### Decide for Yourself

Prepare a brief presentation about diabetes. Have students read the Decide for Yourself feature on page 119. Explain to students what diabetes is, what causes it, and how it can be managed.

**Decide for Yourself Answers**

1. Sample answer: Tanya could try to convince her friend to tell her parents about the problem. If that didn't work, Tanya could tell her friend's parents herself. Tanya could go to another responsible adult for advice, such as her own parents, a school nurse, or a teacher. Another choice is to do nothing. 2. Sample answer: She might be embarrassed about her problem. 3. Sample answer: Tanya should help her friend get the medical help she needs. If Tanya ignores the problem, her friend's condition could worsen.

### Lesson 5 Review Answers

1. When your mouth waters, it produces saliva. 2. The esophagus is a long tube that connects the mouth and the stomach. 3. Enzymes are chemicals that break down foods into a form the body can use. 4. Villi are important because they help the body take in nutrients. 5. If a person's kidneys stopped working properly, his or her body could not get rid of wastes, and the person might die. The body can be damaged severely if wastes are not removed from the blood.

### Portfolio Assessment

Sample items include:
- Health Journal entry
- Summary from Teaching the Lesson
- Flowchart from ELL/ESL Strategy
- Article from Research and Write
- Answers from Decide for Yourself
- Lesson 5 Review answers

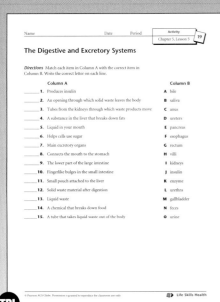

Activity 19      Workbook Activity 19

*I.D.ing Human Body Systems* Chapter 5 119

# Lesson at a Glance

## Chapter 5 Lesson 6

**Overview** This lesson describes the structure and function of the body's skin and includes information about hair and nails.

### Objectives

- To describe the structure of the skin
- To explain how skin protects the body
- To understand the structure of hair and nails

**Student Pages** 120–122

**Teacher's Resource Library**

Activity 20
Alternative Activity 20
Workbook Activity 20

## Vocabulary

| | |
|---|---|
| bacteria | melanin |
| dermis | perspiration |
| epidermis | pore |
| keratin | subcutaneous layer |

Have students write a short paragraph that includes five vocabulary words used correctly in context.

## Background Information

All vertebrates have some sort of digital appendage. Hawks have long, curved claws called talons. Horses have hooves that protect the entire foot. Humans have fingernails and toenails. In each case, the structure probably developed in order to protect the tip of the appendage.

 **Warm-Up Activity**

Have students observe the skin on one of their hands, using a magnifying glass. Help them to identify the epidermis and the pores of the sweat glands. Then have them put on a plastic glove for ten minutes. When they remove the glove, they should observe moisture on their skin. The moisture is perspiration from their sweat glands.

---

## Lesson 6  The Body's Protective Covering

### Objectives

*After reading this lesson, you should be able to*

- describe the structure of the skin
- explain how skin protects the body
- understand the structure of hair and nails

Skin is the body's largest organ. It is part of a system that covers and protects your entire body. The skin is one of your sense organs and is also part of the excretory system. It is a separate system that includes your sweat glands, hair, and nails.

### What Is the Structure of the Skin?

Figure 5.6.1 shows that the skin is made up of three layers. The **epidermis** is the outer layer of skin. The lower part of the epidermis makes new cells that are pushed toward the skin's surface. The layer on the top of the epidermis is made up of dead cells. The dead cells flake off as the new ones take their place.

The **dermis** is the inner layer of skin that is made up of living cells. Nerves and blood vessels are in the dermis. Oil glands in the dermis help keep your hair and skin from cracking.

*Epidermis*
Outer layer of skin

*Dermis*
Inner layer of skin made up of living cells

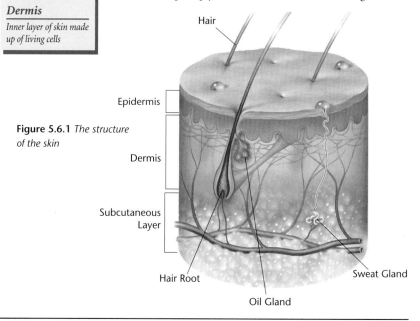

**Figure 5.6.1** *The structure of the skin*

**Subcutaneous layer**
*Deepest layer of skin*

**Bacteria**
*Germs that cause disease*

**Pore**
*Tiny opening in the epidermis*

**Perspiration**
*Sweat*

**Melanin**
*Substance that gives skin its color*

**Keratin**
*A type of protein that makes nails hard*

Below the dermis is the **subcutaneous layer**—the deepest layer of skin. It is made up mostly of fatty tissue. The tissue protects your body from heat and cold. This layer also protects your body from injury.

## How Does the Skin Protect the Body?

Skin protects your body in many ways. First, your skin prevents most **bacteria** from entering your body. Bacteria are tiny germs that cause disease.

Skin also helps control your body's temperature. The epidermis has tiny openings called **pores.** Pores are connected to the sweat glands.

Your body gets rid of **perspiration,** or sweat, through sweat glands. Perspiration helps to cool your body.

The dermis helps protect your skin through cells that produce **melanin.** Melanin is the substance that gives skin its color. When you are in the sun, your skin produces more melanin. This shields the skin from the sun's harmful rays. The increase in melanin makes skin look tanned. Freckles are actually small spots of melanin.

## What Is the Structure of the Hair and Nails?

Hair and nails are two other parts of your body's protective covering. The roots of hair are made up of living cells and grow out of the dermis. A hair root grows out of a small pocket in the dermis that holds the root.

Hair is like the skin in that new cells are pushed up to replace the old cells. The hair on your head and body is made up of dead cells.

Nails are also dead cells. Nails grow out of the skin's epidermis. They contain a type of protein called **keratin,** which makes nails hard. Your nails protect the soft ends of your fingers and toes.

Identifying Human Body Systems Chapter 5

## 2 Teaching the Lesson

Ask students: How does hair protect your body? *(Hair keeps you warm in cold weather and helps protect you from the sun in hot weather. Hair helps keep dirt out of your eyes, ears, and nose. Eyebrows keep sweat out of your eyes.)* Why doesn't it hurt when you cut your hair? *(Hair is made of dead cells.)* Why is it important that nails are located where they are? *(Students might speculate that feet and hands have the most contact with other objects in the environment and, therefore, the protection of toes and fingers is especially important.)*

### TEACHER ALERT

Ask students if they have ever heard someone complain about hot weather by saying, "It's not the heat—it's the humidity." Tell students that although perspiring cools the body, if the humidity of the air around you is high, sweat will not evaporate and cool the body as well as it does when the humidity is low.

## 3 Reinforce and Extend

### ELL/ESL STRATEGY

**Language Objective:** *To teach and practice functional language skills*

Have English language learners look at the words on the vocabulary list. Point out the words *dermis* and *epidermis*. Explain to students that *epi-* is a prefix that means "on" or "over." Help students understand that the epidermis is on or over the dermis.

Now point out the term *subcutaneous layer*. Tell students that cutaneous means "on, or affecting the skin." Explain that *sub-* is a prefix that means "under" or "beneath." Help students understand that the subcutaneous layer is under the skin.

## Lesson 6 Review Answers

1. glands 2. three 3. bacteria 4. The skin would grow thicker and thicker. 5. The person's body temperature could not be controlled. The person would not be able to get cool, and his or her body temperature would continue to rise.

## Portfolio Assessment

Sample items include:
- Paragraph from Vocabulary
- Health in Your Life answers
- Lesson 6 Review answers

## Health in Your Life

Ask a volunteer to read aloud the Health in Your Life feature on page 122. Make sure students understand that in order to work, sunscreen must be applied in sufficient quantity, and it must stay on. Tell them to apply sunscreen a half an hour before going out in the sun and to reapply it every two hours or after swimming or excessive sweating. Sun reflects off water and snow, so tell students to make sure they are wearing sunscreen if they are boating or skiing.

Tell students that other ways to protect their skin from the sun include staying out of the sun between 10 A.M. and 3 P.M., when the sun's rays are most intense; wearing hats and T-shirts to block the sun; and wearing a lip balm with a SPF of 15.

### Health in Your Life Answers

1. Sun protection is important because the sun can damage your skin and cause illnesses. 2. It is a good idea to use sunscreen on your face every day because it is one part of the body that does not get covered with clothing. 3. People who swim outdoors should use sunscreen because the sun's rays can burn them through the water.

122  Unit 2  Personal Health & Family Life

---

**Lesson 6 Review** On a sheet of paper, write the word in parentheses that best completes each sentence.

1. Sweat (glands, pores, cells) help the body remove water and salt.
2. The skin has (two, three, four) layers.
3. Skin protects the body from (bacteria, perspiration, melanin).

On a sheet of paper, write the answers to the following questions. Use complete sentences.

4. **Critical Thinking** What would happen if the dead cells of the epidermis did not flake off as new cells were formed?
5. **Critical Thinking** Suppose that someone's pores were blocked and sweat could not escape. What would happen?

### Health in Your Life

#### Skin Protection

Protecting your skin is important. Sunburn is painful. It can also lead to illnesses, such as skin cancer, when you get older. To keep your skin safe, use a sunscreen with a sun protection factor (SPF) of at least 15. You may need to reapply the sunscreen later because it wears off. Check the bottle for instructions.

It is a good idea to use sunscreen on your face every day. Protect all of your skin when you plan to be out in the sun. Be sure to use a waterproof sunscreen so it will not wear off or wash off.

The sun's rays can burn you through the water of a pool or lake. On a cloudy day, the sun is still shining, and it can burn you. Using sunscreen is an important protection.

On a sheet of paper, write the answers to the following questions. Use complete sentences.

1. Why is sun protection important?
2. Why is using sunscreen on your face every day a good idea?
3. Why should people swimming outdoors wear sunscreen?

122  Unit 2  Personal Health and Family Life

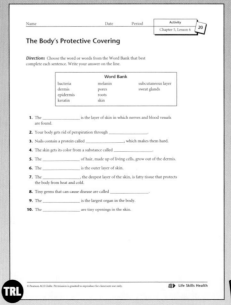

Activity 20

Workbook Activity 20

# Chapter 5  SUMMARY

- The body systems are the skeletal and muscular systems, the nervous system, the endocrine system, the circulatory and respiratory systems, the digestive and excretory systems, and skin.

- The sense organs receive messages from outside your body and send them to the brain. The sense organs are eyes, ears, tongue, nose, and skin.

## Vocabulary

adrenal gland, 111
adrenaline, 111
alveolus, 115
anus, 118
artery, 113
auditory nerve, 108
bacteria, 121
bile, 117
blood pressure, 113
brain stem, 105
bronchus, 114
bronchiole, 114
capillary, 113
carbon dioxide, 112
cerebellum, 105
cerebrum, 104
chest cavity, 115
clotting, 112
contract, 101
cornea, 106
dermis, 120
diaphragm, 115
digestion, 116
eardrum, 108
enzyme, 116
epidermis, 120

esophagus, 117
excretory system, 118
exhale, 115
feces, 118
gallbladder, 117
gland, 110
hormone, 110
inhale, 115
insulin, 117
iris, 107
keratin, 121
kidney, 118
large intestine, 117
lens, 107
ligament, 101
liver, 117
marrow, 101
medulla, 105
melanin, 121
metabolism, 110
nutrient, 100
optic nerve, 107
pancreas, 117
peripheral nervous system, 105

perspiration, 121
platelet, 112
pore, 121
pulse, 112
pupil, 106
receptor cell, 109
rectum, 118
reflex, 106
respiration, 114
retina, 107
saliva, 116
small intestine, 117
spinal column, 105
spinal cord, 105
subcutaneous layer, 121
taste bud, 109
tendon, 102
trachea, 114
ureter, 119
urethra, 119
urinary bladder, 119
urine, 119
vein, 113
villus, 117

## Chapter 5 Summary

Have volunteers read aloud each Summary item on page 123. Ask volunteers to explain the meaning of each item. Direct students' attention to the Vocabulary box on the bottom of page 123. Have them read and review each term and its definition.

## ONLINE CONNECTION

The National Agricultural Library is part of the U.S. Department of Agriculture. One part of its Web site, the Consumer Corner, is presented in cooperation with the University of Maryland. It contains links to a wide variety of sites dealing with nutrition, from special nutritional needs of children with diabetes to activities and recipes for families as a whole:

www.nal.usda.gov/fnic/consumersite/index.html

This Web site, offered by the National Institute of Arthritis and Musculoskeletal and Skin Diseases, provides information on many topics, ranging from acne to vitiligo, in a FAQs format:

www.niams.nih.gov/hi/

This section of the Web site run by the U.S. National Cancer Institute's Surveillance, Epidemiology and End Results (SEER) Program, contains detailed information in easy-to-understand modules on all of the body systems discussed in this chapter, as well as many others:

http://training.seer.cancer.gov/module_anatomy/unit1_1_body_structure.html

## Chapter 5 Review

Use the Chapter Review to prepare students for tests and to reteach content from the chapter.

## Chapter 5 Mastery Test

The Teacher's Resource Library includes two forms of the Chapter 5 Mastery Text. Each test addresses the chapter Goals for Learning. An optional third page of additional critical-thinking items is included for each test. The difficulty level of the two forms is equivalent.

## Review Answers

**Vocabulary Review**

1. melanin 2. vein 3. bile 4. spinal cord
5. medulla 6. reflexes 7. pupil 8. auditory nerve 9. hormones 10. arteries
11. ligament 12. blood pressure 13. small intestine 14. dermis

### TEACHER ALERT

In the Chapter Review, the Vocabulary Review activity includes a sample of the chapter's vocabulary terms. The activity will help determine students' understanding of key vocabulary terms and concepts presented in the chapter. Other vocabulary terms used in the chapter are listed below.

| | |
|---|---|
| adrenal gland | keratin |
| adrenaline | kidney |
| alveoli | large intestine |
| anus | lens |
| artery | liver |
| bacteria | marrow |
| brain stem | metabolism |
| bronchi | nutrient |
| bronchioles | optic nerve |
| capillary | pancreas |
| carbon dioxide | peripheral nervous system |
| cerebellum | |
| cerebrum | perspiration |
| chest cavity | platelet |
| clotting | pore |
| contract | pulse |
| cornea | receptor cell |
| diaphragm | rectum |
| digestion | respiration |
| eardrum | retina |
| enzyme | saliva |
| epidermis | spinal column |
| esophagus | subcutaneous layer |
| excretory system | taste bud |
| exhale | tendon |
| feces | trachea |
| gall bladder | ureter |
| gland | urethra |
| inhale | urinary bladder |
| insulin | urine |
| iris | villi |

124   Unit 2   *Personal Health & Family Life*

---

# Chapter 5  R E V I E W

**Word Bank**
arteries
auditory nerve
bile
blood pressure
dermis
hormones
ligament
medulla
melanin
pupil
reflexes
small intestine
spinal cord
vein

## Vocabulary Review

On a sheet of paper, write the word or phrase from the Word Bank that best completes each sentence.

1. The substance that gives skin its color is _____.
2. A vessel that carries blood back to the heart is a(n) _____.
3. The liquid in the liver that breaks down fats is _____.
4. The central nervous system is made up of the brain and the _____.
5. One part of the brain stem, the _____, controls automatic activities such as breathing.
6. Automatic responses to heat or pain are _____.
7. The _____ in the middle of the eye adjusts its size to let in the right amount of light.
8. The ear picks up vibrations that are transferred to the _____.
9. The endocrine system includes glands that produce _____, which control how the body does its jobs.
10. Blood is carried away from the heart by _____.
11. The tough band of tissue that holds joints together or keeps organs in place is a(n) _____.
12. The force of blood on the walls of the blood vessels is _____.
13. The _____ is where most of the breakdown of food into chemicals takes place.
14. The inner layer of the skin made up of living cells is the _____.

124   Unit 2   *Personal Health and Family Life*

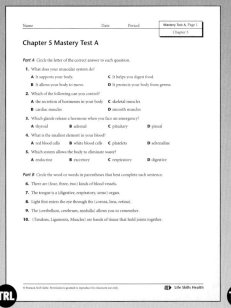

Chapter 5 Mastery Test A, pages 1–3

## Concept Review

On a sheet of paper, write the letter of the answer that best completes each sentence.

15. All of these are types of muscles EXCEPT _____.

    A smooth  C skeletal
    B cardiac  D auditory

16. The body makes nutrients from _____.

    A hair  C hormones
    B food  D bones

17. The air you breathe travels to your lungs through your _____.

    A pancreas  C trachea
    B kidneys  D cornea

## Critical Thinking

On a sheet of paper, write the answers to the following questions. Use complete sentences.

18. Why are healthy food and exercise important to all body systems?

19. Do you think one system of the body is more important than the others? Explain your answer.

20. What do you think is most amazing about how the human body works? Explain your answer.

**Test-Taking Tip** Read and follow each set of directions carefully.

---

## Concept Review
15. D  16. B  17. C

## Critical Thinking
18. Sample answer: All the body systems interact with one another. Healthy food and exercise give the systems what they need to work properly. 19. Sample answer: I think the nervous system is the most important system. Without the brain, nothing else would work. 20. Sample answer: I think the most amazing thing is that all the systems work together to give us the ability to do the things we can do. The systems all do things that help one another.

## ALTERNATIVE ASSESSMENT

Alternative Assessment items correlate to the student Goals for Learning at the beginning of this chapter.

- Alternative Assessment items correlate to the student Goals for Learning at the beginning of this chapter.
- Have students orally describe the purpose of the muscular and skeletal systems.
- Have students write a paragraph explaining the parts of the nervous system and how information comes through the sense organs.
- Have students write a script for a radio health spot that explains how the endocrine glands work with the nervous system.
- Have students draw a diagram that describes the process of respiration and how blood circulates through the body.
- Have students write an encyclopedia article about the digestive and excretory systems.
- Have students build a model that shows the structure of skin, hair, and/or nails. Text accompanying the model should describe the purpose of each element shown.

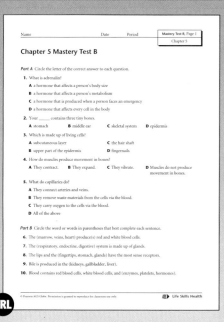
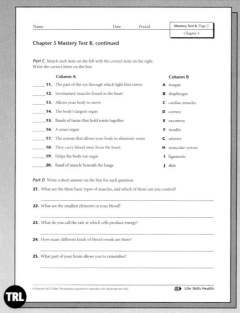

Chapter 5 Mastery Test B, pages 1–3

# Chapter at a Glance

## Chapter 6: Maintaining Personal Hygiene and Fitness
pages 126–147

### Lessons
1. Hygiene for Good Health
   pages 128–133
2. Exercise and Physical Fitness
   pages 134–139
3. Personal Fitness Plan
   pages 140–144

**Chapter 6 Summary** page 145

**Chapter 6 Review** pages 146–147

**Audio CD**

**Skill Track for Life Skills Health**

**Teacher's Resource Library** TRL

   Activities 21–23
   Alternative Activities 21–23
   Workbook Activities 21–23
   Self-Assessment 6
   Community Connection 6
   Chapter 6 Self-Study Guide
   Chapter 6 Outline
   Chapter 6 Mastery Tests A and B
   (Answer Keys for the Teacher's Resource Library begin on page 559 of this Teacher's Edition.)

## Opener Activity

Begin by having a class discussion about physical exercise. Ask students why they think physical exercise is important. Ask students to list physical activities that they enjoy. Ask them if they feel that they get enough physical exercise. If they answer no, ask them to brainstorm ways they could get more exercise.

126  Unit 2  Personal Health & Family Life

# Self-Assessment

## Can you answer these questions?

☑ 1. What is hygiene?

☑ 2. What are three ways to keep your eyes healthy?

☑ 3. What effect does loud music have on your hearing?

☑ 4. What are two basic rules for healthy skin?

☑ 5. What is most important for healthy hair, nails, and teeth?

☑ 6. What are three benefits of exercising?

☑ 7. What is the difference between aerobic and isometric exercise?

☑ 8. What is a personal health plan?

☑ 9. Why is a cooldown a necessary part of exercising?

☑ 10. Why are rest and sleep important parts of personal health?

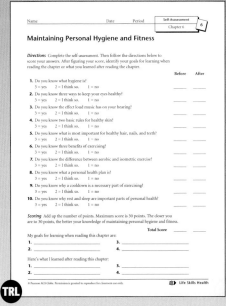

Self-Assessment 6

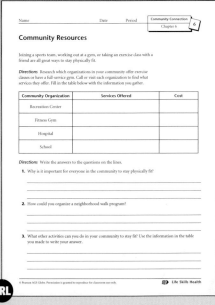

Community Connection 6

# Chapter 6
## Maintaining Personal Hygiene and Fitness

Taking care of yourself helps you feel healthier. Taking care of yourself has other benefits, as well. It can help you look better and feel better about yourself. Taking care of your eyes, ears, skin, hair, nails, and teeth shows you value yourself. Exercising and getting enough sleep are other ways you value yourself.

In this chapter, you will learn about ways to take care of yourself and stay healthy. You will also learn about the importance of exercise, fitness, and rest to your health. You will learn how these actions are important to your social and emotional well-being. Finally, you will have the opportunity to create a fitness plan for yourself.

### Goals for Learning

- ◆ To explain the purpose of basic hygiene
- ◆ To describe ways to protect eyes, ears, skin, hair, nails, and teeth
- ◆ To identify the benefits and parts of a regular exercise program
- ◆ To create a personal fitness plan
- ◆ To explain why your body needs rest and sleep

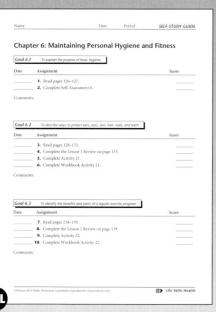

Chapter 6 Self-Study Guide, pages 1–2     Chapter 6 Outline, pages 1–2

## Introducing the Chapter

Have a volunteer read the introductory paragraphs on this page. Ask students to list ways they take care of themselves every day. Ask them to explain how taking care of oneself leads to increased self-esteem. (*When you take care of yourself, you feel better about yourself. You feel confident and proud of the way you look and feel.*)

Have students read the Goals for Learning. Explain that in this chapter students will learn how to take care of themselves and stay healthy.

## Notes and Questions

Ask volunteers to read the notes and questions that appear in the margins throughout the chapter. Then discuss them with the class.

## Self-Assessment

Have students complete the Self-Assessment worksheet before and after reading the chapter. Before reading the chapter, have students fill in the "Before" column. Ask students to identify their goals for learning. To get ideas for setting goals, students might use the chapter introductory material on page 127, the checklist on page 126, or the questions on the Self-Assessment worksheet. Students can use the back of the worksheet if they need more space to write.

Collect the Self-Assessment worksheets and pass them out again at the end of the chapter. Have students fill in the "After" column. Ask them to identify at least four major points they have learned. Again, suggest they use the back of the worksheet if they need more space to write. You may want to collect and review the worksheets, but return them to students so they have a record of their goals and accomplishments.

# Lesson at a Glance

## Chapter 6 Lesson 1

**Overview** This lesson explains the purpose of basic hygiene and what can be done to protect eyes, ears, skin, hair, nails, and teeth.

### Objectives

- To list five ways to take care of your eyes
- To explain risks to your ears and skin
- To describe three ways to protect your teeth

**Student Pages** 128–133

**Teacher's Resource Library**

Activity 21

Alternative Activity 21

Workbook Activity 21

## Vocabulary

| | |
|---|---|
| acne | deodorant |
| allergy | fungus |
| antiperspirant | hygiene |
| athlete's foot | infection |
| dandruff | pimple |

Students have probably heard of most, if not all, of these terms. Students can assess their prior knowledge by writing on a sheet of paper what they think each term means. At the end of the lesson, they can adjust their definitions as appropriate.

## Background Information

When someone has 20/20 vision, this means that the person can see at a distance of 20 feet an object that a person with "normal" vision can see at 20 feet. People with 20/15 vision have better-than-normal vision. An object that a person with normal vision can see clearly at 15 feet can be seen clearly by someone with 20/15 vision at 20 feet. People with 20/100 vision have poorer-than-normal vision. What a person with normal vision can see at 100 feet, people with 20/100 vision can see at 20 feet.

128  Unit 2  Personal Health & Family Life

---

## Lesson 1  Hygiene for Good Health

### Objectives

*After reading this lesson, you should be able to*

- list five ways to take care of your eyes
- explain risks to your ears and skin
- describe three ways to protect your teeth

**Hygiene**
*Practices that promote cleanliness and good health*

**Hygiene** refers to things you do to promote cleanliness and good health. Taking care of your eyes, ears, skin, hair, nails, and teeth are part of hygiene. These are all important parts of your body, so taking care of them is important. Good hygiene habits can have a positive effect on your personal health and your personal and work-related relationships.

### How Can You Take Care of Your Eyes and Sight?

Your body protects your eyes and sight in many ways. The hard bones of your skull surround your eyes. Your eyelids, eyebrows, and eyelashes keep dirt, sweat, and direct light out of your eyes. Blinking protects your eyes from dirt and direct light. Tears wash away tiny objects and help get rid of harmful germs that may get into your eye.

To protect your eyes, avoid touching them with dirty hands. Keep sharp objects away from your eyes. Avoid being in bright sunlight for long periods. If you wear contact lenses, follow instructions for cleaning and wearing them. Wear safety glasses, goggles, or a helmet to shield your eyes from glare, damaging light, chemicals, dirt, and sports injuries. Wear sunglasses in the sun but not indoors.

Table 6.1.1 lists some common vision problems.

| Table 6.1.1 Some Vision Problems | |
|---|---|
| Farsightedness | Difficulty seeing close objects |
| Nearsightedness | Difficulty seeing distant objects |
| Astigmatism | Blurred vision caused by an uneven shape of the cornea or lens |
| Strabismus | Crossed eyes; the eye muscles do not work together; one or both eyes turn inward or outward |
| Color blindness | Inherited lack of ability to see certain colors; usually occurs in males |

128  Unit 2  Personal Health and Family Life

**Infection**
*A sickness caused by a germ in the body*

**Allergy**
*A bad reaction of the body to a food or to something in the air*

You need regular eye exams by an eye doctor to check for eye and vision problems. You know your vision is fine if you can see a sharp, clear picture. If your vision is not fine, your eye doctor can write an order for lenses to correct the problem.

## How Can You Take Care of Your Ears and Hearing?

Your skull protects the most delicate parts of your ears. The wax in your outer ears keeps dirt from going into other parts of your ears. A certain amount of wax in your ears is normal. Sometimes, the wax may build up so you don't hear well. If that happens, a doctor or nurse can remove the wax.

The most common reason for hearing loss is loud noise. If you are around loud noise a lot, you may be at risk for hearing loss. The loss usually happens gradually and without pain. Sometimes, it can happen instantly from a single extremely loud noise. You can almost always avoid noises that cause hearing loss.

Other things that can cause hearing loss are **infection,** heredity, **allergies,** too much wax, or a wound. Infection is a sickness caused by a germ in your body. An allergy is a bad reaction of your body to a food or to something in the air.

A hearing test shows how well you hear. If a hearing test or an ear exam show you have hearing or ear problems, a doctor can tell you what to do.

Protect your hearing and your ears by avoiding loud noise and music. Wear earplugs or earmuffs when you are near loud noise. Clean your ears only with a damp washcloth on your fingertip—never with a cotton swab. Avoid putting objects in your ears. Swim only in clean water and dry your ears after swimming. Wear a head guard or helmet when playing contact sports such as football or rugby.

### Link to >>>
**Biology**

Some hearing and vision problems can begin as early as your 30s. A person's eardrums may become thicker, making sounds less clear. Some older people may need a hearing aid. Changes in the shape of the eye can reduce how sharply a person sees things. Many older people need to wear glasses or contacts. Driving at night may also become harder.

---

 **Warm-Up Activity**

Read the lesson title and ask students to define hygiene *(things you do to promote cleanliness and good health)*. Ask them to share ideas about what might happen if they did not take care of their eyes *(vision problems, loss of vision)*, ears *(earaches, infection, loss of hearing)*, skin *(acne, body odor)*, hair and nails *(dandruff, infections)*, and teeth *(cavities, gum disease)*. Discuss the consequences of poor hygiene.

 **Teaching the Lesson**

Help students organize the material presented in the lesson by having them make a 2-column table. The first column should be titled "Area of Health." The second column should be titled "Health Practices for Care and Protection." Row titles in Column 1 should be "Eyes," "Ears," "Skin," "Hair and Nails," and "Teeth." Students should summarize general ways to take care and protect each area of health in Column 2.

**Reinforce and Extend**

### Link to

**Biology.** Have students read the Link to Biology feature on page 129. Remind students that aging is a natural process. Although the body changes with age, ask students what steps they can take to remain as healthy as possible as they get older. *(They can have regular check-ups with their family doctor, optometrist, and dentist; eat a well-balanced diet; and maintain physical fitness through regular exercise.)*

 **TEACHER ALERT**

Ask students if they enjoy listening to loud music. Stress the importance of avoiding extremely loud music. Encourage students to wear earplugs if they are in a very noisy situation such as a concert or near a construction site.

## GLOBAL CONNECTION

 Point out to students that between two and three million people worldwide develop skin cancer every year. Skin cancer is especially prevalent among people who have light-colored skin and hair, blue or green eyes, and freckles and/or moles. Skin cancer can be cured if it is diagnosed and treated while it is in its early stages. Encourage interested students to find out about the different kinds of skin cancers, what they look like, and what causes them. Ask students to investigate the use of sunscreens and sunblocks as protection against sunburn and skin cancer.

## HEALTH JOURNAL

 Have students write a paragraph in their Health Journals about what steps they take every day to keep their skin healthy. Have them explain what they do to protect their skin when they go out in the sun. Ask them if, after reading the lesson material, there is anything about their skin care practices that they would like to change.

## ELL/ESL STRATEGY

 **Language Objective:** *To practice study skills*

With a partner, have students create note cards to study the vision problems listed in Table 6.1. On one side of the card, students should write the vision problem; on the other side of the card, they should describe the problem. Students may quiz one another on the vision problems. Encourage students to use this strategy to study for tests when there are a number of items to remember.

---

*Deodorant*
A product that covers body odor

*Antiperspirant*
A product that helps control perspiration by closing the pores

*Acne*
Clogged skin pores that cause pimples and blackheads

*Pimple*
An inflamed swelling of the skin

Risk factors are things that may be able to harm your health. How can knowing about risk factors help you stay in good health?

### How Can You Protect Your Skin?

The first rule of good skin care is to keep your skin clean. The best way to keep your skin clean is to wash with soap and water. How often you need to wash depends on your skin type and how active you are.

You may think perspiration has an odor, but it doesn't. Odor results when perspiration mixes with bacteria on your skin. A **deodorant** is a product that can cover up the odor. An **antiperspirant** is a product that helps control perspiration by closing your pores. Neither product, however, removes odor-causing bacteria. Washing with soap and water reduces body odor by removing bacteria.

The second basic rule for skin care is to protect your skin from wind, cold, and sun. Strong wind and cold weather can cause your skin's outer layer to lose moisture quickly. Your skin can crack and split, or become chapped. Staying out too long in the sun can hurt your eyes and give you sunburn. It can wrinkle your skin and cause skin disease. Damage from the sun builds up over the years. It cannot be reversed. The chance of getting sunburned rises when you are skiing or at the beach. This is because the sun's rays bounce off snow, sand, and water. It is possible to get sunburned even on cloudy days.

Here are some other ways you can protect your skin. In windy, cold weather, dress warmly and cover your face and hands. Use a cream or lotion to keep in your skin's natural moisture. Use a sunscreen to protect your skin. A sunscreen with a sun protection factor (SPF) of 15 or more gives the best protection.

### How Can You Take Care of Skin Problems?

**Acne** is the most common skin problem for teens. Acne occurs when pores, the tiny openings in the skin, get clogged. The amount of hormones in your body increases when you are a teen. These hormones cause oil glands to make more oil and may plug up your pores. Then you may get **pimples,** whiteheads, or blackheads. A pimple is an inflamed swelling of the skin. A whitehead forms when oil becomes trapped in a pore. A blackhead is an oil plug that darkens when the air hits it.

**Athlete's foot**
*Itching or cracked skin between the toes*

**Fungus**
*An organism that grows in damp places*

### Health Myth

**Myth:** People who sweat a lot while exercising are not fit.

**Fact:** People who are fit sweat more than those who do not exercise regularly. As you exercise, your body sends blood to the surface of your skin. If you exercise regularly, this happens faster than it does in someone who does not exercise regularly. When the blood reaches the surface of your skin, you give off heat. Your sweat glands then make more sweat to cool you off.

There are many oil glands in your face, neck, and back. Acne usually affects these areas. Acne has no cure, but you can reduce the problem by doing the following:

- Wash with soap and water every morning and evening. If your face is very oily, wash it once more during the day.
- Wash right after exercise to remove sweat and bacteria that can clog your pores. Use a clean washcloth every day. Bacteria grow quickly on a damp washcloth.
- Shampoo your hair often to limit acne on your forehead, neck, and shoulders.
- Eat well-balanced meals and get plenty of rest and exercise.
- Limit time in the sun and avoid tanning booths and sunlamps. Tanning booths damage skin in the same way the sun does.
- Avoid oil-based creams and lotions.
- Do not squeeze or pick pimples and blackheads. This can cause infection or leave a scar on your skin.
- Visit a skin doctor if you have very bad acne.

**Athlete's foot** is itching or cracked skin between the toes. A **fungus** causes it. A fungus is an organism that grows in damp places, such as a locker room. To keep from getting athlete's foot, keep your feet dry. Avoid walking barefoot in locker rooms or showers. Mild athlete's foot can be treated with a special foot powder or ointment found in drugstores. Severe cases of athlete's foot are treated with prescription medication.

### Technology and Society

Lasers can create strong beams of light. Unlike a lightbulb, a laser creates a beam of light that does not spread out. This light can be focused to a small point. Doctors use lasers to close up blood vessels and break up kidney stones. Lasers can help remove scars, whiten teeth, and correct vision problems. Lasers are used in eye surgery to change the eyeball's shape and improve vision.

## LEARNING STYLES

**Visual/Spatial**

Have students create comic strips in which they depict a person practicing some aspect of skin care. They may choose to focus on a particular problem, or they can be more general. Students can share their work with the rest of the class.

### Health Myth

Ask students to read the Health Myth feature on page 131. Explain that sweating is an important part of the body's methods of maintaining its internal temperature. As the body heats up from exercise, it sweats to help release heat and cool itself, keeping internal body temperature the same. If body temperature rises too high, it can be fatal.

### Technology and Society

Have a volunteer read aloud the Technology and Society feature on page 131. Ask students if they know anyone who has had any laser surgery. Ask students if they would choose to undergo laser surgery. Encourage students to research the various types of laser surgery that have been developed. Discuss what the future possibilities of laser surgery might be, as well as the advantages and disadvantages of these types of surgery.

## LEARNING STYLES

 **Interpersonal/Group Learning**

In small groups, have students create a pamphlet geared toward middle school students. The pamphlet should include information on basic hygiene, using the information from this text. After reviewing the pamphlets, you might want to present them to a middle school for inclusion in its library.

## IN THE COMMUNITY

 Because fluoride can help to prevent tooth decay, many communities across the country have added fluoride to the drinking water. Encourage students to call the water department in their town or city and ask if there is fluoride in the drinking water. If there is, ask the students to find out the level of fluoride in the water. Have them discuss the information with their dentists. If there is an insufficient level of fluoride or no fluoride in the drinking water in the community, the students might want to ask their dentists what to do.

## Research and Write

Have students read and conduct the Research and Write feature on page 132. Encourage students to call local hospitals or social service agencies to research this topic. Suggest that students might want to work in pairs or small groups on this project.

---

*Dandruff*
Flaking from the scalp

*Gingivitis*
First stage of gum disease

**Research and Write**

Not everyone has medical insurance or can afford its cost. Find out about medical services where you live. Where can people in your area go if they cannot afford medical care? Create a community medical services booklet.

See your doctor for a checkup and a vision test once a year. Have a dentist clean and check your teeth twice a year.

## How Can You Take Care of Your Hair and Nails?

Shampooing and brushing your hair get rid of dirt and help spread the oils that make your hair shiny and soft. To protect your hair, do not overuse blow dryers and curling irons. Too much heat can damage your hair.

Some **dandruff,** or flaking from your scalp, is normal. Your scalp is skin. It sheds dead cells just as the rest of your skin does. Special shampoos can control dandruff. Check with a doctor if you have a lot of dandruff.

For good nail care, you can do two things regularly: (1) Keep your nails clean, especially underneath. (2) Cut your nails evenly and file ragged edges.

## How Can You Take Care of Your Teeth?

Your teeth play a part in how you feel about yourself and how others think of you. When your teeth are clean and healthy, you can smile with confidence. A bright smile with clean, strong teeth is one of the first things people notice. With proper care, teeth can last a lifetime. You should brush your teeth at least twice a day for 3 to 4 minutes each time. You should also floss at least once a day. Use toothpaste with fluoride, a chemical that helps prevent tooth decay, or use a fluoride rinse. Have regular dental checkups and cleanings. Avoid eating too many sweets. Wear a mouth guard if you play contact sports. Do not chew on hard objects such as pens, pencils, and hard candies.

You should also take care of your gums to avoid gum disease. Gum disease can damage your teeth. The first stage of gum disease is **gingivitis.** Signs of gingivitis are red, swollen gums. The gums may bleed and be painful. Proper brushing and flossing and a healthy diet can prevent gingivitis.

**Word Bank**
allergies
blinking
sunscreen

**Lesson 1 Review** On a sheet of paper, write the word from the Word Bank that best completes each sentence.

1. You can keep dirt out of your eyes by _____.
2. Hearing loss can be caused by _____.
3. To protect your skin from sunburn, use _____.

On a sheet of paper, write the answers to the following questions. Use complete sentences.

4. **Critical Thinking** Why is getting regular medical and dental checkups important?
5. **Critical Thinking** How can taking care of your hair, skin, and teeth improve your self-concept?

### Health in Your Life

#### Noise and Your Hearing

Loud noise can cause hearing loss over time. Strength, or intensity, of sound is measured in decibels (dB). Here are the decibel levels for some common sounds:

| Normal breathing | 10 dB |
| Whisper | 30 dB |
| Normal conversation | 60 dB |
| Truck traffic | 90 dB |
| Rock concert | 120 dB |
| Jet engine at takeoff | 140 dB |

Decibel levels higher than 85 can cause hearing loss. You can protect your hearing from harmful noise levels. Be aware of the noise levels around you. Keep the volume at safe levels on CD players, computers, radios, and TVs.

On a sheet of paper, write the answers to the following questions. Use complete sentences.

1. What are some places where loud noise can cause hearing loss?
2. What can you do to protect your hearing from loud noises?
3. In what way are loud noises a risk factor for good lifelong health?

*Maintaining Personal Hygiene and Fitness* Chapter 6   133

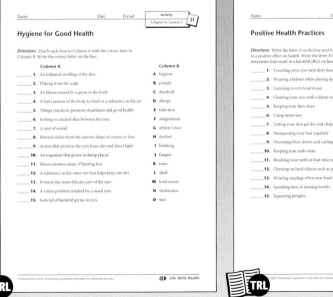

Activity 21

Workbook Activity 21

## Lesson 1 Review Answers

1. blinking 2. loud noise 3. sunscreen 4. Answers should refer to the idea that people should get regular checkups to prevent any existing problems from getting worse, to make sure they are healthy, and to find out what they can do to improve their health. 5. People feel better about themselves and have more confidence when they look their best. Taking care of your hair, skin, and teeth shows you value yourself.

### Portfolio Assessment

Sample items include:
- Definitions from Vocabulary
- Tables from Teaching the Lesson
- Note cards from ELL/ESL Strategy
- Paragraphs from Health Journal
- Comic strips from Learning Styles: Visual/Spatial
- Pamphlets from Learning Styles: Interpersonal/Group Learning
- Booklets from Research and Write

### Health in Your Life

Have students read the Health in Your Life feature on page 133. Explain to students that if you have to shout over a noise to be heard, it is too loud. If your ears buzz or ring when you leave a noisy place, the noise level is too high. And if others can hear your music when you are wearing earphones, the volume is set too high.

#### Health in Your Life Answers

1. Sample answer: Some places where loud noise can cause hearing loss are rock concerts, fireworks displays, and factories. Answers should show an understanding of places with loud noises. 2. You can turn the volume down on radios, televisions, and headsets. You can wear earplugs and avoid standing near loudspeakers. 3. Loud noises can cause permanent hearing loss.

*Maint. Pers. Hygiene/Fitness* Chapter 6   133

# Lesson at a Glance

## Chapter 6 Lesson 2

**Overview** This lesson describes the benefits of healthy exercise and the types of exercise that can be done to promote good health.

### Objectives

- To identify four benefits of exercise
- To explain the different purposes of exercise
- To describe how to determine your pulse rate
- To explain the purpose of warm-up and cooldown activities

**Student Pages** 134–139

**Teacher's Resource Library**

   Activity 22

   Alternative Activity 22

   Workbook Activity 22

## Vocabulary

| | |
|---|---|
| aerobic exercise | isokinetic exercise |
| anaerobic exercise | isometric exercise |
| calorie | isotonic exercise |
| endurance | maximum heart rate |
| flexibility | resistance |

Have volunteers read the definitions of each term as given in the vocabulary boxes throughout the lesson. Then have students describe a situation or an activity that is an example of each term. For example, a student might describe a situation in which someone needs great endurance, a food that has a lot of calories, or a type of isometric exercise.

## Background Information

Exercise can lead to changes in the body including a reduction of body fat, weight loss, and an increase in basal metabolic rate. Moderate exercise can even lead to a decrease in appetite, especially in people who are obese. The amount of exercise each person needs depends on his or her health goals, whether a person is trying to lose weight, and a person's level of physical fitness. Exercise benefits health by reducing the risk of developing cardiovascular disease, helping maintain a healthy weight, and reducing the risk of osteoporosis.

---

## Lesson 2 — Exercise and Physical Fitness

**Objectives**

*After reading this lesson, you should be able to*

- identify four benefits of exercise
- explain the different purposes of exercise
- describe how to determine your pulse rate
- explain the purpose of warm-up and cooldown activities

You are physically fit when your body can meet the demands of daily living. That means having enough energy to do all the things you want to do. Physical fitness is a key part of your overall good health. It affects your emotional, social, and physical well-being.

### What Are Three Parts of Health-Related Fitness?

Three parts of health-related fitness are heart and lung **endurance, flexibility,** and muscular fitness. Endurance is the ability to stay with an activity for a long time. Heart and lung endurance is the fitness of your heart, blood vessels, and lungs. It means that your heart and blood vessels are easily able to move oxygen through your blood to your lungs. Your energy depends mainly on your body being able to take in and use oxygen. To do this, you need a strong heart, clear lungs, and blood vessels free of fat buildup.

Flexibility is the ability to twist, turn, bend, and stretch easily. Flexibility helps keep your muscles from pulling and straining. It also increases the range of motion in your joints.

Muscular fitness is the strength and endurance of your body's muscles. Muscular strength is the amount of force that your muscles put out against **resistance,** or the pushing back of something. Muscular endurance allows you to lift, push, and pull objects without your muscles getting too tired.

### How Does Exercise Help You?

Regular exercise can improve your health-related fitness. It increases your heart and lung endurance. It firms and strengthens your muscles. It improves your flexibility. Exercise gives you more energy. It helps you feel better and look better.

---

**Endurance**
*The ability to stay with an activity for a long time*

**Flexibility**
*The ability to twist, turn, bend, and stretch easily*

**Resistance**
*The pushing back of something*

**Calorie**
*A unit that measures the amount of energy in food*

**Aerobic exercise**
*Activity that raises your heart rate*

**Anaerobic exercise**
*Activity that quickly uses up oxygen in your body*

**Isometric exercise**
*Activity that builds muscle strength by using tension*

**Isotonic exercise**
*Activity that builds muscle strength by using weights and repeated movements*

**Isokinetic exercise**
*Activity that builds muscle strength by using resistance*

Exercise does not have to be hard to be good for you. Cleaning the house and gardening have been shown to be as helpful as biking and walking.

Exercise also can reduce your chances of getting sick. For example, regular exercise reduces the risk of heart disease. Regular exercise can shorten the time needed to get well if you do become sick.

Another reason to exercise is weight control. Exercise burns up extra **calories**. A calorie is a measure of the amount of energy in food. Besides burning up extra calories, exercise speeds up the rate at which your body burns calories.

Finally, exercise helps you reduce anxiety and get rid of stress. Exercise can be fun and is a good way to make friends.

## What Is the Purpose of Exercise?

Different types of exercise have different purposes. Look at Table 6.2.1 to see five types of exercise and their purpose. Exercise is one of the best ways of keeping your mind and body healthy.

| Table 6.2.1 Types of Exercise | | |
|---|---|---|
| Exercise Type | Purpose | Examples |
| Aerobic exercise (steady, continuous activity) | Helps heart and lungs work better by raising the heart rate | Walking, running, bicycling, cross-country skiing, swimming |
| Anaerobic exercise (short spurts of activity) | Improves how the body works when it is working the hardest by quickly using up oxygen | Sprinting, tennis |
| Isometric exercise | Builds muscle strength by tensing muscles | Pushing against a wall |
| Isotonic exercise | Builds muscle strength using weights | Any body movement, push-ups, pull-ups |
| Isokinetic exercise | Builds muscle strength by using resistance | Workout with machines that control force of pushing and pulling |

 **Warm-Up Activity**

Explain to students that the three parts of health-related fitness are heart and lung endurance, flexibility, and muscular fitness. Ask students to explain what they think each of these terms mean. Have them give examples of exercises they think people could do to achieve each type of fitness.

 **Teaching the Lesson**

Help students organize the material on the parts of an exercise program by having them make a flow chart to sequence the steps involved in a good exercise workout. Have them include a brief description of each step as part of the flowchart.

Have students make three separate spider diagrams, one for each type of fitness. The type of fitness should be in the center of the diagram and the surrounding circles should contain examples of the types of exercises that people can do to achieve that type of fitness.

 **Reinforce and Extend**

### IN THE ENVIRONMENT

 Discuss with students that many people engage in jogging and aerobics to increase their physical well-being. Exercise causes the respiratory rate to increase. Any pollutants in the air are, therefore, taken up more rapidly. Cigarette smoke is one such pollutant that can affect breathing rates during exercise. Because there are many people who enjoy exercising outdoors, they come in contact with pollution. Although most restaurants and other public places have designated smoking areas, some people believe that smoking should be banned from all public places, including parks and other recreation areas. Ask students how they feel about this statement.

## TEACHER ALERT

Emphasize the importance of warming up at the beginning of a workout. Ask students to explain the purpose of a warm-up. Ask them to explain what could happen if a person exercised vigorously without warming up. *(People who don't warm up before they exercise are more likely to experience injuries such as sprains or strains than are people who do warm up before they exercise.)*

## LEARNING STYLES

**Body/Kinesthetic**

Show students how to find their maximum heart rate by subtracting their age from 220. Show students how to find their resting heart rate by taking their pulse for 6 seconds and multiplying by 10. A target heart rate is 50 to 75 percent of the maximum heart rate. It is the rate a person wants to reach during aerobic exercise. Have students calculate their target heart rate. Then tell them to run in place for two minutes and find their heart rate again. Ask them to see if their heart rate is within their target heart rate after two minutes of exercise. If they are close to their maximum heart rate, they should slow down when exercising.

## ELL/ESL STRATEGY

**Language Objective:** *To use cooperative learning*

When discussing the importance that exercise plays in maintaining a healthy body, explain the American slang expression "couch potato." Working in cooperative groups, have students write a paragraph describing the habits of a couch potato. Then ask them to design a plan to help the couch potato become a healthier person. Suggest that different groups compare their plans.

---

**Maximum heart rate**
*The number of heartbeats in a minute when one exercises as hard, fast, and long as possible*

Although maximum heart rate is different for everyone, it should be about 220 minus your age. For example, if you are 16, your maximum heart rate should be about 204 beats in a minute.

## What Are the Parts of an Exercise Program?

A good exercise workout includes these five parts:

- a warm-up period
- an aerobic exercise to improve heart and lung endurance
- exercise to improve muscular fitness
- exercise to improve flexibility
- a cooldown period

### Warm-Up

Always begin your exercise workout with a 5-minute warm-up. Warming up makes more blood flow to your muscles. It allows your heart rate to rise gradually. You can get hurt if you push your body into hard exercise too fast. You can warm up by walking and doing stretching exercises.

### Exercises to Improve Heart and Lung Endurance

Your heart rate is the number of times your heart beats in a minute. Your **maximum heart rate** is your heart rate when you exercise as hard, fast, and long as you can.

Pulse rate is another term for heart rate. To determine your pulse rate, do the following:

- Place two fingers of your right hand on the side of your neck just below your jawbone.
- You can also place two fingers of your right hand on the inside of your left wrist. CAUTION: Do not take your pulse with your thumb, because it has its own pulse.
- To locate your pulse, feel the thump against your fingers.
- Count each thump as one heartbeat.
- Use a watch with a second hand to count the number of heartbeats in 10 seconds.
- Multiply this number by 6 to get your heart rate for 1 minute.
- The number of thumps each minute is your pulse rate.

To find out how long you should exercise to improve your heart and lung endurance, do this test:

1. Figure out your maximum heart rate by subtracting your age from 220.
2. Do an aerobic exercise for at least 10 minutes.
3. Immediately after you stop exercising, take your pulse for 10 seconds.
4. Multiply this number by 6 to get your exercising pulse rate per minute.
5. Divide your exercising pulse rate by your maximum heart rate.
6. Multiply the result by 100 to determine the percent of maximum heart rate used.
7. Look at Table 6.2.2 to find the number of minutes of exercise you need to do three times a week.

Here is an example for a 15-year-old with a pulse rate per minute of 123 after walking briskly for 10 minutes:

$220 - 15 = 205$
$123 \div 205 = 0.60$
$0.60 \times 100 = 60$ percent

This person must walk at this pace for 30 to 37 minutes three times a week to improve heart and lung endurance.

| Table 6.2.2 Pulse Rate and Exercise Times | |
|---|---|
| If your maximum pulse rate is: | You can exercise three times a week for: |
| 50 percent | 45–52 minutes |
| 55 percent | 37–45 minutes |
| 60 percent | 30–37 minutes |
| 65 percent | 25–30 minutes |
| 70 percent | 20–25 minutes |
| 75 percent | 15–20 minutes |

## CAREER CONNECTION

A physical therapist works with people whose physical activity has been curtailed due to accidents, surgery, physical handicaps, arthritis, or heart disease. Some physical therapists work with a wide range of problems. Others specialize, working in pediatrics, sports therapy, neurology, or other specialties. Physical therapists teach their patients stretches, exercises, and techniques, as well as how to use specialized equipment. Physical therapists need to be flexible, physically strong, and supportive of their clients. A physical therapist must have at least a master's degree. Discuss with students the pros and cons of choosing this career.

## GLOBAL CONNECTION

Martial arts and yoga are two forms of exercise that are part of many Asian cultures. For example, karate and judo both originated in Japan. Tae kwon do and hapkido are forms of exercise from Korea. Tai chi and kung fu hail from China. Yoga originated primarily in India. Have students work in groups to research and give oral reports on one of these types of exercise. Students should discover how the exercise originated, the benefits of the exercise, and perhaps even how to do some of the moves associated with the exercise.

## LEARNING STYLES

**Logical/Mathematical**
Encourage students to obtain all the information they need in order to calculate the length of time they need to exercise to improve heart and lung endurance. Offer to help students with calculations once they have all of the numbers they need. Remind them that they will not have to share any information with classmates to complete this exercise.

### Research and Write

Have students read the Research and Write feature on page 138. Remind students to conduct their research on reputable Web sites, such as those with .gov or .edu domain names. These are government Web sites and college and university Web sites, respectively.

## AT HOME

Encourage students to develop with their family or friends a list of fitness activities they can do together to improve heart and lung fitness. Invite students to share their lists with the class for comparison. Encourage students to perform some or all of activities on their lists with their family or friends.

**138** Unit 2 Personal Health & Family Life

---

**Research and Write**

You can find lots of information on the Internet about health. What is something about health that interests you? Choose something about health to research. Write a report and prepare an online presentation showing what you learned.

If you want to lower your exercising pulse rate, slow down. If you want to raise it, pick up your pace. How hard or how fast should you exercise? Do not push yourself past the point of being able to talk to someone next to you.

When you exercise to improve heart and lung endurance, you need to do two things:

1. Run, swim, bicycle, walk, or exercise at a steady pace without stopping. Sports such as baseball and softball are not aerobic, because the activity is not steady.

2. At least three times a week, exercise at a level that is 50 to 75 percent of your maximum heart rate for the number of minutes shown in Table 6.2.2 on page 137.

### Exercises to Improve Muscular Fitness

To improve muscular fitness, you must use resistance. Your muscles need to overcome some sort of resistance to become stronger. Exercises such as push-ups, sit-ups, and pull-ups improve muscle strength and endurance. Weight lifting and isometric exercises also improve muscular fitness.

### Exercises to Improve Flexibility

Slow stretching exercises can improve your flexibility. Stretching makes your muscles relax and lengthen. Slow, gradual movements are best for building flexibility. Avoid movements that are fast and bouncy. These make your muscles tighten instead of relax.

### Cooldown

End your workout with a cooldown. Your body needs a chance to slow down gradually. If you stop suddenly, you could become light-headed or even faint. Your blood flow needs time to adjust itself. To cool down, continue to exercise but at a slower pace. Also, stretch to improve your flexibility. A cooldown should take about 5 minutes.

**Figure 6.2.1** End your workout with a cooldown.

138 Unit 2 Personal Health and Family Life

**Word Bank**
aerobic
anaerobic
isotonic

*Lesson 2 Review* On a sheet of paper, write the word from the Word Bank that best completes each sentence.

1. Sprinting and other short spurts of activity are examples of _____ exercise.
2. Lifting weights, doing push-ups, and other muscle-building activities are examples of _____ exercise.
3. Swimming, running, walking, or some other activity that improves the health of your heart and lungs is _____ exercise.

On a sheet of paper, write the answers to the following questions. Use complete sentences.

4. **Critical Thinking** Why should an exercise program include all five parts of a good program?
5. **Critical Thinking** How can regular exercise improve your health?

**Why do some people avoid exercising?**

### Link to ▶▶▶
**Arts**
Ballet dancers leap and turn across the stage. Many of the moves in ballet can help people who play sports. Many athletes study parts of ballet to improve how they move. The moves taught in ballet can be used in sports such as baseball, basketball, ice-skating, and football.

*Maintaining Personal Hygiene and Fitness* Chapter 6  **139**

## Lesson 2 Review Answers

1. anaerobic 2. isotonic 3. aerobic 4. Warm-ups increase your blood flow and get you ready for more exercise. Aerobics build your heart and lung endurance. Strength training builds your muscles. Improving your flexibility allows you a greater range of motion. A cooldown allows your body to slow down gradually. 5. Exercising regularly helps the body fight disease, increases heart and lung endurance, firms and strengthens muscles, improves flexibility, gives you more energy, reduces anxiety, and helps you cope with stress.

## Portfolio Assessment

Sample items include:
- Flow charts and spider diagrams from Teaching the Lesson
- Calculations from Learning Styles: Body/Kinesthetic
- Paragraphs from ELL/ESL Strategies
- Calculations from Learning Styles: Logical/Mathematical
- Reports and online presentations from Research and Write
- Lists from Link to Arts
- Lesson 2 Review answers

### Link to

**Arts.** Ask a volunteer to read aloud the Link to Arts feature on page 139. Have students research and make a list of sports figures who have studied ballet in an effort to improve their skills.

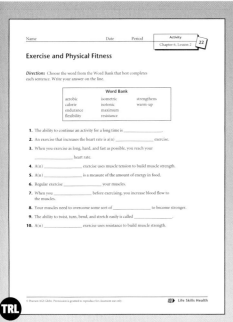
**Activity 22**

**Workbook Activity 22**

*Maint. Pers. Hygiene/Fitness* Chapter 6  **139**

# Lesson at a Glance

## Chapter 6 Lesson 3

**Overview** This lesson describes the parts of a good fitness plan.

### Objectives

- To describe the parts of a good fitness plan
- To explain how rest and sleep are important to fitness
- To explain how fitness can be a part of your life

**Student Pages** 140–141

**Teacher's Resource Library** TRL
- Activity 23
- Alternative Activity 23
- Workbook Activity 23

## Vocabulary

Encourage students to incorporate the vocabulary terms they learned in Lessons 1 and 2 when discussing the material in Lesson 3.

## Background Information

Sleep is essential for good health. While sleeping, the body grows and repairs injuries to muscles, bones, skin, and the brain. If you are well-rested, you have increased concentration and problem-solving abilities. To get a good night's sleep, avoid caffeine four to six hours before bedtime, avoid large meals close to bedtime, get up at the same time every day, and reduce noise and light in your bedroom.

 **Warm-Up Activity**

Have students make a list of what, in their opinions, are the most important parts of a good fitness plan. Have them review and revise their lists after they have completed Lesson 3.

140   Unit 2   Personal Health & Family Life

---

## Lesson 3   Personal Fitness Plan

**Objectives**

*After reading this lesson, you should be able to*

- describe the parts of a good fitness plan
- explain how rest and sleep are important to fitness
- explain how fitness can be a part of your life

A personal fitness plan includes eating well, exercising, and getting enough rest and sleep. The first step in any fitness plan is eating the right foods. You need foods that provide a good balance of nutrients and calories.

Exercising should be a part of your day. It will strengthen your muscles and burn extra calories. Table 6.3.1 shows the number of calories burned for different types of activities.

Talk to your doctor before you start a fitness plan. Make sure there are no limits on the type of activities you can include in your plan. You might want to ask a friend to be a part of your fitness plan. Exercising is always fun to do with a friend.

**Figure 6.3.1** *Jobs such as lawn mowing can be good exercise.*

| Table 6.3.1 Calories Burned for a Person Weighing 150 Pounds ||
|---|---|
| **Activity** | **Calories** |
| Jogging | 675 per hour |
| Soccer | 468 per hour |
| Dancing | 370 per hour |
| Yoga | 360 per hour |
| Brisk walk | 297 per hour |
| Golf (using no golf cart) | 240 per hour |
| Sitting | 81 per hour |
| Watching TV | 72 per hour |
| Sleeping | 45 per hour |
| Mowing lawn | 240 per hour |

140   Unit 2   *Personal Health and Family Life*

Here are a few things to remember as you think about your fitness plan:

- Choose activities that you enjoy.
- Drink lots of water before, during, and after you exercise.
- If you miss a day of exercise, plan for a makeup day. Do not double your exercise time to make up for a lost day.
- Start slowly. Do not overdo it.
- Wear comfortable clothes and shoes.

### What Are the Steps of a Fitness Plan?

To make a fitness plan that will work for you, follow these steps.

**Step 1** Determine your current fitness level. How much do you exercise? What kinds of exercise do you get? Do you eat a variety of healthy foods?

**Step 2** Set realistic goals. Do you want to lose weight or gain weight? Do you want to increase your heart and lung health? What activities will you add to your current exercise level?

**Step 3** Find resources to help you meet your goals. Where will you work out? What kinds of equipment will you need? Will you have friends or family members supporting you?

**Step 4** Get started. After you start exercising, pay attention to your body. Are you feeling dizzy or short of breath? Do your muscles hurt too much? You can always talk with your healthcare professional about how to change your plan to better suit your body.

**Step 5** Keep a record. Keep track of what you did to meet your goals daily or weekly. What exercises did you do? How many minutes did you exercise?

## Teaching the Lesson

Have students write a summary paragraph of material presented in the lesson.

Have students write down any questions they might have about the material. Hold a question-and-answer session during class in which students can ask their questions. Work together with the class to answer each question.

## Reinforce and Extend

### TEACHER ALERT

Encourage students to discuss a new fitness plan with a health care professional before they begin. It is important to make sure the activities and levels of exertion in the plan are appropriate.

### LEARNING STYLES

**Auditory/Verbal**

Have students make a list of fast-paced songs that they would like to listen to as they workout. They should include enough songs for the length of their workout. Student can make mix CDs of their chosen workout music at home by legally purchasing songs from a licensed Web site and burning them onto CDs.

## Decide for Yourself

Have students read and complete the Decide for Yourself feature on page 142. Emphasize the idea that students should listen to their bodies as they exercise. If they overexert themselves on the first day, they will be sore and tired, and this could cause them to not follow through on their plans.

### Decide for Yourself Answers

**1.** A personal fitness plan can improve muscular strength and endurance, heart and lung endurance, and flexibility. Eating well, exercising regularly, and getting enough sleep will improve your physical health. **2.** Often, people enjoy exercising and playing sports together. Sharing physical activity gives a person a sense of belonging to a group. **3.** A personal fitness plan can help relieve anxiety and feeling of depression, help you handle stress, and improve your self-concept.

## HEALTH JOURNAL

Have students record their personal fitness plan in their Health Journals. Have them plan out a week's worth of exercises that will target all three types of fitness. Encourage them to carry out their fitness plans and record how they feel every week.

---

### Decide for Yourself

Before you start on a personal fitness plan, you should do two things. First, decide what your fitness needs are. Next, decide how much time each day you can give to your fitness plan. Then follow these steps in making your personal fitness plan. Remember that eating healthy foods is part of any fitness plan. Be sure to include grains, vegetables, fruits, milk, meat, or beans every day.

**Step 1** Establish a starting point of your fitness level. Write down the answers to these questions:

- How fit is my body? Choose *not at all fit, somewhat fit,* or *very fit.*
- How flexible am I? Choose *not at all flexible, somewhat flexible,* or *very flexible.*
- How much exercise can I do now at one time? Choose *less than 15 minutes, 15–30 minutes,* or *more than 30 minutes.*

**Step 2** Design a written program for yourself based on goals you can actually meet. Write down the answers to these questions:

- Do I just want to lose weight?
- Do I want to increase my endurance and flexibility?
- What activities will be a part of my plan?
- How will my fitness plan fit into my life?
- When will I look at how well I am doing?

**Step 3** Get what you need. For example, does your school or community have a fitness center? Do you have the right kinds of workout clothes and shoes?

**Step 4** Get started. Listen to your body. Pay attention to signs that you are doing too much, such as shortness of breath, dizziness, or feeling sick to your stomach. Make a list of these signs. Share this list with your doctor before you go on with your fitness plan. Be flexible. Change your fitness plan to match what your body can handle.

**Step 5** Every week, figure out how you are doing. Are you sticking to your plan? Do you need to change your activities? How much time do you spend on them? Do you feel more fit? Keep a record of how well you do.

On a sheet of paper, write the answers to the following questions. Use complete sentences.

1. How can a personal fitness plan improve your physical health?
2. How can a personal fitness plan improve your social health?
3. How can a personal fitness plan improve your emotional health?

## Why Are Rest and Sleep Important to Fitness?

Rest is basic to the body's well-being. When you are tired, your body cannot work right. You have a harder time paying attention and often feel more stressed. Your body also is at greater risk of disease and injury when you are tired. Sleep and rest are necessary to feel better and to stay healthy.

Most teens need at least 8 hours of sleep each night. Depending on your level of activity, you might need more rest. Rest does not always have to mean sleep. Relaxing or reducing your level of activity are good ways to rest during the day.

### Health Myth

**Myth:** No pain, no gain!

**Fact:** Being sore after first exercising is common. However, you should not feel pain during or after exercise. If you do, it may mean you are doing something wrong. You may be pushing yourself too hard. You do not need to have pain to improve your muscle strength or endurance. If you continue to exercise when you feel pain, you could injure yourself.

### Health at Work

**Fitness Instructor**

If you like helping others stay fit, you might think about becoming a fitness instructor. Many companies have health and wellness programs. Fitness instructors work with companies to design health programs for their workers. They show workers the correct way to exercise. Fitness instructors also explain how the right exercises help the body. They may work with doctors and physical therapists to help workers get over medical problems. You must have at least a college degree to be a fitness instructor.

The training you receive is similar to the training for physical therapists, physical education teachers, or recreation specialists.

## Health Myth

Have students read the Health Myth feature on page 143. Ask students if they have heard this popular saying. Have them explain what they think it means in their own words. Stress that they should not feel pain while they are exercising; pain is a sign that something is wrong. Encourage students to talk to a health care professional if they are feeling pain during a workout.

## Health at Work

Have students read the Health at Work feature on page 143. Ask students if they think that fitness instructors have an important job and give reasons to support their conclusion. Ask if any students would like to become fitness instructors and discuss the pros and cons of this type of career. You might want students to research the other related careers mentioned in the text, such as physical therapists, physical education teachers, or recreation specialists.

### ELL/ESL Strategy

**Language Objective:** *To practice reading strategies*

As you work through the lesson, have students learning English identify the main idea of each paragraph. Pair students learning English with students who are proficient in English to review the paragraphs together.

# Lesson 3 Review Answers

**1.** C **2.** B **3.** C **4.** Sample answer: Your body needs certain nutrients in order to work properly. Eating healthy foods is important for strong bones and muscles. **5.** Sample answer: Exercising is an important part of a personal fitness plan because it strengthens muscles and burns extra calories. Exercising also helps the lungs work better. Exercising helps people cope with stress.

## Portfolio Assessment

Sample items include:
- Lists from Warm-Up Activity
- Questions from Teaching the Lesson
- Answers from Decide for Yourself
- Personal Fitness Plans from Health Journal
- Lists of songs/mix CDs from Learning Styles: Auditory/Visual
- Main Ideas from ELL/ESL Strategy
- Lesson 3 Review answers

---

*Lesson 3 Review* On a sheet of paper, write the letter of the answer that best completes each sentence.

1. You should decide about _____ before you start a personal fitness plan.
   A school sports
   B a friend's plan
   C your fitness needs
   D local fitness centers

2. Every personal fitness plan should include _____.
   A the use of fitness machines and classes
   B a look at how well you are doing
   C jogging, running, or walking
   D isokinetic exercise

3. Rest and sleep are important for _____.
   A maximum heart rate
   B nutrition
   C well-being
   D calorie intake

On a sheet of paper, write the answers to the following questions. Use complete sentences.

4. **Critical Thinking** Why is eating healthy foods an important part of a personal fitness plan?

5. **Critical Thinking** Why is exercising an important part of a personal fitness plan?

144  Unit 2  *Personal Health and Family Life*

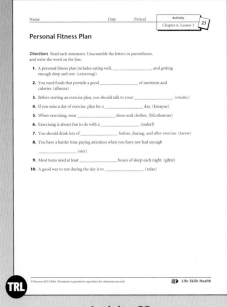

Activity 23

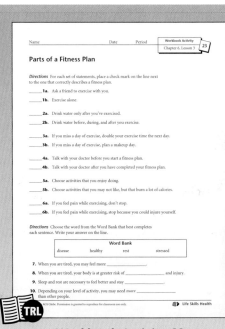

Workbook Activity 23

# Chapter 6 SUMMARY

- Hygiene refers to things you do to have good health. It means taking care of your eyes, ears, skin, hair, nails, and teeth.
- Your body protects your eyes and sight with bones, eyebrows and eyelashes, blinking, and tears. You can protect your eyes in several ways, including having regular eye checkups.
- Your skull protects your ears, and wax keeps dirt out of them. You can protect your ears by avoiding loud sounds and by keeping objects out of them.
- Loud noise is the most common reason for hearing loss. You can almost always avoid hearing loss.
- The two basic rules of good skin care are to keep your skin clean and to avoid strong winds, cold, and sun.
- You can take care of your teeth by brushing after meals, flossing daily, and having regular dental checkups.
- Physical fitness is your body's ability to meet the demands of daily life.
- Three parts of health-related fitness are heart and lung endurance, flexibility, and muscular fitness.
- Regular exercise increases energy, lowers the chances of getting sick, helps control weight, reduces anxiety, and helps get rid of stress.
- Some types of exercise are aerobic, anaerobic, isometric, isotonic, and isokinetic exercise.
- A good exercise program includes a warm-up activity, aerobic activity, exercises to improve muscular fitness and flexibility, and a cooldown activity.
- Lifelong health includes a personal fitness plan that sets reasonable goals.
- Rest and sleep are basic to fitness. The body does not work right without enough sleep.

### Vocabulary

| | | |
|---|---|---|
| acne, 130 | dandruff, 132 | infection, 129 |
| aerobic exercise, 135 | deodorant, 130 | isokinetic exercise, 135 |
| allergy, 129 | endurance, 134 | isometric exercise, 135 |
| anaerobic exercise, 135 | flexibility, 134 | isotonic exercise, 135 |
| antiperspirant, 130 | fungus, 131 | maximum heart rate, 136 |
| athlete's foot, 131 | gingivitis, 132 | pimple, 130 |
| calorie, 135 | hygiene, 128 | resistance, 134 |

## Chapter 6 Summary

Have volunteers read aloud each Summary item on page 145. Ask volunteers to explain the meaning of each item. Direct students' attention to the Vocabulary box on the bottom of page 145. Have them read and review each term and its definition.

### ONLINE CONNECTION

For more information about fitness, have students visit this Web site offered by the President's Council on Physical Fitness and Sports. The Web site explains the President's Challenge, a national awards program that encourages students to participate in physical activity to improve endurance, strength, and flexibility at:

www.fitness.gov/challenge/challenge.html

Students also can visit this Web site offered by the Executive Office of the President and the Department of Health and Human Services. The Web site contains information about physical fitness, nutrition, making healthy choices, diabetes, and obesity at:

www.healthierus.gov/exercise.html

Bam! (Body and Mind), a Web site sponsored by the Centers for Disease Control and Prevention, allows students to make their own fitness plan using an activity calendar. The Web site, which also offers information on nutrition, diseases, and safety, can be found at:

www.bam.gov/sub_physicalactivity/index.html

## Chapter 6 Review

Use the Chapter Review to prepare students for tests and to reteach content from the chapter.

## Chapter 6 Mastery Test

The Teacher's Resource Library includes two forms of the Chapter 6 Mastery Text. Each test addresses the chapter Goals for Learning. An optional third page of additional critical-thinking items is included for each test. The difficulty level of the two forms is equivalent.

## Review Answers

**Vocabulary Review**

1. deodorant  2. maximum heart rate
3. pimple  4. calorie  5. infection  6. allergy
7. acne  8. fungus  9. resistance
10. dandruff  11. endurance  12. flexibility
13. athlete's foot

### TEACHER ALERT

In the Chapter Review, the Vocabulary Review activity includes a sample of the chapter's vocabulary terms. The activity will help determine students' understanding of key vocabulary terms and concepts presented in the chapter. Other vocabulary terms used in the chapter are listed below.

aerobic exercise
anaerobic exercise
antiperspirant
hygiene
isokinetic exercise
isometric exercise
isotonic exercise

146 Unit 2  Personal Health & Family Life

## Chapter 6 REVIEW

**Word Bank**
acne
allergy
athlete's foot
calorie
dandruff
deodorant
endurance
flexibility
fungus
infection
maximum heart rate
pimple
resistance

### Vocabulary Review

On a sheet of paper, write the word or phrase from the Word Bank that best completes each sentence.

1. Using _____ each morning will cover body odor.

2. You can find your _____ after you exercise as hard, fast, and long as you can.

3. A(n) _____ is an inflamed swelling on the skin.

4. A measure of the amount of energy in food is a(n) _____.

5. Some causes of hearing loss are _____ and heredity.

6. A(n) _____ is a bad reaction of the body to a food or to something in the air.

7. Clogged skin pores cause _____.

8. A(n) _____ can cause athlete's foot.

9. The pushing back of something is _____.

10. You can buy shampoos that will prevent _____ on your scalp.

11. People with heart and lung _____ can easily move oxygen through the blood to the lungs.

12. Being able to bend and stretch easily shows you have good _____.

13. If the skin between your toes is itching, you might have _____.

146  Unit 2  Personal Health and Family Life

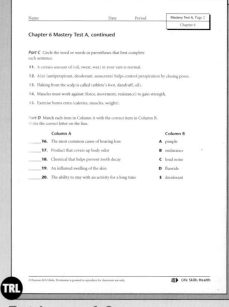

Chapter 6 Mastery Test A, pages 1–3

## Concept Review

On a sheet of paper, write the letter of the answer that best completes each sentence.

**14.** One way to protect your ears is to _____.
- A wear earplugs when there is loud noise
- B clean them daily with a cotton swab
- C keep them wet after swimming
- D avoid head guards

**15.** One thing you can do to take care of your teeth is to _____.
- A avoid fluoride
- B chew on hard objects
- C not eat too many sweets
- D eat mostly soft foods

**16.** A good reason to exercise regularly is that it _____.
- A makes you less flexible
- B improves your hearing
- C gets you out in the fresh air
- D helps control your weight

**17.** Your body needs sleep to _____.
- A work right
- B lose weight
- C improve lung endurance
- D get rid of body odor

## Critical Thinking

On a sheet of paper, write the answers to the following questions. Use complete sentences.

**18.** How can you help others understand the risk factors that might affect the health of your eyes and ears?

**19.** What goals can you set for yourself to stay healthy for the rest of your life?

**20.** Why do you think physical fitness needs to be lifelong?

 **Test-Taking Tip** When taking a test where you must write your answer, read the question twice to make sure you understand what is being asked.

*Maintaining Personal Hygiene and Fitness* Chapter 6 147

## Concept Review

14. A 15. C 16. D 17. A

## Critical Thinking

**18.** Answers could include ways to help others understand the dangers of loud noises and excessive exposure to bright light. For example, you might ask your brother not to play his stereo at a high volume, or you might ask your sister to loan you her sunglasses if you are going to run an errand outdoors on a bright day. You could explain to your family members that you want to protect your hearing and your vision. **19.** Answers should include a minimum of three achievable goals such as maintaining fitness, proper care of teeth, and adequate sleep. **20.** Answers should refer to the benefits of exercise in maintaining heart, lung, and general muscular health throughout one's life. Physical fitness is important for people of all ages: children, teenagers, adults, and senior citizens.

## ALTERNATIVE ASSESSMENT

Alternative Assessment items correlate to the student Goals for Learning at the beginning of this chapter.

- Have students orally describe why it is important to have good hygiene habits.
- Have students make spider diagrams that contain information about how to protect eyes, ears, skin, hair, nails, and teeth.
- Have students design a poster that explains the benefits and parts of a regular exercise program.
- Have students work in groups to create a personal fitness plan for a hypothetical person. Give each group a description of a different hypothetical person to aid them in their work.
- Have students role play a scenario that explains why rest and sleep are important to maintaining good health.

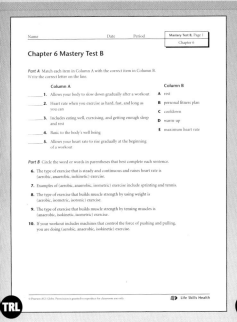
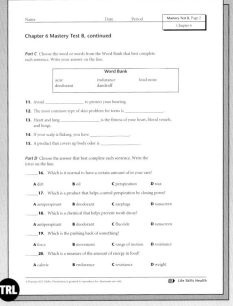

Chapter 6 Mastery Test B, pages 1–3

# Chapter at a Glance

## Chapter 7: The Life Cycle and Human Development
pages 148–173

### Lessons
1. **The Life Cycle and Adolescence**
   pages 150–155
2. **Reproduction**
   pages 156–159
3. **Pregnancy and Childbirth**
   pages 160–166
4. **Heredity and Genetics**
   pages 137–170

### Chapter 7 Summary page 171
### Chapter 7 Review pages 172–173
### Audio CD
### Skill Track for Life Skills Health
### Teacher's Resource Library
   Activities 24–27
   Alternative Activities 24–27
   Workbook Activities 24–27
   Self-Assessment 7
   Community Connection 7
   Chapter 7 Self-Study Guide
   Chapter 7 Outline
   Chapter 7 Mastery Tests A and B
   (Answer Keys for the Teacher's Resource Library begin on page 559 of this Teacher's Edition.)

## Opener Activity

Ask students to bring in magazine photographs of celebrities and their children (adult, teen, or child, but not babies). Ask them also to bring in photographs of celebrities as they looked 10 to 30 years ago and those same celebrities as they look today. Alternately, you or a student volunteer can collect these pictures from the Internet. Discuss whether or not the children look like the parents. Then discuss how people's appearances have changed over the years. (Note: Save the photos for use in Lesson 4.)

148  Unit 2  Personal Health & Family Life

# Self-Assessment

## Can you answer these questions?

1. ☑ What are the eight stages of life?
2. ☑ What are the early stages of human development?
3. ☑ What are some of the changes that occur during puberty?
4. ☑ How does the hormone testosterone affect males during puberty?
5. ☑ How do hormones affect females during puberty?
6. ☑ How can a woman tell if she is pregnant?
7. ☑ How do a human embryo and fetus develop?
8. ☑ What are some healthy behaviors for pregnant women?
9. ☑ How did you inherit traits, or qualities, from your parents?
10. ☑ Why is the history of your family's health important to your own health?

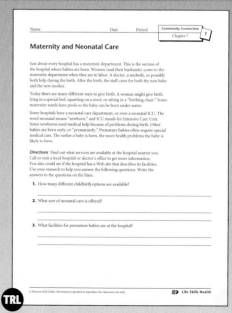

Self-Assessment 7              Community Connection 7

# Chapter 7

## The Life Cycle and Human Development

People pass through many stages during their lifetime. When you were younger, your life was very different. You have grown, developed, and moved through many stages since you were born. Even before you were born, you passed through many stages. At one time, your body was just one tiny cell. One of life's most remarkable events is how that cell developed into the person you are today.

In this chapter, you will learn about the stages in the life cycle. You will learn about the physical changes that happen during your teen years. You will learn how new life forms and develops and what happens during childbirth. Finally, you will learn about how families pass characteristics, or qualities, from parents to their children.

### Goals for Learning

◆ To identify the stages in the life cycle
◆ To explain the changes that take place as young people mature into adults
◆ To describe the male and female reproductive systems
◆ To describe how new life forms and develops
◆ To describe the process of birth
◆ To explain the importance of prenatal care
◆ To explain how characteristics are passed from parents to their children

149

## Introducing the Chapter

Have volunteers read page 149 aloud, including the Goals for Learning. Ask students to think about questions they might have regarding the upcoming topics in the chapter. Have each student write at least two questions anonymously on index cards. Collect the cards. Do this for all your classes and put the cards in a container. Examine the questions in private to check for appropriateness. When time allows at the end of a class, draw a card out of the container. Read a question aloud and answer it. Mark on the card in which class or classes the question has been used so it is not repeated on another day.

### Notes and Questions

Ask volunteers to read the notes and questions that appear in the margins throughout the chapter. Then discuss them with the class.

### TEACHER'S RESOURCE

The AGS Publishing Human Body Systems Transparencies may be used with this chapter. The transparencies add an interactive dimension to expand and enhance the *Life Skills Health* chapter content.

### Self-Assessment

Have students complete the Self-Assessment worksheet before and after reading the chapter. Before reading the chapter, have students fill in the "Before" column. Ask students to identify their goals for learning. To get ideas for setting goals, students might use the chapter introductory material on pages 148–149, the checklist on page 148, or the questions on the Self-Assessment worksheet. Students can use the back of the worksheet if they need more space to write.

Collect the Self-Assessment worksheets and pass them out again at the end of the chapter. Have students fill in the "After" column. Ask them to identify at least four major points they have learned. Again, suggest they use the back of the worksheet if they need more space to write. You might want to collect and review the worksheets, but return them to students so they have a record of their goals and accomplishments.

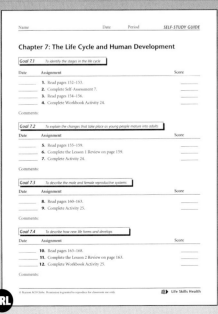

Chapter 7 Self-Study Guide, pages 1–2     Chapter 7 Outline, pages 1–2

# Lesson at a Glance

## Chapter 7 Lesson 1

**Overview** This lesson describes the physical and psychological stages of human development.

### Objectives

- To identify and describe the eight stages of life
- To describe changes that occur during puberty
- To recognize and accept differences in body types among adolescents

**Student Pages** 150–155

**Teacher's Resource Library**

Activity 24

Alternative Activity 24

Workbook Activity 24

### Vocabulary

| | |
|---|---|
| adolescence | reproductive organ |
| estrogen | role confusion |
| genitals | secondary sex |
| inferior | characteristic |
| progesterone | testosterone |
| puberty | |

Suggest students create crossword puzzles using the vocabulary words. They can use the definitions in the margins throughout the lesson to write the clues for the puzzles. Have pairs of students exchange puzzles to solve.

### Background Information

The word adolescence comes from the Latin verb *adolescere*, meaning "to grow up." The French philosopher Jean-Jacques Rousseau (1712–1778) developed the modern western concept of adolescence as a time of physical and emotional stress. In later years, G. Stanley Hall (1844–1924) was regarded as the founder of the scientific study of adolescence. Anthropologist Margaret Mead (1901–1978) made the controversial argument that adolescents in less developed countries have less stress than their counterparts in the West. Therefore, she concluded, they move through their developmental stages more happily and without conflict.

**150** Unit 2 Personal Health & Family Life

---

## Lesson 1  The Life Cycle and Adolescence

### Objectives

*After reading this lesson, you should be able to*

- identify and describe the eight stages of life
- describe changes that occur during puberty
- recognize and accept differences in body types

Many years ago, Doctor Erik Erikson determined that people usually go through eight life stages. Erikson's ideas are called a model of social and psychological development.

Erikson believed that a person could overcome an unpleasant experience in one stage by being successful in a later stage. Look at the overview of these stages in Table 7.1.1 as you read. Keep in mind that different people progress through different stages of the life cycle at different rates.

### What Are the Early Stages of Development?

The first stage is from birth to 12 months. If the basic needs of infants are met during this time, they learn to trust. Infants who are not well cared for may learn to mistrust others.

| Table 7.1.1 Stages of Social and Psychological Development | | | |
|---|---|---|---|
| Stage | Ages | Psychological Task | Description of Task |
| 1 | 0–1 | Trust and mistrust | Baby whose needs are met by caretaker develops a sense of trust in others. |
| 2 | 2–3 | Independence and doubt | Older baby tries to learn independence and self-confidence. |
| 3 | 4–6 | Enterprise and guilt | Young child learns to start his or her own activities. |
| 4 | 7–12 | Ability and inability | Child tries to develop skill in physical, mental, and social areas. |
| 5 | 13–19 | Identity and role confusion | Adolescent tries out several roles and forms a connected, single identity. |
| 6 | 20–40 | Closeness and loneliness | Young adult tries to form close, permanent relationships and to make career commitments. |
| 7 | 41–64 | Caring for children and self-absorption | Middle-aged person tries to contribute to the world through family relationships, work productivity, and artistry. |
| 8 | 65 on | Meaning and purpose and despair | Older person thinks back on life, experiencing satisfaction or disappointment. |

**150** Unit 2 Personal Health and Family Life

**Inferior**
Not as good as others

**Adolescence**
The years between ages 13 and 19

**Role confusion**
Being unsure about who one is and one's goals as an adult

During the second stage, from 2 to 3 years, children begin to do things on their own. They begin to depend less on others. If they are not allowed to learn some independence, they may start to doubt their abilities or feel shame.

Children between ages 4 and 6 are in the third stage of development. Here they learn to explore their world within safe limits. They begin to ask "Why?" They begin to use their imagination. They initiate, or begin, activities instead of simply following someone else. If children are encouraged at this time, they begin to feel able to do things. If not, they may develop feelings of guilt.

Children usually enter the fourth stage between ages 7 and 12. This is when they begin to make things. This is called having a sense of industry. If children hear positive comments about their efforts, they feel good about their abilities. If they are scolded or told negative comments, they may feel **inferior**. Children who feel inferior think they are not as good as others or as good as they should be.

## What Happens from Adolescence to Early Adulthood?

The fifth stage begins at **adolescence**, the years from ages 13 to 19. This is when many young people are trying to discover things about themselves. They are asking themselves questions such as "Who am I?" and "What do I believe in?" At this stage, teens are involved with many groups of people the same age, or peer groups. They sometimes think about themselves as others do. They are trying to figure out their own identity—a sense of who they are. They try to determine their needs, abilities, and values.

If a person had trouble with any of the earlier stages of development, that person may have difficulty dealing with the fifth stage. If they have feelings of mistrust, guilt, or being inferior, they may become confused about their role in life. **Role confusion** is being unclear about one's sense of identity.

### Warm-Up Activity

Have students draw timelines of their lives. They could start at birth, marking what they consider to be highlights in their psychological and physical development. Students might wish to bring in photographs to illustrate their timelines.

### Teaching the Lesson

Help students to understand that all of the ages and age ranges used as benchmarks in this lesson are merely approximations. According to Table 7.1.1, Stage 5 ends at age 19 and Stage 6 begins at age 20. That does not mean there are not some people who are age 22, age 25, or even older who are still in Stage 5.

### Reinforce and Extend

#### GLOBAL CONNECTION

 Erik Erikson's theories help people define stages of social development as seen in the West. However, not all cultures view adolescence this way. For example, in some Asian and African cultures, there is no prolonged period of adolescence. Children are encouraged to mimic adults' work by playing a parental role in younger siblings' lives. Have groups of students research the views of adolescence in different cultures such as the Samoans, the New Guineans, and the Barasona tribe in the Amazon rainforest. Suggest students specifically research the "coming of age" rituals. Have them present their findings to the class.

#### HEALTH JOURNAL

 Have students describe in their journals any memories they might have from each of the Erikson stages through which they have passed.

# ELL/ESL Strategy

**Language Objective:** *Explaining main ideas*

Have English language learners determine the main idea of each paragraph in the lesson. Students should write one sentence for each paragraph that restates the main idea. Then have them do the same thing for each section (What Are the Early Stages of Development?, What Happens from Adolescence to Early Adulthood?, What Are the Later Stages of Development? and What Changes Occur During Puberty?). Finally, have students write one sentence that restates the main idea of the lesson as a whole.

# Learning Styles

**Auditory/Verbal**

After students read through the Erikson stages, ask a student volunteer to read aloud the description of each stage. After each stage is read, have another student volunteer to portray a person in that stage of development. Each volunteer may deliver a monologue or respond to questions from an interviewer—either you or another student. (Students acting out Stages 1 and 2 should assume the "infant" they are portraying has complete speech and vocabulary skills.)

# Health Journal

Ask students to imagine a time in the future when they themselves are parents of teenagers. Have students describe in their journals what they think they might do to help their teenaged children during puberty. Have them address whatever issues they think are most important to adolescents.

---

**Puberty**
*Period at the beginning of adolescence when children begin to develop into adults and are able to have children*

You may be concerned about the changes that happen to you during adolescence. Talking with your parents or other trusted adults can help. Use effective communication skills. Listen without interrupting. If you are embarrassed, be honest. Keep in mind that adults went through puberty at one time. They understand what you are going through.

The sixth stage is early adulthood. This stage usually includes ages 20 to 40. This is often the point in life when people leave their parents' home. They decide how they feel about a close relationship with another person. Again, if they have had problems during earlier stages, this time can be difficult. People who do not feel good about forming close relationships may feel isolated, or alone.

## What Are the Later Stages of Development?

The seventh stage represents middle adulthood. During this stage, adults often turn their attention to their families. They also may help others outside the family. If they do not, they may become too concerned with their personal needs. This is called being self-absorbed. Self-absorbed people may have a difficult time finding satisfaction in life.

The eighth stage is older adulthood. This last stage of life is a time when people look back and think about what they have achieved. If they feel good about what they have done, they experience a sense of completeness. If not, they may feel depressed. Older adults also are at higher risk for diseases and disorders. As a person gets older, the body's immune system does not work as well as it once did. The immune system protects against diseases. Older adults may get the flu more easily, for example. Aging also causes changes in the senses. Parts of the ear and eye weaken or break down. As a result, many older adults have hearing or vision problems.

## What Changes Occur During Puberty?

The beginning of adolescence is called **puberty**. Puberty is the time in life when children begin to develop into adults and are able to have children. Puberty begins at different ages for different people. Girls may begin puberty at ages 8 to 13. Boys may begin puberty at ages 10 to 15. During puberty, the body changes size and shape. These changes are called growth spurts. They occur at different rates and at different times for each person.

Sometimes adolescents worry if they have not reached puberty at the same time as their peers. Others are embarrassed because they feel awkward. During a growth spurt, for example, hands and feet often grow first. Until the arms and legs catch up, a person may feel clumsy. It is normal for adolescents to have different body types and different rates of growth. You can promote your self-esteem and that of others by accepting these differences. You can focus on strengths and abilities rather than differences.

### Health in Your Life

**Good Health Habits in Adolescence**

Each stage of life is important. Yet the fifth stage, or adolescence, has some special challenges. As teens become more independent, they may take more risks. For example, they may decide to use alcohol or other drugs. An adolescent's brain is not fully developed. Alcohol damages a part of the brain that is still growing. It can cause memory and learning problems.

Drinking alcohol is a behavior to avoid. Other behaviors are healthy and help your body develop properly. During adolescence, your body is growing and changing. You need at least 8 hours of sleep each night. Sleep helps you grow and handle stress better. Exercise and good eating habits also help your body develop. You need at least 60 minutes of activity every day. This helps you keep a healthy weight and build muscles and bones.

You can be a spokesperson for good health habits during adolescence. Write a public service announcement (PSA) that encourages your classmates to practice healthy behaviors.

On a sheet of paper, write the answers to the following questions. Use complete sentences.

1. Teens who practice healthy behaviors are more likely to continue them when they are adults. Why do you think this is true?
2. What is a healthy behavior that you are not practicing now? How can you make it a daily habit?
3. As you gain independence, you also become more responsible. What is one way you can show you are responsible for your health?

## PRONUNCIATION GUIDE

Use the pronunciation shown here to help students pronounce difficult words in the lesson. Refer to the pronunciation key that appears in the Glossary at the back of the Teacher's Edition for the sounds of the symbols.

endocrine (en´ dō krən)

pituitary (pə tü´ ə ter ē)

## Decide for Yourself

Ask a volunteer to read aloud the Decide for Yourself feature on page 154. As the volunteer reads each numbered paragraph, discuss it with the class. Then have students complete the exercise.

### Decide for Yourself Answers

**1. – 5.** Answers will vary. Students should choose a specific goal and explain the steps they will take to reach it.

---

**Reproductive organ**
*Organ in the body that allows humans to mature sexually and to have children*

**Genitals**
*Reproductive, or sex, organs*

**Secondary sex characteristic**
*Feature that signals the beginning of adulthood*

The endocrine system includes glands that produce hormones. The pituitary gland produces hormones that signal many changes during puberty. Some hormones control the amount and timing of growth. Some hormones control the development of the **reproductive organs.** Reproductive organs allow humans to grow up, develop sexually, and have children.

Hormones signal changes in both males and females. Oil and sweat glands become more active. Hair begins to grow in other areas of the body and around the **genitals,** or sex organs. Hormones also signal the body to develop **secondary sex characteristics,** or features that signal the beginning of adulthood. Males develop hair on their faces, their voices change, and their shoulders broaden. Females develop breasts, wider hips, and a narrower waist.

### Decide for Yourself

During adolescence, you start thinking about what you want to do with your life. You are making the change from your role as a child to your role as an adult. Here is an example of how one teen set a goal for a career after high school.

1. **Write down your goal.** The teen wants to work with children in a day care setting.

2. **List the steps needed to reach your goal.** Step 1 is to find out the type of degree needed. Step 2 is to identify several schools that offer the degree and the cost of each. Step 3 is to look at ways to pay for school. Step 4 is to apply to study at one or more schools.

3. **Set up a timeline for completing each step.** The teen wants to be ready to start college right after high school graduation.

4. **Identify any problems that might get in your way.** The teen gets a part-time job now to save money for school.

5. **Check your progress as you work toward your goal.** The teen realizes that more time is needed to find out about loans. The teen changes the timeline for Step 3.

On a sheet of paper, write the answers to the following items. Use complete sentences.

1. List a goal you would like to achieve.
2. Go through numbers 2, 3, 4, and 5 above, using your specific goal.

**Testosterone**
Male sex hormone

**Estrogen**
Female sex hormone

**Progesterone**
Female sex hormone

Three hormones signal the body's reproductive system to develop during puberty. These are **testosterone, estrogen,** and **progesterone.** Testosterone is the hormone that signals the development of secondary sex characteristics in males. Estrogen and progesterone control reproductive development in females.

*Lesson 1 Review* On a sheet of paper, write the answers to the following questions. Use complete sentences.

1. What might cause a child to feel inferior?
2. What health problems might an older adult have?
3. What three hormones signal the body's reproductive system to develop?
4. **Critical Thinking** During the fifth stage of development, why is it important for teens to choose peers who share their values?
5. **Critical Thinking** What would you tell a friend who is upset because he is shorter than the other boys in your class?

### Health at Work

#### Medical Technologist

Medical technologists work in hospitals and laboratories. They are like detectives, because they observe and discover things. They collect samples for medical tests. They test for drug levels and chemicals in the blood. They look for germs that cause disease. They check for cells that are not normal that might show health problems. Medical technologists usually need a college degree. They take courses in chemistry, biology, math, and statistics. Technologists must have good judgment and be able to work under pressure. They must pay close attention to detail and be good at problem solving. A medical technician usually has a 2-year degree. Technicians perform simpler tasks and work for a technologist. Most states require both technologists and technicians to be licensed or registered.

### Health at Work

Have students read the Health at Work feature on page 155. Tell students that a medical technologist usually works in the laboratories of hospitals, clinics, or pharmaceutical companies. Have students research other health-related careers such as pathologists, genetic counselors, chemists, pharmacists, and nurses. Encourage students to contact a person in one of these careers and conduct an informal interview. Have students present their findings to the class.

### Lesson 1 Review Answers

**1.** A child might feel inferior if he or she is scolded or receives negative comments. **2.** An older adult is at higher risk for diseases and disorders. Older adults might get the flu more easily because their immune systems are weakened, and they often have problems with hearing and vision. **3.** The three hormones that signal the body's reproductive system to develop are testosterone, estrogen, and progesterone. **4.** Sample answer: During the fifth stage, teens are trying to figure out their identity. Part of this is determining their values. The people they choose to spend time with can influence these decisions. **5.** Sample answer: I would tell him it is normal for people to grow at different rates and that he probably will grow taller in time. I would encourage him to focus on his strengths and not worry about physical differences.

### Portfolio Assessment

Sample items include:
- Timelines from Warm-Up Activity
- Findings from Global Connection application
- Entry from Health Journal
- Sentences from ELL/ESL Strategy
- Health in Your Life answers
- Entry from Health Journal
- Decide for Yourself answers
- Findings from Health at Work
- Lesson 1 Review answers

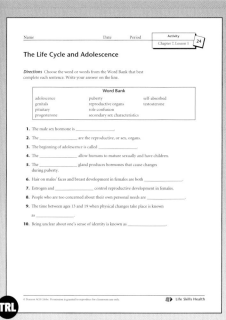

Activity 24     Workbook Activity 24

# Lesson at a Glance

## Chapter 7 Lesson 2

**Overview** This lesson describes changes within the body in males and females during puberty and explains the menstrual cycle and the fertilization process.

## Objectives

- To discuss changes in the reproductive system of males and females during puberty
- To explain the process of menstruation
- To explain how new life forms

**Student Pages** 156–159

**Teacher's Resource Library**

Activity 25
Alternative Activity 25
Workbook Activity 25

## Vocabulary

| | |
|---|---|
| conceive | penis |
| erection | pregnant |
| fallopian tube | reproduction |
| fertilized | sexual intercourse |
| menstrual cycle | sperm |
| menstruation | testes |
| ovaries | uterus |
| ovulation | vagina |

On an overhead projector or on the chalkboard, show simple diagrams of the male and female reproductive systems. Label the systems as volunteers read the vocabulary words.

## Background Information

Menstruation begins around age 12 for most girls. However, it can begin as early as age 8 or as late as age 16. Skipping a month or having a period every three weeks, instead of every 28 days, also can be normal with young girls. Other factors can affect menstruation as well, including overtraining for athletics, eating disorders, stress, anemia, or drastic weight loss. Teens who fall into these categories might stop menstruating or not ever begin. Some girls may get cramps during their period. Some actions that help relieve or prevent cramps are exercise and reducing the amount of caffeine and salt in one's diet.

156    Unit 2    *Personal Health & Family Life*

---

# Lesson 2    Reproduction

### Objectives

*After reading this lesson, you should be able to*

- discuss changes in the reproductive system of males and females during puberty
- explain the process of menstruation
- explain how new life forms

*Reproduction*
The process through which two human beings produce a child

*Testes*
Two glands in males that produce sperm cells

*Sperm*
Male sex cells produced by the testes

The ability to reproduce is one of the most amazing things about the body. **Reproduction** is the process through which two human beings produce a child. The reproductive system is the only body system that differs between males and females.

## What Changes Occur in Males During Puberty?

At puberty, the male hormone testosterone begins to be released into a young man's body. Testosterone is produced in the **testes,** the two main glands in the male reproductive system. Figure 7.2.1 shows the male reproductive system. Another function of the testes is to make **sperm,** or male sex cells. Beginning at puberty, the testes make more than 200 million sperm cells every day. Males are usually able to produce sperm from puberty throughout the rest of their life.

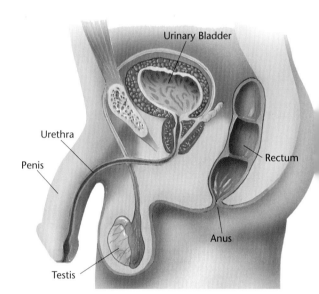

**Figure 7.2.1** *The male reproductive system*

156    Unit 2    *Personal Health and Family Life*

*Ovaries*
Two organs in females that produce egg cells

*Ovulation*
The release of a mature egg cell

*Conceive*
To become pregnant

*Fallopian tube*
Passage through which mature egg cells pass from the ovaries to the uterus

*Uterus*
Female organ that holds a growing baby

*Vagina*
Canal connecting the uterus to the outside of a female's body

## What Changes Occur in Females During Puberty?

When girls reach puberty, hormones begin to act on the **ovaries**. Ovaries are the two organs in a woman's body that produce egg cells. Females are born with more than one million egg cells already present in the body. Hormones cause the egg cells to mature. As they mature, the ovaries release egg cells. This is called **ovulation**. Once a woman begins to ovulate, she is able to **conceive**, or to become pregnant. Each ovary usually releases one egg every other month. Figure 7.2.2 shows the female reproductive system and part of the excretory system.

The released egg then travels into one of the two **fallopian tubes**. The egg cell moves in these tubes to the **uterus**. The uterus is an organ where an egg will grow into a baby if it is joined by a sperm. Otherwise, the egg cell leaves the body through the **vagina**, a canal that leads to the outside of a female's body.

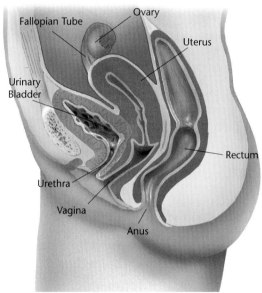

**Figure 7.2.2** *The female reproductive system*

The Life Cycle and Human Development   Chapter 7

## 1 Warm-Up Activity

Have students read about the changes in males during puberty on page 156. Stress the physiological aspects of the reproductive systems in males and females rather than the sexual aspects. Tell students that all living things have a means of reproducing to create other similar living things.

## 2 Teaching the Lesson

Make sure students understand that some of the organs shown in Figures 7.2.1 and 7.2.2 have nothing to do with reproduction. The urinary bladder, urethra, rectum, and anus are parts of the excretory system, not the reproductive system. They are shown in the diagrams simply because they are near the reproductive organs, and it might be confusing to leave them out. If necessary, refer back to Chapter 5. Ask students whether the penis is part of the reproductive system or the excretory system. *(It is part of both systems.)*

## 3 Reinforce and Extend

### ELL/ESL Strategy

**Language Objective:** *Learning vocabulary specific to the content area*

Students learning English may have particular difficulty pronouncing the vocabulary words. Suggest that they keep a notebook and record the names of human body parts in alphabetical order. Tell them to include a brief description of each body part and a phonetic pronunciation for each word. If they have questions about the correct pronunciation of a word, help them look up the word in a dictionary and show them how to use the pronunciation guide at the front of the dictionary.

## TEACHER ALERT

Students have read that a female is born with more than one million egg cells present in her body. Some might think that the end of a woman's fertile stage of life—*menopause*, though they may not know the term—occurs when she has used up all her eggs. In fact, a woman's body will release only a small percentage of her available eggs during her lifetime. As is stated in the last paragraph on page 158, menopause occurs when the body stops producing the required levels of estrogen and progesterone.

## AT HOME

Have female students create a flyer that stresses the importance of practicing good hygiene during menstruation. Students should include the following information:

- Feminine hygiene products are available for a teen's smaller size.
- Studies have shown that high-absorbency tampons increase the risk of toxic shock syndrome (TSS), a rare illness that can cause death.
- Females should change their tampon or pad every four to six hours.
- Females should wear pads or panty shields at night, instead of tampons, to reduce the risk of TSS.
- Females should wash their hands before inserting tampons.

---

**Menstrual cycle**
*The monthly cycle of changes in a female's reproductive system*

**Menstruation**
*The process in females during which blood tissue leaves the lining of the uterus*

The process of releasing an egg is one step in a monthly cycle in women. This process is called the **menstrual cycle.** When the ovary releases an egg cell, the lining of the uterus begins to thicken with blood tissue. If the egg is to become a baby, it attaches to the wall of the uterus. It stays there for 9 months to grow. The blood tissue is the source of nutrients, or food, for a developing baby. Figure 7.2.3 shows the menstrual cycle. FSH and LH are hormones that play important roles in the menstrual cycle.

If the egg does not attach to the wall of the uterus, it leaves the body through the vagina. The blood tissue is not needed. It leaves the body during **menstruation,** a process that takes 4 to 6 days.

The entire menstrual cycle is usually about 28 days long. When one cycle ends, the next cycle begins. The menstrual cycle repeats itself until a woman is between 45 and 55 years old. Then the body produces less and less estrogen and progesterone, and the menstrual cycle stops.

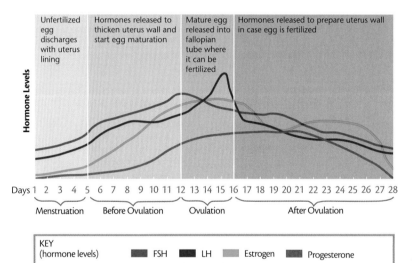

**Figure 7.2.3** The menstrual cycle

**Sexual intercourse**
Insertion of the penis into the vagina

**Penis**
Male reproductive organ

**Erection**
Condition in which the penis becomes hard and larger

**Fertilize**
An egg cell and a sperm join together

**Pregnant**
Carrying a developing baby in the female body

### Health Myth

**Myth:** All girls start menstruating around age 12.

**Fact:** The average age is around 12. However, every girl is different. It is normal for a girl to start as young as age 8 or as old as age 16.

## How Does New Life Form?

Once males and females reach puberty, they are able to reproduce. **Sexual intercourse** is the insertion of the **penis** into the vagina. This allows the male sperm and a female egg cell to join together to form a new life.

During intercourse, sperm pass from the testes through tubes that lead to the penis. The penis is the male reproductive organ. It has many small blood vessels. When these vessels fill with blood, the male has an **erection.** That means the penis becomes hard and larger. During this time, sperm can travel through the penis to enter the female body through the vagina.

During intercourse, millions of sperm are released into the female. If the female is in the right stage of the menstrual cycle, an egg may become **fertilized,** or joined by the sperm. If not, the sperm cells die and leave the female's body. Only one sperm cell is needed to fertilize the female's egg cell. If the sperm and egg unite, the woman becomes **pregnant,** or carries a developing baby.

**Lesson 2 Review** On a sheet of paper, write the word or phrase in parentheses that best completes each sentence.

1. The male hormone is _____ (estrogen, progesterone, testosterone).
2. Males produce _____ (testes, sperm, egg cells) from puberty throughout adult life.
3. The organs that produce eggs are the _____ (ovaries, testes, fallopian tubes).

On a sheet of paper, write the answers to the following questions. Use complete sentences.

4. **Critical Thinking** Explain how egg cells and sperm are different in the numbers that are produced and when they are produced.
5. **Critical Thinking** Suppose a classmate says that menstruation is about 28 days long. What is wrong with this statement?

*The Life Cycle and Human Development* Chapter 7

### TEACHER ALERT

It is possible that a student will ask about the white fluid that is released by the male during intercourse. This fluid is called semen, and it contains sperm (which are too small to be seen with the unaided eye). Semen serves three functions: It provides sperm with a medium for delivery; it protects sperm from the acidic environment of the uterus, and it provides sperm with nutrition.

### Health Myth

Have students read the Health Myth feature on page 159. Tell students that the average age at which girls start menstruating has been dropping for more than 165 years. In 1840, the average age was 15.3 years in France, 16.5 years in England, and 17 years in Norway. Before that, it may have been as high as age 18. In western Europe from 1840 through 1950, the average age dropped four months each decade. In Japan the decline happened later and much more quickly. From 1945 to 1975 in Japan, the average age dropped by nearly a year each decade. Possible reasons are environmental factors and obesity.

### Lesson 2 Review Answers

**1.** testosterone **2.** sperm **3.** ovaries **4.** Females have more than one million egg cells when they are born. Males begin producing more than 200 million sperm cells every day at puberty. **5.** Menstruation is the process in which blood tissue leaves the body. This takes about 4 to 6 days. The menstrual cycle takes about 28 days.

### Portfolio Assessment

Sample items include:
- Steps from Research and Write feature
- Vocabulary notebook from ELL/ESL Strategy
- Lesson 2 Review answers

# Lesson at a Glance

## Chapter 7 Lesson 3

**Overview** This lesson describes the development of a fertilized egg into an embryo and then a fetus, the benefits of good prenatal care, the process of childbirth, and the benefits of good perinatal care.

## Objectives

- To explain how a woman knows she is pregnant
- To describe how an embryo and a fetus develop
- To recognize the importance of prenatal and perinatal care
- To analyze harmful effects of certain substances on the fetus

**Student Pages** 160–166

**Teacher's Resource Library**

  Activity 26
  Alternative Activity 26
  Workbook Activity 26

## Vocabulary

| | |
|---|---|
| cervix | perinatal care |
| cesarean childbirth | placenta |
| embryo | postpartum |
| fetal alcohol syndrome | prenatal care |
| fetus | trimester |
| gestation | ultrasound |
| implantation | umbilical cord |

Have students write the vocabulary words in original sentences. Then ask them to read their sentences in groups. Encourage group members to evaluate each other's use of the vocabulary words and make necessary corrections.

## 1 Warm-Up Activity

Search on the Internet for images of fetal ultrasounds, such as the one on page 160. Show the ultrasounds to the class. Have volunteers point out body parts on the ultrasounds and compare the fetus to the drawings on page 164.

---

## Lesson 3 Pregnancy and Childbirth

### Objectives

*After reading this lesson, you should be able to*

- explain how a woman knows she is pregnant
- describe how an embryo and a fetus develop
- recognize the importance of prenatal and perinatal care
- analyze harmful effects of certain substances on the fetus

### Gestation
*The period of development in the uterus from the time the egg is fertilized until birth; pregnancy*

### Ultrasound
*The use of sound waves to show pictures of what is inside the body*

As you learned in Lesson 2, when an egg cell and sperm join, the egg becomes fertilized. This begins the process called pregnancy, or **gestation**. Gestation lasts about 9 months in humans. This is the period of time needed to allow the fertilized egg to grow and develop.

### How Does a Woman Know She Is Pregnant?

There are several signs of pregnancy. A woman may feel sick or very tired. Her breasts may feel sore. She has probably missed a menstrual cycle, because menstruation stops during pregnancy.

If a woman believes she might be pregnant, she should see a doctor. A blood test at a doctor's office can check for pregnancy as early as 6 to 8 days after ovulation. Some urine tests can tell in as little as 6 days after the egg is fertilized. Both tests look for a certain hormone that is released only during pregnancy. A doctor can determine whether a woman is pregnant in three ways: (1) The doctor can hear the fetus's heartbeat. (2) The fetus shows signs of movement. (3) An **ultrasound** picture shows a developing baby. Ultrasound is the use of sound waves to show what is inside the body. See Figure 7.3.1.

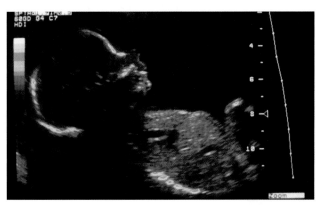

**Figure 7.3.1** This ultrasound picture shows a baby developing in the uterus.

**Prenatal care**
*Health care during a pregnancy*

**Fetal alcohol syndrome**
*A condition caused by drinking alcohol during pregnancy*

Women can test at home for pregnancy by using kits they buy at a drugstore. Most home tests give the right information if done correctly. If a result tells the woman she is not pregnant, she should repeat the test in a few days. If a woman thinks she is pregnant but the test keeps telling her she is not, she should see a doctor.

## What Is Prenatal Care?

**Prenatal care** is health care during a pregnancy. A pregnant woman needs to take special care of her body. She needs to see a doctor many times during the pregnancy. The doctor can tell her about health behaviors that will help her take care of herself and her unborn baby. Here are some ways a woman can take good care of herself and her unborn baby:

- Take a multivitamin that has folic acid. This vitamin helps prevent certain birth defects. A pregnant woman should not take any other drugs or medicines unless a doctor tells her to.
- Eat healthy foods, such as fruits, vegetables, and whole grains.
- Exercise regularly, but only the type and length of time the doctor recommends.
- Get regular checkups. Women who do not have health insurance usually can go to a public health clinic in their community.

Smoking, drinking alcohol, and using other drugs can harm the developing baby. Smoking tobacco can cause a baby to be born too early. The baby might weigh too little at birth. The baby might develop a disease of the lungs called asthma. A baby whose mother drinks alcohol might develop **fetal alcohol syndrome** (FAS). FAS causes unusual facial features, damage to the nervous system, and growth problems. Children who have FAS may have trouble learning and may have other birth defects. Other drugs also can harm the developing baby.

### Link to ➤➤➤

**Environmental Science**

Lead is a mineral that used to be common in paint and gasoline. It also is found in some drinking water and soil. Lead that enters a pregnant woman's body is harmful to the developing baby. It can cause mental slowness, hearing problems, and kidney damage. Pregnant women should avoid breathing lead dust and drinking water from pipes that contain lead.

## Background Information

There are three layers of cells in the developing embryo. The endoderm (innermost layer) will develop into the linings of the digestive and respiratory tracts and into glands, including the liver and pancreas. The mesoderm (middle layer) will develop into the muscles, bones, circulatory and excretory systems, and the dermis. The ectoderm (outermost layer) will develop into the brain, nerves, epidermis, hair, and nails The average gestation period is 40 weeks, or 280 days.

### 2 Teaching the Lesson

As you read page 161, help students see how important the mother's care of her own body during pregnancy is to the health of the unborn baby. As you read page 163, point out that the mother's blood flows through the embryo (and later the fetus). This means that any toxins in her blood also flow through the fetus.

### PRONUNCIATION GUIDE

Use the pronunciation shown here to help students pronounce difficult words in the lesson. Refer to the pronunciation key that appears in the Glossary at the back of the Teacher's Edition for the sounds of the symbols.

asthma (az´ mə)

folic (fō´ lik)

### Link to

Environmental Science. Have a volunteer read aloud the Link to Environmental Science feature on page 161. Tell students that lead in the environment is a health hazard for children and adults as well as unborn children. Students who want to learn more can look for information on the Web sites of the Environmental Protection Agency (www.epa.gov/lead/) and the National Institutes of Health (www.niehs.nih.gov/kids/lead.htm).

### 3 Reinforce and Extend

### ELL/ESL Strategy

**Language Objective:** *To explain how prefixes aid comprehension*

Write the words *prenatal* and *perinatal* on the chalkboard. Draw slashes to separate the prefixes from the root word (pre/natal, peri/natal). Tell English language learners that *pre-* means "before," and *peri-* means "around" or "about." Explain that *natal* comes from the Latin word *natalis*, and it means "having to do with birth." Then ask them to deduce the definitions of the two words on the chalkboard. *(Prenatal means "before birth" and perinatal means "around birth."*) Challenge students learning English to use a dictionary to find three more words that begin with each prefix. Have them write the words and their definitions in a notebook. *(For instance, students might choose* precaution, *which means "care taken beforehand";* preexisting, *which means "existing before";* perimeter, *which means "the outer boundary or the total length of the outer boundary of a figure or area";* pericardium, *"the membranous sac around the heart and the roots of the great blood vessels"; or* periodontics, *"the branch of dentistry concerned with diseases of the bone and tissue supporting the teeth."*)

---

*Implantation*
*Process during which the fertilized egg plants itself in the lining of the uterus*

For example, cocaine can cause early birth. Children whose mothers smoked marijuana during pregnancy may have trouble concentrating.

Substances in the environment also can harm an unborn baby. Pregnant women should avoid chemicals such as paint thinners and products that kill insects. They should read the labels on household products. Most dangerous products have pregnancy warnings.

### What Happens as the Fertilized Egg Grows?

The tiny fertilized egg takes about 4 days to travel from the fallopian tube to the uterus. While it travels, its cells divide many times. It becomes a ball of cells that plants itself in the lining of the uterus. This process, called **implantation,** happens about 6 to 8 days after fertilization. Figure 7.3.2 shows the steps from ovulation to implantation.

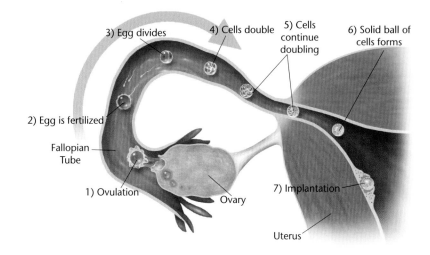

**Figure 7.3.2** *Journey of a fertilized egg from ovulation through implantation*

**Embryo**
*Fertilized egg after implantation*

**Placenta**
*An organ lining the uterus that surrounds the embryo or fetus*

**Umbilical cord**
*Structure that joins the embryo or fetus with the placenta*

**Trimester**
*A period of 3 months*

**Fetus**
*Unborn baby from 8 weeks after fertilization until birth*

### Link to ▶▶▶
**Social Studies**
Through history, most women have had their babies at home without a doctor. Today, 99 percent of babies in the United States are born in a hospital. There, mother and child are more likely to remain healthy. Ultrasound and other tests can check for problems. Doctors and nurses are there to help.

The uterus is normally small enough to fit into the palm of your hand. As the tiny cells grow, the uterus expands to hold the fully developed baby. At this time, the fertilized egg becomes an **embryo.** An embryo is about the size of the dot on the letter *i*.

A special organ called the **placenta** develops from the embryo along the lining of the uterus. The placenta surrounds the embryo. The **umbilical cord** joins the embryo with the placenta. The cord carries nutrients to the embryo and removes its wastes. The cord also carries anything else the mother eats or drinks.

A special fluid protects the embryo in the uterus. The embryo floats in this liquid, which protects the embryo against temperature changes and injury.

## How Does the Embryo Develop?

Figure 7.3.3 on the next page shows the major events that usually occur during each month of pregnancy. The 9 months of pregnancy are divided into three **trimesters.** Each trimester is 3 months long. After 8 weeks, or about 2 months, all the vital organs have started to develop. By this time, the embryo is called a **fetus.** The fetus continues to develop during the rest of the gestation period. At the end of this time, the fetus is considered full term and is ready to be born.

### Technology and Society
An ultrasound picture helps the health of the baby in many ways. The doctor can tell whether the baby has health problems. Sometimes these can be corrected before the baby is born. Ultrasound also can tell how far along the pregnancy is. The doctor can try to prevent labor, or childbirth, if the baby is not yet fully developed.

### TEACHER ALERT
Help students understand the sequence and the meaning of the terms presented here. An *egg* is a single cell. During fertilization, it joins with a sperm. The *fertilized egg* is still a single cell. Once it begins to divide, it is called an *embryo*. After the end of the eighth week of development, the embryo has developed enough to be called a *fetus*.

### Technology and Society
Have students read the Technology and Society feature on page 163. Divide students into small groups and ask them to research other uses for ultrasounds (besides determining whether a woman is pregnant). Have them present their findings to the class. *(Ultrasound also can be used to confirm that a pregnancy is within the uterus, rather than in a Fallopian tube—a dangerous condition known as an ectopic pregnancy; assess the size and growth of an embryo or fetus; determine what week the pregnancy is in; determine the gender of a fetus; determine whether there is one or more fetuses in the uterus; and determine whether the developing fetus has any problems. Some problems can be corrected while the fetus is still in the uterus. Also, knowing about birth defects beforehand helps parents plan appropriate care for a baby after birth.)*

### Link to
Social Studies. Have a volunteer read aloud the Link to Social Studies feature on page 163. Tell students that women who had their babies at home without a doctor were often assisted by a midwife. Today, midwives attend childbirths at many hospitals. Have students research the history of midwives. Questions on which to focus include: What do midwives do? Where do midwives practice? *(Midwives help women give birth, in place of doctors. Midwives also provide care and advice during pregnancy. They work in hospital, clinics, and in their clients' own homes.)*

## LEARNING STYLES

### Interpersonal/Group Learning

Divide the class into groups. Have each group prepare a presentation describing fetal development in each trimester. Students may include illustrations in their presentations, based on illustrations in the text and in other sources, if they like. Illustrations may be drawn, photographed, or scanned and printed. Students also may create multi-media presentations if they have access to the appropriate technology.

## LEARNING STYLES

### Logical/Mathematical

Have students synthesize all the material they have encountered in this chapter so far, including fertilization on page 159, implantation on page 162; fetal development on pages 163–164; and the eight stages of life in Lesson 1, including the changes that occur during puberty, on pages 152–158. Have students make a timeline for the life of an individual, sequencing the events of a life in order.

## IN THE COMMUNITY

Most hospitals have a Neonatal Intensive Care Unit (NICU) that specializes in the care and treatment of premature babies. Help students develop questions about premature babies, such as the percentage of babies born prematurely in the United States; causes and effects of prematurity; special care required for premature babies; long-term health problems of premature babies; and so on. Ask a student volunteer to contact an NICU in the area to get students' questions answered.

9 weeks

11 weeks

15 weeks

36 weeks

| Changes During Pregnancy |||
|---|---|---|
| Trimester/Month | Embryo/Fetus | Mother |
| **First Trimester** First Month | Embryo ¼ inch long, heart is obvious; eyes, nose, and brain appear; arms and legs are small bumps | Menstruation stops; breasts are tender; fatigue; frequent urination; positive pregnancy test |
| Second Month | Embryo 1 inch long; head quite large; fingers, toes appear; nervous system and brain coordinate body functions; mouth opens and closes | Possible nausea and vomiting lasting into third or fourth month |
| Third Month | Embryo becomes a fetus; 4 inches long; weighs about 1 ounce | Breasts possibly swollen; thickened waist; uterus well rounded |
| **Second Trimester** Fourth Month | Fetus 6 inches long; weighs 3½ ounces; sex may be determined; body covered with soft hair | Breasts may discharge liquid; uterus wall more stretched; larger midsection |
| Fifth Month | Fetus 9½ inches long; weighs 10 ounces; skin less transparent | Feels stronger fetal movements; uterus higher; breasts not much larger; midpoint of pregnancy |
| Sixth Month | Fetus 12 inches long; weighs 1½ pounds; eyebrows, eyelashes showing; sucks thumb; hiccups; if born at this time, usually dies | Uterus increases in size; fetal movements may feel sharp |
| **Third Trimester** Seventh Month | Fetus 14 inches long; weighs 2½ pounds; covered with a greasy substance to protect it; responsive to sound and taste; if born at this time, moves actively, cries weakly; if born, can live with expert care | Still feels fetal movements; size of uterus increases; weight gain continues |
| Eighth Month | Fetus 15½ inches long; weighs 3½ pounds; hair on head; skin red and wrinkled; if born, can survive with proper care | Gains some additional weight; uterus extends; fetal movements continue |
| Ninth Month | Fetus 18 inches long; weighs 5½ pounds; gestation period ends; body fat makes figure more round, less wrinkled in face | Top of uterus nears breastbone; possible frequent urination because of pressure on bladder; possible difficulty in walking |

**Figure 7.3.3** *Events in pregnancy. The drawings from top to bottom show the development of the fetus at 9, 11, 15, and 36 weeks.*

**Cervix**
*Narrow outer end of the uterus*

**Cesarean childbirth**
*The delivery of a baby through a cut in the abdomen and uterus*

## What Happens During Childbirth?

Childbirth has three stages. Figure 7.3.4 shows the second stage. The process of birth begins with labor pains that happen when the uterus tightens, or contracts. These contractions begin to pull open the **cervix.** The cervix is the narrow outer end of the uterus. The cervix has been closed to hold the growing baby inside the uterus. The cervix must open, or dilate, to a diameter of about 4 inches.

Dilation of the cervix is the first stage of birth. It lasts an average of 7 to 15 hours. During this time, the woman feels labor pains that happen more often as the contractions continue. The protective placenta around the fetus breaks, and fluid flows out of the woman's body.

When the cervix is dilated, the second stage begins. During this stage, the baby is pushed out, usually head first, from the uterus. The baby is then able to breathe on its own. The second stage of birth usually takes from 15 to 60 minutes.

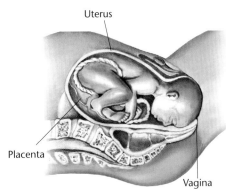

**Figure 7.3.4** *A baby leaves the uterus through the vagina.*

The third stage is called the afterbirth. At this time, the uterus contracts to release the placenta, which leaves the woman's body. This stage usually lasts just a few minutes.

There are different types of childbirth. The process just described is vaginal childbirth. **Cesarean childbirth** is the delivery of a baby through a cut in the abdomen and uterus. A cesarean may be done for many reasons. For example, labor may be lasting too long. The baby's head may be too large to pass through the birth canal. Cesarean births make up about one-fourth of all childbirths in the United States.

### TEACHER ALERT

If students ask, explain that some people think that cesarean childbirth (or cesarean section) derives its name from the legend that Julius Caesar was born that way. But since Caesar's mother reportedly lived many years after his birth, and most women who underwent the procedure before the 1800s died during childbirth or shortly afterwards, Caesar most likely was not delivered via cesarean section.

### CAREER CONNECTION

Doctors and nurses specially trained in obstetrics care for a woman's medical needs prenatally and during labor. However, a woman can enlist the help of a labor coach called a doula while in labor. Most doulas take certification classes in labor and delivery and then teach classes to expectant parents on how to breathe and push during labor and how to care for the newborn after birth. Have students research this career and report back to the class. Reports also might focus on the Lamaze method of childbirth.

### IN THE ENVIRONMENT

A unique form of recycling begins in the delivery room when, after the delivery, the discarded placenta is often sent to a cosmetic manufacturer. Some cosmetics, such as face creams, moisturizers, and cream rinses, are made by the extraction of hormones from the discarded placentas. Invite students to research this practice by checking labels of cosmetics and writing to the companies that produce these products. Have students then report back to the class on the information they received.

### Research and Write

Have students read and conduct the Research and Write feature on page 166. The Central Intelligence Agency (CIA) has compiled a chart showing estimated 2005 IMRs for 226 different countries: www.cia.gov/cia/publications/factbook/rankorder/2091rank.html.

Students also might be interested in seeing a table compiled by the Centers for Disease Control that shows United States infant mortality rates, fetal mortality rates, and perinatal mortality rates, according to race, for selected years from 1950 through 1999: www.cdc.gov/nchs/linked.htm#tabulated.

## Lesson 3 Review Answers

**1.** Gestation lasts about nine months in humans. **2.** The fertilized egg travels from the fallopian tube to the uterus. While it travels, its cells divide many times. **3.** An embryo is called a fetus after 8 weeks, or about 2 months, after fertilization. **4.** If the woman is pregnant, she needs to get checkups and prenatal care to help her baby be healthy. **5.** The umbilical cord joins the embryo with the placenta. The umbilical cord carries nutrients and anything else the mother eats or drinks to the embryo.

## Portfolio Assessment

Sample items include:
- Sentences from Vocabulary
- Words and definitions from ELL/ESL Strategy
- Findings from Technology and Society
- Presentations from Learning Styles: Interpersonal/Group Learning
- Timeline from Learning Styles: Logical/Mathematical
- Reports from Application: Career Connection
- Letters from Application: In the Environment
- Reports from Research and Write
- Lesson 3 Review answers

---

**Perinatal care** Care of the baby around the time of birth and through the first month

**Postpartum** Following birth

### Research and Write

The infant mortality rate (IMR) is the rate at which babies die before they are 1 year old. Use the Internet or the library to find the IMR for the United States and several other countries. Write how the IMR relates to prenatal and perinatal care. Make a table to show your findings.

Why do you think it is important for a baby to have regular doctor visits after it is born?

## What Is Perinatal Care?

**Perinatal care** is care of the baby around the time of birth and through the first month. Here are ways to provide good perinatal care:

- Keep all appointments with the baby's doctor.
- Place the baby on his or her back to sleep on a firm mattress. This helps to prevent crib death, or sudden infant death syndrome (SIDS).
- Consider breast-feeding the baby for at least the first few months. Breast milk is the perfect food for babies. It is easier to digest than cow's milk or baby formula. It helps protect babies from diseases.
- Do not allow anyone to smoke around the baby.

## How Do the Woman's Body and Mind Adjust After Childbirth?

The new life has begun, and it is usually a happy, exciting time for the new parents. The woman's body must adjust after birth. It may take several months to return to normal. Women may have emotional ups and downs as a result of changing hormone levels. Sometimes, women have **postpartum** depression, which can happen any time within the first year after giving birth. It can last from a few weeks to several months. If it lasts longer or is severe, a woman may need to ask her doctor for help.

***Lesson 3 Review*** On a sheet of paper, write the answers to the following questions. Use complete sentences.

1. How long is gestation in humans?
2. What happens to the fertilized egg during the first 4 days?
3. When is an embryo called a fetus?
4. **Critical Thinking** Suppose a woman thinks she is pregnant, but the home test results are negative. Why is it important for the woman to find out for sure?
5. **Critical Thinking** Suppose a woman drinks alcohol while pregnant. How does the alcohol affect the baby?

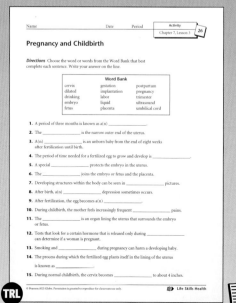

Activity 26

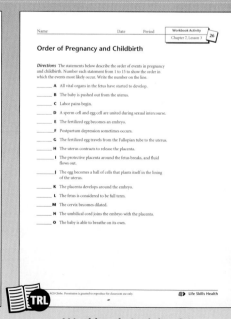

Workbook Activity 26

# Lesson 4: Heredity and Genetics

### Objectives
After reading this lesson, you should be able to
- explain how genes are passed from parents to children
- describe how traits are determined
- identify some genetic disorders

**Genetics**
The science that deals with heredity and inherited characteristics

**Chromosome**
Tiny structure in cells that contains hereditary information

**Gene**
A tiny structure in chromosomes that controls the transfer of traits from parents to children

**Trait**
Quality

**Gender**
The condition of being male or female; sex

Heredity is important because it explains why many children look much like their parents. Heredity also influences our health.

## What Is Genetics?

**Genetics** is the science that studies heredity. Heredity is possible because of tiny structures called **chromosomes.** Every cell in the human body has 46 chromosomes, except sperm and egg cells. These cells have only 23 chromosomes each. When a sperm and an egg unite, the fertilized egg cell receives 23 chromosomes from each parent. The fertilized cell then contains 46 chromosomes. As the cell divides, each chromosome copies itself for a total of 92 chromosomes. When the cell divides into two cells, each cell receives 46 chromosomes. This process in the cells continues throughout life.

Each chromosome is made up of thousands of **genes.** Genes contain chemical codes that determine a person's **traits,** or characteristics. Traits of each parent are passed to the children. Genes determine your hair, eye, and skin color, and even the tone of your voice. Genes can also be the cause of diseases passed from parents to children.

## What Determines Gender?

Two of the 46 chromosomes in a fertilized egg cell determine the **gender,** or sex, of the baby. Females have two copies of an X chromosome. Males have one X and one Y chromosome. Each egg cell has an X chromosome. Each sperm cell has either an X or a Y chromosome. When the two cells join, the chromosome in the sperm determines the sex of the baby. X + X results in a girl. X + Y results in a boy. Look at the following example.

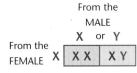

**Figure 7.4.1** Example of chromosome combinations

*The Life Cycle and Human Development* Chapter 7

---

# Lesson at a Glance

## Chapter 7 Lesson 4

**Overview** This lesson discusses heredity and genetics and why they are important to a person's health.

### Objectives
- To explain how genes are passed from parents to children
- To describe how traits are determined
- To identify some genetic disorders

**Student Pages** 167–170

**Teacher's Resource Library**
Activity 27
Alternative Activity 27
Workbook Activity 27

## Vocabulary

| | |
|---|---|
| abnormal | genetic disorder |
| chromosome | genetics |
| dominant | recessive |
| gender | sex-linked |
| gene | trait |
| genetic counselor | |

Divide the class into four groups. Have students study the vocabulary words and the definitions. Write the vocabulary words on the board. Read the words at random. Have each group designate a member to write the definitions the group decides upon on a sheet of paper. Then have the groups read their answers aloud. Give one point for each correct answer. The group with the highest score wins.

## Background Information

Biologist and monk Gregor Mendel (1822–1884) is often considered the father of genetics. Experimenting with pea plants, he found a pattern in the way certain characteristics were passed down from one generation to the next. Another botanist, Carl Correns (1864–1933), also experimented with plants. He discovered that some genes are neither dominant nor recessive. They blend with each other instead. This property is called incomplete dominance. For example, the genes of a red flower blend with the genes of a white flower, forming a pink flower.

*Life Cycle & Human Devel.* Chapter 7

**Warm-Up Activity**

Ask students to bring in family pictures. If students are adopted or are not living with their biological parents, they could bring in magazine photographs of celebrity parents and their children instead. Have students point out ways the children are similar to the parents. How are they different from the parents? What traits of each parent does each child share? Why do the children resemble the parents? Continue discussing to find out what students know about heredity.

 **Teaching the Lesson**

Revisit the photographs of celebrities and their children that you used in the Chapter Opener Activity. Remind students of the similar features they found, and discuss how these features might have been passed from parent to offspring. Remind students that an embryo develops from a single cell. That cell is formed when an egg cell from the female is fertilized by a sperm from the male.

## LEARNING STYLES

 **Visual/Spatial**

Draw this on the chalkboard:

|   | B | b |
|---|---|---|
| b |   |   |
| b |   |   |

Tell students that this is a Punnett square. It describes what results can be expected when dominant and recessive genes combine.

Explain that a dominant gene is depicted by a capital letter and a recessive gene is depicted by a lowercase letter. In this case, "b" represents the gene for blond hair, which is recessive. "B" represents the gene for brown hair, which is dominant. The letters at the top of the square represent one parent. That parent carries the dominant gene for brown hair, and so has brown hair. The letters on the side of the square represent the other parent. That parent carries two genes for blond hair, and so has blond hair.

**Dominant**
Having the most control

**Recessive**
Hidden

**Sex-linked**
Carried by a sex chromosome

**Genetic disorder**
A disease that is caused by abnormal genes or chromosomes

**Genetic counselor**
Health care professional who helps people determine the chance of passing inherited disorders to their children

### How Are Traits Determined?

The chromosomes from each parent determine the traits of a child. All the chromosomes from the mother line up with the chromosomes from the father. The genes for each trait are next to each other in pairs. One or more gene pairs control a trait. In most pairs of genes, one is **dominant,** or has the most control. That gene determines which trait a child will inherit. The other gene is called **recessive,** or hidden. That gene does not determine the trait. For example, suppose the genes a girl received from her father were for dark hair. The genes from the mother were for blonde hair. The genes for dark hair are dominant, so the daughter has dark hair. To have blond hair, she would have to receive a gene for blond hair from both parents.

The Y chromosome is shorter than the X chromosome. The Y chromosome does not have as many genes. Sometimes this can result in health problems for the male. With fewer genes, the male may not have a dominant gene to link with the X chromosome. A male could receive a gene for an inherited disease that could be a recessive gene. Without the dominant gene to link to it, that trait then appears in the male. The gene and the trait it produces are called **sex-linked,** because the sex chromosome carries the gene.

One example of a sex-linked trait is red-green color blindness. A person with this condition cannot tell the difference between these two colors. Color blindness rarely affects females.

 Which of your traits do you think were passed from your father? from your mother?

### What Are Genetic Disorders?

**Genetic disorders** are diseases caused by unusual genes or chromosomes. These genes or chromosomes are abnormal. They are passed from parents to their children. Table 7.4.1 lists some genetic disorders and their causes. Parents with a family history of a genetic disorder may ask for help from a **genetic counselor.** This health care professional can perform tests and advise parents about the chances of passing on a genetic disorder. If the risk is great, the couple may choose not to have children. They may decide to adopt children instead.

---

Show students how to transfer the two letters at the top into each of the boxes below them. Then transfer the two letters on the side into each of the boxes to the right:

|   | B  | b  |
|---|----|----|
| b | Bb | bb |
| b | Bb | bb |

This tells you that if these parents were to have a child, the odds are 2 in 4, or 50 percent, that the child would have brown hair (the left side of the square). The odds are also 2 in 4, or 50 percent, that the child would have blond hair (the right side of the square).

| Table 7.4.1 Genetic Disorders and Their Causes | | |
|---|---|---|
| Disorder | Description | Cause |
| Hemophilia | Blood does not clot, or clump, normally | Recessive abnormal gene |
| Dwarfism | Long bones do not develop properly | Dominant abnormal gene |
| Cystic fibrosis | Abnormally thick mucus, constant respiratory infections | Two recessive genes |
| Sickle cell disease | Abnormally shaped red blood cells, weakness, irregular heart action | Two recessive genes |
| Down syndrome | Mental slowness, slanting eyes, broad skull, broad hands, short fingers | Extra chromosome |
| Spina bifida | Open spine | Many causes |
| Cleft palate | Roof of the mouth does not grow together during development | Many causes |

### How Is Genetics Important to Health?

Genes cause many health traits that tend to run in families. That is why doctors ask about family health histories. For example, genes influence your risk of getting **cancer**. Cancer is a harmful growth in the body that destroys healthy cells. Suppose you know that several family members had or have cancer. This does not mean that you will necessarily develop cancer. You can practice healthy behaviors that decrease your chances of developing cancer. You can eat healthy foods, keep a normal weight, exercise, and avoid smoking.

**Cancer**
*A harmful growth that destroys healthy cells*

> **Health Myth**
>
> **Myth:** Genetics is important only to people planning to have children.
>
> **Fact:** Genes influence many health conditions. Your family's health background can help your doctor predict your risk of certain diseases, such as cancer. Your doctor may do more tests to check for those diseases. He or she may advise you on health behaviors or give you medicine to decrease your risk of getting the disease.

---

## 3 Reinforce and Extend

### ELL/ESL Strategy

**Language Objective:**
*Reading for meaning*

As you work through the lesson, have students learning English take thorough notes. Tell English language learners that as they finish each section ("What Is Genetics?," "What Determines Gender?," etc.), they should look at its heading and see if they can answer the question in their own words. Help any students who cannot answer one or more questions by discussing the specific section or sections with them. Help students to understand the section--and the question in the heading--and to find the vocabulary to answer the question as needed.

### Health Myth

Have students read the Health Myth feature on page 169. Tell students that any doctor seeing a patient for the first time will take a long medical history of that patient. The history includes any illnesses or medical procedures the person might have had. It also includes any illnesses or medical procedures family members—parents, siblings, aunts and uncles, grandparents—might have had. The doctor includes family members in a patient's medical history so that he or she can see what illness or conditions you might be genetically predisposed to inherit.

## Research and Write

Have students read and conduct the Research and Write feature on page 170. A good source for finding the necessary information is the Internet. Show students how to use a search engine and keywords to find helpful Web sites. To research self-exams for breast cancer, you also might suggest that students visit the Web site of the Susan G. Komen Breast Cancer Foundation (www.komen.org/bse/). To research self-exams for testicular cancer, you might suggest that students search the site of the American Cancer Society (www.cancer.org/).

## Lesson 4 Review Answers

**1.** C **2.** B **3.** A **4.** A scientist could tell if a cell was from a male or a female by looking at the chromosomes. If the cell contains two X chromosomes, the person is female. If the cell contains an X and a Y chromosome, the person is male. **5.** The other son may not have inherited the genes that influence high blood pressure.

## Portfolio Assessment

Sample items include:
- Punnett squares from Learning Styles: Visual/Spatial
- Notes from ELL/ESL Strategy
- Lesson 4 Review answers

---

**Research and Write**

Use the Internet or the library to learn the steps for a self-exam for cancer of the testes or breast. Write the steps, laminate the paper, and place it in your bathroom as a reminder.

**Lesson 4 Review** On a sheet of paper, write the letter of the answer that best completes each sentence.

1. The number of chromosomes in a fertilized egg cell is _____.
   A 2   C 46
   B 23  D 92

2. In most pairs of genes, the gene that determines which trait a child will inherit is _____.
   A normal     C recessive
   B dominant   D sex-linked

3. A person who inherits a health trait such as high blood pressure _____.
   A can help keep the condition from developing
   B probably has a sex-linked disorder
   C has two recessive genes for the trait
   D will always develop high blood pressure

On a sheet of paper, write the answers to the following questions. Use complete sentences.

4. **Critical Thinking** Could a scientist tell if a person was male or female by examining a skin cell? Explain your answer.

5. **Critical Thinking** Suppose a woman has high blood pressure. One of her sons develops high blood pressure. The other does not. Besides a healthy lifestyle, what might be another reason the other son does not have high blood pressure?

170  Unit 2  Personal Health and Family Life

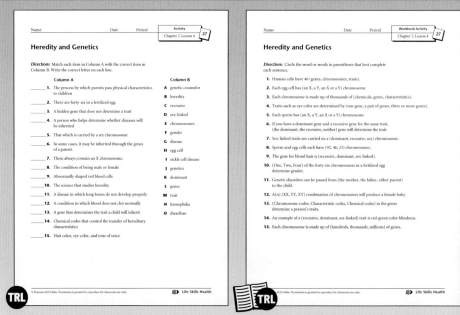

Activity 27     Workbook Activity 27

# Chapter 7 SUMMARY

- Adolescence is a time of many changes. Puberty is the time when children become adults and are able to have children.

- Three hormones signal the body's reproductive system to develop during puberty. They are testosterone for males, and estrogen and progesterone for females.

- The male produces sperm cells, and the female produces egg cells. If these two cells join during sexual intercourse, they form a new life.

- If the egg cell is fertilized, it begins to grow and divide. The uterus expands to hold the growing embryo or fetus. Gestation is about 9 months.

- The three stages of childbirth are the dilation of the cervix, the pushing out of the baby, and the afterbirth.

- Genetics is the science that studies heredity, or the passing of traits from parents to children.

### Vocabulary

| | | |
|---|---|---|
| adolescence, 151 | genetic disorder, 168 | recessive, 168 |
| cancer, 169 | genetics, 167 | reproduction, 156 |
| cervix, 165 | genitals, 154 | reproductive organ, 154 |
| cesarean childbirth, 165 | gestation, 160 | role confusion, 151 |
| chromosome, 167 | implantation, 162 | secondary sex characteristic, 154 |
| conceive, 157 | inferior, 151 | sex-linked, 168 |
| dominant, 168 | menstrual cycle, 158 | sexual intercourse, 159 |
| embryo, 163 | menstruation, 158 | sperm, 156 |
| erection, 159 | ovaries, 157 | testes, 156 |
| estrogen, 155 | ovulation, 157 | testosterone, 155 |
| fallopian tube, 157 | penis, 159 | trait, 167 |
| fertilize, 159 | perinatal care, 166 | trimester, 163 |
| fetal alcohol syndrome, 161 | placenta, 163 | ultrasound, 160 |
| fetus, 163 | postpartum, 166 | umbilical cord, 163 |
| gender, 167 | pregnant, 159 | uterus, 157 |
| gene, 167 | prenatal care, 161 | vagina, 157 |
| genetic counselor, 168 | progesterone, 155 | |
| | puberty, 152 | |

*The Life Cycle and Human Development* Chapter 7

## Chapter 7 Summary

Have volunteers read aloud each Summary item on page 171. Ask volunteers to explain the meaning of each item. Direct students' attention to the Vocabulary box on the bottom of page 171. Have them read and review each term and its definition.

### ONLINE CONNECTION

On MedLinePlus, a Web site offered by the U.S. National Library of Medicine and the National Institutes of Health, there are dozens of articles on prenatal care. The index can be found at: www.nlm.nih.gov/medlineplus/prenatalcare.html.

On the National Women's Health Information Center's Web site, the U.S. Department of Health and Human Services has posted information on prenatal care: www.womenshealth.gov/faq/prenatal.htm.

At the Genetics Home Reference Web site, the National Library of Medicine and the National Institutes of Health provide information on genetics: http://ghr.nlm.nih.gov/.

# Chapter 7 Review

Use the Chapter Review to prepare students for tests and to reteach content from the chapter.

## Chapter 7 Mastery Test

The Teacher's Resource Library includes two forms of the Chapter 7 Mastery Text. Each test addresses the chapter Goals for Learning. An optional third page of additional critical-thinking items is included for each test. The difficulty level of the two forms is equivalent.

## Review Answers

**Vocabulary Review**

1. puberty 2. testosterone 3. ovulation
4. fertilization 5. pregnant 6. gestation
7. ultrasound 8. embryo 9. umbilical cord
10. cervix 11. genetics 12. prenatal care
13. recessive 14. gender

## TEACHER ALERT

In the Chapter Review, the Vocabulary Review activity includes a sample of the chapter's vocabulary terms. The activity will help determine students' understanding of key vocabulary terms and concepts presented in the chapter. Other vocabulary terms used in the chapter are listed below.

| | |
|---|---|
| abnormal | ovaries |
| adolescence | penis |
| cesarean | perinatal care |
| childbirth | placenta |
| chromosome | postpartum |
| conceive | progesterone |
| dominant | reproduction |
| erection | reproductive |
| estrogen | organ |
| fallopian tube | role confusion |
| fetal alcohol syndrome | secondary sex characteristic |
| fetus | sex-linked |
| gene | sexual |
| genetic counselor | intercourse |
| genetic disorder | sperm |
| genitals | testes |
| implantation | trait |
| inferior | trimester |
| menstrual cycle | uterus |
| menstruation | vagina |

172  Unit 2  *Personal Health & Family Life*

# Chapter 7 REVIEW

**Word Bank**
cervix
embryo
fertilization
gender
genetics
gestation
ovulation
pregnant
prenatal care
puberty
recessive
testosterone
ultrasound
umbilical cord

## Vocabulary Review

On a sheet of paper, write the word or phrase from the Word Bank that best completes each sentence.

1. Adolescents are able to have children when they reach _____.

2. The hormone that signals the development of secondary sex characteristics in males is _____.

3. The release of a mature egg cell every month is _____.

4. When an egg cell joins with a sperm cell, _____ has taken place.

5. If a sperm cell and an egg cell unite, a woman becomes _____.

6. Pregnancy, or _____, usually lasts 9 months.

7. The use of sound waves to show a picture of something inside the body is _____.

8. After implantation, a fertilized egg is called a(n) _____.

9. The _____ joins the embryo or fetus with the placenta.

10. Labor pains that pull open the _____ are called contractions.

11. The science that deals with heredity is _____.

12. Part of _____ is seeing a doctor many times during a pregnancy.

13. A trait that is hidden is _____.

14. The _____ of a child is determined by the chromosome that comes from the male.

172  Unit 2  *Personal Health and Family Life*

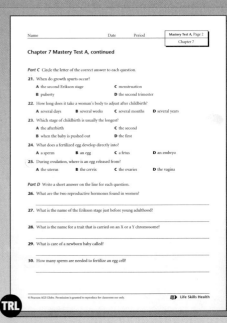

Chapter 7 Mastery Test A, pages 1–3

## Concept Review

On a sheet of paper, write the letter of the answer that best completes each sentence.

**15.** It is important for children to make things to _____.
  **A** avoid being self-absorbed
  **B** determine their identity
  **C** feel they have ability
  **D** reflect on their achievements

**16.** A woman can tell for sure whether she is pregnant if _____.
  **A** a doctor hears the fetus's heartbeat
  **B** a home pregnancy test is done
  **C** her breasts are large and sore
  **D** she always feels sick and tired

**17.** A fertilized egg cell attaches itself to the wall of the _____.
  **A** cervix    **B** ovary    **C** uterus    **D** vagina

## Critical Thinking

On a sheet of paper, write the answers to the following questions. Use complete sentences.

**18.** Someday, you may have children. Which of your physical traits would you want your children to have? Explain your answer.

**19.** What are two goals you plan to reach as an adult to help you feel good about your achievements?

**20.** It takes only one sperm cell to fertilize an egg. Why do you think the body produces so many sperm?

**Test-Taking Tip** Study for a test in short sessions rather than in one long session.

*The Life Cycle and Human Development* Chapter 7

## Concept Review
15. C  16. A  17. C

## Critical Thinking

**18.** Sample answer: I would want them to have my dark hair, because I like the color, and my broad shoulders, because they help carry heavy loads. **19.** Sample answer: I want to raise a family and have my children turn out well. I want to go to school after high school and have a career helping other people. **20.** Sample answer: Many sperm are produced so that there is a good chance of one of them fertilizing the egg cell.

## Alternative Assessment

Alternative Assessment items correlate to the student Goals for Learning at the beginning of this chapter.

- Have students make a chart describing the stages in the life cycle.
- Have students orally explain the changes that take place as young people mature into adults.
- Have students draw diagrams of the male and female reproductive systems.
- Have students write an encyclopedia article describing how new life forms and develops.
- Have students work in groups to create posters that illustrate the process of birth.
- Have students write and act out scripts for an instructional tape that explains the importance of prenatal care.
- Have students draw a diagram showing how characteristics are passed from parents to their children.

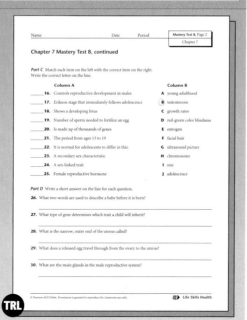

Chapter 7 Mastery Test B, pages 1–3

# Chapter at a Glance

## Chapter 8: The Family
pages 174–195

### Lessons
1. The Family Life Cycle, Dating and Marriage
   pages 176–181
2. Parenting and Family Systems
   pages 182–186
3. Problems in Families
   pages 187–192

**Chapter 8 Summary** page 193

**Chapter 8 Review** pages 194–195

**Audio CD**

**Skill Track for Life Skills Health**

**Teacher's Resource Library**
   Activities 28–30
   Alternative Activities 28–30
   Workbook Activities 28–30
   Self-Assessment 8
   Community Connection 8
   Chapter 8 Self-Study Guide
   Chapter 8 Outline
   Chapter 8 Mastery Tests A and B

(Answer Keys for the Teacher's Resource Library begin on page 559 of this Teacher's Edition.)

## Opener Activity

Ask students for words that they associate with the word *family* and write them on the chalkboard. *(Responses might include words such as parent, brother, sister, home, fun, and together.)* Discuss different family structures with students and ways in which a family is a unique group. *(Spending time together, sharing responsibilities, and so on)*

174  Unit 2  Title

# Self-Assessment

## Can you answer these questions?

☑ 1. How does dating fit into a family's life cycle?

☑ 2. For what reasons do couples marry?

☑ 3. What are some signs of a healthy marriage?

☑ 4. How does life change when a person becomes a parent?

☑ 5. What are parents responsible for?

☑ 6. What are some different ways children learn from parents?

☑ 7. What are some different kinds of families?

☑ 8. What are some signs of a healthy family?

☑ 9. What are some problems that families face?

☑ 10. Where can families get help if they have problems?

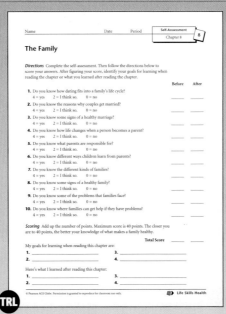

Self-Assessment 8     Community Connection 8

# Chapter 8: The Family

Most people belong to a family and depend on the support they get from their family. There are many different types of families.

In this chapter, you will learn about the family life cycle. You will learn how new families are formed when people marry. You will learn about being a parent and what kinds of things parents are responsible for. You will also learn about several different kinds of families and what makes a family healthy. Finally, you will learn about problems families have and where to get help with these problems.

## Goals for Learning

- To explain the place of dating and marriage in the family life cycle
- To identify some signs of a healthy marriage
- To describe what a parent is and what parents are responsible for
- To identify several different kinds of families
- To describe some signs of a healthy family
- To identify where families can get help for problems

## Introducing the Chapter

Have students read the introductory paragraphs on page 175. Then ask volunteers the following questions. What kinds of support do you think family members should get from their family? *(Emotional, financial, and physical)* What stages do you think the family life cycle consists of? *(Responses might include getting married, buying a house, learning parenting skills, having children, and growing older together.)* What characteristics do you think help make a family healthy? Why do you think so? *(Honesty, good communication, love, feeling safe with your family members, and so on)*

Have students read the Goals for Learning. Explain that in this chapter students will learn about different types of families, what makes a healthy family, and where families can get help for problems.

## Notes and Questions

Ask volunteers to read the notes and questions that appear in the margins throughout the chapter. Then discuss them with the class.

## Self-Assessment

Have students complete the Self-Assessment worksheet before and after reading the chapter. Before reading the chapter, have students fill in the "Before" column. Ask students to identify their goals for learning. To get ideas for setting goals, students might use the chapter introductory material on page 175, the checklist on page 174, or the questions on the Self-Assessment worksheet. Students can use the back of the worksheet if they need more space to write.

Collect the Self-Assessment worksheets and pass them out again at the end of the chapter. Have students fill in the "After" column. Ask them to identify at least four major points they have learned. Again, suggest they use the back of the worksheet if they need more space to write. You may want to collect and review the worksheets, but return them to students so they have a record of their goals and accomplishments.

Chapter 8 Self-Study Guide, pages 1–2        Chapter 8 Outline, pages 1–2

# Lesson at a Glance

## Chapter 8 Lesson 1

**Overview** This lesson discusses socializing, dating, and marriage.

### Objectives

- To discuss some of the difficult parts of dating
- To explore abstinence as a choice
- To describe some healthy reasons for marrying
- To look at the difference between a healthy marriage and an unhealthy one

**Student Pages** 176–181

**Teacher's Resource Library**

Activity 28
Alternative Activity 28
Workbook Activity 28

## Vocabulary

abstinence
collaborative
commitment
divorce
socializing

Have students read the paragraphs in which each vocabulary word appears. Ask if students have any questions about the meaning of the words and discuss them. Ask for examples of the use of the terms in news items.

## Background Information

People have many different reasons for remaining single. Some people may prefer to pursue individual interests or goals. Others may want to concentrate on their careers. Some have many satisfying relationships and friendships without having a marriage partner. Societal pressures to get married are much more relaxed than in the nineteenth and first half of the twentieth centuries, when a woman was expected to marry before age 30.

176 Unit 2 Title

---

## Lesson 1 — The Family Life Cycle, Dating, and Marriage

### Objectives

After reading this lesson, you should be able to
- discuss some of the difficult parts of dating
- explore abstinence as a choice
- describe some healthy reasons for marrying
- look at the difference between a healthy marriage and an unhealthy one

**Socializing**
Getting together with others during free time

Just as a person has a life cycle, a family also has a life cycle. The family life cycle usually includes events such as marriage and the birth of children.

A new family life cycle begins when two people decide to marry. Deciding to get married is one of the most important choices you will make in your life. Many events lead up to getting married and starting a new family.

### What Is Socializing?

Throughout your life, you have grown physically and socially. During adolescence, from ages 12 through 19, you grow into an adult and are usually ready to experience many new things. You are ready to meet and spend time with many different people. You are forming ideas about yourself and are ready to form ideas about others.

**Socializing** is getting together with others to enjoy free-time activities. Socializing includes learning about many male and female friends. By socializing, you learn what you like and dislike about others. You learn more about yourself and the kind of future you hope to have.

One part of socializing is dating. Usually, when you date, you pair up with another person and go out as a couple. Dating is a good way to get to know other people of the opposite sex better. You learn about other people by doing things together. You learn about interests you have in common. For example, you might learn that you both enjoy bike riding. You also learn about interests that are different from your own. You may enjoy seeing movies, while others would rather go dancing. Dating can be a good experience. Going out in groups can also be a positive experience. Going out in groups gives you the chance to get to know other people without the stress that a one-to-one date can cause.

176 Unit 2 Personal Health and Family Life

If two people enjoy being with each other, they may decide to date only each other. This can help you become closer and learn more about each other. However, when you date only one person, you give up the chance to get to know other people.

Two people who date usually have affection for each other. Even early in a relationship, there are many appropriate ways to express this affection. Physically, you might hold hands, hug, and kiss. You can also show affection by paying close attention to another person. Or you can give small gifts or write letters.

### What Are Some Difficult Parts of Dating?

Having a healthy dating relationship means agreeing on certain things. Both people should talk openly about their feelings and values. Sometimes, it can be difficult for two people to agree.

One of the most important parts of all relationships is communication. When you communicate successfully, you express your thoughts and feelings openly and honestly. Talking openly and honestly helps you understand each other. It keeps mix-ups and fights from happening. While it can be hard to express your feelings or thoughts, this is an important part of a relationship. Carefully listening to another person is also an important part of communication.

Another part of dating is deciding who pays for things you do together. Couples should talk this over. They may decide that one person always pays for the date. They may decide to take turns paying, or to "go dutch," with each person paying his or her own way.

Something else to think and talk about is your physical relationship. A healthy relationship is based on many things. Sometimes, relationships are based mainly on physical attraction. This attraction may cause you to begin a sexual relationship. If this happens too early in a relationship, you may have big problems. Some problems include diseases and unexpected pregnancy. Beginning a sexual relationship too early can lead to one person losing respect for the other. It can keep you from getting to know each other in other ways.

**Figure 8.1.1** *Couples should talk about who pays for things they do together.*

## 1 Warm-Up Activity

Ask students what characteristics they think would help a person get along with others. List the suggestions on the chalkboard. Add to or revise the list as you teach the lesson, according to how students' ideas change.

## 2 Teaching the Lesson

Help students process material presented in the lesson by having them write a separate summary paragraph for the material under each heading in the lesson.

## 3 Reinforce and Extend

### LEARNING STYLES

**Interpersonal/Group Learning**

Divide students into groups. Assign each group one of the following dating situations: dating as a group, one-to-one dating, or dating one person exclusively. Have group members discuss the advantages and disadvantages of their assigned situation. Ask volunteers to present each group's ideas to the class.

### ELL/ESL STRATEGY

**Language Objective:** *To identify multiple meanings of a word*

Point out to students who are learning English that the word *date* has several meanings, besides the one used in the lesson. Review with students some other meanings, such as an exact day (June 5) or the fruit of a palm tree.

## AT HOME

Suggest students interview their parents and other older relatives about differences in dating practices between the time they dated and the present. Ask volunteers to share their interviews with the class. Encourage students to discuss similarities and differences in dating practices across generations.

## TEACHER ALERT

Abstinence and sexual relationships are sensitive topics. Encourage students to have a mature attitude when discussing these topics. Suggest they write down any questions they may have on the topics on an index card. The questions can be collected and then read and answered by you to allow students anonymity.

## GLOBAL CONNECTION

Ask students to use library books and/or the Internet to research dating and marriage customs in foreign countries. Have students share their researched information with the class and compare foreign customs to American customs. What do students think about the customs from other countries? Would they be acceptable in American society?

## Health Myth

Have students read the Health Myth on page 178. Then have students list possible problems a couple may encounter in a relationship. For each problem listed, have students discuss a healthy solution for the couple.

---

**Abstinence**
*The choice not to have a sexual relationship*

**Commitment**
*A promise*

Many dating couples choose **abstinence.** Abstinence is the choice not to have a sexual relationship. It allows you to get to know each other in many ways before you begin a sexual relationship. It allows you time to become more grown up, to make better decisions, and to avoid problems. It allows you time to find other ways to be close, such as sharing deep personal thoughts and feelings. Abstinence is the best way to keep from getting pregnant or catching a disease.

Sometimes, a person does not want to have sex but is talked into it on a date. Staying away from places or not doing things that often lead to sex can make abstaining easier. Using alcohol and other drugs, for example, can make you think you have changed your mind. Someone may try to talk you into having sex. Sometimes, friends want you to have sex to fit in. These are all wrong reasons to have sex. You should never feel as if you should have sex. Having sex when you do not really want to do so lowers your self-esteem.

Sometimes, a part of dating is breaking up. You may decide the person you are dating is not right for you after all. Or the person you are dating decides you are not right for him or her. Sometimes, the decision to break up is hard. You may think of it as a loss and be sad. If you feel depressed, look for support from your friends and family.

### Why Do People Marry?

Getting married is a step people may take after dating. Where dating helps you get to know a person, getting married means a **commitment,** or a promise, to that person. This commitment involves promising to support your partner and to care for your partner if he or she gets sick. It also means being a friend to your partner.

> **Health Myth**
>
> **Myth:** If a couple really loves each other, they will not have problems.
>
> **Fact:** All couples have problems. With care and communication, almost all problems can be worked out.

There are many healthy and unhealthy reasons to get married. Some healthy reasons include loving a person and wanting to have a family with that person. If two people understand themselves and each other, they have a good chance for a healthy life together. Being married can help them be happy and satisfied with their lives.

Being married is usually good for couples. Many married people live longer than single people. They have better health and usually have more money. Marriage partners can be best friends. They enjoy talking with each other and working together on their home and family. They enjoy making a social life together with other friends and family members. Getting along well and staying together as partners takes work but is usually worth the effort. Dividing up chores and having good communication help a couple stay together.

Unhealthy reasons for getting married include doing it because other people want you to get married. Some people get married because they are trying to avoid an unhappy home life or loneliness. They think if they get married, their problems will be solved.

Sometimes, people decide to marry because of an unexpected pregnancy. This usually puts a strain on the couple from the start. The decision to have children is an extremely important one. The couple can best make it when they are ready to be responsible for another human being. If a couple has a baby accidentally, they may not be ready for the huge responsibility.

## When Do People Marry?

In the past, people often got married when they were quite young. This is changing for many reasons. One reason is that people live longer now. They do not need to marry young. Also, people have found that they need more education to get the jobs they want. They spend more time in school or in training and wait to get married. Some people wait to marry because they want to earn enough money to support themselves and their families.

## TEACHER ALERT

Explain to students that couples will often get engaged for a period of time before they get married. After a proposal of marriage has been made and accepted, a couple is said to be engaged (until the legal union of marriage occurs). Have students discuss why couples become engaged, what factors may lead to a decision to become engaged, and why a person may break an engagement.

## HEALTH JOURNAL

Have students make lists in their journal of daily and weekly responsibilities that a teenager in a particular situation might have. Have them choose one of the following situations: a married teenager, a married teenager with a child, or an unmarried teenager. Ask volunteers to present their lists to the class and then compare and contrast the lists to those made by other students for the same situation.

## LEARNING STYLES

### Body/Kinesthetic
Divide the class into pairs. Present each pair with a problem that is common in marriage, such as a conflict about money, relatives, raising children, or dividing chores. Ask each pair to role-play the problem and work out a solution.

## HEALTH JOURNAL

Have students write a paragraph in their journal about teenage marriages that includes reasons why they are not highly successful.

---

**Divorce**
The legal end of a marriage

**Collaborative**
Working together as a team

---

Today, the average age at which men in the United States marry is 27. For women, the average age is 25. Studies show that if a couple is older when they marry, they will have a longer life together. For example, 75 percent of all teenagers who get married end up getting a **divorce.** Divorce is the legal end of a marriage.

Some people choose not to get married. Some people choose a career or close friendships over married life. This does not mean that their lives are less happy or emotionally healthy.

### What Are Some Signs of a Healthy Marriage?

Many studies have been done on things that usually lead to happy married life. No one answer is sure to work. Here are some things that are important:

- agreeing on money matters
- having shared interests
- knowing each other well before getting married
- accepting and supporting each other
- agreeing about having children and about how to rear children
- having common goals
- sharing jobs around the house
- having family backgrounds that are somewhat alike
- having good relationships with parents

At some time or another, good and bad things affect a marriage. Moving to a new home or a new job is exciting, but both are stressful. Losing a job or getting sick can be hard. Couples who support each other are better able to cope with these changes.

A happy marriage usually needs more than just love. Couples have to be able to communicate, trust each other, and agree on many things. Couples should encourage each other. They should take a **collaborative** approach to life. This means they approach life as a team. They have to be willing to work out problems.

**Health Myth**

**Myth:** Married people are never lonely.

**Fact:** Even healthy marriages go through hard times when the partners may feel lonely.

Some common problems include handling money, getting along with relatives, and raising children. Some couples have trouble working out a fair way to divide household jobs. Some may not agree about how often to have sex. However, married people are likely to have more sex than single people. Also, married people are more likely to enjoy sex both physically and emotionally.

Marriages are happier when the two people know and understand themselves. Then they are better able to understand their partner and to have a stronger relationship.

*Lesson 1 Review* On a sheet of paper, write the letter of the answer that best completes each sentence.

1. Having sex when you do not want to do so _____ your self-esteem.
   - **A** lowers
   - **B** increases
   - **C** does not harm
   - **D** maintains

2. The best way to keep from getting pregnant is _____.
   - **A** divorce
   - **B** abstinence
   - **C** dating
   - **D** marriage

3. About _____ percent of teenage marriages end in divorce.
   - **A** 33
   - **B** 50
   - **C** 66
   - **D** 75

On a sheet of paper, write the answers to the following questions. Use complete sentences.

4. **Critical Thinking** What are some healthy reasons for getting married?

5. **Critical Thinking** What signs might show you that a marriage is healthy?

## Lesson 1 Review Answers

**1.** A **2.** B **3.** D **4.** Sample answer: Healthy reasons include loving a person and wanting to have a family with that person. **5.** Sample answer: In a healthy marriage you might find:
- agreement on money matters.
- similar interests.
- knowing each other well.
- acceptance and support of each other.
- agreement about children and how to discipline them.
- common goals.
- shared household tasks.
- similar family backgrounds and good relationships with parents.

## Portfolio Assessment

Sample items include:
- Research from Global Connection
- Paragraph from Health Journal
- Lesson 1 Review Answers

### Health Myth

Have students read the Health Myth feature on page 181. Then have them define the word *lonely*. Discuss with students the difference between feeling lonely and being alone.

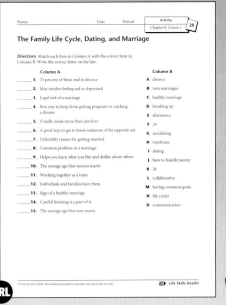

Activity 28

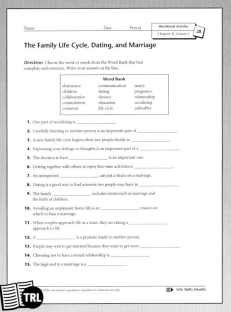

Workbook Activity 28

## Lesson at a Glance

### Chapter 8 Lesson 2

**Overview** This lesson identifies responsibilities of parents, describes types of families, and characterizes healthy families.

### Objectives

- To describe what a parent is responsible for
- To identify four different kinds of families
- To describe signs of a healthy family

**Student Pages** 182–186

**Teacher's Resource Library**

Activity 29
Alternative Activity 29
Workbook Activity 29

### Vocabulary

| | |
|---|---|
| adoptive family | discipline |
| blended family | extended family |
| consequence | foster family |
| custodial parent | nuclear family |
| custody | single-parent family |

Have students read through the definitions of the vocabulary words. Then ask them for examples of the different kinds of families defined by most of the vocabulary words. Examples may be students' own families, families they know, celebrity families, or families on TV shows.

### Background Information

According to the U.S. Census in 2000, over 2 million grandparents were living with and the primary caregivers for their grandchildren. Many of these grandparents are caring for children under age 5. About 34% of the grandparents who are primary caregivers make up a "skipped generation" household, in which the biological parents are not present.

### Technology and Society

Have students read the Technology and Society feature on page 182. Then have them research more information about what causes infertility in men and women. Ask volunteers to present the results of their research to the class.

182  Unit 2  Title

---

## Lesson 2  Parenting and Family Systems

**Objectives**

*After reading this lesson, you should be able to*
- describe what a parent is responsible for
- identify four different kinds of families
- describe signs of a healthy family

*Why is it usually harder for teens to be parents than it is for adults?*

A major event in the family life cycle is having children. Deciding to have children is an important one and can be healthy or unhealthy.

### Why Do Couples Decide to Have a Baby?

When couples purposely decide to make their family larger, they are likely to feel good about what they decided. Sometimes, couples have a baby for other reasons. They may be talked into it by friends or parents. They may want to pass on the family name. Some couples have children because they believe it will make their marriage stronger. Often, the opposite happens. If a marriage already has problems, the stress of having a baby can make the problems worse.

Some couples have a baby that they did not plan to have. This also causes stress because being a parent means being responsible for many things.

### Technology and Society

About 10 percent of U.S. couples face infertility. Infertility is a condition in which a woman cannot get pregnant. Couples can be infertile because of physical problems in either the man or the woman. One way to treat infertility is assisted reproduction, in which doctors try to help couples overcome infertility.

First, fertility drugs are often used. Doctors may be able to fix a problem in the man or woman with surgery. Another option is to collect eggs from the woman. Sperm is collected from the man. The sperm fertilizes the egg in a laboratory. Then the fertilized eggs are placed in the woman. In 2003, more than 40,000 babies were born in the United States through assisted reproduction. However, assisted reproduction is very expensive. It also can be hard on the couple physically and emotionally.

182  Unit 2  *Personal Health and Family Life*

**Consequence**
*Something that happens as a result of something else*

**Discipline**
*A fair and safe way to teach a child how to behave correctly*

### Link to ➤➤➤
**Language Arts**

The book *Little Women* by Louisa May Alcott is a story about an American family. The story is set in the 1860s during the U.S. Civil War. The father is a minister. He goes away to help soldiers in the war. The four daughters and their mother struggle to handle their daily lives. Everyone in the family loves, trusts, and encourages one another. This story provides a good example of how a healthy family behaves.

## What Are Parents Responsible For?

Parents are responsible for the health, safety, and well-being of their children. When you have a baby, your life changes. There are many jobs that you must do, many times a day, every day. Babies need to be fed, held, and have their diapers changed often. Someone must always be in the home when a baby is present. Babies often cry in the middle of the night and must be comforted and fed. Babies sometimes cry for no reason you can see. This is extremely stressful. Spending time with children reduces the time you have to spend with your partner. You also must pay the costs of having a child, including food, clothing, and medical care. It is important for you to talk to your partner to work out plans to take care of the child.

You need to learn parenting skills. You may want to go to classes or read books or articles on child care. You need to learn how to take care of the physical health of your child. You also need to learn how to help your child grow emotionally. You need to help and encourage your children to begin to do things for themselves.

You need to set fair limits for your child. Children should learn at an early age that they must follow certain rules. If children break the rules, you need to present a **consequence**. A consequence is what happens as a result of something else. For example, if a child does something wrong, the consequence may be to sit alone for a few minutes. Using a consequence is one way to **discipline** children. Discipline is a way to teach children how to behave correctly. Discipline should be fair and never cause harm.

Finally, you are an important teacher for your children. Children learn from you in many ways. They may learn from stories you read or tell them. They learn life skills. Children also learn by watching you. They copy the things you do. You must set a good example so your children can learn healthy behaviors.

 **Warm-Up Activity**

Have students role-play common disagreements that occur between parents and their children. Discuss how role-playing can help students understand the responsibilities of parenting.

 **Teaching the Lesson**

Have students make a table to organize information about the different types of families. Columns in the table should be titled "Family Type" and "Description."

Have students make a word web that contains circles for each of the six signs of a healthy family.

 **Reinforce and Extend**

**Pronunciation Guide**

Use the pronunciation shown here to help students pronounce difficult words in the lesson. Refer to the pronunciation key that appears in the Glossary at the back of the Teacher's Edition for the sounds of the symbols.

infertility (in f r til t )

### Link to

**Language Arts.** Have students read the Link to Language Arts feature on page 183. Obtain the book *Little Women* and read passages to the class that exemplify behaviors of a healthy family. After reading each passage, ask students to explain why it is a good example of how a healthy family behaves.

## LEARNING STYLES

### Body/Kinesthetic

Through the following activity, students can learn about the responsibilities of parenting. Students can do this individually or with a partner. Tell students that they must care for a child, represented by a 5-pound sack of sand or flour. Each day for one week, they must dress and watch over the child. To dress the child, they must make a container that protects the child from harm. To watch over the child at all times, one parent must always have the child nearby. Suggest students add their own rules to better simulate parental responsibilities. At the end of a designated time, have students record their thoughts about the aspects of parenting they have simulated and report to the class.

## Decide for Yourself

Ask volunteers to take turns reading the paragraphs in the Decide for Yourself feature on page 184 aloud. Have students research how much day care in the local area costs on a daily, weekly, or monthly basis.

**Decide for Yourself Answers**

**1.** Families that have little money coming in usually spend about one-fourth of what they earn on child care. **2.** Sample answer: Neither the state nor federal government has enough money to give to people who need child care help. **3.** Sample answer: I think the government should increase its funding for child care because every child deserves good child care. Proper child care will help a child grow both physically and emotionally.

---

*Nuclear family*
Family made up of a mother, a father, and their children

*Blended family*
Family made when parents live with children from an earlier marriage or marriages

*Single-parent family*
Family that includes a child or children and one adult

When a couple decides to have a child and be responsible for it, they form what is called a **nuclear family.** This is one of many different types of families.

### What Are Some Family Types?

The most common type of family today is a family with only two people—a couple without children. The second most common type is the nuclear family—parents and children.

Another family type is the **blended family.** In this family, one or both parents have been married or in a relationship before. When children from an earlier marriage or relationship live with the new couple, they become a blended family.

In a **single-parent family,** just one adult raises the children. This family type often results from a divorce or the death of a partner. Sometimes, the adult has never married. Mothers are most often the head of these families. Sometimes, the father, a grandparent, an aunt, or an uncle heads a single-parent family.

---

**Decide for Yourself**

When you have children, you must decide who is going to take care of them. When both parents have jobs, someone must take care of the children while the parents are at work. Many parents must use day care centers. However, this can be expensive. For many families, the three biggest expenses they have are housing, food, and child care. Most families spend about one-fourth of what they do make on child care. Many families cannot afford it.

National and state governments provide some help to families. However, neither have much money to spend. Both levels of government spend more than they take in. Because of this, many families find themselves on waiting lists for help.

Many states have made it harder to qualify. Some states have decided to give out less money for help.

On a sheet of paper, write the answers to the following questions. Use complete sentences.

1. About how much money does a family with little money spend on child care?
2. Why do families who need child care help end up on waiting lists?
3. Do you think the government should spend more money on child care? Explain your answer.

**Extended family**
*Family that includes parents, children, and other relatives*

**Adoptive family**
*Family that includes parents and an adopted child or children*

**Foster family**
*Family that cares for children who are in need of short-term parenting from people other than their birth parents*

**Custodial parent**
*The parent who is responsible for a child after a divorce*

**Custody**
*The legal right and responsibility to care for a child*

A fourth family type is the **extended family.** An extended family is a nuclear family plus other relatives, such as grandparents, aunts, uncles, and cousins. Grandparents often support a family by babysitting, helping with household chores, or giving money. Extended families can be valuable support systems.

Families may form and change in other ways. For example, when children are adopted, they become part of an **adoptive family.** Some same-sex couples have adopted children. Other children may join a **foster family** for a while. A foster family is one that cares for children who are in need of short-term parenting from people other than their birth parents.

Sometimes, children have two families. The parent who is responsible for children after a divorce is the **custodial parent.** In blended and some single-parent families, children live with the custodial parent. They may spend time with the noncustodial parent in another home. This is true when parents have joint **custody** of children. Both parents are responsible for raising and caring for their children.

Any of these types of families can be healthy or unhealthy. What is important is how the family members work together.

## How Is the Family Changing?

Family life is changing. Many families are smaller today than they were many years ago. Parents may decide to have fewer children because of the cost or because of their jobs. They may believe that they can provide a healthy home for only two rather than four or five children.

Many families want or need extra money. To help support the family, both parents work. This may be hard in some ways. For example, with both parents working outside the home, deciding how to divide the household chores may be harder for them. Also, many parents must leave their children in day care centers while the parents work.

The number of single-parent families is increasing in the United States. A single-parent family may have a harder time because only one adult is responsible for the family.

### Research and Write

Use the Internet to find Web sites with information to support parents who have jobs. Write some helpful hints for parents who are struggling to balance working and raising a family.

## ELL/ESL Strategy

**Language Objective:** *To orally describe a situation*

Students who are learning English and are immigrants may be living in an extended family. If so, invite students to describe the makeup of their family and some of the advantages for them of living in their extended family.

## Health Journal

Have students list the chores they do every week in their home to help their family. Then have them list some additional chores they could do. Encourage students to share their list with their family and to incorporate all the chores into a weekly schedule.

## Research and Write

Have students read the Research and Write feature on page 185. Remind them to use reputable Web sites for their research. You might suggest that students provide an illustration for each hint they write.

## Career Connection

Invite an employee from a child care center to speak to your class. Encourage students to ask the speaker questions about job responsibilities and qualifications as well as educational requirements. Ask the speaker to tell students what the most rewarding aspects of her or his job are as well as the most difficult parts of the job.

## Lesson 2 Review Answers

1. Parents must provide food, shelter, and medical care. They must also educate their children and help them with their problems. 2. Answers should list any four types of families: nuclear family, blended family, extended family, single-parent family, adoptive family, and foster family. 3. A blended family occurs when one or both of the parents have children from a previous marriage and these children live with the newly married couple. 4. Answers should list three characteristics. Members of healthy families support, encourage, trust, and value each other. 5. Sample answer: I would find it hardest to get up with a baby in the night, because I am a heavy sleeper. Also, I usually feel tired the next day if I don't get a good night's sleep.

## Portfolio Assessment

Sample items include:
- Research from Technology and Society
- Answers from Decide for Yourself
- Lesson 2 Review answers

### LEARNING STYLES

**Logical/Mathematical**
Make packs of two to six index cards, each pack representing a different family unit. (For example, the cards in one pack may represent a mother, a grandparent, a teenager, a 10-year-old, and a 5-year-old; another pack might include a father, a mother, and an infant.) Distribute the packs to groups of students of corresponding size. Have groups decide how to share and divide daily chores such as preparing meals, washing clothes, or taking out the trash. Then have groups share with the class how they determined the daily responsibilities of family members.

## Research and Write

Have students read the Research and Write feature on page 186. Students may want to visit the library for history books about ancient Rome. Encourage students to be creative when making their posters.

---

No matter what type of family you live in, you must talk about and work out problems for all the relationships to be healthy.

**Research and Write**

Research family life in ancient Rome. Find out about food, houses, and jobs in ancient Rome. What would a teenager have done on a normal day in ancient Rome? Make a poster that shows the differences in families then and now.

How must family members work together to make a family healthy?

### What Are the Signs of a Healthy Family?

Any family can be healthy or unhealthy. In a healthy family, members value, encourage, and trust one another. All family members help with the household chores. A good thing about doing chores is that you feel better about yourself. When you do more than you are told to do, you learn to be self-disciplined and dependable. This raises your self-esteem. Also, you know that you are helping and supporting your family. Healthy families also play together.

Two people who have studied families are Nick Stinnet and John DeFrain. They found that healthy families have these qualities:

- **Commitment**—Family members support and encourage one another.
- **Appreciation**—Family members appreciate one another.
- **Communication**—Everyone talks with and listens to everyone else.
- **Time**—Family members spend time together either working or having fun.
- **Beliefs**—Family members use their common beliefs to give them strength and purpose.
- **Coping ability**—Everyone is better able to handle problems because they get help and support from one another.

*Lesson 2 Review* On a sheet of paper, write the answers to the following questions. Use complete sentences.

1. Name two ways parents are responsible for their children.
2. What are four types of families?
3. What is a blended family?
4. **Critical Thinking** Describe three characteristics of healthy families.
5. **Critical Thinking** Of all the things that parents are responsible for, which do you think would be hardest? Explain your answer.

186  Unit 2  Personal Health and Family Life

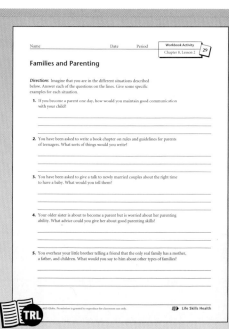

# Lesson 3   Problems in Families

**Objectives**

*After reading this lesson, you should be able to*

- describe how separation and divorce can affect a family
- explain how aging can affect a family
- describe how illness and death can affect a family
- explain how abuse can affect a family
- explore ways for a troubled family to get help

**Separation**
*A couple's agreement, or a court decision made for the couple, to stop living together*

Families change for many reasons. Some changes are hard for families, such as divorce, illness, and death. However, troubled families can get help.

## How Do Separation and Divorce Affect a Family?

Sometimes, when a couple decides they cannot live together, they separate. **Separation** may be an agreement between the two people, or it may be a court decision. During the separation, couples usually think about their marriage. They decide if they can make things better. If they believe they cannot, they may decide to divorce.

Separation and divorce can be hard for children. They may worry or feel depressed. They miss the parent who is gone or feel they are losing a parent. They may believe that they caused their parents' problems, but children are almost never the cause.

Children may feel sad if one parent tries to blame the other parent. Children should not be put in the middle of the adults' problems. A child cannot solve this kind of problem. Parents have a responsibility to work on their conflicts without blaming their children.

## How Does Aging Affect a Family?

People are living longer than ever before. Many older people want to live on their own. Sometimes, however, an older person in a family may have special needs. For example, as people age, they may have a hard time getting around. They may need to use a cane, a walker, or a wheelchair. They may have vision or hearing loss or get sick more often. Sometimes, older adults begin to lose their memory. They become confused and cannot remember simple things. They may have a harder time taking care of their basic needs. They may have to give up driving.

*The Family  Chapter 8*  **187**

---

## Lesson at a Glance

### Chapter 8 Lesson 3

**Overview** This lesson describes how divorce, aging, illness, death, violence, and abuse affect families and explains how troubled families can get help.

### Objectives

- To describe how separation and divorce can affect family
- To explain how aging can affect a family
- To describe how illness and death can affect a family
- To explain how abuse can affect a family
- To explore ways for a troubled family to get help

**Student Pages** 187–192

**Teacher's Resource Library**
Activity 30
Alternative Activity 30
Workbook Activity 30

### Vocabulary

| | |
|---|---|
| abuse | separation |
| domestic violence | sexual abuse |
| marriage counselor | violence |
| neglect | |

Have students read the definitions of the vocabulary words to themselves. Then state several behaviors that are examples of some of the words. Let students try to match the behavior to the word. For example, you might say, "One adult is yelling at and belittling another adult." Students could say this is an example of abuse or domestic violence. Point out that more than one vocabulary word can be used to describe the behavior.

### Background Information

Domestic violence includes physical, sexual, or emotional abuse against a current or former partner or spouse. Domestic violence has effects on all family members. Teens involved in relationships in which there is domestic violence have a higher rate of substance abuse. Increased rates of depression and other health problems are correlated with domestic violence in adults.

*The Family  Chapter 8*  **187**

## Warm-Up Activity

Have students write a paragraph about the type of pressure and stress that various events and situations put on a family and how the family can get help Encourage students to use specific situations as examples in their paragraphs.

## Teaching the Lesson

Have students make a Cause and Effect chart for the problems that can occur in families. On the left side of the chart, under "Cause," a problem should be listed. On the right side of the chart, under "Effect," the effects of the problem on family members should be listed.

## Reinforce and Extend

### IN THE COMMUNITY

Ask students to find out about services for senior citizens in your community. Discuss with students the kind of care that might be needed for elderly family members as they become less able to care for themselves. Encourage interested students to find out about the availability and cost of day care for the elderly in your community.

### IN THE ENVIRONMENT

Ask students to interview an elderly family member, friend, or neighbor. Have students ask for examples of how the way people lived years ago affected the environment positively and negatively. (Positive examples might include: People used fewer disposable items, fewer cars were driven. Negative examples might include: People cut down many forests; garbage and trash were not disposed of in a sanitary manner.) Instruct students to write about their findings. Ask volunteers to read their reports to the class.

---

Since family members do not automatically know what you are thinking or feeling, you are responsible for telling them.

---

Sometimes these age-related problems cause changes in a family. A home may have to be adapted to meet the needs of an older adult. Some family members may need to spend more time taking care of an older adult. A grandparent may need to live with a family if he or she needs more support.

A parent who takes care of a grandparent has less time to take care of his or her own needs. This parent also has less time to spend with children and less energy to help with their problems. When an older person comes to live with a family, everyone in the family has to adjust. Sometimes, all members must take on more chores. Everyone needs to be patient and considerate. Having an older person come to live with a family can be a positive experience. The family gets ideas from another point of view. Older people often have a lot of wisdom to share. Being with an older person is helpful to children, because it gives them another adult to learn to relate to.

### How Do Illness and Death Affect a Family?

When a family member gets sick, other members must adjust. They may have to help by doing more chores. They may have to give more support or get support from others. If an adult is sick for a long time, he or she may not be able to work. That may mean less money coming in and high medical bills.

Another difficult change is when someone in the family dies. For most people, a sudden death is harder to deal with than one that is expected. It is often easier to accept the death of someone who has been sick for a long time. When a young person dies suddenly, that death may be especially hard to accept. When anyone dies, each person deals with it in his or her own way. However, everyone goes through the five stages of grief—denial, anger, bargaining, depression, and acceptance. Everyone gets through these stages at their own speed and in their own order. Sometimes people repeat stages.

As with other changes, family members adjust to their loss in time. They may look for support from others or talk with one another about their grief.

## How Does Abuse Affect a Family?

Problems such as illness and death cannot be avoided. However, **violence** is a problem that can be avoided. Violence is actions or words that hurt people. It can destroy relationships and cause physical and emotional pain. Violence should never be accepted in a family.

Violence in families has many causes. Sometimes, members become angry and hit others. People who use alcohol or other drugs may become violent. Violence can also result from stress or be a learned behavior.

**Abuse** is a form of violence. Many kinds of abuse can affect families and be a serious problem. Both children and adults can be abused. Abuse can be physical or emotional. Physical abuse can be so bad that it results in bruises, broken bones, or even death. Emotional abuse also causes many problems. Most often, people do not show signs of emotional abuse on the outside, such as cuts and bruises. Harsh words and threats, however, can affect a person on the inside for a lifetime. People who are emotionally abused are often depressed and afraid, avoid others, and have low self-esteem.

Abuse is usually not just a one-time thing. It happens over and over. A child who is abused often is afraid to ask for help. The abuser may threaten the child. The child believes that something even worse will happen if he or she tries to get help.

Another form of abuse is **neglect,** or failure to take care of a person's basic needs. It may include leaving a young child alone for long periods at a time. It may mean the adult does not provide food, clothing, or a safe place to live. A child who is neglected may become ill or die.

**Sexual abuse** is any sexual contact that is forced on someone. Sexual abuse includes unwanted touching and talk as well as forcing a child to have sex. This kind of abuse affects many children and teens. Sexually abusing any child is illegal. It should be reported to someone such as a teacher, a counselor, or a police officer.

---

*Violence*
Actions or words that hurt people

*Abuse*
Physical or emotional mistreatment; actions that harm someone

*Neglect*
Failure to take care of a person's basic needs; regular lack of care

*Sexual abuse*
Any sexual contact that is forced on a person, including unwanted talk and touching

---

### TEACHER ALERT

Have students discuss the impact on their family of having a seriously ill family member that needs care in their home. Ask students to explain what things they could do to show their care and concern toward the ill person, even if they can't actually participate in the health care of the ill person.

### ELL/ESL STRATEGY

**Language Objective:**
*To classify contents of an article*

Have students form pairs composed of a student learning English and a student with strong English skills. Present students with an article about a family from a newspaper or a magazine. Challenge students to classify the article into one of the following categories: an article that illustrates family communication and support; an article about a family in crisis; or an article about a family coping with a crisis or receiving help in dealing with a crisis.

### GLOBAL CONNECTION

Have students choose a culture different from their own and research how the people in that culture deal with abusive behavior in a family. Ask for volunteers to share their findings with the class.

## LEARNING STYLES

**Auditory/Verbal**
Have students interview the school counselor to learn how to find professional support for a family problem. Students should write out their questions in advance. For example, they may want to find out what kind of professional help is available through the school or school system. Students could request information about the names and addresses of agencies in the community that serve teenagers and their families. They could find out which services are free and which have a charge. They could find out exactly what happens during a therapy session or a family counseling session.

## Link to

**Arts.** Have students read the Link to Arts feature on page 190. Ask students if they have seen the movie *Good Will Hunting*. Invite students who saw the movie to explain how learning about himself helped Will make a job choice. Ask students to discuss how a psychologist might be helpful to any person who needs to solve a problem.

---

**Domestic violence**
*Violence that happens mostly at home*

**Marriage counselor**
*A person trained in helping couples work on their relationship with each other*

### Link to >>>
**Arts**
The 1998 film *Good Will Hunting* tells about Will, a young man who works as a university janitor. A professor discovers that Will is a math genius. He wants to work with Will, who has been a victim of abuse. Will can work with the professor only if he has counseling. With the help of a psychologist, Will learns about himself. Then Will has to choose. Should he stay in the same job he is used to, or should he try to do something in which he can use his abilities?

---

Some people who are abused physically, emotionally, or sexually may not seek help. They may be afraid to report it. Some people try to hide abuse because they feel shame, even though it is never their fault.

Sometimes, older people are abused or neglected. In an unhealthy marriage, one partner may hurt or abuse the other. This is called **domestic violence,** because it happens mostly at home. People can get seriously hurt and die from domestic violence.

With any abuse, the people who get hurt, the abusers, and the other family members need to get professional help. People who are abused need to know that the abuse is not their fault. They did not in any way deserve, ask for, or encourage the abuse. The abused person is not wrong. The abuser's behavior is wrong in all cases.

### How Can a Troubled Family Find Help?

Problems in a family can affect the health of each member. Sometimes, families need help. Dealing with a problem is better than trying to avoid it. For example, a teenager may run away to avoid a problem. That usually causes more problems. Some problems are so serious that people may even try to kill themselves.

Family members can do several things to get help. Children can talk to teachers or school counselors. Families can go to special counseling centers at little or no cost. People at these centers give advice about dealing with the problem. The counselors help families figure out how to avoid these same problems in the future.

A couple whose marriage is in trouble can talk with a **marriage counselor.** Marriage counselors are trained to help couples improve or save their marriage.

Family counselors work with problems in families. They talk with family members and offer suggestions to help solve these problems.

Many communities have several kinds of support groups. These are groups of people who share a common problem and offer one another encouragement in dealing with it. Many support groups help people who have had a death in the family. Other support groups help people deal with alcohol or other drug problems.

### Health in Your Life

**Family Belief Systems**

Although people may not realize it, all families have specific ideas that guide how they behave. These ideas are about how family members are expected to behave, what the family sees as being important, and the things they believe to be true. These ideas are handed down in families from generation to generation. Figuring out these ideas is important so that family members can change their behavior if needed. Generally, when adolescents are having problems, they are either acting upon or questioning these ideas.

For example, many parents have strong feelings about education. They expect that their child will do well in school and go to college. A teen who is questioning his or her family's ideas might fail in school or skip classes. A counselor would have to uncover the family's ideas about education to help the family solve this problem.

In some families, abusing others may be part of how that family has always dealt with anger. An abusive parent may somehow believe that this is how parents should behave. For example, a great-grandfather abused his son, who abused his son, who abuses his son. This is how abuse is passed down through families.

To find out what ideas the family has, a counselor listens to the stories family members tell. The counselor looks at how the stories are told and pays attention to how family members describe one another. The counselor asks questions about how the family members act and react toward one another. Sometimes, the counselor encourages family members to change. Other times, the problem may be in school or in peer relationships.

On a sheet of paper, write the answers to the following questions. Use complete sentences.

1. What makes up the ideas that families believe in?
2. Is the family always the reason a teen has problems? Explain your answer.
3. Name one idea that is part of your family's group of ideas.

### LEARNING STYLES

**Visual/Spatial**

Bring pictures to class of families participating in activities together. Pictures can depict recreational activities, sports, sharing chores, talking, and so on. Invite students to discuss how doing things with other family members contributes to the development of a healthy family.

### HEALTH JOURNAL

Discuss with students how a child's plans to continue education after high school affect a family. Have students write a paragraph in their journal about their plans for after high school. Encourage students to talk with their parents about their plans.

### Health in Your Life

Have students read and complete the Health in Your Life feature on page 191. Ask students why it may be hard for individuals or families to seek counseling when they need help (*May not be aware that there are services available, pride, lack of money, think it won't help*)

**Health in Your Life Answers**

1. The key ideas that families believe in are shaped by expectations, values, and assumptions. Families have beliefs and values that influence how they see themselves. 2. No, problems can also occur in school and with peers. 3. Samples answer: My family believes holidays are supposed to be spent with the extended family. My family devotes a great deal of time, effort, and money to travel hundreds of miles for holiday gatherings.

## Lesson 3 Review Answers

**1.** worried **2.** less time and energy **3.** neglect **4.** Sample answer: I would go see my English teacher. I feel she really understands me and cares for me. I also think she is wise and could help. If I could not find my English teacher, then I would ask the school nurse for help. **5.** Sample answer: I would probably have to give up my room. Also, I would have to help out more around the house.

### Portfolio Assessment

Sample items include:
- Report from In the Environment
- Answers to Health in Your Life
- Lesson 3 Review answers

### Health at Work

Have students read the Health at Work feature on page 192. Emphasize that help is available for teenagers who need to solve problems. Then ask students what qualities might help a person be an effective counselor for adolescents.

---

If you are experiencing any form of abuse, get help. Talk with a trusted adult, such as a teacher, school counselor, or police officer.

Support groups can sometimes be found by looking in the phone book. Many groups advertise in the newspaper or appear in the newspaper's calendar section. They can also be found by searching the Internet. School counselors can recommend support groups. Finding help for a problem in your school, community, or home will help you feel better. It will also help you stay happier and healthier.

**Lesson 3 Review** On a sheet of paper, write the word or phrase in parentheses that best completes each sentence.

1. When there is a divorce, children may feel (worried, happy, content).
2. A parent who has to care for a grandparent has (less time and energy, more time and energy, less time but more energy) for his or her children.
3. When parents fail to take care of a child, they are guilty of (violence, sexual abuse, neglect).

On a sheet of paper, write the answers to the following questions. Use complete sentences.

4. **Critical Thinking** If you had a problem in your family, where would you go for help? Explain your answer.
5. **Critical Thinking** If a grandparent came to live with your family, what adjustments would you have to make?

### Health at Work

**Adolescent Counselor**

Many teens face problems such as alcohol and other drug abuse, eating disorders, depression, violence, and AIDS. Adolescent counselors help teens with these problems. They may also be asked to find safe places for runaway and homeless teens. Sometimes, adolescent counselors can help teens prepare for jobs. Adolescent counselors must be able to understand how teenagers feel. They must be able to form good relationships with young people. Requirements are a master's degree and a state license or certification.

192  Unit 2  Personal Health and Family Life

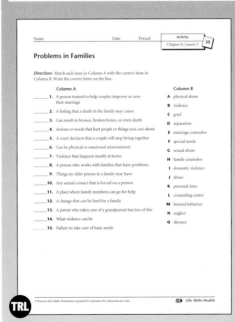

Activity 30

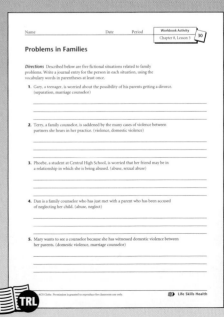

Workbook Activity 30

# Chapter 8 SUMMARY

- Socializing and dating may lead to marriage. A new family life cycle begins when two people get married.
- Dating couples must agree on their physical or sexual relationship. A sexual relationship can lead to an unexpected pregnancy or a disease.
- Often couples who are dating decide to stop dating. This can leave a person feeling unhappy or depressed.
- Most people in the United States marry at some time. People get married for both healthy and unhealthy reasons.
- Coming from the same background, being educated, and having good jobs usually lead to healthy marriages.
- People have a baby for both healthy and unhealthy reasons.
- Parents are responsible for the health, safety, and well-being of their children.
- Parents need to learn certain skills to help their children grow physically and emotionally. They must set rules and provide discipline.
- There are several types of families. These are nuclear, blended, single-parent, extended, adoptive, and foster families.
- Any type of family may be healthy or unhealthy.
- Separation, divorce, aging, illness, and death can affect families.
- Abuse may be physical or emotional. It may be directed at children or adults. Violence and abuse should never be accepted in a family.
- Married couples and families with problems can get help.

| Vocabulary | | |
|---|---|---|
| abstinence, 178 | custody, 185 | neglect, 189 |
| abuse, 189 | discipline, 183 | nuclear family, 184 |
| adoptive family, 185 | divorce, 180 | separation, 187 |
| blended family, 184 | domestic violence, 190 | sexual abuse, 189 |
| collaborative, 180 | extended family, 185 | single-parent family, 184 |
| commitment, 178 | foster family, 185 | socializing, 176 |
| consequence, 183 | marriage counselor, 190 | violence, 189 |
| custodial parent, 185 | | |

The Family   Chapter 8   **193**

## Chapter 8 Summary

Have volunteers read aloud each Summary item on page 193. Ask volunteers to explain the meaning of each item. Direct students' attention to the Vocabulary box on the bottom of page 193. Have them read and review each term and its definition.

## ONLINE CONNECTION

For more information about family statistics in the United States, have students visit the United States Census Bureau Web site that contains recent information at: www.census.gov/population/www/socdemo/hh-fam/cps2004.html

A Web site sponsored by the United States Department of Health and Human Services has information of families, domestic violence, and aging at: www.hhs.gov

## Chapter 8 Review

Use the Chapter Review to prepare students for tests and to reteach content from the chapter.

## Chapter 8 Mastery Test

The Teacher's Resource Library includes two forms of the Chapter 8 Mastery Text. Each test addresses the chapter Goals for Learning. An optional third page of additional critical-thinking items is included for each test. The difficulty level of the two forms is equivalent.

## Review Answers

**Vocabulary Review**

1. abstinence 2. consequence 3. nuclear family 4. single-parent family 5. abuse 6. marriage counselor 7. divorce 8. collaborative 9. extended family 10. emotional abuse 11. neglect 12. sexual abuse 13. domestic violence 14. foster family

### TEACHER ALERT

In the Chapter Review, the Vocabulary Review activity includes a sample of the chapter's vocabulary terms. The activity will help determine students' understanding of key vocabulary terms and concepts presented in the chapter. Other vocabulary terms used in the chapter are listed below.

adoptive family
blended family
commitment
custodial parent
custody
discipline
emotional abuse
separation
socializing
violence

194    Unit 2    Title

---

## Chapter 8 REVIEW

**Word Bank**
abstinence
abuse
collaborative
consequence
divorce
domestic violence
emotional abuse
extended family
foster family
marriage counselor
neglect
nuclear family
sexual abuse
single-parent family

### Vocabulary Review

On a sheet of paper, write the word or phrase from the Word Bank that best completes each sentence.

1. Avoiding a sexual relationship, or _____, will help you avoid an unexpected pregnancy.

2. Parents need to set fair rules and provide a(n) _____ if children break the rules.

3. A married couple living with their own children is a(n) _____.

4. One adult raising children alone is a(n) _____ family.

5. Treating someone badly, either physically or emotionally, is _____.

6. A(n) _____ helps married couples work out their problems.

7. Couples who choose not to live together after separating may choose to get a(n) _____.

8. A family that works together as a team is using a(n) _____ approach to life.

9. A nuclear family, plus aunts, uncles, and grandparents, is a(n) _____.

10. Saying mean and hurtful things to children is _____.

11. Failing to care for a child's basic needs is _____.

12. Unwanted touching and talk are part of _____.

13. Violence that takes place in the home is _____.

14. A(n) _____ cares for children who are in need of short-term parenting.

194    Unit 2    Personal Health and Family Life

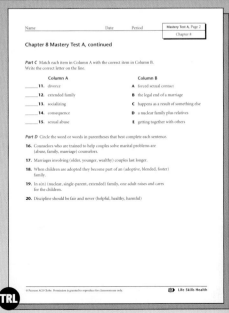

Chapter 8 Mastery Test A, pages 1–3

## Concept Review

On a sheet of paper, write the letter of the answer that best completes each sentence.

15. One sign of an unhealthy marriage is _____.
    A companionship
    B communication
    C lack of appreciation
    D working together

16. One way families are changing is that people _____.
    A are getting married younger
    B are waiting to get married so that they can get more education
    C are working less after they get married
    D have stopped earning money to support the family

17. Having a baby always adds _____ to a marriage.
    A ease
    B strength
    C discipline
    D stress

## Critical Thinking

On a sheet of paper, write the answers to the following questions. Use complete sentences.

18. Age-related problems can be hard for older adults. How could you help an older adult?

19. Why is it better for a family to solve problems than to avoid them? Explain your answer.

20. What signs of a healthy marriage are most important to you? Explain your answer.

**Test-Taking Tip** Before you begin a test, look it over quickly to see how much time you may need for each section. Try to set aside enough time to complete each section.

*The Family* Chapter 8  195

# HEALTH IN THE WORLD

Have students read the Health in the World feature on page 196. Have students relate the idea of overpopulation to limited resources such as clean water, food, and living space. Ask them to explain how overpopulation can lead to greater competition for resources and how this competition can affect a person's health. *(Sample answer: If a person lacks resources, he or she might not be getting a balanced diet, which can lead to deficiencies and malnutrition. Not having enough clean water can spread disease. Living in cramped living spaces might affect a person's mental and emotional health.)*

**Health in the World Answers**

1. Sample answer: The population grows particularly fast in developing nations, and this issue is related to education. Some developing nations lack good systems of education. Improving education helps control population growth because people who are educated about health issues can make informed decisions about birth control. Ignorance about family planning issues contributes to overpopulation. People who are educated about the way their bodies work can make decisions to control the size of their families.

2. Sample answer: Population growth puts a strain on health care systems because medical professionals have to treat an exceptionally large number of patients. In these situations, a doctor may not have time to see all the patients that he or she needs to see each day. Population growth also causes some medical facilities to run out of important medical supplies.

3. Sample answer: The problem of global malnutrition will only be solved when many countries (including the United States) pool their resources and make a strong commitment to solve the problem. Some of the keys for solving malnutrition are improved agricultural methods, improved methods for predicting the weather, and improved methods for quickly transporting food to areas affected by famine. Another key is helping farmers around the world to become self-sufficient.

# HEALTH IN THE WORLD

## Overpopulation and Health

Every year, the world's population grows by about 76 million people. Currently, world population is more than 6.3 billion. This growth is faster in some places than in others. Population is growing six times faster in developing nations than it is in the nations of Europe, North America, Oceania, and Japan.

More than 500 million people around the world do not have enough clean water to drink. This leads to diseases such as cholera, typhoid fever, and dysentery. The more the population grows, the less land and water there are for everyone. Also, higher populations put heavier strains on food supplies and health care systems.

About 8 million children die every year from malnutrition or other diseases caused by not getting enough food. Clearly, overpopulation has negative effects on health.

On a sheet of paper, write the answers to the following questions. Use complete sentences.

1. Why do you think the population grows faster in developing nations than it does in the nations of Europe and North America?

2. How does population growth put a strain on health care systems?

3. What are some ways to solve the serious problem of malnutrition around the world?

# Unit 2 SUMMARY

- The bones in the skeletal system give your body its shape and protect your organs.
- The muscular system allows your body to move. The body has smooth, skeletal, and cardiac muscles.
- The central nervous system controls your body's activities. It includes your brain and spinal cord.
- The circulatory system includes your blood, heart, and three kinds of blood vessels. Blood carries food and oxygen to every cell. Your heart pumps blood throughout your body.
- The respiratory system allows you to breathe in oxygen and breathe out carbon dioxide. Cells need oxygen to break up nutrients.
- Good hygiene means taking care of your eyes, ears, skin, hair, nails, and teeth.
- Physical fitness is your body's ability to meet the demands of daily life.
- Three parts of health-related fitness are heart and lung endurance, flexibility, and muscular fitness.
- Regular exercise improves fitness, increases energy, lowers the chances of getting sick, helps control weight, reduces anxiety, and reduces stress.
- Three hormones signal the body's reproductive system to develop during puberty. They are testosterone for males, and estrogen and progesterone for females.
- In the reproductive system, the male produces sperm cells, and the female produces egg cells. If these two cells join during sexual intercourse, they form a new life.
- The female's ovaries release egg cells in a monthly cycle. If the egg cell is not fertilized, it passes out of the body.
- If the egg cell is fertilized, it begins to grow and divide, and it plants itself in the uterus. The uterus expands to hold the growing embryo or fetus.
- Genetics is the science that studies heredity, or the passing of traits from parents to children.
- There are several types of families. These are nuclear, blended, single-parent, extended, adoptive, and foster families.
- Some qualities of healthy families are commitment, appreciation, communication, common beliefs, and being able to cope. Healthy families spend time together.
- Abuse may be physical or emotional. People who suffer any kind of violence or abuse need to get help.

## Unit 2 Summary

Have volunteers read aloud each Summary item on page 197. Ask volunteers to explain the meaning of each item.

## Unit 2 Review

Use the Unit Review to prepare students for tests and to reteach content from the unit.

## Unit 2 Mastery Test

The Teacher's Resource Library includes two forms of the Unit 2 Mastery Test. An optional third page of additional critical-thinking items is included for each test. The difficulty level of the two forms is equivalent.

## Review Answers

### Vocabulary Review

1. iris  2. estrogen  3. socializing  4. hygiene
5. uterus  6. domestic violence  7. tendon
8. abstinence  9. epidermis  10. chromosome
11. endurance  12. metabolism

### Concept Review

13. A  14. D  15. D  16. B

---

# Unit 2 REVIEW

**Word Bank**
- abstinence
- chromosome
- domestic violence
- endurance
- epidermis
- estrogen
- hygiene
- iris
- metabolism
- socializing
- tendon
- uterus

## Vocabulary Review

On a sheet of paper, write the word or phrase from the Word Bank that best completes each sentence.

1. The part of an eye that gives an eye its color is the _____.

2. One of the hormones that control reproductive development in females is _____.

3. Getting together with people during your free time is _____.

4. Protecting and caring for your ears and hearing are important parts of _____.

5. The female organ where an egg grows into a baby is the _____.

6. One form of abuse that sometimes takes place between family members is _____.

7. A strong set of fibers joining muscle to bone or muscle to muscle is a(n) _____.

8. When two people choose _____, they are choosing not to have a sexual relationship.

9. The outer layer of skin is the _____.

10. A tiny structure inside a cell that contains hereditary information is a(n) _____.

11. The ability to keep doing a physical activity for a long time is _____.

12. Cells in the human body produce energy at a rate that is called a person's _____.

198  Unit 2  Personal Health and Family Life

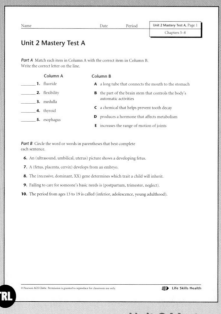
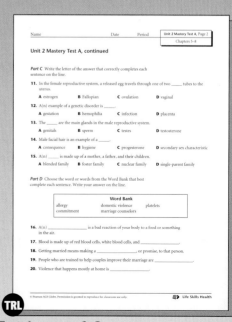

Unit 2 Mastery Test A, pages 1–3

## Concept Review

On a sheet of paper, write the letter of the answer that best completes each sentence.

13. The three parts of blood are red blood cells, white blood cells, and _____.
    A platelets
    B bronchioles
    C ligaments
    D villi

14. Exercise that builds muscle strength by using resistance from a machine is _____.
    A aerobic
    B anaerobic
    C isometric
    D isokinetic

15. A male is usually able to produce sperm from puberty through _____.
    A adolescence
    B early adulthood
    C about age 60
    D the rest of his life

16. An extended family includes _____.
    A only adopted children
    B parents, children, and other relatives, such as grandparents
    C one or two parents who have been married before
    D just one adult who raises the children

## Critical Thinking

On a sheet of paper, write the answers to the following questions. Use complete sentences.

17. Why are a person's reflexes important? Explain your answer.

18. Why is it dangerous to begin exercising without doing a warm-up first?

19. What is the difference between a dominant gene and a recessive gene? Explain your answer.

20. What are the qualities of a healthy family? How is a healthy relationship with a relative similar to a healthy relationship with a friend?

## Critical Thinking

**17.** Sample answer: Reflexes are automatic responses that help and protect people. If you put your hand on a hot radiator, you will automatically pull your hand away. This reflex prevents you from seriously damaging the skin and muscles of your hand. Another reflex is feeling cold when the temperature drops outdoors. This is a way for the peripheral nervous system to send a message to the brain. The message that it is cold outside causes a person to do something to get warm, such as putting on a jacket. This reflex prevents the skin and muscles from being damaged by cold air. **18.** Sample answer: A warm-up makes blood flow to the muscles. This helps prepare the body for a longer period of exercise. A warm-up also allows the heart rate to rise gradually. If a person began exercising without doing a warm-up first, he or she could strain a muscle. It is not healthy for the heart to go from a slower rate to a faster one very quickly. A warm-up helps the heart to start beating faster in a gradual fashion. This is particularly important for people with heart problems who are following an exercise program. **19.** Sample answer: The dominant gene in a pair of genes is the one that determines which trait a child will inherit. The dominant gene has the most control. A recessive gene is sometimes called a hidden gene because it does not determine the trait. For example, the gene for the eye color brown is dominant, and the gene for the eye color blue is recessive. In order for a child to have blue eyes, he or she would need to receive recessive genes from both parents. **20.** Sample answer: Healthy families spend time together, and they have these qualities: commitment, appreciation, good communication, common beliefs, and coping abilities. A healthy relationship with a relative is very similar to a healthy relationship with a friend because in both types of relationships, the two people support one another, encourage one another, appreciate one another, and listen to one another. Honest, open communication is important for both families and friendships.

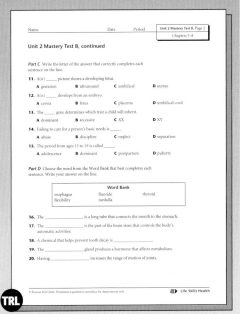

Unit 2 Mastery Test B, pages 1–3

# Unit 3

## Planning Guide
## Nutrition

| | Student Pages | Vocabulary | Health in the World | Review | Critical-Thinking Questions | Chapter Summary |
|---|---|---|---|---|---|---|
| **Chapter 9** The Role of Diet in Health | 202–227 | ✔ | | ✔ | ✔ | 225 |
| Lesson 1 A Healthy Diet | 204–209 | ✔ | | ✔ | ✔ | |
| Lesson 2 Carbohydrates, Fats, and Protein | 210–215 | ✔ | | ✔ | ✔ | |
| Lesson 3 Vitamins, Minerals, and Water | 216–220 | ✔ | | ✔ | ✔ | |
| Lesson 4 Special Dietary Needs | 221–224 | ✔ | | ✔ | ✔ | |
| **Chapter 10** Choosing Healthy Foods | 228–245 | ✔ | | ✔ | ✔ | 243 |
| Lesson 1 Healthy Eating Patterns and Food Choices | 230–232 | ✔ | | ✔ | ✔ | |
| Lesson 2 How the Media Influences Eating Patterns | 233–238 | ✔ | | ✔ | ✔ | |
| Lesson 3 Food Labels and Food Additives | 239–242 | ✔ | 246 | ✔ | ✔ | |

### Unit and Chapter Activities

**Student Text**
Unit 3 Summary

**Teacher's Resource Library**
Home Connection 3
Community Connections 9–10

**Teacher's Edition**
Opener Activity, Chapters 9–10

### Assessment Options

**Student Text**
Chapter Reviews, Chapters 9–10
Unit 3 Review

**Teacher's Resource Library**
Self-Assessment Activities 9–10
Chapter Mastery Tests A and B, Chapters 9–10
Unit 3 Mastery Tests A and B

**Teacher's Edition**
Chapter Alternative Assessments, Chapters 9–10

| | Student Text Features | | | | | | | | Teaching Strategies | | | | | | Learning Styles | | | | | Teacher's Resource Library | | | | |
|---|---|---|---|---|---|---|---|---|---|---|---|---|---|---|---|---|---|---|---|---|---|---|---|
| | Self-Assessment | Health at Work | Health in Your Life | Decide for Yourself | Link to | Research and Write | Health Myth | Technology and Society | Background Information | ELL/ESL Strategy | Health Journal | Applications (Home, Career, Community, Global, Environment) | Online Connection | Teacher Alert | Auditory/Verbal | Body/Kinesthetic | Interpersonal/Group Learning | Logical/Mathematical | Visual/Spatial | Activities | Alternative Activities | Workbook Activities | Self-Study Guide | Chapter Outline |
| | 202 | | | | | | | | | | | | 225 | 226 | | | | | | 31–34 | 31–34 | 31–34 | ✓ | ✓ |
| | | | | | 207 | | | | 204 | 208 | 206 | 207, 209 | 206 | | | | 207 | | | 31 | 31 | 31 | | |
| | | | 214 | | | 213 | | 210 | 210 | 213 | 215 | 211, 214 | | 212 | | | | | 212 | 32 | 32 | 32 | | |
| | | | 219 | | 220 | 218 | 218 | | 216 | 219 | | 218 | | | | | 217 | 218 | | 33 | 33 | 33 | | |
| | | 223 | | | | | 222 | | 221 | 223 | | 224 | 225 | 224 | 225, 226 | | | | | 34 | 34 | 34 | | |
| | 228 | | | | | | | | | | | | 243 | | | | | | | 35–37 | 35–37 | 35–37 | ✓ | ✓ |
| | | | | | 232 | 230 | | | 230 | 231 | 231 | 231, 232 | | | | | | | | 35 | 35 | 35 | | |
| | | 238 | 235 | 237 | | 235 | 234, 236 | | 233 | 237 | 236 | 234, 236 | | 234, 236 | 238 | | 237 | 236 | | 36 | 36 | 36 | | |
| | | | | | 240 | | | 242 | 239 | | | | | | | | | | | 37 | 37 | 37 | | |

## Alternative Activities

The Teacher's Resource Library (TRL) contains a set of lower-level worksheets called Alternative Activities. These worksheets cover the same content as the regular Activities but are written at a second-grade reading level.

## Skill Track

Use Skill Track for *Life Skills Health* to monitor student progress and meet the demands of adequate yearly progress (AYP). Make informed instructional decisions with individual student and class reports of lesson and chapter assessments. With immediate and ongoing feedback, students will also see what they have learned and where they need more reinforcement and practice.

## Unit at a Glance

**Unit 3:**
**Nutrition**
pages 200–249

**Chapters**

9. **The Role of Diet in Health**
   pages 202–227
10. **Choosing Healthy Foods**
    pages 228–245

**Unit 3 Summary** page 247

**Unit 3 Review** pages 248–249

**Audio CD**

**Skill Track for**
**Life Skills Health**

**Teacher's Resource Library** TRL

Home Connection 3

Unit 3 Mastery Tests A and B

(Answer Keys for the Teacher's Resource Library begin on page 559 of this Teacher's Edition.)

### Other Resources

#### Books for Teachers

*Consumer Reports Guide to Diet, Health & Fitness.* New York: Time Inc. Home Entertainment, 2005. (product rankings, advice, and action plans)

Ronzio, Robert A. *The Encyclopedia of Nutrition and Good Health.* New York: Facts on File, 2003.

#### Books for Students

Hovius, Christopher. *The Best You Can Be: A Teen's Guide to Fitness and Nutrition.* Philadelphia: Mason Crest Publishers, 2005. (food guides, balancing diets, exercise)

Salmon, Margaret B. *Food Facts for Teenagers: A Guide to Good Nutrition for Teens and Preteens.* Springfield, IL: Charles C. Thomas, 2002. (quick methods for calculating vitamin and calorie content of foods, recipes)

#### CD-ROM/Software

*Body Management: Nutrition Power.* Learning Multi-Systems (1-800-362-7323). (multimedia activities for nutritional and weight loss information)

#### Videos and DVDs

*Food Fight: Childhood Obesity and the Fast Food Industry* (23 minutes). ABC News, 2003 (1-800-257-5126). (food industry marketing, healthy food initiatives)

*Nutrition and Diet* (30 minutes). Teen Health Video Series. Wynnewoood, PA: Schlessinger, 1996 (1-800-843-3620). (nutritionist and dietician advise teens)

*Winning Sports Nutrition 2000* (30 minutes). Champaign, IL: Human Kinetics Publishers, 2000. (calculate daily values of nutrients in foods for peak performance)

#### Web Sites

www.nutrition.gov (dietary guidance, food-related diseases)

www.health.gov/dietaryguidelines (current and past federal guidelines, related links)

http://teamnutrition.usda.gov/parents.html (links to school-based nutrition education, tips for families to get involved)

# Unit 3: Nutrition

Did you have breakfast this morning? Eating breakfast is important. Students who do not eat breakfast usually score lower on tests than those who have had a healthy morning meal. Consider food as fuel for your body. Think about how a radio sounds when the batteries are low. Without food, your body fades like music on a radio when its batteries are low.

People are becoming more concerned about the importance of eating properly. A poor diet can play a major role in causing heart attacks. It can also be one of the reasons for strokes and some types of cancer. Even with serious health warnings, many people still choose unhealthy diets. In this unit, you will learn how to make healthy food choices.

### Chapters in Unit 3

Chapter 9: The Role of Diet in Health ....................... 202
Chapter 10: Choosing Healthy Foods ....................... 228

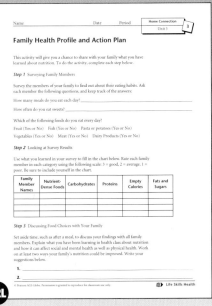

Home Connection 3

## Introducing the Unit

Have students read the text on page 201. Ask students if they ate a healthy breakfast this morning. Ask them how they feel later in the morning on days when they have skipped breakfast. *(Possible answers: tired, less able to concentrate or pay attention)* Ask students what changes they could make to make sure they eat a healthy breakfast in the morning. *(Possible answers: Leave a little more time for breakfast, plan what you will have the night before)* Tell students they will learn more about how their eating habits affect their health as they read this unit.

### HOME CONNECTION

The Home Connection unit activity gives students practical experience with concepts taught in the *Life Skills Health* student text. Students complete the Home Connection activity outside the classroom with the help of family members. These worksheets appear on the Life Skills Health Teacher's Resource Library (TRL) CD-ROM.

### CAREER INTEREST INVENTORY

The AGS Publishing Harrington-O'Shea Career Decision-Making System-Revised (CDM) may be used with the chapters in this unit. Students can use the CDM to explore their interests and identify careers. The CDM defines career areas that are indicated by students' responses on the inventory.

## Chapter at a Glance

### Chapter 9: The Role of Diet in Health
pages 202–227

### Lessons

1. **A Healthy Diet**
   pages 204-209
2. **Carbohydrates, Fats, and Protein**
   pages 210-215
3. **Vitamins, Minerals, and Water**
   pages 216-220
4. **Special Dietary Needs**
   pages 221-224

**Chapter 9 Summary** page 225

**Chapter 9 Review** pages 226–227

**Audio CD**

**Skill Track for Life Skills Health**

**Teacher's Resource Library**

   Activities 31–34
   Alternative Activities 31–34
   Workbook Activities 31–34
   Self-Assessment 9
   Community Connection 9
   Chapter 9 Self-Study Guide
   Chapter 9 Outline
   Chapter 9 Mastery Tests A and B
   (Answer Keys for the Teacher's Resource Library begin on page 559 of this Teacher's Edition.)

## Opener Activity

Bring in several menus from restaurants that include nutritional information. Give copies to students and ask them to discuss or write down their thoughts as a response to these topics:

- What are calories, and how are they important to your diet?
- What types of foods have the most calories? What types of foods have the least calories?
- What types of foods have carbohydrates, fats, and/or proteins?
- What makes a healthy diet?

202 Unit 3 Nutrition

## Self-Assessment

### Can you answer these questions?

☑ 1. Which food groups are the healthiest?

☑ 2. What is an example of a healthy meal menu?

☑ 3. Which foods contain saturated fat?

☑ 4. Why is it important to clean utensils, dishes, and your hands before eating?

☑ 5. How can you read and understand a nutrition label?

☑ 6. Which foods have natural sugar and fiber in them?

☑ 7. How much sugar is in a soft drink?

☑ 8. Why is protein important to your diet?

☑ 9. What foods give you calcium to strengthen your bones?

☑ 10. What is the connection between your diet and your health?

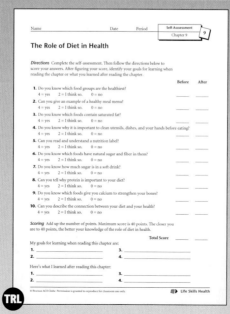

Self-Assessment 9

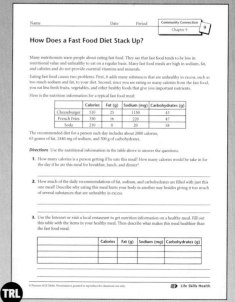

Community Connection 9

# Chapter 9
# The Role of Diet in Health

Food is fuel. Food provides the nutrients your body's systems need to work right. Deciding which foods to eat is important. Good food choices help you stay healthy. Poor food choices can have a bad effect on your life. Eating too much or not enough food can also have a bad effect on your health. The food you eat not only affects your health. It can also affect the quality of your life.

In this chapter, you will learn what a healthy diet is. You will discover how health guidelines can help you choose a healthy diet. You will learn about the nutrients your body needs. You will also learn about special dietary needs and how diet relates to health problems.

## Goals for Learning

◆ To describe how healthy eating affects the body
◆ To explain how food provides calories and nutrients
◆ To study the importance of the six essential nutrient classes and name which foods contain them
◆ To name special dietary needs
◆ To understand how diet can affect health

## Introducing the Chapter

Have a volunteer read aloud the introductory information on page 203. Discuss the importance of eating a balanced diet and the meaning of the expression "You are what you eat" before studying this chapter. Have students reconsider the meaning in a follow-up discussion after reading the chapter.

Have students read the Goals for Learning. Explain that in this chapter students will learn how a proper diet can nourish and strengthen your body, while an improper diet can lead to health problems.

## Notes and Questions

Ask volunteers to read the notes and questions that appear in the margins throughout the chapter. Then discuss them with the class.

## Self-Assessment

Have students complete the Self-Assessment worksheet before and after reading the chapter. Before reading the chapter, have students fill in the "Before" column. Ask students to identify their goals for learning. To get ideas for setting goals, students might use the chapter introductory material on pages 202–203, the checklist on page 202, or the questions on the Self-Assessment worksheet. Students can use the back of the worksheet if they need more space to write.

Collect the Self-Assessment worksheets and pass them out again at the end of the chapter. Have students fill in the "After" column. Ask them to identify at least four major points they have learned. Again, suggest they use the back of the worksheet if they need more space to write. You may want to collect and review the worksheets, but return them to students so they have a record of their goals and accomplishments.

Chapter 9 Self-Study Guide, pages 1–2    Chapter 9 Outline, pages 1–2

# Lesson at a Glance

## Chapter 9 Lesson 1

**Overview** This lesson provides dietary guidelines and shows students how to use those guidelines to plan a healthy diet.

### Objectives

- To explain the correct meaning of *diet*
- To recognize how food gives your body what it needs
- To understand how to make a healthy menu

**Student Pages** 204–209

**Teacher's Resource Library**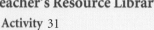

Activity 31

Alternative Activity 31

Workbook Activity x31

## Vocabulary

**essential nutrient**  **nutrient-dense**
**Food Guide Pyramid**

Ask a volunteer to read aloud the definitions of the vocabulary words. Tell students to write a few sentences describing how these words are related. Ask for volunteers who will read their sentences aloud to the class.

## Background Information

In 1774, Antoine Lavoisier performed experimental work demonstrating that combustion, or burning of a material, results from chemicals combining with oxygen. This research led Lavoisier to believe that the breaking down of food by the human body results in the release of the energy used to maintain life processes. As a result of his work, Lavoisier is sometimes called "the father of modern chemistry" and "the father of the science of nutrition."

---

## Lesson 1: A Healthy Diet

### Objectives

*After reading this lesson, you should be able to*

- explain the correct meaning of *diet*
- recognize how food gives your body what it needs
- understand how to make a healthy menu

**Food Guide Pyramid**
*A chart that helps you decide how much and what kinds of food to eat*

The word *diet* does not mean just a way of losing or gaining weight. A diet is made up of the foods you eat and drink. Diet can affect the way you look, feel, and perform. A well-balanced diet gives you a healthy appearance. It makes your hair shine and helps keep your skin clear. It gives you energy to do everything you need or want to do. Eating healthy food also adds to your emotional health. It gives you energy to deal with stress. Your diet affects your general health. Clearly, choosing healthy food to eat is important to your overall health.

Look at Tables C.2 and C.3 in Appendix C in this book. Gather information about what you should weigh based on your age and how tall you are. Your doctor or school nurse can help you learn about healthy ways to maintain or change your weight. A healthy diet and exercise are both important for maintaining or changing your weight.

### What Is a Healthy Diet?

A healthy diet includes different kinds of foods. The U.S. government has set up guidelines to help people choose a healthy diet. These guidelines are listed here.

**Dietary Guidelines**

- Eat lots of different kinds of foods.
- Balance the food you eat with physical activity.
- Choose a diet with plenty of grains, vegetables, and fruits.
- Choose a diet low in fat, saturated fat, and cholesterol.
- Choose a diet low in sugar.
- Choose a diet low in salt and sodium.

### The Food Guide Pyramid

The government developed the **Food Guide Pyramid** (see Figure 9.1.1 on page 205). It is a chart that helps you decide how much and what kinds of food to eat. It groups foods according to the number of servings you should eat each day.

204   Unit 3   Nutrition

The six bands on the pyramid represent five food groups plus oils. The five groups are grains, vegetables, fruits, milk, and meat and beans. The wider food group bands, such as grains, show that you should eat more of these foods than of foods in narrower bands. The narrow yellow band stands for oils. It shows that you should eat very few fats and oils.

Notice that each food group band of the pyramid is wider at the base than at the top. The wide base stands for foods that are low in fats and added sugars. People should choose more of these foods. The narrow tops of the bands stand for foods with more fats and added sugars. People should eat fewer of these foods. However, the more you exercise, the more you can choose foods that are higher in fat. The person climbing the steps of the pyramid reminds people of the importance of exercise.

**Figure 9.1.1** *The Food Guide Pyramid.*  Source: U.S. Department of Agriculture

 **Warm-Up Activity**

Ask students to compare their bodies to a car. Can a car work well using any kind of fuel? Remind students of the different grades of gasoline at a gas station, in addition to other types of fuel such as propane, diesel, and heating oil. Some of these fuels make a car operate poorly, and some won't allow a car to operate at all. Make the analogy with the human body—you cannot put just any kind of food in your body and expect it to work the way it should. Tell students that people take care to give their cars the fuel they need; they should do the same for their bodies. Ask students to write a paragraph explaining their ideas about good nutrition. After they have read Lesson 1, have them review their paragraphs and discuss any misconceptions about nutrition they might have.

 **Teaching the Lesson**

Point out examples of different types of people. Ask students if everyone needs the same amount of food each day. Explain that adolescents need more food during their growing stage. Older adults need less food. Even within age groups, calorie requirements vary from person to person. Have students consult the Food Guide Pyramid on page 205 and consider why it is important to eat a variety of foods. *(No one food supplies the body with all the nutrients it needs.)*

### Pronunciation Guide

Use the pronunciation shown here to help students pronounce difficult words in the lesson. Refer to the pronunciation key that appears in the Glossary at the back of the Teacher's Edition for the sounds of the symbols.

nutrient (nü′ trē ənt)

pyramid (pir′ ə mid)

# 3 Reinforce and Extend

## Health Journal

Have students write down every food they eat for one day, including serving sizes. The next day, have them identify the types of food according to the Food Guide Pyramid chart and estimate how many calories each food contained. Using their journal entries, students can see the types of foods and the amount of calories they consume in a typical day.

## Teacher Alert

The servings listed might be smaller than the size students expected. Point out that more than one serving of a food can be included in a meal. For example, a spaghetti dinner may include a full cup of pasta, or two servings. Ask students to think about the last meal they ate. How many servings of each food did they eat? (Students might need help visualizing some of the measurements. Have ounce and cup measures available.)

---

**Essential nutrient**
*Chemical in foods that the body cannot make*

The chart that goes along with the Food Guide Pyramid (see Figure 9.1.2 below) gives details about the pyramid. The chart lists some of the foods to choose in each food-group band. It gives information about how much of each food to eat each day. Information about finding a balance between food and physical activity is also in the chart. Facts about how to choose and limit fats, sugars, and salt are also listed.

### How Does Food Provide What Your Body Needs?

Food contains nutrients. Nutrients the body gets from food are called **essential nutrients.** Essential nutrients are chemicals in foods that the body cannot make. Scientists have discovered about 50 of these nutrients. They are divided into six essential nutrient classes you will read about in Lessons 2 and 3. These nutrient classes are carbohydrates, fats, proteins, vitamins, minerals, and water. No one food contains all the nutrients your body needs. A healthy diet must include different kinds of foods. Some food sources for the six essential nutrient classes can be found in Table B.1 of Appendix B.

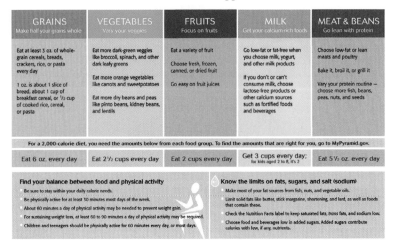

**Figure 9.1.2** *Chart for the Food Guide Pyramid. Source: U.S. Department of Agriculture*

206  Unit 3  Nutrition

**Nutrient-dense**
*Having a good source of nutrients but being low in calories*

Foods vary in the amount of calories and nutrients they contain. Some foods are high in calories but low in nutrients. For example, a teaspoon of sugar has about 16 calories. A teaspoon of green beans has about 1 calorie. Green beans have more nutrients and fewer calories than sugar has. This is important for two reasons. First, your body needs nutrients to stay healthy. Second, it is important to know which foods contain the most nutrients. Your body also needs a specific number of calories. Foods that are good sources of nutrients but are low in calories are **nutrient-dense.** A healthy diet includes many nutrient-dense foods.

Food provides the calories and nutrients your body needs. A calorie is a unit that measures the amount of energy in food. Your body uses the calories for heat, movement, growth, and repair. What should you think about to figure out how many calories you need each day? Whether you are a male or female, how fast you grow, and how active you are all play a role. Here's one example. A person who is tall, muscular, and active needs more calories than a small, inactive person.

Look at serving size when reading nutrition labels. A label may show only 100 calories and 8 grams of fat. This information is for just one serving. A package may contain more than one serving. What if you eat all the food in the package? You will have to multiply the fat and calories by the number of servings you eat.

### Link to ➤➤➤
**Math**

All the information you need about the Food Guide Pyramid is on the United States Department of Agriculture's (USDA's) Web site *(www.MyPyramid.gov)*. You can go to this site to create a healthy eating plan just for you. Under "My Pyramid Plan," you can type in your age, sex, and physical activity level. Then the site gives you the amounts and kinds of foods you should eat each day.

*The Role of Diet in Health   Chapter 9*

## ELL/ESL Strategy

**Language Objective:**
*To describe by example*

Describe or act out an example of a 16-year-old male who weighs 150 pounds, is 5 feet 10 inches tall, and takes in about 2,300 calories each day. He complains about being tired all the time. Ask students to consult Table 9.1.1 and describe what they think the problem might be. *(According to Table 9.1.1, he needs to eat more based on his age, weight, and height. Students should be aware, however, that individuals may have specific dietary or calorie requirements, such as people who are trying to lose weight or people who have diabetes. People with such requirements should consult a doctor before beginning any diet plan.)*

## How Can You Plan a Healthy Diet?

Use the Food Guide Pyramid as your guide to plan a healthy diet. You will be able to make healthy choices that include plenty of different kinds of foods. The pyramid can help you plan your diet. It shows the six food groups and how many servings you need each day. Table 9.1.1 shows sample food plans for 13-year-olds who are physically active 30 to 60 minutes a day in addition to their normal routines.

| Table 9.1.1 Sample Daily Food Plans for Moderately Active 13-Year-Olds | | |
|---|---|---|
| Food Group | Females | Males |
| Grains | 6 ounces | 7 ounces |
| Vegetables | 2.5 cups | 3 cups |
| Fruits | 2 cups | 2 cups |
| Milk | 3 cups | 3 cups |
| Meat and beans | 5.5 ounces | 6 ounces |
| Oils | 6 teaspoons | 6 teaspoons |
| Other fats and sugars | No more than 265 calories | No more than 290 calories |

Go to the USDA Food Pyramid Web site (www.MyPyramid.gov) and enter your information. Use your results to plan what you will eat for the next week. The Web site also offers a meal tracking worksheet you can print out to help you plan what you will eat. At the end of each day, check to see how many of the planned foods you actually ate.

**Word Bank**
essential nutrients
Food Guide Pyramid
nutrient-dense

*Lesson 1 Review* On a sheet of paper, write the phrase from the Word Bank that best completes each sentence.

1. Foods that are a good source of nutrients and are low in calories can be described as _____.
2. The _____ is a way to plan how much and what kind of food to eat.
3. Food provides calories and _____ for your energy needs.

On a sheet of paper, write the answers to the following questions. Use complete sentences.

4. **Critical Thinking** What would happen if you ate the number of calories of someone who is much taller and more active than you are?
5. **Critical Thinking** Suppose you are planning your meals for the day. Which should you have more of—brown rice or cheese? Explain your answer.

The Role of Diet in Health   Chapter 9   209

## Lesson 1 Review Answers

1. nutrient-dense 2. Food Guide Pyramid 3. essential nutrients 4. Sample answer: I would probably gain weight. 5. Sample answer: I should plan to eat more brown rice. The Food Guide Pyramid shows that you should eat more foods from the grain group than from the milk, yogurt, and cheese group.

## Portfolio Assessment

Sample items include:
- Sentences from Vocabulary
- Paragraph from Warm-Up Activity
- Recording of one day's worth of food and calories from Health Journal
- Individualized Food Guide Pyramid plan from Link to Math
- Lesson 1 Review answers

### CAREER CONNECTION

Some students may have part-time jobs in fast-food restaurants. Ask them, as well as other students, if they would ever consider managing a fast-food restaurant. Perhaps they could use their experience to open their own fast-food restaurant. If so, what kinds of changes would they make to a typical fast-food menu? *(Students might include healthier choices, such as more fruits and vegetables, meatless burgers, leaner meats, and so on.)*

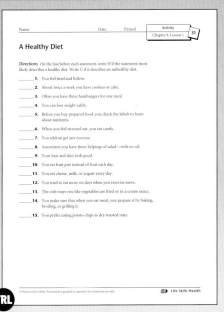

Activity 31    Workbook Activity 31

# Lesson at a Glance

## Chapter 9 Lesson 2

**Overview** This lesson describes three of the six essential nutrients—carbohydrates, fats, and protein—and explains which foods contain them.

### Objectives

- To compare carbohydrates, fats, and protein
- To state the difference between simple and complex carbohydrates
- To understand how the body uses carbohydrates, fats, and protein

**Student Pages** 210–215

**Teacher's Resource Library**

Activity 32
Alternative Activity 32
Workbook Activity 32

## Vocabulary

| | |
|---|---|
| added sugar | fructose |
| amino acid | glucose |
| blood cholesterol | glycogen |
| carbohydrate | lactose |
| cholesterol | natural sugar |
| complex carbohydrate | protein |
| dietary cholesterol | saturated fat |
| essential amino acid | simple carbohydrate |
| fat | |
| fatty acid | sucrose |
| fiber | unsaturated fat |

For each vocabulary word, have students find two sentences from the text that contain that word. Have them read the sentences at least twice and then write an original sentence using that word.

## Background Information

Approximately 3,500 calories of energy are stored in 1 pound of body weight. If people were directed by their doctors to lose weight and subtracted 500 calories per day from their diets, their resulting weight loss should be about 1 pound per week. Many doctors consider a safe rate of weight loss to be no more than 2 pounds per week.

---

## Lesson 2: Carbohydrates, Fats, and Protein

### Objectives

*After reading this lesson, you should be able to*

- compare carbohydrates, fats, and protein
- state the difference between simple and complex carbohydrates
- understand how the body uses carbohydrates, fats, and protein

**Carbohydrate**
*Chemical in foods that provides starches and sugars*

**Fat**
*Stored energy*

**Protein**
*A material in food needed for growth and repair of body tissues*

**Complex carbohydrate**
*Carbohydrate that contains fiber; the best source of dietary carbohydrates*

**Simple carbohydrate**
*A sugar the body uses for quick energy*

Your diet must provide your body with six essential nutrients. The nutrients that provide energy in the form of calories are **carbohydrates, fats,** and **protein.** Carbohydrates are starches and sugars that come mainly from plant food. Fats are stored energy. Protein helps build muscle and repair all body tissues. Carbohydrates, fats, and protein are all nutrients your body needs. Not enough or too much of these nutrients may result in health problems. Eating the right amount of many foods gives your body the nutrients it needs. This creates a healthy, balanced diet.

### What Are Carbohydrates?

Carbohydrates give your body much of the energy it needs each day. There are two kinds of carbohydrates—complex and simple. **Complex carbohydrates** are found in foods such as potatoes, pasta, and bread. Table 9.2.1 shows examples of foods with high amounts of complex carbohydrates. Most of your calories should come from complex carbohydrates. **Simple carbohydrates** are found in sugar, fruits, jelly, and syrup. Many carbonated beverages contain simple carbohydrates.

> **Technology and Society**
>
> Learning what your body mass index (BMI) is can help you decide if you have health risks related to your weight. Web sites can help you find your BMI. Use a safe, reliable site from a government or school agency. Enter the search term *body mass index.* You will not need to enter your name or contact information. As you learned earlier, sites that end in *.gov, .edu,* or *.org* are usually the most reliable.

210 Unit 3 Nutrition

---

## Technology and Society

Ask a volunteer to read aloud the Technology and Society feature on page 210. Explain to students that the body mass index is a much better tool than a scale for determining a healthy weight. The body mass index takes into account the height and weight of a person. Give students an example by asking a question like: "Is 180 pounds a healthy weight for a 30-year-old man?" After they answer, ask them if they would change their answer if they knew the man was 5 feet 2 inches tall or 6 feet 8 inches tall.

**Fiber**
*The coarse material in fruits, vegetables, and grains that helps you digest your food*

**Sucrose**
*Table sugar*

**Fructose**
*The natural sugar found in fruit*

**Lactose**
*The natural sugar found in milk*

**Added sugar**
*Sugar that is added to a food*

**Natural sugar**
*Sugar that exists in a food in nature*

Complex carbohydrates are important to the body. They are the best source of carbohydrates to eat. Foods with complex carbohydrates make you feel satisfied and full. They also contain other nutrients that add **fiber** to your diet. Fiber is not a nutrient, but it is important in helping move food through your digestive system. It is a coarse material in fruits, vegetables, and grains. Table B.2 in Appendix B lists the fiber content of specific foods. Studies show that some types of fiber in the diet help protect against certain health risks.

Simple carbohydrates, or sugars, have many different names. For example, **sucrose** is table sugar. **Fructose** is the natural sugar found in fruit. **Lactose** is the natural sugar found in milk.

Your body uses sugar for quick energy. Your body easily digests sugar and it enters your bloodstream quickly. However, the energy boost you get from sugar wears off quickly. You then feel tired. This is because sugar has no other nutrients.

Sugars found in foods are either **added sugar** or **natural sugar**. Natural sugar is already present in a food as the food grows in nature. It is not added when it goes to the factory or store.

| Table 9.2.1 Foods High in Complex Carbohydrates | | | |
|---|---|---|---|
| | Total Carbohydrates (grams) | Simple Carbohydrates (grams) | Complex Carbohydrates (grams) |
| Bread, 1 slice | 13 | 1 | 12 |
| Corn flakes cereal, 1 ounce (low sugar) | 24 | 2 | 22 |
| Pasta or rice, $\frac{1}{2}$ cup (cooked) | 20 | 0 | 20 |
| Beans, 1 cup (cooked) | 40 | 0 | 40 |
| Potatoes, corn, or peas, 1 cup | 30 | 6 | 24 |
| Carrots or beets, 1 cup | 12 | 6 | 6 |
| Broccoli, 1 cup (cut up) | 7 | 0 | 7 |

### Warm-Up Activity

Write the headings *Carbohydrates*, *Fats*, and *Protein* on the chalkboard. Ask students to name different foods they eat and discuss the carbohydrate, fat, and protein content of each item. Accept all suggestions at this time. Write the food item under the appropriate heading(s) on the chalkboard. Refer back to this chart after studying the lesson. Allow students to make changes to the chart based on their new knowledge.

### Teaching the Lesson

Ask students if they have heard the statement "Bread is the staple of life." The statement reflects the fact that bread is a main component of many diets. Ask them if this is reflected in the Food Pyramid. *(yes)* Would it be wise to eat only bread as a source of carbohydrates? *(No, foods contain more than one kind of nutrient, and by eating a variety of foods that contain carbohydrates, you are providing other essential nutrients to your body.)*

### Reinforce and Extend

#### GLOBAL CONNECTION

Throughout history, starches have been—and continue to be—an important dietary component. Many Native Americans obtained starch from corn. In Asia, the main source of starch is rice. In the Caribbean and in African countries, starch comes primarily from cassavas and yams. Many other cultures depend on varying forms of corn, wheat, rice, beans, and tubers as a basic component of people's daily diet. Ask students to identify sources of starch that may be popular in their own culture or in the homeland of their ancestors. Ask volunteers to bring in samples of some of the foods for the class.

## TEACHER ALERT

Students might think that all fats are bad for them. Discuss with them why it is important to have some fats in your body. *(Fats help protect internal organs from temperature change, protect the body from external blows, help protect nerves, and are a source of energy.)* Tell them that saturated fats are the fats they should avoid because they can raise your cholesterol level, which can lead to cardiovascular disease. Explain that to maintain a diet low in saturated fats, they should limit their intake of foods such as red meats, whole milk, and butter.

## LEARNING STYLES

### Visual/Spatial

Have students cut three squares (about 6 inches by 6 inches) of brown paper from a bag. Rub shortening on one square, egg white on the second square, and a slice of raw potato on the third square. Let the pieces of paper dry. Explain to students that light will pass through paper that has fats on it. Then have students hold each square up to a bright light and compare the amount of light that passes through each piece. From these observations, ask students to draw conclusions about which foods contain fats and which do not. *(Shortening contains fats, while egg whites and potatoes do not contain fats.)* Invite students to use this process to test other foods for fats.

---

*Glucose*
The main sugar in blood; the major energy source for your body

*Glycogen*
Glucose that is stored in your body

*Fatty acid*
Acid that includes saturated and unsaturated fats

*Saturated fat*
A material in food that comes from animal products and that can lead to health problems

Eating fruit is usually healthier than drinking fruit juice. You will feel satisfied longer if you eat the fruit, because it has more fiber. Fruit has lots of fiber. Most of the fiber is lost when the juice is squeezed out of fruit. Also, many fruit juices contain added sugars.

---

Added sugar is put into a food at the factory or plant where it is made. For example, an apple contains the natural sugar fructose. The sugar is part of the apple as it grows. A candy bar contains added sugar. The sugar is put into the candy bar when it is being made. Some foods with added sugar have little or no nutrients. For example, a 12-ounce can of sweetened cola contains 9 teaspoons of sugar. It contains few nutrients. Too much added sugar in your diet can cause health problems.

Many foods that are a source of sugar also have other nutrients. Fruits, vegetables, and bread have carbohydrates that give you enough sugar for your body.

### How Does Your Body Use Carbohydrates?

Carbohydrates are a mixture of chemicals. Your body changes carbohydrates into **glucose,** the main sugar in blood. Glucose moves through your body to give your cells fuel for energy. Extra glucose is stored in your liver and muscle tissue as **glycogen,** or starch. When your body needs more glucose, it changes the glycogen back into glucose. Glucose is the main nutrient your brain needs to help you think clearly and pay attention to what you are doing.

What happens when you eat too many carbohydrates and do not get enough physical activity? You do not use all of the sugar and starch that the carbohydrates give you. Then your body stores the extra sugar and starch as fat.

### What Are Fats?

Like carbohydrates, fats give your body energy, but fats are stored energy. Fats are part of all body cells. They help protect your internal organs from temperature changes. They protect your body if you bump into or are hit by anything and are part of what protects your nerves.

Fats in your diet are broken down in your body into **fatty acids.** These are either saturated or unsaturated. Examples of foods that have a lot of **saturated fat** are butter, meat, ice cream, and chocolate. Eating too much fat can lead to health problems.

**Cholesterol**
*A waxy, fatlike material in your body cells that helps with certain body functions*

**Blood cholesterol**
*Cholesterol in the bloodstream*

**Dietary cholesterol**
*Cholesterol found in some foods*

**Unsaturated fat**
*Material in food from vegetable and fish oils that helps lower cholesterol*

Eating too many saturated fats can cause you to have too much **cholesterol.** Cholesterol is a waxy, fatlike material that is important in making hormones. It also helps digestion. This good kind of cholesterol, **blood cholesterol,** is carried in your bloodstream. But some cholesterol can build up in blood vessels and clog them. Then your blood cannot flow freely. If this happens, you could have serious health problems.

Another kind of cholesterol, **dietary cholesterol,** is found in some foods. One way to lessen your chances of having high blood cholesterol is to stay away from foods that have a lot of saturated fats. Instead, eat foods containing **unsaturated fats.** For example, you can eat less red meat and more fish. Fish contains unsaturated fat. You can also use vegetable oils instead of animal fats and butter. You can drink skim milk instead of whole milk.

Fats are important to your body. However, too many fats can lead to a greater chance of getting some diseases, such as clogged arteries. Eating too many fats can also cause you to gain weight. Table C.1 in Appendix C shows some exercises to do to keep your body weight at a good level.

## What Is Protein?

Protein helps build muscle and repair all your body tissues. It is part of every cell in your body. Your muscles, bones, blood, and skin all contain protein. Since cells are always being replaced, your body needs protein throughout your life. Carbohydrates, fats, and proteins are all sources of energy. When does your body use protein as an energy source? This happens only when it does not get enough calories from carbohydrates and fats.

### Research and Write

Why might people in a poorer community find it hard to eat healthy foods? Work with a partner to discover the answer. Visit a library and gather information from newspapers, magazines, and the Internet. Then write an editorial to your local newspaper. Explain ways that people might overcome this problem and suggest ways in which a poorer community might get more healthy foods to eat.

### Research and Write

Have students read and complete the Research and Write feature on page 213. With the permission of students and their parents, mail or email the letters they write to local newspapers. Monitor the newspapers to see if any of your students' letters are printed. Share the printed letters with the class.

### ELL/ESL Strategy

**Language Objective:** *Learning vocabulary specific to the content area*

Provide pictures of some common foods on index cards. Ensure that there is a diversity of foods across many cultures. Have the students sort the cards into piles that represent carbohydrates, fats, or proteins. If the food pictured on a card fits into more than one category, ask students to explain why. (There is a good chance that many of the foods on the cards will fit into more than one category.) Ask volunteers to work with their parents at home to create some of the dishes pictured on the cards. Volunteers can share these dishes with the class.

## GLOBAL CONNECTION

Have students investigate which is commonly known as "the Mediterranean diet"—the basis for the typical diet of people who live in Mediterranean countries. Students could compare it with their own diet or with a typical American diet. They will discover that the Mediterranean diet contains many fruits, vegetables, breads and other grains, beans, and nuts, as well as moderate amounts of dairy products, eggs, fish, and poultry. It focuses on unsaturated fats (such as olive oil) rather than saturated fats. Therefore, the diet contains very little red meat.

## AT HOME

Ask each student to work with his or her parents or guardians to take an inventory of the food their family eats at home. Students and their families can make some generalizations about the carbohydrate, fat, and protein content of their diets. Tell students that they can work with their families to make dietary changes, if appropriate, that will result in a healthier balance of the three nutrients. With parents' permission, have volunteers share their conclusions with the class.

## Decide for Yourself

Have students read the Decide for Yourself feature on page 214. Ask them to think about what kinds of foods make healthy snacks. As an assignment, have them bring to class at least one recipe for a healthy snack. Ask students to explain why their recipe is healthy.

### Decide for Yourself Answers

**1.** Sample answer: I should consider the nutritional value of the food, the amount of saturated and unsaturated fats, the calorie content, and the cost. **2.** Baked apples have fiber, are low in calories, and are low in fat and cholesterol. **3.** Sample answer: I could suggest carrot sticks, pear slices, and a nutrition bar.

---

**Amino acid**
A small chemical unit that makes up protein

**Essential amino acid**
An amino acid the body cannot make by itself but that can be obtained from food

Why is not skipping meals important?

Proteins are made up of chains of building blocks called **amino acids.** There are many kinds of amino acids. Your body is able to make many proteins. It is unable to make nine proteins called **essential amino acids.** You must get these amino acids from your diet. Your body can make new proteins from these nine essential amino acids.

The foods that include all nine essential amino acids are fish, meat, eggs, milk, and poultry. Some foods provide only a few of the essential amino acids. These are plant foods such as grains, seeds, peas, and beans. Combining certain foods can provide all nine amino acids. Two such combinations are peanut butter and wheat bread or macaroni and cheese.

As with other nutrients, your body needs only a specific amount of protein. If it gets more than it needs, protein becomes a source of extra calories. Foods give you different levels of nutrients. A food high in protein may also be high in fat. For example, eggs are a good source of protein. They also have a lot of fat. It is important to choose foods with protein carefully. Do not eat more than your body needs.

### Decide for Yourself

You just got home from school. Dinner will not be until 7:30. What kind of snack should you eat? Consider time, taste, cost, nutrition, and your goals for a healthy diet. Here is a recipe you might try.

**Baked Apples**
4 apples
1 teaspoon brown sugar (optional)
2 cups unsweetened apple juice
1 teaspoon allspice or ground cloves
1 teaspoon cinnamon

Preheat oven to 350°F. Use a knife or apple corer to remove the core of each apple. Place the apples in a baking dish. Add apple juice so the liquid covers halfway up the apples. Bake for 40 to 50 minutes.

Remove apples. Drain juice into a small saucepan. Add cinnamon, allspice or cloves, and brown sugar. Boil liquid until it turns syrupy. Pour over apples. Serve warm.

Yield: 4 servings
Per Serving: 126 calories, 4 g fiber, 1 g fat, 0 mg cholesterol

On a sheet of paper, write the answers to the following questions. Use complete sentences.

1. What should you consider when deciding what to eat?
2. Why are baked apples a healthy snack?
3. What are three other healthy snacks you could suggest to someone?

***Lesson 2 Review*** On a sheet of paper, write the answers to the following questions. Use complete sentences.

1. How does glucose help your body?
2. What are two foods that have saturated fat and two that have unsaturated fat?
3. What are essential amino acids?
4. **Critical Thinking** What might happen if you did not eat any fat? Explain your answer.
5. **Critical Thinking** How can you use the Food Guide Pyramid to plan your meals?

The Role of Diet in Health   Chapter 9   215

## Lesson 2 Review Answers

**1.** Glucose helps the body by giving it energy. **2.** Answers should include two foods that contain saturated fat, such as red meat, butter, and whole milk. Answers also should include two foods that contain unsaturated fat, such as vegetable oil, skim milk, and fish. **3.** Essential amino acids are the amino acids your body cannot make by itself, but must get through the diet. **4.** Sample answer: I would probably have little energy and might have damage to nerves. Fat protects nerves and provides energy. **5.** Sample answer: The Food Guide Pyramid can help me make healthy choices about my diet. It reminds me to keep track of the number of servings I have from the five main food groups.

## Portfolio Assessment

Sample items include:
- Sentences from Vocabulary
- Letters from Research and Write
- Diet comparisons from Application: Global Connection
- Food inventories from Application: At Home
- Lesson 2 Review answers

### HEALTH JOURNAL

Now that students have read Lesson 2 and have more information about nutrition, ask them to reconsider the foods they ate and recorded in the Health Journal activity in Lesson 1. Have them write an evaluation of their diet for that day, and ask them to include suggestions on how to improve their diet.

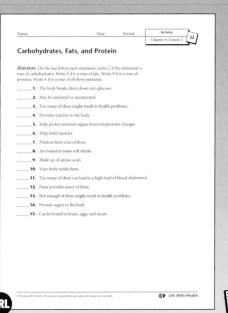

Activity 32

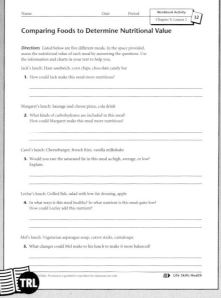

Workbook Activity 32

The Role of Diet in Health   Chapter 9   215

# Lesson at a Glance

## Chapter 9 Lesson 3

**Overview** This lesson describes the remaining essential nutrients—vitamins, minerals, and water—and the importance of each.

### Objectives

- To explain why vitamins, minerals, and water are important
- To recognize the healthiest way to prepare vegetables
- To identify ways in which you can get different types of vitamins
- To understand where to find calcium

**Student Pages** 216–220

**Teacher's Resource Library**

Activity 33

Alternative Activity 33

Workbook Activity 33

## Vocabulary

calcium  sodium
mineral  vitamin
phosphorus

Discuss the definitions of the words with the class. Then have students create a matching game by writing the vocabulary words on the left side of a sheet of paper and, in random order, clues for the words on the right side of the sheet of paper. Tell students to trade papers and match the words with the clues.

## Background Information

Many vitamins were first discovered when their role in preventing diseases caused by dietary deficiencies became known. Scurvy is a disease caused by a lack of vitamin C. British sailors often had scurvy because most food carried on long voyages did not contain enough vitamin C. In 1774, British doctor James Lind proved that supplementing shipboard diets with lemons supplied the needed vitamin. By 1795, all British navy ships carried limes (thought to be better at preventing scurvy because they contained more acid), and from that time on British sailors became known as "limeys." (Tell students that today "limey" is considered a derogatory word.)

216  Unit 3  Nutrition

---

## Lesson 3  Vitamins, Minerals, and Water

### Objectives

After reading this lesson, you should be able to
- explain why vitamins, minerals, and water are important
- recognize the healthiest way to prepare vegetables
- identify ways in which you can get different types of vitamins
- understand where to find calcium

**Vitamin**
*A material needed in small amounts for growth and activity*

**Mineral**
*A natural material needed for fluid balance, digestion, and other bodily functions*

Carbohydrates, fats, and proteins are three essential nutrients that give your body energy. The other three essential nutrients are **vitamins, minerals,** and water. A vitamin is a material your body needs. Minerals are materials formed in the earth. Vitamins, minerals, and water do not give you energy. They help your body change food into energy. Your body needs smaller amounts of vitamins and minerals than it does carbohydrates, fats, and protein. Your body also needs enough water to replace the amount it loses every day.

### Why Are Vitamins Important?

Your body needs vitamins so you can grow and stay healthy. Table 9.3.1 shows why you need certain vitamins and where you can get them.

When you cook, your food loses a number of vitamins. For example, the B vitamins and vitamin C dissolve in water. What happens when you cook vegetables that contain these vitamins? Many of the vitamins mix with the water. When you pour off the water, you lose the vitamins. To keep this from happening, use only a small amount of water to cook vegetables. Also, do not cook them too long. You get more vitamins from uncooked vegetables than from cooked vegetables. To stay healthy, you should eat foods with vitamins every day. Your body does not store vitamins from one day to the next.

Vitamins A, D, E, and K dissolve in fat. These vitamins can build up and be stored in your body's fatty tissues. These vitamins are found in meat, milk, eggs, and other foods.

As with other nutrients, vitamins help keep your body healthy. It is possible, however, to have too many vitamins. Extra vitamins such as vitamins A and E can be bad for your body. Some people can harm their bodies by taking too many vitamin supplements, or pills. You need vitamins in small amounts. If you eat healthy foods, you are probably able to get all the vitamins you need from your diet.

216  Unit 3  Nutrition

**Calcium**
*Mineral important for keeping bones and teeth strong*

**Phosphorus**
*Mineral that works with calcium to keep bones and teeth strong*

## Why Are Minerals Important?

Your body needs small amounts of many minerals to stay healthy. Minerals are needed to help you digest your food and keep your bodily fluids balanced.

One important mineral is **calcium.** Calcium is found in dairy products and green leafy vegetables. It is important for controlling how parts of your body behave and for keeping your bones and teeth strong. Another important mineral is **phosphorus,** which mixes with calcium to keep your bones firm. Phosphorus is found in peas, beans, milk, meat, and other foods. Your body needs larger amounts of calcium and phosphorus than of other minerals.

### Table 9.3.1 Essential Vitamins

| Vitamin | Purpose | Sources |
|---|---|---|
| Vitamin A | Helps skin, hair, eyes, lining of nose and throat | Milk, egg yolk, beef liver, carrots, sweet potatoes, yellow squash, spinach, other greens |
| B Vitamins: | | |
| Niacin | Protects skin and nerves, aids digestion | Beef, chicken, turkey, liver, whole wheat, milk, cereals, mushrooms |
| Thiamin | Protects nervous system, aids appetite and digestion | Pork, sunflower seeds, whole grains, cereal, green beans, peanuts, liver |
| Riboflavin | Increases resistance to infection, prevents eye problems | Milk, milk products, pork, liver, eggs, breads, rolls, crackers, green leafy vegetables |
| Vitamin C | Helps form bones and teeth, increase iron absorption, resist infection and stress | Tomatoes, most citrus fruits, kiwis, potatoes, fruit juices, green pepper |
| Vitamin D | Helps form strong bones and teeth | Fish oils, milk, sunlight |
| Vitamin E | Helps maintain cell health, has possible role in reproduction | Vegetable oils, margarine, peaches |
| Vitamin K | Aids in blood clotting | Green leafy vegetables, soybeans, bran, peas, green beans, liver |

## Warm-Up Activity

Before reading the lesson, encourage students to name examples of foods that contain vitamins and minerals. Students will likely name a variety of fruits, vegetables, meats, dairy products, grains, and so on. List these items on the chalkboard.

## Teaching the Lesson

Have the students find the foods they named in the Warm-Up Activity in Table 9.3.1 on page 217. Ask them which vitamins are present in those foods. Tell students that as people eat fewer fresh fruits and vegetables, they do not get the vitamins they need. In response, an increasing number of companies are adding vitamins and minerals to their foods. It can be important for people to take vitamin supplements when they are suffering from an illness, since vitamins and minerals are important in the body processes that maintain health.

## Reinforce and Extend

### LEARNING STYLES

**Interpersonal/Group Learning**

Bring in a variety of packaged foods, including breakfast cereal, for students to examine. Have groups of students read the nutritional labels to each other and note the presence of vitamins and minerals. Tell them that many of the vitamins and minerals listed are added artificially—that is, the vitamins and minerals listed do not exist naturally in the ingredients of the packaged foods. Rather, companies add them during processing.

## Health Myth

Have students read the Health Myth feature on page 218. Show students the warning label on the side of a vitamin bottle. Tell students that some health myths about vitamins are persistent, but untrue. For instance, taking massive doses of Vitamin C will not ward off the common cold. It might give you diarrhea, however, and could lead to other complications as well.

## LEARNING STYLES

### Logical/Mathematical

To demonstrate the amount of water contained in fresh vegetables, begin by weighing fresh sliced carrots or celery. Then dry the vegetable pieces on a flat tray in an oven at a very low temperature. After the vegetables have completely cooled, weigh them again. Subtract the dried weight from the fresh weight to demonstrate how much water was in the fresh vegetables. The results will show that vegetables are mostly water, some as much as 85 percent.

## IN THE ENVIRONMENT

Students have learned that water is one of the six nutrients that humans need to live. Tell them that in some parts of the world, millions of people do not have access to clean water. Unclean water contributes to many diseases, including malaria, hepatitis, ringworm, and cholera, and other conditions such as lead poisoning and diarrhea, both of which can be fatal.

## Research and Write

Have students read and conduct the Research and Write feature on page 218. Show students how to use a search engine and keywords to find helpful Web sites. You might want to have students work with partners or in small groups to complete this activity.

218   Unit 3   Nutrition

---

**Sodium**
*Mineral that controls fluids in the body*

**What are crash diets? Why are they dangerous?**

**Health Myth**

**Myth:** You should take as many vitamins as you can.

**Fact:** Taking more than the suggested dose of vitamins can have serious health effects. Some problems from taking too many vitamins are bone pain, itching, and involuntary movements of the muscles.

Your body uses only small amounts of the mineral iron. Iron is found in liver, beef, dried fruits, whole-grain foods, and green leafy vegetables.

**Sodium** is another mineral used in small amounts. It is necessary for controlling the amount of fluids in your body. You need only a small amount of sodium in your diet. Too much sodium may give you high blood pressure. Sodium is found in table salt and many other foods. You probably do not have to add salt to your foods.

Minerals cannot be stored in the body. It is important to get enough minerals every day to keep your body working as it should. A balanced diet includes all the minerals your body needs.

### Why Is Water Important?

Water helps each cell in your body do its work. About 65 percent of your body's weight is water. Water carries nutrients, hormones, and waste products to and from the cells in your body. It helps you digest and use the food you eat. Water also helps control your body's temperature.

Water does not contain calories, but it is an important nutrient. Each day, your body loses about eight glasses' worth of water. This is why you should drink at least eight 8-ounce glasses of water every day. You can drink plain water or get water in soups and juices and other healthy drinks.

### Research and Write

Do some research to learn more about nutrition labels. Visit a grocery store and use the Internet to learn about the information shown on these labels. Make a set of posters to teach people in your community about healthy eating. Show large nutrition labels on the posters. Explain what each part of the label means. Then explain which parts of the labels show healthy choices.

218   Unit 3   Nutrition

## Health in Your Life

### Avoiding Germs

There is more to good health than a healthy diet. You also need to keep your hands clean and free of germs. Your hands pick up germs throughout the day. If you do not wash your hands before you touch your food, the germs get on your food, and you are eating the germs!

Wash your hands thoroughly. Use soap and warm water, not hot or cold. Use soap on both sides of your hands. Wash your wrists and the areas between your fingers. Do not forget to get soap under your fingernails. Wash for about 30 seconds. To measure 30 seconds, slowly count to 10 three times. Then rinse your hands well.

You may not have the chance or time to wash your hands before lunch at school. There is a solution. You can carry a small bottle of hand sanitizer. It does not require water or soap. You can easily carry it in your pocket, lunch bag, purse, or backpack.

On a sheet of paper, write the answers to the following questions. Use complete sentences.

1. Where can you wash your hands at school before eating breakfast or lunch? Describe a way in which you can remind yourself to wash your hands before eating at school.
2. Why is it important for waiters and waitresses to wash their hands when they are working?
3. Name two jobs in which employees not only wash their hands frequently but also wear disposable rubber gloves. Explain why these employees wear and throw away several pairs of rubber gloves throughout the day.

---

## Health in Your Life

Ask students to read the Health in Your Life feature on page 219. Tell them that many colds and other illnesses are transferred between humans through some contact with hands. Most doctors recommend washing hands regularly as the best way to avoid getting sick.

**Health in Your Life Answers**

1. Sample answer: I can wash my hands in the restroom or in the locker room.
2. Sample answer: It is very important that waiters and waitresses wash their hands when they are working because they handle silverware, napkins, plates, and sometimes food. If a restaurant employee has dirty hands, he or she will spread germs to food that customers will eat.
3. Sample answer: Doctors and nurses frequently wash their hands when they are working, and they wear disposable rubber gloves on the job. Since doctors and nurses physically interact with ill people, it is very important that they don't spread germs and disease-causing bacteria.

### ELL/ESL STRATEGY

**Language Objective:** *To learn vocabulary specific to the content area*

Have students create a concept map to compare the six essential nutrients. They should include the purpose of each nutrient and examples of sources of each nutrient in the concept map. After they construct their maps, have students test their organization by using the map to explain the nutrients to one another. Allow them to make adjustments to their maps.

### Link to

**Social Studies.** Ask a volunteer to read aloud the Link to Social Studies on page 220. Students can learn more about vitamins and minerals at the Web site for MedlinePlus, a service of U.S. National Library of Medicine and the National Institutes of Health: www.nlm.nih.gov/medlineplus/vitaminsandminerals.html.

### Lesson 3 Review Answers

**1.** D **2.** A **3.** C **4.** Sample answer: I will probably have weak bones. **5.** Sample answer: My body temperature would probably rise or fall a great deal. I might have trouble digesting food because I would be dehydrated.

### Portfolio Assessment

Sample items include:
- Matching Game from Vocabulary
- Lesson 3 Review answers

---

**Lesson 3 Review** On a sheet of paper, write the letter of the answer that best completes each sentence.

1. The healthiest way to prepare vegetables is to _____.
   A reheat them over high heat
   B boil them in a lot of water
   C bake them in a hot oven
   D cook them in a little water

2. A good source of calcium is _____.
   A milk           C water
   B oil            D cola

3. Calcium and phosphorus are examples of _____.
   A fluids         C minerals
   B vitamins       D vegetables

On a sheet of paper, write the answers to the following questions. Use complete sentences.

4. **Critical Thinking** What might happen if your body does not get enough calcium and phosphorus?

5. **Critical Thinking** What might happen to your body temperature if you do not drink water for a long time?

### Link to >>>

**Social Studies**

Fresh vegetables and orange or grapefruit juice are good sources of vitamin C. What happens when a person does not get vitamin C for a long time? He or she might get sick with scurvy. Scurvy is a disease that affects the skin, gums, teeth, and bones. A person can die from scurvy. Scurvy was a big problem for sailors during the 1700s and 1800s. They could not keep fresh vegetables on their ships when they were gone for a long time. Later, people realized that eating fruits such as lemons and limes could prevent scurvy.

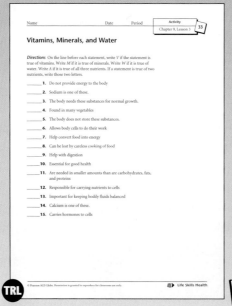

Activity 33

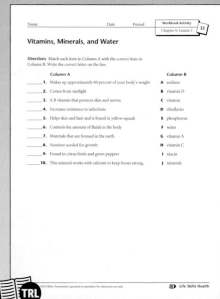

Workbook Activity 33

# Lesson 4: Special Dietary Needs

**Objectives**

*After reading this lesson, you should be able to*

- understand what teenagers need to eat
- recognize what athletes need to eat
- explore the importance of diet in weight control
- discuss how a poor diet affects health

The Food Guide Pyramid gives general guidelines to follow so people can choose a healthy diet. You must adjust your diet to meet your own needs. For example, you must know how many calories you need each day. This is based on your size, your activity level, and whether you are male or female. You also must know which foods give you the nutrients you need. As you get older, your calorie needs will change.

## What Special Dietary Needs Do Teenagers Have?

Because you are still growing and changing, you need large quantities of calories, calcium, and iron. Since boys are usually larger and grow more than girls do, they need more calories.

Calcium is an important mineral in a teenager's diet. You need calcium for bone growth. If your diet does not contain enough calcium, your body takes calcium from your bones. Over time, if your body keeps taking the calcium from your bones, it can cause them to break easily. The best sources of calcium are milk, yogurt, and cheese (see Table B.3 in Appendix B). Your body is able to absorb calcium from dairy products because they contain vitamin D.

Iron is important for getting oxygen into your blood and cells. It is found in liver and other meats, dried fruits, and whole-grain foods. It is also found in green leafy vegetables. Foods with vitamin C in them help your body absorb iron. For example, a meal with meat and potatoes is a healthy choice. It provides iron and vitamin C to help your body absorb the iron. If you are a teenage girl, you need more iron to make blood cells to replace the ones you lose during menstruation.

*The Role of Diet in Health  Chapter 9  221*

---

# Lesson at a Glance

## Chapter 9 Lesson 4

**Overview** This lesson discusses special dietary needs for teenagers and athletes and the importance of diet in weight control.

## Objectives

- To understand what teenagers need to eat
- To recognize what athletes need to eat
- To explore the importance of diet in weight control
- To discuss how a poor diet affects health

**Student Pages** 221–224

**Teacher's Resource Library**
Activity 34
Alternative Activity 34
Workbook Activity 34

## Vocabulary

**deficiency**          **malnutrition**

Have a volunteer read the definitions of *deficiency* and *malnutrition* and discuss how the words are related. Then ask students to write one sentence using deficiency to define malnutrition and another sentence using malnutrition to describe the effects of a dietary deficiency. Ask volunteers to read their sentences aloud.

## Background Information

Teens need a lot of calcium in their diets. Many people think that the only foods that contain calcium are milk and other dairy products. If teenagers have an allergy to dairy products, as some people do, or if teenagers don't like dairy products, where can they get the calcium they need? Luckily, calcium is found in other foods, such as legumes (beans, peas, lentils, and peanuts); green, leafy vegetables (like broccoli and kale); and soybean products.

*The Role of Diet in Health  Chapter 9  221*

### 1 Warm-Up Activity

Have students read page 221. Ask them if they think they are getting enough calcium, iron, and calories in their diets. For those who say that they don't think they are getting enough calcium, iron, and/or calories in their diets, ask them to think about how they could. Ask the class for suggestions and write them on the board. *(Students might say that they could drink a glass of milk with breakfast, use whole-grain bread to make sandwiches, and so on.)*

---

**PRONUNCIATION GUIDE**

Use the pronunciation shown here to help students pronounce difficult words in the lesson. Refer to the pronunciation key that appears in the Glossary at the back of the Teacher's Edition for the sounds of the symbols.

dietary (dī′ ə ter ē)

menstruation (men strü a′ shən)

---

### 2 Teaching the Lesson

Point out that the Food Guide Pyramid provides general guidelines to help people choose a healthy diet. Stress the word *guidelines*. Ask why the word guidelines is used instead of the word rules. *(Individuals have their own needs, and the information in the Food Guide Pyramid has to be adjusted to account for those needs.)*

### 3 Reinforce and Extend

**Health Myth**

Ask a volunteer to read aloud the Health Myth feature on page 222. Point out that this myth probably persists because when most people stop working out, they not only tend to lose muscle tone but also gain weight due to the decrease in exercise. Tell students that the best way to stay in shape is to eat nutritious foods in moderate amounts and to get an hour of moderate to vigorous exercise every day.

---

**Malnutrition**
*Condition you get from a diet that lacks some or all nutrients*

**Health Myth**

**Myth:** Your muscles will turn into fat if you do not exercise.

**Fact:** Fat and muscle are two different kinds of tissue. One cannot turn into the other. If you stop exercising, your muscles can lose some of their muscle tone. People often mistake the loss of muscle tone for fat.

---

### What Special Dietary Needs Do Athletes Have?

If you play sports, you need more calories. The extra calories give you the extra energy you need. What if you play sports and eat a balanced diet? Studies show that you do not need extra protein, vitamins, or minerals.

If you play sports, you should eat at least 3 hours before your game or race. That way, your stomach has time to digest the food. Your meal should be low in fat and easy to digest. It should include three glasses of water or another drink. You need to replace the water you lose when you sweat before, during, and after your game or race.

Over the years, athletes have tried many different diets to try to improve how they play. Sometimes, these diets can be harmful. It is safest for people who play sports to eat a healthy, balanced diet. If you play sports, make sure you get enough calories every day.

### How Is Diet Important in Weight Control?

Eating foods with too many calories can cause you to gain weight. Not getting enough calories can cause you to lose weight. A healthy diet with the right amount of physical activity can help you maintain a healthy weight.

If you want to make a big change in your weight, talk to a health care professional first. Some plans can be unhealthy and cause other problems. A healthy and balanced diet is the best way to keep a healthy weight.

### How Does a Poor Diet Affect Health?

If your diet lacks the six essential nutrients, you may have **malnutrition**. Malnutrition comes from a diet that lacks one or more of the essential nutrients. Some people eat enough food but still do not get enough nutrients.

Some signs of malnutrition are tiredness, frequent headaches, stomachaches, or depression. Malnutrition may lead to serious health problems such as softened bones.

*Deficiency*
*A lack of something*

Another problem that comes from a poor diet is a **deficiency**. A deficiency is a lack of something. For example, people without enough iron in their diet may have an iron deficiency. Deficiencies may lead to more serious health problems. These include damage to the nervous system or heart. You can keep from getting malnutrition and deficiencies by eating a balanced diet.

## How Are Diet and Disease Related?

Several diseases are linked to diet. If your diet is unhealthy, your risk is greater for some diseases. These include high blood pressure, heart disease, and cancer. For example, a person may have too much salt in his or her diet. The extra salt makes the body hold extra fluid. This affects the blood vessels and can cause higher blood pressure.

### Health at Work

**Nutrition Technician**

Hospitals and retirement homes need workers who know about nutrition. Nutrition technicians follow guidelines to prepare and serve meals to patients. Nutrition technicians change recipes based on specific needs of patients. Some record keeping is also required. An additional duty might be keeping track of what supplies are on hand and what supplies are needed.

Nutrition technicians can be supervisors. A supervisor manages and trains other kitchen workers. A supervisor might also create work schedules.

A nutrition technician needs a high school diploma. A college degree is generally not needed.

---

## ELL/ESL Strategy

**Language Objective:** *To discuss an assigned topic as part of a small group*

Divide students into small groups and ask them to discuss what they would suggest to a parent or guardian whose child can't drink milk or eat dairy products (because of allergies) and who also refuses to eat foods such as broccoli, beans, and tofu. Have each group explain to the class its suggestions. (*Students might suggest that the child's parents give him or her fruit juices fortified with calcium, soymilk, soy yogurt, or almonds. They also might try using calcium supplements or "disguising" calcium-rich foods the child doesn't like. However, the parent or guardian should consult a physician before using calcium supplements.*)

### Health at Work

Have students read the Health at Work feature on page 223. Ask them why it is important to employ a Nutrition Technician in places that regularly serve food to large numbers of people. (*If there is no one to coordinate the nutrition of the food served, people who eat there regularly might not get all the nutrients they need, and they could suffer health problems.*) Ask students if they would want someone to plan their meals for them every day. Ask them if their families or others they know have ever used menus provided by a nutritionist, an organization, or a computer software program. Discuss whether the programs were helpful and effective.

## Lesson 4 Review Answers

**1.** calories **2.** calcium **3.** high blood pressure **4.** Sample answer: You might not get enough nutrients and calories through the planned diet. You might have personal health issues that need to be addressed. A health care professional can give good advice because he or she is educated and objective. **5.** Sample answer: Person A is more likely to get sick because the diet is not balanced. Person A would not get necessary nutrients. The human body is better able to fend off disease when a person eats a healthy, balanced diet.

### Portfolio Assessment

Sample items include:
- Sentences from Vocabulary
- Lesson 4 Review answers

### TEACHER ALERT

Encourage students to think of all the ways salt gets into their diet. Point out that many processed foods already contain salt. People might add more salt when they cook and even more salt at the dinner table. Ask students how they can reduce salt in their diets. *(Buy low-salt or low-sodium versions of foods; cut down on the amount of salt used in cooking and at the table.)*

### IN THE COMMUNITY

Ask students if they have heard of community food-delivery programs such as Meals on Wheels. Discuss how such programs are helpful for the community. *(They provide balanced meals for people who are homebound, usually people who are senior citizens, people who are disabled, or people who are ill.)* Have students investigate whether any food-delivery programs exist in their community. How do these programs ensure the delivery of nutritious meals? Are there opportunities for young people to volunteer? Encourage interested students—with the permission of their parents or guardians—to volunteer for a food-delivery program.

---

Your body systems need the different nutrients that specific foods give you. If your diet does not include these foods, your systems may not work well. As a result, you may get a disease. You can also have too many of some nutrients, including salt, sugar, and fat. This condition also may cause you health problems.

A healthy diet cannot keep you from getting every disease. A healthy diet can often lower your chances of getting sick. Nutrition is one thing you can control.

***Lesson 4 Review*** On a sheet of paper, write the word or phrase in parentheses that best completes each sentence.

1. Teenage boys usually need more (water, calories, iron) than girls need.
2. Teenagers usually need more (salt, calcium, iron) in their diets for bone growth.
3. Too much sodium in your diet can lead to (deficiency, high blood pressure, depression).

On a sheet of paper, write the answers to the following questions. Use complete sentences.

4. **Critical Thinking** Name three reasons you should talk to a health care professional before you make plans to change your weight.
5. **Critical Thinking** Person A eats mostly candy bars, grilled cheese sandwiches, and soft drinks. Person B eats mostly fish, vegetables, and yogurt. Which person is more likely to get sick? Explain your answer.

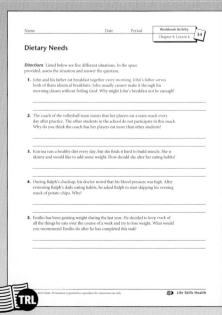

# Chapter 9 SUMMARY

- The U.S. government's Food Guide Pyramid shows how much and what kinds of foods to eat.

- Your body needs six types of essential nutrients. Carbohydrates, fats, and protein give your body energy. Vitamins, minerals, and water help your body change food into energy.

- Carbohydrates are starches and sugars that provide much of your energy.

- Fats are stored energy. Fats protect your organs and nerves. Too many saturated fats can lead to high levels of cholesterol in your blood. Some kinds of cholesterol can clog your blood vessels.

- Protein builds muscle and repairs tissue. Animal products and some plants give you protein.

- Your body needs vitamins in small amounts for normal growth and activity.

- Minerals are important in small amounts for fluid balance and digestion. Calcium, phosphorus, and iron are especially important minerals.

- Water carries nutrients, hormones, and waste products to and from your body's cells.

- Teenagers have special dietary needs, including more calcium and iron.

- A healthy diet is an important tool in weight control.

- Some diseases are linked to diet. A healthy diet can lower your chances of getting some diseases.

### Vocabulary

| | | |
|---|---|---|
| added sugar, 211 | essential nutrient, 206 | mineral, 216 |
| amino acid, 214 | fat, 210 | natural sugar, 211 |
| blood cholesterol, 213 | fatty acid, 212 | nutrient-dense, 207 |
| calcium, 217 | fiber, 211 | phosphorus, 217 |
| carbohydrate, 210 | Food Guide Pyramid, 204 | protein, 210 |
| cholesterol, 213 | fructose, 211 | saturated fat, 212 |
| complex carbohydrate, 210 | glucose, 212 | simple carbohydrate, 210 |
| deficiency, 223 | glycogen, 212 | sodium, 218 |
| dietary cholesterol, 213 | lactose, 211 | sucrose, 211 |
| essential amino acid, 214 | malnutrition, 222 | unsaturated fat, 213 |
| | | vitamin, 216 |

## Chapter 9 Summary

Have volunteers read aloud each Summary item on page 225. Ask volunteers to explain the meaning of each item. Direct students' attention to the Vocabulary box on the bottom of page 225. Have them read and review each term and its definition.

### ONLINE CONNECTION

The Food and Nutrition Information Center—which is a part of the U.S. Department of Agriculture (USDA)—offers a Web site that provides information on topics such as dietary guidelines, food supplements, and food safety: www.nal.usda.gov/fnic/

The Web site of the U.S Food and Drugs Administration (FDA) contains up-to-date information on weight-loss programs, food safety, and food product recalls: www.fda.gov

### LEARNING STYLES

**Auditory/Verbal**

Have students work in groups of three or four. Group members should bring in packaged food items, including at least one popular snack food, such as potato chips. Have students read the food labels to find out how much salt, sugar, and fat are in each of the foods. As a class, discuss students' findings. Were they surprised by what they found? What surprised them most? *(They might be surprised that so many foods contain large amounts of salt.)*

## Chapter 9 Review

Use the Chapter Review to prepare students for tests and to reteach content from the chapter.

## Chapter 9 Mastery Test

The Teacher's Resource Library includes two forms of the Chapter 9 Mastery Text. Each test addresses the chapter Goals for Learning. An optional third page of additional critical-thinking items is included for each test. The difficulty level of the two forms is equivalent.

## Review Answers
**Vocabulary Review**

1. deficiency 2. unsaturated fat
3. carbohydrates 4. glucose 5. lactose
6. sodium 7. protein 8. amino acids
9. vitamins 10. calcium 11. saturated fat
12. cholesterol 13. fiber 14. malnutrition

### TEACHER ALERT

In the Chapter Review, the Vocabulary Review activity includes a sample of the chapter's vocabulary terms. The activity will help determine students' understanding of key vocabulary terms and concepts presented in the chapter. Other vocabulary terms used in the chapter are listed below.

added sugar
blood cholesterol
complex carbohydrate
dietary cholesterol
essential amino acid
essential nutrient
fat
fatty acid
Food Guide Pyramid
fructose
glycogen
mineral
natural sugar
nutrient-dense
phosphorus
simple carbohydrate
sucrose

226  Unit 3  Nutrition

---

## Chapter 9  REVIEW

**Word Bank**
amino acids
calcium
carbohydrates
cholesterol
deficiency
fiber
glucose
lactose
malnutrition
protein
saturated fat
sodium
unsaturated fat
vitamins

### Vocabulary Review
On a sheet of paper, write the word or phrase from the Word Bank that best completes each sentence.

1. If you lack something, you are said to have a(n) _____.
2. The kind of fat that comes from vegetable and fish oils is _____.
3. Potatoes, pasta, and bread are good sources of _____.
4. Carbohydrates are broken down into blood sugar, or _____.
5. The natural sugar that is found in milk is _____.
6. A mineral found in table salt is _____.
7. Muscles, bones, blood, and skin all contain _____.
8. The body needs nine _____ that are the building blocks of protein.
9. Water soaks up some _____, which means you may lose their health value.
10. An important mineral found in dairy products is _____.
11. The kind of fat that comes from animal products is _____.
12. If you eat too much fat, you can get a buildup of _____, which can clog your blood vessels.
13. The coarse material found in food that helps in digestion is called _____.
14. A condition you might get if your diet lacks some or all nutrients is called _____.

226  Unit 3  Nutrition

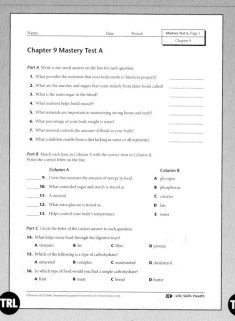
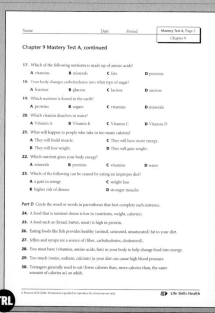

Chapter 9 Mastery Test A, pages 1–3

## Concept Review

On a sheet of paper, write the letter of the answer that best completes each sentence.

**15.** All of these foods give you complex carbohydrates EXCEPT_____.
- **A** bread
- **B** soft drinks
- **C** pasta
- **D** rice

**16.** Extra sugar and starch in your body is _____.
- **A** turned into fat
- **B** used for energy
- **C** passed out as waste
- **D** stored to repair body tissues

**17.** Calcium is a _____.
- **A** fat
- **B** vitamin
- **C** carbohydrate
- **D** mineral

## Critical Thinking

On a sheet of paper, write the answers to the following questions. Use complete sentences.

**18.** What are four foods you would choose to make a balanced dinner? Explain your reasoning.

**19.** How could you change your own diet to make it healthier?

**20.** Imagine you are a health care worker. How does the advice you would give a teenaged athlete differ from the advice you would give a teenager with an average activity level?

> **Test-Taking Tip** Always read directions more than once. Underline words that tell how many examples or items you must provide.

*The Role of Diet in Health* Chapter 9 227

## Concept Review
15. B  16. A  17. D

## Critical Thinking

**18.** Sample answer: I would serve fish, broccoli, rice, and skim milk. This would provide low-fat protein, vitamins, minerals, and fiber. **19.** Sample answer: I would stop eating so much sugar. I would eat less fat, and I would add more calcium. **20.** Sample answer: I would tell the athlete to drink plenty of water and eat more calories than a teenager of average activity level.

## Alternative Assessment

Alternative Assessment items correlate to the student Goals for Learning at the beginning of this chapter.

- Have students write a paragraph describing how healthy eating can affect a person's body.
- Have pairs of students take turns explaining a food nutrition label to each other, with an emphasis on the calories and nutrients provided by the sample food.
- Have students give an example of a type of food that provides each of the six essential nutrient classes.
- Have students work on projects that describe some special dietary needs people might have.
- Have students create a poster or brochure that describes how diet choices can affect health.

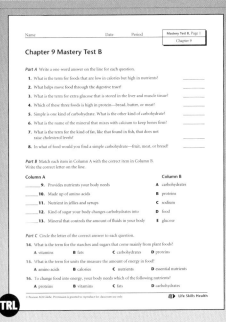
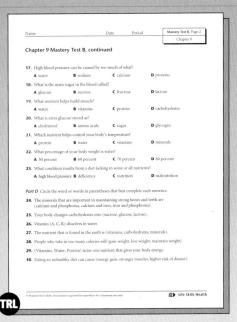

Chapter 9 Mastery Test B, pages 1–3

*The Role of Diet in Health* Chapter 9 227

## Chapter at a Glance

### Chapter 10: Choosing Healthy Foods
pages 228–245

### Lessons
1. Healthy Eating Patterns and Food Choices
   pages 230–232
2. How the Media Influences Eating Patterns
   pages 233–238
3. Food Labels and Food Additives
   pages 239–242

**Chapter 10 Summary** page 243

**Chapter 10 Review** pages 244–245

**Audio CD**

**Skill Track for Life Skills Health**

**Teacher's Resource Library**
   Activities 35–37
   Alternative Activities 35–37
   Workbook Activities 35–37
   Self-Assessment 10
   Community Connection 10
   Chapter 10 Self-Study Guide
   Chapter 10 Outline
   Chapter 10 Mastery Tests A and B

(Answer Keys for the Teacher's Resource Library begin on page 559 of this Teacher's Edition.)

## Opener Activity

Have students brainstorm methods food companies use to try to convince teens to buy their food products. Possible ways include running commercials during TV programs that teens watch, offering coupons for free or reduced-price food, and catering to food fads. Have students discuss whether they find these methods effective and whether they are influenced by them.

Have volunteers write and act out a TV commercial for a food product. Have the class discuss whether the ad made them want to buy the product, and if so, why.

## Self-Assessment

### Can you answer these questions?

1. What is a healthy eating pattern?
2. Why do you choose the kinds of foods you eat?
3. How can stress affect food choices?
4. How can advertising affect food choices?
5. How can fads affect diet?
6. How has the fast-food industry affected the American diet?
7. What does the government do to make sure food is safe?
8. What information is found on a food label?
9. How do you read a food label?
10. Why is knowing about food additives important?

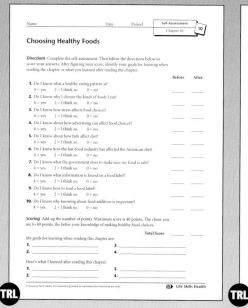

Self-Assessment 10

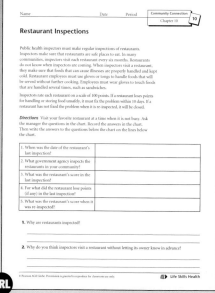

Community Connection 10

228  Unit 3  Nutrition

# Chapter 10
## Choosing Healthy Foods

Food is an important part of life. For most people, mealtimes are times to talk with family members or friends. Social events often include food. Many things affect the food choices you make. Realizing why you choose the foods you do is important. Developing an eating pattern that will help you keep eating healthy foods is also important.

In this chapter, you will learn about healthy eating patterns. You will also learn about how advertising and the fast-food industry affect how you eat. You will find out what is done to make sure foods are safe. Finally, you will learn about reading food labels and about things that are added to foods.

### Goals for Learning

◆ To identify healthy eating patterns
◆ To describe what influences food choices
◆ To describe how government agencies make sure food is safe
◆ To explain how to read food labels in order to make healthy food choices
◆ To understand how things that are added to foods may change them

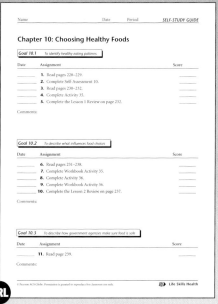

## Introducing the Chapter

Bring in some common packaged foods such as cereals and canned vegetables. Have students look at the labels on the packages and read the nutrients listed. Discuss the major sources of various nutrients in the foods they eat each day. Explain that this chapter will discuss the importance of choosing a healthy diet. Ask how reading food labels can help them make good choices. *(Sample answer: The nutritional information on food labels helps you plan a healthy and balanced diet and helps you avoid too much of any one nutrient.)*

Have students read the Goals for Learning. Explain that in this chapter, students will learn about healthy eating patterns, what influences eating patterns, what the government does to make sure foods are safe, and what students can do to make healthy food choices.

### Self-Assessment

Have students complete the Self-Assessment worksheet before and after reading the chapter. Before reading the chapter, have students fill in the "Before" column. Ask students to identify their goals for learning. To get ideas for setting goals, students might use the chapter introductory material on pages 228–229, the checklist on page 228, or the questions on the Self-Assessment worksheet. Students can use the back of the worksheet if they need more space to write.

Collect the Self-Assessment worksheets and pass them out again at the end of the chapter. Have students fill in the "After" column. Ask them to identify at least four major points they have learned. Again, suggest they use the back of the worksheet if they need more space to write. You may want to collect and review the worksheets, but return them to students so they have a record of their goals and accomplishments.

# Lesson at a Glance

## Chapter 10 Lesson 1

**Overview** This lesson describes healthy eating patterns and discusses influences on food choices.

## Objectives

- To explain how eating patterns affect your diet
- To list four factors that influence food choice
- To describe three ways stress affects eating patterns

**Student Pages** 230–232

**Teacher's Resource Library** TRL
Activity 35
Alternative Activity 35
Workbook Activity 35

## Vocabulary

**association**

Have students look up the vocabulary word in the dictionary and write a short paragraph that includes the word *association* used correctly in context. Encourage students to write about a food choice they have made that they associate with a personal experience.

## Background Information

Each world culture can be distinguished by its foods. For example, many Asian cultures are noted for dishes containing tofu or soy sauce, and in Asian diets, rice is often the main source of complex carbohydrates. Foods that have been important in American culture since the country's founding are beef, chicken, turkey, pork, potatoes, corn, greens, wheat, beans, and apples.

## Research and Write

Have students read and conduct the Research and Write feature on page 230. Have students work with partners or in small groups to complete this activity. Students can share their charts with the class.

---

## Lesson 1 — Healthy Eating Patterns and Food Choices

**Objectives**

*After reading this lesson, you should be able to*

- explain how eating patterns affect your diet
- list four factors that influence food choice
- describe three ways stress affects eating patterns

**Research and Write**

About 2.5 percent of Americans are vegetarians. Vegetarians do not eat meat or other animal products. Research the different types of vegetarian diets, such as the vegan diet and the lacto-ovo vegetarian diet. Create a chart that shows which foods are not eaten in each diet. Then write the answer to this question: How can vegetarian diets provide enough daily nutrition for teens?

Your lifestyle affects your diet. Sometimes a busy lifestyle results in a poor diet. Knowing more about what affects the foods you choose can help you improve your diet.

### How Does Your Eating Pattern Affect Your Life?

People usually have a set time for when they eat. For example, their normal lunchtime may be noon. Eating at a set time is helpful. It gives your body food for energy at regular times each day. Sitting down to a balanced meal each morning, midday, and evening will help you set a healthy eating pattern.

Sometimes, following a healthy eating pattern is difficult. Today, many people try to fit mealtimes into very busy lives. People may not have time to prepare and eat meals at home. They may rely on fast-food restaurants or have high-calorie snacks. Some people may skip breakfast or other meals.

Setting a healthy eating pattern is important. For example, eating a good breakfast every morning helps give you the energy you need for the day. In some ways, breakfast is more important than an evening meal. You are usually more active in the earlier part of the day. You use up more calories and energy during that time, so you need a nutritious breakfast. In the evening, you may be less active and need fewer calories. So the evening meal could be smaller. Breakfast foods should include complex carbohydrates to give your body the energy it needs. Avoid foods high in sugar. They give you an energy burst that your body uses up quickly and then leaves you feeling tired.

Snacks can be a part of a healthy diet, as long as they do not take the place of regular meals. Nutritious snacks can add protein, vitamins, and minerals to your diet. For example, you can add vitamin C by eating fruit. You can add vitamin A by eating carrots or green peppers. Avoid high-calorie foods such as candy and cookies. Also, avoid late-night snacking, which can disturb your sleep and cause you to gain weight.

230 Unit 3 Nutrition

---

## Pronunciation Guide

Use the pronunciation shown here to help students pronounce difficult words in the lesson. Refer to the pronunciation key that appears in the Glossary at the back of the Teacher's Edition for the sounds of the symbols.

complex carbohydrate (kəm pleks´ kär bō hī´ drāt)

nutritious (nü trish´ əs)

**Association**
Something that reminds a person of something else

## What Affects Your Food Choices?

Many things affect the foods you choose to eat. These may include where you live, the time of year, your religion, or your cultural background. For example, if you live close to a body of water, you are more likely to eat fish and other seafood. Some foods are in season only at certain times during the year. For example, you may find fresh strawberries at the grocery store during June but not during December.

Cultural background or religion often affect your food choices. Some cultural groups include more rice, beans, or spicy foods in their diets than others do. Some religions have rules about foods, such as not eating certain kinds of meat. Eating foods that are common to other cultures can enrich your diet and be fun to try.

Several other things can affect what you choose to eat. The cost of food can affect what you buy. Checking prices and knowing the nutritional value of food will help you get the best food for the best price. Your family and friends can also influence your food choices. For example, you may have dinner at a friend's house and discover a food that you have never tried before.

**Association** also influences what you choose to eat. Association means that the food reminds you of something else. You may strongly like or strongly dislike certain foods because of association. For example, you may enjoy toasted marshmallows because they remind you of a picnic that you enjoyed with friends. You may dislike marshmallows because you once got sick eating them.

If you think about why a certain food is a part of your diet, you may decide to make some changes. For example, you may choose to skip the high-sugar cereal at breakfast and have whole-grain toast instead.

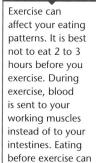

Exercise can affect your eating patterns. It is best not to eat 2 to 3 hours before you exercise. During exercise, blood is sent to your working muscles instead of to your intestines. Eating before exercise can cause you to have cramps and feel sick. A slow walk right after you eat, however, should not bother you.

**Figure 10.1.1** How much you can afford to spend affects what you buy.

Choosing Healthy Foods   Chapter 10   **231**

 **Warm-Up Activity**

Ask each student to make a list of all the snacks he or she has eaten in the last 24 hours. Then, ask students how they can modify their lists to make their snacks more nutritious. *(They can substitute fruit or vegetables for junk food.)* Ask them to name some nutritious snack foods they might like and encourage them to try those foods.

 **Teaching the Lesson**

Discuss the concept of eating patterns with students. Be sure they understand what an eating pattern is. Then ask students to write short paragraphs describing their eating patterns. They should include their eating schedule, whether they take their time to eat or eat on the go, and whether they eat at regular times each day. Ask volunteers to share their paragraphs with the class and discuss whether they feel they have healthy eating habits. Ask if they would like to make any changes in their daily eating routine.

 **Reinforce and Extend**

### Career Connection

Ask students who they think plans the menus for their school lunch program, as well as the menus for airlines, hospitals, corporate office cafeterias, and restaurants. Tell them that these menu decisions are made with the help of dietitians. Have students use reference material and/or the Internet to find out how dietitians choose foods to includes on the menus and how they are especially helpful to people with special dietary needs.

### ELL/ESL Strategy

**Language Objective:** *To practice using oral interaction skills*

Have students form pairs. One student should be learning English. Have the student learning English describe how the foods from his or her culture are different from American foods. The students with strong English skills should correct errors in grammar, offer encouragement, and ask questions about the various foods the ELL students are describing.

### Health Journal

Have students describe in their journals how their eating patterns change when they are under stress. Do they eat more or less? Do they indulge in sweets or salty foods such as potato chips? Ask them to consider what they could do to relieve stress in other, healthier ways.

*Choosing Helathy Foods*   Chapter 10   **231**

## Lesson 1 Review Answers

1. A healthy eating pattern is one that includes nutritious meals at regular times. 2. Sample answer: Four factors that affect food choices are cost, association, culture, and the influence of friends. Two other factors are religion and the time of the year. 3. Stress can cause people to skip meals, make poor food choices, or overeat. 4. Sample answer: Three strategies I can use to avoid a diet of unhealthy foods are establishing a healthy eating pattern, choosing healthy snacks, and suggesting ideas for healthy foods for parties or other events. 5. Sample answer: I can improve my eating patterns by taking time to eat a nutritious breakfast before school.

## Portfolio Assessment

Sample items include:
- Paragraphs from Vocabulary
- Charts and answers from Research and Write
- Modified lists from Warm-Up Activity
- Paragraphs from Teaching the Lesson
- Pictures from ELL/ESL Strategy
- Entry from Health Journal
- Paragraphs from Application: At Home
- Lesson 1 Review answers

### Link to

**Social Studies.** Have a volunteer read aloud the Link to Social Studies feature on page 232. Discuss whether students like to eat fish and how they like it prepared. If students do not eat fish often, ask whether they would consider eating it because it could help prevent illness. Discuss other ways people might get the benefits of fish without eating it twice a week (taking fish-oil supplements, for example.)

### AT HOME

Ask students to interview the person in their families who does most of the cooking. Suggest the following questions: Do you plan menus for each meal in advance? What nutritional quality do you keep in mind when you select menu items? Do you by generic brands? Do you read the list of ingredients on food labels? Have students write several paragraphs summarizing the answers they received.

---

### Link to
**Social Studies**

North Americans get 10 times as many heart attacks as do people who live in Greenland or Japan. People who live in places such as these eat more fish than North Americans do. Many North Americans eat a great deal of meat. Meat has cholesterol that can clog blood vessels. The oils in fish contain cholesterol that helps blood flow quickly. Many nutrition experts suggest eating fish at least twice a week.

At times, making healthy food choices may be difficult, such as at a party where only high-calorie and high-fat foods are offered. Keep in mind the nutritional value of food. Choose foods that are the best for your health. You might make suggestions to people planning parties or other events about healthy food to serve.

### How Does Stress Affect Eating Patterns?

Stress often affects the food you choose to eat. If you feel a great deal of mental stress, you may forget about your diet. You may not realize that during times of stress you are using more energy and need more nutrients. You may ignore your feelings of hunger and may not eat enough. By not getting the nutrients you need, you may have even greater physical stress.

Stress can sometimes cause you to eat too much. When you are stressed, you may choose foods that have too much sugar or salt. Remember that a healthy diet and exercise may help cut down the effects of stress. When you are stressed, choose foods that are high in vitamins, minerals, and protein. Good food choices would be nuts, orange juice, or bananas. Stay away from coffee, tea, cola, chocolate, and other foods that have a lot of caffeine. The caffeine can make you feel more nervous and make sleeping difficult. Choosing food wisely is a good way to help manage your stress.

**?**
Think about a time when you felt stress. How did the stress affect your eating patterns?

**Lesson 1 Review** On a sheet of paper, write the answers to the following questions. Use complete sentences.

1. What is a healthy eating pattern?
2. What are four factors that affect your food choices?
3. What effect can stress have on healthy eating patterns?
4. **Critical Thinking** What are three strategies you can use to avoid a diet of unhealthy foods?
5. **Critical Thinking** What can you do to improve your eating patterns?

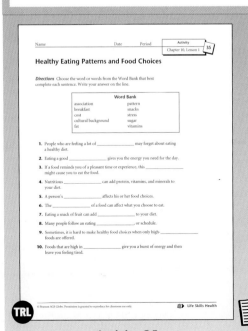

# Lesson 2: How the Media Influences Eating Patterns

### Objectives

After reading this lesson, you should be able to

- explain how the media affects the foods you choose
- describe how fads affect your diet
- identify three ways fast food affects healthy eating patterns

**Media**
*Ways people can communicate information: TV, Web sites, newspapers, radio, magazines, and so on*

**Product**
*Item for sale made by someone or something*

**Advertise**
*Communicate to draw attention to a product*

The **media** can have a big effect on what people do. The media is made up of the many ways we communicate information: TV, Web sites, newspapers, radio, magazines, and so on. Companies that make **products**, or items to be sold, use the media to reach people who may buy the products. Understanding how the media influences your food choices will help you make better ones.

## How Does Advertising Affect the Food You Choose?

Many foods and drinks are **advertised** in the media. Advertising is communicating to draw attention to a product. The idea is to make the product appealing to people so they will buy it.

Food products are advertised often. Many times, people choose foods simply because they have heard about them. Food companies spend large amounts of money to encourage people to buy their products.

Some phrases companies use to try to get you to buy their food products are explained below.

- *New product*—Packages may include the words *new* or *improved*. Advertisers use these words to interest you in something that may be a little different or easier to use. Sometimes, the product may be almost the same, but the packaging is designed to look new.

- *Old-fashioned*—Some products say the food is prepared as it was many years ago. Adults may associate certain foods with happy memories from childhood. They may buy a food because it reminds them of a relative or an old friend. For example, a company may say its mashed potatoes are "just like Grandma used to make."

*Choosing Healthy Foods* Chapter 10  233

---

# Lesson at a Glance

## Chapter 10 Lesson 2

**Overview** This lesson describes how advertising, the fast-food industry, and fads affect your food choices.

### Objectives

- To explain how the media affects the foods you choose
- To describe how fads affect your diet
- To identify three ways fast food affects healthy eating patterns

**Student Pages** 233–238

**Teacher's Resource Library**
Activity 36
Alternative Activity 36
Workbook Activity 36

### Vocabulary

advertise  organic food
Daily Values (DV)  product
media  trans fat

Have students use index cards to make flash cards for the vocabulary words, writing a word on one side of a card and the definition on the other side. Have students write sentence using the words.

### Background Information

More than 65 million Americans have tried diets that promise fast weight loss. One diet fad is the liquid diet. Many obesity specialists are concerned about people who begin an over-the-counter liquid diet program without consulting a doctor or following directions. Most of these dieters are looking for a fast way to lose weight, and there is no way to accomplish this safely and consistently. Researchers are currently looking at diet alternatives. Future diet aids may include powerful appetite suppressants, fat-absorption blockers, medicines to increase metabolism, or pills designed to block the pleasure receptors in the brain so that food will not taste as good.

*Choosing Helathy Foods* Chapter 10  233

## Warm-Up Activity

Ask students to name famous people and the food products that they advertise. Use a game format for this activity. Having students correctly match the product with the celebrity can be used as a test of how effective the ad is. Ask students if they are influenced by the fact that a celebrity is advertising the product.

## Teaching the Lesson

Have students go through old magazines and newspapers to find advertisements. By asking them to do this, you can make sure students understand the formats in which print media advertisements appear. Sometimes it is easy to confuse advertisements with legitimate magazine articles, because advertisers cleverly make the ads look like articles.

### Health Myth

Have students read the Health Myth feature on page 234. Tell them that some vitamins, such as vitamins B and C, are eliminated in the urine when they are in excess in the body. Others, such as vitamins A, D, and E, are stored in the liver and in adipose tissue (fat that accumulates within muscles, under the skin, and around internal organs). Serious health problems can occur if these vitamins are in excess in the body.

## Reinforce and Extend

### TEACHER ALERT

Be sure that students understand that trans fats are artificial, not natural. They are formed when vegetable oils are processed into solid fats such as margarine or shortening. Manufacturers use margarine and shortening to reduce costs, improve flavor and texture, and extend the shelf-life of food products.

**Organic food**
Food raised without pesticides, unnatural fertilizers, antibiotics, or growth hormones

**Health Myth**

**Myth:** You can lose weight if you stop eating and take lots of vitamins.

**Fact:** At first, not eating will cause you to lose weight. Soon, however, your body adjusts to less food and begins to burn calories at a slower rate. When you begin to eat again, you will gain weight quickly. You can get very sick if you take large amounts of certain vitamins. Always take the recommended amount of any vitamin.

- *Expensive and the best*—These advertisements make people feel that the more something costs, the more they should want it. Usually, products advertised as "the best" cost more. For example, advertisers say their mustard "may cost a little bit more, but it's worth it."

- *Inexpensive and a good deal*—Some companies try to convince people they are getting the most for their money. Buyers must look at the price and number of servings to figure out if this is true.

- *Quick and easy to prepare*—These advertisements try to interest people who have little time to prepare meals. Most often, the meals are heated in a microwave oven. Technology and busy lifestyles have made a demand for many new food products that are simple to make.

- *Contains no cholesterol*—People may buy a product simply because it says that the product has no cholesterol. For example, an ad for peanut butter may state "This product has no cholesterol." Yet, no peanut butter contains cholesterol. This is because peanuts come from plants, and cholesterol is found only in animal products.

- *Fat free or low carb*—Many food packages use terms that appeal to people who are following a certain kind of diet. People who want to avoid fat may buy products that are advertised as fat free. People who carefully count carbohydrates may buy products that are advertised as being low in carbohydrates or "carbs." A product may be low in carbohydrates and fat but high in calories.

- *Organic foods*—Food products must meet government standards before they can be called **organic foods.** The word *organic* also means *natural.* Yet not all products called "natural" are organic. Some advertisers would like people to think that natural and organic foods are the same in every case. However, foods advertised as "natural" do not have to meet the same government standards as foods labeled "organic."

234  Unit 3  Nutrition

### IN THE ENVIRONMENT

Many food products are advertised as being environmentally friendly. Ask students what they can do to choose products that help the environment. Suggestions include buying foods that use the least packaging, using cloth napkins, and using a reusable thermal lunch bag. Other suggestions may include buying fruits and vegetables that are grown organically—without the use of pesticides and herbicides—and buying foods that do not use artificial preservatives or other chemicals.

### PRONUNCIATION GUIDE

Use the pronunciation shown here to help students pronounce difficult words in the lesson. Refer to the pronunciation key that appears in the Glossary at the back of the Teacher's Edition for the sounds of the symbols.

saturated fat (sach´ ə rā tid fat)

**Research and Write**

Today, about 25 percent of children in the United States are overweight. This can lead to lifelong health problems. Use the Internet to research these health problems. Write a report on your findings. Include a list of the health problems with descriptions of them.

Companies use all these ways to encourage you to buy products. You must look closely at the foods you buy to be sure you are choosing foods that are nutritious and priced fairly.

## How Does the Fast-Food Industry Affect Your Diet?

Today, many people want food that is fast and easy to prepare and eat. Almost one-third of all children in the United States eat fast food at some point each day. Studies have shown that fast-food companies target much of their advertisements at children. Children who watch TV ads for fast food are more likely to ask for fast food. Fast-food ads also have a strong effect on adults.

Fast foods have been linked to the growth in weight problems in America. Many fast foods are high in fat. Your body needs a certain amount of fat for growth and good health. Fat also provides energy for your body in the form of calories. However, taking in too much fat is unhealthy. Fat should make up only 30 percent of your total daily calories. A fast-food burger can contain many of your daily calorie needs. Half or more of those calories in the burger can come in the form of fat.

---

**Health in Your Life**

**A Look at Food Advertising**

Today, many foods are advertised as *light, low fat, nonfat, low calorie,* or *fat free*. The use of these terms is regulated by the FDA. Words in ads and on food labels are written to get you to choose a certain food. Many people think these foods will help them stay healthy.

Some fast-food restaurants advertise reduced-fat menu items, such as salads and low-fat sandwiches. Yet foods that are low in fat can still be less healthy because they are high in sugar. Many fast-food restaurants have information on the nutritional value of their menu items.

Request this information from the restaurant's manager, or find it on the Internet. Then you can decide if you should eat the food.

On a sheet of paper, write the answers to the following questions. Use complete sentences.

1. Why do advertisers of food use the words they do?
2. In what way can fast food be part of a healthy eating pattern?
3. What are three strategies you can use to make sure you have a healthy eating pattern?

---

**Research and Write**

Have students read and conduct the Research and Write feature on page 235. Introduce and explain the term *childhood obesity* as a term that students can use to start their Internet research. Students can work with a partner to do the research, but they should write individual reports. Students' reports should mention diabetes as a health problem that stems from being overweight. Students can share their reports with the class.

**Health in Your Life**

Find a food advertisement from a magazine or newspaper. The ad should include one or more words that advertisers use to convince people to choose a certain food. Look for the terms mentioned in the feature. Also find the actual food item advertised.

Have students read the Health in Your Life feature on page 235 and answer the questions. Bring the ad and the food item to class on the day students will read the feature. Challenge students to evaluate the truthfulness of the ad. Is the ad misleading when compared with the information on the food label?

**Health in Your Life Answers**

1. Sample answer: They choose "healthy" words so you will buy their products.
2. Sample answer: Fast food can be part of a healthy eating pattern if the foods chosen are high in fiber and low in fat and sugar. Some fast-food restaurants offer healthy items, such as salads, fruit, and bottled water.
3. Sample answer: Three strategies I can use to have a healthy eating pattern are limiting my intake of saturated and trans fats, avoiding fast foods and convenience foods, and understanding the influence of advertising.

## Health Myth

Have students read the Health Myth feature on page 236. Tell them that fats contain 9 calories per gram; carbohydrates and proteins contain 4 calories per gram. If you eat the same amount of fat as protein or carbohydrates, you will take in more than twice as many calories from fat as you did from protein or carbohydrates. Taking in more calories than you expend is what causes you to gain weight, not eating foods that contain fat.

### HEALTH JOURNAL

Have students describe some food fads and explain whether these diets are healthy. Examples might include the grapefruit diet, cabbage soup diet, and liquid protein diets. These diets are not healthy if they restrict you to specific foods or food groups. Diets need to contain a balance of all the essential nutrients.

### IN THE COMMUNITY

A popular way to provide nutrition and health information is with a health fair. As a class, plan a health and nutrition fair for your school or community. Determine the type of information you want to provide, such as the effectiveness and safety of fad diets, how to make healthful food choices, how advertising influences food choices, and any other topics from this chapter. Use local restaurants, hospitals, and colleges as resources. You might invite representatives from a restaurant to attend and discuss how their menu is influenced by fad diets, for example.

***

**Trans fat**
Type of fat from processed vegetable oil

**Daily Values (DV)**
Section of a nutrition label that provides information about the percentage of nutrients in a product

### Health Myth

**Myth:** Eating fatty foods makes you overweight.

**Fact:** Your body changes the fat you eat into calories that give you energy. Fat has twice as many calories as carbohydrates or protein. If you eat a lot of fat, you are eating a lot of calories. A little bit of fat should be a part of your daily diet. However, if you eat too much fat, you will gain weight.

***

Many of the foods you eat contain fat. Saturated fat is found in meat, poultry, milk, butter, eggs, and palm oil. Unsaturated fat is found in vegetables, nuts, seafood, and certain oils such as olive oil. Another kind of fat, called **trans fat,** is made from processed vegetable oil. Trans fat is added to food to lengthen the time it can stay on the shelf without spoiling. Trans fat is found in many foods, including most vegetable shortening, crackers, cookies, and snack foods.

Knowing about the three different types of fat in food is important. For good health, the U.S. Food and Drug Administration (FDA) recommends you eat a diet that is low in saturated and trans fats.

The **Daily Values (DV)** are ways of measuring the ingredients in food. The Daily Values are printed on nutrition labels. The DV percentage for fat tells how much fat is in the food compared with the total amount you need each day. A Daily Value of 5 percent or less of saturated fat is low. A DV of 20 percent or more is high. Table 10.2.1 below lists the saturated fats contained in several foods.

| Table 10.2.1 Saturated Fats per Serving ||||
|---|---|---|---|
| Product | Common Serving Size | Saturated Fat (grams) | % DV for Saturated Fat |
| Butter | 1 tablespoon | 7 | 35% |
| Candy bar | 1 (40 grams) | 4 | 20% |
| Doughnut | 1 | 4.5 | 23% |
| Fast-food French fries | Medium (147 grams) | 7 | 35% |
| Potato chips | Small bag (42.5 grams) | 2 | 10% |
| Skim milk | 1 cup | 0 | 0% |
| Whole milk | 1 cup | 4.5 | 23% |

Source: U.S. Food and Drug Administration

***

### TEACHER ALERT

Be sure that students understand what the percentage of Daily Value means. It tells how much of the nutrient is found in the food compared with the total amount of the nutrient that people need each day. Percent Daily Values are based on a diet containing 2,000 calories per day. However, teen girls usually need 2,200 calories per day and teen boys usually need 2,800 calories per day, depending on how active they are. In that case, the Percent Daily Value would be lower for any given food.

### LEARNING STYLES

**Logical/Mathematical**
Have students calculate the number of calories that come from saturated fat in one serving of each food listed in Table 10.2.1. Students should use the value of 9 calories per gram of fat for their calculations. They should multiply the number of grams of saturated fat in a serving by 9 calories per gram. This will tell them how many calories from fat that they receive from one portion of each food.

## How Do Fads Affect Your Diet?

A fad is something that becomes popular for a short time. When you believe a food fad works, you may change your diet. For example, you might have heard that you should eat grapefruit six times a day for 2 weeks to lose weight. Grapefruit can be a healthy food, but eating too much of it is not healthy. Food fads usually focus on one food or on an unhealthy eating pattern. Your body needs nutrients from a variety of foods. Food fads can cause serious health problems.

Another example is the low-carb diet in which people reduce the number of carbohydrates they eat. Your body needs carbohydrates, so low-carb diets can be unhealthy.

There are many things to think about when choosing foods for your diet. Choosing good food and setting a healthy eating pattern will help your body get the nutrients it needs.

### Decide for Yourself

Suppose you are planning a snack. Read the nutrition labels. Compare the information so you can choose the healthier snack.

On a sheet of paper, answer the following questions. Use complete sentences.

1. Which snack would you choose?
2. Why would you choose this snack? Explain your choice, using the nutritional information in both labels.

**Wheat Crackers**

**Nutrition Facts**
Serving Size 14 Crackers (22g)
Servings Per Container about 6

Amount Per Serving
Calories 110    Calories from Fat 10
Total Fat 1g
  Saturated Fat 0g
  Polyunsaturated Fat 0.5g
  Monounsaturated Fat 1.5g
Cholesterol 0mg
Sodium 110mg
Total Carbohydrate 21g
  Dietary Fiber 2g
  Sugars 2g
Protein 3g

Vitamin A 0%    Vitamin C 0%
Calcium 2%      Iron 6%

**Nut Mix**

**Nutrition Facts**
Serving Size 1 ounce
Servings Per Container about 20

Amount Per Serving
Calories 400    Calories from Fat 126
Total Fat 14g
  Saturated Fat 5g
  Polyunsaturated Fat 0.5g
  Monounsaturated Fat 2.5g
Cholesterol 0mg
Sodium 200mg
Total Carbohydrate 15g
  Dietary Fiber 5g
  Sugars 8g
Protein 6g

Vitamin A 0%    Vitamin C 3%
Calcium 5%      Iron 3%

---

### Decide for Yourself

Have students read the Decide for Yourself feature on page 237 and answer the questions. When they have finished, make a list on the board that compares the information from the nutrition labels of the two snacks. Ask students for their input while you are making the list. Have students recall the Food Guide Pyramid from Chapter 9 and decide where each snack should be placed in the pyramid.

**Decide for Yourself Answers**

1. Sample answer: I would choose the wheat crackers. 2. Sample answer: The wheat crackers contain 3 grams of protein and only 1 gram of fat. They also contain 2 grams of dietary fiber and quite a bit of iron. They contain no saturated fat. I would not choose the nut mix. This nut mix contains saturated fat.

### LEARNING STYLES

**Interpersonal/Group Learning**

Have students work in groups to design an advertisement for an original food product. They should try to be honest and accurate in their advertisement yet be creative enough to encourage the public to consider the product. Let groups present their ads to the class.

### ELL/ESL STRATEGY

**Language Objective:** *To understand different purposes of writing*

Pair an ELL student with an English-speaking student. Have the ELL students examine a magazine advertisement for a food product and explain which advertising techniques are used. Have the English-speaking student ask questions about how much the ad has influenced the other student.

## Lesson 2 Review Answers

1. low fat  2. unsaturated fat  3. Daily Value  4. Sample answer: I could work with a teacher to start an "eat healthy" campaign at school. This campaign could include a plan to display "eat healthy" posters in the school cafeteria and on a bulletin board.  5. Sample answer: I could use the Internet to find information about a particular food fad.

### Portfolio Assessment

Sample items include:
- Flash cards and sentences from Vocabulary
- Reports from Research and Write
- Advertisements from Learning Styles: Interpersonal/Group Learning
- Entries from Health Journal
- Plans for health fair from Application: In the Community
- Calculations from Learning Styles: Logical/Mathematical
- Lesson 2 Review answers

### Health at Work

Have students read the Health at Work feature on page 238. Ask students if they think this job is important. If there were no dietary aides, what effect would that have on people? *(Students should realize that the quality of patient dietary care would be affected.)*

### LEARNING STYLES

**Auditory/Verbal**

Present the following topic for group discussion. A food advertised as "low-calorie" must have fewer than 40 calories per serving. How might this food still have more calories than what you might think? *(The serving size listed on the food label might be tiny. If a normal serving for you is four or five times the size listed on the food label, you might be taking in more calories than if you ate a food that was not "low-calorie.")*

---

**Lesson 2 Review** On a sheet of paper, write the word or phrase in parentheses that best completes each sentence.

1. Advertisers use words such as (low fat, high sugar, high calorie) to try to get you to buy their products.
2. Three types of fat are saturated fat, (unsaturated fat, vegetable oil, reduced fat), and trans fat.
3. One way to measure the amount of fat in food is to know its (calorie percent, Daily Value, weight).

On a sheet of paper, write the answers to the following questions. Use complete sentences.

4. **Critical Thinking** How would you advertise the value of a healthy diet to students at your school?
5. **Critical Thinking** How can you find out more about the health risks of a food fad?

### Health at Work

**Dietary Aide**

Dietary aides work in places such as hospitals and nursing homes. They work with dieticians to set good nutritional standards for patients. They may help prepare food for residents. They may also help plan menus, talk to patients about what they choose to eat, and keep diet records. Many dietary aides work on a tray line. They follow directions about what should go on each patient's meal tray. Then they deliver the trays. After a meal, they pick up the trays and return them to the kitchen. They record the amount of food that patients eat. Some dietary aides may help sort recyclables and trash, wash dishes, and do other clean-up tasks. Dietary aides are sometimes called food service workers or nutrition assistants.

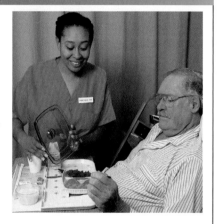

Dietary aides usually receive on-the-job training. Some may be asked to take special certification classes.

# Lesson 3: Food Labels and Food Additives

### Objectives

After reading this lesson, you should be able to
- describe how the government makes sure food is safe
- explain what a food label tells you
- identify food additives

You know which nutrients your body needs, but you also need to know which foods contain these nutrients. To adjust your diet to your needs, you should know how many calories are in one serving of the foods you eat. This information is found on food labels. Food labels also list other things that are added to the foods you eat.

## How Does the Government Make Sure Food Is Safe?

The government has agencies to make sure the foods you buy are labeled correctly and are safe to eat. These agencies set rules that companies must follow when they make food. The agencies have rules for how safe food must be, how clean the companies must be, how the companies advertise, how foods are labeled, and how the food is packaged.

The United States Department of Agriculture (USDA) is in charge of all programs that deal with making food. It conducts national food surveys to find out eating patterns in the United States. It is responsible for checking and grading meat and poultry.

The U.S. Food and Drug Administration (FDA) controls food and drug safety. It figures out which ingredients in the food are safe. The FDA also controls how food is labeled.

The Food and Nutrition Board of the National Research Council performs studies on the nutrients people need. It also suggests ways to help people meet those needs.

## What Does a Food Label Tell You?

The government has set guidelines for what information must appear on a food label. The information includes what is in the package and how to store the food. Labels give the weight and lot number of the product. A lot number identifies a group of packages. Some labels say the product should be used by a certain date to be sure the product is fresh.

---

## Lesson at a Glance

### Chapter 10 Lesson 3

**Overview** This lesson explains how the government ensures that food is safe, tells what a food additive is, and explains how students can use food labels to make healthy food choices.

### Objectives
- To describe how the government makes sure food is safe
- To explain what a food label tells you
- To identify food additives

**Student Pages** 239–242

**Teacher's Resource Library**
Activity 37
Alternative Activity 37
Workbook Activity 37

### Vocabulary
additive          preservative
Nutrition Facts

After discussing the meanings of the vocabulary words, have a variety of empty food packages available for students to examine. Ask volunteers to choose a package, examine it, and give a short description of the product, using all of the vocabulary words.

### Background Information

The USDA was established by Abraham Lincoln in 1862. At first, it issued research bulletins and reports on agricultural crops. In 1869, the USDA published the first analysis of a food product, corn. By 1883, methods that detected adulteration of food were developed. These methods led to enactment of the Pure Food and Drug Act. During the 1890s, the USDA began to fund research in human nutrition and studies of nutrients. Today, we take federal funding of nutrition research for granted. We also have come to expect safe food in the marketplace.

*Choosing Helathy Foods* Chapter 10  **239**

# Warm-Up Activity

Show students two loaves of bread, one packaged from a supermarket and the other fresh from a bakery. Tell students that the bakery bread has no preservatives. Ask them which bread should be eaten first. What will happen to the two loaves if they are not eaten in a week? *(Mold will grow on the bakery bread, which has no preservatives. Nothing will happen to the packaged bread.)* Their predictions can be tested by sprinkling a slice of each loaf with water and placing the slices in separate self-seal bags. Place in a warm, dark location for several days.

# Teaching the Lesson

Encourage students to bring in labels from food products that they eat regularly. Have them share the label information with the class and discuss the various additives that they consume by eating these foods. Ask students their feelings about the amount of additives found in foods. Are they concerned about the possible health risks these ingredients pose? Encourage interested students to research possible dangers of food additives.

# Reinforce and Extend

## Link to

**Math.** Have students read the Link to Math feature on page 240. Ask a student to come to the board and show how 60 calories is equal to about 3 percent of the Daily Value for calories.

*(2000 Calories ÷ 60 = 0.03 × 100 = 3%).*

## LEARNING STYLES

**Visual/Spatial**

Have each student bring a food package with its label to class. Ask why it is important to check the serving size of a food on the food label. *(Amounts of nutrients are based on the serving size. If you eat a smaller or larger serving, you are getting a different amount of nutrients from what is stated on the label.)* Measure out the serving sizes listed on several food labels to show students different serving sizes.

---

**Nutrition Facts**
*Section of a food label that provides information about the product*

Figure 10.3.1 shows a sample food label. Food labels also must include information about the nutrients and contents of the product. These appear under the heading **Nutrition Facts**. Nutrition Facts tell the size of one serving and how many servings are in the package. Nutrition Facts also list the number of calories per serving, as well as the number of calories from fat.

The Nutrition Facts label also lists the Daily Value (DV) for each nutrient found in the food. The DV is based on a diet of 2,000 calories per day. For example, the DV for sodium is 2,400 milligrams (mg). The label in Figure 10.3.1 shows that the product has 250 mg of sodium per serving. Then it shows that the food has 10 percent of the DV for sodium. The Daily Value information helps you figure out how many nutrients you are getting in one serving of that food.

Under Daily Values, a food label must give the amounts of total fat, cholesterol, sodium, total carbohydrate, and protein. Amounts are given in grams (g) or milligrams (mg). Next to the amount in grams or milligrams, the label shows the actual percent of Daily Value.

**Figure 10.3.1** *Food label*

## Link to

**Math**

Figure 10.3.1 shows that a serving of tuna has 60 calories. This is about 3 percent of the Daily Value (DV) for calories. Which nutrients in the tuna provide more than 3 percent of the DV?

The Nutrition Facts label also must include the amount of vitamin A, vitamin C, calcium, and iron in the product. The label may list other vitamins and minerals. This information also is listed as a percent.

Finally, the label must list everything that is in the product, or the ingredients. The ingredients are listed in order by weight, from the most to the least. For example, on a can of soup, the first thing listed is water. That means there is more water in the soup than anything else. Some people must check the ingredients listed on labels carefully. If they have an allergy to a food, they can tell from the ingredients whether that food is included.

## TEACHER ALERT

Tell students that artificial food colorings are among the most common additives. Many of the foods students eat have been color-enhanced. For example, strawberry sherbet is not naturally a rich pink color. It has been enhanced with red food coloring.

Nine certified colors are approved for use in foods in the United States. All are artificially made. Each batch of color is tested by both the manufacturer and the FDA to ensure the safety, quality, consistency, and strength of the additive before it is used in foods.

**Additive**
*A substance added in small amounts to foods that changes them in some way*

**Preservative**
*A substance added to foods to prevent them from spoiling*

Why is it important to be able to read and understand a food label?

All the information on the food label can be helpful in choosing food. If you are trying to limit something in your diet, such as sodium, look at the label. It will tell you what percent of sodium is in one serving. By using the information on the label, you can tell if the product is priced fairly. You can look at the price and the number of servings and decide how much each serving costs. You can then look at other brands to see which product is the best value.

## What Are Food Additives?

A food **additive** is something added to food that changes it in some way. Additives change color, flavor, or texture. Additives such as sugars or salt are sometimes added to help prevent the food from spoiling. Sometimes, nutrients are added to foods. For example, vitamin C is added to some fruit drinks. Calcium is added to some fruit juices.

Some people question whether food additives are safe. The U.S. government has passed laws about food additives. The laws state that additives must be tested and must not be found to cause cancer in animals or humans. Food producers must test the additives and give their results to the FDA. The FDA decides whether the additive is safe.

**Preservatives** are also added to food. They keep bacteria from growing and spoiling the food. Preservatives such as sugar and salt are necessary, because many foods spoil quickly. Preservatives help by keeping canned, packaged, or frozen foods safe to eat.

In general, food additives are used in very small amounts. A person would have to eat a large amount of a food with additives to reach a dangerous level. Because foods are being processed more, it is a good idea to find out which additives are in the foods you eat often. You may decide to try to add more fresh foods to your diet.

## GLOBAL CONNECTION

Have students research the U.S. Food and Drug Administration (FDA) and the United States Department of Agriculture (USDA) to discover how these agencies set rules for food producers to follow. Ask students to research whether other countries have similar organizations, and if so, to find out what their guidelines and recommendations are.

## ELL/ESL STRATEGY

**Language Objective:** *To learn content area vocabulary*

Have ELL students examine a food label and make a list of the words on the label that they do not understand. Have ELL students look up each word and write it and its definition in a notebook. It would be helpful to the students if they also wrote in the notebook the translations of the words in their native languages.

## LEARNING STYLES

**Visual/Spatial**
Have students visit a grocery store and examine three different brands of a type of cereal, such as corn flakes or puffed rice. Ask students to compare the cost and the nutritional information for each brand and answer the following questions. Does each brand use the same serving size? Are the ingredients in all brands the same? Which brand provides the most nutritious breakfast? Why?

## LEARNING STYLES

**Body/Kinesthetic**
Have students create their own label for a unique food product. The label should include nutritional information as well as a list of ingredients, including any additives and/or preservatives. Refer students to Figure 10.3.1 in the text to make sure they include all the required information.

## Technology and Society

Ask a volunteer to read aloud the Technology and Society feature on page 242. Ask students to use the Internet to plan a dinner that is both interesting and healthful. Encourage them to try recipes from other cultures. Have students write out a menu for their dinner. Students can give oral reports to the class about their menus, discussing why they chose the foods they did and explaining the nutritional value of the foods. The menus can be displayed in the classroom. Encourage students to prepare the foods on their menus for their families.

## Lesson 3 Review Answers

**1.** D **2.** C **3.** A **4.** Sample answer: I can use the nutrition and calorie information on food labels to plan a healthy diet. Reading food labels can help me avoid snacks that have very little nutritional value. **5.** Sample answer: I would use the Internet to learn about food additives. I would then be able to back up my beliefs with facts about what the additives are and how they are used. I could use these facts to support my view that the average American consumes far too many food additives each day.

## Portfolio Assessment

Sample items include:

- Reports from Application: Global Connection
- Words and definitions from ELL/ESL Strategy
- Cereal research from Learning Styles: Visual/Spatial
- Menu from Technology and Society
- Food label from Learning Styles: Body/Kinesthetic
- Lesson 3 Review answers

242  Unit 3  Nutrition

---

The number of calories you need each day depends on how much physical activity you do. The more active you are, the more calories you need.

***Lesson 3 Review*** On a sheet of paper, write the word or phrase from the Word Bank that best completes each sentence.

**1.** The government agency that controls food labeling is the _____.
  **A** Department of Agriculture
  **B** Food and Nutrition Board
  **C** Department of Education
  **D** Food and Drug Administration

**2.** Ingredients on a food label are listed according to _____.
  **A** calories          **C** weight
  **B** Daily Value       **D** saturated fat

**3.** Daily Value percents are based on a diet of _____ per day.
  **A** 2,000 calories
  **B** three meals and a snack
  **C** 2,000 carbohydrates and fats
  **D** 500 grams

On a sheet of paper, write the answers to the following questions. Use complete sentences.

**4. Critical Thinking** How can you use labels to develop and track a personal health plan?

**5. Critical Thinking** How would you explain your point of view about food additives to someone who disagrees with you?

### Technology and Society

People lead busy lives. Planning meals and grocery shopping can be difficult. Healthy recipes for snacks, quick meals, or special diets can be found on the Internet. You can also find recipes from other cultures. Some people shop for groceries, special ingredients, or special cooking utensils on the Internet. Using the Internet can make meal planning faster and fun.

242  Unit 3  Nutrition

Activity 37

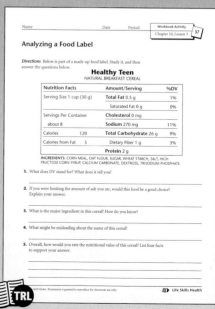

Workbook Activity 37

## Chapter 10 SUMMARY

- A healthy eating pattern includes regular, balanced meals and nutritious snacks. A healthy eating pattern is important.
- Many things affect how you choose food. These include where you live, the time of year, the cost of food, and your cultural background. Association and stress also affect how you choose food.
- Companies use advertising to get people to buy their products.
- Fast food has become part of many people's regular eating patterns.
- Fad diets are diets that usually focus on a certain food. These diets can cause serious health problems.
- The U.S. government has set up agencies to make sure foods are labeled correctly and are safe to eat. They are the USDA, the FDA, and the Food and Nutrition Board of the National Research Council. The FDA is responsible for food labeling.
- Food labels provide information about the product. Labels include Nutrition Facts, Daily Value information, and a list of the product's ingredients. Food labels are helpful in choosing foods for your diet.
- Nutrition Facts tell serving sizes, number of servings in the package, and number of calories per serving.
- Additives are things that are added to foods to change them in some way. Some additives are preservatives that help prevent the food from spoiling. Additives are included in very small amounts.

### Vocabulary

| | | |
|---|---|---|
| additive, 241 | media, 233 | preservative, 241 |
| advertise, 233 | Nutrition Facts, 240 | product, 233 |
| association, 231 | organic food, 234 | trans fat, 236 |
| Daily Values (DV), 236 | | |

*Choosing Healthy Foods* Chapter 10

## Chapter 10 Summary

Have volunteers read aloud each Summary item on page 243. Ask volunteers to explain the meaning of each item. Direct students' attention to the Vocabulary box on the bottom of page 243. Have them read and review each term and its definition.

### ONLINE CONNECTION

For more information about the work and responsibilities of the Center for Food Safety and Applied Nutrition, a branch of the FDA, have students go to this Web site. They can access information about diverse topics such as food additives and supplements, food safety, and food-borne illnesses: www.cfsan.fda.gov/list.html

This Web site, offered by the American Heart Association, discusses quick-weight loss, or fad, diets. The site will help students understand why such diets are not healthy and gives recommendations for a healthful plan of weight loss: www.americanheart.org/presenter.jhtml?identifier=4584

This Web site, offered by the U.S. Department of Health and Human Services, has an interactive menu planner. Students will enjoy planning menus that balance calorie counts and items from various food groups: http://hin.nhlbi.nih.gov/menuplanner/menu.cgi

*Choosing Helathy Foods* Chapter 10

# Chapter 10 Review

Use the Chapter Review to prepare students for tests and to reteach content from the chapter.

## Chapter 10 Mastery Test

The Teacher's Resource Library includes two forms of the Chapter 10 Mastery Text. Each test addresses the chapter Goals for Learning. An optional third page of additional critical-thinking items is included for each test. The difficulty level of the two forms is equivalent.

## Review Answers

**Vocabulary Review**

1. organic food 2. additive 3. advertises 4. association 5. preservative 6. Daily Value 7. Nutrition Facts 8. media 9. product 10. trans fat

# Chapter 10 REVIEW

**Word Bank**
- additive
- advertises
- association
- Daily Value
- media
- Nutrition Facts
- organic food
- preservative
- product
- trans fat

## Vocabulary Review

On a sheet of paper, write the word or phrase from the Word Bank that best completes each sentence.

1. Food grown without pesticides or unnatural fertilizers can be labeled as _____.
2. Something that changes the color of a food is a(n) _____.
3. A company _____ an item to draw attention to it.
4. A(n) _____ happens when a food reminds you of something you once did.
5. Something added to food to keep bacteria from growing is a(n) _____.
6. The _____ of a food is the percent of a nutrient that is found in a food.
7. You can find how much protein there is in a can of soup by looking at the _____.
8. The _____ includes TV, radio, the Internet, and magazines.
9. The purpose of advertising is to sell a(n) _____.
10. Processed vegetable oil is mostly made of _____.

## Concept Review

On a sheet of paper, write the letter of the answer that best completes each sentence.

11. An ad that shows a child eating popcorn while watching a movie is using _____ to sell popcorn.
    A fast food
    B association
    C claims of superior food
    D peer pressure

244 Unit 3 Nutrition

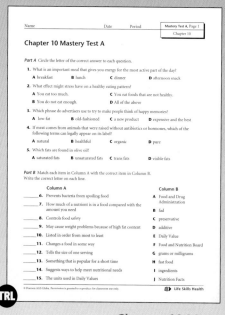
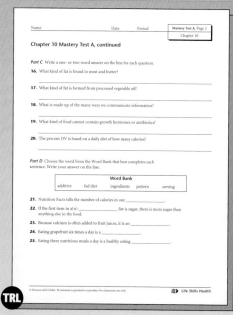

Chapter 10 Mastery Test A, pages 1–3

12. The USDA is responsible for _____.
    A inspecting meat
    B inspecting health food stores
    C regulating the fast-food industry
    D suggesting new products

13. Nutrition facts on a food label show the ingredients and the _____ of the Daily Value of each nutrient.
    A kilogram  B percent  C milliliter  D calorie

14. Eating only cabbage soup for 2 weeks is an example of a(n) _____ diet.
    A trans fat  B nutritious  C fad  D organic

15. Two things that may affect which foods you choose are your cultural background and _____.
    A how old you are      C your interests
    B the weather          D the time of year

16. The most important meal of the day is _____.
    A dinner  B breakfast  C lunch  D a snack

## Critical Thinking

On a sheet of paper, write answers to the following questions. Use complete sentences.

17. What is your definition of a healthy eating plan?

18. For a healthy diet, why should people know the nutritional value of fast foods?

19. How can food labels help someone with food allergies?

20. Your friend wants you to go on a fad diet with him or her. What would you tell your friend? Why?

**Test-Taking Tip** When you read test directions, say them in your own words to yourself to make sure you understand them.

## Concept Review

11. B  12. A  13. B  14. C  15. D  16. B

**Critical Thinking**

17. Sample answer: A healthy eating plan includes regular and nutritious meals and snacks. Foods included in a healthy eating plan would be low in fat and sugar and high in vitamins and fiber. 18. Sample answer: Fast food is a regular part of many people's diets. Knowing the nutritional value of the food will help people make healthy diet decisions. 19. Sample answer: Reading the ingredients on food labels will help people with food allergies stay away from the foods that make them sick. For example, a person who has an egg allergy should not buy a box of dried pasta that lists eggs as an ingredient. 20. Sample answer: I would tell my friend that being on a fad diet is not good for you because you need the nutrients from a variety of foods.

## ALTERNATIVE ASSESSMENT

Alternative Assessment items correlate to the student Goals for Learning at the beginning of this chapter.

- Have students write a paragraph describing how their lifestyle affects their diet and how they could change their lifestyle to create a healthier eating pattern.

- Have students orally describe factors that influence food choices.

- Have students work in small groups to make a list of ways that government agencies make sure food is safe.

- Have students make a poster showing a food label and pointing out how it helps them make healthy food choices.

- Have students point out the additives and preservatives in the ingredients list on a food label and explain how things that are added to foods may change them.

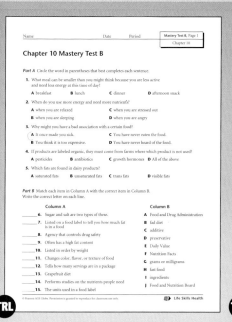
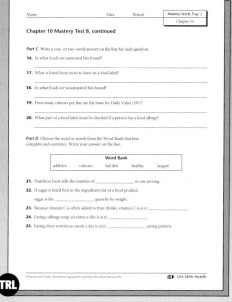

Chapter 10 Mastery Test B, pages 1–3

# HEALTH IN THE WORLD

Have students read the Health in the World feature on page 246. Ask students to explain how a natural disaster could affect people's physical, emotional, and mental health. (Sample answer: It could affect physical health because people could be injured during the disaster; if they are unharmed, they might not have access to clean water or food for an extended period of time. The stress and problems brought on by a natural disaster could be very hard on people's emotional health. People could become depressed about the losses caused by the disaster.)

## Health in the World Answers

**1.** Sample answer: After a natural disaster, the International Red Cross works with other organizations because this cooperation results in more people being helped more quickly. Each relief organization has different strengths, and by partnering with other relief organizations, the International Red Cross can help ensure that the most number of people get help in the least amount of time. **2.** Sample answer: During a natural disaster, volunteers help the International Red Cross locate people who are stranded or who need medical attention. Volunteers are often local residents who know an area very well. For example, in 2004 a tsunami struck parts of Indonesia, India, and Sri Lanka. In the aftermath of this natural disaster, volunteers helped the International Red Cross because they were very knowledgeable about what the areas looked like before the tsunami struck. **3.** Sample answer: Red Cross workers helped people in Louisiana and Mississippi after Hurricane Katrina in 2005. Red Cross workers also helped people in Indonesia, India, and Sri Lanka after the tsunami of 2004. In both situations, Red Cross workers helped injured people get medical care, they helped survivors get bottled water, and they helped survivors reunite with their families.

# HEALTH IN THE WORLD

## International Red Cross

The International Red Cross is part of the world's largest network for helping people. This network is made up of 181 countries and includes many organizations.

The goal of the International Red Cross is to bring relief to people following a natural disaster. Its mission is to protect life and health for all people of the world. Each member country has equal status and carries equal responsibilities. Help is given without regard for political opinion, race, or religious beliefs.

Natural disasters come in many forms: crop failures, earthquakes, hurricanes, tsunamis, mudslides, and others. Natural disasters can affect thousands of people at the same time. Often they strike with little or no warning. Communities are destroyed. Often many lives are lost. Families are separated in the confusion brought on by the disaster. With little or no food, clean water, or medical supplies, malnutrition and disease are a constant threat to victims of natural disasters.

International Red Cross volunteers work within their own countries to bring relief to victims of natural disasters. Volunteer youth groups are an important part of this effort. The International Red Cross provides financial and technical support for relief efforts in individual countries.

This includes supplying natural disaster victims with additional volunteers, shelter, water, and food supplies. These organizations also help to reconnect families separated by a natural disaster. Often relief efforts last several months.

On a sheet of paper, write the answers to the following questions. Use complete sentences.

1. Why do you think the International Red Cross works with other organizations to help people after a natural disaster?

2. How are volunteers an important part of the International Red Cross?

3. Can you think of a recent natural disaster in which the Red Cross helped? Describe what the Red Cross did to help.

# Unit 3 Summary

- The U.S. government's Food Guide Pyramid shows how much and what kinds of foods to eat.

- Your body needs six essential nutrients. Carbohydrates, fats, and protein are three nutrients that give your body energy. Vitamins, minerals, and water are three nutrients that help your body change food into energy.

- Carbohydrates are starches and sugars that come mainly from plants. They provide much of the body's energy.

- Fats are stored energy. Fats protect your organs and nerves. Too many saturated fats can lead to high levels of cholesterol in your blood. Some kinds of cholesterol can clog blood vessels.

- Protein builds muscle and repairs tissue. Animal products and some plants give you protein. Protein is made of amino acids.

- Your body needs vitamins in small amounts for normal growth and activity. Some vitamins can be stored in the body. Others cannot be stored.

- Minerals are important in small amounts for fluid balance and digestion. Calcium, phosphorus, and iron are especially important minerals.

- A healthy diet includes foods with all six essential nutrients. Not enough or too much of these nutrients can lead to health problems.

- A healthy eating pattern includes balanced meals and nutritious snacks.

- Many things affect how you choose food. These include where you live, the time of year, the cost of food, and your cultural background.

- Advertisers use several ways to get people to buy their products.

- Fad diets are diets that usually focus on a certain food. These diets can cause serious health problems.

- The U.S. government has set up agencies to make sure foods are labeled correctly and are safe to eat. They are the USDA, the FDA, and the Food and Nutrition Board of the National Research Council.

- Food labels include Nutrition Facts, Daily Value information, and a list of the product's ingredients. Nutrition Facts tell serving sizes, number of servings in the package, and number of calories per serving.

- Additives are things that are added to foods to change them in some way. Some additives are preservatives that help prevent the food from spoiling.

## Unit 3 Review

Use the Unit Review to prepare students for tests and to reteach content from the unit.

## Unit 3 Mastery Test

The Teacher's Resource Library includes two forms of the Unit 3 Mastery Text. An optional third page of additional critical-thinking items is included for each test. The difficulty level of the two forms is equivalent.

## Review Answers

**Vocabulary Review**

1. carbohydrates 2. protein 3. deficiency 4. association 5. calcium 6. trans fat 7. additive 8. preservative 9. unsaturated fat 10. Daily Values (DV) 11. sodium 12. cholesterol

**Concept Review**

13. D 14. C 15. B 16. A

---

# Unit 3 REVIEW

## Vocabulary Review

On a sheet of paper, write the word or phrase from the Word Bank that best completes each sentence.

**Word Bank**
additive
association
calcium
carbohydrates
cholesterol
Daily Values (DV)
deficiency
protein
preservative
sodium
trans fat
unsaturated fat

1. Your body gets energy from _____, the starches and sugars that come mainly from plant food.
2. A nutrient that helps build muscle is _____.
3. People who do not eat a balanced diet are at risk of having a(n) _____.
4. A connection between a food and a memory is a(n) _____.
5. A mineral found in dairy products that helps keep your bones strong is _____.
6. Vegetable shortening and many cookies contain _____, which is made from processed vegetable oil.
7. Food producers must test a(n) _____ before it can be added to a food.
8. A(n) _____ is used to keep canned foods from spoiling.
9. Eating fish instead of red meat will replace saturated fat with _____.
10. The _____ are printed on food labels.
11. A mineral that helps control the amount of fluids in your body is _____.
12. A waxy, fatlike material that is important in making hormones is _____.

## Concept Review

On a sheet of paper, write the letter of the answer that best completes each sentence.

13. Table sugar is _____.
    A fructose   B glucose   C lactose   D sucrose

248  Unit 3 Nutrition

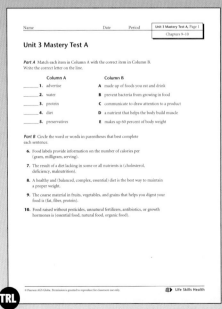
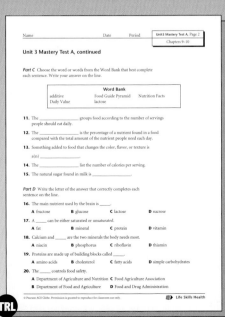

Unit 3 Mastery Test A, pages 1–3

14. The proteins that you must get from your diet are _____.

   A  vitamins
   B  minerals
   C  essential amino acids
   D  complex carbohydrates

15. Many children want to eat fast food because they have seen _____ for the food.

   A  Nutrition Facts
   B  TV advertisements
   C  lot numbers
   D  serving sizes

16. The agency that checks and grades meat and poultry is the _____.

   A  United States Department of Agriculture (USDA)
   B  U.S. Environmental Protection Agency (EPA)
   C  U.S. Food and Drug Administration (FDA)
   D  Food and Nutrition Board of the National Research Council

## Critical Thinking

On a sheet of paper, write the answers to the following questions. Use complete sentences.

17. Many companies that sell vitamins use the word "daily" on the labels of vitamin pills. Why?

18. Suppose someone is recovering from a heart attack. Why might a doctor tell that person to eat less red meat?

19. Advertisements can affect a person's food choices. Describe a situation in which you or someone you know chose to eat something after seeing an advertisement. Was this a healthy choice? Explain your answer.

20. The Daily Values (DV) are based on a diet of 2,000 calories per day. Why is it important to include this amount (2,000 calories) on a Nutrition Facts label?

Unit 3 Review  249

## Critical Thinking

**17.** Sample answer: There are two key reasons why manufacturers of vitamin pills encourage their customers to take vitamin pills daily. First of all, the human body does not store vitamins for long periods of time, so the body needs a new supply of vitamins daily. Secondly, these manufactures want to sell as many vitamins as possible, so if a customer takes vitamins daily, then he or she will frequently need to buy bottles of vitamin pills. **18.** Sample answer: Having a healthy diet is very important for someone who is recovering from a heart attack. A doctor might tell someone to avoid red meat because it contains a lot of saturated fat. Eating a diet that is high in saturated fats can cause cholesterol to build up in blood vessels and clog them. When blood cannot flow freely in the body, the person can have serious health problems, including heart problems. **19.** Sample answer: Last week, my younger brother bought a bag of potato chips because he saw an advertisement for the chips in a sports magazine. This was not a healthy choice because he later learned that these chips are high in fat, sodium, and calories. My family tries to avoid eating unhealthy snacks. **20.** Sample answer: It is important to include this information on a Nutrition Facts label because some people eat more than 2,000 calories per day, and some people eat less than 2,000 calories per day. The percentages for the Daily Values are based on a daily diet of 2,000 calories. Including this information on the label helps people make informed decisions about the food choices they make. If a person is trying to reduce the amount of sodium in his or her diet, the Daily Values can help the person choose whether or not to eat a certain food.

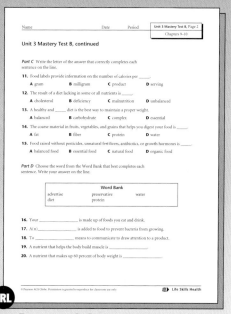

Unit 3 Mastery Test B, pages 1–3

Nutrition  Unit 3  249

# Unit 4

## Planning Guide
## Preventing and Controlling Diseases and Disorders

| | Student Pages | Vocabulary | Health in the World | Review | Critical-Thinking Questions | Chapter Summary |
|---|---|---|---|---|---|---|
| **Chapter 11** Disease—Causes and Protection | 252–267 | ✔ | | ✔ | ✔ | 265 |
| Lesson 1 Causes of Disease | 254–258 | ✔ | | ✔ | ✔ | |
| Lesson 2 How the Body Protects Itself from Disease | 259–264 | ✔ | | ✔ | ✔ | |
| **Chapter 12** Preventing AIDS and Sexually Transmitted Diseases | 268–283 | ✔ | | ✔ | ✔ | 281 |
| Lesson 1 AIDS | 270–274 | ✔ | | ✔ | ✔ | |
| Lesson 2 Sexually Transmitted Diseases | 275–280 | ✔ | | ✔ | ✔ | |
| **Chapter 13** Recognizing Common Diseases | 284–309 | ✔ | | ✔ | ✔ | 307 |
| Lesson 1 Cardiovascular Diseases and Problems | 286–290 | ✔ | | ✔ | ✔ | |
| Lesson 2 Cancer | 291–297 | ✔ | | ✔ | ✔ | |
| Lesson 3 Diabetes | 298–301 | ✔ | | ✔ | ✔ | |
| Lesson 4 Arthritis, Epilepsy, and Asthma | 302–306 | ✔ | 310 | ✔ | ✔ | |

### Unit and Chapter Activities

**Student Text**
Unit 4 Summary

**Teacher's Resource Library**
Home Connection 4
Community Connections 11–13

**Teacher's Edition**
Opener Activity, Chapters 11–13

### Assessment Options

**Student Text**
Chapter Reviews, Chapters 11–13
Unit 4 Review

**Teacher's Resource Library**
Self-Assessment Activities 11–13
Chapter Mastery Tests A and B, Chapters 11–13
Unit 4 Mastery Tests A and B

**Teacher's Edition**
Chapter Alternative Assessments, Chapters 11–13

250A

| | Student Text Features | | | | | | | | Teaching Strategies | | | | | | | Learning Styles | | | | | Teacher's Resource Library | | | | |
|---|---|---|---|---|---|---|---|---|---|---|---|---|---|---|---|---|---|---|---|---|---|---|---|---|---|
| | Self-Assessment | Health at Work | Health in Your Life | Decide for Yourself | Link to | Research and Write | Health Myth | Technology and Society | Background Information | ELL/ESL Strategy | Health Journal | Applications (Home, Career, Community, Global, Environment) | Online Connection | Teacher Alert | Auditory/Verbal | Body/Kinesthetic | Interpersonal/Group Learning | Logical/Mathematical | Visual/Spatial | Activities | Alternative Activities | Workbook Activities | Self-Study Guide | Chapter Outline |
| 252 | | | | | | | | | | | | | 265 | 266 | | | | | | 38–39 | 38–39 | 38–39 | ✔ | ✔ |
| | | | 257 | 256 | 258 | 258 | | 254 | 256 | 256 | 255, 256, 257 | | 255 | | | 255 | | 257 | 38 | 38 | 38 | | |
| | 264 | 261 | | 259 | 262 | 260 | 263 | 259 | 260 | 262 | 262, 264 | | 262 | 264 | 260 | | 263 | | | 39 | 39 | 39 | | |
| 268 | | | | | | | | | | | | 281 | 282 | | | | | | 40–41 | 40–41 | 40–41 | ✔ | ✔ |
| | 274 | | | 271, 273 | 273, 274 | 271, 273 | | 270 | 271 | 273 | 272, 273 | | 273 | 272 | | | 272 | | 40 | 40 | 40 | | |
| | | 280 | 279 | | | 276 | 275 | 276 | 278 | 278 | | 277, 279 | | 277 | 279 | 278 | | | 41 | 41 | 41 | | |
| 284 | | | | | | | | | | | | 307 | 308 | | | | | | 42–45 | 42–45 | 42–45 | ✔ | ✔ |
| | | | 289, 290 | 290 | 289 | 288 | 286 | 287 | 289, 290 | 281–291 | 290 | | 290 | | | 288 | | 42 | 42 | 42 | | | |
| | | 297 | | 296 | 294 | | 292 | 293 | | 294–296 | | 295 | | 294 | 295 | | | 43 | 43 | 43 | | | |
| | | | 300 | | | | 298 | 299 | | | | 299 | | | | | | 44 | 44 | 44 | | | |
| | 305 | | | | 304 | | 302 | 304 | 303 | 305 | | | | | | 303 | | 45 | 45 | 45 | | | |

## Alternative Activities

The Teacher's Resource Library (TRL) contains a set of lower-level worksheets called Alternative Activities. These worksheets cover the same content as the regular Activities but are written at a second-grade reading level.

## Skill Track

Use Skill Track for *Life Skills Health* to monitor student progress and meet the demands of adequate yearly progress (AYP). Make informed instructional decisions with individual student and class reports of lesson and chapter assessments. With immediate and ongoing feedback, students will also see what they have learned and where they need more reinforcement and practice.

## Unit at a Glance

### Unit 4:
### Preventing and Controlling Diseases and Disorders
pages 250–313

### Chapters

11. Disease—Causes and Protection
    pages 252–267
12. Preventing AIDS and Sexually Transmitted Diseases
    pages 268–283
13. Recognizing Common Diseases
    pages 284–309

**Unit 4 Summary** page 311

**Unit 4 Review** pages 312–313

**Audio CD**

**Skill Track for Life Skills Health**

**Teacher's Resource Library** TRL
  Home Connection 4
  Unit 4 Mastery Tests A and B
  (Answer Keys for the Teacher's Resource Library begin on page 559 of this Teacher's Edition.)

## Other Resources

### Books for Teachers

Kennedy, Michael. *A Brief History of Disease, Science, and Medicine: From the Ice Age to the Genome Project.* Mission Viejo, CA: Asklepiad Press, 2004.

Kiple, Kenneth F., ed. *The Cambridge World History of Human Disease.* New York: Cambridge University Press, 1993.

### Books for Students

*American Medical Association Complete Guide to Your Children's Health.* New York: Random House, 1999. (written for parents but accessible to teens)

Powell, Michael, and Oliver Fischer. *101 Diseases You Don't Want to Get.* New York: Thunder's Mouth Press, 2005. (humorous pocket guide to scary diseases)

### CD-ROM/Software

*Emerging and Re-Emerging Infectious Diseases.* Colorado Springs, CO: BSCS and Videodiscovery, Inc., 1999. (http://science.education.nih.gov/) (major concepts in diseases and their impact on society)

### Videos and DVDs

*Sexually Transmitted Diseases: The Silent Epidemic* (29 minutes). Dartmouth, NH: Dartmouth-Hitchcock Medical Center, 2000 (1-800-257-5126). (prevention, diagnosis, and treatment of STDS)

### Web Sites

www.hsls.pitt.edu/guides/chi/diseases (information on common diseases)

http://www.ncbi.nlm.nih.gov/books/bv.fcgi?call=bv.View..ShowSection&rid=gnd.preface.91 (overview of the relationship between genetics and diseases)

www.mic.ki.se/Diseases (diseases, health conditions, and disorders)

# Unit 4: Preventing and Controlling Diseases and Disorders

When was the last time you were sick? Has it been so long ago that you cannot remember? This could be due to wise decisions you have been making about your health. Young adulthood is usually a time of excellent health. It is generally a time when your body is in peak physical form. However, young adulthood does not automatically mean good health. It is important for people of all ages to make healthy decisions. All people should do what they can to prevent disease.

Diseases affect how you feel in a number of ways. You can prevent many diseases by taking steps to protect yourself. In this unit, you will learn about causes and signs of disease. You will also discover how you can try to keep yourself from getting sick.

### Chapters in Unit 4

Chapter 11: Disease—Causes and Protection..............252
Chapter 12: Preventing AIDS and Sexually Transmitted Diseases ................268
Chapter 13: Recognizing Common Diseases..............284

## Introducing the Unit

Have volunteers read aloud the introductory paragraphs on page 251. Ask students what things they do every day to prevent themselves from getting sick. *(Possible answers include washing hands before eating or after using the bathroom, not drinking from the same glass as others, and staying away from others who are coughing and sneezing.)* Ask students to list some diseases that they can help prevent themselves from getting. *(Possible answers include colds, flu, and sexually transmitted diseases.)*

### HOME CONNECTION

The Home Connection unit activity gives students practical experience with concepts taught in the *Life Skills Health* student text. Students complete the Home Connection activity outside the classroom with the help of family members. These worksheets appear on the Life Skills Health Teacher's Resource Library (TRL) CD-ROM.

### CAREER INTEREST INVENTORY

The AGS Publishing Harrington-O'Shea Career Decision-Making System-Revised (CDM) may be used with the chapters in this unit. Students can use the CDM to explore their interests and identify careers. The CDM defines career areas that are indicated by students' responses on the inventory.

Home Connection 4

## Chapter at a Glance

**Chapter 11:**
**Disease—Causes and Protection**
pages 252–267

### Lessons

1. **Causes of Disease**
   pages 254–258
2. **How the Body Protects Itself from Disease**
   pages 259–263

**Chapter 11 Summary** page 265

**Chapter 11 Review** pages 266–267

**Audio CD**

**Skill Track for Life Skills Health**

**Teacher's Resource Library**

   Activities 38–39
   Alternative Activities 38–39
   Workbook Activities 38–39
   Self-Assessment 11
   Community Connection 11
   Chapter 11 Self-Study Guide
   Chapter 11 Outline
   Chapter 11 Mastery Tests A and B

(Answer Keys for the Teacher's Resource Library begin on page 559 of this Teacher's Edition.)

### Opener Activity

Ask students to name some common diseases. *(Answers may include the common cold, the flu, and cancer.)* What are some uncommon diseases? *(Uncommon diseases might include cirrhosis, Parkinson disease, and muscular dystrophy.)* Encourage students to share any knowledge they have about an uncommon disease with the class.

252  Unit 4  Prev. & Contr. Diseases

## Self-Assessment

### Can you answer these questions?

1. What is an infectious disease?
2. What causes infectious diseases?
3. Why should you wash your hands frequently?
4. What are five ways that germs are spread?
5. How does AIDS affect the body?
6. What causes inherited diseases?
7. How do your skin, tears, and saliva help keep you from getting sick?
8. Why do people get some diseases only once?
9. Why does vaccination prevent some diseases?
10. Why is it important to be up to date on your vaccinations?

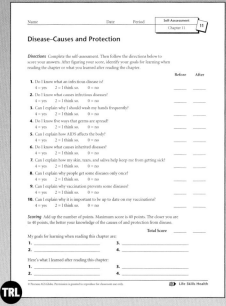

Self-Assessment 11

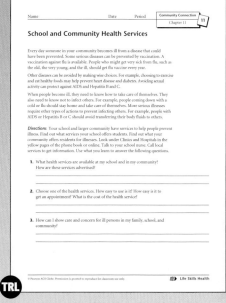

Community Connection 11

# Chapter 11
# Disease—Causes and Protection

Have you ever wondered how you came down with the flu or got a cold? Have you wondered what your body does when you get sick? Your body has an amazing way of preventing many diseases. If you do get sick, your body fights the cause of the disease. When you know how diseases spread, you can take action to prevent getting or giving them.

In this chapter, you will learn the causes of diseases. You will also learn how the body fights diseases.

## Goals for Learning

◆ To explain the causes of acquired diseases and inherited diseases
◆ To identify some of the body's barriers to infection
◆ To explain how the immune system works
◆ To recognize the importance of vaccinations

## Introducing the Chapter

Have students read the chapter opener on page 253. Then discuss the contents of the paragraphs with students as well as the Goals for Learning. Ask them what they think are some ways to protect themselves from disease. Keep a list of students' responses so that they can compare them with their new knowledge after completing the chapter.

## Notes and Questions

Ask volunteers to read the notes and questions that appear in the margins throughout the chapter. Then discuss them with the class.

## Self-Assessment

Have students complete the Self-Assessment worksheet before and after reading the chapter. Before reading the chapter, have students fill in the "Before" column. Ask students to identify their goals for learning. To get ideas for setting goals, students might use the chapter introductory material on page 253, the checklist on page 252, or the questions on the Self-Assessment worksheet. Students can use the back of the worksheet if they need more space to write.

Collect the Self-Assessment worksheets and pass them out again at the end of the chapter. Have students fill in the "After" column. Ask them to identify at least four major points they have learned. Again, suggest they use the back of the worksheet if they need more space to write. You may want to collect and review the worksheets, but return them to students so they have a record of their goals and accomplishments.

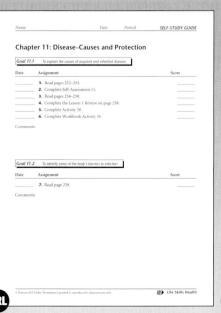

Chapter 11 Self-Study Guide, pages 1–2    Chapter 11 Outline, pages 1–2

# Lesson at a Glance

## Chapter 11 Lesson 1

**Overview** The lesson explains some causes of acquired and inherited diseases and the stages of infectious disease.

### Objectives

- To explain some causes of acquired diseases
- To identify the stages of infectious disease
- To recognize some causes of inherited diseases

**Student Pages** 254–258

**Teacher's Resource Library**
  Activity 38
  Alternative Activity 38
  Workbook Activity 38

## Vocabulary

| | |
|---|---|
| acquire | infectious |
| antibiotic | influenza |
| communicable | inherit |
| deformity | pathogen |
| immune system | virus |
| incubation period | |

Divide the students into pairs. Ask the pairs to find the vocabulary words in the Glossary and to discuss the meaning(s) of each word. Then ask them to write a sentence for each word.

## Background Information

Controlling disease during the past century has led to a change in life expectancy from 47 in 1900 to 77 in the early 2000s. Understanding how diseases spread is essential to minimizing their spread. Diseases such as HIV/AIDS and hepatitis B and C threaten young adults who may be more susceptible to the diseases because they often participate in risky behaviors.

 **Warm-Up Activity**

Have students think about a time when they have been ill. Ask them whether they were willing to make changes in their lives to help keep them from getting ill again. Explain that by learning about the causes of diseases, they can learn how to avoid becoming ill.

254  Unit 4  Prev. & Contr. Diseases

---

## Lesson 1  Causes of Disease

### Objectives

*After reading this lesson, you should be able to*

- explain some causes of acquired diseases
- identify the stages of infectious disease
- recognize some causes of inherited diseases

**Inherit**
*To receive something passed down from parents*

**Acquire**
*To get after being born*

**Infectious**
*Passed from a germ carrier to another living thing*

**Communicable**
*Infectious*

**Influenza**
*A common illness known as the flu*

Disease is a condition that harms or stops the normal workings of the body. A person can **inherit** or **acquire** a disease. An inherited disease is passed physically from parents to children. Acquired diseases are those that people get after they are born.

### What Are Some Causes of Acquired Diseases?

Many diseases are acquired. Infections, human behaviors, or the environment cause an acquired disease.

When infection causes an acquired disease, the disease is **infectious.** Infectious means passed from a germ carrier to another living thing. Infectious diseases are often caused by germs that pass from person to person. They also can spread from food, water, or animals that carry germs. Another word for infectious is **communicable.** Infectious diseases such as **influenza,** or the common flu, often spread very easily. Here are the ways germs can be spread:

- through direct physical contact with an infected person
- by drops that an infected person coughs into the air
- by contact with an object that an infected person has used
- by contact with food or water with germs in it
- through the bites of infected animals, including insects

An acquired disease can also result from a person's behaviors. For example, eating healthy foods and getting regular exercise may help prevent heart disease. People who do not choose healthy behaviors are more likely to have heart disease. People may also tend toward heart disease because of heredity. For these people, choosing a healthy lifestyle is especially important.

An acquired disease can also come from the environment. For example, children living in older homes who eat lead-based paint chips can get lead poisoning. Nonsmokers who breathe others' cigarette or cigar smoke may get lung diseases, including cancer.

254  Unit 4  *Preventing and Controlling Diseases and Disorders*

**Pathogen**
Germ that causes infectious diseases

**Virus**
Small germ that reproduces only inside the cells of living things

**Incubation period**
Time it takes after a person is infected for symptoms of a disease to appear

**Immune system**
Combination of body defenses that fight pathogens

Why do you think people often get a cold or the flu within a couple of days after flying on an airplane?

## What Are the Stages of Infectious Disease?

When a person acquires an infectious disease, it usually passes through certain stages. Figure 11.1.1 shows these stages. Germs that cause infectious diseases are **pathogens.** Pathogens include bacteria, fungi, and **viruses.** Viruses are small germs that reproduce only inside the cells of living things. Viruses cause the common cold, chickenpox, and influenza.

After a pathogen infects a person, the disease's first stage is the **incubation period.** This is the time it takes after infection for symptoms of the disease to appear. This period may be days, weeks, months, or even years.

The second stage is illness or getting well. If the body is able to defend, or fight off, the pathogens during incubation, a person may never become ill. The combination of body defenses that fights pathogens is the body's **immune system.**

Sometimes the body's defenses cannot fight off pathogens during the incubation period. Then the person becomes ill. The immune system continues to fight the illness. The third stage is recovery or death. The immune system successfully fights most infectious diseases and the person gets well. For example, if you get a cold, you are usually well again in about 10 days.

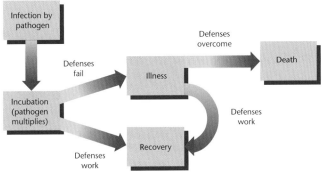

**Figure 11.1.1** *Stages of infectious disease*

Disease—Causes and Protection   Chapter 11   **255**

### Teaching the Lesson

Before students read about the causes of disease, ask them to look at Figure 11.1.1 on page 255. Focus their attention on the step prior to the first stage of an infectious disease—Infection by pathogen. Ask students to identify the ways that pathogens are spread as they read page 254.

Ask students to look at the first stage of an infectious disease—incubation. During this stage of a disease, an infected person usually does not know that his or her body is trying to fight off infection.

Point out that after the incubation period, the person either recovers or becomes ill. During the illness itself, the body continues to try to fight the disease. If the body fights the disease effectively, the person recovers. If not, the disease takes over.

### Reinforce and Extend

#### LEARNING STYLES

**Interpersonal/Group Learning**
Invite students to work in small groups to read and discuss the contents of the lesson. Ask group members to work together to produce a graphic organizer with the main ideas and important details for the three subhead questions in the lesson.

#### GLOBAL CONNECTION

A flu pandemic is a global outbreak that occurs when a new influenza virus causes serious human illness and spreads easily from person to person. Have students research influenza pandemics at the World Health Organization on their Epidemic and Pandemic Response page of their Web site at www.who.int/csr/disease/influenza/en/index.html.

### TEACHER ALERT

Inform students that a person can get infected with a disease by eating food that has been contaminated with bacteria or other pathogens. In many places with hot climates, where food can spoil quickly, the people use a lot of hot spices in their food. Some people believe that these spices can kill bacteria and prevent the food from becoming contaminated. However, the spices simply mask the taste of food that has spoiled.

## HEALTH JOURNAL

Ask students to write a plan in their journal for making decisions that will help them avoid the risk of being infected by HIV/AIDS and hepatitis B and C. Explain that the plan needs to include the following points: 1) goal, 2) behaviors to avoid, 3) refusal skills, 4) behaviors to practice, and 5) evaluation of progress.

## ELL/ESL STRATEGY

**Language Objective:** *To read for meaning*

Pair a student with strong English skills with a student learning English. Have the student with the strong skills read the second paragraph on page 256, about how disease can affect the immune system, aloud. The student learning English should orally summarize the contents of the paragraph, assisted by his or her partner as necessary.

## Link to

**Environmental Science.** Have students read the Link to Environmental Science feature on page 256. Then have them create a cause and effect chain to show how people are affected by disease in an environment with standing water.

## IN THE ENVIRONMENT

Have students find out how pollution causes or worsens disease. For example, air pollution worsens respiratory diseases such as asthma and bronchitis. Water pollution can cause dysentery and other infectious diseases. Suggest students present their information on a poster. They might draw the human body, and then add labels to describe diseases that affect the labeled parts of the body.

---

**Antibiotic**
*Substance that can destroy the growth of certain germs*

Some diseases can be spread through contact with blood or other body fluids. AIDS and hepatitis B and C are serious diseases that can be spread through the exchange of body fluids during sexual activity or other contact.

For some diseases, such as the common cold, there is no cure. Rest and plenty of liquids help a person get over a cold more quickly. **Antibiotics,** substances that can destroy the growth of certain germs, may be given for some diseases that bacteria cause. Antibiotics do not work against viruses. Sometimes bacteria change, and antibiotics do not work against them. This is one reason you may want to keep track of antibiotics you are treated with. You can determine whether they work.

A disease can affect the immune system itself. As a result, a person's defenses are weak or are not able to protect against other diseases. Even germs that normally are not pathogens can cause disease. People may then die from severe infections. In some people, disease of the immune system is inherited. In others, the disease is acquired. For example, AIDS is an acquired disease of the immune system.

## What Are Some Causes of Inherited Diseases?

An inherited disease passes through the genes from parents to children. You cannot get an inherited disease any other way. There are three main causes for inherited diseases.

1. A single abnormal gene can cause diseases such as hemophilia. In this disease, blood does not clot normally. The person bleeds more than normal.

> **Link to** ➤➤➤
>
> **Environmental Science**
>
> The environment can affect the rates of some infectious diseases. For example, mosquitoes carry several different types of infectious germs, such as West Nile virus. Mosquitoes breed in standing water. When rainfall in an area increases, the number of mosquitoes goes up. Diseases caused by germs carried by mosquitoes may increase.

**Deformity**
*Abnormally formed body structure*

2. A combination of many abnormal genes and environment can cause diseases such as diabetes and some cancers. **Deformities** also are caused this way. Deformities are abnormally formed body structures, such as cleft palate. In this disorder, the plates in the roof of an unborn baby's mouth do not close.

3. Abnormal chromosomes cause genetic disorders such as Down syndrome, a mental impairment. A chromosome may break or fail to divide properly. Genes can also be placed incorrectly on the chromosomes.

---

### Decide for Yourself

#### Home or School?

*Define the problem.* You wake up one morning with a bad cold. Should you stay home or go to school? You have a big test today.

*Come up with possible solutions.* You can go to school. You can stay home and get plenty of liquids and rest.

*Consider the consequences of each solution.* If you go to school, you might feel worse, and you will spread pathogens. You might perform badly on your test. If you stay home, you will get well quicker and not spread pathogens. However, you will have to take a makeup test.

*Put your plan into action.* You decide to stay home and rest.

*Evaluate the outcome.* You feel much better in a couple of days. You do well on the makeup test. You feel good about yourself because you did not make other people sick.

On a sheet of paper, write answers to the following questions. Use complete sentences.

1. Have you ever been faced with a similar problem? If so, what did you decide to do?
2. What was your outcome?
3. Would you make the same decision in the future? Why?

---

## LEARNING STYLES

### Visual/Spatial

Give students differently colored beads and pipe cleaners to make a model of a healthy chromosome. Ask students to arrange the beads on two pipe cleaners to make an X. Point out that the beads represent strings of genes on the four "arms" of a chromosome. Explain that disorders can occur when genes are in the wrong position on an arm, if the entire arm is attached in the wrong place, if there are extra chromosome arms, or if an arm is missing.

### Decide for Yourself

Have students read and complete the Decide for Yourself feature on page 257. Discuss options that students might have for deciding to go to school with a bad cold on the day of a test. Have them role play the conversations they might have with the teacher if they were able to call in advance to explain their decision or a message they might leave for the teacher regarding their illness and absence.

**Decide for Yourself Answers**

1. Sample answer: Yes, I made the same decision—to stay home and rest.
2. Sample answer: In two days, I was well enough to go back to school. It wasn't hard to make up the work I had missed.
3. Sample answer: Yes, it turned out to be a good idea. I was more comfortable at home, my dad was glad I stayed home, and I didn't pass my pathogens to anyone at school.

## IN THE COMMUNITY

Obtain permission to visit a health care facility in your community. Ask the public relations officer at the facility to help students become more aware of how the facility reaches out to the community and provides for its needs. If a field trip is not possible, invite the public relations officer from the facility to make a presentation at your school.

# Lesson 1 Review Answers

1. normal 2. acquired 3. deformity
4. Sample answer: I would tell him to cover his mouth and nose when he sneezed and to wash his hands afterward because drops from his sneeze can infect others. 5. People with inherited diseases cannot prevent them because they are born with these diseases. The diseases are passed to them through genes from their parents.

## Portfolio Assessment

Sample items include:
- Personal health goal and the plan from Health Journal
- Decide for Yourself answers
- Lesson 1 Review answers
- Table of common infectious diseases from Research and Write

## Research and Write

Have students read the Research and Write feature on page 258. Show them how to do an Advanced Search on the Internet to access only governmental sites, which generally have reliable statistics. Also point out that the Centers for Disease Control and Prevention have an appropriate Web site for this assignment at www.cdc.gov.

## Health Myth

Invite students to read and discuss the Health Myth on page 258. You might wish to point out that scientists are currently studying whether antibacterial cleaning and hygiene products, including use of antibacterial hand soap, are linked to drug resistant bacteria.

---

**Research and Write**

Work in small groups to research the five most common infectious diseases in your state. Also find out ways to prevent each disease. Use the Internet or the library. Prepare a table to show your findings.

*Lesson 1 Review* On a sheet of paper, write the word or phrase in parentheses that best completes each sentence.

1. Disease harms (abnormal, immune, normal) body function.
2. A disease that one person passes to another is (immune, inherited, acquired).
3. A combination of abnormal genes and environment can cause a (deformity, pathogen, virus).

On a sheet of paper, write the answers to the following questions. Use complete sentences.

4. **Critical Thinking** You have a friend who has a cold. He does not cover his mouth and nose when he sneezes. What advice would you give your friend?
5. **Critical Thinking** Is it possible for people with inherited diseases to have prevented them? Explain why or why not.

### Health Myth

**Myth:** Antibacterial soap is better than regular soap at preventing diseases.

**Fact:** Washing the hands physically removes pathogens from the skin. Regular soap does this just as well as antibacterial soap. In addition, viruses cause the most common diseases. Antibacterial soap does not work against viruses.

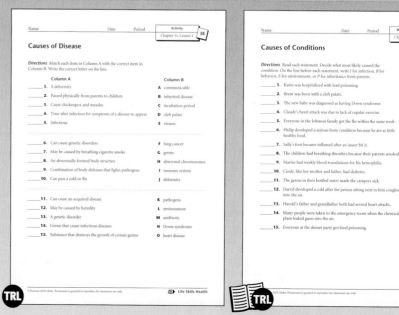

# Lesson 2: How the Body Protects Itself from Disease

### Objectives

After reading this lesson, you should be able to
- explain how the body fights infection
- describe how the immune system works
- recognize the importance of vaccinations

**Barrier**
Something that blocks passage

**Mucous membrane**
The moist lining of the mouth and other body passages

**Mucus**
The fluid that mucous membranes produce

The human body has many ways to protect itself from disease and infection. Germs that are washed off the skin never enter the body. The body's physical and chemical **barriers** protect the body from infection. A barrier is something that blocks passage. The immune system also defends the body against disease.

## How Does the Body Fight Infection?

Physical and chemical barriers are the first way your body protects itself from infection. These are the body's first line of defense. The skin and **mucous membranes** are physical barriers to infection. Mucous membranes are the moist linings of your mouth and other body passages. These barriers prevent pathogens from getting through your body's surface. The tiny hairs in your breathing passages are other physical barriers. These hairs whip rapidly back and forth to sweep pathogens and dust out of your body. Your body also has chemical barriers to infection. Saliva, tears, and sweat clear pathogens from your body's surface. Your stomach produces strong acids that kill pathogens that enter the digestive system. Urine carries pathogens out of your body. Tears, saliva, and **mucus** contain substances that fight pathogens. Mucus is the fluid that mucous membranes produce.

### Link to ➢➢➢

**Social Studies**

In 1900, the number one cause of death in the United States was pneumonia. Pneumonia is a disease of the lungs. Today, the number one disease is heart disease. In 1900, 30 percent of all deaths in the United States occurred in childhood. Today, about 1 percent of all deaths occur in childhood. Throughout the 1900s, infectious diseases decreased because of antibiotics, childhood vaccinations, and public health improvements such as clean drinking water. Life expectancy has increased from 47 years in 1900 to 77 years today.

*Disease—Causes and Protection* Chapter 11 **259**

---

# Lesson at a Glance

## Chapter 11 Lesson 2

**Overview** The lesson describes the body's natural defenses against disease and how vaccinations provide protection against disease.

### Objectives
- To explain how the body fights infection
- To describe how the immune system works
- To recognize the importance of vaccinations

**Student Pages** 259–264

**Teacher's Resource Library**
- Activity 39
- Alternative Activity 39
- Workbook Activity 39

### Vocabulary

antibody                mucous membrane
barrier                 mucus
immunization            vaccination
inflammation

Discuss the vocabulary words and their meanings with the class. Have students write a sentence for each word that illustrates its meaning.

### Background Information

The U.S. Department of Health and Human Services considers how vaccines have reduced infectious diseases a great public health success story. The smallpox vaccination has wiped out smallpox. The polio virus has nearly been stamped out. Vaccines can prevent death and disability. They have saved billions of dollars in health care costs every year.

### 1 Warm-Up Activity

Have students imagine that they are building a house in the woods. They want to keep out insects, wild animals, and dirt. What are some things they would design into their house to accomplish this? *(Students may mention such things as screened doors and windows, walls and a roof without any holes or gaps, and insect traps.)* Tell students that their body has parts that keep unwanted things from entering it.

### Link to

**Social Studies.** Ask students to read and discuss the Link to Social Studies feature on page 259. Ask them to identify some of the changes that have occurred in the past hundred years that led to the thirty year increase in life expectancy in the United States *(antibiotics, vaccinations, clean drinking water).*

**Teaching the Lesson**

Draw the following flowchart on the board and use it to explain to students how the immune system works.

**Immune System**

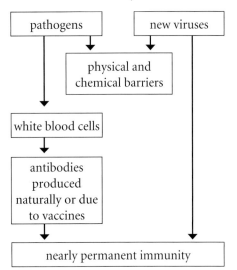

 **Reinforce and Extend**

### Health Myth

Invite students to read and discuss the Health Myth on page 260. Point out that economics is the biggest factor in countries that have low vaccination rates. Explain that the health organizations in developing countries do not have enough money to vaccinate many people. As a result, diseases that are at bay in the United States are prevalent in poor countries. Even in wealthy countries, people who are not vaccinated are vulnerable to infection.

### ELL/ESL Strategy

 **Language Objective:** *To learn vocabulary specific to the content area*

Have students work in pairs to create two word banks for vocabulary from the lesson with the topics "diseases" and "the immune system". When a word related to either topic is mentioned during class discussions, students can add it to the appropriate word bank. In this way, they are listening for the words, analyzing the words, and classifying them.

---

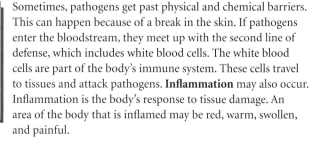

*Inflammation*
Redness, soreness, and swelling after tissue damage

*Antibody*
A protein that kills a particular pathogen

You can practice many behaviors to keep your immune system healthy. Get the recommended amount of vitamins and minerals. Exercise regularly. Keep stress to a level you can manage. Get enough sleep.

Sometimes, pathogens get past physical and chemical barriers. This can happen because of a break in the skin. If pathogens enter the bloodstream, they meet up with the second line of defense, which includes white blood cells. The white blood cells are part of the body's immune system. These cells travel to tissues and attack pathogens. **Inflammation** may also occur. Inflammation is the body's response to tissue damage. An area of the body that is inflamed may be red, warm, swollen, and painful.

### How Does the Immune System Work?

The immune system is the body's final line of defense against infection. If pathogens enter the bloodstream, the blood makes **antibodies.** Antibodies are proteins that kill particular pathogens. If the correct antibodies are in the blood when an infection occurs, they can begin to fight it. They may knock out the infection during the incubation period. A person may never become sick.

Antibodies form in the blood in a number of ways. If you do become sick, antibodies that work against a pathogen develop during the course of the illness. These antibodies help fight the illness so that recovery is speeded up.

Some antibodies remain in the body after a person has recovered. These antibodies give you immunity so you will not get the disease again. For example, once you have mumps, you will not get the disease again. You are immune.

> **Health Myth**
>
> **Myth:** Vaccines are not necessary because the diseases they prevent are rare.
>
> **Fact:** The reason some diseases are rare is because of vaccines. If vaccines were not given, more people would get the diseases. Countries that have low vaccination rates have high rates of diseases such as polio, measles, and diphtheria.

### Learning Styles

 **Body/Kinesthetic**

Let students use 2 paper towel tubes, 10 straight pins, and 2 small balls to create a model of how hairs in the lungs can keep out germs. Have partners stick the 10 pins randomly into one tube, pressing them in as far as they will go. Remind students to take safety precautions when using the pins. Then have them roll a small ball through each tube. Ask them to compare the tubes to determine which was easier to roll the ball through and why.

**Immunization**
*The process of making the body immune to a disease*

**Vaccination**
*A means of getting dead or weakened pathogens into the body*

Antibodies may also be in the blood as a result of **immunization,** or the process of making the body immune to a disease. A **vaccination** is a means of getting dead or weakened pathogens into your body. Vaccinations also can contain parts of organisms. A vaccination is usually an injection, or shot. A vaccination will not make you sick from the disease. However, your immune system will make antibodies against the bacteria or virus in the vaccine. For example, a vaccination against measles contains the weakened virus that causes measles. When you get a measles vaccination, your body makes antibodies against the measles virus. Then you are immune to measles.

### Health in Your Life

**Vaccinations—Not Just for Little Kids**

Are you too old for vaccinations? No. Some vaccines are recommended for young people and adults. Adolescents need to make sure they receive the vaccinations shown in the list at right. The diseases these vaccines prevent can be very serious in teens and adults.

About 20 percent of children are not up to date on their vaccinations. They are at risk of getting diseases. Ask your family to help you make a list of all the vaccinations you have had and when you had them. If they are not up to date, see your doctor or local health department. Your school nurse can also help you locate places that offer vaccines.

On a sheet of paper, write the answers to the following questions. Use complete sentences.

1. Have you received all the vaccinations listed? Do you plan to get them? Why?
2. Use Internet or print sources to find more information about the diseases listed. What is tetanus? What is a common symptom of rubella?
3. How can you communicate the importance of getting vaccinations to students in your school?

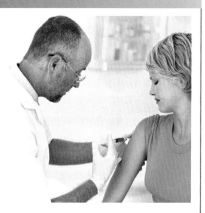

**Vaccines Teens Need**
- Varicella (chickenpox)
- Measles-Mumps-Rubella (MMR)
- Hepatitis B
- Tetanus-Diphtheria

## TEACHER ALERT

Point out that tuberculosis, often called consumption, was also a deadly disease that was effectively brought under control in the 1900s with the introduction of antibiotics. Stress that overuse of antibiotics, however, threatens an increase in tuberculosis. The bacteria that cause tuberculosis may become resistant to some antibiotics. Tell students that when taking antibiotics, they must follow the doctor's instructions carefully.

## HEALTH JOURNAL

Invite students to write in their journal about a recent time when they have been ill. Ask them to write about how and where they might have gotten the illness, how they felt during the illness, how their body might have fought it, and how long it took to feel well again.

## AT HOME

Ask students if they know whether they have had the vaccinations teens need mentioned in the chart on page 262. Which vaccinations are they sure about? Which ones are they unsure about? Stress the importance to students of finding out this information. Suggest students ask a parent or guardian about their vaccinations. Each member of a family should have a card that shows when she or he received vaccinations. Tell students that a family physician keeps information about patients' vaccinations. Also tell them that public health departments offer vaccinations.

Why do you think washing your hands is the best way to keep from getting pathogens?

Another way people get immunity is from their mothers during pregnancy. Babies are protected from the same infectious diseases as their mothers are. This is because antibodies pass from the mother to the baby through the placenta. Antibodies also pass to babies through breast milk. Babies have these antibodies until they are 6 to 12 months old. Then babies' own immune systems begin to take over.

Without the immune system, people would have many more diseases and infections. Your immune system is an important defense against disease.

**Figure 11.2.2** *Babies get immunity from their mothers during pregnancy.*

### Research and Write

Find out where free or low-cost vaccines are available in your community. Use the Internet or contact your local health department. Then make a brochure that explains what vaccines are, why they are important, and where people can get them. Place the brochure in your school library.

### Research and Write

Have students read and conduct the Research and Write feature on page 262. Point out that the address and telephone number of the local health department is listed in the government pages of the local telephone directory. Encourage students to find needed information on the health department's Web site. Check brochures for accuracy before making them available to the school library.

**Lesson 2 Review** On a sheet of paper, write the letter of the answer that best completes each sentence.

1. An example of a chemical barrier is _____.
   A the flow of urine
   B the outer layer of skin
   C strong acids in the stomach
   D tiny hairs in breathing passages

2. If you are immune to a disease, you _____.
   A have antibodies against the disease
   B must have been ill from the disease
   C will probably get the disease again
   D must have received a vaccination against the disease

3. One of the first ways that the body deals with pathogens is to _____.
   A make antibodies
   B become red, warm, and swollen
   C clear pathogens with saliva and tears
   D kill the pathogens during the incubation period

On a sheet of paper, write the answers to the following questions. Use complete sentences.

4. **Critical Thinking** Why is it important for school children to have certain vaccinations?

5. **Critical Thinking** Suppose you have the flu. Your brother gets it, too, but your sister does not. Why might your sister stay well?

### Technology and Society

Some people dislike getting vaccines because of the needle. Scientists are working on other ways to give vaccines. The day is not far off when people might get their vaccines from vegetables or skin patches. An influenza vaccine is now available that squirts up the nose. A liquid vaccine that was swallowed was used at one time against polio.

*Disease—Causes and Protection* Chapter 11  263

## Lesson 2 Review Answers

1. C 2. A 3. C 4. Sample answer: School children are around a lot of other people during the day; therefore, they could easily pick up and spread pathogens. 5. Sample answer: Her immune system was able to fight off the flu germs before they made her sick.

### Portfolio Assessment

Sample items include:
- Vaccines brochure from Research and Write
- Health Journal entry about a recent illness
- Health in Your Life answers
- Lesson 2 Review answers

### Technology and Society

After students read the Technology and Society feature on page 263, encourage them to search on the Internet for "new technologies in administering vaccinations." Let students share their findings.

### LEARNING STYLES

**Logical/Mathematical**
Let students create a flow chart to answer item 5 in the Lesson 2 Review, such as the following:

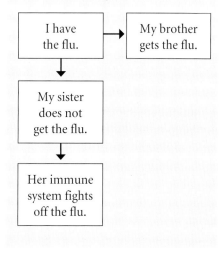

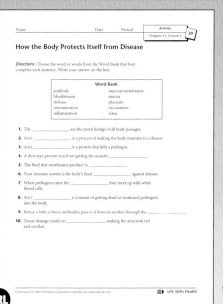

Activity 39 | Workbook Activity 39

*Disease–Causes & Protection* Chapter 11  263

## Health at Work

Have students read the Health at Work feature on page 264. Ask them what they think might happen if a medical facility lacked a health unit coordinator. (Patients might not receive the correct treatment, tests, or medication.) Help students name some jobs that a coordinator might deal with. *(doctor, nurse, nursing aide, laboratory technician, X ray technician, dietician, pharmacist, physical therapist, and so on)* If possible, invite a guidance counselor to speak to the class about how to identify and evaluate their personal interests and goals to make plans about personal and career choices.

### CAREER CONNECTION

Inform students that an infectious disease (ID) specialist is a doctor with advanced training in the diagnosis and treatment of infectious diseases. Suggest that students who might be interested in pursuing a career as an ID specialist research what such a doctor does and report their findings to the class. Also, suggest that they create and revise a plan related to the career.

### LEARNING STYLES

**Auditory/Verbal**
Have groups of students role play a scenario in which a health care coordinator is the major participant. The scenario should include communication between a coordinator and two or more departments in a health care facility.

---

## Health at Work

### Health Unit Coordinator

Health unit coordinators work in hospitals, clinics, and nursing homes. They perform nonnursing duties. They handle forms when patients enter or leave health care facilities. They enter doctors' orders on a computer. They communicate with other departments to order special diets, drugs, equipment, supplies, laboratory tests, and X rays. They may serve as receptionists on patient floors. Health unit coordinators may be called ward or unit secretaries, ward or unit clerks, or hospital service coordinators.

To become a health unit coordinator, a person needs a high school diploma and a 6-month to 1-year training program. Some employers require certification.

Health unit coordinators need excellent communication skills. They must be pleasant and businesslike when dealing with the public and coworkers.

# Chapter 11 SUMMARY

- Diseases can be acquired or inherited.
- An acquired disease is caused by infection, human behaviors, or the environment.
- An acquired disease is infectious when infection causes it.
- Pathogens are the germs that cause infectious diseases.
- The first stage of infectious disease is the incubation period, which can last days, weeks, months, or years.
- If the body's immune system can fight off a pathogen, a person may never become ill.
- If the body's immune system cannot fight off a pathogen, a person becomes ill. The immune system continues to fight the illness, and a person usually recovers.
- The immune system fights pathogens by producing antibodies.
- Diseases of the immune system may cause people to become ill with organisms that are not normally pathogens.
- The body has physical and chemical barriers to infection. Some physical barriers are the skin and mucous membranes. Some chemical barriers are saliva, tears, mucus, and stomach acids.
- Antibodies get in the blood in a number of ways. Diseases or vaccinations build antibodies in the blood to fight infections. A baby gets antibodies through its mother's placenta and breast milk.
- A person may get an inherited disease through a single abnormal gene, a combination of many abnormal genes and environment, or an abnormal chromosome.

| Vocabulary | | |
|---|---|---|
| acquire, 254 | immune system, 255 | inherit, 254 |
| antibiotic, 256 | immunization, 261 | mucous membrane, 259 |
| antibody, 260 | incubation period, 255 | mucus, 259 |
| barrier, 259 | infectious, 254 | pathogen, 255 |
| communicable, 254 | inflammation, 260 | vaccination, 261 |
| deformity, 257 | influenza, 254 | virus, 255 |

## Chapter 11 Summary

Have volunteers read aloud each Summary item on page 265. Ask volunteers to explain the meaning of each item. Direct students' attention to the Vocabulary box on the bottom of page 265. Have them read and review each term and its definition.

### ONLINE CONNECTION

The fact sheet about HIV/AIDS in youth is published by the U.S. Department of Health and Human Services Centers for Disease Control and Prevention: www.cdc.gov/hiv/pubs/facts/youth.htm

The World Health Organization Web site includes an Epidemic and Pandemic Alert and Response (EPR) page with information about disease outbreaks: www.who.int/csr/disease/en/

The U.S. Department of Health and Human Services National Vaccine Program Office Web site provides information about childhood, adolescent, and adult immunizations: www.hhs.gov/nvpo/

## Chapter 11 Review

Use the Chapter Review to prepare students for tests and to reteach content from the chapter.

## Chapter 11 Mastery Test

The Teacher's Resource Library includes two forms of the Chapter 11 Mastery Text. Each test addresses the chapter Goals for Learning. An optional third page of additional critical-thinking items is included for each test. The difficulty level of the two forms is equivalent.

## Review Answers

**Vocabulary Review**

1. infectious 2. pathogens 3. incubation period 4. immune system 5. mucous membranes 6. antibodies 7. vaccination 8. mucus 9. acquired 10. communicable 11. deformity 12. inflammation 13. immunization 14. virus

### TEACHER ALERT

In the Chapter Review, the Vocabulary Review activity includes a sample of the chapter's vocabulary terms. The activity will help determine students' understanding of key vocabulary terms and concepts presented in the chapter. Other vocabulary terms used in the chapter are listed below.

antibiotic   influenza
barrier

---

## Chapter 11 REVIEW

**Word Bank**
acquired
antibodies
communicable
deformity
immune system
immunization
incubation period
infectious
inflammation
mucous membranes
mucus
pathogens
vaccination
virus

### Vocabulary Review

On a sheet of paper, write the word or phrase from the Word Bank that best completes each sentence.

1. An acquired disease is _____ when infection causes it.
2. Germs that cause infectious diseases are _____.
3. The first stage of infectious disease is the _____.
4. The body's _____ produces antibodies.
5. The skin and _____ are physical barriers to infection.
6. Particular pathogens are killed by _____.
7. A shot of weakened or dead pathogens is a(n) _____.
8. Tears, saliva, and _____ are chemical barriers to pathogens.
9. A disease caused by infection, human behaviors, or the environment is _____.
10. Another word for infectious is _____.
11. A combination of many abnormal genes and environment can cause a person to be born with a(n) _____.
12. An area of the body that is red, warm, swollen, and painful is showing signs of _____.
13. After weakened pathogens are injected into the body, the process of _____ begins.
14. A germ that reproduces only within cells of living things is a(n) _____.

**266** Unit 4 Preventing and Controlling Diseases and Disorders

## Concept Review

On a sheet of paper, write the letter of the answer that best completes each sentence.

**15.** An inherited disease is caused by _____.
  A pathogens
  B the environment
  C personal behaviors
  D genes from parents

**16.** A person can get an acquired disease from _____.
  A an infection
  B a single abnormal gene
  C abnormal chromosomes
  D many abnormal genes and the environment

**17.** The body's first line of defense includes _____.
  A antibodies
  B vaccinations
  C stomach acids
  D white blood cells

## Critical Thinking

On a sheet of paper, write the answers to the following questions. Use complete sentences.

**18.** What do you think you could do to protect yourself from getting a cold from another person?

**19.** Why do you think making sure you get vaccinations is important?

**20.** Why are diseases of the immune system especially dangerous?

**Test-Taking Tip** Be sure you understand what the test question is asking. Reread it if necessary.

## Concept Review

**15.** D **16.** A **17.** C

## Critical Thinking

**18.** Sample answer: To protect myself from getting a cold from another person, I could avoid being around the person and wash my hands if I touch anything the person touches. **19.** Sample answer: It is important to make sure I get vaccinations so that I am protected against some diseases. **20.** Sample answer: Diseases of the immune system are especially dangerous because the body cannot defend itself from pathogens.

## ALTERNATIVE ASSESSMENT

Alternative Assessment items correlate to the student Goals for Learning at the beginning of this chapter.

- Ask students to write two separate paragraphs describing the causes of acquired and inherited diseases.
- Have student create a word web identifying some of the body's barriers to infection.
- Invite small groups of students to orally explain how the immune system works.
- Let students create posters about the importance of vaccinations.

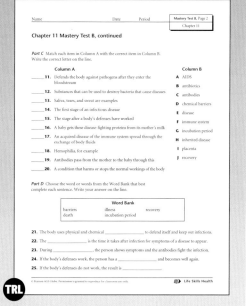

Chapter 11 Mastery Test B, pages 1–3

## Chapter at a Glance

### Chapter 12: Preventing AIDS and Sexually Transmitted Diseases
pages 268–283

### Lessons
1. AIDS
   pages 270–274
2. Sexually Transmitted Diseases
   pages 275–280

### Chapter 12 Summary  page 281

### Chapter 12 Review  pages 282–283

### Audio CD 🎧

### Skill Track for Life Skills Health 🖱

### Teacher's Resource Library 🆃🆁🅻
- Activities 40–41
- Alternative Activities 40–41
- Workbook Activities 40–41
- Self-Assessment 12
- Community Connection 12
- Chapter 12 Self-Study Guide
- Chapter 12 Outline
- Chapter 12 Mastery Tests A and B

(Answer Keys for the Teacher's Resource Library begin on page 559 of this Teacher's Edition.)

## Opener Activity

Collect a variety of newspaper and magazine articles about AIDS. Clip these articles and post them in a place in the classroom where students will pass by them when they enter the room. Instruct students to take a few moments at the beginning of class to skim the articles before you begin the chapter. Ask students which articles most caught their attention and why.

268  Unit 4  Prev. & Contr. Diseases

## Self-Assessment

### Can you answer these questions?

☑ 1. What is AIDS?

☑ 2. What causes AIDS?

☑ 3. Why are sexually transmitted diseases serious?

☑ 4. What is gonorrhea?

☑ 5. What is chlamydia?

☑ 6. What is syphilis?

☑ 7. What is genital herpes?

☑ 8. What are genital warts?

☑ 9. How can sexually transmitted diseases be prevented?

☑ 10. Which sexually transmitted diseases cannot be cured?

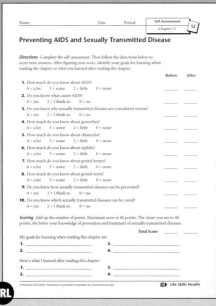

Self-Assessment 12

Community Connection 12

# Chapter 12
## Preventing AIDS and Sexually Transmitted Diseases

Communicable diseases that are passed through sexual contact can be serious. A pregnant mother who is infected with a sexually transmitted disease can pass the disease to her baby. Some of these diseases can cause cancer or sterility. Some of them can be cured or treated with medicine. While some of these diseases cannot be cured, all of them can be prevented.

In this chapter, you will learn what causes AIDS, how it is acquired, and its symptoms. You will learn about infections that result when the immune system is weakened. You will learn about symptoms, treatments, and prevention of some common sexually transmitted diseases.

### Goals for Learning

◆ To describe the causes of AIDS and how it is acquired

◆ To identify symptoms of AIDS and infections it causes

◆ To describe the symptoms and treatment of common sexually transmitted diseases

◆ To identify help or resources for people with sexually transmitted diseases

◆ To explain how sexually transmitted diseases can be prevented

269

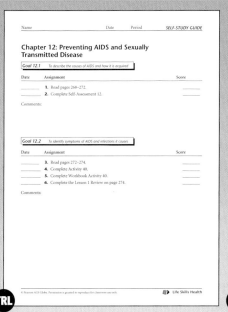

Chapter 12 Self-Study Guide, pages 1–2

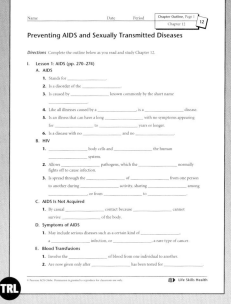

Chapter 12 Outline, pages 1–2

## Introducing the Chapter

When two people are attracted to each other, they might decide to become closer by engaging in a sexual relationship. Such a relationship has serious implications. This chapter discusses one of those implications—the transmission through sexual contact of communicable diseases such as AIDS, genital herpes, syphilis, chlamydia, and gonorrhea. Invite students to tell what they know about the transmission and prevention of these diseases. Discuss what they have heard about AIDS in the media and from friends and family.

Have students read the Goals for Learning. Explain that in this chapter students will learn about the prevention, causes, symptoms, and treatment of AIDS and other sexually transmitted diseases.

### Notes and Questions

Ask volunteers to read the notes and questions that appear in the margins throughout the chapter. Then discuss them with the class.

### Self-Assessment

Have students complete the Self-Assessment worksheet before and after reading the chapter. Before reading the chapter, have students fill in the "Before" column. Ask students to identify their goals for learning. To get ideas for setting goals, students might use the chapter introductory material on pages 268–269, the checklist on page 268, or the questions on the Self-Assessment worksheet. Students can use the back of the worksheet if they need more space to write.

Collect the Self-Assessment worksheets and pass them out again at the end of the chapter. Have students fill in the "After" column. Ask them to identify at least four major points they have learned. Again, suggest they use the back of the worksheet if they need more space to write. You might want to collect and review the worksheets, but return them to students so they have a record of their goals and accomplishments.

*Prevent. AIDS & STDs* Chapter 12   269

## Lesson at a Glance

### Chapter 12 Lesson 1

**Overview** In this lesson, students will learn about AIDS, including how it is acquired and how it affects the world.

### Objectives

- To know the difference between HIV and AIDS
- To describe how AIDS is acquired
- To discuss the safety of the blood supply

**Student Pages** 270–274

**Teacher's Resource Library**
   Activity 40
   Alternative Activity 40
   Workbook Activity 40

### Vocabulary

**Acquired immunodeficiency syndrome (AIDS)**
**Human immunodeficiency virus (HIV)**
**Kaposi's sarcoma**
**opportunistic pathogen**
**pneumonia**
**transfusion**

Have students volunteer to take turns reading aloud the words and definitions. Then have students write a short paragraph that includes the six vocabulary words used correctly in context.

### Background Information

AIDS was not discovered and recognized until the early 1980s in the United States. Although it is likely that some people died of the disease before the early 1980s, it wasn't until 1981 that the Centers for Disease Control began to link increasing numbers of deaths from a rare form of pneumonia and Kaposi sarcoma among gay men. By mid-1982, the disease had been named AIDS and linked to blood transmission.

### 1 Warm-Up Activity

Ask students to go through the lesson and read the subheads. Each subhead is in question form. Have students write out each question and answer it before they read the lesson. Have them add to their lists any other questions they might have.

270  Unit 4  Prev. & Contr. Diseases

---

## Lesson 1  AIDS

### Objectives

*After reading this lesson, you should be able to*

- know the difference between HIV and AIDS
- describe how AIDS is acquired
- discuss the safety of the blood supply

**Acquired immunodeficiency syndrome (AIDS)**
*A disorder of the immune system*

**Human immunodeficiency virus (HIV)**
*A virus that infects body cells and severely damages the human immune system, causing AIDS*

**Opportunistic pathogen**
*Germ that the body would normally fight off but that takes advantage of a weakened immune system*

A serious communicable disease is **acquired immunodeficiency syndrome (AIDS)**. AIDS is a disorder of the immune system. The first cases of AIDS in the United States were identified in 1981. Since then, the disease has spread rapidly in the United States and throughout the world. Many people have died from it. Figure 12.1.1 shows the number of reported AIDS cases in the United States in 2003.

### What Causes AIDS?

**Human immunodeficiency virus (HIV)** causes AIDS. HIV infects body cells and severely damages the human immune system. As a result, the body becomes open to **opportunistic pathogens**. These pathogens are germs that the body normally fights off. When the immune system is weakened, they cause infection.

**Figure 12.1.1** *Estimated numbers of people living with AIDS in the United States at the end of 2003*

270  Unit 4  Preventing and Controlling Diseases and Disorders

### Link to >>>
**Arts**

When AIDS first came to public awareness, many people were afraid of those who had it. People thought they could get AIDS just from being near someone who had it. The movie *Philadelphia* tells the story of a lawyer whose firm fires him because he has AIDS. He sues the firm. The film shows how people were misinformed about AIDS and how it is acquired.

### Health Myth

**Myth:** HIV affects only homosexual men and drug users.

**Fact:** HIV can affect anyone. Babies, heterosexual couples, and seniors have all been infected with HIV.

## How Has AIDS Affected the World?

In the United States, more than half a million people with AIDS have died. Each year since 1989, there have been about 40,000 new cases of AIDS. An estimated 929,985 cases in total have been in the United States. Worldwide, more than 20 million people have died of AIDS since 1981. Young people from 15 to 24 years old account for half of all new HIV infections. Around the world, more than 6,000 young people become infected with HIV every day. In 2004, about five million adults and children became infected with HIV.

## How Is AIDS Acquired?

Like all illnesses caused by a virus, AIDS is a communicable disease. HIV is spread in these ways:

1. Body fluids—HIV can be transmitted through blood, semen, vaginal secretions, and breast milk of infected people. It is spread through exchange of these fluids from one person to another. It has not been transmitted through fluids such as saliva, sweat, or urine.

2. Sexual activity—AIDS can be spread through sexual activity with an infected partner, including oral sex. Oral sex is any mouth-to-genital contact. Each new partner in turn can infect others.

3. Sharing needles among drug users—Any needle that is used for injection of a drug or medicine into the body has blood on it. Drug users who share needles risk getting HIV.

4. Mother to child—HIV can be passed from a mother to a baby before, during, or after birth. It can also be passed to a baby when the baby drinks the mother's breast milk.

A blood test often shows whether a person is infected with HIV or has AIDS. If people are not aware that they are infected with HIV, they do not realize that they may get or already have AIDS.

## How Is AIDS Not Acquired?

AIDS is a serious disease that cannot be cured. Because of this, some people have fears about how it is acquired. These people are afraid to be around people with AIDS.

*Preventing AIDS and Sexually Transmitted Diseases* Chapter 12 **271**

---

### 2 Teaching the Lesson

Have students generate a list of ways that communicable diseases are spread. Write the list on the chalkboard. Items may include shaking hands, sharing cups or silverware, sexual contact, sharing needles, sneezing, and so on. Then ask students to identify which of the ways on the list they think AIDS can be transmitted. After reading the lesson, have students revisit the list and make any necessary changes. Stress that AIDS is not spread through casual contact such as hugging, shaking hands, or sharing food.

### 3 Reinforce and Extend

#### PRONUNCIATION GUIDE

Use the pronunciation shown here to help students pronounce difficult words in the lesson. Refer to the pronunciation key that appears in the Glossary at the back of the Teacher's Edition for the sounds of the symbols.

saliva (sə lī′ və)

#### ELL/ESL STRATEGY

**Language Objective:** *To practice conversation using lesson material*

Have students form small groups. Include both students learning English and students who are proficient in English in each group. Ask the groups to engage in an informal conversation about the topics they have learned about in the lesson. For instance, students can talk about any misconceptions they might have had about the ways AIDS is transmitted. Allow students to feel comfortable speaking about the topics of AIDS and sexually transmitted diseases.

---

### Health Myth

Ask a volunteer to read aloud the Health Myth feature on page 271. Tell students that the concept of AIDS as a disease that affects only homosexuals is dangerous, because it can cause people to expose themselves to AIDS through unsafe behavior. Conversely, the ways that AIDS is spread are very specific, and following basic safety practices can help protect people from the disease.

### Link to

**Arts.** Have students read the Link to Arts feature on page 271. Tell students that *Philadelphia* was released in December 1993. It was the first Hollywood movie to tackle the subject of AIDS for a mainstream audience. Tom Hanks won his first Best Actor Oscar for his performance as the lawyer who was fired because he had AIDS. Bruce Springsteen also won an Oscar and a Grammy award for his song "Streets of Philadelphia," which was featured on the movie's

## LEARNING STYLES

**Visual/Spatial**

Refer to Figure 12.1.1 on page 270. Identify the states with the highest and lowest number of reported AIDS cases. Discuss how this number might be related to the size of each state's population. Have students visit the U.S. Census Bureau's Web site at www.census.gov and use the "Population Finder" feature on the right side of the home page to find state population estimates for 2004. Then show students how to use the data in Figure 12.1.1 and the state population figures to make a double-bar graph. Displaying the data in this manner will provide a clear visual representation of the relationship between population and AIDS cases and AIDS cases compared with the total U.S. population.

## GLOBAL CONNECTION

Tell students that AIDS is thought to have occurred in Africa as early as 1962. Had the international community noticed the initial occurrences of this disease, epidemiologists (people who study the spread and control of epidemics) might have gained a head start on learning how AIDS is transmitted and how it can be prevented.

## IN THE COMMUNITY

Many communities organize annual AIDS walks to raise awareness of the disease and to raise money for research. Suggest that students investigate whether such walks or other AIDS fund-raising projects exist in their community. Encourage students to participate in these projects.

---

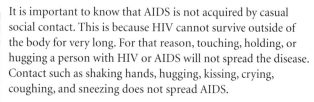

**Pneumonia**
A lung infection

**Kaposi's sarcoma**
A rare type of cancer

**Transfusion**
Transfer of blood from one person to another

It is important to know that AIDS is not acquired by casual social contact. This is because HIV cannot survive outside of the body for very long. For that reason, touching, holding, or hugging a person with HIV or AIDS will not spread the disease. Contact such as shaking hands, hugging, kissing, crying, coughing, and sneezing does not spread AIDS.

AIDS is also not spread by sharing cups, dishes, and other eating utensils. Bathing in the same pool or hot tub has not been shown to spread AIDS. No evidence shows that bites from insects such as mosquitoes spread AIDS.

### What Are the Symptoms of AIDS?

Most teens with HIV or AIDS were exposed to the virus through sexual activity or by sharing needles among drug users.

AIDS is an illness that can have a long incubation period. This means that a person infected with HIV may not show outward signs of AIDS for 6 to 10 years or longer. During that time, however, HIV is weakening the person's immune system.

When symptoms of AIDS appear, they are like symptoms of infection with HIV. Some of these symptoms are a feeling of weakness, chills and fever, night sweats, dry cough, and shortness of breath. Other symptoms are fatigue, stiff neck, headache, weight loss, skin rashes, diarrhea, and swollen glands. A person may also have sores in the mouth. The entire body may be swelled with fluid.

Because HIV weakens the immune system, people who are infected with it also can have other serious diseases. These are caused by the opportunistic pathogens that the body cannot fight against. For example, a person can have a certain kind of **pneumonia.** Pneumonia is a lung infection. Another serious disease can be **Kaposi's sarcoma,** a rare type of cancer. Still other serious problems can result from the attack of AIDS on the nervous system. The disease can also cause brain damage.

### How Safe Is the Blood Supply?

**Figure 12.1.2** AIDS is not acquired by casual social contact.

People sometimes need **transfusions.** A transfusion is the transfer of blood from one person to another.

272   Unit 4   Preventing and Controlling Diseases and Disorders

---

## LEARNING STYLES

**Auditory/Verbal**

Some students might benefit from hearing Lesson 1 on the Audio CD. As students listen, have them write down in a notebook at least five major points they should remember. Students might need to listen to the lesson as a whole first, and then listen again to write down the key points. After the students have written down what they think are the lesson's key points, discuss as a class what important information they need to remember from the lesson and have them check what they wrote to make sure those points are included.

## IN THE ENVIRONMENT

Inform students that, in order to control the spread of AIDS and other diseases, medical personnel dispose of hazardous biochemical wastes in safe ways. Perhaps students have noticed special waste containers in doctors' examining rooms and hospital rooms. Biomedical waste includes needles, bandages, urine and stool containers, and other items that come in contact with body fluids. Suggest to interested students that they find out which waste disposal companies haul away this waste and what is done with it.

### Research and Write

In May of 2003, President George W. Bush signed the HIV/AIDS Act of 2003. The official name of this law is the United States Leadership Against Global HIV/AIDS, Tuberculosis and Malaria Act of 2003. Use the Internet to find a copy of this act. Then write a letter to the editor of a newspaper expressing your opinion of this act.

AIDS can be acquired through infected blood. In 1985, a test was developed to detect HIV in blood. Since that time, all blood given for transfusions is tested. Blood can be used for transfusions only when the test for HIV is negative. Today, the U.S. blood supply is almost always safe. Some people who had blood transfusions before 1985 have been infected with HIV. These infections are now prevented by the HIV blood test.

Sometimes, people know in advance they need a transfusion. For example, they may have a surgery scheduled. They can give their own blood. Then the blood is set aside until it is needed.

### Are Health Care Workers Safe from Acquiring AIDS?

Doctors, dentists, and other health care workers can be at risk for acquiring AIDS because they work with body fluids. These people may care for patients infected with HIV. Small cuts on the skin can provide a pathway for HIV to enter a health care worker's bloodstream.

Health care workers protect themselves through preventive measures such as wearing gloves and masks when working with body fluids. They treat every patient as though he or she may carry HIV or other pathogens.

### Health Myth

**Myth:** An AIDS diagnosis means that you will die soon.

**Fact:** New drugs and cancer-fighting therapies are helping people live longer and more comfortably with HIV and AIDS.

### Link to >>>

**Social Studies**

An area in Africa south of the Sahara Desert is the part of the world affected most by HIV and AIDS. More than 25 million people in the area are living with HIV. In 2004, AIDS killed about 2.3 million people there. Around 2 million children under the age of 15 have HIV. More than 12 million children have been orphaned by AIDS. Levels of economic well-being, education, and life expectancy in the area have always been lower than those in more developed nations. Since the AIDS outbreak, school enrollment has declined. AIDS is erasing much of the progress that had been made on life expectancy. Many workers have died. This has slowed economic growth. This makes it even harder for Africa to deal with the AIDS epidemic.

---

### Research and Write

Have students read and complete the Research and Write activity on page 273. Look over students' letters and make any suggestions or corrections you might have. Tell students to incorporate your suggestions and corrections, print a final copy of their letters, and then mail them. Check local newspapers to see if any of your students' letters are printed. Post in the classroom letters that are printed in newspapers.

### Health Myth

Ask a volunteer to read aloud the Health Myth feature on page 273. Tell students that new drugs and technology have made a huge difference in an AIDS prognosis. When AIDS first surfaced in the United State in the early 1980s, patients often lived only a few years after being diagnosed with the HIV virus. Faster diagnosis and new antiviral drugs and drug combinations now allow patients to live much longer after diagnosis—some as long as 10 or even 20 years.

### Link to

**Social Studies.** Have students read the Link to Social Studies feature on page 273. Tell students that while AIDS is a huge problem for the United States, it is much more devastating to the people of Africa. The drugs that are used to treat AIDS are very expensive, and many Africans cannot afford treatment. Have students do some research on the AIDS epidemic in Africa and present their findings to the class.

---

### HEALTH JOURNAL

Have students write in their journals a description of the health status of someone who has been diagnosed with AIDS. Have them include a description of the testing, treatment, and prognosis of the patient.

### AT HOME

Hospice programs allow a person with AIDS to be cared for in a homelike setting. Have students discuss with family members their feelings on hospice programs for people with terminal illnesses such as cancer, liver disease, or AIDS. Have students express their own wishes, fears, and concerns regarding this type of medical care.

### TEACHER ALERT

The tests that are used to detect HIV work by looking for antibodies in the bloodstream. When the body is attacked by a virus, including HIV, it produces antibodies to fight off the infection. An antibody test looks for specific proteins that are present only in the antibodies that are produced as a reaction to an HIV infection.

### Research and Write

Have the students read and complete the Research and Write feature on page 274. Tell students that when they use the Internet to do research, they must be careful to use reliable sources. The most reliable Web sites generally have *gov* or *edu* domain names.

### Health at Work

Have students read the Health at Work feature about laboratory assistants on page 274. Ask the students to use the Internet to find job listings for laboratory assistants. What qualifications and educational level do the posting require? Have students report their findings to the class.

## Lesson 1 Review Answers

1. Human immunodeficiency virus (HIV) causes AIDS. 2. AIDS can be acquired through the exchange of body fluids, through sexual contact, through sharing needles, and from a mother to a baby during birth or through breast milk. 3. Two diseases (caused by opportunistic pathogens) that a person with AIDS can get are pneumonia and Kaposi's sarcoma. 4. AIDS cannot be acquired through casual social contact because HIV cannot live outside the body. 5. Sample answer: Health care workers take precautions with all patients because many people do not know they are infected with HIV.

## Portfolio Assessment

Sample items include:
- Paragraphs from Vocabulary
- Q and A from Warm-Up Activity
- Graphs from Learning Styles: Visual/Spatial
- Outlines from Learning Styles: Auditory/Verbal
- Letters from Research and Write
- Research from Link to Social Studies
- Reports from Research and Write
- Entries from Health Journal
- Research findings from Health at Work
- Lesson 1 Review answers

---

### Research and Write

HIV sometimes becomes resistant to drugs. Drugs need to be tested to determine whether they will work. Use the Internet to research HIV drug resistance and how it is tested. Write a two-page report on your results.

### What Is the Future for People with AIDS?

At this time, AIDS is a disease with no vaccine and no cure. Some treatments increase the quality and length of life for people with AIDS. Eventually, however, all people with AIDS will die of the disease unless a cure is found. Scientists are researching many possible cures, but it may be years before they find one that works. For now, prevention is the only choice. Education has helped people slow down the spread of AIDS. AIDS infection can be prevented by avoiding contact with HIV-infected blood or other body fluids. Abstinence from sexual activity is one way to avoid this contact.

***Lesson 1 Review*** On a sheet of paper, write the answers to the following questions. Use complete sentences.

1. What causes AIDS?
2. What are four ways that AIDS is acquired?
3. What are two diseases caused by opportunistic pathogens that a person with AIDS can get?
4. **Critical Thinking** Why is AIDS not acquired by casual social contact?
5. **Critical Thinking** Why do health care workers take precautions with all patients?

---

### Health at Work

**Laboratory Assistant**

Laboratory assistants work with medical technicians. Assistants look at samples of human urine, blood, and saliva. They look for organisms that cause disease. They report all organisms that are not normal. Laboratory assistants usually work in hospitals, clinics, and doctors' offices. They help to maintain equipment. If anything is not working properly, they report the problem.

Some 1-year programs are available, but a 2-year degree is usually needed.

Activity 40

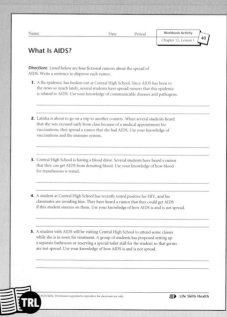

Workbook Activity 40

## Lesson 2: Sexually Transmitted Diseases

**Objectives**

*After reading this lesson, you should be able to*

- name and describe five sexually transmitted diseases
- explain which sexually transmitted diseases can be cured
- explain how to avoid sexually transmitted diseases

**Sexually transmitted disease**
*Any disease that is spread through sexual activity*

**Gonorrhea**
*A sexually transmitted disease*

Any disease that is spread through sexual activity is a **sexually transmitted disease**. For example, AIDS is a sexually transmitted disease because it can be spread through sexual activity. In the United States, one in four people under age 21 has had a sexually transmitted disease. These diseases can be serious because they can cause other diseases if they are not treated. Millions of people around the world will have a sexually transmitted disease at some point in their lifetime. Even one sexual experience can pass a sexually transmitted disease from one person to another.

### What Is Gonorrhea?

**Gonorrhea** is a sexually transmitted disease. One quarter of the people who have this infectious disease are between 10 and 19 years of age. It can also be transmitted to babies as they pass through the birth canal of an infected mother. Gonorrhea can cause an eye infection in babies that can lead to blindness.

In males, symptoms are a white discharge from the urethra and burning while urinating. From 20 to 40 percent of males, however, have no symptoms and do not know they are infected. Females may have a discharge from the vagina and some swelling and redness in the genital area. Most females, however, also have no symptoms and do not know they are infected. The gonorrhea bacteria can also infect the rectum and throat.

Gonorrhea used to be easy to treat with antibiotics. An antibiotic is a substance that can destroy the growth of certain germs.

Recently, some strains of gonorrhea have become resistant to some antibiotics. However, if a person thinks he or she may have been infected, clinics can still quickly diagnose and begin treating gonorrhea. All newborn babies are treated with a special eye medicine immediately after birth to prevent the eye infection.

*Preventing AIDS and Sexually Transmitted Diseases* Chapter 12  **275**

---

## Lesson at a Glance

### Chapter 12 Lesson 2

**Overview** This lesson discusses five common sexually transmitted diseases and their symptoms and treatment, plus ways to prevent their spread.

### Objectives

- To describe gonorrhea, chlamydia, syphilis, genital herpes, and genital warts
- To explain which sexually transmitted diseases can be cured
- To explain how to avoid sexually transmitted diseases

**Student Pages** 275–280

**Teacher's Resource Library**

Activity 41
Alternative Activity 41
Workbook Activity 41

### Vocabulary

| | |
|---|---|
| antibiotic | gonorrhea |
| chancre | sexually transmitted |
| chlamydia |    disease |
| chronic | sterility |
| genital herpes | syphilis |
| genital warts | |

Have students write a sentence for each vocabulary word. Then ask them to write the sentences again, substituting the definition of the word for the word itself. Students will see that their definitions for some of the words will have to be more specific so that the sentences with substituted definitions do not read exactly the same.

### Background Information

Some STDs are caused by bacteria; they can be treated with antibiotics, such as penicillin. However, some STDs are caused by viruses and cannot be treated with antibiotics, such as human papillomavirus (HPV). Some types of HPV cause genital warts, and other types have been linked to cervical cancer. HPV is a growth that can occur on the cervix, vagina, vulva, mouth, anus, rectum, or penis. The warts may be burned or frozen off or removed with laser surgery. Despite these treatments, the virus stays in a person's body for life, although it may be dormant.

*Prevent. AIDS & STDs* Chapter 12  **275**

**Warm-Up Activity**

Ask students what they already know about sexually transmitted diseases (STDs). Ask volunteers to contact a local community health agency for information on one of the STDs described in the lesson. Ask them to report back to the class with the information they have acquired.

 **Teaching the Lesson**

Have students make a chart using each of the diseases covered in the lesson as column headers. As they read the lesson, have them fill in the column below each header with words, phrases, or other notes.

Make sure students understand that some STDs—such as HPV, viral hepatitis, trichomoniasis, gonorrhea, and chlamydia—do not always produce symptoms in people who have them. People can be infected with STDs and not know it, which is one reason why they are so dangerous.

 **Reinforce and Extend**

### PRONUNCIATION GUIDE

Use the pronunciation shown here to help students pronounce difficult words in the lesson. Refer to the pronunciation key that appears in the Glossary at the back of the Teacher's Edition for the sounds of the symbols.

urethra (yü rē´ thrə)

### Technology and Society

Have the students read the Technology and Society feature on page 276. Tell students that some drug trials split the patients into two groups. Half the patients receive the test drug and half receive a sugar pill, or placebo. The patients in the study do not know if they are taking the drug or the placebo. This helps researchers determine if the drug being tested is actually working, although some concerns have been raised about whether it is ethical to withhold treatment from patients who receive only the placebo.

276  Unit 4  Prev. & Contr. Diseases

---

**Sterility**
Inability to produce a child

**Chlamydia**
A sexually transmitted disease

**Syphilis**
A sexually transmitted disease

**Chancre**
A hard sore that is the first sign of syphilis

### Technology and Society

Doctors are working to develop new drugs that will help people with sexually transmitted diseases. A new drug is tested in a clinical trial. The main purpose of a clinical trial is to make sure that the drug is safe and effective. A clinical trial helps doctors decide what information to put on the label for a drug.

If gonorrhea is not treated in males, it can lead to **sterility,** which is the inability to produce a child. If it is not treated in females, it can spread throughout the reproductive system. Untreated gonorrhea and other sexually transmitted diseases can cause an advanced infection that can lead to sterility in females.

Gonorrhea can spread to the bloodstream. Once there, the bacteria can infect other body parts. This can lead to heart valve problems and other diseases such as arthritis, a painful inflammation in the joints.

Gonorrhea is an important public health concern. Because many people do not have symptoms, they do not know they are infected and can spread the disease.

### What Is Chlamydia?

The symptoms of **chlamydia,** another sexually transmitted disease, are similar to those for gonorrhea. For some females, symptoms may be pelvic pain and a discharge from the vagina. Usually, however, females have no symptoms. Males may have a discharge from the urethra and pain when they urinate. Chlamydia can be treated with an antibiotic.

Chlamydia can lead to sterility. Like gonorrhea, chlamydia can infect a baby as it passes through the birth canal. It can cause an eye infection that results in blindness.

### What Is Syphilis?

**Syphilis** is a sexually transmitted disease that progresses in three stages if left untreated. The first stage of syphilis is a **chancre,** a hard sore with a small amount of yellow discharge. A chancre may be painful. The chancre is usually on the penis, anus, or rectum in men. It appears on the cervix and genital areas in women. It can also appear on the lips, tongue, tonsils, fingers, and mucous membranes. In this stage, the lymph glands may be swollen but not painful.

The second stage of syphilis appears 17 days to 6½ months after infection, often as a rash that does not itch. A person has a feeling of discomfort, headache, aching bones, loss of appetite, and other symptoms.

276  Unit 4  Preventing and Controlling Diseases and Disorders

---

### ELL/ESL STRATEGY

 **Language Objective:** *Explaining main ideas*

Have English language learners determine the main idea of each paragraph in the lesson. Students should write one sentence for each paragraph that restates the main idea. Have students write one sentence that restates the main idea of the lesson as a whole.

**Genital Herpes**
*A sexually transmitted disease*

**Chronic**
*Lasting for a long time*

During the third stage of syphilis, the symptoms disappear for many years. In about one-third of people with the disease, it progresses and affects the heart and brain.

Most people with syphilis become infected by sexual contact. Syphilis can also be passed from mother to baby through the placenta.

Many states require people to have a blood test for syphilis before they receive a marriage license. The test identifies people with the disease so that they can be treated with an antibiotic. Treatment also prevents the spread of infection to babies at birth.

## What Is Genital Herpes?

Another sexually transmitted disease is **genital herpes**. It is a **chronic,** or lasting, disease. The main symptom of genital herpes is clusters of small, painful blisters in the genital area. The blisters break, heal, and then come back.

Genital herpes is spread by contact with someone who has it. Genital herpes can be spread even when blisters are not present. Avoiding contact with the pathogen that causes genital herpes can prevent this communicable disease. Contact can be avoided by abstaining from sex, including oral sex.

Genital herpes can cause other problems. Sometimes, repeated infections in females involve the cervix, which may lead to cancer of the cervix. Genital herpes can infect a baby passing through the birth canal, and can cause brain damage in the baby.

The disease has no cure. Medicines can control the symptoms of genital herpes. They speed up healing but do not get rid of the infection. Over time, the infections tend to be less severe, but genital herpes never goes away.

**Figure 12.2.1** *Many states require that couples get tested for syphilis before they can get a marriage license.*

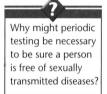

Why might periodic testing be necessary to be sure a person is free of sexually transmitted diseases?

## HEALTH JOURNAL

Together with students, skim the first page or two of the chapter and show students how to make an outline in their health journals. Discuss how to identify and summarize the main point(s) of each paragraph and how to organize the information into an outline format. Have students continue the process on their own for the rest of the chapter. Emphasize how helpful an outline can be as a study guide for reviewing material and preparing for a test.

## CAREER CONNECTION

Tell students that the Centers for Disease Control (CDC) in Atlanta, Georgia, works to prevent and control infectious diseases, including AIDS and other STDs. Have students research the CDC and create a chart that describes the jobs of the many different kinds of health experts that work there.

## LEARNING STYLES

**Logical/Mathematical**
Have students research the occurrences of STDs in Amsterdam. Since the start of the 1960s, Amsterdam has been considered a liberal city for sexual behavior. One may expect, therefore, that the Netherlands would have an extraordinarily high number of STDs. Actually, the Netherlands has is one of the few countries in which the number of STDs has decreased. Authorities credit much of this decrease to massive advertising campaigns, especially those focused on teens.

---

## What Are Genital Warts?

**Genital warts** are the result of an infection with human papilloma virus (HPV). These soft bumps appear in the genital area weeks or months after infection. They sometimes appear in clusters that resemble cauliflower. They can be raised or flat, small or large. About 5.5 million new cases of sexually transmitted HPV are reported every year. At least 20 million people in the United States are currently infected.

Genital warts are spread by skin-to-skin contact during sexual activity with an infected partner. There may be no symptoms of infection. People who are infected but have no symptoms can still spread the disease to a partner. Some types of HPV can cause cancer.

HPV has no cure. Further, there is no way to predict whether the warts will grow or disappear. They can be treated with medication, freezing, burning, or surgery. The only way to prevent an HPV infection is by avoiding direct contact with the virus.

*Genital warts*
Bumps caused by a virus spread through sexual contact

## What Help Is Available?

Except for genital herpes, genital warts, and AIDS, most sexually transmitted diseases can be cured when they are detected early. For diseases without a cure, medical treatment may make people more comfortable.

Many clinics concentrate on diagnosing and treating people who suspect they have a sexually transmitted disease. Health care workers know that the possibility of having a sexually transmitted disease can be difficult and embarrassing. They appreciate a person's strength and good judgment in seeking medical advice. They protect the identity of their patients.

You can get information about sexually transmitted diseases from doctors, local health departments, and family planning clinics. Also, the American Social Health Association (ASHA) gives free information. It also keeps lists of clinics and doctors who provide treatment for people with sexually transmitted diseases.

> Many people with sexually transmitted diseases have no symptoms. The only way a person can know for sure that he or she does not have a sexually transmitted disease is by being tested regularly.

Are you at risk for a sexually transmitted disease if you have only one sexual experience?

## How Are Sexually Transmitted Diseases Prevented?

There are no vaccines that prevent sexually transmitted diseases. The body does not build up immunity to sexually transmitted diseases. Like all communicable diseases, they can be prevented by avoiding contact with the pathogens that cause them. The best prevention for sexually transmitted diseases is choosing abstinence from sexual activity. If two people are uninfected and have sexual activity only with their partners, they are not at risk for sexually transmitted diseases.

### Decide for Yourself

There are ways to avoid a sexually transmitted disease. One is to be in a mutually faithful relationship with a partner who has been tested and found free of sexually transmitted diseases. The only sure way is to choose abstinence from sex.

There are many pressures on teens to have sex. Peer pressure can be strong. Teens may think that having sex will help them fit in and be popular. This pressure can make saying no hard. However, there is nothing cool about getting a sexually transmitted disease.

In addition to peer pressure, many messages in the media make casual sexual activity seem appealing. These messages are found in movies, TV shows, magazines, and advertisements. All of these show people having sex outside committed relationships. The problem is that these messages do not include the real danger of sexually transmitted diseases.

Here are some ways to say no to sexual activity:

1. Say no and repeat it as many times as needed to make your point.
2. Practice how to say no if someone pressures you. For example, if someone says "Everybody does it," respond with "I don't have to do what everyone else does."
3. If you get into a troublesome situation, walk away and stay away.
4. Stand tall, speak clearly, and look the other person in the eye. Your body language will express your choice.

On a sheet of paper, write the answers to the following questions. Use complete sentences.

1. What sometimes makes choosing abstinence hard?
2. Why should you practice saying no?
3. Why is using body language important to help express your choices?

### Decide for Yourself

Have students read the Decide for Yourself feature on page 279. Have students work in groups to develop skits that show various ways of saying "no" to something they do not want to do. The skit scripts should focus on situations that are not sexual, such as saying "no" to drugs, alcohol, driving with a friend who drives dangerously, partaking in vandalism, and so on. Skits may include a variety of ways to say "no" through words and actions. One way should be to avoid putting oneself in the situation in the first place. After students present their skits, discuss how the ways of saying "no" can be applied to sexual situations. For example, students could avoid a potentially harmful situation by avoiding being alone with someone whom they do not trust.

**Decide for Yourself Answers**

1. Sample answer: Peer pressure, the media, and the person's partner may all be saying yes. 2. Sample answer: Practicing saying no can make it much easier to say no in an actual situation in which someone is pressuring you to have sex. 3. Sample answer: If your words say no and your body language does not, people may think you do not mean no.

## LEARNING STYLES

### Interpersonal/Group Learning

Divide the class into small groups. Assign each group a different STD to research: gonorrhea, syphilis, herpes, AIDS, chlamydia, pediculosis (pubic lice), HPV, and others. Have students create a Public Service Announcement (PSA) for the prevention of the diseases based on their research. If they are not sure what a PSA is, remind them of the "Just Say No to Drugs" ads they have undoubtedly seen on TV. Students then can present the PSAs to the class.

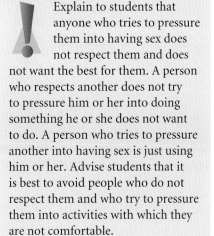

### TEACHER ALERT

Explain to students that anyone who tries to pressure them into having sex does not respect them and does not want the best for them. A person who respects another does not try to pressure him or her into doing something he or she does not want to do. A person who tries to pressure another into having sex is just using him or her. Advise students that it is best to avoid people who do not respect them and who try to pressure them into activities with which they are not comfortable.

## Health in Your Life

Have students read the Health in Your Life feature on page 280. Tell students that "informed consent" (the third bullet point in the feature) involves patients giving their consent to undergo a medical procedure. For a person to give consent, his or her doctor should explain the patient's diagnosis, what would happen if no treatment was attempted, the pros and cons of a proposed treatment and what it would accomplish, and alternative treatments and their risks and benefits.

### Health in Your Life Answers

1. Sample answer: All health problems are personal, so privacy is important to most patients. 2. Sample answer: Any person who might die from a serious health problem needs to be treated immediately by doctors. A person should not be turned away simply because he or she does not have insurance or the money to pay for emergency services. 3. Sample answer: Informed consent would be very important to me. If I did not know what all of my treatment options were, I probably would not be able to choose the one that is best for me.

## Lesson 2 Review Answers

1. sterility 2. gonorrhea 3. syphilis
4. Sample answer: Short-term problems that genital herpes causes are clusters of painful blisters and the possibility of spreading the disease to a partner.
5. Sample answer: Long-term problems that genital herpes causes are continuing future outbreaks and the need to tell a (current or future) partner about the disease.

## Portfolio Assessment

Sample items include:
- Sentences from Vocabulary
- Charts from Teaching the Lesson
- Sentences from ELL/ESL Strategy
- Outlines from Health Journals
- Charts from Career Connection
- Skit scripts from Decide for Yourself
- PSAs from Learning Styles: Interpersonal/Group Learning
- Research from Learning Styles: Logical/Mathematical
- Lesson 2 Review answers

280  Unit 4  Prev. & Contr. Diseases

---

**Word Bank**
gonorrhea
sterility
syphilis

**Lesson 2 Review** On a sheet of paper, write the word or phrase from the Word Bank that best completes each sentence.

1. Gonorrhea can lead to _____.
2. The symptoms for chlamydia are similar to the symptoms for _____.
3. The first stage of _____ is a chancre.

On a sheet of paper, write the answers to the following questions. Use complete sentences.

4. **Critical Thinking** What are some short-term problems that genital herpes causes?
5. **Critical Thinking** What are some long-term problems that genital herpes causes?

### Health in Your Life

**A Patient's Basic Rights**

Everyone who wants medical treatment has certain basic rights. These include:

- You have the right to all information known about your illness or injury.
- You have the right to know about all the different treatment choices.
- Caregivers must have your informed consent to any treatment. This means that you understand what will happen before, during, and after the treatment.
- You have the right to privacy. This means that only caregivers who need to know about you will see your information.
- You have the right to be treated with dignity and respect.
- You have the right to use emergency services when needed.
- You have the right to refuse treatment.

On a sheet of paper, write the answers to the following questions. Use complete sentences.

1. Why is the right to privacy important?
2. Why is the right to use emergency services important to a person with no medical insurance?
3. How important would informed consent be to you if you were being treated in a hospital? Explain your answer.

280  Unit 4  Preventing and Controlling Diseases and Disorders

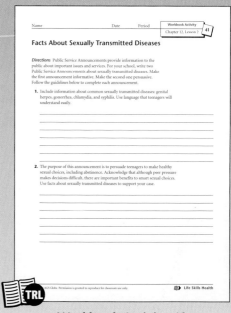

Activity 41          Workbook Activity 41

# Chapter 12 SUMMARY

- Acquired immunodeficiency syndrome (AIDS) is a disorder caused by the human immunodeficiency virus (HIV). HIV weakens the immune system.
- AIDS is spread through contact with certain body fluids, through sexual activity, by drug users sharing needles, and from mother to child.
- AIDS is not spread through casual social contact.
- The blood supply in the United States is safe because of a test for HIV. The test is used on all blood that is given for transfusions.
- Health care workers who are around body fluids and needles use preventive measures to protect themselves and patients.
- No cure is now available for AIDS. It can be prevented by avoiding contact with body fluids that may be infected with HIV.
- Any disease spread through sexual contact is a sexually transmitted disease.
- Gonorrhea has few symptoms. Antibiotics cure it, but untreated gonorrhea can lead to sterility.
- Chlamydia has symptoms and treatment similar to gonorrhea.
- Syphilis can progress in three stages if it is not treated. Antibiotics are used to treat syphilis.
- Genital herpes is chronic and has no cure.
- Genital warts are the result of an infection with human papilloma virus (HPV). HPV has no cure.
- People with sexually transmitted diseases can get treatment at clinics that protect their identity.
- All sexually transmitted diseases can be prevented by avoiding contact with the pathogen. The most reliable way to do this is by abstaining from sex, including oral sex.

### Vocabulary

| | | |
|---|---|---|
| acquired immunodeficiency syndrome (AIDS), 270 | genital warts, 278 | pneumonia, 272 |
| chancre, 276 | gonorrhea, 275 | sexually transmitted disease, 275 |
| chlamydia, 276 | human immunodeficiency virus (HIV), 270 | sterility, 276 |
| chronic, 277 | Kaposi's sarcoma, 272 | syphilis, 276 |
| genital herpes, 277 | opportunistic pathogen, 270 | transfusion, 272 |

## Chapter 12 Summary

Have volunteers read aloud each Summary item on page 281. Ask volunteers to explain the meaning of each item. Direct students' attention to the Vocabulary box on the bottom of page 281. Have them read and review each term and its definition.

### ONLINE CONNECTION

The Centers for Disease Control and Prevention, a branch of the U.S. Department of Health and Human Services, has an extensive section about HIV/AIDS on its Web site: www.cdc.gov/hiv/

The U.S. National Library of Medicine and the National Institutes for Health operate MedLinePlus, a Web site with up-to-date information on sexually transmitted diseases. Students can take an online tutorial here: www.nlm.nih.gov/medlineplus/sexuallytransmitteddiseases.html

## Chapter 12 Review

Use the Chapter Review to prepare students for tests and to reteach content from the chapter.

## Chapter 12 Mastery Test

The Teacher's Resource Library includes two forms of the Chapter 12 Mastery Text. Each test addresses the chapter Goals for Learning. An optional third page of additional critical-thinking items is included for each test. The difficulty level of the two forms is equivalent.

### Review Answers

**Vocabulary Review**

1. transfusion 2. human immunodeficiency virus (HIV) 3. opportunistic pathogens 4. genital warts 5. AIDS 6. genital herpes 7. Kaposi's sarcoma 8. sexually transmitted disease 9. sterility 10. antibiotics 11. syphilis 12. chlamydia 13. chronic 14. chancre

---

### TEACHER ALERT

In the Chapter Review, the Vocabulary Review activity includes a sample of the chapter's vocabulary terms. The activity will help determine students' understanding of key vocabulary terms and concepts presented in the chapter. Other vocabulary terms used in the chapter are listed below.

gonorrhea     pneumonia

---

# Chapter 12 REVIEW

### Word Bank
AIDS
chancre
chlamydia
chronic
genital herpes
genital warts
human immunodeficiency virus (HIV)
Kaposi's sarcoma
opportunistic pathogens
pneumonia
sexually transmitted disease
sterility
syphilis
transfusion

### Vocabulary Review

On a sheet of paper, write the word or phrase from the Word Bank that best completes each sentence.

1. The U.S. blood supply is tested for HIV, so a(n) _____ is safe.

2. AIDS is caused by _____.

3. When a person's immune system is weakened, it is open to _____, such as HIV.

4. Small bumps caused by a virus spread through sexual contact are _____.

5. A disorder of the immune system is _____.

6. Blisters that appear, break, heal and come back are symptoms of _____.

7. Two opportunistic infections that can appear as AIDS progresses are pneumonia and _____.

8. One in four people under age 21 in the United States has had a(n) _____.

9. Untreated gonorrhea can lead to _____, or the inability to have children.

10. A lung infection that often develops in AIDS patients is _____.

11. The sexually transmitted disease that progresses in three stages is _____.

12. Females with _____ have almost no symptoms.

13. Genital herpes is a(n) _____ infection because it never goes away.

14. The first symptom of syphilis is a(n) _____.

**282**   Unit 4   Preventing and Controlling Diseases and Disorders

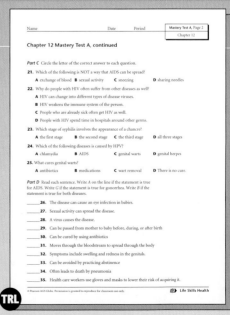

Chapter 12 Mastery Test A, pages 1–3

## Concept Review

On a sheet of paper, write the letter of the answer that best completes each sentence.

15. A sexually transmitted disease that has no cure is _____.

   A gonorrhea
   B chlamydia
   C syphilis
   D genital herpes

16. Before 1985, a person could get _____ from a blood transfusion.

   A gonorrhea
   B AIDS
   C genital herpes
   D syphilis

17. Babies can get a serious eye infection that is caused by _____.

   A gonorrhea
   B Kaposi's sarcoma
   C HIV
   D pneumonia

## Critical Thinking

On a sheet of paper, write the answers to the following questions. Use complete sentences.

18. Why are sexually transmitted diseases serious?

19. How do you think you could show caring and concern for a person infected with AIDS?

20. What would be the best way for you to avoid a sexually transmitted disease? Explain your answer.

**Test-Taking Tip** If you do not know the answer to a question, put a check mark beside it and go on. Then when you are finished, go back to any checked questions and try to answer them.

*Preventing AIDS and Sexually Transmitted Diseases* Chapter 12 283

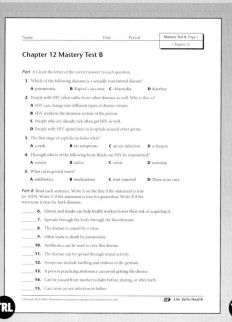

Chapter 12 Mastery Test B, pages 1–3

## Concept Review

15. D  16. B  17. A

## Critical Thinking

18. Sample answer: Sexually transmitted diseases are serious because they can lead to sterility, cancer, or even death. Also, some sexually transmitted diseases have no cure. 19. Sample answer: Because AIDS is not spread through casual social contact, someone can show caring and concern for AIDS patients in a safe manner by hugging, talking, listening, and helping out with simple tasks. 20. Sample answer: The most reliable way to avoid sexually transmitted diseases is to abstain from sex. This is because not everyone who has a sexually transmitted disease knows it. Also, some people will not talk about having a sexually transmitted disease, even to a potential partner. You cannot be sure that someone does not have a sexually transmitted disease just because he or she does not have symptoms.

## ALTERNATIVE ASSESSMENT

Alternative Assessment items correlate to the student Goals for Learning at the beginning of this chapter.

- Have students write an encyclopedia article describing the causes of AIDS and how it is acquired.
- Have students make a graphic organizer that identifies the symptoms of AIDS and infections it causes.
- Have students orally explain the symptoms and treatment of five common sexually transmitted diseases.
- Have students work in groups to identify help or resources for people with sexually transmitted diseases.
- Have students write and act out a script that explains how sexually transmitted diseases can be prevented.

# Chapter at a Glance

## Chapter 13: Recognizing Common Diseases
pages 284–309

### Lessons
1. Cardiovascular Diseases and Problems
   pages 286–291
2. Cancer
   pages 292–297
3. Diabetes
   pages 298–301
4. Arthritis, Epilepsy, and Asthma
   pages 302–306

### Chapter 13 Summary  page 307
### Chapter 13 Review  pages 308–309
### Audio CD
### Skill Track for
Life Skills Health

### Teacher's Resource Library TRL
Activities 42–45
Alternative Activities 42–45
Workbook Activities 42–45
Self-Assessment 13
Community Connection 13
Chapter 13 Self-Study Guide
Chapter 13 Outline
Chapter 13 Mastery Tests A and B
(Answer Keys for the Teacher's Resource Library begin on page 559 of this Teacher's Edition.)

## Opener Activity

Divide students into pairs. Have one student in each pair choose a common disease and list its symptoms. These students can use the Internet to help them identify the symptoms of the disease. Steer students toward a chronic disease and away from an infectious disease such as the flu. The other student in a pair should try to "diagnose" his or her partner's disease from the list of symptoms.

When the student pairs have finished their task, have a class discussion of some common diseases. Students can contribute some of the information they learned from the activity.

284  Unit 4  Prev. & Contr. Diseases

# Self-Assessment

## Can you answer these questions?

☑ 1. Why is it a good idea to be tested for high blood pressure?

☑ 2. How can you keep from getting diseases of the arteries?

☑ 3. What causes a heart attack?

☑ 4. What causes a stroke?

☑ 5. Which diseases are you more likely to get as you get older?

☑ 6. What are some risk factors and treatments for cancer?

☑ 7. How can you do a self-exam to check for cancer?

☑ 8. What are the symptoms of diabetes?

☑ 9. What causes epilepsy?

☑ 10. What are symptoms and treatments for asthma?

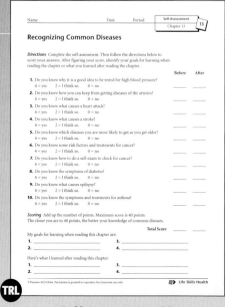

Self-Assessment 13

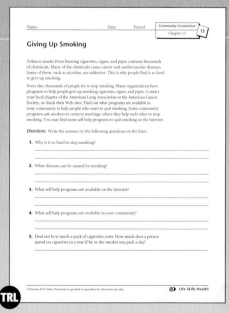

Community Connection 13

# Chapter 13
## Recognizing Common Diseases

You can compare a healthy body to a musical band. Each system of your body is like an instrument that plays its own part. Each system is like part of the mix of music made by the band as a whole. Sometimes, though, a body system hits a "bad note." When a disease affects a body system, your body's "music" is interrupted. The disease can affect your body's overall health.

In this chapter, you will learn about heart and blood vessel diseases and why people get them. You will learn what causes cancer. You will discover the warning signs of this disease. You will also find out how cancer is treated. Finally, you will learn about diabetes, arthritis, epilepsy, and asthma, and how they are treated.

### Goals for Learning

◆ To describe some cardiovascular diseases and what causes them

◆ To explain the causes, symptoms, warning signs, and treatments for cancer

◆ To explain the types of diabetes and treatment for the disease

◆ To describe arthritis, epilepsy, and asthma, as well as treatments for these diseases

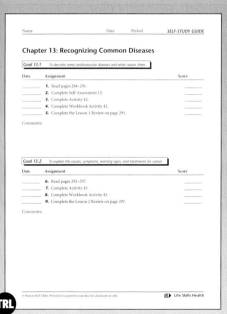

Chapter 13 Self-Study Guide, pages 1–2    Chapter 13 Outline, pages 1–2

## Introducing the Chapter

Encourage volunteers to name some diseases they or people they know have had. List on the chalkboard as many diseases as students can name. Then challenge students to rank the diseases according to how common they think the diseases are.

Have students read the chapter introduction on page 285. Ask them how a person can maintain a healthy body. *(Answers will vary but should include exercising, eating healthy meals, and getting enough rest.)* Why might a person not have complete control over his or her health? *(Genetics can cause some diseases or determine a tendency toward certain health conditions.)*

Have students read the Goals for Learning. Explain that in this chapter, students will learn about some common diseases including their causes, symptoms, and treatments.

### Notes and Questions

Ask volunteers to read the notes and questions that appear in the margins throughout the chapter. Then discuss them with the class.

### Self-Assessment

Have students complete the Self-Assessment worksheet before and after reading the chapter. Before reading the chapter, have students fill in the "Before" column. Ask students to identify their goals for learning. To get ideas for setting goals, students might use the chapter introductory material on page 285, the checklist on page 284, or the questions on the Self-Assessment worksheet. Students can use the back of the worksheet if they need more space to write.

Collect the Self-Assessment worksheets and pass them out again at the end of the chapter. Have students fill in the "After" column. Ask them to identify at least four major points they have learned. Again, suggest they use the back of the worksheet if they need more space to write. You may want to collect and review the worksheets, but return them to students so they have a record of their goals and accomplishments.

# Lesson at a Glance

## Chapter 13 Lesson 1

**Overview** This lesson describes cardiovascular diseases and problems, including treatment, prevention, and risk factors associated with them.

### Objectives

- To explain what causes high blood pressure
- To describe cardiovascular diseases and what might cause them
- To discuss how a heart attack happens
- To identify causes of a stroke

**Student Pages** 286–291

**Teacher's Resource Library**

Activity 42

Alternative Activity 42

Workbook Activity 42

## Vocabulary

| | |
|---|---|
| abnormality | heart attack |
| angina pectoris | hypertension |
| arterial | risk factor |
| arteriosclerosis | stroke |
| atherosclerosis | tissue |
| cardiovascular | surgery |
| chronic | |

Write the following vocabulary words on the chalkboard: cardiovascular, hypertension, arteriosclerosis, atherosclerosis, angina pectoris, stroke, and heart attack. Ask students what parts of the body the words relate to. Keep probing until students conclude that the words relate to the heart and arteries.

## Background Information

Coronary arteries are the blood vessels that supply blood, which contains oxygen, to the heart muscle. When these arteries are narrowed by the accumulation of fatty deposits called plaque, coronary heart disease results. Blood flow to the heart decreases, and the heart does not get enough oxygen to beat properly. This is the most common cause of heart disease, affecting about 13.2 millions Americans each year.

**286** Unit 4 Prev. & Contr. Diseases

---

## Lesson 1  Cardiovascular Diseases and Problems

**Objectives**

*After reading this lesson, you should be able to*

- explain what causes high blood pressure
- describe cardiovascular diseases and what might cause them
- discuss how a heart attack happens
- identify causes of a stroke

**Cardiovascular**
*Relating to the heart and blood vessels*

**Arterial**
*Relating to the arteries*

**Hypertension**
*High blood pressure*

**Abnormality**
*A condition that is not usual*

The group of diseases that affect your heart and blood vessels is called **cardiovascular** disease. More people die in the United States from these diseases than from anything else. Appendix C lists the numbers of deaths from cardiovascular disease. Some common cardiovascular diseases are high blood pressure and **arterial** diseases. Arterial diseases are diseases of the arteries. Arteries are blood vessels that carry blood from the heart.

You can decide to do things that will help prevent some of these diseases. These things include eating low-fat, low-sodium diets, exercising, and not smoking.

### What Is High Blood Pressure?

Blood pressure is the force of your blood against your artery walls when your heart pumps. Blood pressure normally goes up when you exercise and down when you sleep or relax. If your blood pressure always remains too high, you have **hypertension**, or high blood pressure.

Artery walls become thicker if you have high blood pressure. If more pressure is in your arteries, your heart pumps harder than it normally does. If hypertension is not treated, it leads to other kinds of cardiovascular disease.

The exact cause of most hypertension is unknown. It can affect adults and children. For adults, stress, tobacco use, and a high-salt or high-fat diet raise the chances of having hypertension. Heredity also can affect high blood pressure. If one of your parents has high blood pressure, you might be likely to get it, too. For children, causes include kidney disease or an **abnormality** of blood vessels that the child is born with. An abnormality is an unusual condition.

Most people with hypertension do not know it. Its symptoms are not easy to see. The only way to find out if you have hypertension is to have your blood pressure checked by a health care professional.

**286** Unit 4 *Preventing and Controlling Diseases and Disorders*

**Arteriosclerosis**
*Hardening of the arteries because of a buildup of fat*

**Atherosclerosis**
*A form of arteriosclerosis in which the large and medium arteries become narrow from a fat buildup along their walls*

**Heart attack**
*A condition in which the blood supply and the nutrients to the heart are greatly reduced*

**Tissue**
*Material in the body that is made up of many cells that are somewhat alike in what they are and what they do*

High blood pressure cannot be cured. It can usually be controlled with medicine and diet. The medicine helps lower blood pressure. Eating a diet low in salt, sugar, cholesterol, and saturated fats can also help control blood pressure. Regular exercise is another measure that is good for lowering blood pressure.

## What Are Some Diseases of the Arteries?

When you are born, the lining of your arteries is smooth. Over many years, fat can build up on artery walls. Usually, this happens because of foods you eat or heredity. Eventually, this fat buildup can harden like cement. This process is called **arteriosclerosis,** or hardening of the arteries. It is a chronic disease. This means it lasts for a long time. A chronic disease often causes major changes in your body. When your arteries harden, the arterial walls thicken and cannot move easily. When that happens, your blood cannot move around your body as it should.

Arteriosclerosis can cause many other diseases, including one called **atherosclerosis.** In atherosclerosis, your large and medium arteries get narrow because fat has built up along their walls. These narrowed arteries slow the flow of your blood to your heart. Your blood then gets too thick. You may get a blood clot. The fat deposits or a clot can block a blood vessel. If this happens, your heart does not get enough oxygen, or air. This is called a **heart attack.** Eating healthy foods can often prevent this buildup.

## What Is a Heart Attack?

Like every muscle in your body, your heart muscle needs nutrients and oxygen. A heart attack occurs when the blood supply and nutrients to your heart are reduced or stopped. Suppose your blood supply is cut off for a life-threatening amount of time. Then muscle **tissues** get hurt and die. Tissue is body material made up of many cells. These cells are somewhat alike in what they are and what they do. How bad a heart attack is depends on how much tissue dies.

 **Warm-Up Activity**

Give groups of students a short piece of hose, a small lump of clay, water, and a dishpan. Discuss with groups how they can use the materials to model the development of atherosclerosis. *(If some material (clay) builds up on the inside wall of an artery (hose), the material obstructs the flow of blood (water) through the artery. As more material builds up, much of the blood flow is blocked.)* Allow groups to manipulate their materials so that they can observe the decreased flow of "blood" through the "artery."

 **Teaching the Lesson**

Invite volunteers to tell what they already know about heart attacks, strokes, and hypertension. Encourage them to relate experiences family members, such as grandparents, may have had with these conditions. Then have students read about the cause and treatment of hypertension on pages 286 and 287.

 **Reinforce and Extend**

### ELL/ESL Strategy

 **Language Objective:** *To learn vocabulary specific to the content area*

Some of the terms in this lesson can be difficult for all students to pronounce and remember. Have students look up *arteriosclerosis*, *atherosclerosis*, and *angina pectoris* in the dictionary and discuss the word origins. For example, *angina* comes from *angere*, which means "to choke." In angina pectoris, the supply of oxygen-rich blood is "choked off" from the heart.

In addition, have students use the dictionary to differentiate between *arteriosclerosis* and *atherosclerosis*. Ask them to explain in their own words how the words are different. *(Arteriosclerosis is the thickening of the arterial lining, which causes the disease atherosclerosis.)*

## Technology and Society

Have students read the Technology and Society feature on page 288 about portable defibrillators. Have students find out what places in the community, such as churches and senior citizen centers, are likely to have portable defibrillators. Students can report their findings to the class.

### GLOBAL CONNECTION

Inform students that heart attack is the number one cause of death in the United States. Have students research how this compares with deaths caused by heart attack in other countries. Let students hypothesize reasons for higher or lower heart attack death rates in different countries. Suggest they do further research to check their hypotheses.

### LEARNING STYLES

**Visual/Spatial**

Make a simple drawing of the heart on the chalkboard. Add a branching coronary artery. Have a volunteer add to the drawing to show how a heart attack occurs. *(The student should show a blockage in the artery.)* Tell the student to shade the area that might be affected by the blockage of the artery. *(The student should shade an area that is below the blocked part of the artery.)* Have another volunteer add to the drawing to show how blood vessels can be attached to the heart and allow blood flow to bypass the blocked vessel.

### PRONUNCIATION GUIDE

Use the pronunciation shown here to help students pronounce difficult words in the lesson. Refer to the pronunciation key that appears in the Glossary at the back of the Teacher's Edition for the sounds of the symbols.

defibrillator (dē fib′ rə lā tər)

sphygmomanometer (sfig mō mə nom′ ə tər)

288   Unit 4   *Prev. & Contr. Diseases*

---

**Surgery**
Opening up the patient's body

There are two kinds of cholesterol, LDL and HDL. LDL cholesterol is often called "bad" cholesterol. It can build up in your arteries and make a substance that leads to atherosclerosis. HDL cholesterol is often called "good" cholesterol. High levels of HDL cholesterol keep you from getting a heart attack by carrying cholesterol away from your arteries.

If a heart attack is mild, the heart can still function. Other arteries will do the work that the blocked arteries did. Scar tissue may form where the other tissue was hurt. Scar tissue does not move easily and does not contract when the heart contracts. That means the heart may never work again as well as it once did.

Machines can help people who have had a heart attack. For example, one machine can keep track of the sounds and electrical activity of the heart. Another machine can figure out how much the heart has been hurt.

Doctors can do **surgery** to open a blocked artery or make the blood go around it. During surgery, a doctor opens up a patient's body.

To replace a diseased heart, doctors can use the healthy heart of a person who has died and donated the heart. Sometimes artificial, or human-made, heart valves can take the place of diseased valves. A valve is the part of the heart that stops the return flow of blood. It does this by closing or folding.

Another option is to place a pacemaker next to the heart. A pacemaker is a small machine that makes the heart beat regularly.

There are other treatments besides these to help people who have had a heart attack. Medicine is one treatment. Changes in diet, exercise, and smoking are also part of the treatment.

### Technology and Society

What if someone has a heart attack far away from a hospital? The heart attack victim may be saved by a portable defibrillator. This is a machine that can be carried or stored on airplanes, in places around the community, or in homes. Paddles or electrodes from the defibrillator are placed on the victim's chest. Small electronic shocks are sent to the heart. This helps get back a heartbeat rhythm. Then the blood can flow again.

288   Unit 4   Preventing and Controlling Diseases and Disorders

**Angina pectoris**
*Severe pain that happens when the heart does not receive enough blood*

**Stroke**
*A cardiovascular disease that happens when the blood supply to the brain is suddenly stopped*

### Health Myth

**Myth:** All people die soon after having a heart attack.

**Fact:** Many people live for many years after having a heart attack. Having a heart attack may make them change their diet and exercise plans and try to keep stress low.

## What Is Angina Pectoris?

People with atherosclerosis may get **angina pectoris,** also called angina. It happens when not enough oxygen-rich blood reaches the heart. Often, painful angina happens before a heart attack.

You can have angina pain without actually having a heart attack. The pains, however, can seem like those of a heart attack. If you have angina, you can take medicine to get rid of the pain when it starts. You can also get rid of the pain by rest and relaxation. If you have angina, you may need to be less active at times. You also might need surgery to make your blood flow more easily.

## What Is a Stroke?

A **stroke** happens when the blood supply to the brain is suddenly stopped. Arterial disease or a clot that is blocking an artery can stop the blood supply. High blood pressure may be one of the reasons a person has a stroke.

When the blood supply to your brain is stopped, some of your brain cells cannot work right. If these cells cannot get the blood and nutrients they need, they die. After a stroke, the part of your body that the dead brain cells controlled does not work any more. A stroke affects people differently. The effect depends on the part of your brain that is hurt. Some people notice almost no difference, while others may never be able to do certain things well again. It can take years to learn ways to make up for the damage that a stroke has caused.

Some new medicines can help lessen or do away with the effects of a stroke. A person must be treated within 6 hours of the stroke.

### Link to >>>
**Social Studies**

A sphygmomanometer is a device used to check blood pressure. Early sphygmomanometers of the 1800s did not have a cuff. Later sphygmomanometers had cuffs, but the cuffs were too thin to be used to get a correct reading. Today's sphygmomanometers have a wide cuff that is placed around the middle of the upper arm. Health care professionals use the devices with a stethoscope to listen to how the blood is flowing.

*Recognizing Common Diseases  Chapter 13  **289***

### Health Myth

Have students read the Health Myth feature on page 289. Tell them that there are many programs—such as mall walking programs—designed to rehabilitate heart attack patients. Invite a person who has had a heart attack to visit the class and discuss how he or she has made life style changes.

### Link to

**Social Studies.** Have students read the Link to Social Studies feature on page 289. Then explain to students how a sphygmomanometer works. *(The cuff is inflated until an artery is pressed closed. When the pressure is slowly released, a whooshing sound is heard through a stethoscope applied to the artery in the lower arm as blood flow starts again. This is the top blood pressure number. The cuff pressure is further released until the sound is no longer heard. This is the bottom number.)*

### CAREER CONNECTION

People who have had a heart attack usually go through a rehabilitation program, which includes exercise. Have students get brochures from a hospital that describe a rehab program for people with heart disease. Students should create a comic strip describing how physical and occupational therapists help these heart patients return to a state of improved health. Comics should include exercise, diet, and stress management.

### HEALTH JOURNAL

Have students respond in their journal to the following statement: Your uncle says he doesn't need to make any changes in his lifestyle because he's taking medicine to prevent himself from having another heart attack. *(Students should recognize that lifestyle habits such as smoking and lack of regular exercise counteract the medicinal treatment and that another heart attack is more likely if the uncle does not change his lifestyle.)*

## Research and Write

Have students read and conduct the Research and Write feature on page 290. You may want to have students work in small groups to research myths about heart disease and make their posters. Students can share their posters with the class, or they can be displayed in the classroom or school hallways.

### TEACHER ALERT

Have students compare controllable and uncontrollable cardiovascular risk factors. Ask students to list both kinds of factors and identify those that apply to them.

### AT HOME

Suggest that students interview family members to form a better picture of their medical history. Tell students to gather such information as names, ages, diseases, surgeries, and causes of deaths. Students can analyze this information to see if certain cardiovascular diseases run in the family.

### HEALTH JOURNAL

Have students answer the following questions in their journal by addressing the four risk factors for cardiovascular disease that can be controlled. Do you think you are at risk for getting a cardiovascular disease later in life? What specific things can you do now that will help prevent you from getting a cardiovascular disease?

---

**Risk factor**
*A habit or trait known to increase the chances of getting a disease*

### Research and Write

Find and research the American Heart Association's Web site to learn more about heart disease. Make a poster that lists myths about heart disease. Explain why each one is a myth and what the truth is. Tell how people can guard against heart disease. Include drawings of a blocked artery and an artery that works properly.

## What Are Some Cardiovascular Risk Factors?

A **risk factor** is a habit or trait known to increase your chances of getting a disease. Serious diseases, such as cardiovascular disease, have many risk factors. Some risk factors cannot be changed. Risk factors for cardiovascular problems that usually cannot be changed are:

- Family medical history—You may inherit the chance to have cardiovascular disease from your parents. You cannot change your genes. Knowing your family's medical history, however, can help you and your doctor figure out how high your risk is.

- Being male—Men have a greater risk of getting a heart attack before middle age than women do. After middle age, a woman's risk goes up.

- Aging—As you get older, the chance of having a heart attack is greater. More than half of all heart attacks happen to people who are age 65 or older.

- Race—Some races have a high risk of cardiovascular disease. For example, African Americans have an unusually high rate of hypertension.

## How Can Cardiovascular Problems Be Prevented?

You can change the following risk factors to lessen your chances of having cardiovascular problems:

- Smoking—A person can lower the risk for cardiovascular problems by choosing not to smoke. Smoking can lead to hypertension and other cardiovascular diseases. A smoker is twice as likely to have a heart attack as a nonsmoker is. Death from a heart attack is also twice as likely for a smoker as for a nonsmoker. When a person quits smoking, his or her risk of cardiovascular disease quickly lessens.

290    Unit 4   Preventing and Controlling Diseases and Disorders

---

### LEARNING STYLES

**Auditory/Verbal**

Have students listen to the Audio CD recording of pages 290 and 291. Suggest they take notes while they listen. Have students explain in their own words how they can reduce their chances of getting cardiovascular disease.

### Link to Math

Do you want to check your pulse? Use the tips of your first two fingers on one hand. Do not use your thumb because it has a pulse of its own. Place your fingertips on the inside of the wrist on your other hand. Press lightly. Count the number of beats for 10 seconds. Multiply by 6. The answer is your heart rate. Ask your health care professional to help you figure out your target exercising heart rate.

**Word Bank**
angina pectoris
arteriosclerosis
stroke

- **High cholesterol**—High blood cholesterol is another major risk factor for cardiovascular problems. Eating less saturated fats helps lower blood cholesterol levels. That means you should eat small amounts of animal fat, such as eggs, butter, and meat. You should exercise regularly, especially if cardiovascular disease runs in your family.

- **Being overweight**—Too much weight strains the heart and adds to cardiovascular problems. Also, it can raise blood cholesterol levels and blood pressure. Planning a balanced diet helps you keep to a healthy weight.

- **Physical inactivity**—Being inactive can make handling a sudden change in your blood pressure hard. Physical activity helps you handle these changes. Aerobic activities such as walking, running, swimming, or bicycling for 20 minutes every day can help. These activities help raise blood flow to your heart and strengthen it. Physical activity can also reduce mental stress, which may add to cardiovascular problems.

*Lesson 1 Review* On a sheet of paper, write the word or phrase from the Word Bank that best completes each sentence.

1. The terrible pain that comes with a short supply of blood to the heart is _____.

2. A cardiovascular disease that happens when the blood supply to the brain is stopped is a(n) _____.

3. Hardening of the arteries because of fat buildup is _____.

On a sheet of paper, write the answers to the following questions. Use complete sentences.

4. **Critical Thinking** Imagine you are a doctor. A young adult has come to you for medical advice. Both of the young adult's parents have hypertension. What would you tell the young adult?

5. **Critical Thinking** Write descriptions of the backgrounds and lifestyles of two people, one person who is at high risk for heart disease and one person who is at low risk for heart disease. Tell whose lifestyle is more like yours.

*Recognizing Common Diseases* Chapter 13 **291**

---

### Link to

**Math.** Have students read the Link to Math feature on page 291. Tell them that their pulse can be taken in many places besides the wrist. It may be easier for students to take their carotid pulse by placing the same two fingers along the outer edge of their trachea (windpipe). This pulse may be easier to find than the radial pulse in the wrist.

### AT HOME

Studies have shown that African-Americans have a higher incidence of hypertension than the general population. Have students review their own cultures and family ancestry to see if they can determine any other cultural links that might promote or help prevent cardiovascular disease. For example, are the foods and the preparation of those foods in that culture particularly healthy or unhealthy regarding cardiovascular disease?

### Lesson 1 Review Answers

**1.** angina pectoris **2.** stroke **3.** arteriosclerosis **4.** Sample answer: I would tell the young adult that hypertension can be avoided. I would advise staying away from tobacco and eating a low-salt, low-fat diet. I would also advise following a program of regular exercise and avoiding stress as much as possible. **5.** Sample answer: A person with high risk is a male who smokes, eats fatty foods, and has a family history of heart disease. A person with low risk is a young woman who eats a low-fat diet and doesn't smoke. I am at low risk. I eat healthy foods and exercise regularly, and my parents don't have heart disease.

### Portfolio Assessment

Sample items include:

- Atherosclerosis model from Warm-Up Activity
- Comic strip from Career Connection
- Poster from Research and Write
- Lesson 1 Review answers

# Lesson at a Glance

## Chapter 13 Lesson 2

**Overview** This lesson discusses causes, symptoms, risk factors, and treatments of cancer.

### Objectives

- To discuss causes and symptoms of cancer
- To tell the difference between malignant and benign tumors
- To understand risk factors for cancer
- To perform a self-exam to check for cancer

**Student Pages** 292–297

**Teacher's Resource Library**

Activity 43
Alternative Activity 43
Workbook Activity 43

## Vocabulary

| | |
|---|---|
| benign | malignant |
| carcinogen | mutation |
| chemotherapy | radiation |
| Hodgkin's disease | toxic |
| lymph | tumor |

Have students provide definitions of each of the vocabulary words based on their current knowledge. Then review the definitions in the textbook and have students compare the definitions with their previous ideas. Finally, suggest students write a question they have about each word. For example: Does chemotherapy get rid of cancer?

## Background Information

Chemotherapy works by attacking dividing cells. Some drugs attack a cell while it is dividing into two cells. Other drugs attack a cell prior to dividing. Because cancer cells divide much more rapidly than normal cells do, they are affected by these drugs to a greater degree. Radiation damages all cells that are actively growing and dividing. Because cancer cells are less well-organized than healthy cells, they are less able to repair the damage and recover.

292  Unit 4  Prev. & Contr. Diseases

---

## Lesson 2 Cancer

**Objectives**

*After reading this lesson, you should be able to*

- discuss causes and symptoms of cancer
- tell the difference between malignant and benign tumors
- understand risk factors for cancer
- perform a self-exam to check for cancer

Cancer is the second leading cause of death in the United States. For this reason, it can seem frightening to many people. Knowing about the causes and risk factors for cancer can help reduce people's fears.

### What Causes Cancer?

An abnormal and harmful growth of cells in your body causes cancer. There are more than 100 different types of cancers. Some cancers are more harmful than others.

Cancer cells grow in your body in a wilder and more uncontrolled way than normal cells do. Cancer may invade normal tissues. This means the cancer gets in and spreads to other parts of your body. When cancer takes over normal tissues, those organs cannot work properly. Sometimes, the abnormal growth of cells forms a mass, or pile, of tissue called a **tumor.**

If the cells keep growing, the tumor is **malignant.** This means it is harmful to your health. When cancer gets in normal tissues and spreads to other organs, it becomes more harmful. Some tumors are not malignant. They are **benign.** They do not spread and are not harmful to your health. More than 90 percent of tumors are benign. For example, some tumors in the breast are benign. They are actually collections of cells filled with liquid. They are not cancerous and are usually harmless.

**Tumor**
*A mass of tissue made from the abnormal growth of cells*

**Malignant**
*Harmful to health*

**Benign**
*Not harmful to health*

### Where Does Cancer Develop?

Cancer can develop in any tissue of your body. It usually develops in some areas more than others. Table C.4 in Appendix C shows the most common places where people get cancer.

Once you get cancer, it can spread to other parts of your body. It can even spread to places in your body that are far from the original tumor. Cells from a malignant tumor can break away. They can get into your bloodstream.

292  Unit 4  Preventing and Controlling Diseases and Disorders

These loose cancer cells reach other places in your body. There, they can grow into more tumors. Some cancers do not travel through your bloodstream. Instead, they spread by attaching themselves to nearby organs and getting inside them.

**Lymph**
*A colorless liquid that feeds tissue*

## What Are the Symptoms of Cancer?

The symptoms of cancer depend on where it is in the body. For example, a person with lung cancer may have a cough that does not get better with treatment. A person with breast cancer may feel a lump in the breast. If the cancer has spread to the bones, a person may have back pain.

If cancer has moved into your blood, it may cause unusual bleeding. Cancer may attack blood. It may attack tissues that produce **lymph.** Lymph is a colorless liquid that feeds tissues. When your blood and lymph tissues are attacked, your immune system does not work properly. Then you can get other infections.

Not all symptoms for a particular cancer are easy to recognize. You may feel somewhat sick but may not have specific symptoms. Too much of a weight loss too quickly and a poor appetite may be possible symptoms of cancer.

Sometimes, changes in skin and in the way your body works can be warning signs of cancer. For example, blood in your feces may be a sign of cancer in the intestines. If the cancer grows, it may cause a change in your bowel habits.

### The Seven Warning Signs of Cancer

- **C** hange in bowel or bladder habits
- **A** sore that will not heal
- **U** nusual bleeding or discharge
- **T** hickening or lump in the breast or elsewhere
- **I** ndigestion or a hard time swallowing
- **O** bvious change in a wart or mole
- **N** agging cough or hoarseness

## What Are the Warning Signs of Cancer?

Usually, cancer gives warning signs that can help in detecting the disease early. Read about the seven early warning signs in the box on the left. One sign, or a combination of signs, means there might be cancer.

*Recognizing Common Diseases Chapter 13* **293**

## Warm-Up Activity

Have students examine Table C.4 in Appendix C. Have them answer the following questions. What is the most common type of cancer among men? *(prostate)* Among women? *(breast)* What is the second most common type of cancer in both men and women? *(lung)* Who gets cancer of the urinary bladder more often, men or women? *(men)*

## Teaching the Lesson

Say the word *cancer* aloud as you write it on the chalkboard. Ask students to verbalize some of the feelings they associate with this word. Record their feelings on the chalkboard. Students may express fear, sadness, confusion, anger, and worry. Have students read what cancer is and what causes it on page 292.

## Reinforce and Extend

### ELL/ESL STRATEGY

**Language Objective:**
*To speak in front of a small group*

Divide students into groups of seven. Have each group make a set of seven posters, one per student. Each poster should feature one of the seven warning signs of cancer. The purpose of the posters should be to educate younger students in the school. When the posters are finished, each group member should give a short talk about his or her warning sign in front of the group. Encourage group members to ask questions of the presenter. At the end of the talks, group members will have discussed all seven warning signs. The posters can then be hung in the school hallways.

## Health Myth

Have students read the Health Myth feature on page 294. Tell them that mammograms are X-rays of the breast tissue that detect any abnormalities. Mammograms can be done on both men and women. Doctors suggest that women have mammograms as a preventive test to check for irregular tissue growth or the production of calcium deposits that can clump and form a cancerous mass. Sometimes ultrasound is used as well to diagnose fibrous cysts or fluid-filled pockets in the breast tissue.

## AT HOME

Contact your local public health office or a gynecologist's office for brochures on self-exam of the breasts. Also contact a urologist's office for information on testicular cancer and obtain brochures on self-exam of the testes. Distribute the brochures to students and tell them that a self-exam is the first step in preventative health care. Stress to students that members of a family should do a self-exam at least once a month to ensure early detection of any abnormality.

## LEARNING STYLES

### Body/Kinesthetic

Acquire demonstration models of a breast and scrotum from a science supply company. These models contain "tumors" so students can feel what a lump would be like if they found one while doing a self-exam. You may also wish to contact Inventive Products Inc. (1-800-356-6911) regarding a self-exam pad called Sensor Pad®. When this pad is placed over the skin, it magnifies any irregular tissue so that it can be felt more easily.

---

Why is it important to do self-exams and pay attention to the warning signs of cancer?

Finding cancer early is the most important thing for long-term survival. Talking about warning signs with a health care professional means you are more likely to be cured.

## How Can You Do Self-Exams for Warning Signs of Cancer?

You can examine yourself for changes that may be early warning signs of cancer.

### Females: Doing a Breast Self-Exam

Females should do a self-examination after each menstrual period, or once a month for older women. Breasts should be examined in front of a mirror, in the shower, or lying down. Females can stand in front of a mirror and look for any unusual shapes, dips, or lumps in the breasts.

For a breast exam in the shower or lying down, females can use the left hand to examine the right breast. They can then use the right hand to examine the left breast. The pads of the first three fingers should be used. Each breast is checked in a widening circle around the nipple until the entire breast is covered. Areas to check are the nipple, beneath the nipple, and the area between the armpit and breast. Look for unusual changes.

### Males: Doing a Testes Self-Exam

Cancer of the testes is one of the most common cancers in males 20 to 35 years of age. Most testicular cancers are first found during self-examination. Males should do a self-examination of their testicles once a month. The best time for self-examination of the testes is immediately after a bath or shower. Gently roll each testicle between thumb and fingers, looking for any abnormal lumps.

### Health Myth

**Myth:** Only women get breast cancer.

**Fact:** Men can also get breast cancer—though fewer men than women get the disease. Men have breast tissue and small amounts of the hormones women have. Men are more likely to die from breast cancer than women are. This is because men often do not consider or treat symptoms.

**Hodgkin's disease**
*A lymph gland cancer that often affects young adults*

**Chemotherapy**
*A treatment that involves taking drugs to kill cancer cells*

**Radiation**
*A cancer treatment that involves sending energy in the form of waves*

**Carcinogen**
*Material that causes cancer*

**Toxic**
*Harmful to human health or to the environment*

### How Is Cancer Treated?

Now more than ever before, scientists are finding cures for cancer. For example, **Hodgkin's disease** is a lymph gland cancer that often affects young adults. With treatment, up to 90 percent of people with Hodgkin's disease can be cured.

There are as many treatments for cancer as there are types of cancers. Usually, one or more of three treatments is used to treat cancer. One common treatment for many kinds of cancers involves taking drugs that kill the cancer cells. This kind of treatment is called **chemotherapy.** Each kind of cancer may need a different type of drug.

A second kind of treatment is surgery to remove the abnormal growth of cells. The third kind of treatment for cancer is **radiation.** This involves sending energy in the form of waves. Radiation destroys cancerous tissue.

Often, cancer treatments give people side effects that make them very sick. For example, a person receiving chemotherapy may get an upset stomach. Even so, these treatments allow many people to survive cancer.

### What Are the Risk Factors for Cancer?

Risk factors may cause cancer. If you know what the risk factors are, you can work to stay away from them.

- Cigarette and cigar smoke—Materials that cause cancer are called **carcinogens.** Carcinogens that cause the greatest number of deaths in America are found in cigarette and cigar smoke. As a result, smokers have a much higher risk for cancer than nonsmokers do. However, even nonsmokers can develop cancer from secondhand smoke.

- **Toxic** chemicals—Toxic chemicals are materials that are harmful to people or to the environment. Many chemicals used in industry, on farms, and in homes have been identified as toxic and are carcinogens.

- Radiation—While radiation is used to treat cancer, too much radiation can cause it. It is important to get X-rays only when you need them.

## IN THE ENVIRONMENT

 Newspapers often have articles about people whose cancers are thought to be caused by environmental pollution such as secondhand smoke and toxic chemical wastes from nearby factories or old waste dumps. Have students investigate information about their local environment for possible carcinogens. They should report their findings to the class in the form of an oral report, a poster, a newspaper article, a play, or a news broadcast. The reports should include the potential danger, how it affects the community, how it can be prevented, who the key players are, and what the key issues are.

## TEACHER ALERT

Invite a chemist to talk to the class about toxic chemicals. Students may not recognize that many everyday chemicals can be toxic. Chemists now believe that many chemicals that were once considered to be safe are, in fact, toxic.

## LEARNING STYLES

 **Interpersonal/Group Learning**
When the class completes the In the Community activity, you might extend it by having some students role-play city council members who must vote on a bill to ban smoking in all public buildings. Let these students decide on the roles. Suggest that other students role-play citizens who attend the council meeting to voice their opinions before the council votes. You might suggest that council members act as an advisory board rather than a voting council. Then each student citizen can have a vote.

## Research and Write

Have students read and conduct the Research and Write feature on page 296. Show students how to use a search engine and keywords to find helpful hospital Web sites. You may want to have students work with partners or in small groups to complete this activity.

## In the Community

Have students research smoking ordinances for public and private buildings in your area. Discuss the reasons for smoking bans and ask for students' opinions about the need for the bans. Ask if there are other places in the community where students would like to see no smoking ordinances enforced. Are such ordinances fair to everyone?

## Health in Your Life

Have students read the Health in Your Life feature on page 297 and answer the questions. Tell them that vermiculite is a mineral that often contains asbestos. Vermiculite has the unusual property of expanding into wormy pieces when heated. The expanded vermiculite is a lightweight, fireproof, absorbent and odorless material. These properties allow vermiculite to be used to make numerous products, including attic insulation, packing material, and garden products.

**Health in Your Life Answers**

**1.** Sample answer: There were more smokers in the past because people did not know that smoking could cause serious lung problems and cancer.
**2.** Sample answer: Modern society has many environmental carcinogens because we have made many things that contain carcinogens, such as home insulation and electrical insulation. **3.** Sample answer: Pollution probably has a big effect on the amount of carcinogens in food sources. Polluted water can poison the fish people eat.

---

**Mutation**
*Change*

- Sunlight—Sunlight is a source of natural radiation that can cause skin cancer. For example, people who sunbathe or use tanning booths are a target of harmful radiation. That gives them a high risk for malignant melanoma, the most common form of skin cancer. See Appendix C.
- Heredity—The same genes that control heredity also control the growth and work of cells. Many people have cancers that run in their families. For example, daughters of women who have breast cancer have a higher risk of getting breast cancer.
- Viruses—Viruses can cause cancer in almost every living creature. A virus can get its genes into human chromosomes. This causes a **mutation,** or change, that leads to cancer. No cancer caused by a virus, however, can spread from person to person. Also, not everyone is open to cancers caused by viruses. Some people have a weakened immune system from AIDS, which is caused by HIV. Their systems are more open to cancer.

### How Can You Avoid the Risk Factors for Cancer?

You can prevent many forms of cancer. You can stay away from the things and behaviors that are known to cause it. For example, keep away from cigarette smoke. Apply sunscreen before you go outdoors to help protect your skin. Use protective clothing such as long-sleeved shirts, wide-brimmed hats, and sunglasses. Try to stay away from toxic chemicals and X rays. If you have an inherited tendency to get certain kinds of cancer, see your doctor regularly.

**Research and Write**

Research hospital Web sites for disease prevention strategies. Learn how to prevent high blood pressure. Learn about pap smears and colorectal exams. Make a brochure to tell people about the importance of these preventive steps. Explain the age at which people should begin taking each step. Tell how often each step should be done.

**Lesson 2 Review** On a sheet of paper, write the answers to the following questions. Use complete sentences.

1. What is the difference between a malignant tumor and a benign tumor?
2. What are the seven warning signs of cancer?
3. What are the three types of treatments for cancer?
4. **Critical Thinking** Which patient would be more likely to get Hodgkin's disease—a 20-year-old or a 40-year-old. Why?
5. **Critical Thinking** What would you tell people in a presentation about how to avoid risk factors for cancer? How would you present the information?

### Health in Your Life

#### Environmental Carcinogens

Things in the environment that can cause cancer are called environmental carcinogens. Scientists figure out whether something in the environment causes cancer by working in labs and doing studies.

One environmental carcinogen is secondhand smoke. It contains thousands of dangerous chemicals.

Asbestos is another known carcinogen. It was once used as insulation, a way to keep buildings warm or cool. Asbestos should immediately be removed by a professional if it is found in your home or school.

Mercury is another environmental carcinogen. Mercury is found in some kinds of fish from certain locations.

You can learn about many environmental carcinogens by checking the American Cancer Society (ACS) Web site.

You can also do more research in books and articles.

On a sheet of paper, write the answers to the following questions. Use complete sentences.

1. Before scientists studied the effects of smoking, there were many more smokers. Why do you think this is true?
2. Why do you think modern society has so many environmental carcinogens?
3. How do you think pollution affects the amount of carcinogens in food sources such as fish?

*Recognizing Common Diseases* Chapter 13 **297**

## Lesson 2 Review Answers

1. A malignant tumor is harmful to health. A benign tumor is not. 2. The seven warning signs of cancer are (a) a change in bowel or bladder habits, (b) a sore that will not heal (c) unusual bleeding or discharge, (d) thickening or a lump in the breast or elsewhere, (e) indigestion or difficulty swallowing, (f) an obvious change in a wart or mole, and (g) a nagging cough or hoarseness. 3. Sample answer: The three types of treatment for people with cancer are chemotherapy, surgery, and radiation. 4. Sample answer: Someone who is 20 would be more likely to get Hodgkin's disease because it usually affects young adults. 5. Sample answer: I would tell people not to smoke. I would advise people to use sunscreen. I would tell people to stay away from toxic chemicals. I would present the information with posters, and I would have health care professionals speak during this presentation.

## Portfolio Assessment

Sample items include:
- Posters from ELL/ESL Strategy
- Reports, posters, newspaper articles, plays, or news broadcasts from In the Environment
- Lesson 2 Review answers

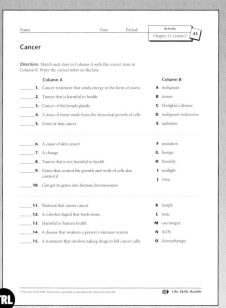

Activity 43

Workbook Activity 43

# Lesson at a Glance

## Chapter 13 Lesson 3

**Overview** This lesson discusses the causes and symptoms of diabetes and treatments for the disease.

### Objectives
- To explain type I and type II diabetes
- To recognize symptoms of diabetes
- To discuss treatment for diabetes

**Student Pages** 298–301

**Teacher's Resource Library**
- Activity 44
- Alternative Activity 44
- Workbook Activity 44

## Vocabulary

| | |
|---|---|
| coma | type I diabetes |
| diabetes | type II diabetes |
| obesity | |

Instruct students to write questions for which the vocabulary words are the answers. Caution students to write a question so that only one word is the correct answer. For example: What is the type of diabetes you could acquire when you are an adult? *(type II)* Have pairs of students exchange papers and answer each other's questions.

## Background Information

When people hear of diabetes, they may associate it with taking daily injections of insulin. Today, there are alternative, more modern ways of getting insulin. One way uses a pump that infuses a continuous flow of insulin into the body. The pump is worn much like a beeper. The insulin enters the body through a tiny needle inserted into the abdomen. Another treatment involves transplanting a pancreas or cells from a pancreas into a patient. Stress to students that these are examples of how continued medical research benefits people.

---

# Lesson 3 Diabetes

### Objectives

*After reading this lesson, you should be able to*
- explain type I and type II diabetes
- recognize symptoms of diabetes
- discuss treatment for diabetes

**Diabetes**
*A disease caused by lack of insulin in the body or when the body cannot use insulin effectively*

**Type I diabetes**
*A condition in which insulin is needed to stay alive*

**Coma**
*A condition in which a person loses consciousness*

**Diabetes** is a chronic disease. It comes from not having enough insulin in your body. As you learned in Chapter 5, insulin is a hormone made in your pancreas. You can also get diabetes if your body is not using insulin properly. Insulin is important for growth because it is needed for the storage and release of energy.

Your digestive system turns carbohydrates into glucose. With the help of insulin, your body's cells use glucose for the energy they need. Without insulin, the cells cannot use the glucose. Then the sugar you eat stays in your bloodstream.

As diabetes gets worse, it makes changes in the blood vessels. Because of these changes, people with diabetes may lose their eyesight. Diabetes may hurt their kidneys and cause cardiovascular disease. It can also cause strokes or lessen the movement of the blood into the hands and feet. The two most common types of diabetes are type I and type II.

## What Is Type I Diabetes?

**Type I diabetes** is also known as insulin-dependent diabetes. You have to take insulin to stay alive. This kind of diabetes usually starts in childhood. Your body may not be able to use glucose. This problem gives you high levels of glucose in your blood. When your high blood glucose level is not controlled, it can lead to a diabetic **coma**. When this happens, you pass out, or lose consciousness. A person who goes into a diabetic coma must be treated at once.

People with type I diabetes seem to have a defect in their immune system. Their immune systems kill the cells of the pancreas, which makes insulin. Type I diabetes symptoms include frequent urination and extreme thirst. Other symptoms are rapid weight loss, fatigue, and extreme hunger.

**Type II diabetes**
A condition in which the pancreas makes insulin that the body cells cannot use correctly

**Obesity**
Condition of being extremely overweight

## What Is Type II Diabetes?

**Type II diabetes** is also called noninsulin-dependent diabetes. It usually happens in adulthood after the age of 40. This is the most common type of diabetes. In type II diabetes, your pancreas makes insulin, but your body cells cannot use it right. Type II diabetes is milder than type I diabetes is. Type II diabetes has been linked to heredity, lack of activity, and **obesity,** or being very overweight.

Often, if you have type II diabetes, you do not know that your blood glucose level is high. Your symptoms may be blurry vision, slow-healing sores, and sleepiness. You may also have a tingling feeling in your hands or feet.

## How Do You Know If You Have Diabetes?

There are two ways to check for diabetes. During a regular medical checkup, your health care professional can test your blood for glucose. If your blood glucose level is high, you may have diabetes. Your health care professional can also check your urine for glucose to test for diabetes. Normally, urine does not contain glucose.

## What Is the Treatment for Diabetes?

If you have diabetes, you may get treated with insulin. You learn to measure the proper amount and give yourself insulin.

Some people with diabetes do not need to take insulin. A change in diet may keep their blood glucose at a safe level. For people who are obese, weight loss and exercise may treat the diabetes.

An important part of treatment for diabetes is learning which foods are safe to eat. If you have diabetes, foods with a lot of sugar can make your glucose level too high. If you have diabetes, you also must eat regularly, especially after taking insulin. You may need to eat nutritious snacks throughout the day. Insulin makes your blood sugar level go down. If you have an empty stomach, a shot of insulin could be dangerous.

**Figure 13.3.1** This teenager is using a glucose meter to check blood glucose levels.

---

 **Warm-Up Activity**

People with diabetes often wear ID tags or bracelets indicating this condition. Ask students why it is a good idea for a person with diabetes to wear this ID. *(If a diabetic coma or another emergency occurs, medical personnel will know what treatment to give or not to give the person.)*

 **Teaching the Lesson**

Review with students the role that hormones play in helping the body function. Refer to pages 110 and 111 in Chapter 5. Point out the location of the pancreas in Figure 5.3.1. Then have students read about the hormone insulin on page 298. Have them write short reports on the functions of insulin.

 **Reinforce and Extend**

### TEACHER ALERT

 Ask students why they think it would be especially important for a person with diabetes to avoid cuts, sores, and blisters on the feet. *(Circulation to the feet decreases as diabetes progresses, making it difficult for cuts, sores, and blisters to heal.)*

### ELL/ESL STRATEGY

 **Language Objective:**
*To read for meaning*

Have ELL students read pages 298 and 299. Ask the following questions to assess students' language proficiency: What is the relationship between the immune system and type I diabetes? *(Type I diabetes results from a defective immune system that destroys the cells of the pancreas.)* If a neighbor's child has diabetes, which type do you think it is? *(type I because this type commonly begins during childhood)* What are two ways type II diabetes differs from type I? *(Type II usually begins in adulthood, and it is usually milder than type I.)*

### Decide for Yourself

Have students read the Decide for Yourself feature on page 300 and answer the questions. Lead a class discussion by asking the following questions: Why do people with diabetes need to watch their intake of glucose? *(The pancreas cannot produce enough insulin to use glucose effectively. Therefore, diets that are low in sugar prevent excess glucose from accumulating in the blood.)* Suppose a person with diabetes plans on hiking for three hours. What would you expect the person to take along? Why? *(The person should take along food or juice to maintain proper levels of glucose in the body.)*

**Decide for Yourself Answers**

1. Sample answer: No, he's probably not right. He has type I diabetes, and he won't outgrow this because diabetes is a chronic disease. He may have a defect in his immune system. 2. Sample answer: I would tell my friend that many people have diabetes. I would explain that having diabetes is nothing to be embarrassed about. I would encourage him to have discussions with his doctor regarding various methods of testing his blood that could be done in private. I would tell him that diabetes is serious, and therefore is not to be taken lightly. I would remind my friend that I do not view him any differently now than I did before I learned that he had diabetes. 3. Sample answer: Yes, I would tell the nurse at school. My friend's life could be in danger.

---

It could give you very low levels of glucose. The sugar that is found in orange juice or hard candy can quickly return your blood glucose level to normal.

The American Diabetes Association lists these as risk factors:

- being overweight or obese
- getting little or no exercise during the day
- being age 45 or older
- being a woman who has had a baby weighing more than 9 pounds at birth
- having a sister, brother, or parent with diabetes

If you have two or more of these risk factors, you are more likely to develop diabetes than the average person.

### Decide for Yourself

Your friend has just learned that he has diabetes. His doctor has told him to test his blood during the day at school. He takes blood from his fingertip. Then he uses a small machine to tell if his glucose level is safe. He also has to eat a controlled diet.

Your friend has told you he is too embarrassed to test his blood at school. He does not want anyone to know he has diabetes. He does not plan to stop eating sugary foods. He says he has always loved these foods. He tells you that everyone will know something is wrong with him if he no longer eats them. He has begged you not to say anything to anyone. He thinks he will outgrow the problem.

On a sheet of paper, write the answers to the following questions. Use complete sentences.

1. Do you think your friend is right about outgrowing diabetes? Explain your answer.
2. What could you tell your friend to help him feel more comfortable about having diabetes? How could you help him deal with the things he must do?
3. Should you tell someone about your friend's problem? Explain your answer.

**Lesson 3 Review** On a sheet of paper, write the letter of the answer that best completes each sentence.

1. Diabetes comes from too little _____ or from the body not being able to use it.
   - **A** urine
   - **B** insulin
   - **C** oxygen
   - **D** fat

2. If a person with diabetes does not control the blood glucose level, the person might experience a(n) _____.
   - **A** heart attack
   - **B** asthma attack
   - **C** seizure
   - **D** coma

3. All of these are symptoms of type I diabetes EXCEPT _____.
   - **A** lack of thirst
   - **B** extreme hunger
   - **C** fatigue
   - **D** rapid weight loss

On a sheet of paper, write the answers to the following questions. Use complete sentences.

4. **Critical Thinking** A patient goes to a health care professional with a number of symptoms. These include sleepiness, tingling in the feet, blurred vision, and a sore that will not heal. Do you think this patient has type I or type II diabetes? What treatment do you think the health care professional will recommend? Explain your answer.

5. **Critical Thinking** Could a person with diabetes be on a swim team or play another team sport? Explain your answer.

*Recognizing Common Diseases* Chapter 13   **301**

## Lesson 3 Review Answers

1. B  2. D  3. A  4. Sample answer: These are symptoms of type II diabetes, which is also called noninsulin-dependent diabetes. The health care professional might recommend that the patient exercise regularly, lose weight, and eat a carefully controlled diet. The patient may need to visit a doctor regularly because his or her body cells are not using insulin properly. 5. Sample answer: Yes, a person who has the correct treatment for diabetes usually can lead a normal life. The person may have to monitor his or her diet carefully and test his or her blood regularly, but this person can play a team sport.

## Portfolio Assessment

Sample items include:
- Reports from Teaching the Lesson
- Lesson 3 Review answers

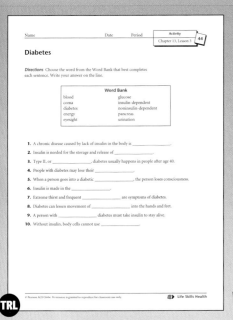

**Activity 44**

**Workbook Activity 44**

# Lesson at a Glance

## Chapter 13 Lesson 4

**Overview** This lesson examines the characteristics and treatment of the three common chronic diseases of arthritis, epilepsy, and asthma.

## Objectives

- To discuss arthritis and its treatment
- To understand epilepsy and its treatment
- To recognize what to do when someone has a seizure
- To discuss symptoms of asthma and how it is treated

**Student Pages** 302–306

**Teacher's Resource Library**

Activity 45
Alternative Activity 45
Workbook Activity 45

## Vocabulary

| | |
|---|---|
| arthritis | osteoarthritis |
| asthma | petit mal seizure |
| cartilage | rheumatoid |
| epilepsy | arthritis |
| grand mal seizure | |

Have students classify the vocabulary words into two categories: familiar words and unfamiliar words. For unfamiliar words, have students write ideas of word meanings based on the parts of the words. For example, students may be able to figure out that a *petit mal seizure* is not as bad as a *grand mal seizure*. Encourage students to compare their ideas of what words mean with their actual definitions as they read the lesson.

## Background Information

Rheumatic diseases affect joints, tendons, ligaments, bones, or muscles. Arthritis is characterized by inflammation of joints. About one in every three adults is affected by arthritis. Each year, arthritis results in 750,000 hospitalizations and 36 million outpatient visits. Arthritis is not just an "old person's disease". Nearly two-thirds of people with arthritis are younger than 65 years of age.

---

# Lesson 4 — Arthritis, Epilepsy, and Asthma

**Objectives**

*After reading this lesson, you should be able to*

- discuss arthritis and its treatment
- understand epilepsy and its treatment
- recognize what to do when someone has a seizure
- discuss symptoms of asthma and how it is treated

**Arthritis**
*A group of diseases that causes swelling of the joints*

**Rheumatoid arthritis**
*A condition in which the joints swell badly and become stiff and weak*

**Osteoarthritis**
*A condition in which the cartilage in the joints wears away*

**Cartilage**
*A cushion, or padding, in the joints*

You have read about diabetes, which is a chronic disease. Most chronic diseases have no cure. With medical care, however, a chronic disease can be controlled. This lesson tells about three common chronic diseases.

## What Is Arthritis and How Is It Treated?

**Arthritis** is a group of very painful diseases. They can cause your joints to swell and rub on the bones around them. Sometimes, your joints will not work if you have arthritis. There are two main types of arthritis. Together, they affect about 40 million Americans.

The first and most serious type of arthritis is **rheumatoid arthritis.** It is a harmful inflammation, or swelling, of your joints. In rheumatoid arthritis, your joints become stiff, swollen, and tender. If you have this disease, you may not be able to move your joints. You may feel weak and tired. Rheumatoid arthritis can include other tissues in your body and may cause them to be deformed, or misshapen. Usually, the treatment for rheumatoid arthritis includes drugs that help with the inflammation and pain.

The second and most common type of arthritis is **osteoarthritis.** In this kind of arthritis, the **cartilage,** or cushioning in your joints, wears away. When this happens, your bones rub against one another. This wear can change the makeup and shape of your joints. It can make your joints ache and feel sore. It can also make moving them hard. The same drugs that are used to treat rheumatoid arthritis have been used to treat osteoarthritis. New studies have shown that these drugs may cause problems. They may affect your body's natural way of fighting osteoarthritis.

302  Unit 4  *Preventing and Controlling Diseases and Disorders*

**Epilepsy**
*A chronic disease that is caused by disordered brain activity and that causes seizures*

**Seizure**
*A physical reaction to an instance of mixed-up brain activity*

**Grand mal seizure**
*A type of epileptic seizure in which a person passes out*

**Petit mal seizure**
*A type of epileptic seizure in which a person does not pass out*

---

Why is knowing what to do important when someone is having a seizure?

---

When arthritis badly harms your joints, you may be able to have a joint replacement. This surgery gives you an artificial joint in place of the old one. Artificial joints work just like your body's own joints.

## What Is Epilepsy and How Is It Treated?

**Epilepsy** is a chronic disease that affects 2.5 million Americans. It is caused by disordered brain activity. This is brain activity that sometimes gets mixed up. **Seizures** are usually a part of epilepsy. A seizure is a physical reaction to the mixed-up brain activity. There are two main kinds of epileptic seizures.

The most common kind of seizure is a **grand mal seizure.** If you have a grand mal seizure, you usually pass out and fall down. Your muscles get stiff, and your body shakes uncontrollably. During a grand mal seizure, you may lose control of your bladder and urinate. A grand mal seizure usually lasts about 2 to 5 minutes. Afterward, you may act strangely or be sleepy for a while. Often you do not remember the seizure.

The other kind of seizure is a **petit mal seizure.** If you have a petit mal seizure, you do not pass out. You may simply stare at something for a moment. You might drop something or act confused. A petit mal seizure lasts less than 30 seconds. Others nearby may not notice anything different. Only your own thinking is affected during a petit mal seizure.

Not all seizures are from epilepsy. A person is said to have epilepsy only after having two seizures without knowing why. While seizures may have many other causes, the cause of most seizures is not known.

The right medical care and prescribed medicine can help control seizures from epilepsy. In all other ways, a person with epilepsy can live a normal, active life.

Recognizing Common Diseases   Chapter 13   **303**

---

 **Warm-Up Activity**

Ask students if they or someone they know has asthma. Have students share their experiences with asthma with the rest of the class.

 **Teaching the Lesson**

Ask students if they know of anyone with arthritis or epilepsy. Invite students to tell what they know of each disease or how it is treated. Make a list on the chalkboard of main ideas students present. Students can correct or refine their ideas as they read the lesson.

 **Reinforce and Extend**

### LEARNING STYLES

 **Logical/Mathematical**
After students have read page 303, start to ask a question. Then suddenly stop, stare straight ahead, and drop a pencil or book from your hand. Do nothing for about 10 seconds. Then start talking again. Ask what type of seizure students might have just witnessed? *(petit mal)* How do they know? *(The seizure lasted less than 30 seconds.)* Ask students to write a paragraph reflecting on whether and how their relationship with a person would change if they found out that the person has epilepsy. *(Some students might suggest that they would be embarrassed if they were with the person during a seizure.)*

### HEALTH JOURNAL

 Have students describe in their journal what should be done if a person sees someone having a grand mal seizure. *(Descriptions could include: call for help, turn the person's head to the side, loosen clothing especially around the neck, place a cushion under the head, and do not try to pry open the mouth.)*

## ELL/ESL Strategy

**Language Objective:** *To learn vocabulary specific to the lungs*

Ask ELL students the following questions to assess their understanding of vocabulary specific to the lungs: What are bronchioles? *(Refer students to Figure 5.4.1 on page 114 to review the location and function of the bronchioles, the tiny air tubes in the lungs.)* What happens to the lungs in an asthma attack? *(The lining of the bronchioles produces mucus and contracts. These actions block the passage of air through the bronchioles.)* What kind of a sound is wheezing? *(a whistling-type of sound that happens when breathing is hard)*

## Health Myth

Have students read the Health Myth feature on page 304. Tell them that there are over 16,000 summer camps in the United States where children and teens with asthma can have fun and actively engage in sports under medical supervision. The American Lung Association runs many of these camps.

---

**Asthma**
*A respiratory disease that makes breathing difficult*

If someone has a seizure, follow these guidelines:

- Move nearby objects to protect the person.
- Loosen clothing such as buttoned collars or scarves around the person's neck.
- Place a cushion under the person's head.
- Turn the person's head to the side.
  - Do not put an object in the person's mouth.
  - Call for help if the seizure lasts more than 5 minutes.

### What Is Asthma and How Is It Treated?

**Asthma** is a disease that makes breathing difficult. It is a respiratory, or breathing, disease. Many things cause asthma to act up. One cause is allergies. These may be allergies to mold, pollen (powdery material from plants), animal fur, or dust. Asthma affects some people during some seasons of the year more than other seasons. Strong feelings about something may also start an asthma attack.

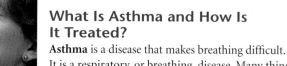

**Figure 13.4.1** *Allergies can cause you to sneeze and can cause an asthma attack.*

During an asthma attack, you may cough, wheeze, and be short of breath. Wheezing is a whistling-type sound that happens when breathing is hard. The linings of your bronchioles may make a great deal of mucus. Then the bronchioles tighten, and you have a hard time breathing. You may get scared about not being able to breathe and become even more stressed. People with asthma may have an attack as often as once every few hours or as rarely as once every few years.

> **Health Myth**
>
> **Myth:** People who have asthma should not exercise.
>
> **Fact:** Exercise can be good for people who have asthma. Exercise helps them stay in good shape. Health care professionals tell patients with asthma to take their medicine before they exercise. Then it is safe for them to exercise.

You can get asthma at any age, but this disease often affects children. More than 20.3 million Americans have asthma, and about 6.1 million of those are under age 18. The overall number of people who have asthma is getting higher.

Many different medicines are used to treat asthma. Usually, you would breathe asthma medicine directly into your air passages. The medicine can quickly relieve your breathing problem. Sometimes stronger medicines are needed. Without treatment, asthma can be life threatening.

### Health at Work

**Ultrasound Technician**

Using an ultrasound exam is a common way to find disease. An ultrasound technician performs an ultrasound exam. Sound waves are sent into the body. They bounce off body tissues and organs. This creates echoes. The echoes create a picture that shows if the tissues and organs are normal or abnormal. An ultrasound can help find many kinds of problems. These include problems related to heart disease, stroke, and cancer.

An ultrasound technician is trained to use ultrasound equipment and perform ultrasound exams. The technician makes and records the information from ultrasound exams. Then the information is given to a doctor. The technician does not tell patients what has been discovered in exams.

Usually, an ultrasound technician must complete a training program that lasts from 2 to 4 years.

This training may be given through a college or a specialty school. Some hospitals or other health care clinics require that students get a special certification or license.

### Health at Work

Have students read the Health at Work feature on page 305. After they have read about an ultrasound technician, ask them what parts of the body they think can be seen with ultrasound. Tell students that a technician can specialize in performing ultrasound exams of the female reproductive system, including the fetus; the organs of the abdomen (liver, kidneys, gallbladder, spleen, and pancreas); the brain; the eyes; and the heart and blood vessels. Students may have seen a sonogram of a fetus. If not, try to bring one in to show the class.

### AT HOME

Suggest that students make a health contract with themselves, listing what they will do if they become ill or notice unusual symptoms. Tell them to provide information about the kind of diet they will follow, exercises they will do, and how much sleep they will get. Have students share this contract with their parents and encourage their parents to make a similar contract.

# Lesson 4 Review Answers

1. joints 2. asthma 3. cartilage 4. Sample answer: She might not have another asthma attack for years, or she might start having attacks frequently. She should see a health care professional because without proper treatment, asthma can be life threatening. 5. Sample answer: The student does not necessarily have epilepsy. Other conditions can cause seizures. A person usually loses consciousness during a grand mal seizure, which typically lasts 2 to 5 minutes. The body shakes uncontrollably during this type of seizure. A person does not pass out during a petit mal seizure, which lasts less than 30 seconds. Only a person's thinking is affected during this type of seizure.

## Portfolio Assessment

Sample items include:
- Paragraph from Learning Styles Logical/Mathematical
- Contract from At Home
- Lesson 4 Review answers

---

Use these search terms to find Internet sites with information about chronic diseases: "National Institutes of Health," "American Diabetes Association," "National Heart, Lung, and Blood Institute," "Arthritis Foundation," "Epilepsy Foundation," and "Asthma and Allergy Foundation of America."

*Lesson 4 Review* On a sheet of paper, write the word in parentheses that best completes each sentence.

1. Arthritis is a disease that swells the (blood vessels, joints, bronchioles) and may change how well they work.
2. An allergy to mold, pollen, animal fur, or dust can trigger an (asthma, epilepsy, osteoarthritis) attack.
3. The body has (bone, cartilage, muscle), which serves as a cushion in joints.

On a sheet of paper, write the answers to the following questions. Use complete sentences.

4. **Critical Thinking** One of your friends had an asthma attack for the first time last week. Now she is afraid she will have one every week. What can you tell her about the likelihood of having an asthma attack every week?
5. **Critical Thinking** A student just had a seizure and has passed out. Can you be sure this student has epilepsy? Explain your answer. Describe the two types of seizures in your explanation.

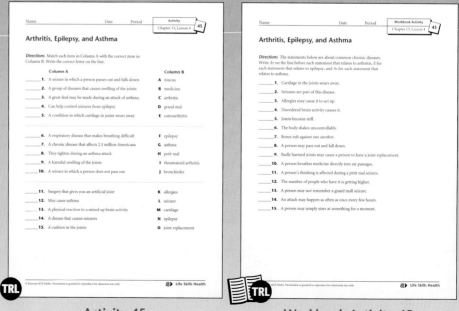

# Chapter 13 SUMMARY

- Cardiovascular disease affects the heart and blood vessels. It is the leading cause of death in the United States.
- Hypertension, or high blood pressure, happens when your blood pressure regularly stays too high.
- Heart attacks happen when narrowed blood vessels do not send enough oxygen and nutrients to the heart.
- A stroke happens when the blood supply to the brain is cut off. Some brain cells may die. This affects the work of the part of the body these brain cells control.
- An abnormal, harmful growth of cells causes cancer and usually forms a tumor. Some tumors are malignant, or harmful. Others are benign, or not harmful.
- Cancer's symptoms depend on the body part that is involved. Most treatment involves chemotherapy, surgery, or radiation. Some risk factors for cancer can be avoided.
- In type I diabetes, a person depends on insulin to stay alive. In type II diabetes, the pancreas makes insulin, but the body cells cannot use it the proper way.
- Rheumatoid arthritis is a painful swelling of the joints. In osteoarthritis, cartilage wears away, causing the bones to rub against one another.
- Epilepsy comes from disordered, or mixed-up, brain activity. It often gives people seizures.
- Asthma makes breathing difficult. An allergy usually triggers an asthma attack, which can be controlled with medicine.

### Vocabulary

| | | |
|---|---|---|
| abnormality, 286 | coma, 298 | petit mal seizure, 303 |
| angina pectoris, 289 | diabetes, 298 | radiation, 295 |
| arterial, 286 | epilepsy, 303 | rheumatoid arthritis, 302 |
| arteriosclerosis, 287 | grand mal seizure, 303 | risk factor, 290 |
| arthritis, 302 | heart attack, 287 | seizure, 303 |
| asthma, 304 | Hodgkin's disease, 295 | stroke, 289 |
| atherosclerosis, 287 | hypertension, 286 | surgery, 288 |
| benign, 292 | lymph, 293 | tissue, 287 |
| carcinogen, 295 | malignant, 292 | toxic, 295 |
| cardiovascular, 286 | mutation, 296 | tumor, 292 |
| cartilage, 302 | obesity, 299 | type I diabetes, 298 |
| chemotherapy, 295 | osteoarthritis, 302 | type II diabetes, 299 |

## Chapter 13 Summary

Have volunteers read aloud each Summary item on page 307. Ask volunteers to explain the meaning of each item. Direct students' attention to the Vocabulary box on the bottom of page 307. Have them read and review each term and its definition.

### ONLINE CONNECTION

The home page of the Centers for Disease Control and Prevention leads to a wealth of statistics, graphs, and charts for many diseases. For example, students can go to this Web site to access information about different kinds of cancer and their incidence. Maps show "hot spots" for different kinds of cancer: www.cdc.gov/

This site, offered by the National Institute of Arthritis and Musculoskeletal and Skin Diseases, discusses these diseases, their treatment, recent press releases, and new research. NIAMS is a branch of the National Institutes of Health: www.niams.nih.gov/index.htm

This site, from the U.S. National Library of Medicine and the National Institutes of Health, offers illustrations of how asthma affects lungs and bronchioles. Diagrams of asthma triggers are also shown: www.nlm.nih.gov/medlineplus/ency/article/000141.htm

# Chapter 13 Review

Use the Chapter Review to prepare students for tests and to reteach content from the chapter.

## Chapter 13 Mastery Test

The Teacher's Resource Library includes two forms of the Chapter 13 Mastery Text. Each test addresses the chapter Goals for Learning. An optional third page of additional critical-thinking items is included for each test. The difficulty level of the two forms is equivalent.

## Review Answers

**Vocabulary Review**

1. hypertension  2. arteriosclerosis
3. heart attack  4. stroke  5. malignant
6. lymph  7. chemotherapy  8. tumors
9. type I diabetes  10. chronic
11. rheumatoid arthritis  12. grand mal seizure  13. asthma  14. cartilage

---

### TEACHER ALERT

In the Chapter Review, the Vocabulary Review activity includes a sample of the chapter's vocabulary terms. The activity will help determine students' understanding of key vocabulary terms and concepts presented in the chapter. Other vocabulary terms used in the chapter are listed below.

| | |
|---|---|
| abnormality | mutation |
| angina pectoris | obesity |
| arterial | osteoarthritis |
| arthritis | petit mal seizure |
| atherosclerosis | radiation |
| benign | risk factor |
| carcinogen | seizure |
| cardiovascular | surgery |
| coma | tissue |
| diabetes | toxic |
| epilepsy | type II diabetes |
| Hodgkin's disease | |

---

# Chapter 13 REVIEW

**Word Bank**
- arteriosclerosis
- asthma
- cartilage
- chemotherapy
- grand mal seizure
- heart attack
- hypertension
- lymph
- malignant
- mutation
- rheumatoid arthritis
- stroke
- tumors
- type I diabetes

## Vocabulary Review

On a sheet of paper, write the word or phrase from the Word Bank that best completes each sentence.

1. High blood pressure, or _____, happens when the heart pumps blood too hard.
2. The condition of hardening of the arteries is _____.
3. A _____ comes from a short supply of oxygen and nutrients to the heart.
4. When the supply of blood to the brain is cut off, a person could have a(n) _____.
5. An abnormal growth of cells that is harmful is a(n) _____ tumor.
6. A colorless liquid that feeds cells is _____.
7. A treatment that kills cancer cells is _____.
8. Abnormal cell growth can produce _____.
9. The kind of diabetes for which insulin is needed to stay alive is _____.
10. A virus can invade chromosomes and cause a(n) _____ that leads to cancer.
11. A painful swelling of the joints is _____.
12. A(n) _____ makes a person with epilepsy pass out.
13. An allergy to mold, pollen, animal fur, or dust may trigger a(n) _____ attack.
14. A cushion in the joints is _____.

308  Unit 4  Preventing and Controlling Diseases and Disorders

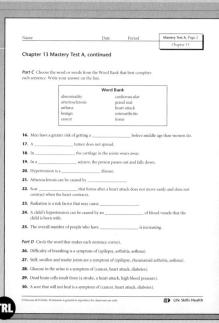

Chapter 13 Mastery Test A, pages 1–3

## Concept Review

On a sheet of paper, write the letter of the answer that best completes each sentence.

**15.** A risk factor for cardiovascular disease that CANNOT be controlled is _____.
- **A** weight
- **B** heredity
- **C** smoking
- **D** physical inactivity

**16.** Materials that cause cancer are called _____.
- **A** carcinogens
- **B** viruses
- **C** mutations
- **D** seizures

**17.** All of these words describe a mass of body tissue EXCEPT _____.
- **A** tumor
- **B** benign
- **C** artificial
- **D** malignant

## Critical Thinking

On a sheet of paper, write the answers to the following questions. Use complete sentences.

**18.** Why is getting help as quickly as possible important for someone who has had a stroke?

**19.** What did you learn about how diet affects your health as you read this chapter? What surprised you the most about this?

**20.** How important do you think it is to know your family's medical history? Explain your answer.

> **Test-Taking Tip** Look for words in each question that tell you the correct form for the answer you should give. For example, some questions may ask for a *paragraph*. Others may ask for a *sentence*, and still others may ask for a *list*.

Recognizing Common Diseases  Chapter 13

---

### Concept Review
15. B  16. A  17. C

### Critical Thinking
**18.** Sample answer: If brain cells cannot get the blood and nutrients they need, the cells die. Getting help quickly for a stroke victim is important so fewer brain cells die and there are fewer negative effects.
**19.** Sample answer: I learned that diet is very important. For example, sugary foods can be dangerous to a person with diabetes. I was surprised to learn that some people who have diabetes do not need to take insulin.
**20.** Sample answer: It is very important. My family's medical history can help me know if I am at risk for diseases such as cancer and diabetes. If I have a higher risk factor, I can take extra steps to protect myself.

## ALTERNATIVE ASSESSMENT

Alternative Assessment items correlate to the student Goals for Learning at the beginning of this chapter.

- Have students write a paragraph describing what they can do now to prevent cardiovascular problems later in life.
- Have students make a poster showing an outline of a male and a female body. They should indicate the five most common places where people get cancer by marking them with numbers 1–5.
- Have students make a chart showing the two types of diabetes, their symptoms, causes, and treatments.
- Have students describe orally how having arthritis, epilepsy, or asthma affects a person's daily life.

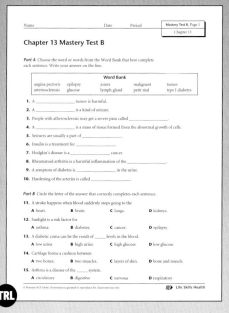
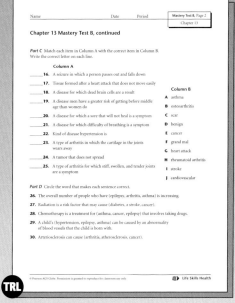

Chapter 13 Mastery Test B, pages 1–3

# HEALTH IN THE WORLD

Have students read the Health in the World feature on page 310. Review with students what vaccines are and how they work to provide immunity from disease. Ask students why it is important for people all over the world to have vaccines available to them. *(Possible answer: Vaccines can keep people from getting diseases and help save lives.)*

**Health in the World Answers**

1. Sample answer: The country must have money to purchase the vaccinations. The country must have a plan for getting the vaccinations to people throughout the country, as well as medical professionals who can administer the vaccinations. The country must have government leaders who think that vaccinations are important. The country must have a campaign to convince people that they should get vaccinated and to explain where they can go to get the vaccinations. 2. Sample answer: Many organizations help people get vaccinations, including UNICEF, the International Committee of the Red Cross (ICRC), the Global Alliance for Vaccines and Immunization (GAVI), individual hospitals and clinics, and pharmaceutical companies. 3. Sample answer: A vaccination program is effective if it prevents people from getting a disease. This effectiveness can be shown with statistics indicating fewer deaths from the disease, fewer hospital admissions due to the disease, and fewer visits to doctors and clinics due to the disease. Vaccinations can lead to a healthier population, which improves the quality of life, improves the economy, and gives people the opportunity to focus on other issues. For example, if many farmers in a country are sick, the country will not produce as much food.

---

# HEALTH IN THE WORLD

## The Need for Vaccinations

In 2003, only 56 cases of measles were reported in the United States. Most of these people with measles had lived or traveled outside the United States. In 2003, about 30 million people got the measles in countries around the world. About 530,000 of these people died. Every year, tetanus kills 300,000 newborn babies. Pertussis, or whooping cough, kills nearly 300,000 people. All it takes to prevent these deadly infectious diseases are vaccinations. Why, then, do so many people get these diseases?

Many people in countries in Asia, Africa, Latin America, and islands near Australia do not get vaccinations because they lack the money. Another reason is that many areas do not have health services. There is no one to deliver and give the vaccinations. In some countries, the government does not think vaccinations are important.

However, this is slowly changing. Many groups around the world have joined together to deliver vaccinations to countries that cannot afford them. Their efforts are starting to work. In 1988, for example, 350,000 children were paralyzed with polio worldwide. In 2004, only about 1,200 people got the disease. From 1999 to 2003, the number of deaths from measles in Africa decreased by half.

The fight against polio and other infectious diseases shows what can happen when people work together. Smallpox, a deadly viral disease, was declared "wiped out" in 1980 because of vaccinations. One day, perhaps we can say the same for other diseases that vaccinations can prevent.

On a sheet of paper, write the answers to the following questions. Use complete sentences.

1. What must a country have in order to provide vaccinations for a large number of people?
2. Use the Internet or print resources to find more information about vaccinations around the world. What are three organizations that help people get vaccinations?
3. What are some ways to show that vaccinations have been effective in a country?

# Unit 4 SUMMARY

- Diseases can be acquired or inherited.
- Pathogens are the germs that cause infectious diseases.
- If the body's immune system can fight off a pathogen, a person may never become ill. If the body's immune system cannot fight off a pathogen, a person becomes ill. The immune system continues to fight the illness, and a person usually recovers.
- The immune system fights pathogens by producing antibodies.
- A person may get an inherited disease through a single abnormal gene, a combination of many abnormal genes and environment, or an abnormal chromosome.
- Acquired immunodeficiency syndrome (AIDS) is a disorder caused by the human immunodeficiency virus (HIV). HIV weakens the immune system, leaving it open to other opportunistic pathogens.
- AIDS is spread through contact with certain body fluids, through sexual activity, by drug users sharing needles, and from mother to child.
- AIDS is not spread through casual social contact.
- Health care workers who are around body fluids and needles use measures to protect themselves and patients.
- No cure is now available for AIDS. It can be prevented by avoiding contact with body fluids that may be infected with HIV.
- All sexually transmitted diseases can be prevented by avoiding contact with the pathogen. The most reliable way to do this is by abstaining from sex.
- Cardiovascular disease affects the heart and blood vessels. It is a leading cause of death in the United States.
- Heart attacks happen when narrowed arteries do not send enough oxygen and nutrients to the heart.
- A stroke happens when the blood supply to the brain is cut off. Some brain cells may die. This affects the work of the part of the body these brain cells control.
- An abnormal, harmful growth of cells forms a tumor. Some tumors are malignant and others are benign.
- Cancer can affect any body part. Its symptoms depend on which part is involved. Most treatment involves chemotherapy, surgery, or radiation.
- In type I diabetes, a person depends on insulin to stay alive. In type II diabetes, the pancreas makes insulin, but the body cells cannot use it the proper way.
- Asthma makes breathing difficult. An allergy usually causes asthma, which can be controlled with medicine.

## Unit 4 Summary

Have volunteers read aloud each Summary item on page 311. Ask volunteers to explain the meaning of each item.

## Unit 4 Review

Use the Unit Review to prepare students for tests and to reteach content from the unit.

## Unit 4 Mastery Test

The Teacher's Resource Library includes two forms of the Unit 4 Mastery Text. An optional third page of additional critical-thinking items is included for each test. The difficulty level of the two forms is equivalent.

## Review Answers

### Vocabulary Review

1. pathogens 2. pneumonia 3. heart attack 4. gonorrhea 5. asthma 6. immune system 7. cardiovascular 8. diabetes 9. incubation period 10. chronic 11. human immunodeficiency virus (HIV) 12. antibiotic

### Concept Review

13. B 14. C 15. D 16. A

### Critical Thinking

17. Sample answer: Hospitals have rules to reduce the likelihood that an infectious disease will be spread from a patient to another person. An infected patient could infect other patients, hospital staff members, or visitors. 18. Sample answer: The polio vaccine worked like any other effective vaccination; after a person had been given the vaccine, he or she became immune to polio.

---

# Unit 4 REVIEW

**Word Bank**
antibiotic
asthma
cardiovascular
chronic
diabetes
gonorrhea
heart attack
human immunodeficiency virus (HIV)
immune system
incubation period
pathogens
pneumonia

## Vocabulary Review

On a sheet of paper, write the word or phrase from the Word Bank that best completes each sentence.

1. Bacteria, fungi, and viruses are all _____.
2. People who have AIDS often get _____, a lung infection.
3. A blocked blood vessel can lead to a(n) _____.
4. A symptom for males infected with _____ is a white discharge from the urethra.
5. A(n) _____ attack may be triggered by allergies.
6. The defenses that fight pathogens make up the body's _____.
7. Diseases that affect the heart and blood vessels are _____ diseases.
8. The chronic disease that comes from not having enough insulin in your body is _____.
9. During the _____, a person is infected but symptoms of the disease have not appeared yet.
10. Another term for "lasts a long time" is _____.
11. Drug users who share needles risk getting _____.
12. A substance that can destroy the growth of certain germs is a(n) _____.

## Concept Review

On a sheet of paper, write the letter of the answer that best completes each sentence.

13. The human body's response to tissue damage is _____.

   A placenta   C type I diabetes
   B inflammation   D influenza

312  Unit 4  Preventing and Controlling Diseases and Disorders

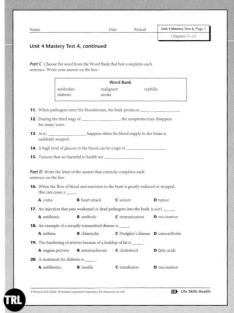

Unit 4 Mastery Test A, pages 1–3

14. HIV often is transmitted from one person to another through _____.
   A saliva   C blood
   B urine   D sweat

15. The best prevention for sexually transmitted diseases is _____.
   A having proper hygiene
   B getting tested often
   C eating a nutritious diet
   D choosing abstinence from sexual activity

16. The cancer treatment that uses waves of energy to destroy cancerous tissue is _____.
   A radiation   C angina pectoris
   B obesity   D hypertension

## Critical Thinking

On a sheet of paper, write the answers to the following questions. Use complete sentences.

17. In hospitals, there are places where patients with infectious diseases are not allowed to go. Why do hospitals have rules such as this?

18. In the early 1900s, thousands of people in the United States had the disease polio. Thanks to a vaccine, polio is extremely rare in the United States today. Why do you think the polio vaccine worked? Explain your answer.

19. About 25 million people in Africa were living with HIV at the end of 2004. How does fighting a disease on one continent help improve the lives of people who live on other continents?

20. Why is it important for people to understand epilepsy and be alert if one of their family members has epilepsy? Explain your answer.

19. Sample answer: The key reason to help people who live on other continents is that it is a kind, humane thing to do. Another reason is that fighting a disease on one continent benefits people living on other continents, because the medical advances that are made can be applied to other medical conditions. 20. Sample answer: People with epilepsy experience seizures. During a seizure, a person could be injured because he or she loses control of his or her body. Family members should be aware of this so they can possibly prevent an injury that could happen during a seizure.

# Unit 5

## Planning Guide
## Use and Misuse of Substances

| | Student Pages | Vocabulary | Health in the World | Review | Critical-Thinking Questions | Chapter Summary |
|---|---|---|---|---|---|---|
| **Chapter 14** Recognizing Medicines and Drugs | 316–343 | ✓ | | ✓ | ✓ | 341 |
| Lesson 1 Medicines | 318–323 | ✓ | | ✓ | ✓ | |
| Lesson 2 Tobacco | 324–327 | ✓ | | ✓ | ✓ | |
| Lesson 3 Alcohol | 328–331 | ✓ | | ✓ | ✓ | |
| Lesson 4 Stimulants, Depressants, Narcotics, and Hallucinogens | 332–337 | ✓ | | ✓ | ✓ | |
| Lesson 5 Other Dangerous Drugs | 338–340 | ✓ | | ✓ | ✓ | |
| **Chapter 15** Dealing with Drug Dependence | 344–361 | ✓ | | ✓ | ✓ | 359 |
| Lesson 1 Drug Dependence—The Problems | 346–351 | ✓ | | ✓ | ✓ | |
| Lesson 2 Drug Dependence—The Solutions | 352–355 | ✓ | | ✓ | ✓ | |
| Lesson 3 Avoiding Drug Use | 356–358 | ✓ | 362 | ✓ | ✓ | |

### Unit and Chapter Activities

**Student Text**
Unit 5 Summary

**Teacher's Resource Library**
Home Connection 5
Community Connections 14–15

**Teacher's Edition**
Opener Activity, Chapters 14–15

### Assessment Options

**Student Text**
Chapter Reviews, Chapters 14–15
Unit 5 Review

**Teacher's Resource Library**
Self-Assessment Activities 14–15
Chapter Mastery Tests A and B, Chapters 14–15
Unit 5 Mastery Tests A and B

**Teacher's Edition**
Chapter Alternative Assessments, Chapters 14–15

| | Student Text Features | | | | | | | | Teaching Strategies | | | | | | Learning Styles | | | | | Teacher's Resource Library | | | | |
|---|---|---|---|---|---|---|---|---|---|---|---|---|---|---|---|---|---|---|---|---|---|---|---|---|
| | Self-Assessment | Health at Work | Health in Your Life | Decide for Yourself | Link to | Research and Write | Health Myth | Technology and Society | Background Information | ELL/ESL Strategy | Health Journal | Applications (Home, Career, Community, Global, Environment) | Online Connection | Teacher Alert | Auditory/Verbal | Body/Kinesthetic | Interpersonal/Group Learning | Logical/Mathematical | Visual/Spatial | Activities | Alternative Activities | Workbook Activities | Self-Study Guide | Chapter Outline |
| | 316 | | | | | | | | | | | | 341 | 342 | | | | | | 46–50 | 46–50 | 46–50 | ✓ | ✓ |
| | | 322 | | | | | | 322 | 318 | 320 | 321 | 320, 322 | | 320, 321 | | | 321 | | | 46 | 46 | 46 | | |
| | | | 327 | | 326 | 325 | 326 | | 324 | 326 | 325 | 326 | | | | | | | | 47 | 47 | 47 | | |
| | | | | 331 | 329 | | 330 | | 328 | 330 | | | | | | | 330 | | | 48 | 48 | 48 | | |
| | | | | | | | | | 332 | 334 | | | | 334, 335 | 335 | | | | 336 | 49 | 49 | 49 | | |
| | | | | | | 340 | | | 338 | 339 | | | 341 | | | | | 339 | | 50 | 50 | 50 | | |
| | 344 | | | | | | | | | | | | 359 | | | | | | | 51–53 | 51–53 | 51–53 | ✓ | ✓ |
| | | | 351 | | 347, 350 | 350 | 348 | | 346 | 347 | 349, 350 | 348–350 | | 348 | | | 349 | | 348 | 51 | 51 | 51 | | |
| | | 355 | | | | | | 353 | 352 | 353 | 354 | 354 | | | 355 | | 354 | 354 | | 52 | 52 | 52 | | |
| | | | | | 358 | | 357 | | 356 | 357 | | 357 | | | 357 | | | | 359 | 53 | 53 | 53 | | |

## Alternative Activities

The Teacher's Resource Library (TRL) contains a set of lower-level worksheets called Alternative Activities. These worksheets cover the same content as the regular Activities but are written at a second-grade reading level.

## Skill Track

Use Skill Track for *Life Skills Health* to monitor student progress and meet the demands of adequate yearly progress (AYP). Make informed instructional decisions with individual student and class reports of lesson and chapter assessments. With immediate and ongoing feedback, students will also see what they have learned and where they need more reinforcement and practice.

## Unit at a Glance

**Unit 5:**
**Use and Misuse of Substances**
pages 314–365

**Chapters**

14. **Recognizing Medicines and Drugs**
    pages 316–343

15. **Dealing with Drug Dependence**
    pages 344–361

**Unit 5 Summary**  page 363

**Unit 5 Review**  pages 364–365

**Audio CD**

**Skill Track for Life Skills Health**

**Teacher's Resource Library**

   Home Connection 5

   Unit 5 Mastery Tests A and B

   (Answer Keys for the Teacher's Resource Library begin on page 559 of this Teacher's Edition.)

## Other Resources

### Books for Teachers

Gahlinger, Paul M. *Illegal Drugs: A Complete Guide to Their History, Chemistry, Use and Abuse.* New York: Plume, 2004. (comprehensive resource on illegal drugs)

Galanter, Marc, and Herbert D. Kleber, eds. *The American Psychiatric Publishing Textbook of Substance Abuse Treatment.* 3d ed. Washington, DC: American Psychiatric Publishing, 2004. (biology of drugs, treatment, outcomes)

### Books for Students

Youngs, Bettie B., Jennifer Leigh Youngs, and Tina Moreno. *A Teen's Guide to Living Drug Free.* Deerfield Beach, FL: HCI Teens, 2003. (advice, commentary and stories)

Denning, Patt. *Over the Influence: The Harm Reduction Guide for Managing Drugs and Alcohol.* New York: Guilford Press, 2004.

### CD-ROM/Software

*Decisions, Decisions 5.0: Substance Abuse.* Watertown, MA: Tom Snyder Productions, 1997 (1-800-342-0236). (responsible decision making to prevent substance abuse)

### Videos and DVDs

*Hooked: Illegal Drugs and How They Got That Way* (2 videos, 200 minutes). New York: A&E Television Network, 2000 (1-888-423-1212). (VHS)

### Web Sites

www.casacolumbia.org (The National Center on Addiction and Substance Abuse at Columbia University reports on substance abuse and risk reduction and studies prevention and treatment programs.)

www.whitehousedrugpolicy.gov (state of drug enforcement and awareness in the United States)

www.child.net/drugalc.htm (substance abuse and help for teens)

# Unit 5: Use and Misuse of Substances

Over-the-counter medicines and medicines given by doctors can help us fight and prevent illnesses. People should learn how to use their medicines wisely and safely. In this unit, you will discover details about drugs that help people. You will also learn about drugs that cause harm. It is important to learn the facts about different types of drugs. Then you can make wise decisions about how you will use them.

What do you think of when you hear the term *drug use*? Most people think only of illegal drugs. However, alcohol and nicotine in tobacco are also drugs—drugs that are legal for adults. You may be surprised to learn that legal drugs cause more deaths than illegal drugs cause.

### Chapters in Unit 5

Chapter 14: Recognizing Medicines and Drugs .........316
Chapter 15: Dealing with Drug Dependence .............344

---

## Introducing the Unit

Have volunteers read aloud the introductory paragraphs on page 315. Ask students if they were surprised to learn that legal drugs cause more deaths than illegal drugs cause. Then ask them to name some of the risks of legal drugs. (*Most students already will know that alcohol and tobacco use can lead to addiction, that smoking can cause cancer, and that excessive alcohol use can cause health problems.*)

### HOME CONNECTION

The Home Connection unit activity gives students practical experience with concepts taught in the *Life Skills Health* student text. Students complete the Home Connection activity outside the classroom with the help of family members. These worksheets appear on the Life Skills Health Teacher's Resource Library (TRL) CD-ROM.

### CAREER INTEREST INVENTORY

The AGS Publishing Harrington-O'Shea Career Decision-Making System-Revised (CDM) may be used with the chapters in this unit. Students can use the CDM to explore their interests and identify careers. The CDM defines career areas that are indicated by students' responses on the inventory.

## Chapter at a Glance

### Chapter 14: Recognizing Medicines and Drugs
pages 316–343

### Lessons

1. **Medicines**
   pages 318–323

2. **Tobacco**
   pages 324–327

3. **Alcohol** pages 328–331

4. **Stimulants, Depressants, Narcotics, and Hallucinogens**
   pages 332–337

5. **Other Dangerous Drugs**
   pages 338–340

**Chapter 14 Summary** page 341

**Chapter 14 Review** pages 342–343

**Audio CD**

**Skill Track for Life Skills Health**

**Teacher's Resource Library**

  Activities 46–50

  Alternative Activities 46–50

  Workbook Activities 46–50

  Self-Assessment 14

  Community Connection 14

  Chapter 14 Self-Study Guide

  Chapter 14 Outline

  Chapter 14 Mastery Tests A and B

  (Answer Keys for the Teacher's Resource Library begin on page 559 of this Teacher's Edition.)

### Opener Activity

Discuss with students such prescription drugs as tranquilizers and pain relievers. Have students do research to determine the beneficial effects of such drugs, as well as their potential for abuse. The MedLinePlus Web site (www.nlm.nih.gov/medlineplus/prescriptiondrugabuse.html), offered by the U.S. National Library of Medicine and the National Institutes of Health, is a good place to start.

## Self-Assessment

### Can you answer these questions?

1. What is a drug?
2. When is a drug thought of as medicine?
3. What are three purposes of medicine?
4. What are three ways medicine affects your body?
5. How does tobacco affect your body?
6. How does alcohol affect your body?
7. What are psychoactive drugs?
8. What effect do psychoactive drugs have on your body?
9. How does marijuana affect your body?
10. Why are anabolic steroids dangerous?

Self-Assessment 14

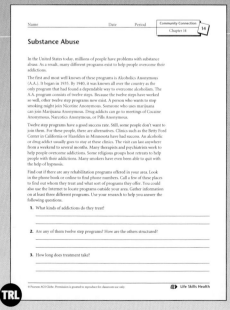

Community Connection 14

# Chapter 14
## Recognizing Medicines and Drugs

Have you ever played a word-association game? Someone says a word, and then you say the first word that comes to your mind. Each person might respond differently to the same word. For example, when you hear the word *drugs*, you might say "medicines." Someone else might say "illegal." Whether a drug is a medicine or an illegal substance, using it involves taking chances with your health. To keep from taking these chances, it is important to know whether a drug is safe. You must know how it can affect your mind and body.

In this chapter, you will learn about drugs that can help keep you from getting sick and drugs that may keep you healthy. You will learn about the harmful drug effects of tobacco and alcohol. You also will learn about the harmful effects of drugs that are illegal.

### Goals for Learning

◆ To explain the types and purposes of medicines

◆ To describe how medicines are taken and possible problems with them

◆ To identify some cautions to follow when taking medicines

◆ To describe the effects of using tobacco and alcohol

◆ To explain the effects of psychoactive and other dangerous drugs

### Introducing the Chapter

Ask students to discuss how some drugs, such as antibiotics, have had a positive effect on society. Then have students discuss how the misuse and abuse of other drugs, such as alcohol and cocaine, affect society in a negative way. Record students' ideas and suggest that they might want to adjust their ideas as they proceed through the lesson.

### Notes and Questions

Ask volunteers to read the notes and questions that appear in the margins throughout the chapter. Then discuss them with the class.

### Self-Assessment

Have students complete the Self-Assessment worksheet before and after reading the chapter. Before reading the chapter, have students fill in the "Before" column. Ask students to identify their goals for learning. To get ideas for setting goals, students might use the chapter introductory material on pages 316-317, the checklist on page 316, or the questions on the Self-Assessment worksheet. Students can use the back of the worksheet if they need more space to write.

Collect the Self-Assessment worksheets and pass them out again at the end of the chapter. Have students fill in the "After" column. Ask them to identify at least four major points they have learned. Again, suggest they use the back of the worksheet if they need more space to write. You might want to collect and review the worksheets, but return them to students so they have a record of their goals and accomplishments.

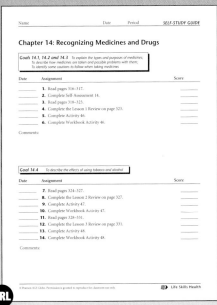

Chapter 14 Self-Study Guide, pages 1–2   Chapter 14 Outline, pages 1–2

## Lesson at a Glance

### Chapter 14 Lesson 1

**Overview** This lesson describes types and purposes of medicines; explains how medicines are taken; discusses effects of medicines; and identifies problems, safeguards, and cautions associated with taking some medicines.

### Objectives

- To compare prescription and over-the-counter (OTC) drugs
- To list five purposes of medicine
- To explain three problems with taking medicines
- To describe how the government keeps medicines safe

**Student Pages** 318–323

**Teacher's Resource Library**

Activity 46
Alternative Activity 46
Workbook Activity 46

### Vocabulary

analgesic
dependent
medicine
over-the-counter (OTC) medicine
pharmacist
prescription
side effect

Write the vocabulary words on the board and ask students what they know about each one before reading the definitions in the textbook. Then have students use all five vocabulary words to create two or three sentences that describe a situation.

### Background Information

The U S. Food and Drug Administration (FDA) is a scientific, regulatory, and public health agency. It oversees food products (other than meat and poultry), human and animal drugs, medical devices, radiation-emitting products, cosmetics, and animal feed.

The agency started as one chemist in the U.S. Department of Agriculture in 1862. The modern era of the FDA began in 1906 with the passage of the Federal Food and Drugs Act. The Act made employees of the FDA regulators, not just scientists.

318 Unit 5 Use & Misuse of Substances

---

## Lesson 1 Medicines

**Objectives**

*After reading this lesson, you should be able to*

- compare prescription and over-the-counter (OTC) drugs
- list five purposes of medicines
- explain three problems with taking medicines
- describe how the government keeps medicines safe

**Medicine**
*A drug that is used to relieve, cure, or keep a person from getting a disease or problem*

**Prescription**
*A written order from a doctor for a medicine*

**Pharmacist**
*Druggist*

**Over-the-counter (OTC) medicine**
*Medicine that can be bought without a prescription*

**Side effect**
*Reaction to a medicine other than what it is intended for*

From time to time, everyone gets sick. You may get simple aches and pains or something more serious. Sometimes, a drug or **medicine** is needed to help you get rid of pain and get you healthy again. Medicines are drugs that can relieve, cure, or keep you from getting a disease. Remember, a drug is something other than food that changes the way your mind and body work. All medicines are drugs, but not all drugs are used for medical reasons.

### What Are the Types of Medicines?

The U.S. Food and Drug Administration (FDA) divides medicines into two types. The first type is **prescription** medicines. These can be gotten only with a written order from a doctor. Only a doctor can write a prescription. Only a **pharmacist** can fill a prescription. A pharmacist reads the prescription and makes sure it is given correctly to the patient. Pharmacists are highly trained druggists who work in drugstores.

The second type of medicines is **over-the-counter (OTC) medicines.** Anyone can buy these medicines without a prescription in drugstores and other stores.

Prescription medicines are generally stronger than OTC medicines. Because prescription medicines are strong, they may have unwanted **side effects.** Side effects are reactions to a medicine other than what it is intended for. Side effects can hurt you or help you. OTC medicines can have unwanted side effects. Because of possible side effects, reading the labels on OTC medicines is important. Reading the information that comes with both OTC and prescription medicines is especially important.

Young children should never take any kind of medicine without adult supervision. Even some of the most common OTC medicines can have side effects that could be serious. Some medicines should not be taken with other medicines. A doctor should always be told the names of all the medicines that a patient takes, including OTC medicines.

318 Unit 5 Use and Misuse of Substances

## What Are Some Purposes of Medicines?

Both prescription and OTC medicines are used for many purposes. Usually, medicines are grouped according to why they are needed. Some of the reasons for the use of medicines are:

*To get rid of pain*—Medicines that a person takes to ease or get rid of pain are called **analgesics**. Some analgesics are mild. They work for minor pains such as a sore muscle. Most mild analgesics, such as aspirin or an aspirinlike drug, are OTC medicines. Other analgesics are powerful. They control serious pain, such as the pain you get after an operation. Powerful analgesics such as morphine are prescription medicines.

*To control or destroy bacteria*—Antibiotics are a kind of medicine that control infection by destroying bacteria. Sulfa drugs also reduce bacteria. Most antibiotics and sulfa drugs are prescription medicines. Some OTC medicines that can be rubbed on small wounds may contain a small amount of an antibiotic.

*To make the heart and blood vessels healthy*—Many different medicines treat problems of the heart and blood vessels. These medicines may slow your heartbeat, control blood pressure, widen veins and arteries, or control your heart's rhythm. These are all prescription medicines. New studies, however, show that aspirin, an OTC medicine, can help prevent heart attacks in some adults. Even so, adults should check with their doctor before taking aspirin for a heart condition.

*To prevent a disease*—Vaccines are medicines that can keep you from getting different diseases by giving your body immunity, or a way of fighting the disease. For example, vaccinations keep you from getting diseases such as measles, mumps, and chicken pox. Vaccines are prescription medicines that you get at a doctor's office or at a health clinic.

*To relieve symptoms of a disease*—Medicines you take for the common cold are examples of medicines that can relieve symptoms. Cold medicines can relieve a sore throat or cough, dry up a runny nose and eyes, or reduce mucus. Most cold medicines are mild and can be bought over the counter.

> **Analgesic**
> A medicine that eases or gets rid of pain

> All drugs have some side effects. With your doctor, talk about ways other than drugs that you may be able to treat an illness.

---

### Warm-Up Activity

Ask students to name medicines with which they are familiar. Challenge students to describe the purpose of each medicine they name. Ask them if they know of any precautions people should take when they are using the medicines. Chances are that they will not know many precautions. Look up the medications they have named on the Internet, or bring in empty medicine or prescription bottles, to show them the labels with indications and contraindications for use. Tell students that it is important to read the labels on medicine or prescription bottles before using them so as to avoid illness or injury.

### Teaching the Lesson

You might want to ask volunteers what medicines they take for such common complaints as headaches or colds. Some might take OTC analgesics such as acetaminophen for headaches, while others might have parents who prefer to limit the use of OTC drugs. Some parents might administer antihistamines or cough syrup for colds, while other parents might prefer herbal supplements such as echinacea. (Tell students that herbal supplements are not regulated by the FDA and therefore manufacturers do not have to prove their safety and effectiveness.) Remind students that they always should talk to their parents and read the label on the bottle before taking an OTC medication.

### PRONUNCIATION GUIDE

Use the pronunciation shown here to help students pronounce difficult words in the lesson. Refer to the pronunciation key that appears in the Glossary at the back of the Teacher's Edition for the sounds of the symbols.

antibiotic (an ti bī ot´ ik)

vaccine (vak sēn´)

asthma (az´ mə)

*To replace body chemicals*—Some prescription medicines replace chemicals that the body cannot produce. For example, insulin is a prescription medicine. People with type I diabetes take insulin because their pancreas cannot produce it.

*To lessen anxiety*—Many medicines can reduce depression, anxiety, nervousness, and other mental health problems. Strong forms of these medicines are available only with a doctor's prescription. Milder forms, such as a mild sleep aid, can be bought over the counter.

### How Are Medicines Taken?

Knowing the purpose of a medicine and how you should take it are important. Usually, a medicine is taken in the way that will start it working the fastest. For example, you may be told to swallow a medicine after a meal. Some medicines may upset your stomach. Taking some medicines with food may keep you from having an upset stomach. At times, you may take your medicine before a meal. This is because food may delay how quickly your body uses the medicine. Because of this, some medicines work faster on an empty stomach.

Sometimes, you receive medicines such as vaccinations as a shot. The medicine goes directly into a blood vessel or a muscle. Creams, lotions, and patches are put on the skin. The medicine in them is absorbed into the body through the skin. Other medicines are breathed in and go through the lungs first. For example, medicines for asthma are breathed in to help respiratory passages.

### How Does a Medicine Affect Your Body?

Your size, weight, age, and how your body works make a difference in how a medicine affects you. For example, a short, thin adolescent may have unwanted side effects from a certain medicine. A tall, heavy adult may have few or no side effects from the same medicine. Your doctor figures out the amount of a medicine that is right for you. Sometimes, your doctor might have to give you more medicine or less so it works right and does not cause unwanted side effects.

---

## 3 Reinforce and Extend

### ELL/ESL Strategy

**Language Objective:** *Using visuals to teach vocabulary*

While you are reading or discussing the section How Are Medicines Taken? with your class, pantomime the various methods that are described. For example, pantomime taking a pill, giving a shot, using an inhaler, and rubbing lotion on your arm. These gestures will help students learning English to connect the action with the words that describe it.

### At Home

Ask students to review with family members the kinds of OTC medicines they have at home. Suggest that families discuss the purpose of each kind of medicine so that medicines are not accidentally misused. Suggest also that students and their families check expiration dates on the medicines and discard those that are outdated. Have students share their experiences with the class.

### Teacher Alert

Tell students that a common side effect of many OTC medications is drowsiness. Many OTC products have a warning on the label telling the user not to drive a car or operate heavy machinery while taking the medication. Tell students they always should talk to their parents and read the labels of OTC medications for warnings before using them.

**Dependent**
*Having developed a need for a medicine or drug*

Because each person's body chemistry is different, a medicine may affect each person differently. For example, most people can take penicillin for a bacterial infection. Some people, though, are allergic to penicillin and cannot take it. A doctor can prescribe another drug to take the place of penicillin.

## What Are Some Problems with Taking Medicines?

The effects of medicine on your body can cause problems. For this reason, you always need to know the specific purpose of a medicine and how it should be taken. You also should know which medicines you cannot take to avoid problems. Four possible problems with taking medicines are having unwanted side effects, developing a strong need for a medicine, mixing medicines, and misusing medicines.

Many medicines cause unwanted side effects. For example, you may get a rash or a headache from taking a medicine. Some of these kinds of side effects are minor and hardly bother you. Other side effects, such as high blood pressure or irregular heartbeat, can be serious.

Sometimes, you can become **dependent** on a medicine. This is when you develop a need for the medicine. For example, a person may take a medicine to help him or her sleep. In time, the person may not be able to sleep without taking the medicine.

Mixing one or more medicines can cause problems. Sometimes, two different medicines cause you to get sick. Always check with your doctor before mixing OTC or prescription medicines with any other medicines.

Misusing medicines is a serious problem. A person can use a medicine incorrectly by taking more or less than the suggested amount. Taking two pills instead of one for a cold, for example, could cause a person to fall asleep while driving. Someone can misuse a medicine by taking it for a longer or shorter time than it should be taken. For example, if a person stops taking an antibiotic too soon, the infection may come back.

## LEARNING STYLES

**Interpersonal/Group Learning**

Divide students into groups. Have each group brainstorm a list of questions students might ask a doctor or pharmacist before taking a prescription medicine. Suggest that one student in each group record the questions and another student read them to the class. Possible questions are: What is this medicine for? Are there any side effects? What should I do if I forget to take a dose? What if I accidentally take too much? Should I take it with food or water? Tell students that they can ask these questions the next time they must take a prescription medication.

## HEALTH JOURNAL

Ask students to record in their journals some common alternative to taking OTC medicines. For example, they could describe gargling with warm salt water for a sore throat, or applying ice or heat to a sore muscle. Ask them to explore in their journals how trying a method that doesn't involve medicine might sometimes be just as effective as taking an OTC medicine.

## TEACHER ALERT

Tell students that many medicines can be dangerous when combined with alcohol; serious illness, injury, or death could result. If a medication should not be combined with alcohol—whether the medicine is prescription or OTC—it will carry a warning telling the user not to drink alcohol while taking it.

## Technology and Society

Have students read the Technology and Society feature on page 322. Explain to students that during a clinical trial, some people take the drug being tested and others take a placebo, or sugar pill. The research participants do not know if they are getting the drug or not. The doctors administering the doses do not know either. This is called a "double-blind" experiment. Ask students why they think it might be important to test drugs this way. *(If people knew if they were taking the drug or not, those that were might imagine they felt better even if they were not getting better, and those that were not taking the drug might feel they were not getting better even if they were.)* Point out that concerns have been raised about whether it is ethical to give some research participants placebos if they are in need of real treatment.

## Global Connection

Ask students to think about medicines they may have taken. Then invite students to discuss whether government regulation of medicines made and sold in the United States affects the level of confidence they feel about medicines. Tell them that in some countries, medicines are not strictly regulated. Ask if they would feel safe taking medicines made and sold in another country that did not have strict government regulations.

## Health at Work

Have students read the Health at Work feature on page 322. Have students research what training or licensing requirements your state has for pharmacy clerks. Then discuss with students whether they think these requirements are too strict or not strict enough, and why. Also ask them what personality characteristics would help make a pharmacy clerk successful. *(Someone who is meticulous and detail-oriented and who enjoys talking to people would do well as a pharmacy clerk.)*

322  Unit 5  *Use & Misuse of Substances*

---

### Technology and Society

Several clinical trials are needed before the FDA approves a drug for sale. A clinical trial is a research study. Trials begin with as few as 20 people. Later trials can involve several thousand people. Special computer technology is used to carefully watch and evaluate the results of each clinical trial.

Another misuse of a medicine is taking it for a purpose other than its intended purpose. Although drugs are usually not used incorrectly on purpose, by misusing them, a person could get hurt or very sick.

## How Does the Government Keep Medicines Safe?

The U.S. government has many laws and agencies to control the safety of medicines. Before a company can sell any medicine, the FDA makes sure the medicine is tested carefully to prove it does what it is supposed to do. The company must give the test results to the FDA. It must also write papers telling the purpose, ingredients, possible side effects, and other information about the medicine. The FDA then decides whether the medicine can be sold to the public. It also decides whether the medicine must be sold only with a doctor's prescription.

The FDA decides how medicines must be labeled. Labels must give the purpose of the medicine, the suggested amount, possible unwanted side effects, and cautions. Reading a label tells you whether taking the medicine is safe and how to keep from having any problems.

### Health at Work

**Pharmacy Technician**

Pharmacy technicians work at hospital and community pharmacies. Pharmacy technicians order supplies. They help label drugs for homebound patients. They keep accurate records of all drugs and supplies. They might type and file reports and answer phones. Under a pharmacist's direction, they measure out or mix prescriptions. They enjoy helping people and working with the public. There is an increasing need for pharmacy technicians. Some pharmacy technicians receive on-the-job training. However, most states require licensing. A 2-year program at a community college can lead to getting a license.

322  Unit 5  *Use and Misuse of Substances*

## What Are Some Cautions When You Take Medicine?

The correct use of medicines can improve your health and life. Misusing medicines can cause serious problems. To be safe, follow these simple guidelines:

- Tell your doctor about any health problems you have. Tell your doctor about any other medicines you are taking before you get a new prescription.
- Follow the directions for taking a medicine exactly as they are given.
- Tell your doctor right away if you have any unwanted side effects from a medicine.
- Take medicines that are prescribed only for you. Be careful not to confuse one medicine with another.
- Store medicines correctly. Throw away old medicines.
- Do not mix medicines with other medicines, unless directed to do so by your doctor.
- Never mix medicines with alcohol.

**Word Bank**
medicine
over-the-counter medicines
side effects

**Lesson 1 Review** On a sheet of paper, write the word or phrase from the Word Bank that best completes each sentence.

1. Stomach upset or rashes are examples of unwanted _____ from some medicines.
2. All _____ can be bought without a prescription.
3. A(n) _____ is used to keep a person from getting a disease or to relieve a problem.

On a sheet of paper, write the answers to the following questions. Use complete sentences.

4. **Critical Thinking** Why is it important to read the label before taking any medicine?
5. **Critical Thinking** Why is knowing about the possible side effects of a medicine important?

*Recognizing Medicines and Drugs* Chapter 14  323

## Lesson 1 Review Answers

1. side effects 2. over-the-counter medicines 3. medicine 4. Sample answer: Labels tell the purpose of the medicine and list the possible side effects. Reading the label can help a person avoid serious health problems. 5. Sample answer: It's important to know possible side effects to avoid medicines that could be harmful. In some situations, the side effects could be worse than the symptoms of the illness.

## Portfolio Assessment

Sample items include:
- Entries from Health Journal
- List of questions from Learning Styles: Interpersonal/Group Learning
- Research from Health at Work
- Lesson 1 Review answers

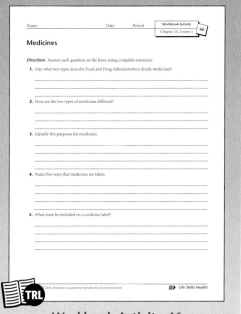

Activity 46    Workbook Activity 46

# Lesson at a Glance

## Chapter 14 Lesson 2

**Overview** This lesson explains why people use tobacco, what the physical effects of tobacco are, how tobacco smoke affects nonsmokers and fetuses, and how people can stop smoking.

## Objectives

- To list five physical effects of tobacco use
- To explain how tobacco smoke affects nonsmokers
- To describe the effects tobacco smoke has on unborn babies

**Student Pages** 324–327

**Teacher's Resource Library**

Activity 47

Alternative Activity 47

Workbook Activity 47

## Vocabulary

| | |
|---|---|
| emphysema | stimulant |
| nicotine | withdrawal |

Read each vocabulary word and its definition aloud. Ask volunteers to provide expanded definitions for each word.

## Background Information

In 1964, U.S. Surgeon General Luther L. Terry reported that smoking cigarettes was harmful to health, news that shocked the nation. At that time, nicotine was described as a habit-forming drug, not an addictive drug. In his 1988 report, "The Health Consequences of Smoking," U.S. Surgeon General C. Everett Koop reported that nicotine is as addictive as cocaine and heroin. Nicotine affects certain nerve receptors in a way that is similar to other addictive drugs. Nicotine is one of the strongest poisons known. In fact, at one time, it was used as an insecticide.

---

# Lesson 2 — Tobacco

### Objectives

After reading this lesson, you should be able to

- list five physical effects of tobacco use
- explain how tobacco smoke affects nonsmokers
- describe the effects tobacco smoke has on unborn babies

**Nicotine**
A chemical in tobacco to which a person can become addicted

**Stimulant**
A drug that speeds up the central nervous system

**Emphysema**
A serious respiratory disease that makes breathing difficult

Some drugs change the way your body works in harmful, rather than helpful, ways. One drug that has harmful physical effects is tobacco.

## Why Do People Use Tobacco?

People use tobacco for many reasons. Many teens and adults smoke to feel they fit in. Some people say smoking makes them feel relaxed. Teens may smoke because their friends smoke or because some TV programs make smoking look cool. They believe they can quit smoking before it harms their bodies. In reality, research shows that the younger people are when they start to smoke, the harder it is for them to stop.

The main reason people keep smoking is because of **nicotine,** a chemical in tobacco. People can get an addiction to nicotine. An addiction is a dependency, or a need for a habit-forming substance. Nicotine is in all forms of tobacco. That includes cigarettes, cigars, pipe tobacco, and smokeless tobacco. Smokeless tobacco is put between the cheek and gums and chewed.

## What Are the Physical Effects of Tobacco?

The nicotine in tobacco is a **stimulant.** A stimulant speeds up your central nervous system. Even though people may believe that smoking relaxes them, nicotine makes the heart rate speed up. It raises blood pressure but narrows the blood vessels. Using tobacco for a long time also causes a buildup of material in your blood vessels. It can give someone a heart attack, stroke, and lung and other cancers. The buildup can cause lasting bronchitis and **emphysema.** Emphysema is a serious respiratory disease that makes breathing difficult.

When tobacco is burned, it creates thousands of other harmful chemicals besides nicotine. These chemicals include tars and other substances that form a thick, brown, sticky material in the lungs. Tars cripple and destroy the tiny parts in the bronchial tubes that help keep the lungs clean.

324   Unit 5   *Use and Misuse of Substances*

## Research and Write

Studies show that most smokers begin smoking as teens. Studies also show that if people have not started smoking by the age of 18, they most likely never will smoke. Using the Internet and print resources, find information on the health risks of smoking during adolescence. What are the connections between smoking as a teenager and having health problems later in life? Create a class presentation that shows your research.

Tobacco smoke also makes gases, including carbon monoxide. This gas takes the place of the oxygen in the red blood cells, giving a person less oxygen. This lack of oxygen makes doing anything physical difficult. Other smoke-produced gases have poisonous chemicals in them that can give someone tumors and cancer.

Smokeless tobacco also can cause health problems. For example, it can rot teeth and cause gums to shrink so the teeth get loose and fall out. Smokeless tobacco can give a person white, leathery patches in the mouth that could turn into cancer.

### How Does Smoke Affect Nonsmokers?

People who do not smoke but who breathe in other people's tobacco smoke also can be harmed. Many studies have shown that healthy people who do not smoke can get sick because of secondhand smoke. People can get respiratory diseases and cancers by breathing in smoke from burning cigarettes or cigars. People who do not smoke get these diseases because they breathe into their lungs many of the same harmful gases that smokers breathe in.

If you do not smoke, you still can get many of the same physical effects that smokers get. For example, you can get heart and lung problems. Your eyes may bother you, you can get headaches, and you might cough from the smoke. If you are sick, secondhand smoke can make you sicker. For example, if you already have a heart condition, breathing in smoke can make it worse.

Many studies have shown the dangers of breathing secondhand smoke. The health concern of secondhand smoke has caused changes in our society. Laws make smoking illegal in public places and on public transportation. Some cities do not let people smoke in restaurants. In other cities, you can ask to sit in nonsmoking areas of restaurants. Also, many workplaces have become smoke free. Each state decides how old someone must be to buy tobacco products. The federal government puts health warnings on the labels of tobacco products. It also has stopped tobacco advertising on television.

##  1 Warm-Up Activity

Discuss the legal battles in recent years in which many states' attorneys general have sued tobacco companies on behalf of citizens who have gotten sick or died from using tobacco. Bring to class various newspaper and magazine articles on the subject. (You can find older articles on newspapers' and magazines' Web sites.) Create a bulletin board display and invite students to add to it. Refer to the articles throughout this lesson.

##  2 Teaching the Lesson

If you are a former smoker, you might want to describe to students how hard it is to overcome the addiction to nicotine. Tell your students that when you were a teenager, you never thought you could get addicted to cigarettes. Answer students' questions and try to be very descriptive as you tell them about the physical effects of smoking. If you are not a former smoker, perhaps you know someone who is. Invite that person to speak to your class.

## 3 Reinforce and Extend

### HEALTH JOURNAL

 For a week, ask students to make a note in their journals each time they are exposed to second-hand smoke. Tell them to note when, where, and for how long they are exposed. Ask them if they can make changes in their lives that would protect them from being exposed to second-hand smoke. For instance, they could stop going to a restaurant that has a smoking section.

### Research and Write

Have students read and conduct the Research and Write feature on page 325. Two reliable Web sites that have information on the health risks of smoking during adolescence are the Centers of Disease Control and Prevention (www.cdc.gov) and the National Institutes of Health (www.nih.gov). You might want to have students work with partners or in small groups to complete this activity.

### PRONUNCIATION GUIDE

Use the pronunciation shown here to help students pronounce difficult words in the lesson. Refer to the pronunciation key that appears in the Glossary at the back of the Teacher's Edition for the sounds of the symbols.

bronchitis (brong kī´ tis)

monoxide (mə nok´ sīd)

## ELL/ESL Strategy

**Language Objective:** To prepare and give a group presentation

Have students work together in groups of three to write a short presentation about the dangers of smoking. Include students learning English and students who are proficient in English in each group. You might want to assign each group a specific topic, such as the dangers of secondhand smoke or the dangers of smoking during pregnancy. Allow students to read from notes or index cards during the presentation. Encourage them to go online to find visuals that can accompany their presentations. If they have access to the appropriate technology, groups can give multimedia presentations.

## Health Myth

Point out to students that the Federal Trade Commission and other government groups no longer believe that low-tar and/or low-nicotine cigarettes are any "healthier" to smoke than "regular" cigarettes. This is because in real life, people tend to compensate for cigarettes that have less tar and nicotine by taking longer, deeper puffs in order to boost the amount of nicotine they're inhaling. They also tend to either knowingly or unwittingly cover small holes in the filters of low-tar and/or low-nicotine cigarettes—holes that are there to dilute the amount of smoke in each puff—which also boosts the amount of nicotine they're inhaling.

## Link to

**Biology.** Ask a volunteer to read aloud the Link to Biology feature on page 326. Students who wish to learn more about neurons, neurotransmitters, and how they are affected by drugs can find information on the University of Texas at Austin Web site: www.utexas.edu/research/asrec/addiction.html

---

**Withdrawal**
*A physical reaction to not having a drug in the body*

**Health Myth**

**Myth:** Low-tar cigarettes are safer to smoke than regular cigarettes.

**Fact:** Using any tobacco product can cause cancer and other health problems. Low-tar cigarettes still contain toxic materials such as lead, arsenic, benzene, and butane gas. No tobacco product is completely safe, or even safer than others.

## How Does Smoke Affect Unborn Babies?

Today, more than 17 percent of women ages 15 to 44 smoke while they are pregnant. Cigarette smoke is poison for unborn babies. Women who smoke are twice as likely as nonsmokers to have smaller babies. Recent research shows that tobacco smoke also affects brain growth. Babies born to mothers who smoked during pregnancy have a higher chance of developing attention deficit disorder (ADD) or of having lower intelligence levels. Babies born to mothers who smoke show a high chance of having upset stomachs. They also show signs of stress and drug **withdrawal**. This is a physical reaction to not having a drug in the body. For all of these reasons, women who are pregnant should not smoke. Pregnant women also should stay away from places with much secondhand smoke.

## How Can People Stop Smoking?

As more people realize the problems that smoking causes, they are choosing to quit smoking. However, the addiction to nicotine is powerful. When people stop smoking, they will probably go through a period of withdrawal. During withdrawal from nicotine, a person may have a headache, be unable to sleep, and have a hard time focusing or concentrating. A person may be irritable and have some anxiety.

Help is available for people who choose to quit smoking. They can get skin patches or a special gum prescription from a doctor. These aids allow smokers to ease away slowly from their addiction to nicotine. Classes, support groups, and professionals also are available to help smokers quit smoking.

### Link to >>>

**Biology**

Neurons, or nerve cells in the brain, pass on information. Chemicals called neurotransmitters carry the information from neuron to neuron. Nicotine attaches itself to one special neurotransmitter. When this happens in the brain, a person gets a feeling of pleasure and relaxation. Many people who smoke have a strong desire to experience these feelings again and again. The same kind of chemical release happens with cocaine. That is why both drugs are highly addictive.

---

## Career Connection

Invite students to discuss the role a health educator might play in helping people to stop smoking. Lead students to talk about how an educator might plan school and community programs as well as develop printed and multimedia materials in order to spread the message about the dangers of smoking. Ask students what qualities and skills might help a person to be a successful health educator. Encourage interested students to find out more about educational requirements for a career as a health educator.

## In the Environment

Divide students into two groups for a debate. Present the following topic: Outlawing smoking in public places provides protection from secondhand smoke for nonsmokers. Assign opposite positions to each group and allow students time to research and prepare an argument supporting their point of view. Then ask volunteers to present each group's opinion. Serve as moderator as groups take turns arguing their case.

*Lesson 2 Review* On a sheet of paper, write the word or phrase in parentheses that best completes each sentence.

1. The (tar, nicotine, oxygen) in tobacco is addictive.
2. People who do not smoke can still have smoke-related problems from (secondhand, primary, filtered) smoke.
3. Nicotine acts as a central nervous system (stimulant, pathogen, barrier).

On a sheet of paper, write the answers to the following questions. Use complete sentences.

4. **Critical Thinking** How does nicotine become addictive?
5. **Critical Thinking** Why should women who are pregnant stay away from places with a lot of smoke?

### Health in Your Life

**Breaking the Nicotine Habit**

Nicotine addictions are hard to break. The effects of nicotine last from 40 minutes to 2 hours. Some people smoke many times during a day because they want to feel these effects. When people try to stop smoking, they may experience irritability, anxiety, depression, and a craving for nicotine. People who decide to quit smoking are taking the first step toward lifelong health. Think about this:

- Within 20 minutes of quitting smoking, blood pressure returns to normal.
- Within 2 to 12 weeks, circulation improves.
- Within 3 to 9 months, coughing and breathing problems get better.
- Within 10 years, chances of getting lung cancer are cut in half.
- Within 15 years, chances of a heart attack are the same as for someone who never smoked.

On a sheet of paper, write the answers to the following questions. Use complete sentences.

1. Why is an addiction to nicotine hard to break?
2. What might a smoker experience when he or she decides to quit smoking?
3. Why is quitting smoking a step toward lifelong health?

*Recognizing Medicines and Drugs* Chapter 14

### Lesson 2 Review Answers

1. nicotine 2. secondhand 3. stimulant 4. Sample answer: People can develop a dependency on nicotine, which is a stimulant. Nicotine is a chemical in tobacco. Nicotine becomes addictive because its effects are short-lived. People smoke again (or use smokeless tobacco again) to feel the effects again. 5. Sample answer: Pregnant women should not smoke, and they should avoid places with a lot of tobacco smoke because both types of smoke can be very harmful to unborn babies. Cigarette smoke is poison for unborn babies. Some babies born to mothers who smoke show signs of drug withdrawal.

### Portfolio Assessment

Sample items include:
- Entries from Health Journal
- Presentations from Research and Write Presentations from ELL/ESL Strategy
- Lesson 2 Review answers

### Health in Your Life

Have students read the Health in Your Life feature on page 327. If students want to find out more about smoking cessation, the National Institutes of Health has a page on its Web site devoted to links to articles about stop-smoking methods, products and studies, as well as articles that address how teenagers and women can quit smoking: www.nlm.nih.gov/medlineplus/smokingcessation.html.

**Health in Your Life Answers**

1. Sample answer: Nicotine addiction is hard to break because nicotine causes chemical changes that last only from 40 minutes to 2 hours. Smokers develop a craving for nicotine because they want to experience those chemical changes.
2. Sample answer: A person who decides to quit smoking might experience headaches, sleeplessness, irritability, depression, a hard time focusing or concentrating, and anxious feelings.
3. Sample answer: People who quit smoking reduce their chances of having a heart attack, lung cancer and other health problems. When people quit smoking, their circulation and breathing improve within months.

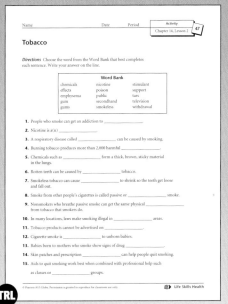

**Activity 47**

**Workbook Activity 47**

# Lesson at a Glance

## Chapter 14 Lesson 3

**Overview** This lesson explains why people use alcohol and how alcohol affects the body and a person' driving abilities. It also defines alcoholism and fetal alcohol syndrome.

## Objectives

- To explain the physical effects of alcohol on the body
- To describe how alcohol affects an unborn baby
- To explain why alcoholism is a disease

**Student Pages** 328–331

**Teacher's Resource Library**

   Activity 48

   Alternative Activity 48

   Workbook Activity 48

## Vocabulary

| | |
|---|---|
| alcoholism | intoxication |
| binge drink | psychoactive |
| depressant | |

Provide students with copies of a magazine ad for alcohol. Have students edit the ads to incorporate the vocabulary words and any other applicable information.

## Background Information

The temperance movement of the 19th century led to Prohibition. The 18th Amendment to the U.S. Constitution prohibited the production and sale of alcohol beverages. The rural population was mostly law-abiding, but in the cities people wanted alcohol, and bootleggers stepped forward to provide illegal beer and liquor. Quarrels between groups of bootleggers brought more violence into cities, and the corruption spread to government agencies and lawmakers. When the U.S. Congress repealed Prohibition in 1933 by passing the 21st Amendment, the bootleggers simply switched to other forms of crime, including narcotics, gambling, prostitution, loan sharking, and extortion, which they continue to run today.

---

# Lesson 3   Alcohol

**Objectives**

*After reading this lesson, you should be able to*

◆ explain the physical effects of alcohol on the body

◆ describe how alcohol affects an unborn baby

◆ explain why alcoholism is a disease

**Depressant**
*A drug that slows down the central nervous system*

**Psychoactive**
*Affecting the mind or mental processes*

Across America, alcohol is served in restaurants, bars, and bowling alleys and at sporting and community events. From the smallest town to the largest city, alcohol is easy to find and buy. It also is easily abused. About 14 million people in the United States abuse alcohol. Abuse of alcohol occurs when people drink too much alcohol or drink it too frequently.

## Why Do People Use Alcohol?

Some people use alcohol to fit in with a particular group. Some people drink to make themselves feel better, to relax, or not to think about how they feel. In reality, alcohol does not help any of these things. Many teens try alcohol because their friends do it or because they have seen their parents drink. Some try it because of the ads they see on TV or in the movies.

You can become both physically and psychologically addicted to ethyl alcohol. This is the kind of alcohol that is found in wine, beer, and liquor. Alcohol is a **depressant,** a type of drug that slows down your central nervous system. It is also a **psychoactive** drug. Psychoactive drugs affect your mind, or mental processes.

It is a mistake to think that some forms of alcohol, such as beer or wine coolers, are safer than liquor. You can become just as addicted to beer or wine as you can to liquor.

## How Does Alcohol Affect Your Body?

Alcohol has a negative effect on every major system in your body. Your body does not digest alcohol. It is absorbed directly from your stomach into your bloodstream. It is then carried throughout the body. How alcohol affects your body depends on several factors:

- the amount of alcohol you drink
- body size—the less you weigh, the more alcohol will affect you
- your sex—females are affected sooner than males

**328**    Unit 5   *Use and Misuse of Substances*

### Link to ▸▸▸

**Math**

Blood alcohol concentration (BAC) is a measure of the percent of alcohol in the blood. A 0.10 BAC means a person has 1 part of alcohol for every 1,000 parts of blood. A 0.20 BAC means there is 1 part of alcohol for every 500 parts of blood. On average, one drink will cause a 0.04 increase in a person's BAC.

- whether your stomach is empty or full—food slows the effects of alcohol
- your previous experiences with alcohol

After entering your body, alcohol combines with oxygen in a process called oxidation. This happens mainly in your liver. Your liver can oxidize only about $\frac{1}{2}$ ounce of alcohol per hour. Drinking any amount of alcohol affects your judgment, vision, reaction time, and muscle control. Your blood alcohol concentration (BAC) begins to rise. Some short-term effects of alcohol are impaired driving ability, dizziness, flushed skin, dulled senses and memory, and a relaxed feeling.

With just one drink, you can rapidly lose muscle coordination and may not be able to perform simple activities such as walking a straight line. You cannot speak or think as clearly and your hearing and vision become less sharp. When you are affected physically in these ways, serious accidents such as car crashes can happen.

Long-term use of alcohol can cause serious diseases such as cancer of the liver, stomach, colon, and mouth. It can also damage the brain, pancreas, and kidneys. Long-term alcohol use can result in high blood pressure, heart attacks, strokes, and diseases of the liver. These health problems have costs to society.

### What Is Fetal Alcohol Syndrome?

Drinking alcohol during pregnancy can cause fetal alcohol syndrome (FAS). Babies who have FAS are born with birth defects. These include unusual facial features and central nervous system disorders. In extreme cases, FAS can cause prenatal death—the baby dies before it is born. Babies who have FAS have difficulty learning and can have vision, hearing, and speech problems. Problems affecting the heart, kidneys, and bones also can occur. Older children with FAS have a greater chance of having problems in school. They are also more likely to be involved in crime.

FAS is a leading cause of birth defects. It is a permanent condition. However, it is entirely preventable. Only babies born to women who drink alcohol have FAS.

 **Warm-Up Activity**

Have students list the brand names of as many alcoholic beverages—beer, wine coolers, hard liquor—as they can. Point out that alcohol is so established in our culture that even students who will not be able to buy these products legally until they are 21 years old are familiar with many brand names.

**Teaching the Lesson**

Some students may think that alcohol can't be "so bad" because it is legal, while other drugs are not. To show them that alcohol is just as dangerous as drugs, you can emphasize some of the facts in sections titled How Does Alcohol Affect Your Body? and How Does Alcohol Affect Driving?

### Link to

**Math.** Have a volunteer read aloud the Link to Math feature on page 329. To help students understand blood alcohol concentration, you might want to write on the chalkboard:

0.10 BAC ➞ 1/1,000

0.20 BAC ➞ 2/1,000 = 1/500

The calculations for one drink are:

0.04 BAC ➞ 0.4/1,000 = 1/2500

### PRONUNCIATION GUIDE

Use the pronunciation shown here to help students pronounce difficult words in the lesson. Refer to the pronunciation key that appears in the Glossary at the back of the Teacher's Edition for the sounds of the symbols.

concentration (kon sən trā´ shən)

oxidation (ok sə dā´ shən)

### 3 Reinforce and Extend

#### ELL/ESL Strategy

**Language Objective:**
*To speak colloquially*

Describe a situation in which a teenager who has been drinking alcohol wants to drive and invites some friends to be passengers in his or her car. Have students work in groups and role play a scene in which they not only refuse to ride in the car with their friend but also find a way to keep their friend from driving. Include students learning English and students who are proficient in English in each group. Have groups write scripts for their skits and practice before they perform for the class.

#### Health Myth

Tell students that most American beers are about 4 percent alcohol by mass, which is roughly 5 percent alcohol by volume. So a 12-ounce can of beer contains 0.6 ounces of alcohol.

$12 \times 0.05 = 0.6$ ounces of alcohol

Wine is roughly 15 percent alcohol by volume, so a 4-ounce glass of wine contains 0.6 ounces of alcohol.

$4 \times 0.15 = 0.6$ ounces of alcohol

Whiskey or other hard liquor that is labeled "80 proof" is 40 percent alcohol by volume. A 1.5-ounce shot of this liquor contains 0.6 ounces of alcohol.

$1.5 \times 0.4 = 0.6$ ounces of alcohol

Any hard liquor that is more than 80 proof will contain more alcohol.

---

**Intoxication**
*Excitement or stimulation caused by use of a chemical material*

**Alcoholism**
*A dependence on alcohol and lack of control in drinking*

**Binge drink**
*To drink alcohol heavily, then stop for a time, then drink heavily again*

#### Health Myth

**Myth:** Beer contains less alcohol than does liquor such as whiskey.

**Fact:** The alcohol in beer is the same as in wine and liquor. The same amount of alcohol is found in 12 ounces of beer, 5 ounces of wine, or one shot ($1\frac{1}{2}$ ounces) of whiskey. Although there is a different amount of liquid in each of these three drinks, the total amount of alcohol is the same.

### How Does Alcohol Affect Driving?

When a person's BAC rises to the point of **intoxication,** it can affect many things, including how you drive. Intoxication is excitement or stimulation caused by use of a chemical material. It means being drunk. In the United States, any person with a BAC of 0.08 is considered legally intoxicated. At this level, it is against the law to drive a car.

Alcohol affects your judgment, lessens your attention level, and causes you to get sleepy. Alcohol also cuts down your reaction time and makes it hard to judge your speed and distances. Drinking and driving can lead to death. In the United States, an average of one alcohol-related vehicle death occurs every 30 seconds. Among teenagers, alcohol-related accidents are the number one cause of death. A person with a BAC of 0.04 is twice as likely as a nondrinking driver to be in a fatal car crash. The chances rise to three times as likely for a person with a BAC of 0.08.

Driving drunk has many negative consequences, including paying heavy fines and losing a driver's license for 6 months to 1 year. People who are stopped for driving drunk may have to spend time in jail and be evaluated for **alcoholism.** Alcoholism is a dependence on alcohol and lack of control in drinking.

### When Is a Person an Alcoholic?

The main sign of alcoholism is that a person cannot control his or her drinking. Alcoholics are physically and mentally dependent on alcohol. They cannot manage stress without alcohol, and they cannot stop drinking each time they start. Some alcoholics drink on a daily basis. Others **binge drink,** or drink heavily and then stop for a while before drinking heavily again.

People who start drinking at an early age have a greater chance of becoming alcoholics. Among teens who use alcohol, about 38 percent said they had tried their first drink by eighth grade. Studies show that more than 43 percent of teenagers who start drinking before the age of 14 become alcoholics. If a member of a family is an alcoholic, other family members have an increased chance of getting the disease.

---

#### Learning Styles

**Body/Kinesthetic**

Divide students into groups and ask them to spontaneously role play a situation involving peer pressure and alcohol. Have one student portray a non-drinker while two or three other students try to pressure him or her into drinking. See how many different things the student can say or do to avoid drinking. *(Students might come up with verbal "come-backs" or simply leave the situation.)*

**In your state, what are two of the penalties for people convicted of drunk driving?**

People can recover from alcoholism if they stop drinking. Often medical and psychological help is needed for an alcoholic to recover from the disease.

*Lesson 3 Review* On a sheet of paper, write the answers to the following questions. Use complete sentences.

1. What are five physical effects of alcohol on the body?
2. Why should women who are pregnant not drink alcohol?
3. What is binge drinking?
4. **Critical Thinking** Do you know anyone who has either driven drunk or been a victim of a drunk driver? Describe what happened.
5. **Critical Thinking** Why is alcoholism thought to be a disease, not just a behavior problem?

### Decide for Yourself

Drinking alcohol can affect your grades, your friendships, and how well you do at sports. Drinking can cause vomiting, severe headaches, and death from accidents or alcohol poisoning. Teens who drink are more likely to have unprotected sex and engage in other high-risk behaviors.

Studies show that one out of every 30 high school students drinks daily. Four out of every 10 sixth-grade students feel pressure to drink alcohol. You must be age 21 to drink legally. If you are caught drinking before you are age 21, you can be fined, lose your driver's license, or be put in jail.

Here are some ways to avoid drinking:

- Become friends with people who choose not to drink.
- Get involved with school activities.
- Leave a party or event if you see people your age drinking.
- Say no when someone offers you an alcoholic beverage. Keep saying no even if you are being pressured to have a drink.
- Do not allow friends to drink and drive. Find a ride home for your friend with someone who has not been drinking.

On a sheet of paper, write the answers to the following questions. Use complete sentences.

1. How much pressure to use alcohol do you think teens experience? Explain your answer.
2. Why is it important to know how to say no to drinking alcohol before you go to a party?
3. What law do you think should be passed that might help stop underage drinking?

*Recognizing Medicines and Drugs* Chapter 14 **331**

### Decide for Yourself

Have students read the Decide for Yourself feature on page 331. Lead a discussion about things students can say to deflect peer pressure when someone is trying to persuade them to drink. Remind students that the best way to resist peer pressure to drink is simply not to go where you know there will be alcohol.

**Decide for Yourself Answers**

**1.** Sample answer: Teens feel a lot of pressure to drink alcohol because young people in movies and on television often drink alcohol. **2.** Sample answer: Practicing saying no builds self-confidence and will help me avoid the pressure to drink at parties. **3.** Sample answer: I think there should be a law stating that anyone who is guilty of underage drinking should have to talk with someone who was seriously hurt because a drunk driver caused an automobile accident.

### Lesson 3 Review Answers

**1.** Sample answer: Alcohol slows down the central nervous system; it affects mental processes; it can damage organs; it can cause cancer of the liver and high blood pressure. **2.** Drinking alcohol can cause fetal alcohol syndrome (FAS). **3.** Sample answer: Binge drinking is drinking a lot in a short time, not drinking for a while, and then drinking a lot again. **4.** Sample answer: I knew a victim of a drunk driver. One of my teachers had a son who was killed by a drunk driver on New Year's Eve. **5.** Sample answer: Alcoholism is considered a disease because a person needs medical and psychological help to quit drinking.

### Portfolio Assessment

Sample items include:

- Edited ads from Vocabulary
- Skit scripts from ELL/ESL Strategy
- Lesson 3 Review answers

Activity 48

Workbook Activity 48

*Recog. Medicines/Drugs* Chapter 14 **331**

## Lesson at a Glance

### Chapter 14 Lesson 4

**Overview** This lesson describes the effects of stimulants, depressants, narcotics, and hallucinogens.

### Objectives

- To list the effects of stimulants
- To list the effects of depressants
- To list the effects of narcotics
- To list the effects of hallucinogens

**Student Pages** 332–337

**Teacher's Resource Library**

Activity 49

Alternative Activity 49

Workbook Activity 49

### Vocabulary

| | |
|---|---|
| amphetamine | methamphetamine |
| barbiturate | narcotic |
| cocaine | overdose |
| crack | PCP |
| ecstasy | sedative-hypnotic |
| flashback | drug |
| hallucinogen | tolerance |
| heroin | tranquilizer |
| LSD | |

Have students read the words and definitions. Then have them begin a chart with each of the types of drugs in a separate column. Tell students to add information to the chart as they go through the lesson. They can use the chart as a study tool at the end of the chapter and as an information sheet to take home.

### Background Information

Cocaine was legal in the United States until 1914, when the Harrison Act banned the importation and non-medical use of cocaine. It also imposed the same criminal penalties for cocaine users as for opium, morphine, and heroin users. This law, along with the emergence of cheaper, legal substances such as amphetamines, made cocaine scarce in the United States. However, use began to rise again in the 1960s. In 1970, Congress classified it as a Schedule II substance. A Schedule II substance has an accepted medical use with severe restrictions, a high potential for abuse, and a potential for severe psychological or physical dependence.

**332** Unit 5 *Use & Misuse of Substances*

---

## Lesson 4: Stimulants, Depressants, Narcotics, and Hallucinogens

**Objectives**

After reading this lesson, you should be able to
- list the effects of stimulants
- list the effects of depressants
- list the effects of narcotics
- list the effects of hallucinogens

Stimulants, depressants, narcotics, and hallucinogens are psychoactive drugs. Each one makes a person behave or think differently. Some psychoactive drugs can be used as prescription medicines under a doctor's direction. Buying these drugs without a prescription is against the law. Some psychoactive drugs have no medical use and are made and bought illegally. Buying or selling these drugs is a felony, a serious crime. If a person is found guilty of a felony, he or she can spend a long time in prison. In some states, people found guilty of a felony lose the right to vote.

### What Are the Effects of Stimulants?

Stimulants are one kind of psychoactive drug. Stimulants speed up the body's central nervous system. They speed up breathing and heart rate, and they raise blood pressure. People can become dizzy and not want to eat. People also may have headaches, blurred vision, sweating, and sleeplessness. In large amounts, stimulants can make the heart beat irregularly and cause other problems. Breathing in or injecting stimulants can raise blood pressure so high that a person has a stroke or high fever. People could have heart failure and die. People can also become physically dependent on stimulants.

**Figure 14.4.1** *People who are dependent on stimulants can feel tired and depressed when the drugs wear off.*

Stimulants change the way the mind works. They provide a feeling of energy and fool people into thinking they are not tired. These feelings can cause someone to become psychologically and physically dependent on stimulants. The dependence begins when the effect of the drug wears off and the person feels tired and depressed. Then the person wants the drug to get back the same feeling of energy again.

The nicotine in tobacco is a stimulant. Descriptions of two other stimulants follow.

**Amphetamine**
A central nervous system stimulant

**Methamphetamine**
A highly addictive, illegal amphetamine

**Cocaine**
A highly addictive illegal stimulant

### Amphetamines

**Amphetamines** are synthetic, or human-made, stimulants. This means they are made from chemicals in a laboratory. These drugs speed up heart rate and breathing. They will make a person anxious and not want to eat or sleep.

Doctors do not prescribe amphetamines as much as they once did. Doctors may still prescribe them for certain diseases.

Illegal amphetamines are used to lose weight, stay awake and alert, or do better in sports. They are also used to cancel out the effects of depressants. *Speed, ice, crank,* and *crystal* are a few of the names used for an amphetamine called **methamphetamine** or *meth*. Methamphetamine is an addictive amphetamine. A person can become hooked on meth the first time he or she uses it. Meth can cause liver, kidney, lung, and brain damage. It can also cause a stroke or heart failure, which can cause death.

### Cocaine

**Cocaine** is a stimulant that comes from the leaves of the coca plant. It is made into a white powder. This highly addictive drug is now illegal, but at one time it was used for medical purposes. Like other stimulants, cocaine gives a person a lot of energy. The feeling of energy is actually increased heart rate, blood pressure, and breathing. In large amounts, it may cause shaking, seizures, and heart and respiratory failure. These problems can be life threatening. Dependence on cocaine happens quickly—sometimes the first time it is used.

Cocaine dependence can greatly harm physical and mental health. Usually, cocaine is inhaled, or snorted, through the nose. It is absorbed through the mucous membranes and enters the bloodstream. The drug tightens the blood vessels in the nose, making it feel dry. Constant use can wreck the wall that separates the nasal passages. People who use cocaine all the time stop eating properly. They often get malnutrition. If a user has a heart problem, cocaine will make it worse. It also can raise the chances of getting a heart attack. People who inject cocaine risk getting AIDS if they share needles with other cocaine users.

### Warm-Up Activity

Ask students to discuss ways in which psychoactive drugs would impair their capacity to perform in school and to think and act responsibly.

### Teaching the Lesson

As the class works through the lesson, put the information it contains on the chalkboard in the form of a table. The table should have four columns, labeled Stimulants, Depressants, Narcotics, and Hallucinogens. In each column, you can list the various drugs that fall into that category, as well as the effects of each of them.

You can use this table to help students construct their own table (see the Vocabulary activity on page 332) and in the ELL/ESL Strategy described on page 334.

## 3  Reinforce and Extend

### ELL/ESL Strategy

**Language Objective:** *Summarizing Information*

Refer to the table you created in Teaching the Lesson. Review the information with students, clarifying the terms and descriptions. Instruct both students learning English and students who are proficient in English to use the information in the table to write a few complete sentences describing the purpose and dangers associated with each of the categories of drugs. For example: Barbiturates, minor tranquilizers, and major tranquilizers are all depressants. Depressants make a person calm. An overdose of depressants can cause death.

### Teacher Alert

If students ask, explain that the legal definition of *narcotic* varies from place to place and instance to instance. A good general-purpose definition is given in the vocabulary box on page 335. A useful, more precise definition might be, "a drug with addictive potential that reduces pain, alters mood and behaviors, usually induces sleep, and in excessive doses can cause stupor, coma, or death."

---

**Tolerance**
*The need for more of a drug to get the effect that once occurred with less of it*

**Crack**
*A form of cocaine that is smoked*

**Overdose**
*Too much of a drug*

Cocaine also affects mental health. Using cocaine gives a false feeling of confidence for the first few minutes. That feeling is often followed by depression. To keep from getting depressed, people will use more and more of the drug to get the same feeling of confidence. When that happens, people quickly build up **tolerance** to cocaine. Tolerance occurs when larger amounts of a drug are needed to get the same effect a person used to get with less drugs. Cocaine affects people mentally, too. Some people say they have felt as if insects were crawling under their skin. Others get confused and have hallucinations, or imagine things or events that are not really there.

**Crack** is a form of cocaine that is smoked. It is made when cocaine is mixed with baking soda to form lumps that look like rocks. Sometimes, people mix the rocks with ether and then set fire to them so the drug can be smoked. This is called freebasing. A big danger of freebasing is setting off an explosion and uncontrolled fire.

Like other forms of cocaine, the effects of crack last only a few minutes. Like other forms, too, people can become extremely addicted to it. The same health problems from using cocaine are even worse with crack.

### What Are the Effects of Depressants?

Depressants are a second kind of psychoactive drug. They slow down the central nervous system. Depressants sedate, or calm, a person. They lower blood pressure, heart rate, and metabolic rate. Using large amounts of depressants can make a person sleepy, not think clearly, lose muscle control, and nauseous. An **overdose,** or too much, of a depressant may give a person heart, lung, or kidney failure. It can even put people in a coma and kill them. Depressants can create both physical and psychological dependence.

If people use alcohol and other depressants together, they are taking great chances with their health. This combination can severely depress the central nervous system, causing death.

**Sedative-hypnotic drug**
*A prescribed depressant that lessens anxiety or helps a person sleep*

**Barbiturate**
*A category of sedative-hypnotic drugs*

**Tranquilizer**
*A category of sedative-hypnotic drugs*

**Narcotic**
*A pain-killing drug made from the opium poppy and in laboratories*

When a woman uses depressants while she is pregnant, her baby may be born dependent on the drugs. At birth, the baby also may have signs of withdrawal from the depressants. Also, babies can be born with birth defects and have long-term behavioral problems.

Some people use depressants to handle stress, but that is very dangerous. Tolerance to depressants builds up quickly. Within a few weeks of regular use, depressants lose their effectiveness and no longer work. By that time, people may be dependent on the drugs and need more of them. People may feel worse than when they started using the depressants. People who continue to use more depressants may die, depending on the amount of these drugs and the way they are used.

**Sedative-hypnotic drugs** are depressants prescribed by a doctor to lessen anxiety (sedatives) or help a person sleep (hypnotics). There are three categories of sedative-hypnotic drugs—**barbiturates,** minor **tranquilizers,** and major tranquilizers.

### What Are the Effects of Narcotics?

**Narcotics** are a third kind of psychoactive drug. Used as pain relievers, narcotics are made from the opium poppy and in laboratories. Some common medicines that are narcotics are morphine and codeine.

Morphine is given to help people with the pain that they get after an operation. It is also given to patients in the painful later stages of cancer. Codeine is used to help with pain and to control coughing.

Narcotics act as strong depressants on the central nervous system. At first, narcotics produce a feeling of well-being. After that, they cause sleepiness, nausea, and sometimes vomiting. An overdose of narcotics makes the breathing shallow, causes cold, damp skin and seizures. It can result in a coma and even death. Usually, a strong physical dependence and tolerance happen with narcotics. The symptoms of withdrawal are so painful that people continue to use narcotics so they do not have to go through withdrawal.

### TEACHER ALERT

Tell students that withdrawal symptoms for people who are abusing depressants—especially sedative-hypnotic drugs—include anxiety, sweating, restlessness, agitation, and hallucinations. Ask students to imagine what effect those symptoms might have on a newborn or premature baby.

### LEARNING STYLES

**Auditory/Verbal**
Divide students into groups. Ask groups to write a press release about a celebrity who died after abusing cocaine. Make sure they include details about how cocaine affects the body. Then have a volunteer from each group hold a press conference and read the press release. Others in each group can act as reporters by taking notes about the press release. Invite the reporters to ask the speaker questions and to read their notes aloud to check for accuracy.

## LEARNING STYLES

### Visual/Spatial
Have students present the information in this lesson in the form of posters or murals. The visual design should be more creative than simply a table. Links may be delineated between types of drugs, types of effects, or something else entirely. Encourage students to use photographs, their own drawings, or other graphics in their posters and murals.

## PRONUNCIATION GUIDE

Use the pronunciation shown here to help students pronounce difficult words in the lesson. Refer to the pronunciation key that appears in the Glossary at the back of the Teacher's Edition for the sounds of the symbols.

metabolism (mə tab´ ə liz əm)

---

**Heroin**
*An illegal narcotic made from morphine*

**Hallucinogen**
*A psychoactive drug that confuses the central nervous system and makes a person see, hear, feel, and experience things that are not real*

**Flashback**
*Experiencing the effects of a drug without using it again*

**Ecstasy**
*An illegal, psychoactive drug*

How can you use peer pressure in a good way to keep your friends from using drugs?

---

**Heroin,** which is illegal to make or use in the United States, is a narcotic made from morphine. Babies of mothers who use heroin may be born early or dead. Babies who are born addicted to heroin may go through severe withdrawal pain. People who inject heroin with a used or dirty needle can get blood poisoning or HIV. A heroin overdose can cause death.

## What Are the Effects of Hallucinogens?

**Hallucinogens** are a fourth kind of psychoactive drug. These drugs confuse the central nervous system. They change the way the brain handles information from the sense organs, including information about sight, hearing, smell, and touch. Because they change thoughts and moods, hallucinogens make things appear to be what they are not. Physically, they can cause the pupils to dilate and can raise body temperature, blood pressure, and heart rate. A person who uses hallucinogens may feel weak and sick.

The mental dangers of hallucinogens are worse than their physical effects. People may feel anxious, confused, or terrified, and may get panic attacks. Because hallucinogens twist people's thinking, people may do something that hurts either themselves or someone else.

Hallucinogens can cause permanent brain damage. For example, users may have **flashbacks.** That means experiencing the effects of the drug without using it again. Because the drug stays in a person's body, flashbacks can happen at any time for the rest of a person's life.

Hallucinogens are illegal and have no medical use. Some hallucinogens are human-made drugs known as street drugs. These dangerous drugs can leave poisonous materials in the brain and may damage the way it works.

**Ecstasy** is a psychoactive drug much like methamphetamine. It is also much like a hallucinogen. Ecstasy changes a person's mood and metabolism. It is addictive. People using ecstasy get the same harmful effects on their body as they would with cocaine or an amphetamine.

**LSD**
A human-made hallucinogen

**PCP**
A human-made hallucinogen

**LSD,** perhaps the best-known human-made hallucinogen, was originally used in research to treat mental illnesses. LSD did not cure any mental problems, however, but often caused additional ones. LSD is both illegal and dangerous.

**PCP** may be the most dangerous human-made hallucinogen. This is because its effects are different for everyone. It may give one person memory problems, another person depression, and a third person violent behavior. A person who takes a large dose of PCP may have seizures, a coma, and heart and lung failure.

*Lesson 4 Review* On a sheet of paper, write the letter of the answer that best completes each sentence.

1. A drug that speeds up the central nervous system is called a(n) _____.
   A depressant
   B overdose
   C stimulant
   D narcotic

2. A person who uses cocaine quickly builds up a(n) _____ to it.
   A tolerance
   B intoxication
   C withdrawal
   D immunity

3. Narcotics act as a strong _____ on the central nervous system.
   A tolerance
   B withdrawal
   C stimulant
   D depressant

On a sheet of paper, write the answers to the following questions. Use complete sentences.

4. **Critical Thinking** Why are stimulants, such as amphetamines, very harmful to use?

5. **Critical Thinking** Why are the mental effects of hallucinogens especially dangerous?

*Recognizing Medicines and Drugs* Chapter 14 337

## Lesson 4 Review Answers

1. C 2. A 3. D 4. Sample answer: Stimulants are psychoactive drugs. They speed up the central nervous system, causing increased respiratory and heart rates. Stimulants can cause blurred vision, sleeplessness, or a stroke. Some stimulants are highly addictive.
5. Sample answer: Hallucinogens can cause confusion, terror, and panic attacks. Because they distort a person's thinking, physical harm may result. Hallucinogens can cause a person to see something that does not exist. People who take hallucinogens may experience flashbacks at any time in their lives.

## Portfolio Assessment

Sample items include:

- Charts from Vocabulary
- Sentences from ELL/ESL Strategy
- Press releases from Learning Styles: Auditory/Verbal
- Poster or mural from Learning Styles: Visual/Spatial
- Lesson 4 Review answers

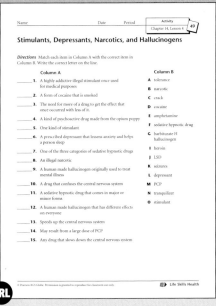
Activity 49

Workbook Activity 49

# Lesson at a Glance

## Chapter 14 Lesson 5

**Overview** This lesson describes the effects of other dangerous drugs, including marijuana, inhalants, designer and look-alike drugs, and anabolic steroids.

## Objectives

- To list five effects of marijuana use
- To explain how inhalants affect the body
- To explain how anabolic steroids are used both medically and illegally

**Student Pages** 338–340

**Teacher's Resource Library**

Activity 50

Alternative Activity 50

Workbook Activity 50

## Vocabulary

| | |
|---|---|
| anabolic steroid | look-alike drug |
| designer drug | marijuana |
| inhalant | THC |

Have students add these vocabulary words to the chart they made for the words in the previous lesson. They may have to create another chart on a separate sheet of paper. Tell them to take notes about each drug under its header.

## Background Information

Scientists first developed steroids in the 1930s to help treat men who were unable to produce enough of the hormone testosterone. During World War II malnourished soldiers were given this artificial testosterone to gain weight. After the war, Olympic and professional athletes and even college and high school athletes began to use steroids. In 1975, the International Olympic Committee banned the use of steroids in Olympic competition. In 2000, the Anabolic Steroid Enforcement Act of 1990 placed certain anabolic steroids on Schedule III of the Controlled Substances Act. (Until then, steroids had been unscheduled and controlled only by state laws.) Today, surveys show that adolescent use of steroids is on the rise and that a great number of adults are actively using them, too.

---

## Lesson 5 Other Dangerous Drugs

**Objectives**

*After reading this lesson, you should be able to*

- list five effects of marijuana use
- explain how inhalants affect the body
- explain how anabolic steroids are used both medically and illegally

Some drugs are dangerous to use because of their effects on your body and your mind. Many of these drugs are also illegal.

### What Is Marijuana?

**Marijuana** is a drug that comes from the hemp plant, called *Cannabis sativa*. It is illegal to grow or use marijuana. Because hemp leaves and buds have an intoxicating effect, some people smoke, drink, or eat them. Marijuana contains more than 400 chemicals, one of which is called **THC**. THC has a strong psychoactive effect that can make a person dependent on it.

THC is harmful because it may remain in the body's tissues for up to 30 days. It stays mainly in the brain, kidneys, liver, lungs, and reproductive organs. Using marijuana for a long time may harm all of these organs. Also, THC harms proper growth and division of cells.

Marijuana affects each person differently. It can act like a central nervous system depressant, a pain reliever, or a hallucinogen. Marijuana can lower body temperature and raise blood pressure and heart and pulse rates. It makes the eyes red and makes a person hungry. Driving and movement skills lessen after a person uses marijuana.

Marijuana affects how a person functions mentally by weakening memory, concentration, and judgment. As a result, a person's schoolwork may suffer. People can experience mood changes with the first use of marijuana. People who use marijuana may not care about things they once did.

Studies show that using marijuana has serious health problems, including lung disease. For example, marijuana smokers breathe in the smoke deeply and try to hold it in their lungs for a long time. This can damage the lungs. In addition, marijuana smoke can give a person cancer and may make the immune system less effective when fighting disease and infection.

**Marijuana**
*A drug from the hemp plant,* Cannabis sativa

**THC**
*The psychoactive chemical in marijuana*

**Inhalant**
*A chemical that is breathed in*

**Designer drug**
*A drug with a slightly different chemical makeup from a legal drug*

**Look-alike drug**
*A drug made from legal chemicals to look like a common illegal drug*

## What Are Inhalants?

**Inhalants** are chemicals that people breathe into their bodies. Some inhalants may be used to help with pain. Other inhalants are used for a hallucinogenic effect. Chemicals with a hallucinogenic effect are generally found in products that are not used as drugs. Some examples of these products are glues, gasoline, spray paints, markers, hair spray, nail polish, air freshener, and other products. Using an inhalant for a different purpose than it is meant for is considered illegal drug use.

Using inhalants can harm a person's mind and body by depressing the central nervous system. People who use inhalants may have slurred speech, poor judgment, confusion, dizziness, headaches, and hallucinations. They may even pass out. Some physical effects of using inhalants are sneezing, coughing, sniffing, and nosebleeds. Inhalants can cause more serious damage, including permanent brain damage. People who use a lot of inhalants can die immediately because they suffocate, or stop breathing.

## What Are Designer and Look-Alike Drugs?

A **designer drug's** chemical makeup is like the chemical makeup of a legal drug. Drug dealers, however, make designer drugs slightly different from the original drug. They are usually many times stronger. Many designer drugs act like narcotics. An example of a designer drug is china white, which imitates heroin.

**Look-alike drugs** are made from legal chemicals to look like other illegal drugs. They may contain anything. For example, people who think they are buying an amphetamine may be getting a depressant.

Making and using all designer and look-alike drugs is illegal. Because people really do not know what is in these drugs, someone using these drugs could easily overdose. Even small doses can raise blood pressure and heart rate. These drugs can act like poison in the body. People who mix look-alike drugs with other drugs have a good chance of dying.

*Recognizing Medicines and Drugs   Chapter 14   339*

### Warm-Up Activity

Have students write what they know about each of the drugs discussed in this lesson. Tell them to save their papers and compare their thoughts to the information they learn as they go through this lesson.

### Teaching the Lesson

Explain to students that there is one more reason why illegal drugs are dangerous. Because they are illegal, there is no supervision of their production, transportation, or sale. Drug dealers can and do dilute their drugs with almost anything—sometimes even with substances that are toxic.

### Reinforce and Extend

#### LEARNING STYLES

**Logical/Mathematical**
Ask students to imagine they have just inherited a small country. Their first duty is to develop the country's drug laws. Have students prepare a document that describes the laws pertaining to all the drugs identified in Chapter 14, as well as the penalties or consequences for each offense.

#### ELL/ESL STRATEGY

**Language Objective:** *To summarize subject material*

As you work through the lesson with the class, ask students learning English to create an outline. Tell them to include as much information from the lesson as possible. Students' finished outlines will have a similar form to the Chapter 14 Outline in the TRL, but will be much more complete. Ask students who are proficient in English to answer any questions ELL students might have as they create their outlines.

*Recog. Medicines/Drugs   Chapter 14   339*

# Lesson 5 Review Answers

1. THC is stored in the body's fat tissues for up to 30 days. 2. Inhaling and holding marijuana smoke is harmful because it can damage the lungs. 3. Sample answer: Inhalants can cause slurred speech, poor judgment, headaches, nosebleeds, and hallucinations. 4. Sample answer: Look-alike drugs may look like narcotics, but they could contain anything, including substances that could cause death instantly. 5. Sample answer: Doctors can prescribe anabolic steroids to help people produce more testosterone. Some athletes abuse anabolic steroids to improve their athletic performance. Some anabolic steroids are legal with a prescription, but others are illegal substances that have no medical use.

## Portfolio Assessment

Sample items include:
- Charts from Vocabulary
- Notes from Warm-Up Activity
- Document from Learning Styles: Logical/Mathematical
- Outlines from ELL/ESL Strategy
- Posters from Research and Write
- Presentation from Application: In the Community
- Lesson 5 Review answers

## Research and Write

Have students read and complete the Research and Write feature on page 340. You might want to ask the school librarian to select appropriate and useful books and magazines in advance and provide them to your class. Allow students to spend time in the library using the selected materials.

---

**Anabolic steroid**
A human-made drug that is like the natural male sex hormone, testosterone

**Research and Write**
Drugs can affect a person in many ways. Using print resources at a library, list the negative effects of alcohol, tobacco, and other drugs. Make a poster that presents the facts. Include this poster in a class presentation.

## What Are Anabolic Steroids?

**Anabolic steroids** are human-made drugs that are like the natural male sex hormone, testosterone. When used medically, anabolic steroids help people who do not make enough testosterone naturally. Major League Baseball is one of many organizations that has strict rules banning anabolic steroids. Athletes sometimes abuse anabolic steroids to do better at their sport. By abusing these drugs, athletes may get bigger and stronger.

Anabolic steroids can have both physical and mental effects on the body. They can cause high blood pressure, heart and kidney disease, or life-threatening liver cancer. If a person uses these drugs as an adolescent, the bones may stop growing and the person might not reach normal adult height. Males may get bald, have their testes shrink, and make less sperm. In females, breasts may shrink, facial hair may grow, the voice may deepen, and head hair may thin. Both males and females may never be able to have children after using steroids.

Although steroids generally are not used to change mood, they can have bad mental effects. In both sexes, steroids can cause violence. Other mental effects are anxiety, depression, sudden mood swings, and hallucinations.

*Lesson 5 Review* On a sheet of paper, write the answers to the following questions. Use complete sentences.

1. For how long does THC remain in the body after use and where is it stored?
2. Why is inhaling and holding marijuana in your lungs harmful?
3. What effect do inhalants have on the body?
4. **Critical Thinking** Why are look-alike drugs more dangerous than the drugs they copy?
5. **Critical Thinking** How can anabolic steroids be both helpful and harmful?

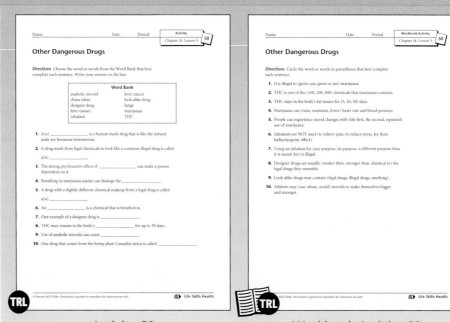

# Chapter 14 SUMMARY

- A drug is something other than food that changes the way the mind and body work. All medicines are drugs, but not all drugs are used for medical reasons.
- Medicines are bought with a doctor's prescription or over the counter without a prescription.
- Medicines are grouped according to their purpose.
- Body size, weight, age, and how your body works influence how a medicine affects you.
- Taking medicines as directed will help you avoid unwanted side effects and dependence.
- The U.S. government controls the safety of medicines.
- The nicotine in tobacco is an addictive stimulant.
- Nonsmokers who breathe in secondhand smoke can have the same health problems that smokers have.
- Alcohol is an addictive central nervous system depressant and psychoactive drug.
- Alcoholism is a disease that comes from dependence on alcohol.
- Using tobacco or alcohol can result in serious diseases involving the heart, lungs, and other body organs.
- Stimulants, depressants, narcotics, and hallucinogens are all addictive psychoactive drugs. Stimulants speed up the central nervous system, and depressants and narcotics slow it down. Hallucinogens confuse the central nervous system.
- Marijuana, inhalants, designer drugs, look-alike drugs, and anabolic steroids are dangerous and addictive drugs.

### Vocabulary

| | | |
|---|---|---|
| alcoholism, 330 | flashback, 336 | over-the-counter (OTC) medicine, 318 |
| amphetamine, 333 | hallucinogen, 336 | PCP, 337 |
| anabolic steroid, 340 | heroin, 336 | pharmacist, 318 |
| analgesic, 319 | inhalant, 339 | prescription, 318 |
| barbiturate, 335 | intoxication, 330 | psychoactive, 328 |
| binge drink, 330 | look-alike drug, 339 | sedative-hypnotic drug, 335 |
| cocaine, 333 | LSD, 337 | side effect, 318 |
| crack, 334 | marijuana, 338 | stimulant, 324 |
| dependent, 321 | medicine, 318 | THC, 338 |
| depressant, 328 | methamphetamine, 333 | tolerance, 334 |
| designer drug, 339 | narcotic, 335 | tranquilizer, 335 |
| ecstasy, 336 | nicotine, 324 | withdrawal, 326 |
| emphysema, 324 | overdose, 334 | |

*Recognizing Medicines and Drugs* Chapter 14 **341**

## Chapter 14 Summary

Have volunteers read aloud each Summary item on page 341. Ask volunteers to explain the meaning of each item. Direct students' attention to the Vocabulary box on the bottom of page 341. Have them read and review each term and its definition.

### ONLINE CONNECTION

Ohio State University's Web page that describes how to read the labels of OTC medications and how to properly use them is aimed at senior citizens, but most of the information applies to younger people as well: http://ohioline.osu.edu/ss-fact/0130.html

The U.S. Drug Enforcement Administration's Web site has current news articles, as well as links to articles on everything from the DEA's history to its mission to drugs and drug trafficking in general: www.dea.gov

### IN THE COMMUNITY

 Have groups of students brainstorm activities that would be appropriate for a community-wide drug and alcohol awareness week. Suggest that students contact community officials or ask for help from community volunteer organizations in choosing and planning appropriate activities. Students might contact a local newspaper to publicize their efforts. Then have students plan an informative presentation for community members about the drug and alcohol awareness week.

## Chapter 14 Review

Use the Chapter Review to prepare students for tests and to reteach content from the chapter.

## Chapter 14 Mastery Test

The Teacher's Resource Library includes two forms of the Chapter 14 Mastery Text. Each test addresses the chapter Goals for Learning. An optional third page of additional critical-thinking items is included for each test. The difficulty level of the two forms is equivalent.

## Review Answers
**Vocabulary Review**

1. pharmacist 2. analgesic 3. depressants
4. nicotine 5. emphysema 6. intoxicated
7. alcoholism 8. overdose 9. inhalants
10. withdrawal 11. flashbacks 12. THC
13. anabolic steroids

### TEACHER ALERT

In the Chapter Review, the Vocabulary Review activity includes a sample of the chapter's vocabulary terms. The activity will help determine students' understanding of key vocabulary terms and concepts presented in the chapter. Other vocabulary terms used in the chapter are listed below.

amphetamine
barbiturate
binge drink
cocaine
crack
dependent
designer drug
ecstasy
hallucinogen
heroin
look-alike drug
LSD
marijuana
medicine
methamphetamine
narcotic
over-the-counter medicine
PCP
prescription
psychoactive
sedative-hypnotic drug
side effect
stimulant
tolerance
tranquilizer

342  Unit 5  Use & Misuse of Substances

---

# Chapter 14 R E V I E W

## Vocabulary Review

On a sheet of paper, write the word or phrase from the Word Bank that best completes each sentence.

**Word Bank**
alcoholism
anabolic steroids
analgesic
depressants
emphysema
flashbacks
inhalants
intoxicated
nicotine
overdose
pharmacist
THC
withdrawal

1. Only a(n) _____ can fill a prescription written by a doctor.
2. A medicine that helps get rid of pain is a(n) _____.
3. Drugs that slow down the central nervous system are _____.
4. All forms of tobacco have _____ in them.
5. A serious disease that affects the lungs and can come from smoking is _____.
6. A person who has a blood alcohol concentration of 0.08 is legally _____.
7. People who cannot control their drinking have _____.
8. People can _____ on a drug, which means they have taken too much.
9. Glue, markers, gasoline, and hair spray are examples of _____.
10. Babies born addicted to heroin go through _____ when they are born.
11. People who use hallucinogens may have _____ for the rest of their lives.
12. The _____ in marijuana remains in the brain and other body organs for up to 30 days after use.
13. Adolescents who use _____ may stop growing early.

342  Unit 5  Use and Misuse of Substances

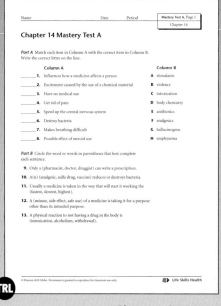

Chapter 14 Mastery Test A, pages 1–3

## Concept Review

On a sheet of paper, write the letter of the answer that best completes each sentence.

**14.** Three purposes of medicine include getting rid of pain, keeping disease away, and _____.
  **A** replacing body chemicals   **C** causing withdrawal
  **B** causing side effects       **D** creating mood swings

**15.** Smokeless tobacco is dangerous because it can lead to _____.
  **A** kidney problems            **C** a slower heart rate
  **B** carbon monoxide poisoning  **D** cancer

**16.** Alcohol and tobacco are both psychologically and physically _____.
  **A** stimulants                 **C** depressants
  **B** addictive                  **D** psychoactive

**17.** Anabolic steroids are human-made drugs that are like _____.
  **A** depressants                **C** hallucinogens
  **B** narcotics                  **D** testosterone

## Critical Thinking

**18.** How do alcohol, tobacco, and other drugs affect the central nervous system?

**19.** Why do you think some people never misuse drugs and others become dependent on them?

**20.** Why do you think alcoholism or any drug dependence could be thought of as a family disease?

 **Test-Taking Tip** Read test questions carefully to pick out those questions that are asking for more than one answer.

*Recognizing Medicines and Drugs* Chapter 14 343

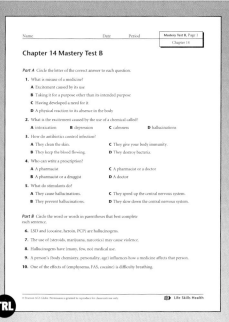

Chapter 14 Mastery Test B, pages 1–3

## Concept Review
14. A  15. D  16. B  17. D

## Critical Thinking

**18.** Sample answer: Tobacco (like all stimulants) speeds up the central nervous system. Alcohol (like all depressants) and narcotics slow down the central nervous system. Hallucinogens confuse the central nervous system. Using inhalants can harm a person by depressing the central nervous system. **19.** Sample answer: Some people misuse drugs because they think that drugs will make them more popular, better at school, or better at sports. Other people understand the extremely serious risks involved with drug use; therefore, these people never misuse drugs. Many people are not tempted to experiment with drugs because they want to do everything possible to have good physical, mental, and emotional health. **20.** Sample answer: Alcoholism or any drug dependence affects everyone in the family in some way. Because of this, it takes an effort by the entire family to help the person stop using drugs. If a member of a family is an alcoholic, other family members have an increased chance of getting the disease.

## ALTERNATIVE ASSESSMENT

Alternative Assessment items correlate to the student Goals for Learning at the beginning of this chapter.

- Have students orally explain several different types and purposes of medicines.
- Have students design a poster that shows how medicines are taken and possible problems with them.
- Have students write and perform a script for an instructional video describing some cautions to follow when taking medicines.
- Have students work in groups to write the script for a public service announcement describing the effects of using alcohol and tobacco.
- Have students write an encyclopedia article describing the effects of psychoactive and other dangerous drugs.

*Recog. Medicines/Drugs* Chapter 14 343

## Chapter at a Glance

### Chapter 15: Dealing with Drug Dependence
pages 344–361

### Lessons

1. Drug Dependence—
   The Problems
   pages 346–351

2. Drug Dependence—
   The Solutions
   pages 352–355

3. Avoiding Drug Use
   pages 356–358

**Chapter 15 Summary** page 359

**Chapter 15 Review** pages 360–361

**Audio CD**

**Skill Track** for
Life Skills Health

**Teacher's Resource Library** TRL

   Activities 51–53

   Alternative Activities 51–53

   Workbook Activities 51–53

   Self-Assessment 15

   Community Connection 15

   Chapter 15 Self-Study Guide

   Chapter 15 Outline

   Chapter 15 Mastery Tests A and B

(Answer Keys for the Teacher's Resource Library begin on page 559 of this Teacher's Edition.)

### Opener Activity

Have students brainstorm a list of drugs. Write each drug on the board as it is mentioned. If alcohol and tobacco are not on the list, ask if they are drugs. Elicit the response that both are drugs, although their use is legal. Ask students to discuss why teenagers start smoking in spite of the known health risks involved. Ask if they think health warnings in advertisements and on cigarette packages influence teenagers not to smoke.

344  Unit 5  *Use & Misuse of Substances*

## Self-Assessment

### Can you answer these questions?

☑ 1. How can friends and family members express concern about a person's drug use?

☑ 2. What is the cost to society of drug dependence?

☑ 3. How can you refuse when pressured to use drugs?

☑ 4. What are the signs of drug dependence?

☑ 5. How could you identify drug treatment programs in your community?

☑ 6. What activities can young people do instead of using drugs?

☑ 7. How does drug dependence by a family member affect other family members?

☑ 8. How does stress contribute to drug use?

☑ 9. How can you help friends avoid drug use?

☑ 10. What is the drug dependence recovery process?

Self-Assessment 15

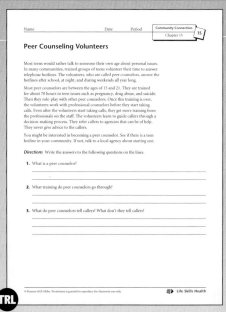

Community Connection 15

# Chapter 15
# Dealing with Drug Dependence

Everyone is faced with choices and problems in life. People who choose drugs usually are struggling with problems. They may see drug use as a quick solution. However, drug use never provides the solutions to life's problems. In fact, drug use adds to them.

Understanding the problems that drug dependence brings is important. Knowing about solutions for drug dependence is equally important. Perhaps most important is knowing about healthy choices, or alternatives, to drug use.

## Goals for Learning

- ◆ To explain physical and psychological drug dependence
- ◆ To identify the main pattern and signs of drug dependence
- ◆ To describe how drug dependence costs the family and society
- ◆ To explain the steps in recovery from drug dependence
- ◆ To identify resources for help overcoming drug dependence
- ◆ To use refusal skills when pressured to use drugs
- ◆ To describe healthy alternatives to drug use

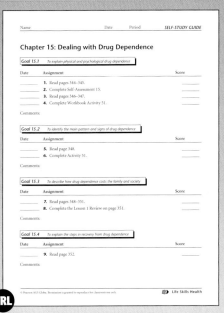

Chapter 15 Self-Study Guide, pages 1–2         Chapter 15 Outline, pages 1–2

## Introducing the Chapter

Have students read the introductory paragraphs on page 345. Ask students why someone who has problems might start using drugs. Have students brainstorm a list of situations that might be stressful for teenagers. Examples might include communication problems with family members or friends, academic pressures, social pressures, and financial pressures. Then ask students to discuss healthy ways to deal with such problems. Refer them to methods discussed in Chapters 2, 3, and 8.

Have students read the Goals for Learning. Explain that in this chapter students will learn about drug dependence, how it affects families and society, and how to recover from it, as well as strategies for dealing with peer pressure and healthy alternatives to drug use.

## Notes and Questions

Ask volunteers to read the notes and questions that appear in the margins throughout the chapter. Then discuss them with the class.

## Self-Assessment

Have students complete the Self-Assessment worksheet before and after reading the chapter. Before reading the chapter, have students fill in the "Before" column. Ask students to identify their goals for learning. To get ideas for setting goals, students might use the chapter introductory material on page 345, the checklist on page 344, or the questions on the Self-Assessment worksheet. Students can use the back of the worksheet if they need more space to write.

Collect the Self-Assessment worksheets and pass them out again at the end of the chapter. Have students fill in the "After" column. Ask them to identify at least four major points they have learned. Again, suggest they use the back of the worksheet if they need more space to write. You might want to collect and review the worksheets, but return them to students so they have a record of their goals and accomplishments.

# Lesson at a Glance

## Chapter 15 Lesson 1

**Overview** This lesson describes physical and psychological drug dependence, identifies patterns and signs of drug dependence, and explains how drug dependence affects individual families and society as a whole.

### Objectives

- To explain physical and psychological drug dependence
- To identify the main pattern and signs of drug dependence
- To recognize how drug dependence affects the family
- To describe how drug dependence costs society

**Student Pages** 346–351

**Teacher's Resource Library**
   Activity 51
   Alternative Activity 51
   Workbook Activity 51

## Vocabulary

**enabling**

After reading the vocabulary word and its definition, ask students to give several examples or descriptions of the term. For example, lending someone money to buy drugs would be considered enabling.

## Background Information

The United States has one of the highest rates of teenage drug abuse of any industrialized nation. More than half of high school seniors have used drugs—mostly marijuana. Most teenagers who use marijuana daily first tried it before the tenth grade. In 2002, more than 14 million Americans over the age of 12 used marijuana at least once a month. Each year from 1995-2001, there were an estimated 2.6 million new marijuana users. However, the prevalence of marijuana use is decreasing. A 2003 study of students in grades 8, 10, and 12 showed both a decrease in use of marijuana and an increase in perceived risk for regular marijuana use.

346  Unit 5  *Use & Misuse of Substances*

---

# Lesson 1  Drug Dependence—The Problems

### Objectives

*After reading this lesson, you should be able to*

◆ explain physical and psychological drug dependence
◆ identify the main pattern and signs of drug dependence
◆ recognize how drug dependence affects the family
◆ describe how drug dependence costs society

Dependence on alcohol and other drugs is a serious problem. It affects not only individuals but also families and society. Understanding more about this dependence and its costs is important.

## What Is Drug Dependence?

Drug dependence is a need for a drug that results from using the drug once in a while or for a long time. Other names for drug dependence are *addiction* and *substance abuse disorder*. Drug dependence can be physical, psychological, or both. Psychological drug dependence affects the mind and behavior.

### Physical Drug Dependence

Physical drug dependence happens when the body develops tolerance to a drug and goes into withdrawal when drug use is stopped. The body begins to develop tolerance as a person uses more of a drug. If people develop tolerance for one drug, they may develop tolerance for other drugs as well. When people who have developed tolerance suddenly stop taking a drug, they go into withdrawal. Withdrawal is the physical reaction to the sudden absence of a drug in the body. Generally, symptoms of withdrawal are the opposite of the effects of the drug. Suppose a person is dependent on alcohol or another depressant. If the person suddenly stops using the drug, the central nervous system speeds up. Suppose a person who is dependent on a stimulant, such as an amphetamine, suddenly stops taking it. The central nervous system slows down.

Heredity may influence whether people become physically dependent on drugs. For example, the risk for alcoholism runs in families. Research studies have proved that the tendency to become addicted to alcohol is genetic. However, this does not mean that people with these genes will always become addicted. For example, people who know that alcoholism runs in their families can avoid drinking alcohol.

346  Unit 5  *Use and Misuse of Substances*

The drugs that produce the strongest physical dependence are narcotics, barbiturates, and alcohol. Withdrawal from narcotics is extremely painful. A person in withdrawal from alcohol or barbiturates can die without medical care.

**Psychological Drug Dependence**
Psychological dependence on a drug happens when people believe they need the drug to feel good or avoid the pain of personal problems. The dependence usually begins when people have personal problems. People may have low self-esteem, for example, or get along poorly with their family members. When they use a drug, they notice that its effects lessen their emotional pain.

When the effect of the drug wears off, the emotional pain comes back. To avoid this pain, people use the drug again to escape their feelings. These people may then begin to have a mental desire to use the drug. Even if they stop using the drug, they may still feel a psychological need for it.

Generally, drugs that produce a psychological dependence are psychoactive drugs. Examples are alcohol, marijuana, cocaine, heroin, and amphetamines. These drugs change the way people think. They can also produce physical dependence.

A drug can result in just physical or just psychological dependence. Usually, however, both types result. The symptoms of dependence are different for each drug. All drugs that produce dependence have one thing in common. They create a need to use more of the drug to feel good or avoid pain.

> **Link to** ➢➢➢
> **Language Arts**
> Writer Langston Hughes wrote the poem "Junior Addict," published in 1961. He describes an African American boy who uses drugs perhaps because of how he feels African Americans have been treated in America. The boy may find strength and hope if he can feel pride in his heritage.

## 1 Warm-Up Activity

Have students work in groups to write skits about a teen who starts using an addictive drug when family problems arise. Skits should follow the teen through physical or psychological dependency. Students may want to perform their skits for the class. The skits will expose misconceptions that students may have about drug addictions. If so, be sure to correct them.

## 2 Teaching the Lesson

Ask students if they have ever seen a television show in which a character is addicted to a drug or is struggling to overcome an addiction. Also, many sitcoms have episodes devoted to drug abuse. Have students describe the behaviors drug-dependent characters exhibit and tell if they are accurate. Invite students to discuss how drug dependence would adversely affect a person's life. Have students read about physical and psychological drug dependence on pages 346 and 347.

## 3 Reinforce and Extend

### ELL/ESL Strategy

**Language Objective:**
*To understand different purposes of writing*

Ask students what a cause-effect purpose of writing is. Explain that a cause-effect paragraph clearly states how one action causes another or proves that a certain action will lead to a certain behavior. Ask them which paragraph on page 346 is a cause-effect paragraph. *(the third paragraph on the page)* Review words and phrases that can be used to express cause-effect relationships. Examples include: if/then, because, since, therefore, consequently, and as a result. Have both ELL students and students with a firm grasp of English write a cause-effect paragraph about the pattern and signs of drug dependence. (Information about the pattern and signs of drug dependence is on page 348.)

---

## Pronunciation Guide

Use the pronunciation shown here to help students pronounce difficult words in the lesson. Refer to the pronunciation key that appears in the Glossary at the back of the Teacher's Edition for the sounds of the symbols.

amphetamine (am fet´ ə mēn)

barbiturate (bär bich´ ər it)

psychological (sī kə loj´ ə kəl)

### Link to

**Language Arts.** Have students read the Link to Language Arts feature on page 347. Locate Hughes' poem in the library or access it on the Internet at http://sbacari.tripod.com/poetry/jraddict.htm. Read the poem to the class and have a class discussion. Do students think the poem is positive or negative? How does it make them feel? Why do they think Hughes wrote the poem?

## Health Myth

Have students read the Health Myth feature on page 348. Tell that methamphetamine also can be fatal after one use. It increases heart rate and blood pressure and can cause the cardiovascular system to collapse. It also can damage blood vessels in the brain, causing strokes.

## TEACHER ALERT

Discuss with students how a person's experiences as a young child might affect the decisions the person makes about drugs as a teenager. Challenge students to discuss how others can help a young child have positive experiences and develop strong self-esteem.

## LEARNING STYLES

### Visual/Spatial

Have students bring in newspaper and magazine advertisements for alcoholic beverages. Display the ads in the classroom and discuss the following questions about each ad: To whom is the ad appealing? What is the ad suggesting about the use of the alcoholic beverages? How realistic is the ad? What are consequences of using alcohol that the ad does not show?

## CAREER CONNECTION

Ask students to research and write a report about the educational requirements and job opportunities for pharmacists. Then have students interview a pharmacist to find out how customers can abuse legal drugs and what drugs are abused. Students should ask about the precautions and guidelines that are followed by the pharmacist to identify and prevent such abuse.

---

**Health Myth**

**Myth:** Drugs are not harmful the first time you use them.

**Fact:** Drugs such as cocaine and inhalants can kill you the first time you use them. Other drugs such as meth may cause addiction the first time you use them.

## What Are the Main Pattern and Signs of Drug Dependence?

The main pattern and signs of drug dependence seem to be the same for all drugs that cause dependence. The main pattern is that people's lives center on the drug. When people start using drugs, their behavior and personalities change. They spend their time thinking about the drug and how to get it. They spend more time using the drug and recovering from its effects. After they recover, the pattern starts again. Over time, the pattern is set. People become dependent on the drug.

Some signs of drug dependence appear in Table 15.1.1. People who have only one or two of the signs may not have a drug problem. If they show many of these signs, however, they may have a drug problem. Lesson 2 describes ways to get help.

| Table 15.1.1 Signs of Drug Dependence | |
|---|---|
| changes in how a person looks | reduced energy and ambition |
| major changes in behavior | loss of interest in favorite activities or hobbies |
| changes in choice of friends | borrowing money |
| sudden mood changes | trouble with the police |
| angry or aggressive behavior | drop in performance at school or work |
| loss of memory and concentration | many school or work absences and tardiness |
| lying, cheating, and stealing | becoming angry or resistant when discussing drug use |

### How Does Drug Dependence Affect the Family?

Being able to recognize signs of drug dependence may be especially important for family members. As people become more dependent on drugs, their behavior and personalities change. These changes greatly affect families.

**Enabling**
*Actions of a friend or family member that help a person continue to use drugs*

Family members cannot predict the behavior of the people who are drug-dependent. They sometimes try to protect the dependent family members. For example, family members may tell other people that someone in the family is sick when the person actually is drunk. They may make excuses for a drug user's actions. They may lie for the person or provide money to buy drugs. These actions actually help the person to continue using drugs. This is called **enabling.** The family keeps the drug user from having to face the effects of his or her behavior.

Enabling usually happens in an attempt to protect the family image. In some cases, a family member is afraid of the drug user. However, enabling winds up hurting everyone. The person using drugs continues getting deeper into drug dependence.

Enabling also delays getting help for the drug user. As the problem goes on, the emotional well-being of family members is affected. Children growing up in families with drug dependence often feel stress and are abused.

### How Does Drug Dependence Cost Society?

Drug dependence affects not only the family but also society. Alcohol and other drug dependence costs the United States billions of dollars each year. Expenses include lost work time, lost jobs, car crashes, crime, and health problems. As a result, everyone must pay higher taxes. They also must pay for higher insurance and health care costs.

Drug dependence costs more than money—it costs lives. Car crashes in which alcohol is involved are the leading cause of death among teens.

**Figure 15.1.1** *Sudden mood changes or changes in behavior can be a sign of drug dependence.*

## HEALTH JOURNAL

Have students describe in their journals how they would respond to an alcoholic friend or family member who asks them to tell a teacher or an employer that he or she is ill with the flu when actually the person is drunk. *(Students might say that they should refuse to lie about the alcoholic's condition.)*

## AT HOME

Discuss how talking about matters of personal importance with family members might help a person stay emotionally healthy. Have students write down four things that are important to them. The items might be in the realm of social interaction, personal achievement, values, future goals, or another area. Suggest that students share this information with family members through a family discussion.

## LEARNING STYLES

**Interpersonal/Group Learning**

Present the following statement: "What I do with my body is nobody else's business. I should be free to use drugs if that is what I choose to do." Divide the class into small groups. Have each group develop an argument to refute the statement. Then call on volunteers to share their arguments and discuss how they would respond to a friend who made such a statement.

### Research and Write

Have students read and conduct the Research and Write feature on page 350. Suggest that they begin their research for the necessary information on the Internet. If necessary, show students how to use a search engine and keywords to find helpful Web sites. Have students continue their research by visiting community organizations.

### Link to

**Math.** Have students read the Link to Math feature on page 350. Ask them what they think this money covers. *(substance-abuse treatments, prevention costs, other healthcare costs, costs associated with reduced job productivity and lost earnings, and other costs to society such as crime and social welfare)* Government agencies bear much of these costs.

### IN THE ENVIRONMENT

Ask students to discuss the various ways they have seen smokers dispose of cigarette butts. Some ways might include emptying car ash trays on the ground, stepping on cigarettes on the ground, and throwing cigarettes into bodies of water. Ask how disposal of cigarettes and cigarette butts has a negative effect on the environment *(Incorrectly disposing of cigarette butts is littering, and cigarette butts that are incorrectly disposed of might start fires.)*

### GLOBAL CONNECTION

Have students research and write reports on how other countries deal with drug problems. Problems include the production of drugs in a country, the shipping of those drugs to another country, drug dealing with a country, drug addiction and treatment in a country, and other healthcare and societal costs. Students should evaluate how effective those approaches are and whether they would be effective in the United States.

### Link to

**Math**

The use of alcohol, tobacco, and other drugs cost American society about $432 billion in 2001. That is about $1,535 for every person.

Drug dependence costs not only present lives but also future lives. Pregnant women who use drugs run the risk of giving birth to drug-dependent babies. Women who drink alcohol when they are pregnant may have babies who are smaller than normal. These babies also may have defects that cannot be reversed. They may be mentally slow or have deformities. Mothers who use cocaine during pregnancy may give birth to babies who are dependent on cocaine. These babies do not like to be held. They fail to respond to other humans. They are likely to have physical and behavior problems throughout their lives.

Drug dependence can increase the risk of acquiring AIDS and other sexually transmitted diseases. Young people who use drugs are more likely to have sex than people who do not use drugs. Many drug users report using alcohol or other drugs when they have sex. In addition, people who inject drugs and share needles have a higher risk for acquiring AIDS.

Drug dependence can have criminal consequences. Being convicted for selling or using drugs can have serious and long-term effects. A person who has been convicted of selling or using drugs may be sent to prison and pay large fines. A convicted drug user cannot get loans or grants for college. In some states, a person who has been convicted of drug use may lose the right to vote or hold office. The person may not be able to train as a nurse, pharmacist, teacher, or other professional. Getting employment may be difficult.

### Research and Write

Work in small groups to locate sources of information on the Internet and in your community on alcohol and other drugs. Determine whether these sources are reliable. Then make a brochure that tells other teens where they can find reliable information on alcohol and other drugs. Place the brochure in the school library.

### HEALTH JOURNAL

Have students write an essay either agreeing or disagreeing with a nationwide minimum drinking age of 21. Students should discuss the maturity levels of people under and over the age of 21 and whether the fact that many 21-year-olds are graduating from college and entering the work force should have any effect on the drinking age.

**Lesson 1 Review** On a sheet of paper, write the answers to the following questions. Use complete sentences.

1. When does physical drug dependence happen?
2. When does psychological drug dependence happen?
3. What are three signs of drug dependence?
4. **Critical Thinking** A friend tells you that she suspects her brother is drinking alcohol. She is doing his homework for him. Are the friend's actions helpful? Explain.
5. **Critical Thinking** A classmate says drugs should be legal because they do not hurt anyone but the person using the drugs. What would you say to this classmate?

### Health in Your Life

**Media, Culture, and Drug Use**

You are watching sports on TV. A commercial shows people drinking beer. The beer company wants you to think that you will have fun and fit in if you drink its beer. Every day, you are surrounded by media messages that try to influence you to use drugs. Movies and TV programs show people using alcohol and other drugs.

Culture also influences drug use. If a person's community or peers accept drug use, a person is more likely to use drugs.

When you resist media and cultural influences, you do not let them control you. You can avoid movies or events that promote drug use. You can use positive peer pressure to help others avoid drug use. You can be a role model, remind your friends of future goals, and choose fun activities that do not involve drugs.

On a sheet of paper, write the answers to the following questions. Use complete sentences.

1. How might talking about future goals help friends avoid drug use?
2. How can you resist negative cultural influences on drug use?
3. A friend wants you to go with her to a party where beer will be served. You are not old enough to drink alcohol. You do not want to drink. How can you use positive peer pressure in this situation?

*Dealing with Drug Dependence* Chapter 15  351

## Health in Your Life

Have students read the Health in Your Life feature on page 351 and answer the questions. Ask them what type of behavior television programs and advertisements often show to represent people having fun at parties. Ask them to suggest some things they can do to have a great party without alcohol. *(Answers might include music, food, structured activities, choosing carefully the people you invite, and so on.)*

### Health in Your Life Answers

**1.** Sample answer: Talking about goals with friends might help them see that drug abuse can interfere with their future. **2.** A person can be selective about entertainment and peers and can practice resistance and refusal. **3.** Sample answer: You can tell your friend that you choose not to go because you want to avoid drinking alcohol, which is illegal at your age.

### Lesson 1 Review Answers

**1.** Physical drug dependence happens when the body develops tolerance to a drug. **2.** Psychological drug dependence happens when people believe they need a drug to feel good or to avoid the pain of personal problems. **3.** Three signs of drug dependence are changes in choices of friends, sudden mood changes, and borrowing money. **4.** Sample answer: The friend's actions are not helpful because she is trying to cover up her brother's drinking for him. **5.** Sample answer: I would tell my classmate that using drugs does hurt other people. It hurts the family. It can cause car accidents and lost jobs.

## Portfolio Assessment

Sample items include:

- Skit scripts from Warm-Up Activity
- Paragraphs from ELL/ESL Strategy
- Reports from Career Connection
- Entry from Health Journal
- Brochures from Research and Write
- Reports from Global Connection
- Health in Your Life answers
- Essays from Health Journal
- Lesson 1 Review answers

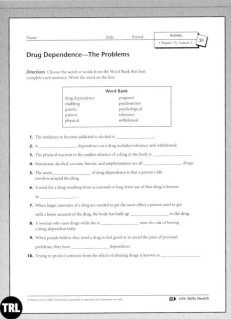
**Activity 51**

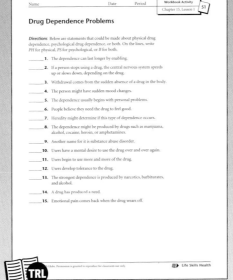
**Workbook Activity 51**

*Dealing w/Drug Dependence* Chapter 15  351

# Lesson at a Glance

## Chapter 15 Lesson 2

**Overview** This lesson explains how people can get help and recover from drug dependence.

### Objectives
- To explain the steps in recovering from drug dependence
- To identify resources for help with drug dependence

**Student Pages** 352–355

**Teacher's Resource Library**
- Activity 52
- Alternative Activity 52
- Workbook Activity 52

## Vocabulary

**detoxification**
**halfway house**
**outpatient treatment center**
**residential treatment center**

Pronounce the vocabulary word *detoxification* and ask students if they have ever heard it. Then have students attempt to define the word based on the different parts of the word. Read the definition so that students can assess their ideas. Do the same thing with *halfway house*. Have students use the words *outpatient* and *residential* in sentences in a context other than drug treatment. Be sure students understand the difference between the words.

## Background Information

Alcoholics Anonymous is an international group of people who have had problems with alcoholism. The only requirement for membership is a person's desire to stop drinking alcohol. Many health professionals include alcoholism in the general category of substance abuse or chemical dependency. They sometimes introduce people who do not have problems with alcohol to A.A. and encourage them to attend meetings. People who do not have problems with alcohol may attend open meetings, in which members describe their experiences with alcohol.

352  Unit 5  Use & Misuse of Substances

---

# Lesson 2 Drug Dependence—The Solutions

**Objectives**

*After reading this lesson, you should be able to*
- explain the steps in recovering from drug dependence
- identify resources for help with drug dependence

**Detoxification**
*Removing addictive drugs from the body*

People who are dependent on drugs can take positive steps to recover. Many programs are available to help people stop using drugs.

## How Do People Recover from Drug Dependence?

The process of recovery from drug dependence includes three main steps. The first step requires a person to recognize that he or she has a problem with drugs.

The second step in recovery is **detoxification**, or removing addictive drugs from the body. Detoxification should take place in a hospital under medical care to avoid serious problems.

The third step in recovery is learning to live day to day without drugs. This is a long-term process. Treatment counselors help a drug user understand his or her behavior and attitude. The person becomes healthier emotionally. The person learns healthy ways to handle problems through decision-making and problem-solving skills.

The process of recovery never stops, because drug dependence cannot be cured in the way that some diseases can be cured. However, drug dependence can be treated successfully so that people can lead healthy lives. Abstinence is the goal of recovery. This means that a person never uses drugs again. Sometimes, the recovering drug user goes back to using drugs. Even so, the person can still achieve recovery and the goal of abstinence.

### Technology and Society

In research studies, scientists studied the effects of alcohol on animals' brains. The animals' brain cells did not respond normally to a brain chemical that produces feelings of pleasure. The animals drank alcohol to produce those feelings. Gene therapy caused the cells to respond normally to the chemical. As a result, the animals drank less alcohol. In time, this therapy may be used to help treat people with alcoholism.

352  Unit 5  Use and Misuse of Substances

---

## Technology and Society

Have students read the Technology and Society feature on page 352. Explain gene therapy to students. Gene therapy is a technique for correcting missing or damaged genes responsible for causing diseases. Sometimes a normal gene is introduced into the cells of a person with a missing or damaged gene. The introduced gene then begins to function to correct the symptoms of the disease caused by the missing or damaged gene. Sometimes techniques are used to fix damaged genes in the cells, rather than introducing new genes. In the research described in the feature, gene therapy caused the animals' brain cells to respond to the brain chemical in a normal way.

### Health Myth

**Myth:** People should be able to stop using drugs by themselves. If they cannot, they are weak.

**Fact:** At some point after people start using drugs, the drugs cause changes in the brain. These changes cause drug dependence. Some drug users cannot quit by themselves.

What are some of the support groups that meet in your community?

## Where Can Individuals and Families Get Help?

Successful treatment programs set up a plan to help each individual. Each person may respond differently to treatment. For this reason, trained professionals use several different methods to help a person. All treatment programs help a person overcome drug dependence. The places where a person gets help may depend on the drug he or she used. It also depends on how long the person used it and on his or her desire for help. Some types of treatment resources or centers are described here.

### Counseling

Most schools have counselors or a student assistance program. Students can talk over concerns about drug use—their own, a friend's, or a family member's. Counselors at community mental health centers can help by directing people to treatment centers. They also can provide help and encouragement after treatment or for family members.

### Support Groups

A common place for treatment is support groups. These are groups in which members help one another. Alcoholics Anonymous (AA) was the first organization to offer support groups. The goal of treatment for AA members is never to drink again. AA believes that people with alcoholism are responsible for managing their disease day by day. The group is based on the buddy system. That means that each person in AA has a sponsor. The sponsor is available around the clock to help the person deal with the desire to drink.

AA has three other groups for family members of people with alcoholism. Al-Anon is a support group for adult family members. Alateen is for 12- to 18-year-olds. Alatot is for 6- to 12-year-olds. Members help one another deal with the problems of living with someone who has alcoholism.

Two other support groups are Cocaine Anonymous and Narcotics Anonymous. You can locate all of these support groups by searching online or the phone directory.

*Dealing with Drug Dependence* Chapter 15

## 1 Warm-Up Activity

Ask students how they can find out about support groups relating to drug dependence in their community. *(looking in a phone book or local newspaper, asking teachers or members of the clergy, searching the Internet, etc.)* Have students examine a phone book and make a list of local organizations and support groups relating to drug dependence.

## 2 Teaching the Lesson

Help students to understand that the functions of support groups, counselors, and treatment centers are different. Support groups bring together addicted people so they can help each other emotionally. Counselors are available to help people discuss concerns and to direct them to appropriate treatment. Treatment centers use a medical approach to stopping drug dependency.

## 3 Reinforce and Extend

### Health Myth

Have students read the Health Myth feature on page 353. Remind them that addiction to a drug is a physical disease that involves changes in brain function. Research has shown that long-term drug use causes major changes in brain function that continue long after the person stops using drugs. These drug-induced physical changes might cause many behavioral changes.

### ELL/ESL STRATEGY

**Language Objective:** *To discuss a topic as a small group*

Divide the class into small groups. Within each group, ask students to discuss what kinds of situations in a teenager's life might be helped by talking with a trusted person. *(family problems, drug use, and so on)* Have them suggest who that trusted person might be. Have one person in each group act as a recorder to make a list of situations and people whom they can trust.

### PRONUNCIATION GUIDE

Use the pronunciation shown here to help students pronounce difficult words in the lesson. Refer to the pronunciation key that appears in the Glossary at the back of the Teacher's Edition for the sounds of the symbols.

alcoholism (al′ kə hôl iz əm)

anonymous (ə non′ ə məs)

## GLOBAL CONNECTION

Some countries, such as England and Sweden, have tried to fight drug addiction by making heroin available legally. Epidemics of drug addiction followed legalization of the drug. These countries have since reversed their laws and made heroin illegal again. Have a class discussion about whether students think these facts support the argument not to legalize other drugs.

## LEARNING STYLES

### Body/Kinesthetic
Have students work in small groups to create a board game about drugs. Games can include cards with questions, points for answers, a path to the end of the game, and clear directions. Instruct students to give the game an interesting name and to make the game challenging and interesting for their age group. Encourage students to try playing each group's game.

## HEALTH JOURNAL

Have students address the following questions in their journals: Did you ever break a habit? Were you able to break it easily, or was it very hard for you to break the habit? What helped you break the habit? Are there some habits you cannot seem to break?

## LEARNING STYLES

### Interpersonal/Group Learning
Instruct groups of students to plan and create a presentation for middle school students on the benefits of being drug-free. Depending on the types of presentation, roles could include video camera operator, script writer, researcher, performer, or art director.

**354** Unit 5 Use & Misuse of Substances

---

**Residential treatment center**
A place where individuals live while being treated for drug dependence

**Halfway house**
A place for recovering individuals to get help easing back into society

**Outpatient treatment center**
A place where individuals go for drug-dependence treatment while living at home

Many organizations and businesses have programs to reduce drug use. They may have policies that drug users can lose their jobs. Employers may conduct drug testing. Many programs provide information about drug use. They also refer employees who use drugs to treatment and support services.

### Residential Treatment Centers
An individual can go to a regular hospital, a mental health hospital, or a center. The individual lives in such a center, called a **residential treatment center,** during treatment. The person receives help from health professionals who are trained in treating drug dependence. Different centers may focus on certain age groups or steps in treatment. For example, some centers may treat only adolescents.

The doctors, social workers, and counselors in these centers use the medical approach to treatment. That means they slowly take the drug-dependent person off a drug. Sometimes, these professionals use medicines to help individuals get through withdrawal.

Some treatment centers are run by state or local governments. Others are privately owned. Centers that the government runs may be less expensive than privately owned centers. In either case, a person's health insurance plan may cover treatment in these centers.

### Halfway Houses
Sometimes, when people leave a treatment center, they go to a **halfway house.** These houses provide shelter and food for recovering individuals as they ease back into society. People may stay in a halfway house to adjust to drug-free living. They must stay off drugs while staying there.

### Outpatient Treatment Centers
Some individuals receive treatment in an **outpatient treatment center.** These centers are for people who can function without drugs while recovering from their drug dependence. Usually, people live at home while they get treatment for a few hours each day at the center. This kind of treatment generally lasts longer than residential treatment.

**354** Unit 5 Use and Misuse of Substances

*Lesson 2 Review* On a sheet of paper, write the word or phrase in parentheses that best completes each sentence.

1. The first step in recovery from drug dependence is (detoxification, recognizing a problem, learning to live without drugs).

2. A place where a recovering drug user can live while easing into society is a (halfway house, support group, medical hospital).

3. Removing addictive drugs from the body is (counseling, detoxification, abstinence).

On a sheet of paper, write the answers to the following questions. Use complete sentences.

4. **Critical Thinking** Why is learning healthy ways to deal with problems important for recovering drug users?

5. **Critical Thinking** Whom could you talk to if you thought a friend had a problem with drugs?

### Health at Work

**Substance Abuse Counselor**

Substance abuse counselors work with people who are dependent on alcohol or other drugs. The counselor's goal is to break the dependence. The counselor helps people identify behaviors and problems related to the dependence. Counselors may work with an individual, the individual and the family, or groups of people who need counseling. Substance abuse counselors may work in halfway houses, hospitals, residential treatment centers, and state and local government agencies. Some counselors are self-employed. A substance abuse counselor must have more than a 4-year degree to be licensed or certified. A person with a 4-year degree can work as a counseling aide.

## Health at Work

Have students read the Health at Work feature on page 355. Invite them to discuss how helping a person change day-to-day behavior patterns and deal with practical problems such as managing pressures at school would be an important part of a substance-abuse counselor's job. Ask students to discuss what roles substance-abuse counselors could play at different types of treatment resources.

## Teacher Alert

Have a class discussion on the advantages and disadvantages of receiving treatment at a center that focuses on a certain age group. Students might have differing opinions. For example, some students might consider it an advantage to receive support from people in the same age group. Other students might consider it an advantage to have the point of view of those in different age groups.

## Lesson 2 Review Answers

**1.** recognizing a problem **2.** halfway house **3.** detoxification **4.** People sometimes use drugs as a way to deal with problems. If they learn more healthy ways to deal with problems, it might be easier for them to avoid using drugs. **5.** Sample answer: You could talk to the school counselor, a teacher, or a parent.

## Portfolio Assessment

Sample items include:
- Lists from Warm-Up Activity
- Board game from Learning Styles: Body/Kinesthetic
- Entries from Health Journal
- Lesson 2 Review answers

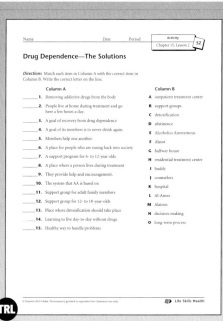

**Activity 52**

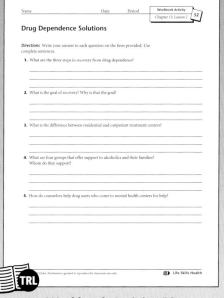

**Workbook Activity 52**

# Lesson at a Glance

## Chapter 15 Lesson 3

**Overview** This lesson discusses ways to avoid drug use and healthy alternatives to drug use.

### Objectives

- To use resistance skills to avoid peer pressure to use drugs
- To identify sources of help to avoid drug use
- To describe healthy alternatives to drug use

**Student Pages** 356–358

**Teacher's Resource Library**

Activity 53
Alternative Activity 53
Workbook Activity 53

---

## Vocabulary

**resistance skill**

Have students look in the dictionary to find the meaning of the root word in *resistance*. Ask them to write sentences using the vocabulary term *resistance skill* and other words containing the root word *resist*.

---

## Background Information

Resistance skills must be practiced if they are to be effective. One way to practice is for the student to learn how to ask clarifying questions. These questions identify the specific activity or behavior that a person wants the student to do. An example might be, "What do you want me to do at the party with you?" Asking the right questions helps students to confirm exactly what the person wants them to do. The questions also serve to shift peer pressure by causing the other person to name the exact behavior in question, which the other person might not want to do.

### 1 Warm-Up Activity

Have students work in groups to develop a public service announcement (PSA) for television or radio about ways to avoid and refuse drugs. Invite each group to present its PSA to the rest of the class. Then classmates should write a summary of the message and evaluate its effectiveness.

356 Unit 5 Use & Misuse of Substances

---

## Lesson 3 Avoiding Drug Use

**Objectives**

*After reading this lesson, you should be able to*

- use resistance skills to avoid peer pressure to use drugs
- identify sources of help to avoid drug use
- describe healthy alternatives to drug use

**Resistance skill**
*A way to resist negative peer pressure*

Eating a balanced diet every day increases energy levels and maintains normal brain chemistry. A person who has normal brain chemistry can cope better with stress and other problems.

There are many ways to avoid ever starting to use drugs. You can refuse offers from people trying to get you to use drugs. You can choose healthy alternatives to drug use.

### How Do You Refuse Drugs?

One main reason adolescents use drugs is peer pressure. **Resistance skills** are ways to resist negative peer pressure. One way for anyone feeling pressure to use drugs is to say no and walk away. Another way is to avoid places where you know or suspect drugs are being used. For example, if you know there will be drugs at a party, you simply do not go.

Here are some other refusal techniques.

- Change the subject. Refuse an offer of drugs and then talk about something else.
- Give honest reasons for not using drugs. For example, you could say you want to be in control of your life rather than having drugs control your life.
- Keep repeating that you do not want to use drugs.
- Give facts about why the drugs will harm you and why you will not use them.
- Choose to spend time only with people who do not use drugs.

### How Can You Get Help?

Getting help for stress and other problems can help people avoid drug use. This is especially important for people with a family member who uses drugs. Those people need to know that they are not responsible for the drug user's behavior. They also need to realize that they are responsible for their own behavior.

People can get help to avoid drug use by talking with a trusted, trained person. Sometimes, supportive friends can help. Sources of help are parents, school counselors or nurses, teachers, and other adult family members.

356 Unit 5 Use and Misuse of Substances

**Research and Write**

Interview students in your school. Ask them their favorite alternative activities to using drugs. Then make a poster that describes and illustrates some of these ideas. Place the poster where many students can see it, such as in the gym or cafeteria.

Many hotlines are also available, such as the National Drug and Alcohol Treatment Hotline run by the Center for Substance Abuse. This service can link people with treatment in their own community.

## What Are Some Alternatives to Drug Use?

Using drugs to deal with problems is a risky choice that never solves the problems. Here are some activities that can help you be part of a group and have fun. These activities can increase your self-esteem. You can make them part of your lifestyle.

- Do something physically active. Exercise can improve mental and physical fitness, reduce stress, and increase energy. It also gives you a sense of accomplishment. You might try a dance or martial arts class.

- Get involved in an important cause that helps others and adds purpose to your life. You might volunteer at a hospital or for an environmental cleanup.

- Join an interest or activity group with people who share common concerns. You might start a new hobby, work on the school newspaper, or help organize a walk for charity.

- Work at a part-time job. You will get a paycheck and the pleasure of doing a job well. A job can help increase your feelings of self-worth, self-esteem, and self-confidence.

- Choose entertainment that promotes mental and physical health. You might choose movies, books, and magazines that do not feature violence or drug use.

- Support efforts by your school and community to prevent drug use. Join a committee to develop your school's drug policy. Find out how you can participate in community prevention programs and activities.

**Figure 15.3.1** *Getting help for drug dependence can start with a call to a substance abuse hotline.*

### 2 Teaching the Lesson

Ask students to discuss things that are easier to do because of not using drugs. The discussion can include personal, academic, and social situations. Have them give examples of extracurricular activities offered at school or in the community. Ask how these activities might help keep someone drug-free.

### 3 Reinforce and Extend

**Research and Write**

Have students read and conduct the Research and Write feature on page 357. Ask students if they think there is a correlation between people who abuse drugs and people who have too much free time, and why. *(Some students might think that people who keep busy have less time to get into trouble and abuse drugs.)*

#### IN THE COMMUNITY

 Discuss with students how calling a hotline might be valuable for people who are dealing with drug or alcohol dependence themselves or who have relatives or friends with drug or alcohol problems. Share the following hotline phone numbers with students: Al-Anon/Alateen–1-888-425-2666. American Council for Drug Education–1-800-488-3784.

#### LEARNING STYLES

 **Auditory/Verbal**
Have pairs of students take turns role-playing a conversation between someone answering a phone at a drug hotline and a teenager calling for advice about a friend who might be using drugs. Encourage students to listen carefully and take notes about the conversations they hear.

#### ELL/ESL STRATEGY

 **Language Objective:** *To read for meaning*

Have students learning English read page 357. Ask them to give an example that is not mentioned in the text for each of the six bulleted alternatives to drug use. Ask them if they participate in any of these alternatives. Have them write a short paragraph describing an activity in which they participate or would like to participate.

### Decide for Yourself

Have students read the Decide for Yourself feature on page 358 and answer the questions. Ask the following questions: What are other things you can say if someone pressures you to take drugs? *(If you were really my friend, you wouldn't push me to do something I don't want to do.")* Should you tell an adult about the drug use at the party? *(Tell students that the right thing to do is to tell an adult, because friends could be seriously injured or die if they abuse drugs.)*

#### Decide for Yourself Answers

**1.** There are five resistance skills. They are saying no, giving honest reasons for not using drugs, giving facts about why the drugs will harm you, repeating the refusal, and walking away. **2.** Other resistance skills that could be used are changing the subject, choosing to spend time only with people who do not use drugs, and avoiding the situation in the first place. **3.** Mine is the most reasonable argument. I'm arguing against taking a health risk, but my friend is arguing for taking a health risk.

### Lesson 3 Review Answers

**1.** Two techniques for refusing drugs are to change the subject and to keep repeating that you do not want to use drugs. **2.** Three alternatives to drug use are doing something physically active, joining an interest group, and working at a part-time job. **3.** One main reason adolescents use drugs is peer pressure. **4.** Sample answer: I do not want to use drugs because I want to do well in school and stay on the soccer team. **5.** Sample answer: I could turn to my dad and my coach.

### Portfolio Assessment

Sample items include:

- PSAs, summaries, and evaluations from Warm-Up Activity
- Posters from Research and Write
- Notes and evaluations from Auditory/Verbal
- Paragraphs from ELL/ESL Strategy
- Presentations from Interpersonal/Group Learning
- Essays from Logical/Mathematical
- Decide for Yourself answers
- Lesson 3 Review answers

---

**?** Who is someone you admire who does not use illegal drugs?

**Lesson 3 Review** On a sheet of paper, write the answers to the following questions. Use complete sentences.

1. What are two techniques for refusing drugs?
2. Name three alternatives to drug use.
3. What is one main reason adolescents use drugs?
4. **Critical Thinking** What is a reason you would give someone for not using drugs?
5. **Critical Thinking** Who are two people you could turn to if you needed help to avoid drug use?

### Decide for Yourself

The best way to avoid the negative effects of drugs is to avoid situations where drugs are being used. However, you may not always be able to do so.

Suppose you are at a party at a friend's house. Your friend says, "Hey, I found these pain killers in my parents' medicine cabinet. You can get a high with no bad effects because they're not illegal."

You say, "No, I don't want to take them."

He replies, "Well, everyone else here is taking them. What's the matter with you?"

You respond, "It doesn't matter to me what other people are doing. I don't want to take them. I want to be in control of my mind and body. These pills are prescription medicine. They could make me vomit or go into a coma. I don't want to get sick."

Your friend keeps pressuring you, and you keep saying no. You decide to leave the party.

On a sheet of paper, write the answers to the following questions. Use complete sentences.

1. How many resistance skills can you identify in the situation? Name them.
2. What other resistance skills could you use in this situation?
3. Who is making the most reasonable argument here—you or your friend? Explain your answer.

---

**358** Unit 5 *Use and Misuse of Substances*

**Activity 53** — Avoiding Drug Use

**Workbook Activity 53** — Alternatives to Drug Use

# Chapter 15 SUMMARY

- Drug dependence can be physical, psychological, or both.

- In physical drug dependence, the body develops tolerance to a drug and then withdrawal when the drug use is stopped. Heredity may influence whether people become physically dependent on drugs.

- In psychological drug dependence, people believe they need a drug to feel good or avoid the pain of personal problems. Psychoactive drugs produce psychological drug dependence.

- Drug use usually results in physical and psychological drug dependence.

- The main pattern of drug dependence is that a person's life centers on a drug. The more signs of drug dependence a person shows, the greater are the chances the person is drug dependent.

- Enabling is the effort of friends or family members to protect a drug user from the results of drug use. Enabling delays admitting the problem and getting help. It hurts everyone.

- Drug dependence costs society in dollars, present lives, and future lives.

- Drug dependence can increase the risk of acquiring AIDS and other sexually transmitted diseases. Using drugs can have criminal consequences.

- People can recover from drug dependence, although they can never be cured. The three main steps in recovery are admitting the problem, detoxification, and learning to live every day without drugs.

- Treatment in support groups involves members helping one another stay drug free. Support groups help people who live with or are close to a drug-dependent person.

- Some people live in residential treatment centers during treatment. Others live in halfway houses for a short time after treatment. Some live at home when going to outpatient treatment.

- People can avoid drugs by using resistance skills. They can talk with trusted adults when they are having problems or feel stress. There are many healthy alternatives to drug use.

- Hotlines provide information about sources of help for drug dependence.

| Vocabulary | | |
|---|---|---|
| detoxification, 352 | halfway house, 354 | residential treatment center, 354 |
| enabling, 349 | outpatient treatment center, 354 | resistance skill, 356 |

## Chapter 15 Summary

Have volunteers read aloud each Summary item on page 359. Ask volunteers to explain the meaning of each item. Direct students' attention to the Vocabulary box on the bottom of page 359. Have them read and review each term and its definition.

### ONLINE CONNECTION

This site, offered by the National Institute on Drug Abuse, discusses common drugs of abuse and other drug-related concerns: www.nida.nih.gov

This site, offered by the American Council for Drug Education, offers a 20-question quiz about legal and illegal drugs:
http://www.acde.org/youth/quiz.htm

A Web site provided by the Center for Substance Abuse Prevention and the Substance Abuse and Mental Health Services Administration presents suggestions, activities, and teaching tools for teaching resistance skills to students:
http://p2001.health.org/VOL01/MOD7TR.htm#teaching

### LEARNING STYLES

**Logical/Mathematical**
Have students evaluate their feelings as they answer this question in a short essay: How do you feel when you stand up for yourself, when you assert yourself by refusing to do something that is unhealthy, unsafe, or against your values? *(Most students probably will respond that asserting themselves in this way helps them feel better about themselves.)*

# Chapter 15 Review

Use the Chapter Review to prepare students for tests and to reteach content from the chapter.

## Chapter 15 Mastery Test

The Teacher's Resource Library includes two forms of the Chapter 15 Mastery Text. Each test addresses the chapter Goals for Learning. An optional third page of additional critical-thinking items is included for each test. The difficulty level of the two forms is equivalent.

## Review Answers

**Vocabulary Review**

1. residential treatment center 2. enabling 3. detoxification 4. resistance skills 5. halfway house 6. outpatient treatment center

---

# Chapter 15 REVIEW

**Word Bank**
detoxification
enabling
halfway house
outpatient treatment center
residential treatment center
resistance skills

## Vocabulary Review

On a sheet of paper, write the word or phrase from the Word Bank that best completes each sentence.

1. Individuals who live in the place where they are treated are in a(n) _____.
2. Helping someone to keep using drugs is _____.
3. Getting addictive drugs out of the body is _____.
4. Saying no to drug use is one of several _____.
5. A(n) _____ provides help for recovering individuals as they ease back into society.
6. Individuals who live at home while they are treated go to a(n) _____.

## Concept Review

On a sheet of paper, write the letter of the answer that best completes each sentence.

7. Support groups help _____.
   A children only   C family members
   B drug users only   D counselors

8. All of these actions enable drug users EXCEPT _____ them.
   A lying for   C making excuses for
   B getting treatment for   D giving money to

9. Another name for drug dependence is _____.
   A withdrawal   C substance abuse disorder
   B detoxification   D tolerance

10. A good alternative to drug use is to _____.
    A avoid exercise   C keep to yourself
    B try a new hobby   D stop watching TV

360   Unit 5   Use and Misuse of Substances

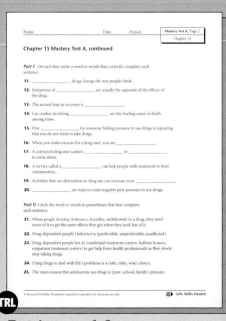

Chapter 15 Mastery Test A, pages 1–3

11. A drug user's parent _____ his behavior.
    A may feel stress from    C is able to control
    B can predict             D is responsible for

12. If you use resistance skills, you _____.
    A are physically active   C refuse offered drugs
    B resist getting help     D give in to peer pressure

13. Treatment for a drug user can happen at a _____.
    A halfway house           C drug user's home
    B support group           D residential treatment center

14. All drug-dependent people _____.
    A can be cured            C have enabling families
    B have stable personalities  D have a drug tolerance

15. The effects of drugs on unborn babies _____.
    A are rarely serious      C may include deformities
    B can usually be reversed D last only a few years

## Critical Thinking

On a sheet of paper, write the answers to the following questions. Use complete sentences.

16. A friend confides in you that her mother has a drinking problem. What would you say to this friend?

17. How might a drug-free person end up paying for drug use?

18. Suppose a friend told you he would not be your friend any longer unless you drank alcohol. What would you say?

19. What are some benefits of a drug-free lifestyle?

20. How does being involved in a loved one's treatment for drug dependence benefit family members?

**Test-Taking Tip** Try to answer all questions as completely as possible. When asked to explain your answer, do so in complete sentences.

## Concept Review

7. C  8. B  9. C  10. B  11. A  12. C  13. D  14. D  15. C

## Critical Thinking

16. Sample answer: I would tell her to talk to a trusted family member or the school counselor or to go to an Alateen support group. 17. Drug use causes higher taxes and higher health insurance and health care costs. 18. Sample answer: I would say no and tell the person he is not a true friend if he is trying to get me to do something that will harm my health. 19. Sample answer: Some benefits of a drug-free lifestyle are being healthy, being able to think clearly, being able to focus on school, friends, and family instead of drugs, and not worrying about being arrested. 20. Sample answer: The family can learn how to support the loved one's recovery efforts. They also might learn new skills, such as how to improve communication within the family.

## ALTERNATIVE ASSESSMENT

Alternative Assessment items correlate to the student Goals for Learning at the beginning of this chapter.

- Have students orally explain physical and psychological drug dependence.
- Have students make a poster identifying the main pattern and signs of drug dependency.
- Have students write and act out a skit describing how drug dependence costs the family and society.
- Have students discuss the steps in recovery from drug dependence.
- Have students work in groups to create brochures that identify resources for help overcoming drug dependence.
- Have students perform a skit showing how to use refusal skills when pressured to use drugs.
- Have students make a graphic organizer that describes healthy alternatives to drug use.

# HEALTH IN THE WORLD

## Drug Trafficking

Drug trafficking is the buying and selling of illegal drugs. It is a big business. Out of about 185 million drug users in the world, about 25 million are in the United States. Cocaine, heroin, marijuana, and methamphetamine are some of the illegal drugs that come into the country by air, land, and sea. Some illegal drugs also are produced in the United States.

Many U.S. government agencies are responsible for detecting the movement of illegal drugs across American borders. The U.S. Coast Guard watches over drug trafficking by air and sea. The U.S. Customs Service searches the country's borders. It also monitors people and luggage coming into the country on commercial flights. One of the most successful customs programs is the Canine Enforcement Program. Dogs such as Labrador retrievers, golden retrievers, and German shepherds are trained to detect narcotics. These dogs found almost 2 million pounds of narcotics in 2004.

Though drug trafficking is still a huge problem, it has been decreasing in the United States. The United States works with many other countries to try to stop drug production.

However, focusing on drug trafficking alone cannot solve the problem. People must be educated about the dangers and consequences of drug use. Efforts must continue to help drug users never use drugs again. When the demand for drugs decreases, the supply decreases.

On a sheet of paper, write the answers to the following questions. Use complete sentences.

1. How would the amount of drugs coming into the United States change if fewer people used drugs? Explain your answer.

2. Suppose you are going to write a letter to your state representatives. You want to urge them to support funding to reduce drug trafficking from other countries. What would you write in your letter?

# Unit 5 SUMMARY

- A drug is something other than food that changes the way the mind and body work. All medicines are drugs. However, not all drugs are used for medical reasons.

- Taking medicines as directed will help you avoid unwanted side effects and dependence.

- Using tobacco or alcohol can result in serious diseases involving the heart, the lungs, and other body organs.

- The nicotine in tobacco is a stimulant. People can become addicted to nicotine.

- Nonsmokers who breathe in secondhand smoke can have the same health problems that smokers have.

- Alcohol is an addictive depressant, and it is a psychoactive drug. Psychoactive drugs affect the mind.

- Alcoholism is a disease. Alcoholism is a dependence on alcohol.

- Stimulants, depressants, narcotics, and hallucinogens are all addictive psychoactive drugs. Stimulants speed up the central nervous system. Depressants and narcotics slow down the central nervous system. Hallucinogens confuse the central nervous system.

- Marijuana, inhalants, designer drugs, look-alike drugs, and anabolic steroids are dangerous and addictive drugs.

- In physical drug dependence, the body develops tolerance to a drug and then withdrawal when the drug use is stopped. Heredity may influence physical dependence.

- In psychological drug dependence, people believe they need the drug to feel good or avoid the pain of personal problems. Psychoactive drugs produce psychological drug dependence.

- The main pattern of drug dependence is that a person's life centers on a drug. The more signs of drug dependence a person shows, the greater are the chances the person is drug dependent.

- Enabling is the effort of friends or family members to protect a drug user from the results of drug use. Enabling delays admitting the problem and getting help. Enabling hurts everyone.

- Drug dependence can increase the risk of acquiring AIDS and other sexually transmitted diseases. Using drugs can have criminal consequences.

- People can recover from drug dependence, although they can never be cured. The three main steps in recovery are admitting the problem, detoxification, and learning to live every day without drugs.

- People can avoid drugs by using resistance skills.

## Unit 5 Summary

Have volunteers read aloud each Summary item on page 363. Ask volunteers to explain the meaning of each item.

## Unit 5 Review

Use the Unit Review to prepare students for tests and to reteach content from the unit.

## Unit 5 Mastery Test

The Teacher's Resource Library includes two forms of the Unit 5 Mastery Text. An optional third page of additional critical-thinking items is included for each test. The difficulty level of the two forms is equivalent.

## Review Answers

**Vocabulary Review**

1. over-the-counter medicine 2. side effect
3. nicotine 4. stimulant 5. psychoactive
6. inhalant 7. tolerance 8. hallucinogen
9. detoxification 10. enabling 11. outpatient treatment center 12. resistance skill

**Concept Review**

13. A 14. D 15. C 16. A

---

# Unit 5 REVIEW

**Word Bank**
detoxification
enabling
hallucinogen
inhalant
nicotine
outpatient treatment center
over-the-counter medicine
psychoactive
resistance skill
side effect
stimulant
tolerance

## Vocabulary Review

On a sheet of paper, write the word or phrase from the Word Bank that best completes each sentence.

1. Anyone can buy a(n) _____ without a prescription.
2. A reaction to a medicine other than what it is intended for is a(n) _____.
3. An addictive chemical in tobacco is _____.
4. A drug that speeds up the central nervous system is a(n) _____.
5. A(n) _____ drug is one that affects your mental processes.
6. A type of chemical that is breathed in is a(n) _____.
7. If a person needs more cocaine to get the same effect that once occurred with less cocaine, the person has developed a(n) _____ to the drug.
8. A drug that confuses the central nervous system and can make the user hear things that are not real is a(n) _____.
9. In the process of recovery from drug dependence, _____ should take place in a hospital.
10. Someone who lies to protect a drug user is _____ the drug addiction.
11. A(n) _____ is a place for people who can function without drugs while recovering from their drug dependence.
12. An effective way to resist negative peer pressure is a(n) _____.

364  Unit 5  Use and Misuse of Substances

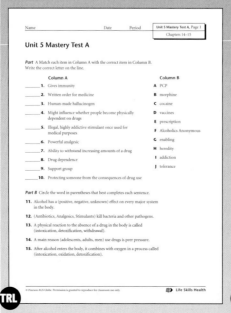
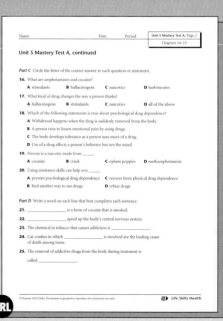

Unit 5 Mastery Test A, pages 1–3

## Concept Review

On a sheet of paper, write the letter of the answer that best completes each sentence.

13. A type of psychoactive drug that slows down the central nervous system is a(n) _____.
    A depressant
    B hallucinogen
    C amphetamine
    D anabolic steroid

14. Aspirin and other over-the-counter medicines that get rid of pain are _____.
    A hallucinogens
    B anabolic steroids
    C inhalants
    D analgesics

15. The physical reaction to the sudden absence of a drug in the body is _____.
    A tolerance
    B emphysema
    C withdrawal
    D intoxication

16. A(n) _____ dependence exists when an alcohol addict believes that she must drink to feel good.
    A psychological
    B look-alike
    C physical
    D enabling

## Critical Thinking

On a sheet of paper, write the answers to the following questions. Use complete sentences.

17. Why is a person who is addicted to alcohol at risk for having liver problems? Explain your answer.

18. Why do some employers require job seekers to take a drug test? Do you think this is fair?

19. The word *anonymous* describes something that is not named or identified. Why do you think the support group Alcoholics Anonymous uses that name?

20. In your school or community, what are some healthy activities that are alternatives to drug use?

## Critical Thinking

17. Sample answer: Compared to people who do not drink, alcoholics are at a greater risk for having liver problems. Alcohol combines with oxygen in a process called oxidation, which takes place mainly in the liver. The liver can oxidize only about one-half ounce of alcohol per hour. People who abuse alcohol also abuse their body organs, including the liver. Abusing alcohol often leads to liver problems. 18. Sample answer: Drug testing is a fair practice because employers want to find the best possible person to fill a job opening. No employer wants to hire someone who is addicted to drugs. People who buy and use illegal drugs are criminals. Employers want their employees to be competent, reliable, and law-abiding. Drug testing helps employers identify people who might not be reliable and law-abiding. Many employers believe that people who break the law in their private activities might break the law while they are at work. If a job applicant believes that drug testing is unfair, then he or she can apply for a job at a company that does not use drug testing. 19. Sample answer: Support groups (such as Alcoholics Anonymous) use the word anonymous in their titles to emphasize that the group never reveals to the public the identities of its members. Anonymity is very important for people who are recovering from alcohol or drug dependence. People in support groups wish to remain anonymous for a variety of reasons. Some people are embarrassed about their addiction. Some people don't want their employers or their neighbors to discover their addiction. 20. Sample answer: In my community, there are several organizations that offer fun, healthy activities for people my age. My school has athletic teams, a debate team, a chess club, and a photography club. My community has a recreation center where people can take classes in martial arts, yoga, painting, and Web site design. My community also has a learning center, where people can sign up for courses in gymnastics, tennis, sign language, and pottery.

# Unit 6

## Planning Guide
## Injury Prevention and Safety Promotion

| | Student Pages | Vocabulary | Health in the World | Review | Critical-Thinking Questions | Chapter Summary |
|---|---|---|---|---|---|---|
| **Chapter 16** Reducing Risks of Injury | 368–393 | ✔ | | ✔ | ✔ | 391 |
| Lesson 1 Reducing Risks at Home | 370–375 | ✔ | | ✔ | ✔ | |
| Lesson 2 Reducing Risks Away from Home | 376–380 | ✔ | | ✔ | ✔ | |
| Lesson 3 Reducing Risks on the Road | 381–384 | | | ✔ | ✔ | |
| Lesson 4 Safety During Natural Disasters | 385–390 | ✔ | | ✔ | ✔ | |
| **Chapter 17** Applying First Aid to Injuries | 394–415 | ✔ | | ✔ | ✔ | 413 |
| Lesson 1 First Aid Basics | 396–400 | ✔ | | ✔ | ✔ | |
| Lesson 2 First Aid for Life-Threatening Emergencies | 401–406 | ✔ | | ✔ | ✔ | |
| Lesson 3 First Aid for Poisoning and Other Problems | 407–412 | ✔ | | ✔ | ✔ | |

*(Unit Planning Guide is continued on next page.)*

### Unit and Chapter Activities

**Student Text**
Unit 6 Summary

**Teacher's Resource Library**
Home Connection 6
Community Connections 16–18

**Teacher's Edition**
Opener Activity, Chapters 16–18

### Assessment Options

**Student Text**
Chapter Reviews, Chapters 16–18
Unit 6 Review

**Teacher's Resource Library**
Self-Assessment Activities 16–18
Chapter Mastery Tests A and B, Chapters 16–18
Unit 6 Mastery Tests A and B

**Teacher's Edition**
Chapter Alternative Assessments, Chapters 16–18

| Student Text Features | | | | | | | | Teaching Strategies | | | | | | Learning Styles | | | | | Teacher's Resource Library | | | | |
|---|---|---|---|---|---|---|---|---|---|---|---|---|---|---|---|---|---|---|---|---|---|---|---|
| Self-Assessment | Health at Work | Health in Your Life | Decide for Yourself | Link to | Research and Write | Health Myth | Technology and Society | Background Information | ELL/ESL Strategy | Health Journal | Applications (Home, Career, Community, Global, Environment) | Online Connection | Teacher Alert | Auditory/Verbal | Body/Kinesthetic | Interpersonal/Group Learning | Logical/Mathematical | Visual/Spatial | Activities | Alternative Activities | Workbook Activities | Self-Study Guide | Chapter Outline |
| 368 | | | | | | | | | | | | 391 | | | | | | | 54–57 | 54–57 | 54–57 | ✓ | ✓ |
| | | 375 | 371 | | 374 | | | 370 | 372 | | 373, 374 | | 372 | 371 | | | | 374 | 54 | 54 | 54 | | |
| | 376 | | | | 380 | | | 376 | 377 | | 379 | | 377 | | 378 | 379 | | | 55 | 55 | 55 | | |
| | | | 384 | 382 | 384 | 383 | | 381 | 382 | 383 | | | | | | | 383 | | 56 | 56 | 56 | | |
| | | | | | | | 386 | 385 | 389 | 390 | 386, 388–9 | | | | 387 | | | 387, 388 | 57 | 57 | 57 | | |
| 394 | | | | | | | | | | | | 413 | 414 | | | | | | 58–60 | 58–60 | 58–60 | ✓ | ✓ |
| | 400 | 399 | | 397 | 396 | | | 396 | 398 | 399 | 398 | | 399 | | | 398 | | | 58 | 58 | 58 | | |
| | | | | | | | 405 | 401 | 403, 405 | 403 | 402, 406 | | 405 | 402 | | 404 | | 404 | 59 | 59 | 59 | | |
| | | 412 | 407 | 412 | | 410, 411 | | 408 | 409 | | 408, 410–11 | | 409 | 410 | | | 411 | | 60 | 60 | 60 | | |

## Alternative Activities

The Teacher's Resource Library (TRL) contains a set of lower-level worksheets called Alternative Activities. These worksheets cover the same content as the regular Activities but are written at a second-grade reading level.

## Skill Track

Use Skill Track for *Life Skills Health* to monitor student progress and meet the demands of adequate yearly progress (AYP). Make informed instructional decisions with individual student and class reports of lesson and chapter assessments. With immediate and ongoing feedback, students will also see what they have learned and where they need more reinforcement and practice.

# Unit 6

## Planning Guide

## Injury Prevention and Safety Promotion (continued)

| | Student Pages | Vocabulary | Health in the World | Review | Critical-Thinking Questions | Chapter Summary |
|---|---|---|---|---|---|---|
| **Chapter 18** Preventing Violence and Resolving Conflicts | 416–437 | ✔ | | | | 435 |
| Lesson 1 Defining Violence | 418–423 | ✔ | | | | |
| Lesson 2 Causes of Violence | 424–428 | ✔ | | | | |
| Lesson 3 Preventing Violence | 429–434 | ✔ | 438 | | | |

| | Student Text Features | | | | | | | | Teaching Strategies | | | | | | Learning Styles | | | | | Teacher's Resource Library | | | | |
|---|---|---|---|---|---|---|---|---|---|---|---|---|---|---|---|---|---|---|---|---|---|---|---|---|
| | Self-Assessment | Health at Work | Health in Your Life | Decide for Yourself | Link to | Research and Write | Health Myth | Technology and Society | Background Information | ELL/ESL Strategy | Health Journal | Applications (Home, Career, Community, Global, Environment) | Online Connection | Teacher Alert | Auditory/Verbal | Body/Kinesthetic | Interpersonal/Group Learning | Logical/Mathematical | Visual/Spatial | Activities | Alternative Activities | Workbook Activities | Self-Study Guide | Chapter Outline |
| | 416 | | | | | | | | | | | | 435 | | | | | | | 61–63 | 61–63 | 61–63 | ✓ | ✓ |
| | | 423 | 422 | | 419, 421 | 421 | | | 418 | 420 | | 420 | | 421 | 420 | | | | | 61 | 61 | 61 | | |
| | | | | | | | 424, 426 | | 424 | 426 | | 425, 428 | 426, 427 | | | | | | 428 | 62 | 62 | 62 | | |
| | | | | 432 | | 430 | | 433 | 429 | 431 | 431, 433 | 433, 434 | | 432 | | 430 | 431 | 430 | | 63 | 63 | 63 | | |

## Unit at a Glance

### Unit 6:
### Injury Prevention and Safety Promotion
pages 366–441

### Chapters

16. Reducing Risks of Injury
    pages 368–393

17. Applying First Aid to Injuries
    pages 394–415

18. Preventing Violence and Resolving Conflicts
    pages 416–437

**Unit 6 Summary** page 439

**Unit 6 Review** pages 440–441

**Audio CD**

**Skill Track for Life Skills Health**

**Teacher's Resource Library** TRL
  Home Connection 6
  Unit 6 Mastery Tests A and B

(Answer Keys for the Teacher's Resource Library begin on page 559 of this Teacher's Edition.)

## Other Resources

### Books for Teachers

Shock, Susan. *Saving Children: A Guide to Injury Prevention.* New York: Oxford University Press, 2005. (developmental approach to understanding injury)

Thygerson, Alton. *First Aid.* 4th ed. Boston: Jones and Bartlett Publishers, 2004. (illustrated workbook including procedures for various injuries)

### Books for Students

*Johns Hopkins Children's Center: First Aid for Children Fast.* Rev. ed. New York: DK Publishing, 2002. (step-by-step instructions for dealing with injuries)

Hipp, Earl. *Understanding the Human Volcano: What Teens Can Do About Violence.* Center City, MN: Hazelden, 2000.

### Videos and DVDs

*Home Safe, Not Sorry!* (45 minutes). Springfield, MO: Opfer Productions, 1996 (1-417-864-5000). (improve safety and remove hazards at home)

*Surviving High School* (90 minutes). Teen Health Video Series. Wynnewoood, PA: Schlessinger, 2001 (1-800-843-3620).

### Web Sites

www.chp.edu/besafe/ (safety tips for a wide range of activities from the Children's Hospital of Pittsburgh)

http://kidshealth.org/parent/firstaid_safe/index.html (first aid and safety in many different environments from the Nemours Foundation's Center for Children's Health Media)

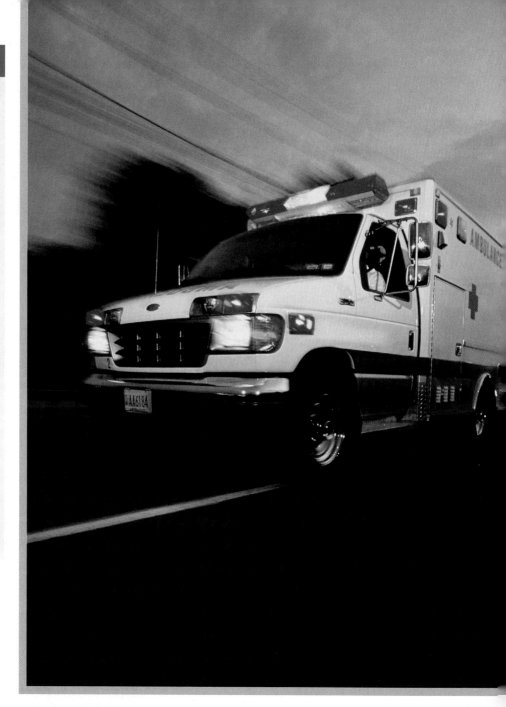

# Unit 6: Injury Prevention and Safety Promotion

The last time you rode a bicycle, did you wear a helmet? In some places, people are ticketed for not wearing a helmet when riding a bike. How about seatbelts? Do you always wear a seatbelt while riding or driving in a vehicle? Wearing a helmet and fastening your seatbelt are two ways to reduce risk of injury. There are many more.

Many Americans might find life today riskier than ever before. Yet the number of years we are expected to live continues to rise. This is because we have discovered new ways to stay safe and live longer. Through laws, education, and technology, communities are working to provide a safe environment. What will you learn in this unit? You will discover things you can do to help yourself, your family, and your community be safe. You will learn ways to reduce risks of injuries at home and away.

### Chapters in Unit 6

Chapter 16: Reducing Risks of Injury ........................... 368
Chapter 17: Applying First Aid to Injuries ................... 394
Chapter 18: Preventing Violence and
　　　　　　Resolving Conflicts ................................. 416

## Introducing the Unit

Have volunteers read aloud the introductory paragraphs on page 367. Have students answer the questions—about wearing bicycle helmets and seat belts—that are posed in the first paragraph. Ask students why most Americans feel that life today is riskier than ever before. (*Students' opinions might include the idea that more cases of extreme violence, such as murder, seem to be in the news today. Students also might think that awareness of violence is greater because of efforts to reduce it.*) Tell students that in this unit they will learn a variety of ways to prevent and treat injuries.

### HOME CONNECTION

The Home Connection unit activity gives students practical experience with concepts taught in the *Life Skills Health* student text. Students complete the Home Connection activity outside the classroom with the help of family members. These worksheets appear on the Life Skills Health Teacher's Resource Library (TRL) CD-ROM.

### CAREER INTEREST INVENTORY

The AGS Publishing Harrington-O'Shea Career Decision-Making System-Revised (CDM) may be used with the chapters in this unit. Students can use the CDM to explore their interests and identify careers. The CDM defines career areas that are indicated by students' responses on the inventory.

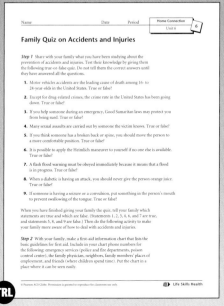

Home Connection 6

## Chapter at a Glance

### Chapter 16: Reducing Risks of Injury
pages 368–393

### Lessons

1. **Reducing Risks at Home**
   pages 370–375
2. **Reducing Risks Away from Home**
   pages 376–380
3. **Reducing Risks on the Road**
   pages 381–384
4. **Safety During Natural Disasters**
   pages 385–390

**Chapter 16 Summary** page 391

**Chapter 16 Review** pages 392–393

**Audio CD**

**Skill Track for Life Skills Health**

**Teacher's Resource Library** TRL

- Activities 54–57
- Alternative Activities 54–57
- Workbook Activities 54–57
- Self-Assessment 16
- Community Connection 16
- Chapter 16 Self-Study Guide
- Chapter 16 Outline
- Chapter 16 Mastery Tests A and B

(Answer Keys for the Teacher's Resource Library begin on page 559 of this Teacher's Edition.)

## Opener Activity

Hold a classroom discussion with the students about the injuries they have experienced in their lives. For each experience, have the student describe how the injury occurred, the environment it occurred in, and what she or he would do to prevent the injury in the future. Tell students that they will learn how injuries commonly occur in this chapter as well as how to prevent them.

## Self-Assessment

### Can you answer these questions?

☑ 1. Who is at the greatest risk for injuries from falls?

☑ 2. How can electrical outlets cause fires?

☑ 3. What are two fire safety devices?

☑ 4. How can the risk of poisoning be reduced?

☑ 5. What is OSHA?

☑ 6. What is survival floating?

☑ 7. How can the number of vehicle crashes be reduced?

☑ 8. Why are bicyclists and motorcyclists at greater risk for injury than riders in cars?

☑ 9. How can people prepare for a hurricane?

☑ 10. How can people prepare for a tornado?

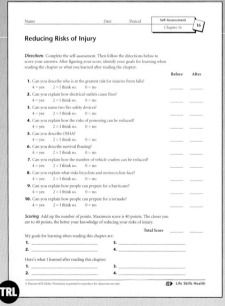

Self-Assessment 16

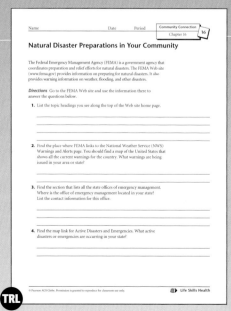

Community Connection 16

368 Unit 6 Injury Prev. & Safety Promo.

# Chapter 16
# Reducing Risks of Injury

Everyone faces risks of injury every day, both at home and away from home. These injuries range from falls to motor vehicle crashes. People today are more aware of risks in their lives than ever before. They are also more aware of the steps they can take to reduce the risks. Laws, education, and technology have all helped to increase the safety of the individual, the family, and the community. Identifying some common risks and learning what can be done to reduce those risks also increases safety. With that knowledge, you can lower both the chances and the results of injuries to yourself and others.

In this chapter, you will learn how to reduce common risks at home, away from home, and on the road. You will also learn what you can do to prepare for and stay safe during natural disasters.

## Goals for Learning

◆ To describe ways to reduce the risks of falls, fire, poisoning, and electrical shock at home

◆ To identify ways to reduce risks at work and during recreational activities

◆ To explain how to reduce the risks of vehicle crashes and increase bicycle and motorcycle safety

◆ To describe ways to stay safe during common natural disasters

369

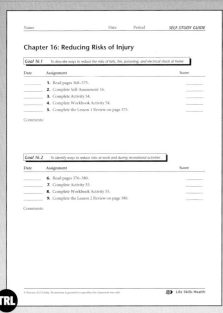

Chapter 16 Self-Study Guide, pages 1–2    Chapter 16 Outline, pages 1–2

## Introducing the Chapter

Have students read the introductory paragraphs on page 369. Then ask them to name certain activities they take part in that are more likely to provide risks for accidents than others. *(Students might mention bicycling, skateboarding, in line skating, contact sports, and so on.)* Ask students to suggest precautions they could take to limit the likelihood of accidents occurring during the activities. *(Suggestions might include wearing helmets and shin guards, bicycling in safe areas, and so on.)*

## Notes and Questions

Ask volunteers to read the notes and questions that appear in the margins throughout the chapter. Then discuss them with the class.

## Self-Assessment

Have students complete the Self-Assessment worksheet before and after reading the chapter. Before reading the chapter, have students fill in the "Before" column. Ask students to identify their goals for learning. To get ideas for setting goals, students might use the chapter introductory material on page 369, the checklist on page 368, or the questions on the Self-Assessment worksheet. Students can use the back of the worksheet if they need more space to write.

Collect the Self-Assessment worksheets and pass them out again at the end of the chapter. Have students fill in the "After" column. Ask them to identify at least four major points they have learned. Again, suggest they use the back of the worksheet if they need more space to write. You may want to collect and review the worksheets, but return them to students so they have a record of their goals and accomplishments.

*Reducing Risks of Injury* Chapter 16   369

# Lesson at a Glance

## Chapter 16 Lesson 1

**Overview** This lesson explains how to reduce the risks of falls, discusses various aspects of fire safety, and explains how to reduce the risks of poisoning and electrical shock.

### Objectives
- To identify safety hazards in the home
- To identify ways to reduce the risk of falls
- To discuss ways to reduce the risk of fire, poisoning, and electrical shock

**Student Pages** 370–375

**Teacher's Resource Library**
- Activity 54
- Alternative Activity 54
- Workbook Activity 54

## Vocabulary

electrical shock　　hazard
fire extinguisher　　smoke detector

Challenge small groups of students to write copy for a TV news story in which each vocabulary word is used. For example, a group might write about how members of a family escaped from a fire because they were prepared. When the news stories are completed, invite volunteers to read them to the class.

## Background Information

The National Safety Council published a report about safety in the home in 2001. They found that in the United States a fatal injury occurs in the home every 16 minutes and a disabling injury occurs every 4 seconds. The leading cause of death was poisoning, and the 25 to 44 age group had the highest death rate.

370　Unit 6　*Injury Prev. & Safety Promo.*

---

## Lesson 1   Reducing Risks at Home

### Objectives

*After reading this lesson, you should be able to*
- identify safety hazards in the home
- identify ways to reduce the risk of falls
- discuss ways to reduce the risk of fire, poisoning, and electrical shock

Everyone would like to think that home is a safe place, but accidents do happen there. People make nearly 21 million medical visits each year because of accidental injuries at home. Nearly 18,000 people die from their injuries. Fortunately, a large number of accidents at home can be avoided.

Most injuries that occur at home are the result of falls. Other causes of injuries at home are fires, poisoning, and electrical shock. Many of these injuries can be prevented. In each case, you can take steps to reduce the risks involved.

### How Can the Risk of Falls Be Reduced?

Everyone can trip and fall, but young children and older people are especially at risk. Stairs are a dangerous place for falls. To make stairways safer, they should be well lit, in good repair, and free of litter. Tightly secured handrails, carpeting, and nonskid strips all can help prevent falls on stairs.

Bathtubs and showers are common places where people fall. To make tubs and showers safe, grip bars can be installed to provide a handle. Nonskid mats or stickers on the bottom of bathtubs also can help prevent falls.

Spills or objects scattered on floors cause people to slide or trip and fall. Spills should be cleaned up when they happen. In addition, fastening throw rugs in place can prevent people from tripping and falling. Keeping toys and other objects off floors and all pathways is another preventive measure.

People often fall from unstable chairs, boxes, or piles that they stand on to reach high places. A safe stepladder or step stool is a better choice than a wobbly chair or box.

### How Can the Risk of Fire Be Reduced?

Making sure electrical outlets are not overloaded is one way to prevent electrical fires. Another way is to check for bad wiring or damaged electrical cords. A licensed electrician can fix these problems.

> Put together a first aid kit for your home. Put it in a place that you can reach quickly.

370　Unit 6　*Injury Prevention and Safety Promotion*

**Smoke detector**
*A device that sounds loudly when it senses smoke or high amounts of heat*

**Fire extinguisher**
*A portable device containing chemicals that put out small fires*

Keep fuel or fire starters away from sources of heat and out of the reach of children. Store all flammable liquids in tightly closed containers away from heat, including a fireplace. Also, keep matches or lighters out of the reach of small children.

Fires in fireplaces and chimneys are often the cause of house fires. To prevent this, place a screen in front of a fireplace when a fire is burning. Also, a fire in a fireplace should be attended at all times. Fires in chimneys can be prevented by having them cleaned each year.

Space heaters and furnaces can cause fires in homes. Clean all heating units regularly and check to be sure they are operating safely. Only use extension cords that match the requirements of a space heater.

Loose-fitting clothing and many materials can easily catch fire near an open flame. Be careful around burning candles, fireplaces, and grills when people wear such clothing. Someone should always watch something cooking on the stove.

### What Are Some Fire Safety Devices?

If a fire does start, a smoke detector and a fire extinguisher can help. A **smoke detector** is a device that sounds loudly when it senses smoke or high amounts of heat. It warns people so they can take action immediately. A house should have at least one smoke detector on every level and in each bedroom. To make sure the batteries work, test smoke detectors every month. Change the batteries twice a year.

A **fire extinguisher** is a portable device containing chemicals that put out small fires. Keep an extinguisher in the kitchen, since many fires can start from spills on the stove or in the oven. Almost everyone in a family can learn when and how to use a fire extinguisher.

### What Can Be Done If a Fire Starts?

Use a fire extinguisher only to put out small fires. A small fire is one that stays in the area where it started. For example, a fire extinguisher can put out a fire in a wastebasket, furniture cushion, or small appliance.

**Link to >>> Social Studies**

In 2004, employers spent $38 billion for injuries that their workers had at home. The costs included health insurance, life insurance, sick leave pay, and disability. Costs for hiring and training new employees also play a part. An injury that causes a hospital stay can cost an employer nearly $20,000.

---

 **Warm-Up Activity**

Ask students to name ways people might get hurt at home. Explain that in order for homes to be safe places, people need to act in safe ways. Tell students that this lesson will point out ways people can act to stay safe.

 **Teaching the Lesson**

As students read through the lesson, have them make a safety checklist for their home. A checklist might include an entry such as the following: (1) Stairway to bedrooms is well lit, in good repair, and free of litter. Suggest students take their checklist home and find areas in the home that need safety improvement.

 **Reinforce and Extend**

**Link to**

**Social Studies.** Have students read the Link to Social Studies feature on page 371. Ask them why it would make sense for a company to spend money on extra safety training for its employees. (*Even one accident can be very costly for a company. A small amount of money spent on training can save a lot of money by reducing accidents.*)

### LEARNING STYLES

**Auditory/Verbal**
Divide the class into two teams. Have each team compile a list of home safety rules. Volunteers from each team should read one rule at a time. If the other team also has that rule, each team should cross the rule off their list. When both teams have finished reading their list, give one point for each rule left on the lists. The team with the most points wins.

## TEACHER ALERT

Many students may believe that the main danger from fire is being burned. Tell students that by far the biggest cause of death in a fire is from smoke inhalation. Many deaths are caused because people underestimate how deadly smoke can be in a fire.

## ELL/ESL STRATEGY

**Language Objective:** *To explain by acting out actions that can be done ahead of time to prepare for a fire*

Divide the students into small groups. Include both students proficient in English and those learning English in a group. Tell the groups to read the actions on page 372 that can be done to prepare ahead of time for a fire. Suggest that the students proficient in English act out any preparatory action that the students learning English do not understand.

## TEACHER ALERT

Point out to students that when you call for help, you need to give certain information. The person you call will need to know what the emergency is and where it is. He or she will ask you questions about the emergency. You need to stay calm so you can answer the questions clearly and correctly. Also, you should not hang up the telephone until you are instructed to do so.

Does your school have regular fire drills? What else could your school do to make sure your school is safe from fire?

A person should know how to use a fire extinguisher correctly. The person's back should always be to an escape route in order to leave quickly if necessary.

Sometimes, small fires can be put out in other ways. For example, baking soda can put out a small grease fire on a stove or in an oven. Putting a tight lid on a pan that has a grease fire in it cuts off oxygen to the fire. Do not use water on a grease fire, because it makes the fire spread.

People should not attempt to fight fires that they cannot easily extinguish. In case of a large fire that cannot be extinguished, it is best to be prepared ahead of time by taking these actions:

- Have regular fire drills, including a drill as if everyone were asleep in bed.
- Have a signal that alerts everyone to a fire drill. For example, someone could press the "test" button on the smoke detector.
- Drop to the floor on hands and knees and crawl to a door. Stay close to the floor where there would be less smoke and more oxygen.
- Feel the door and doorknob to check for warmth. If a door is hot, everyone should take a second escape route, such as a window.
- Stop, drop, and roll on the ground if clothing has caught fire.
- Get out of the house at once without stopping to take belongings. Then go to a designated meeting place.
- Call the fire department from a neighbor's home.

### How Can the Risk of Poisoning Be Reduced?

A poison causes injury, illness, or even death when it enters the body. Most poisonings happen with babies and young children, because they put so many things in their mouth. Sometimes, two different medicines or drugs interact and cause poisoning. Taking too much of a drug is the main cause of poisoning in 15- to 24-year-olds.

**Electrical shock**
*A flow of electric current through the body*

Most risks of poisoning can be reduced in these ways:

- Store medicines and household products in their original containers out of children's reach.
- Never run a car or other engine powered by gasoline or oil in a closed garage. These engines release poisonous gas.
- Inspect the yard for such plants as poison ivy, poison oak, and poison sumac. Find out how to remove them safely.
- Keep the poison control center number by the phone.

Another type of poisoning is food poisoning. This occurs when a person eats food containing certain types of living things, including some kinds of bacteria. Most cases of food poisoning are mild and last only a few days. Other cases require medical care. Some forms of food poisoning can be prevented by cooking and handling food properly.

## How Can the Risk of Electrical Shock Be Reduced?

**Electrical shock** is a flow of electric current through the body. Electrical shock may cause serious burns, injuries to organs, and even death.

Here are some ways to avoid electrical shocks:

- Never use an electric appliance if the floor, your body, or your clothing is wet.
- Pull on the plug, not the cord, when disconnecting appliances.
- Stay away from power lines that are down.
- Never touch a power line, and make sure that no object, such as a ladder, touches a power line.
- Cover electrical outlets with safety plugs.
- Have only a licensed electrician repair appliances and wiring.
- Unplug electrical appliances that are not working properly.

## CAREER CONNECTION

Have students investigate the roles of firefighters during emergencies. Students can talk with a public information officer at their local fire department to gather information. Some questions might focus on the qualifications for becoming a firefighter. *(Some fire departments require only a high school diploma or GED while others require a two-year college degree. Many colleges offer programs in fire science technology.)* Students could also find out in what areas firefighters can specialize. *(Some specialties are water rescue, high-angle rescue with the use of ropes and rappelling, rescue from confined spaces, urban rescue such as from collapsed buildings, hazardous materials, fire investigation, and public education and information.)* Students might be surprised to find out that many firefighters have another job, a trade, or a college education on which to rely because firefighting becomes significantly more difficult past the age of 40, when back problems and heart conditions increase dramatically in the profession.

## PRONUNCIATION GUIDE

Use the pronunciation shown here to help students pronounce difficult words in the lesson. Refer to the pronunciation key that appears in the Glossary at the back of the Teacher's Edition for the sounds of the symbols.

sumac (sü´ mak)

### Health Myth

Have students read the Health Myth feature on page 374 together. Ask them what safety measures a gun owner should take to prevent injuries with children. *(Gun owners should store the gun and ammunition separately in locked containers. Guns should be stored unloaded with locked trigger guards.)*

### LEARNING STYLES

**Logical/Mathematical** Provide students with samples of manuals, users' guides, or other directions that come with electrical appliances. Have students search for safety information in this material. Tell students to note if any symbols are used to indicate cautions or danger. Do the symbols effectively communicate the information? Challenge students to design better symbols if necessary. Have them exchange the symbols with classmates and interpret them.

### AT HOME

Have students find the telephone numbers for the following community services: emergency operator, police station, fire department, ambulance, hospital emergency room, and poison control center. Have students prepare cards with the numbers that can be placed near their telephones at home. Suggest that students add to the list telephone numbers of relatives, friends, and neighbors to contact in an emergency.

---

*Hazard*
Something that is dangerous

**Health Myth**

**Myth:** Keeping a gun in your home will keep you safe.

**Fact:** A person with a gun in the home is nearly three times more likely to be shot and killed than someone in the same neighborhood who does not have a gun in the home.

## What Are Some Other Household Dangers?

Some parts of homes are dangerous and are common sources of injuries. These parts are safety **hazards.** You can act to prevent injuries. Doorknob covers and locks can keep children away from hazardous areas such as swimming pools and bathrooms. Window guards and safety netting can prevent falls from windows. Injuries from sharp edges of furniture and fireplaces can be reduced by the use of corner and edge bumpers. Window blind cords should be cut short enough that children cannot reach them. Children can be choked or strangled by the cords.

Special devices placed on faucets and showerheads can prevent burns. Doorstops and door holders can help prevent hand injuries. Burning candles can be fire hazards. They should be watched carefully and never left unattended.

If firearms are in the home, they should be kept in a locked place. Guns should be unloaded and locked in a separate place from the ammunition.

### Safety When Caring for Children

Watching young children carries many responsibilities for safety. These guidelines can help avoid risks when you are watching young children.

- Know how to contact parents and know when they will return.
- Have emergency and neighbors' phone numbers nearby.
- Learn family rules for playing inside.
- Keep your full attention on the children.
- Keep children away from appliances, matches, cleansers, soap, medicine, and bodies of water.

*Lesson 1 Review* On a sheet of paper, write the answers to the following questions. Use complete sentences.

1. Where are two places that people commonly fall in the home?
2. What are two fire safety devices that can help keep you safe if a fire starts in your home?
3. For what kind of fires should a fire extinguisher be used?
4. **Critical Thinking** How can you help your family reduce injuries from a fall? How can your family help you?
5. **Critical Thinking** How can you help your family reduce the danger of a fire? How can your family help you?

### Health in Your Life

**Planning Escape Routes**

Do you know how you would get out of your home in case of a fire? Follow these guidelines to make a plan of escape from every room of your home.

Draw a floor plan of your home. Show escape routes from each room to use in case of fire. Include an emergency meeting place that your family has chosen outside your home.

Decide who will assist anyone in your family with special needs during the escape.

Post the floor plan where every family member can review it once in a while. Make sure each family member is familiar with the routes and the emergency meeting place.

Beneath the floor plan, write the names of two people whose homes you can go to in case of emergency.

On a sheet of paper, write the answers to the following questions. Use complete sentences.

1. How can your family members help you make these escape plans?
2. Where will you post these escape plans?
3. What can your family do to make sure all family members can use the plan?

## Lesson 1 Review Answers

1. Sample answer: People commonly fall on stairs or in bathtubs. 2. Two fire safety devices are a fire extinguisher and a smoke detector. 3. A fire extinguisher should be used for a small fire, which is one that stays in the area where it started. 4. Sample answer: I can keep objects off the floors and wipe up spills. My family can provide handrails on stairs and nonskid mats or stickers in the bathtub. 5. Sample answer: I can use care in handling matches, lit candles, and flammable liquids and be careful when wearing loose-fitting or lightweight clothing. I can learn to use a fire extinguisher. My family can install smoke detectors and fire extinguishers and make sure electrical wiring and space heaters are in good condition.

### Health in Your Life

Have the students read the Health in Your Life feature on page 375. Point out to students that making a plan of action in case of a fire can help safe their life and the life of others in their family. Encourage students to follow the guidelines to make an escape plan from their home. Suggest that students share their plan with their family.

**Health in Your Life Answers**

1. Sample answer: My family members can help by making suggestions for escape routes, drawing the plan, and rehearsing it. My family members can discuss the safest way to help my sister (who is a toddler) during a fire. 2. Sample answer: We will post the escape plans on the kitchen bulletin board. 3. Sample answer: The adults in my family should take the plan seriously. This will help young children understand how important it is. Everyone should practice it with a daytime drill and a nighttime drill.

### Portfolio Assessment

Sample items include:
- Safety checklist from Teaching the Lesson
- Home fire escape plan from Health in Your Life
- Lesson 1 Review answers

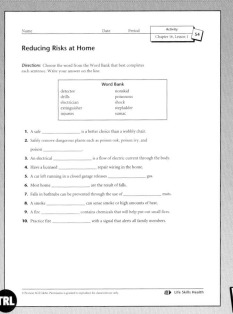

**Activity 54**

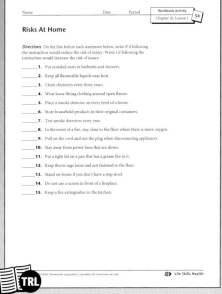

**Workbook Activity 54**

# Lesson at a Glance

## Chapter 16 Lesson 2

**Overview** This lesson explains how to reduce risks of injury away from home.

### Objectives
- To identify ways to avoid injuries in the workplace and at school
- To identify ways to avoid drowning
- To discuss water and sports safety

**Student Pages** 376–380

**Teacher's Resource Library** TRL
- Activity 55
- Alternative Activity 55
- Workbook Activity 55

## Vocabulary

**survival floating**

Write the term *survival floating* on the board. Then direct students' attention to the picture on page 378. Ask them to describe the activity pictured.

## Background Information

By federal law, all boats must carry a personal flotation device (PFD), such as an approved life jacket or life preserver, at all times. For boats shorter than 16 feet, one flotation device must be available for each person on board. For boats longer than 16 feet, one wearable personal flotation device must be available for each person and one throwable personal flotation device must be available for emergencies.

### 1 Warm-Up Activity

Encourage students to describe work or school situations that might have obvious safety hazards. *(Situations could include those associated with construction work, heavy machinery, dangerous chemicals, or a science lab.)* Then ask students to name some safety hazards in work or school situations that might not be obvious. *(Hazards might include malfunctioning equipment, pollution that cannot be seen, spills on the floor, and broken chairs.)*.

**376** Unit 6 *Injury Prev. & Safety Promo.*

---

## Lesson 2 Reducing Risks Away from Home

### Objectives

*After reading this lesson, you should be able to*
- identify ways to avoid injuries in the workplace and at school
- identify ways to avoid drowning
- discuss water and sports safety

Injuries occur not only at home but also in the workplace and during recreational activities. You can reduce your risk for these injuries.

### How Can Injuries at Work Be Reduced?

If you have a job, you may already know about job safety. Every year, nearly four million U.S. workers have disabling injuries on the job. More than 5,000 workers die. In addition, millions of work hours are lost every year because of work-related injuries and illnesses. Expenses for injuries and lost productivity cost employers hundreds of billions of dollars each year.

Many occupational injuries and illnesses can be prevented. A U.S. government agency that seeks to protect the safety and health of American workers is the Occupational Safety and Health Administration (OSHA).

### Health at Work

**Industrial Safety Specialist**

Industrial safety specialists work for OSHA. OSHA's mission is to save lives, prevent injuries, and protect the health of American workers. Industrial safety specialists make sure companies correct unsafe conditions. Safety specialists might discover dangerous asbestos materials in a building. They may convince workers to practice industrial safety activities. All workers need to understand how to be safe. Industrial safety is challenging because of the many rules. Industrial safety specialists must be concerned with waste disposal, environmental safety, pollution, and health hazards.

A 4-year college degree is usually required to become an industrial safety specialist. Continuing education also is necessary to keep up with changes on the job.

**376** Unit 6 *Injury Prevention and Safety Promotion*

OSHA sets standards for industrial safety and provides education about safety procedures. For example, OSHA sets standards for noise and fire safety, protective clothing and safety gear, and safe levels of certain substances. It then works with employers and employees to make sure that these standards are met.

Employers are responsible for reducing risks of injury to employees by informing them of possible dangers on the job. They also must remove dangers in the workplace, train workers properly, and make employees aware of safety regulations.

Employees also can reduce the risks of injuries on the job. Getting necessary rest helps employees be alert at work and avoid injuries. Employees need to follow all safety procedures, learn how to use equipment properly, and wear any required protective clothing or devices.

### How Can Risks of Injuries at School Be Reduced?

Injuries can happen at school, though many of them are preventable. For example, falls can happen at school. To prevent them, report spills and slippery places, such as worn stair edges, to a teacher or janitor. Do not run in the halls, especially when it is raining outside and the floors are wet.

Fires can break out at school. Pay attention during fire drills. Know where stairways and exits are located. Do not use elevators during a fire. Try to help people with disabilities get out of the building.

Your backpack can cause an injury. Do not use it as a portable locker. Backpacks should weigh no more than 15 to 20 pounds. Use both straps when carrying your backpack. Readjust the straps every time you put on your backpack.

### How Can Recreational Injuries Be Reduced?

Everyone enjoys some kind of recreation. Every day, however, people are injured or die while engaging in recreational activity. Water and sports safety are special concerns.

## 2 Teaching the Lesson

Before students begin reading the lesson, have them quickly scan the lesson pages, looking at the section headings. Have students write down one or two questions that they would like answered by the material under each heading. Encourage them to answer their questions, if possible, as they read through the sections.

### Health at Work

Have students read the Health at Work feature on page 376. Tell them to check out the OSHA Web site (www.osha.gov) to see what news items and materials are currently being provided by OSHA.

## 3 Reinforce and Extend

### ELL/ESL Strategy

**Language Objective:** *To prepare and give a group presentation*

Have students form pairs composed of a student learning English and a student with strong English skills. The students should create and then present a skit in which the manager of a certain company and an industrial safety specialist are inspecting working conditions in the company.

### Teacher Alert

Be sure students understand that an employee who refuses to follow safety procedures on her or his job may be subjected to disciplinary action such as attending a safety class or even suspension from the job.

## Learning Styles

**Body/Kinesthetic**
Have students mimic as closely as possible the technique of survival floating. Help students coordinate arm and head movements with breathing.

### Water Safety

Drowning is the third most common cause of accidental death in the United States. Drowning most often occurs among people between the ages of 15 and 24. It usually results from swimming and boating accidents, often when people have been drinking alcohol.

Knowing how to swim can prevent drowning. Swimming classes usually are offered through schools, at community pools, or through community recreation programs.

A technique that can help prevent drowning is called **survival floating.** This technique helps victims in warm water conserve their energy while they wait to be rescued. Figure 16.2.1 shows how survival floating works.

> **Survival floating**
> A technique that can help prevent drowning

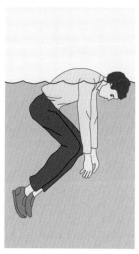 

Drowning can be prevented in other ways. Before going out in a boat, people should check the weather conditions and make sure the boat is in good repair. No one in a boat should ever drink alcohol. Everyone in a boat should wear an approved life jacket or life preserver at all times. If a boat overturns, passengers should stay with the boat until help arrives.

**Figure 16.2.1** *Survival floating*
Left: In the float position, rest face down for several seconds. Let arms and legs dangle.
Right: Slowly tilt head back so it clears the water. Exhale and take a new breath while pressing arms down and bringing legs together. Return to the float position.

Here are some other water safety guidelines that can help prevent drowning:

- Swim only in places where a lifeguard is on duty and follow any posted water safety rules.
- Dive only where you know water is a safe depth.
- If a strong current pulls you away from shore, swim at a diagonal angle toward the shore and try not to panic.
- If you see a swimmer in trouble, yell for help immediately. Do not attempt to rescue someone who is in danger of drowning unless you have had special training.
- Heed any warning signs to stay off thin ice. Signs should be posted if walking or skating on the ice is safe.
- If ice begins to crack while you are on it, lie down and crawl away from the crack to safety.

**Sports and Recreational Safety**

Injuries to people engaging in sports activities are very common. In fact, many doctors treat only sports injuries. You can prevent many sports accidents by following these simple guidelines:

- Know how to use recreational or sports equipment properly and safely as it was intended to be used. Make sure all equipment is working properly.
- Wear the protective gear that is required or recommended for a sport.
- Learn the skills needed to do the activity.
- Learn and follow any recommended safety guidelines, including proper warm-up and cooldown exercises.
- Do not try to do more than you are able or know how to do.
- If you fall while skating or biking, try to curl up into a ball and roll as you fall. This will decrease the chance of injuries.
- When camping, always make sure someone knows where you will be. Take plenty of food and water. Never camp alone.

## GLOBAL CONNECTION

Have students research popular sports in other countries. Students should find out how such sports are played and what safety rules should be followed by people playing the sport. Example sports include cricket, Jai Alai (a Basque sport resembling handball), and rugby.

## LEARNING STYLES

**Interpersonal/Group Learning**

Arrange students into groups. Assign to each group a certain location where people swim, such as the ocean, a river, a swimming pool, a lake, and so on. Ask each group to make a poster of safety precautions appropriate for their swimming situation. Ask groups to share their posters with the class. Compare and contrast the precautions for each different setting.

## Research and Write

Have the students read the Research and Write feature on page 380. Tell the students to look for materials that are written by someone with good qualifications. Have them list the author's qualifications on the poster. Does this person have a degree or other background that suggests that they have studied how stretching and other exercising activities can affect the body?

## Lesson 2 Review Answers

1. C 2. A 3. B 4. Sample answer: If used improperly, some equipment can cause injury to employees or other people. A saw is an example of a work tool that can cause injury if it is not used properly. 5. Sample answer: A person would first float with arms and legs dangling and face in the water. Then the person would slowly tilt the head back out of the water, exhale, and take a new breath. The person would then return to the float position. Survival floating helps people conserve energy while they wait to be rescued.

## Portfolio Assessment

Sample items include:
- Group safety poster for Learning Styles (Interpersonal/Group Learning)
- Stretching poster for Research and Write
- Lesson 2 Review answers

---

**Research and Write**

With a partner, go to a library and research how stretching before exercise can reduce injury. Make a poster that describes some of the stretches. The poster should include the names of the muscles that become more flexible during a stretching session. Working as a team, demonstrate some of the stretches to the class.

**Lesson 2 Review** On a sheet of paper, write the letter of the answer that best completes each sentence.

1. Backpacks should not weigh more than _____.
   A 5 to 10 pounds
   B 10 to 15 pounds
   C 15 to 20 pounds
   D 20 to 25 pounds

2. A U.S. government agency that seeks to protect the health of workers is the _____.
   A Occupational Health and Safety Administration
   B Department of Justice
   C Department of Health and Human Services
   D Council on Environmental Quality

3. The third most common cause of accidental death in the United States is _____.
   A strangulation
   B drowning
   C falling
   D fire

On a sheet of paper, write the answers to the following questions. Use complete sentences.

4. **Critical Thinking** Why is it important for employees to learn how to use equipment properly?

5. **Critical Thinking** Suppose a person fell out of a boat and did not know how to swim. How could that person keep from drowning?

380  Unit 6  Injury Prevention and Safety Promotion

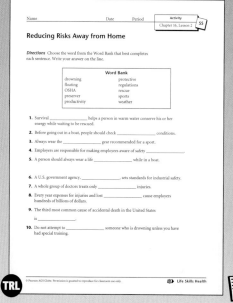

Activity 55

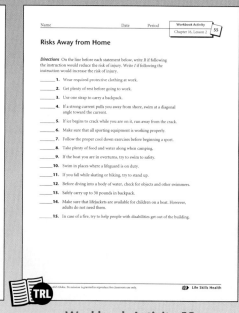

Workbook Activity 55

## Lesson 3: Reducing Risks on the Road

### Objectives

After reading this lesson, you should be able to
- discuss the danger of road injuries
- describe ways to prevent road injuries
- describe safety procedures and equipment for bicyclists and motorcyclists

Motor vehicle crashes cause the highest number of deaths among 16- to 24-year-olds in the United States. Fortunately, safe driving practices can prevent many of these crashes.

### How Can Vehicle Crashes Be Reduced?

The best prevention for vehicle crashes is to take a driver's training course to learn safe driving skills and habits. Students in these courses learn road rules and defensive driving skills. They learn to watch out for drivers who make mistakes. They learn to adjust for unsafe road conditions.

A critical safe driving habit is observing and maintaining the speed limit. When road conditions are wet or icy, traveling below the posted speed allows safe handling of a vehicle. Reducing speed on dirt roads and on damaged pavement is also important.

Wearing a seat belt at all times can reduce injuries if an accident does occur. Both drivers and riders should buckle their seat belts securely. Small children should always be buckled into car seats that meet specific state requirements.

Never use alcohol or other drugs and then drive. Never ride with a driver who is under the influence of alcohol or other drugs.

To avoid crashes:

- Maintain the vehicle properly.
- Maintain good vision by clearing moisture and ice from windows. Make sure windshield wipers and defroster fans work properly.
- To see and be seen, use headlights from dusk until just after sunrise and in rain, snow, and fog.
- Leave enough distance between you and the vehicle in front of you so you can stop quickly and safely.
- Avoid driving when your judgment might be clouded because of frustration, stress, anger, or other strong emotions.

---

## Lesson at a Glance

### Chapter 16 Lesson 3

**Overview** This lesson explains how safety on the road with motor vehicles, bicycles, and motorcycles can be increased.

### Objectives

- To discuss the danger of road injuries
- To describe ways to prevent road injuries
- To describe safety procedures and equipment for bicyclists and motorcyclists

**Student Pages** 381–384

**Teacher's Resource Library**

Activity 56

Alternative Activity 56

Workbook Activity 56

### Background Information

Some estimates of injuries related to bicycling indicate that 85 percent of them could have been prevented with the use of bicycle helmets. Surveys suggest that even though most cyclists believe that helmets effectively protect the head from injury, fewer than half of them actually wear helmets.

### Warm-Up Activity

Find pictures of passengers in a moving car from 20 to 30 years ago. Have the students assess the safety of the passengers. *(Students should realize that automobile safety concerns have changed dramatically over the last few decades. Seatbelts, child safety seats, and airbags are now standard items in a car. Young children must now ride only in a car's back seat.)*

**Teaching the Lesson**

Brainstorm with students some safety guidelines to prevent car, bus, bicycle, and motorcycle accidents and injuries. Make a list of guidelines for each kind of vehicle on the board. As students read through the lesson, have volunteers add guidelines, as necessary, to the appropriate list.

 **Reinforce and Extend**

### ELL/ESL Strategy

 **Language Objective:** *To illustrate vehicular safety rules*

Pair a student learning English with a student proficient in English. Have pairs create an illustrated safety guide for car, bus, bicycle, or motorcycle safety. They should illustrate unsafe behavior as well as the proper, safe method.

### Health Journal

 Ask students how they think buses have become safer during the last ten years. Then ask students what safety devices they think might be added to buses during the next 10 years. Have them write some descriptions of the devices they imagine in their journal.

### Link to

**Arts.** Have students read the Link to Arts feature on page 382. Then have them look for news stories or articles on people who have made major accomplishments after losing the ability to walk. Share or post these stories for the class.

---

 Should school buses be equipped with seat belts?

## How Can You Stay Safe on a School Bus?

Here are some tips for staying safe while riding a bus to or from school.

- When waiting for the bus, stay away from traffic and do not play in the street.
- Wait until the bus has stopped and the door has opened before stepping into the street.
- When on the bus, stay quietly in your seat. Loud talking or other noise can distract the bus driver.
- Never put your head, arms, or hands out of the window.
- Keep the aisles clear.
- Wait for the bus to stop completely before leaving your seat to get off.
- If you have to cross the street in front of the bus, walk ahead of the bus until you can see the driver before crossing. Make sure the driver can see you. Do not cross the center line of the road until it is safe for you to do so.
- Stay away from the bus's wheels at all times.

**Figure 16.3.1** *Wearing seat belts reduces injuries in car crashes.*

### Link to >>>

**Arts**

The movie *Passion Fish* is about a soap opera star who is injured in a car accident. She goes home to Louisiana to deal with her life in a wheelchair. One of her nurses has problems of her own. The movie shows how these two women become friends and help each other heal.

**Health Myth**

**Myth:** If your car has air bags, you do not need to wear a seat belt.

**Fact:** Motorists can slide under an air bag and become injured if they are not wearing a seat belt. Air bags may not help in a rear impact or a rollover crash. Seat belts help keep passengers inside the car during such crashes.

## How Can Bicycle and Motorcycle Safety Be Increased?

Bicyclists and motorcyclists are hard to see and are less protected than people riding in closed vehicles. These reasons and human error increase their risk for accidents. Hundreds of thousands of people are injured or die each year in bicycle and motorcycle accidents.

Both bicyclists and motorcyclists should wear the proper safety helmets. They should obey all traffic rules. They should never grab a moving vehicle and should always keep a safe following distance. In addition, bicyclists should follow these guidelines:

- Make safe turns and cross intersections with care. Signal all turns with the left arm straight out for a left turn and the forearm up for a right turn.
- Some roads have bicycle lanes. Use these lanes when possible.
- Ride single file as far to the right as possible with the flow of traffic, not against it.
- Wear bright clothing to be seen in daylight and reflective clothing to be seen at night.
- Use the right safety equipment and make sure the bike is in good repair.

Motorcyclists should follow these safety tips:

- Follow the rules for getting a motorcycle permit or license in your state.
- Learn to drive a motorcycle properly and safely before taking it on the road.
- Be sure that the motorcycle has working lights, reflectors, and rearview mirrors.
- Make sure the motorcycle is in good repair and have it inspected regularly.
- Wear proper clothing for protection and to avoid getting it caught in moving parts.
- Be especially careful driving on wet surfaces.

---

## Health Myth

Have a student read the Health Myth feature on page 383 aloud in class. Tell students that an air bag is designed to work together with a seat belt. A person who is in an accident without a seat belt on has a greater risk of injury or death than a person wearing a seat belt. Many students may also believe that not wearing a seat belt will enable them to be thrown clear of an accident. Tell students that a person thrown from a car in an accident typically suffers grievous injury or dies.

## LEARNING STYLES

**Logical/Mathematical**

Show students examples of international road signs and discuss reasons for their use. Invite students to evaluate whether the designs and symbols clearly show what they need to show. Then have students create their own road signs for common road conditions, road safety hazards, or other information pertinent to a driver. Remind students that their signs need to be easily understood by people who speak any language.

## Research and Write

Have students read and conduct the Research and Write feature on page 384. Students should be able to find the Web sites of organizations dedicated to head injuries on the internet.

## Lesson 3 Review Answers

1. driving course 2. reduce 3. protected
4. Sample answer: The driver of a school bus should observe the same guidelines for avoiding vehicle crashes. I can help prevent accidents by waiting until the bus stops and the door opens before stepping into the street. 5. Sample answer: If you don't ride single file, one rider could be too far out into the lane and in the path of vehicles.

## Portfolio Assessment

Sample items include:
- Illustrated safety guide for ELL/ESL Strategy
- Report on consequences of nonfatal head injury for Research and Write
- Lesson 3 Review answers

## Decide for Yourself

Have students read the Decide for Yourself feature on page 384 and answer the questions. Demonstrate the proper use of a helmet and other safety equipment for the students.

### Decide for Yourself Answers

1. Wearing a helmet is important because it protects your skull and brain in case of a fall or collision. 2. A good fit keeps the helmet in place during a fall. 3. Sample answer: I would say, "It is very dangerous to ride a bike at night without lights or reflectors. I will remove my bike's headlight and taillight and loan them to you. I can also loan you my reflectors."

---

**Research and Write**

Car crashes can cause head injuries. So can falling from a bicycle or motorcycle. Investigate the possible consequences of a nonfatal head injury. Write a two-page report on your findings.

**Lesson 3 Review** On a sheet of paper, write the word or phrase in parentheses that best completes each sentence.

1. The best prevention for vehicle crashes is a (driving course, reflector, safe distance).
2. It is important to (increase, maintain, reduce) speed with dangerous road conditions.
3. Bicyclists and motorcyclists are less (noisy, alert, protected) than people in closed vehicles.

On a sheet of paper, write the answers to the following questions. Use complete sentences.

4. **Critical Thinking** How do the tips for preventing vehicle crashes apply to school buses? What can you do to make your school bus ride safe?
5. **Critical Thinking** Why is it important for bicyclists to ride single file?

### Decide for Yourself

In the United States, more than 500,000 people are injured riding bicycles each year. In 1999, 750 of these people died. More than 25 percent of the deaths were of children ages 5 to 15. More than 95 percent of those killed were not wearing helmets. About 140,000 children visit emergency rooms each year for head injuries they receive while bicycling.

Helmets are the most important safety equipment you can wear while bicycling. A helmet protects your skull and brain if you fall. Helmets vary in quality, however. The most important thing in choosing one is to find one that fits you properly. The strap should keep the helmet from sliding to one side or to the back.

Other safety equipment includes gloves to protect your hands if you fall. Pads can protect your knees and elbows.

Putting a flag and electric lights on your bike will help drivers of cars and other vehicles see you. Reflective materials on the bike or on clothing help, but lights are the best.

On a sheet of paper, write the answers to the following questions. Use complete sentences.

1. Why is wearing a helmet important?
2. Why is finding a helmet that fits important?
3. Suppose a friend plans to ride a bike home from your house after dark without lights or reflectors. What would you say to your friend?

384  Unit 6  Injury Prevention and Safety Promotion

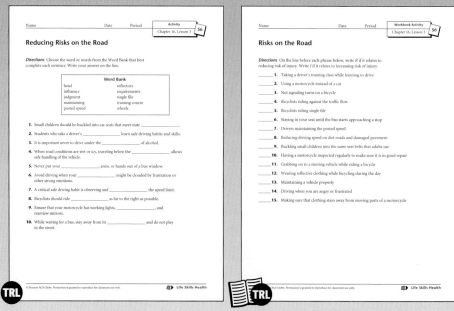

# Lesson 4: Safety During Natural Disasters

**Objectives**

After reading this lesson, you should be able to
- explain what hurricanes, tornadoes, blizzards, floods, and earthquakes are
- describe what to do before, during, and after natural disasters

**Hurricane**
A severe tropical storm with heavy rains and winds above 74 miles per hour

**Tornado**
A funnel-shaped column of wind with wind speeds up to 300 miles per hour

Natural disasters are sudden emergencies that result from causes in nature. Some natural disasters can be predicted. Others cannot be predicted. People who have no warning can be injured or killed in natural disasters, which cause millions of dollars in damage. With warning and preparation, however, people can survive these disasters with few or no injuries.

## How Can People Prepare for a Hurricane?

A **hurricane** is a severe tropical storm with heavy rains and winds above 74 miles per hour. Most hurricanes can be predicted. They affect the southern and eastern coasts of the United States. When a storm begins to develop, the National Weather Service sends out a hurricane watch or warning. A watch means that a hurricane may occur. A warning means that one is expected within the next 24 hours. People can use radios, TVs, or the Internet to get updates and instructions. In dangerous situations, people are told to quickly leave certain areas.

To prepare for a hurricane, tape or board windows and bring in items from outside. If advised to leave your home, go to a designated place until conditions are safe again. Before leaving home, shut off water, gas, and electric power. Avoid any downed power lines to prevent being killed by electric shock. If the hurricane is close, do not drive. Stay indoors on the downwind side of the house or building.

## How Can People Prepare for a Tornado?

Another natural disaster involving high winds and heavy rain is a **tornado**. A tornado is a funnel-shaped column of wind with wind speeds up to 300 miles per hour. Usually, tornadoes happen in the central and southern parts of the United States. These dangerous storms can be predicted. The National Weather Service sends out a tornado watch when weather conditions make it possible for tornadoes to form. A tornado warning is issued when a tornado is formed and is threatening a certain area.

### 1  Warm-Up Activity

Review the safety procedures for natural disasters common in your area with the class. Ask students what they should do if the area experiences a particular disaster while they are in school.

### 2  Teaching the Lesson

Lead a class discussion about students' experiences with an adverse weather condition. Ask if the condition posed a danger to them and to their family. Did they expect the condition? How did they prepare for it? Tell students that as they read the lesson, they may learn ways to prepare for a particular weather condition that they were previously unaware of.

### 3  Reinforce and Extend

**Technology and Society**

Have students read the Technology and Society feature on page 386. Tell them that the popularity of cell phones has added a complication to 911 phone services. People who dial 911 from a wired phone can be tracked to their location even if they cannot speak. However, many cell phones cannot be tracked by emergency personnel. New technology is being added to cell phones to make them easy to locate, but at this time cell phones are less reliable for summoning help than a wired phone.

#### IN THE ENVIRONMENT

Wetlands are nature's means of flood control. When wetlands are destroyed for development, much of the rain that would have soaked into the wetland runs into low lying developed areas or into nearby streams that may overspill their banks. Invite students to survey their community for any wetlands and find out what kinds of wildlife live there. Have students find out how the wetlands in their area protect against flooding.

Figure 16.4.1 *Tornadoes can cause serious damage and injury.*

If you are in the area of an approaching tornado, take shelter at once. Go to the basement and stay there. If you do not have a basement, go to the center of the house on the lowest floor. Crouch or lie flat in a closet or bathtub and cover your head. People who live in mobile homes should go to a tornado shelter. If there is no shelter, they should leave their homes and head for low, protected ground.

If you are outside and you have time, move away from a tornado at right angles to its path. If the tornado is too close, lie facedown in a ditch or the nearest low spot.

**Technology and Society**

People in many areas of the country can dial 911 to speak with emergency fire and police services. New technology is being developed to alert people to natural disasters. The technology enables the National Weather Service to make more than 1,000 calls per minute to people in targeted areas.

**Flood**
*A natural disaster that occurs when a body of water overflows and covers normally dry land*

**Flash flood**
*A dangerous flood that occurs suddenly without warning because of heavy rains*

**Earthquake**
*The shaking caused when the major slabs of rock in the earth's crust shift and move*

## How Can People Prepare for a Flood?

A **flood** is a natural disaster that occurs when a body of water overflows and covers normally dry land. Floods may result from severe rainstorms or from unusually heavy winter snowmelts in the spring. Most floods can be predicted hours to days in advance. Occasionally, however, very heavy rains cause dangerous **flash floods** suddenly and without warning.

If people live in flood areas, they need to know where to go if they are ordered to move to high ground. If they have time before leaving, they should turn off gas, electricity, and water in their homes. They should also move their belongings to an upper floor if they have one.

During a flood, do not try to walk across a stream where water is at or above knee level. Do not drive over flooded roads or land. If your car stalls because of floodwater, leave it and go to higher ground at once.

After a flood, do not touch electrical equipment in wet areas. Throw away any foods or liquids that the floodwater touched. Drink only bottled or boiled water until you are told the water supply is safe again.

## How Can People Prepare for an Earthquake?

Another type of natural disaster is an **earthquake.** An earthquake happens when plates—major slabs of rock in the earth's crust—shift and move, causing the earth to shake. Most earthquakes in the United States occur along the West Coast. Scientists cannot yet predict when an earthquake will occur. Even though earthquakes cannot be predicted, people in earthquake areas can prepare emergency plans.

To prepare for an earthquake, store emergency supplies such as food, water, flashlights, extra batteries, and a first aid kit. Learn where the main gas valve is in a house or building and know how to shut it off. Either move or secure pictures and mirrors so they will not fall off the walls.

---

## LEARNING STYLES

### Body/Kinesthetic

Present a model of how the earth's plates shift to produce an earthquake. Hold up two board erasers side by side. Each eraser represents a plate. Press the plates against each other, building up stress. Let the plates move suddenly a little past each other. This is the earthquake. You might place a few piles of sugar cubes on top of the erasers so that students can see how the resultant shaking damages buildings and other structures on the earth's surface. After this demonstration, let students conduct their own demonstration by snapping their fingers. The thumb and middle finger represent two plates, or two other smaller slabs of rock. As students press their thumb and finger together, stress builds up. Eventually, enough stress builds up that the thumb and finger slide suddenly past each other, releasing energy in the form of sound waves—a snap. These sound waves are like the waves that travel out from the origin of an earthquake and cause damage.

## LEARNING STYLES

### Visual/Spatial

Have students design a magazine ad to sell an emergency supply kit. Ads should include drawings or pictures of equipment that would be included in the kit. Students' ideas for equipment might include flashlights, batteries, bottled water, sealed containers of food, can opener, battery operated radio, blanket, scissors, rubber gloves, pocket mask or face shield, first aid supplies, and so on.

## LEARNING STYLES

**Visual/Spatial**
Bring to class a photograph of a tree that has been struck by lightning. Most books on weather and many encyclopedias have such pictures. Ask students why it is unsafe to seek shelter under a tree during a thunderstorm. *(Lightning usually strikes tall objects such as trees.)* Give students a chance to respond; then show them the picture. It should serve as a reminder of the dangers of seeking shelter beneath a tree. Tell students that when lightning strikes a tree, the sap within the tree heats up so quickly and intensely that it explodes from the tree, shattering the wood.

## GLOBAL CONNECTION

Have students research a type of natural disaster named in this lesson that occurred in another part of the world. Encourage students to find information about the event from a variety of library resources as well as the Internet. Ask students to write a report about the disaster and how people dealt with it. A map showing the location of the disaster should be included with the report. You may wish to have volunteers present their report to the class.

---

During an earthquake, duck under a strong table or desk and cover your face and head if you are indoors. Hold this position until the shaking stops. Stay away from glass and do not use elevators or stairs. If you are driving, move to the shoulder of the road, stop, and stay in the car. Do not stop under bridges, overpasses, or power lines. If outside, stay in a clear space away from buildings, walls, trees, and power lines.

After an earthquake, turn off the gas valve if you smell gas to avoid a fire from a broken gas line. Stay away from downed power lines and keep out of damaged buildings.

### How Can People Prepare for a Blizzard?

A natural disaster that occurs in winter is a **blizzard.** A blizzard is a violent snowstorm with winds above 35 miles per hour that lasts for at least 3 hours. The temperature during a blizzard is at or below 20°F, and there is almost no visibility. The National Weather Service can predict blizzards, which occur mainly in the northern plains of the United States.

During a blizzard, stay in your home or a warm shelter. Do not travel even short distances. If there is a medical emergency, call 911 or alert local authorities for help. If the heat goes out in your home, dress in layers of warm clothing and sleep under several blankets. Turn off water main and open taps to avoid frozen water pipes.

If you live in an area where blizzards occur, carry emergency gear in your vehicle during the winter. Emergency gear includes blankets, flashlights, snacks, warm mittens and hats, extra socks, and warm boots. If your vehicle becomes stuck during a blizzard, stay in it. Run the vehicle only a few times each hour for heat. Clear away snow from the tailpipe to keep car exhaust out of the vehicle. Keep moving and stay awake to keep from freezing.

**Blizzard**
*A violent snowstorm with winds above 35 miles per hour that lasts for at least 3 hours*

Keep a supply of new batteries for radios and flashlights on hand at all times.

**Lightning**
*An electrical discharge that commonly occurs during thunderstorms*

### How Can People Prepare for Lightning Strikes?

Another natural disaster is a **lightning** strike. Lightning is an electrical discharge that commonly occurs during thunderstorms. Lightning is dangerous because it can strike and kill people. The National Weather Service can predict thunderstorms, which occur in certain seasons in most areas of the United States.

During a thunderstorm, stay indoors or in a closed vehicle. Unplug electrical equipment and use telephones with cords only in an emergency. (Cell phones are safe during thunderstorms.) If you are outdoors, find shelter away from trees, tall objects, and water. That is because lightning usually strikes tall objects or water. Lie down in the lowest possible place if you cannot get to shelter. Whether indoors or outdoors, stay away from metal objects during a thunderstorm.

### How Can Communities Prepare for Disasters?

Usually, communities have plans to follow in case of natural disasters. For example, if a tornado or severe storms are approaching, sirens sound to alert people to take shelter. People who make these plans usually include police, hospital, fire, and public health representatives. The planners may also include local Red Cross, United Way, and the Federal Emergency Management Agency (FEMA) representatives. FEMA is part of the U.S. Department of Homeland Security. FEMA's purpose is to help people before, during, and after a disaster.

The agencies these people represent can teach people how to prevent accidents, as well as how to respond in disasters. Many of these agencies offer courses to train people for emergencies. For example, the Red Cross offers courses in first aid and lifesaving.

## IN THE COMMUNITY

Assign students to find out what plans your community has for responding to natural disasters. Some questions to guide students in their research include: What agencies are involved? Where do the agencies get their information? What kind of warnings are given? How adequate are the warnings? Do people know what to do when the warnings are given? Is enough being done to help them prepare? What else could be done to help them prepare?

## ELL/ESL STRATEGY

**Language Objective:**
*To understand safety procedures*

English language learners may have difficulty understanding class safety procedures in case of a natural disaster. Have these students work with a partner who is proficient in English to create either a set of procedure steps in their native language or an illustrated procedure guide.

## HEALTH JOURNAL

Have students write a story in their journal that describes the occurrence of a natural disaster in your area. They should include a narrative that describes the actions taken by the people that experience the disaster, including the preparations they have made.

## Lesson 4 Review Answers

1. flash flood 2. lightning 3. blizzard
4. Sample answer: Hurricanes and tornadoes both have high winds and cause destruction. A tornado has higher wind speeds. Hurricanes affect the southern and eastern coasts of the United States; tornadoes usually happen in the central and southern parts of the United States. 5. Sample answer: You need to be prepared for earthquakes at all times because they cannot be predicted. Hurricanes can be predicted and may involve evacuation in advance.

## Portfolio Assessment

Sample items include:
- Magazine ad from Learning Styles
- Disaster Report from Global Connection
- Lesson 4 Review answers

**Word Bank**
blizzard
flash flood
lightning

***Lesson 4 Review*** On a sheet of paper, write the word or phrase from the Word Bank that best completes each sentence.

1. A _____ occurs suddenly and is caused by heavy rains.
2. To avoid injury by _____, you should use telephones with cords only in an emergency during a thunderstorm.
3. A _____ lasts longer than 3 hours.

On a sheet of paper, write the answers to the following questions. Use complete sentences.

4. **Critical Thinking** How are hurricanes and tornadoes alike and different?
5. **Critical Thinking** How is hurricane preparation different from earthquake preparation?

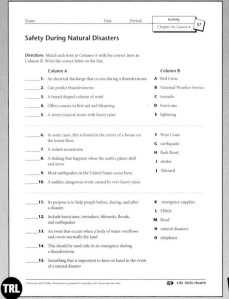

Activity 57

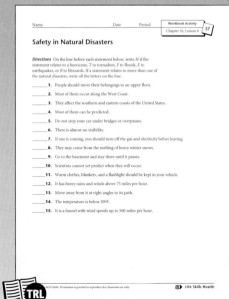

Workbook Activity 57

# Chapter 16 SUMMARY

- Serious falls can happen at home. Installing simple equipment and using some preventive measures can help you avoid falls.

- Fires caused by electrical problems, fuel and fire starters, space heaters and furnaces, and open flames can all be prevented.

- Every home should have one or more smoke detectors and fire extinguishers.

- A few simple measures can prevent poisoning or electrical shocks.

- The Occupational Safety and Health Administration (OSHA) sets standards for safety in the workplace. Employers are responsible to reduce and remove dangers in the workplace. Employees can reduce risks by following procedures and requirements.

- Learning to swim can prevent drowning. Survival floating and following other water safety guidelines also can prevent drowning.

- Sports injuries can be prevented by using equipment properly, wearing protective gear, and learning necessary skills.

- Learning safe driving skills and habits, maintaining the speed limit, and driving according to conditions all help prevent vehicle crashes. Wearing a seat belt can reduce injuries if a crash occurs. Following other tips can help avoid crashes.

- Bicyclists and motorcyclists can be seriously injured in an accident because they are less protected than people in larger vehicles. Wearing helmets, following the rules of the road, and taking other preventive actions can reduce risks of accidents.

- Natural disasters cannot be prevented and may or may not be predictable. People can be prepared and take protective steps in case of hurricanes, tornadoes, floods, earthquakes, blizzards, or lightning strikes.

## Chapter 16 Summary

Have volunteers read aloud each Summary item on page 391. Ask volunteers to explain the meaning of each item. Direct students' attention to the Vocabulary box on the bottom of page 391. Have them read and review each term and its definition.

## ONLINE CONNECTION

The Automobile Safety Foundation posts information about car safety hazards and new technology to improve automobile safety: www.carsafe.org

The United States Department of Homeland Security provides information on planning for natural disasters: www.dhs.gov/dhspublic/

| Vocabulary | | |
|---|---|---|
| blizzard, 388 | flash flood, 387 | lightning, 389 |
| earthquake, 387 | flood, 387 | smoke detector, 371 |
| electrical shock, 373 | hazard, 374 | survival floating, 378 |
| fire extinguisher, 371 | hurricane, 385 | tornado, 385 |

## Chapter 16 Review

Use the Chapter Review to prepare students for tests and to reteach content from the chapter.

## Chapter 16 Mastery Test

The Teacher's Resource Library includes two forms of the Chapter 16 Mastery Text. Each test addresses the chapter Goals for Learning. An optional third page of additional critical-thinking items is included for each test. The difficulty level of the two forms is equivalent.

### Review Answers

**Vocabulary Review**

1. fire extinguisher  2. electrical shock
3. hazard  4. survival floating  5. smoke detector  6. flood  7. flash flood  8. blizzard
9. tornado  10. hurricane  11. earthquake
12. lightning

# Chapter 16 REVIEW

**Word Bank**
blizzard
earthquake
electrical shock
fire extinguisher
flash flood
flood
hazard
hurricane
lightning
smoke detector
survival floating
tornado

## Vocabulary Review

On a sheet of paper, write the correct word or phrase from the Word Bank that best completes each sentence.

1. A(n) _____ is a portable device containing chemicals that put out small fires.
2. A flow of electric current through the body is _____.
3. An unprotected electrical outlet is a safety _____.
4. A technique to conserve energy in warm water while waiting to be rescued is _____.
5. When there is smoke or excess heat, a(n) _____ sounds.
6. A(n) _____ occurs when water covers land that is normally dry.
7. A(n) _____ is a flood that occurs suddenly.
8. A violent snowstorm is a(n) _____.
9. A(n) _____ is a funnel-shaped column of wind.
10. A(n) _____ has winds above 75 miles per hour.
11. Preparation for a(n) _____ includes anchoring heavy items that might fall.
12. An electrical discharge that occurs during a thunderstorm is _____.

## Concept Review

On a sheet of paper, write the letter of the answer that best completes each sentence.

13. The federal government agency in charge of disaster aid is _____.
   A  OSHA
   B  the FBI
   C  FEMA
   D  the National Weather Service

**392**  Unit 6  *Injury Prevention and Safety Promotion*

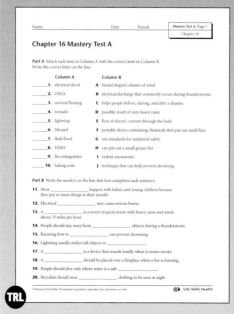
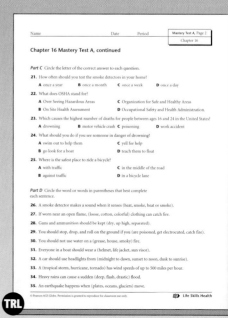

Chapter 16 Mastery Test A, pages 1–3

**14.** Most injuries that occur at home are the result of _____.
   A drowning   C poisoning
   B falls      D fire

**15.** A natural disaster in which people must seek higher ground is a _____.
   A flood      C hurricane
   B blizzard   D tornado

**16.** A natural disaster in which people must go to basements or lower floors is a _____.
   A flood      C hurricane
   B blizzard   D tornado

**17.** Electrical fires can be caused by _____.
   A burning candles   C overloaded outlets
   B gas stoves        D dirty chimneys

## Critical Thinking

On a sheet of paper, write the answers to the following questions. Use complete sentences.

**18.** Which home safety tips are your responsibility? Which are the responsibility of adults in your home?

**19.** You may have heard the caution that you "should not fool with nature." Why do you think this is especially true during natural disasters?

**20.** Which parts of school and school bus safety can you control? Which are under someone else's control?

> **Test-Taking Tip** After you have completed a test, reread each question and answer. Ask yourself: Have I answered the question completely?

Reducing Risks of Injury  Chapter 16  393

## Concept Review
13. C  14. B  15. A  16. D  17. C

## Critical Thinking

**18.** Sample answer: I can take precautions during my own activities, learn an evacuation plan, and avoid creating hazards. Adults must provide safety equipment, repair things that can cause hazards, and do things such as installing handrails, tightening the stair rail, and fixing wiring. **19.** Sample answer: This is especially true during natural disasters because they can destroy all the human-made things in their path, such as houses and buildings. The best way to deal with natural disasters is to get out of their way. People get hurt when they underestimate the severity of a natural disaster. **20.** Sample answer: As a student, I can pay attention during fire drills, I can avoid running in the halls, and I can wait until it is safe to exit the bus. School administrators must keep the school free of hazards and ensure that the buses are in good repair. School administrators must ensure that the school has an adequate number of smoke detectors, sprinklers, and fire extinguishers. The school's bus drivers must drive safely.

## Alternative Assessment

Alternative Assessment items correlate to the student Goals for Learning at the beginning of this chapter.

- Have students create a pamphlet that describes ways to reduce the risks of falls, fire, poisoning, and electrical shock at home.
- Have students orally identify ways to reduce risks at work and during recreational activities.
- Have students write a paragraph that explains how to reduce the risks of vehicle crashes and increase bicycle and motorcycle safety.
- Have students create a poster that illustrates and describes ways to stay safe during common natural disasters.

Chapter 16 Mastery Test B, pages 1–3

Reducing Risks of Injury  Chapter 16  393

# Chapter at a Glance

## Chapter 17: Applying First Aid to Injuries
pages 394–415

### Lessons
1. First Aid Basics
   pages 396–400
2. First Aid for Life-Threatening Emergencies
   pages 401–406
3. First Aid for Poisoning and Other Problems
   pages 407–412

**Chapter 17 Summary** page 413

**Chapter 17 Review** pages 414–415

**Audio CD**

**Skill Track for Life Skills Health**

**Teacher's Resource Library** TRL
- Activities 58–60
- Alternative Activities 58–60
- Workbook Activities 58–60
- Self-Assessment 17
- Community Connection 17
- Chapter 17 Self-Study Guide
- Chapter 17 Outline
- Chapter 17 Mastery Tests A and B

(Answer Keys for the Teacher's Resource Library begin on page xxx of this Teacher's Edition.)

## Opener Activity

Ask students if they ever had a medical emergency requiring first aid. Possibilities include a broken limb, a bad cut, a sprained ankle, passing out from excessive heat, etc. Invite students to tell what first aid treatments, if any, were provided at the scene. Ask students to explain if the treatment kept the injury or illness from worsening.

# Self-Assessment

## Can you answer these questions?

1. What are the basic guidelines for first aid?
2. When should you call 911 and what information should you provide?
3. What laws protect people who help others in emergencies?
4. What should be placed in a first aid kit?
5. How can you help someone who is choking?
6. How can you help someone who has stopped breathing?
7. How can you help someone who is in shock?
8. How can you help someone who is bleeding heavily?
9. How can you help someone who has been poisoned?
10. For what types of injuries do you know how to give first aid?

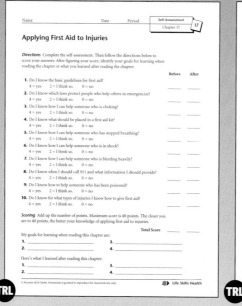

394  Unit 6  Injury Prev. & Safety Promo.

# Chapter 17
## Applying First Aid to Injuries

In an emergency, it is important to closely follow certain rules for giving first aid. In some emergencies, first aid skills can mean the difference between life and death. If you know first aid, you can spot the signs of a serious illness or injury. Then you will be able to do the proper things to help someone. Giving the right first aid can help make sure that these illnesses or injuries do not become more serious.

In this chapter, you will learn the basic guidelines for first aid. You will also find out about laws that protect you if you help someone who needs first aid. You will learn the signs of choking, respiratory failure, cardiovascular failure, shock, and serious bleeding. You will learn how to give first aid for life-threatening emergencies, poisoning, and common injuries.

### Goals for Learning
- To identify basic guidelines for first aid
- To describe first aid steps in five life-threatening emergencies
- To explain first aid for poisoning and common injuries

395

## Introducing the Chapter

Have a volunteer read the introductory paragraphs on page 395. Have students explain why it is important to know basic first aid techniques. *(You can help yourself and others before emergency technicians arrive at an emergency scene.)* Ask students why it is important to administer first aid properly. *(Proper first aid ensures that an illness or injury does not become more serious.)*

Have students read the Goals for Learning. Explain that in this chapter students will learn more about first aid techniques and how to respond to certain emergency situations.

### Notes and Questions

Ask volunteers to read the notes and questions that appear in the margins throughout the chapter. Then discuss them with the class.

### Self-Assessment

Have students complete the Self-Assessment worksheet before and after reading the chapter. Before reading the chapter, have students fill in the "Before" column. Ask students to identify their goals for learning. To get ideas for setting goals, students might use the chapter introductory material on pages 395, the checklist on page 394, or the questions on the Self-Assessment worksheet. Students can use the back of the worksheet if they need more space to write.

Collect the Self-Assessment worksheets and pass them out again at the end of the chapter. Have students fill in the "After" column. Ask them to identify at least four major points they have learned. Again, suggest they use the back of the worksheet if they need more space to write. You may want to collect and review the worksheets, but return them to students so they have a record of their goals and accomplishments.

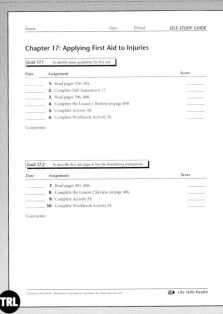

Chapter 17 Self-Study Guide, pages 1–2

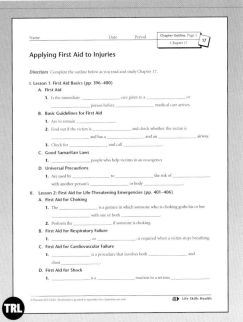

Chapter 17 Outline, pages 1–2

## Lesson at a Glance

### Chapter 17 Lesson 1

**Overview** This lesson explains the basic guidelines for providing first aid and reviews some precautions that protect the first aid provider.

### Objectives

- To explain the basic guidelines for first aid
- To understand how to help someone who needs first aid
- To discuss Good Samaritan laws
- To describe universal precautions

**Student Pages** 396–400

**Teacher's Resource Library**

Activity 58

Alternative Activity 58

Workbook Activity 58

### Vocabulary

dressing
first aid
Good Samaritan laws
latex
universal precautions

Write the vocabulary words on the board and ask students to provide definitions based on prior knowledge. Then read the definitions from the textbook. Have students compare their previous ideas with the definitions.

### Background Information

The 911 Universal Emergency Telephone Number system was first introduced to the American public in 1968. The numbers 911 were chosen because they were not being used in the United States or Canada as the first three numbers of any phone number or area code.

### Research and Write

Have students read the Research and Write feature on page 396. Have students work in groups to complete the research and the report. Students can use the American Red Cross Web site at www.redcross.org to help with research. Have students share their reports with the class.

**396** Unit 6  *Injury Prev. & Safety Promo.*

---

### Lesson 1  First Aid Basics

**Objectives**

*After reading this lesson, you should be able to*

- explain the basic guidelines for first aid
- understand how to help someone who needs first aid
- discuss Good Samaritan laws
- describe universal precautions

**First aid**
*Immediate emergency care given to a sick or injured person before professional medical care arrives*

What can you do at home to prepare for a medical emergency? List what every family member should know, such as the location of a first aid kit.

**First aid** is immediate emergency care. It is given to a sick or injured person before professional medical care arrives. Performing the proper emergency steps can lessen the pain and seriousness of someone's injuries. It can even save a life.

#### What Are the Basic Guidelines for First Aid?

The most important thing to do for someone in an emergency is to remain calm. Then you will be better able to follow these guidelines:

1. Look around at your immediate surroundings to see if there are any dangers to either one of you. Move the person only in case of danger, such as drowning or fire. Make sure you will be safe while helping the person.

2. Find out if the person is conscious, or awake. Tap on the person's shoulder. You can also loudly ask if the person is all right. If he or she is not conscious, do not leave. Ask someone else to call for help. If you are alone, quickly make the call. Then immediately return to take care of the person.

3. Check whether the person is breathing. Make sure he or she has a pulse and open airway. If not, try to clear the airway and perform rescue breathing. Rescue breathing is described in Lesson 2 of this chapter.

**Research and Write**

Research the American Red Cross. Find out who started this organization and for what reason. Write a report that includes how many people the Red Cross helps each year. Include the kinds of emergencies in which the Red Cross helps. Also, include information about how the Red Cross is run and the people who work for the Red Cross.

**396**  Unit 6  *Injury Prevention and Safety Promotion*

**Latex**
*A kind of rubber used to make thin gloves for health care professionals*

Practice knowing when to use 911 with younger brothers or sisters. Ask: "What happens if the adult at home is hurt and cannot tell you what is wrong?" Explain: "Dial 911. Tell your name. Tell our address. Tell what is wrong. Don't hang up."

4. If the person is conscious, ask permission to give help. Check the person's whole body for any other injuries. Check for emergency medical identification, such as a tag, bracelet, or card. This identification will tell any medical problems the person may have. These problems may include a heart condition, diabetes, or allergies to medicines.

5. Check for bleeding. If you have **latex** gloves, put them on. Latex gloves are very thin gloves made of a special type of rubber. You have probably seen them in a doctor's office or hospital. The latex gloves will help protect you. After you put them on, press down on the wound. This will help stop the bleeding. If you do not have latex or plastic gloves, you can use several layers of cloth or a sheet of plastic to protect yourself from the blood.

6. If no one has called for help, call 911 or 0 for the operator. Explain that you need emergency medical help. Be prepared to say where you are and tell important details about the person you are helping. Answer any questions you are asked. Listen carefully to everything you are told to do. Stay on the line until the operator ends the call.

After you have done these things, stay with the person. Explain that more help is on the way. Do your best to keep the person from getting too cold or too warm. Shade him or her from sunlight. Loosen any tight clothing around the neck area if you can, but be careful not to move the person's spine or neck. Use common sense and do only what you have been trained to do. Make sure your help will not hurt the person.

### Link to >>>
**Social Studies**

In 1797, French doctor Dominique-Jean Larrey went to Italy to treat injured French soldiers. He used "flying ambulances." These were wagons pulled by horses. The wagons could quickly take the wounded from the battlefield to get medical care. Larrey's ideas led to the kind of ambulances in use today.

 **Warm-Up Activity**

Ask students what they think are some things they should do if they found a person lying unconscious on the floor. Write their responses on the board. *(Possible answer: Check for pulse and breathing, call 911, and check the area for other possible dangers.)* Ask students to list some things they should not do. *(Possible answer: Do not panic or move the person unnecessarily.)* Have students read pages 396-397 to find correct, complete answers to these questions.

 **Teaching the Lesson**

Have each student make a flow chart that sequences the steps of what to do in an emergency situation. Creating the flow charts will help students organize the material in this lesson. Also have each student make a web diagram that contains information about steps to take for universal precautions.

 **Reinforce and Extend**

### Link to

**Social Studies.** Ask a volunteer to read aloud the Link to Social Studies feature on page 397. Hold a discussion on the importance of ambulances in emergency situations today. What are the advantages of using ambulances? *(They can arrive more quickly than people can reach the hospital on their own; people might not be able to get to the hospital on their own; and in an ambulance, first aid can be started before people reach a hospital.)*

## ELL/ESL Strategy

**Language Objective:**
*To practice functional language skills*

Have students learning English practice saying in English the information they would need to tell a 911 dispatcher. Pair students learning English with students who have a strong grasp of English. Assign the role of 911 dispatcher to students who have a strong grasp of English. Students learning English should learn the kinds of things the 911 dispatcher will say to them, know what those phrases mean, and know how to respond.

## Career Connection

Emergency Medical Technicians (EMTs) respond to medical emergencies, evaluate injuries, initiate basic first aid, and transport victims to a hospital. Another emergency medical professional is a paramedic, whose education is similar to but more advanced than the education of an EMT. Invite an EMT and a paramedic to speak to your class and answer questions about their professions. Make sure to ask them about the similarities and differences between their careers.

## Learning Styles

**Interpersonal/Group Learning**

Have students work in groups to practice the steps in the basic first aid guidelines. Have students role play different parts, including a person who has a head injury, a person who is giving first aid, a bystander, and one person who will check off the list of basic first aid guidelines to make sure everything is done in the right order.

---

*Good Samaritan laws*
Laws that protect people who help others in emergencies

*Universal precautions*
Methods of self-protection that prevent contact with a person's blood or body fluids

*Dressing*
Layers of cloth and other materials that protect a wound

## What Are Good Samaritan Laws?

To protect people who help others in emergencies, many states have **Good Samaritan laws.** These laws state that rescuers use common sense and skills within their training. You can get this training in first aid classes. These classes are offered through groups such as the American Red Cross and the American Heart Association. You are not expected to risk your own life when you help someone else.

## What Are Universal Precautions?

Medical workers protect themselves by using **universal precautions.** Universal precautions keep workers from having contact with another person's blood or other body fluids. These actions help lessen the risk of passing on diseases such as AIDS.

Here are some universal precautions:

- Workers wear latex gloves and other protective gear, such as masks.
- Workers are careful to properly throw away any materials that have come in contact with body fluids. These materials may include bandages and **dressings.** Dressings are layers of cloth and other materials that protect a wound.
- Workers carefully wash their hands after they finish.

If you ever give first aid, you should follow these rules. They protect both you and the person you are helping.

### Health in Your Life

**Protecting Your Family**

Here is a list of ways to protect family members during emergencies:

- Keep cleaning items and medicines in the boxes and bottles they came in. Keep them in separate places. Never store gasoline or other poisonous items in soft-drink bottles.
- Keep all medicines and cleaning items out of the reach of children.
- Mark all poisonous items with a sticker that shows children should stay away. This includes cleaning items. Poison control centers may have these stickers available at no charge.
- Place nonskid mats in bathtubs.
- Do not give or take medicine if you just woke up and the room is still dark. Wait until you turn on the lights and are fully awake.
- Put together a first aid kit. The American Red Cross recommends that you put the following items in your kit: activated charcoal (use only if told to by a poison control center); adhesive tape; antiseptic ointment; bandages (several sizes); blanket; cold pack; disposable latex gloves; gauze pads and roller gauze (several sizes); hand cleaner; plastic bags; scissors and tweezers; small flashlight and extra batteries; syrup of ipecac (use only if told to by a poison control center).

On a sheet of paper, write the answers to the following questions. Use complete sentences.

1. Why do you think it is important to keep cleaning items and medicines in the boxes and bottles they came in?
2. Why do you think it is important to place nonskid mats in bathtubs?
3. Why do you think it is a good idea to have disposable latex gloves in a first aid kit?

---

### Health in Your Life

Have students read and complete the Health in Your Life feature on page 399. Have students explain why it is not a good idea to take medicine immediately after you have woken up in the middle of the night. *(You might not take the right medicine; you might not realize how much you are taking; you might not measure properly; you might accidentally take something that is not medicine.)*

**Health in Your Life Answers**

**1.** Sample answer: If cleaning items and medicines aren't in the boxes or bottles they came in, people could mistake them for something else and could accidentally misuse them. **2.** Sample answer: Bathtubs are slippery, so it is important to use nonskid mats to keep from slipping and falling. **3.** Sample answer: It is a good idea to have disposable latex gloves in case someone is bleeding. Latex gloves protect the victim and the rescuer.

### HEALTH JOURNAL

Have each student use the information in the Health in Your Life feature to create a checklist for his or her home. Have each student complete the checklist and report how he or she improved emergency preparedness in his or her home.

### TEACHER ALERT

Remind students that making a false emergency call to 911 could be a crime. In many states placing a false 911 call is a misdemeanor, and if a person is found guilty of the charge, he or she can be fined or be required to serve jail time. Have students research the law regarding false 911 calls in your state and report to the class on what they find.

# Lesson 1 Review Answers

1. The most important thing to do in an emergency is to stay calm. 2. To find out if a person is conscious, you should tap the person on the shoulder or ask loudly if he or she is all right. 3. Good Samaritan laws protect people who help others in an emergency. 4. Sample answer: You might need a medical professional to help an unconscious person. You might need to leave the person alone for a moment in order to call 911. If you stay with the person and do not call 911, medical professionals might not arrive in time to help the person. 5. Sample answer: The purpose of using universal precautions is to keep rescuers from having contact with another person's blood and body fluids. This reduces the likelihood of spreading diseases such as AIDS.

## Portfolio Assessment

Sample items include:
- Reports from Research and Write
- Flow charts and web diagrams from Teaching the Lesson
- Health in Your Life answers
- Checklist from Health Journal
- Lesson 1 Review answers

## Health at Work

Have students read the Health at Work feature on page 400. Have students explain why having good communication skills would be important to a 911 operator. *(A 911 operator must be able to talk with people who are hurt or under stress or who are possibly panicking. An operator might have to calm someone down enough to get useful information from him or her. An operator would have to be calm and reassuring during a crisis.)*

---

**Lesson 1 Review** On a sheet of paper, write the answers to the following questions. Use complete sentences.

1. What is the most important thing to do in an emergency?
2. How can you find out if a person is conscious?
3. What are Good Samaritan laws?
4. **Critical Thinking** Suppose you are helping someone who is not conscious. What is one reason you might leave the person alone?
5. **Critical Thinking** What is the purpose of using universal precautions?

### Health at Work

#### 911 Operator

A 911 operator answers calls for people who need emergency services. The operator asks questions to find out information.

In an emergency, the operator types the information into a computer. Then emergency help is sent.

It is important for a 911 operator to think fast and make decisions quickly. The operator can offer support until help arrives.

Usually, a 911 operator must have a high school diploma. Knowing how to work with computers and do office work sometimes is needed. Often, training is done on the job. A special certification might also be needed.

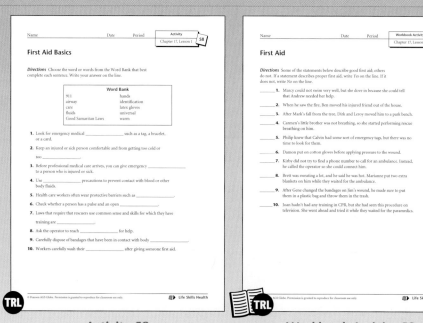

# Lesson 2  First Aid for Life-Threatening Emergencies

## Objectives

*After reading this lesson, you should be able to*

- understand how to give first aid for choking
- understand how to give first aid when someone is not breathing
- know how to give first aid to someone whose heart has stopped beating
- know how to give first aid to people in shock or who are bleeding

**Universal distress signal**
*A sign made by grabbing the throat with the hand or hands to show a person is choking*

**Heimlich maneuver**
*A way of giving firm upward thrusts below a person's rib cage to force out an object from the throat*

In first aid, it is important to treat life-threatening conditions before treating anything else. Choking, inability to breathe, heart attack, shock, and serious bleeding can all be life threatening. Minutes can mean the difference between life and death for people with these conditions. Knowing what to do before help arrives can make that difference.

## What Is First Aid for Choking?

Thousands of children and adults choke to death each year. Often, food or some other object blocks the airway or throat. It is important to recognize the **universal distress signal** for choking. This signal is known in many different places by people who speak many different languages. Someone who grabs the throat with the hand or hands is choking. Look at Figure 17.2.1. Other signs of choking include gasping, a weak cough, and inability to speak. A person may get pale or blue skin and pass out.

Choking is sometimes mistaken for a sign of a heart attack. Before giving first aid, always ask if a person is choking. If the person choking can still speak or cough, allow him or her to try to cough up the object. Never pound on the person's back to try to force out the object. This can make the problem worse.

**Figure 17.2.1** *Someone using the universal distress signal for choking needs help immediately.*

If the person choking cannot speak, cough, or breathe, perform the **Heimlich maneuver.** Firm, upward thrusts, or squeezes, just below the rib cage force air from the lungs. This pushes the object out of the throat.

## Lesson at a Glance

### Chapter 17 Lesson 2

**Overview** This lesson reviews basic first aid techniques such as the Heimlich maneuver, rescue breathing, and cardiopulmonary resuscitation (CPR). It also describes basic procedures to treat shock and severe bleeding.

### Objectives

- To understand how to give first aid for choking
- To understand how to give first aid when someone is not breathing
- To know how to give first aid to someone whose heart has stopped beating
- To know how to give first aid to people in shock or who are bleeding

**Student Pages** 401–406

**Teacher's Resource Library**
Activity 59
Alternative Activity 59
Workbook Activity 59

### Vocabulary

cardiac arrest
cardiopulmonary resuscitation (CPR)
chest compression
Heimlich maneuver
mouth-to-mouth resuscitation
pressure point
rescue breathing
respiratory failure
shock
sterile
universal distress signal

Have students take turns reading the vocabulary words and definitions aloud. Discuss each word and have volunteers use each word correctly in a sentence. Then challenge students to write a fictitious news article that describes an emergency situation and that uses as many of the vocabulary words as possible. Ask volunteers to read their articles to the class.

## Background Information

During rescue breathing, the victim is receiving another person's exhaled air. The air that a person inhales contains about 20 to 21 percent oxygen. The lungs use only 5 percent of that oxygen and exhale the other 15 to 16 percent. So during rescue breathing, the air that the victim is taking in contains about 15 to 16 percent oxygen, which is more than enough to keep vital organs alive and functioning.

**Warm-Up Activity**

Some students may have learned first aid techniques through the Girl Scouts, the Boy Scouts, local park districts, or other organizations. Ask these students to describe their experiences and to tell what goes on in a typical first aid lesson.

 **Teaching the Lesson**

Have students make a table to organize information about the first aid responses for different situations. Table column titles should include "Situation," "Description," and "Actions to Take." Tell students to use the table to study for tests.

 **Reinforce and Extend**

### LEARNING STYLES

 **Body/Kinesthetic**

Ask for volunteers to demonstrate the Heimlich maneuver. Be careful that the "rescuer" does not actually thrust into the abdomen of the "victim" so that no one is injured. After a few demonstrations, have students work in pairs to practice the correct position of the body and hands but, again, caution students against actual thrusting. As an alternative, students could demonstrate the thrusting motion on a pillow.

### IN THE COMMUNITY

 Teens often provide child care for younger children. Point out that these teens are responsible for the children. Discuss how knowledge of first aid techniques and other emergency procedures can be helpful and can provide peace of mind for the parents of the child. Mention that community organizations such as park districts, libraries, fire departments, or hospitals usually offer classes for teens on basic first aid. Have students research which organizations in their community offer these classes.

---

**Respiratory failure**
*A condition in which breathing has stopped*

The Heimlich maneuver is shown in Figure 17.2.2A:

1. Stand behind the person choking and wrap your arms around the person's waist.
2. Make a fist with one hand.
3. Place the thumb side of your fist against the middle of the person's abdomen above the navel but below the ribs.
4. Use your other hand to grab your fist. Give quick, upward thrusts into the abdomen.
5. Repeat this until the object is forced out.

If the blockage is not removed and the person passes out, lower the person face up on the floor. Open the person's mouth and check the throat for the object. Use the finger sweep shown in Figure 17.2.2B. To do the finger sweep:

1. Grasp the person's tongue and lower jaw between your thumb and fingers and lift the jaw.
2. Use your other index finger in a hooking action at the back of the tongue. Sweep the object out of the airway.
3. If this does not work, begin rescue breathing, which is described on the next page.

If you are alone and begin to choke, you can give yourself the Heimlich maneuver. Put your hands in the position shown in Figure 17.2.2A. You also can lean over with your stomach pushing into a firm object, such as a chair.

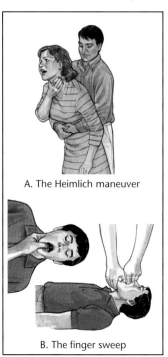

**Figure 17.2.2** *First aid for choking*

### What Is First Aid for Respiratory Failure?

Choking is one reason people stop breathing. Other reasons include electrical shock and drowning. **Respiratory failure,** or not breathing, is a life-threatening emergency. Signs that a person is not breathing are no noticeable breathing movement and blue fingernails, lips, and tongue.

**Rescue breathing**
Putting oxygen from a rescuer's lungs into an unconscious person's lungs to help the person breathe

**Mouth-to-mouth resuscitation**
Rescue breathing

**Rescue breathing** is needed when a person stops breathing, but the person's circulation and pulse continue. Rescue breathing puts oxygen from a rescuer's lungs into an unconscious person's lungs. Figure 17.2.3 shows the steps in rescue breathing, which is also called **mouth-to-mouth resuscitation.** The best way to learn how to do this is in your health class or another training class.

## What Is First Aid for Cardiovascular Failure?

When a person's heart or circulatory system stops working, it is a life-threatening emergency. The cardiovascular system may fail because of electrical shock, poisoning, drug overdose, or heart disease. It may also fail because a blood vessel is blocked, as in a stroke.

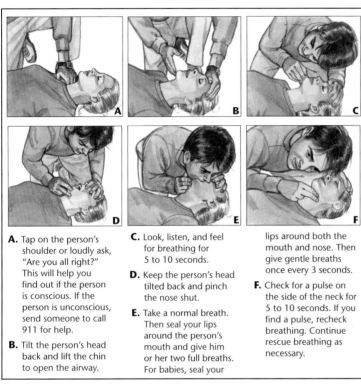

**A.** Tap on the person's shoulder or loudly ask, "Are you all right?" This will help you find out if the person is conscious. If the person is unconscious, send someone to call 911 for help.

**B.** Tilt the person's head back and lift the chin to open the airway.

**C.** Look, listen, and feel for breathing for 5 to 10 seconds.

**D.** Keep the person's head tilted back and pinch the nose shut.

**E.** Take a normal breath. Then seal your lips around the person's mouth and give him or her two full breaths. For babies, seal your lips around both the mouth and nose. Then give gentle breaths once every 3 seconds.

**F.** Check for a pulse on the side of the neck for 5 to 10 seconds. If you find a pulse, recheck breathing. Continue rescue breathing as necessary.

**Figure 17.2.3** The steps in rescue breathing

## HEALTH JOURNAL

Have students draw the universal distress signal for choking. Then have them list the steps involved in performing the Heimlich maneuver correctly.

## ELL/ESL STRATEGY

**Language Objective:**
*To practice oral interaction skills*

Have students learning English practice asking a person who is giving the universal signal for distress if he or she is choking. Then have them practice asking the person if they can help him or her. Make sure that students learning English also understand the responses of the person who might be choking. The responses might be nonverbal, as a person who is choking likely would not be able to speak.

## LEARNING STYLES

**Visual/Spatial**

Have a student volunteer help you demonstrate the correct positioning of rescuer and victim for the compressions used in CPR. If you have used a CPR dummy or mannequin earlier in the lesson, students can practice CPR techniques on it. To avoid injuries, do not let students practice compressions on each other.

## LEARNING STYLES

**Interpersonal/Group Learning**

Divide the class into groups of three or four students. Assign each group an accident situation. Examples include a choking person in a restaurant; an elderly person having a heart attack; a teen who was rescued from a pool who is not breathing but has a pulse; and a car accident victim with mouth injuries who is not breathing. Have students act out the proper first aid techniques for each scene. Have other classmates take notes while each group performs. Afterward, discuss the correctness of the actions performed in each situation.

---

**Cardiac arrest**
*A condition in which the heart has stopped beating and there is no pulse*

**Cardiopulmonary resuscitation (CPR)**
*A procedure for cardiovascular failure*

**Chest compression**
*When doing CPR, the act of pressing against the chest over and over*

Suppose you need to give CPR. You are nervous. How can you calm down?

**Figure 17.2.4** *Cardiopulmonary resuscitation (CPR)*

---

Signs of cardiovascular failure include a hard time breathing, an irregular pulse, a damp face, and pale or blue skin. The person may stop breathing altogether. If a person's heart has stopped beating and there is no pulse, he or she is in **cardiac arrest.** If a person shows these signs, call 911 immediately.

As you wait for help to arrive, you can perform **cardiopulmonary resuscitation (CPR).** This involves both rescue breathing and **chest compressions,** or presses. The goals of CPR are to open the person's airway and restore breathing and circulation. To perform CPR, follow these basic steps, which are shown in Figure 17.2.4:

1. Find the proper hand position.
   - Locate the notch at the lower end of the person's breastbone.
   - Place the heel of your other hand on the breastbone, next to your fingers.
   - Remove your hand from the notch and place it on top of your other hand.
   - Use only the heels of your hands, keeping your fingers off the person's chest.

2. Give compressions to the person in the following way:
   - Position your shoulders over your hands.
   - Compress, or press down on, the person's breastbone $1\frac{1}{2}$ to 2 inches. Do 30 compressions in about 10 seconds.
   - Compress down and up smoothly, keeping your hand in contact with the person's chest at all times.
   - After every 30 compressions, give the person two full breaths.

The best way to learn and practice CPR is in your health class or another training class.

**Shock**

*Failure of the circulatory system to provide enough blood to the body*

## What Is First Aid for Shock?

**Shock** can be a life-threatening emergency. It is a physical reaction to a serious injury, such as bleeding or a heart attack. When a person is in shock, the circulatory system fails to send enough blood to the body. Because shock can lead to death, steps should be taken at once to treat shock.

To tell if a person is in shock, look for specific signs. Signs of shock include a very fast or slow pulse, fast or slow breathing, and large pupils. Other signs are weakness, thirst, and an upset stomach. They may include cold and damp skin, as well as pale or blue skin.

If a person is in shock, call 911 for help at once. Then use these first aid steps for shock until help arrives:

1. Keep the person lying down. Raise the person's legs about 12 inches to help return blood to the heart. Do not raise the legs, however, if you think the person's neck or spine is injured.

2. Maintain the person's normal body temperature. Use blankets or other warm covers.

3. Do not give the person any food or water.

## What Is First Aid for Serious Bleeding?

Because serious bleeding can cause shock or death, it must be controlled quickly. If you are going to help someone who is seriously bleeding, you must do four things:

1. Stop the bleeding.
2. Protect the wound from infection.
3. Treat the person for shock.
4. Get emergency help immediately.

If you are helping someone with serious bleeding, you should first take steps to protect yourself. Latex gloves, plastic gloves, or a sheet of plastic can give you this protection. Several layers of cloth also offer protection.

*Applying First Aid to Injuries  Chapter 17*

### Technology and Society

Technology has made it possible to save lives inside ambulances. There are now air ambulances, as well as ambulances on the ground. Air and ground ambulances contain many kinds of equipment. Most have electrocardiograph machines and electronic defibrillators. The electrocardiograph machine shows heart activity. The electronic defibrillators can shock the heart back to a normal rhythm.

---

### Technology and Society

Have students read the Technology and Society feature on page 405. Have students research other materials or machines that can save a person's life that are carried in ambulances. Have students present reports on their findings to the class. If students have access to the appropriate technology, they can create multimedia reports—including graphics—to share with the class.

### PRONUNCIATION GUIDE

Use the pronunciation shown here to help students pronounce difficult words in the lesson. Refer to the pronunciation key that appears in the Glossary at the back of the Teacher's Edition for the sounds of the symbols.

defibrillator (di fib´ rə lā tər)

electrocardiograph
(i lek trō kär´ dē ə graf)

### ELL/ESL STRATEGY

**Language Objective:** *Identifying multiple meanings*

Students who are learning English may be confused by the different meanings of the word *shock*. Distinguish the medical use of the word from other common meanings such as electrical shock or surprise.

### TEACHER ALERT

Remind students again that they should take universal precautions when they give first aid, especially when they might come into contact with body fluids such as blood or saliva. Stress that universal precautions are meant to protect both the rescuer and the victim. Take a moment to review with students what they learned about universal precautions in Lesson 1.

*Applying First Aid to Injuries  Chapter 17*

## Lesson 2 Review Answers

1. respiratory failure 2. choking 3. shock 4. Sample answer: Thousands of people choke to death every year. A doctor might not be nearby when someone is choking. Lives could be saved if more people knew how to do the Heimlich maneuver. 5. Sample answer: I could keep a card or other reminder in my wallet. I could review the procedures periodically and practice on a doll.

### Portfolio Assessment

Sample items include:
- Articles from Vocabulary
- Tables from Teaching the Lesson
- Entries from Health Journal
- Reports from Technology and Society
- Lesson 2 Review answers

### GLOBAL CONNECTION

 Three organizations that help train people throughout the world in first aid techniques include the International Red Cross, the World Health Organization (WHO), and the United Nations Children's Emergency Fund (UNICEF). Have each student research and write a report about one of these organizations. Have students focus on finding out more about how the organizations work to improve the health of people worldwide. Suggest to students that in their research they focus on one person or incident to relate to the class as an example of the work that people in that particular organization do.

---

**Sterile**
Free from bacteria

**Pressure point**
A place on the body where an artery can be pressed against a bone to stop bleeding in an arm or a leg

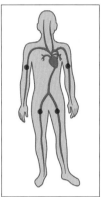

**Figure 17.2.5** Pressure points for the arms and legs

**Word Bank**
choking
respiratory failure
shock

After protecting yourself, follow the steps below to control serious bleeding.

1. Cover the wound with a **sterile** bandage or clean cloth. A sterile bandage is free from bacteria.

2. Use the palm of your hand to put direct pressure on the wound. If blood soaks through the first dressing, leave it there and add more dressings. Continue doing this until the bleeding stops.

3. Elevate, or raise, the wound above the level of the person's heart.

Usually, bleeding can be stopped with direct pressure and elevation. If bleeding continues, you may need to apply pressure at a **pressure point.** This is a place where an artery can be pressed against a bone. The purpose of using pressure is to control the flow of blood and stop bleeding in an arm or a leg. To apply pressure on one of these points, use your fingers or the heel of your hand. Press the artery toward the nearest bone and hold it there until help arrives. This will prevent the person from losing more blood. Figure 17.2.5 shows where the pressure points are for the arms and legs.

**Lesson 2 Review** On a sheet of paper, write the word or phrase from the Word Bank that best completes each sentence.

1. Rescue breathing is the kind of help you give when someone has _____.

2. The Heimlich maneuver is used to help someone who is _____.

3. Signs of _____ include large pupils, very fast or slow pulse, damp skin, and thirst.

On a sheet of paper, write the answers to the following questions. Use complete sentences.

4. **Critical Thinking** Why should people who are not doctors know the Heimlich maneuver?

5. **Critical Thinking** What might you do to help you remember how to give CPR?

406  Unit 6  Injury Prevention and Safety Promotion

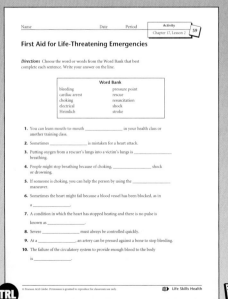

Activity 59

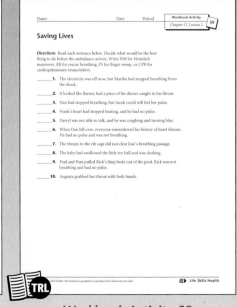

Workbook Activity 59

# Lesson 3: First Aid for Poisoning and Other Problems

## Objectives

After reading this lesson, you should be able to
- recognize the correct first aid steps to take when someone has been poisoned
- understand first aid treatment for a variety of problems
- discuss emergency treatment needed when a person has been in extreme temperatures

**Oral poisoning**
Poisoning that happens after a harmful material is swallowed

**Inhalation poisoning**
Poisoning caused by breathing in harmful fumes

**Fume**
Gas

Someone trained in first aid can recognize signs of poisoning, fainting, burns, and other problems. Then the right help can be offered.

## What Is First Aid for Poisoning?

Poisons may be swallowed, breathed in, or absorbed through the skin. In any case, medical help should be found quickly after first aid is given. Some poisoning can be life threatening.

### Oral Poisoning

**Oral poisoning** happens when a harmful material, such as a household cleaner, is swallowed. Signs of oral poisoning include sudden severe abdominal pain, upset stomach, and vomiting. The person may become sleepy and lose consciousness. The person may have chemical burns on the lips and a chemical odor on the breath. A container of a poisonous material may be nearby.

If a person has swallowed a poison, call your local poison control center immediately. Give as much information as you can, including the person's age. Also tell how much you think the person weighs. Give the name of the poison you think was swallowed. The poison control expert will tell you what to do and send emergency workers if they are needed.

### Inhalation Poisoning

**Inhalation poisoning** happens when someone breathes in harmful **fumes**. Fumes are gases. Signs of inhalation poisoning include headache and dizziness followed by unconsciousness.

### Link to >>>

**Biology**

When a person swallows poison, how does it get into the bloodstream? The poison enters the throat. It is pushed into the stomach by the muscles and moves to the small intestine. Villi absorb chemicals from the small intestine. The villi contain small blood vessels that are connected to the rest of the bloodstream. Poisonous chemicals move from the villi into the bloodstream.

---

## Lesson at a Glance

### Chapter 17 Lesson 3

**Overview** This lesson reviews basic first aid techniques for poisoning, fainting, bites and stings, burns, bone and joint injuries, heat exhaustion, and hypothermia.

### Objectives

- To recognize the correct first aid steps to take when someone has been poisoned
- To understand first aid treatment for a variety of problems
- To discuss emergency treatment needed when a person has been in extreme temperatures

**Student Pages** 407–412

**Teacher's Resource Library**

Activity 60

Alternative Activity 60

Workbook Activity 60

### Vocabulary

| | |
|---|---|
| contact poisoning | hypothermia |
| fainting | inhalation |
| fracture | poisoning |
| frostbite | oral poisoning |
| fume | rabies |
| heat exhaustion | splint |
| heatstroke | sprain |

Write the vocabulary words on the board. Ask students to look at the words and decide which pairs of words they think are related. For example, students may think fracture and splint are related because they both deal with broken bones. Have students review the definitions in the textbook to check their ideas.

### Link to

**Biology.** Ask a volunteer to read aloud the Link to Biology feature on page 407. Show students a diagram of the human digestive system and identify the structures mentioned in the feature as the volunteer reads it aloud.

## Background Information

Many of the world's plants contain substances that are toxic. Some of these substances can cause illnesses and can even kill if people eat them in large or even small amounts. Some of the most poisonous are plants with fruit or flowers that can attract animals or young children. Some common poisonous plants are philodendron, daffodil, narcissus, milkweed, mistletoe, lily of the valley, azalea, rhododendron, and iris.

###  Warm-Up Activity

Ask students to describe other situations, besides those in Lesson 2, which would require first aid treatment. Then ask if anyone has broken an arm or a leg, sprained a muscle, been bitten by a dog or a cat, or been stung by a bee. Have students explain what first aid measures, if any, were taken at the time.

###  Teaching the Lesson

Have students make a table to organize information about the first aid responses for the different situations in this lesson. Table column titles should include "Situation," "Possible Dangers," and "First Aid Procedure."

###  Reinforce and Extend

### IN THE ENVIRONMENT

 Many poisons are household items such as paints, mineral spirits, household cleaners, and oils. These poisons can be harmful if they are disposed of with other trash and end up in a regular landfill. Many communities offer either curbside pickup or a central location to which such poisons can be brought for proper disposal. Have students investigate their local community waste management and recycling programs to find out how they can dispose of these poisons properly.

---

**Contact Poisoning**
*Poisoning that happens when harmful materials are absorbed through the skin*

**Fainting**
*Short-term loss of consciousness caused by too little blood to the brain*

**Rabies**
*A serious disease that affects the nervous system*

---

Someone who has breathed in poison should get fresh air or be moved out of the area immediately. If the person is unconscious, check for a pulse and breathing. If the person is not breathing, perform rescue breathing or cardiopulmonary resuscitation (CPR). Be careful not to breathe in any harmful fumes yourself. If your own safety is in danger, do not try to help the person. Instead, call for help at once.

### Contact Poisoning

**Contact poisoning** happens when harmful materials are absorbed through the skin. Poisons may be absorbed from plants such as poison ivy and from household cleaning products. Poisons may also be absorbed from lawn and farm chemicals, such as fertilizers. Signs of contact poisoning include a bad rash, swelling, blisters, itching, and burning. Immediately take off any clothes that have come in contact with a poison. Wash the skin with soap and large amounts of water. Applying calamine lotion may get rid of the itching for a while.

### What Other Problems Need First Aid?

Fortunately, most problems that need first aid are less serious than those described in Lesson 2. First aid can ease the pain of these problems. It can prevent them from becoming more serious. You can read about a number of these problems below.

### Fainting

**Fainting** is a short-term loss of consciousness caused by too little blood to the brain. Signs that a person is about to faint include pale skin, perspiration, and dizziness. Usually, the person regains consciousness in a minute or two. To give first aid to a person who has fainted, have the person lie flat. Then elevate the person's feet. Another way to help is to bend the person over so the head is lower than the heart. Loosen any tight clothing and sponge the face with cool water. If the person does not regain consciousness, call 911.

### Bites and Stings

*Animal bites*—Bites from dogs and other animals can give diseases such as **rabies** to humans. Rabies is a serious disease that affects a person's nervous system.

**Fracture**
*A broken bone*

**Splint**
*An unbendable object that keeps a broken arm, leg, or finger in place*

Some people think that minor dog and cat bites should not be treated. This is not true. Even small bites can easily get infected. It is important to treat them in the same way a more serious bite would be treated.

To help, first control any bleeding from the animal bite. Then wash the wound with soap and water. Finally, apply a dressing. If possible, have animal control workers catch the animal. Then it can be tested for rabies. Do not try to catch the animal yourself. Follow up later by calling your local health department to find out if the animal had rabies. This will tell you whether the person needs treatment for rabies or other diseases.

*Snakebites*—Only four kinds of snakes in the United States are poisonous. These are rattlesnakes, copperheads, water moccasins, and coral snakes. Snakebites can cause serious problems and even death. Snakebites should be treated at once. Get the person with a snakebite to a hospital right away or call 911. In the meantime, keep the person calm, still, and lying down. This keeps the venom, or the animal's poison, from traveling throughout the body. Keep the affected part of the body still. Make sure it is lower than the person's heart.

*Insect stings*—Some people are allergic to stings from bees, wasps, and hornets. These people may have a life-threatening reaction to a sting. Use the edge of something firm or your fingernail to scrape away the stinger. Do not pull out the stinger. Pulling out the stinger could force the poison under the skin. Wash the area with cold water. Apply vinegar or a paste of baking soda and water. These applications help lessen the swelling and pain. Watch for signs such as vomiting, upset stomach, a hard time breathing, or an irregular heartbeat. If you see any of these signs, call 911 immediately. If the person stops breathing, begin rescue breathing.

### Bone and Joint Injuries

If you think someone has a broken bone, or **fracture,** check for crookedness, swelling, bruising, loss of movement, and serious pain. Keep the body part in the same position in which you found it. Make a **splint.** A splint is an unbendable object that keeps a broken arm, leg, or finger in place. If the bone has broken through the skin, cover the wound with a sterile dressing. Be careful not to press on the bone. If you think the person has a broken neck or back, do not move him or her. Call 911 at once.

## TEACHER ALERT

There are many ideas about what should be done to treat a snakebite, including trying to cut into the bite and suck the venom out. Medical professionals now feel that this may not be an effective treatment and may cause even more harm. The American Red Cross recommends washing the bite with soap and water, immobilizing the bitten area and keeping it lower than the heart, and, most importantly, seeking medical help immediately. Even waiting a few hours can mean the difference between an effective treatment and a treatment that will not be effective.

## ELL/ESL Strategy

**Language Objective:**
*To model reading/study strategies*

Have students select each vocabulary word from the boxes in the margins of the lesson pages. Then have them use their finger to scan the lesson to locate the word. Point out that scanning is a quick way to find the answer to a specific question if students know the key words for which to scan. Compare this study skill to scanning for a key word when using a search engine on the Internet.

## LEARNING STYLES

### Auditory/Verbal
Have students review the first aid procedures they have learned in this lesson by working with a partner. Partners should take turns in which one student verbally identifies a situation in which first aid is needed and the other verbally lists the steps that should be taken in response. This exercise can help students review for a test or a quiz.

## Health Myth

Have students read the Health Myth on page 410. Burns can be caused by heat and fire, chemicals, electricity, radiation, and sunlight. Home remedies including butter, creams, oils, or ointments should not be applied to a burn as part of first aid because they might cause infection or further damage to the skin. Using ice can cause frostbite, which will further damage the skin.

## GLOBAL CONNECTION

People in different countries around the world have adapted to extremely hot or cold conditions. Have students conduct research to find out how people in extreme climates guard against heatstroke or hypothermia. For example, what clothing helps guard against these conditions? What practices in people's daily routines guard against these conditions? Do certain foods or drinks help people maintain the correct body temperature? Have students present their findings to the class.

---

**Sprain**
*The tearing or stretching of tendons or ligaments that connect joints*

A **sprain** occurs when the tendons or ligaments that connect joints together get torn or stretched. The joints are parts of the body where bones are connected. Usually, sprains occur at wrists, knees, or ankles. A person with a sprain should avoid moving that part of the body. To lessen swelling, the injured part of the body should be raised. An ice pack or a cold pack should be placed on the injury. Then it is important to get medical attention.

### Burns

There are three types, or degrees, of burns. First-degree burns are the least serious. They hurt only the outer layer of the skin. Some sunburns are first-degree burns. Second-degree burns affect the top layers of skin. Bad sunburns may cause second-degree burns. Third-degree burns go through all the layers of skin to the tissues underneath.

First aid for first-degree and second-degree burns is the same. First, stop the burn by getting the person away from its source. Next, cool the burn with large amounts of cool water. Then without cleaning the burn, cover it. As a cover, use a dry, loose, sterile dressing or a sheet. This will help keep the burn from getting infected. For third-degree burns, call 911 immediately. Stay with the person and watch for signs of shock.

### Objects in the Eye

If there is an object in the eye, do not rub. Flush the eye by pouring large amounts of water over it. Start the flushing at the corner of the eye near the nose. This may remove the object. If the object will not come out, get medical help.

**Figure 17.3.1** *A person with a sprain should place an ice pack or a cold pack on the injured part.*

### Health Myth

**Myth:** In a burn emergency, you should put butter or ice on the burn.

**Fact:** Butter does not help to treat a burn. Ice can actually be harmful. It can create a cold burn in addition to the burn the person already has.

*Heat exhaustion*
A condition that comes from hard physical work in a very hot place

*Heatstroke*
A condition that comes from being in high heat too long

*Hypothermia*
A serious loss of body heat that comes from being in the cold too long

*Frostbite*
A tissue injury caused by being in extreme cold

### Health Myth

**Myth:** If someone has a nosebleed, pinch the nose and lean the head back.

**Fact:** Leaning the head back does not help. It can cause blood to run down the throat. Then the blood can enter the stomach. This could make the person vomit.

### Nosebleeds

For a small nosebleed, have the person sit and lean forward. The person should breathe through the mouth. Direct pressure should be placed on the bleeding nostril for 10 minutes. A cold cloth across the bridge of the nose may help. If these steps do not stop the bleeding, get medical help.

### Exposure to Extreme Heat and Cold

**Heat exhaustion** usually comes from hard physical work in a very hot place. Signs of heat exhaustion include weakness, heavy sweating, muscle cramps, headache, and dizziness. To give first aid, remove the person from the heat. Loosen the person's clothing, cool him or her with wet cloths, and offer cool water to sip. Call 911 if the person becomes unconscious.

**Heatstroke** may happen when a person stays in high heat too long. The main sign of heatstroke is a lack of sweating. Other signs include high body temperature, red skin, vomiting, confusion, a very fast pulse, and sudden unconsciousness. Call 911 immediately and follow the same steps for treating heat exhaustion.

**Hypothermia** is a serious loss of body heat that comes from being in the cold too long. It happens most often in cold, damp weather. It often happens when a person is tired and is not dressed warmly. Signs of hypothermia include shivering, slurred speech, and below-normal body temperature.

Hypothermia can cause death. If the person is not breathing, perform rescue breathing. Have someone call 911 immediately. Then remove wet clothing and dry the person. Slowly warm the person with blankets or other heat sources. Give the person warm liquids to drink. Watch the person until help arrives.

**Frostbite** is a tissue injury that comes from being out in extreme cold without proper clothing. It usually affects the hands, feet, ears, or nose. Skin that has frostbite looks gray or yellowish and feels numb, cold, and doughy. To treat frostbite, slowly rewarm the affected area by placing it in warm water. Do not rub the area. Cover it with clean, dry bandages and get medical help.

## LEARNING STYLES

**Logical/Mathematical**
Students are most likely familiar with body temperatures given in degrees Fahrenheit. However, many areas of the world use degrees Celsius as a temperature scale. Have students convert the following important body temperatures from Fahrenheit to Celsius using this formula:

$$\frac{(°F - 32)}{1.8}$$

Normal Body Temperature = 98.6°F (37°C)

Hypothermia occurs when body temperature drops below 95°F (35°C).

Heatstroke can occur when body temperature rises above 104°F (40°C).

## AT HOME

Have students prepare a checklist for items that should be in the home to provide basic first aid based on what they have read in this chapter. Items on the list might include adhesive bandages in various sizes, antibiotic creams, cotton bandages, instant ice packs, and so on. After students make their checklists, have them check their home for these items. They should check off each item they find on the list.

### Health Myth

Have students read the Health Myth on page 411. If you keep your head above your heart (in other words, sit or stand but do not lie down), your nose will bleed less. You also should avoid bending over or blowing your nose for several hours after a nosebleed stops, because those actions could cause it to start again.

# Lesson 3 Review Answers

1. A  2. C  3. B  4. Sample answer: The object could cut the eye.  5. Sample answer: Someone playing sports for a long time on a hot summer day might have heat exhaustion. This might be prevented by playing during a cooler part of the day.

## Portfolio Assessment

Sample items include:
- Tables from Teaching the Lesson
- Checklists from Application: At Home
- Findings from Application: Global Connection
- Lesson 3 Review answers

## Research and Write

Have students read and conduct the Research and Write feature on page 412. Tell students that, to gather information, they can call or visit local hospitals and talk to an employee in the public relations department, who will get them the information they need.

## Decide for Yourself

Have students read and complete the Decide for Yourself feature on page 412. Discuss how students might convince the friend to seek medical help immediately.

**Decide for Yourself Answers**

1. Sample answer: No, he does not have a good plan for treatment. The popping sound might indicate that he has broken a bone, in which case he would need treatment from a health care professional. 2. Sample answer: Yes, I should convince my friend to get medical treatment. I would convince my friend to go to a clinic or hospital so that a health care professional could examine his injury. 3. Sample answer: No, I would not call 911 because this is not an emergency situation that requires an ambulance.

---

### Research and Write

Some people are afraid to go to an emergency room. They do not know what to expect when they get there. Research to find out what people can expect when they go to an emergency room. Create a brochure that explains what to expect.

### Lesson 3 Review
On a sheet of paper, write the letter of the answer that best completes each sentence.

1. Inhalation poisoning happens when someone _____ a harmful material.
   A breathes in   C spits out
   B touches or holds   D chews and swallows

2. A _____ burn extends through all layers of skin.
   A first-degree   C third-degree
   B second-degree   D fourth-degree

3. For a bee sting, you should NOT _____.
   A wash the sting with cold water
   B pull out the stinger quickly
   C apply a paste of baking soda
   D watch for an upset stomach

On a sheet of paper, write the answers to the following questions. Use complete sentences.

4. **Critical Thinking** If something is in the eye, why should the eye not be rubbed?

5. **Critical Thinking** Give an example of a time that someone might get heat exhaustion. What are two ways the heat exhaustion might have been prevented?

### Decide for Yourself

You and a friend are walking through the neighborhood on your way to a party. Your friend steps on a wet rock, his foot slides off the rock, and he falls down. When he falls, his ankle twists and makes a popping sound. You help your friend stand up and ask him if he is hurt. He says, "No, I'm fine. It's no big deal." Your friend begins to limp, but he still wants to attend the party. He says that he will put an ice pack on his ankle when he returns home after the party.

On a sheet of paper, write the answers to the following questions. Use complete sentences.

1. Does your friend have a good plan of treatment for his injury? Explain your answer.
2. Should you convince your friend to get medical treatment? If so, what would the treatment be? Explain your answer.
3. Should you call 911? Should you call anyone else? Explain your answer.

412   Unit 6   Injury Prevention and Safety Promotion

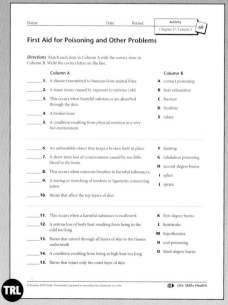

Activity 60

Workbook Activity 60

# Chapter 17 SUMMARY

- In an emergency, stay calm, find out whether the person is conscious and breathing, look for injuries, and call 911 for help.

- Good Samaritan laws protect people who help others in emergencies.

- Following universal precautions helps rescuers from coming in contact with blood and other body fluids and protects wounds from dirt and infections.

- Choking, failure to breathe, cardiovascular failure, shock, and serious bleeding all can be life-threatening emergencies.

- The Heimlich maneuver is used when someone is choking to force an object out of the throat.

- Rescue breathing puts oxygen from the rescuer's lungs into an unconscious person's lungs.

- A person in cardiovascular failure has a hard time breathing and an irregular pulse. A person whose heart has stopped beating and has no pulse is in cardiac arrest. Cardiopulmonary resuscitation (CPR) is needed for both.

- Shock happens when the circulatory system fails to get enough blood to the body. First aid goals are to return blood to the heart and maintain a normal body temperature.

- Immediate direct pressure over a wound is the first step in first aid for serious bleeding.

- The local poison control center can give instructions to treat poisoning.

- First aid can treat fainting, bites and stings, bone and joint injuries, burns, nosebleeds, heat exhaustion, and frostbite.

### Vocabulary

| | | |
|---|---|---|
| cardiac arrest, 404 | Good Samaritan laws, 398 | pressure point, 406 |
| cardiopulmonary resuscitation (CPR), 404 | heat exhaustion, 411 | rabies, 408 |
| chest compression, 404 | heatstroke, 411 | rescue breathing, 403 |
| contact poisoning, 408 | Heimlich maneuver, 401 | respiratory failure, 402 |
| dressing, 398 | hypothermia, 411 | shock, 405 |
| fainting, 408 | inhalation poisoning, 407 | splint, 409 |
| first aid, 396 | latex, 397 | sprain, 410 |
| fracture, 409 | mouth-to-mouth resuscitation, 403 | sterile, 406 |
| frostbite, 411 | oral poisoning, 407 | universal distress signal, 401 |
| fume, 407 | | universal precautions, 398 |

## Chapter 17 Summary

Have volunteers read aloud each Summary item on page 413. Ask volunteers to explain the meaning of each item. Direct students' attention to the Vocabulary box on the bottom of page 413. Have them read and review each term and its definition.

### ONLINE CONNECTION

For more information about first aid, including booklets on babysitting, first aid, and pet first aid, have students visit the American Red Cross Web site: www.redcross.org

Students also can browse the medical encyclopedia on the Web site of MedlinePlus, which is offered by the U.S. National Library of Medicine and the National Institutes of Health. The medical encyclopedia will help students find information related to specific health emergencies: www.nlm.nih.gov/medlineplus/encyclopedia.html

# Chapter 17 Review

Use the Chapter Review to prepare students for tests and to reteach content from the chapter.

## Chapter 17 Mastery Test

The Teacher's Resource Library includes two forms of the Chapter 17 Mastery Text. Each test addresses the chapter Goals for Learning. An optional third page of additional critical-thinking items is included for each test. The difficulty level of the two forms is equivalent.

## Review Answers

**Vocabulary Review**

1. rabies 2. universal precautions
3. rescue breathing 4. universal distress signal 5. heatstroke 6. shock 7. pressure points 8. oral poisoning 9. inhalation poisoning 10. cardiac arrest
11. hypothermia 12. frostbite 13. fainting
14. fracture

### TEACHER ALERT

In the Chapter Review, the Vocabulary Review activity includes a sample of the chapter's vocabulary terms. The activity will help determine students' understanding of key vocabulary terms and concepts presented in the chapter. Other vocabulary terms used in the chapter are listed below.

| | |
|---|---|
| cardiopulmonary resuscitation (CPR) | heat exhaustion |
| chest compression | Heimlich maneuver |
| contact poisoning | latex |
| dressing | mouth-to-mouth resuscitation |
| first aid | respiratory failure |
| fume | splint |
| Good Samaritan laws | sprain |
| | sterile |

---

# Chapter 17 REVIEW

**Word Bank**
- cardiac arrest
- fainting
- fracture
- frostbite
- heatstroke
- hypothermia
- inhalation poisoning
- oral poisoning
- pressure point
- rabies
- rescue breathing
- shock
- universal distress signal
- universal precautions

## Vocabulary Review

On a sheet of paper, write the word or phrase from the Word Bank that best completes each sentence.

1. A disease a person can get when bitten by an infected animal is _____.
2. To lessen their risk of catching diseases, medical workers use _____.
3. The transfer of someone's breath to a person who cannot breathe is _____.
4. Grabbing the throat with the hand or hands is the _____ for choking.
5. The main sign of _____ is a lack of sweating.
6. When the circulatory system does not get enough blood to the body, a person may experience _____.
7. People can press on a(n) _____ to stop someone's bleeding.
8. Chemical burns on lips and chemical odor on the breath are often signs of _____.
9. A person who has breathed in harmful fumes has _____.
10. A person who is in _____ has no heartbeat and no pulse.
11. A serious loss of body heat from being outside too long in cold, damp weather is _____.
12. To treat _____, slowly warm the affected area.
13. A temporary loss of consciousness, or _____, can happen when the brain gets too little blood.
14. Crookedness, swelling, bruising, loss of movement, and pain are signs of a(n) _____.

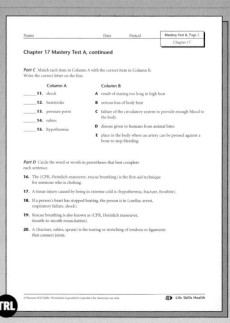

Chapter 17 Mastery Test A, pages 1–3

## Concept Review

On a sheet of paper, write the letter of the answer that best completes each sentence.

**15.** Always ask before helping if you think someone is choking because choking is sometimes mistaken for _____.
  A frostbite
  B fainting
  C a heart attack
  D hypothermia

**16.** The condition LEAST likely to be life threatening is _____.
  A contact poisoning
  B shock
  C cardiovascular failure
  D a nosebleed

**17.** To treat heat exhaustion, do NOT _____.
  A loosen the person's clothing
  B apply warm cloths to the person's neck
  C give the person a glass of water to sip
  D take the person inside or into the shade

## Critical Thinking

On a sheet of paper, write the answers to the following questions. Use complete sentences.

**18.** How can you prepare to give emergency first aid?

**19.** What did you learn about emergency treatment in this chapter that surprised you the most? What might you do as a result? Explain your answer.

**20.** Why might you be less likely to be afraid to help in an emergency now? Explain your answer.

**Test-Taking Tip** After you have taken a test, review it and ask yourself "Do my answers show that I understood the question?"

---

### Concept Review
15. C   16. D   17. B

### Critical Thinking
**18.** Sample answer: I can take courses in first aid and CPR. I can carry information about how to give CPR. I can remember how to give the Heimlich maneuver.
**19.** Sample answer: I was surprised to learn that thousands of people die each year from choking. I had not realized that this was so common. I also was surprised to find out that I can do the Heimlich maneuver on myself. I will be sure I know the right way to do the Heimlich maneuver. I want to be able to help if I ever see anyone choking. I want to be able to help myself if I am ever choking and no one is nearby.
**20.** Sample answer: Now that I know what to do, I will be less likely to be afraid to help others. Learning this information has given me more confidence, and it has made me realize that in an emergency situation, I could save someone's life.

## ALTERNATIVE ASSESSMENT

Alternative Assessment items correlate to the student Goals for Learning at the beginning of this chapter.

- Have students make a diagram that sequences and identifies basic guidelines for first aid.
- Have students make a poster that describes first aid steps in five life-threatening emergencies.
- Have students orally explain first aid for poisoning and common injuries.

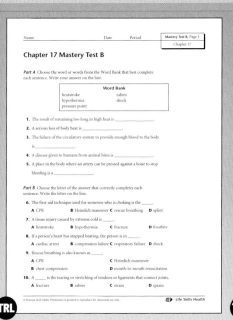
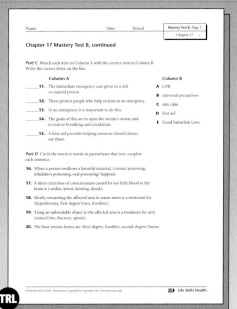

Chapter 17 Mastery Test B, pages 1–3

# Chapter at a Glance

## Chapter 18:
## Preventing Violence and Resolving Conflicts
pages 416–437

### Lessons
1. **Defining Violence**
   pages 418–423
2. **Causes of Violence**
   pages 424–428
3. **Preventing Violence**
   pages 429–434

**Chapter 18 Summary** page 435

**Chapter X Review** pages 436–437

**Audio CD**

**Skill Track for Life Skills Health**

**Teacher's Resource Library** TRL
   Activities 61–63
   Alternative Activities 61–63
   Workbook Activities 61–63
   Self-Assessment 18
   Community Connection 18
   Chapter 18 Self-Study Guide
   Chapter 18 Outline
   Chapter 18 Mastery Tests A and B
   (Answer Keys for the Teacher's Resource Library begin on page 559 of this Teacher's Edition.)

## Opener Activity

Have students leaf through the local daily newspaper. Ask them to point out articles about a conflict of any sort. Then have them go through the paper again, this time looking for articles about acts of violence. Discuss which articles belong in both categories, as well as which articles don't belong in both categories, and why. Ask students how often conflict led to violence in that day's newspaper. Discuss how nonviolent solutions to conflicts are the best solutions. Point out that it sometimes takes a lot of work to come to a nonviolent solution, and that is why violence so often seems to follow conflict.

416  Unit 6  Injury Prev. & Safety Promo.

# Self-Assessment

## Can you answer these questions?

1. ✓ What is violence?
2. ✓ How can conflict lead to violence?
3. ✓ How does violence affect people?
4. ✓ What are three causes of violence?
5. ✓ What types of violence are teens most concerned with?
6. ✓ What are five ways to manage anger?
7. ✓ What does a mediator do?
8. ✓ How can people avoid conflict?
9. ✓ What does a mediator do?
10. ✓ What are three ways communities can fight crime?

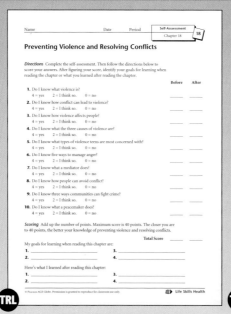

Self-Assessment 18

Community Connection 18

# Chapter 18
# Preventing Violence and Resolving Conflicts

You see violence on TV. You read about it in newspapers or on the Internet. Even the words to some songs are violent. You may ask why violence is in your health book, too. The answer is simple. Violence has become a health topic. It is a growing danger for young people. We all need to understand violence. Then we can prevent it, keep ourselves healthy, and help protect our friends and loved ones.

In this chapter, you will learn about the causes of violence. You will think about how violence affects its victims. You will read about the high cost of violent acts. You will learn how people can avoid violence. You will learn how victims of violence can get help. Finally, you will gain an understanding of how communities can fight crime.

## Goals for Learning

◆ To define *violence* and describe its costs
◆ To identify warning signs of conflict
◆ To identify the causes of violence
◆ To explain ways to prevent violence and find answers to conflicts

## Introducing the Chapter

Invite students to describe what they think about when they hear the word *violence*. Responses may be as diverse as hitting, family abuse, weapons, alcohol and drugs, television shows, movies, and so on. Then ask students to give examples of violence that they have seen or have heard about. Point out to students that violence and violent images surround us every day, and that so much exposure to violence can lead people to think that it is normal and even acceptable, when really it is not.

Have students read the Goals for Learning. Explain that in this chapter students will learn the definition of violence, its costs to society, and its causes and warning signs, as well as ways to prevent violence and find peaceful solutions to conflicts.

## Notes and Questions

Ask volunteers to read the notes and questions that appear in the margins throughout the chapter. Then discuss them with the class.

## Self-Assessment

Have students complete the Self-Assessment worksheet before and after reading the chapter. Before reading the chapter, have students fill in the "Before" column. Ask students to identify their goals for learning. To get ideas for setting goals, students might use the chapter introductory material on page 417, the checklist on page 416, or the questions on the Self-Assessment worksheet. Students can use the back of the worksheet if they need more space to write.

Collect the Self-Assessment worksheets and pass them out again at the end of the chapter. Have students fill in the "After" column. Ask them to identify at least four major points they have learned. Again, suggest they use the back of the worksheet if they need more space to write. You may want to collect and review the worksheets, but return them to students so they have a record of their goals and accomplishments.

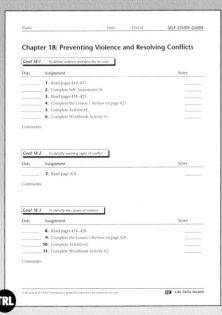

Chapter 18 Self-Study Guide, pages 1–2

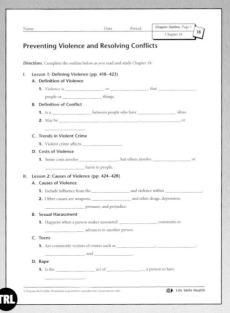

Chapter 18 Outline, pages 1–2

# Lesson at a Glance

## Chapter 18 Lesson 1

**Overview** This lesson defines violence and conflict, explores the effects of violence, describes people affected by violence, and discusses the costs of violence.

### Objectives

- To explain the short-term and long-term effects of violence
- To look at how conflict and violence are alike and different
- To identify the most likely victims of violence
- To describe the cost of violence to victims and society

**Student Pages** 418–423

**Teacher's Resource Library**

Activity 61

Alternative Activity 61

Workbook Activity 61

### Vocabulary

**conflict**        **victim**

Write the word *conflict* on the board. Read the definition from the textbook and ask students for examples of conflict. *(Some examples are disagreements over how much money a friend owes you, an umpire's call in a baseball game, and the actions of a motorist who cuts you off on the road.)* Do the same with the word *victim*. *(Some examples are a person who is robbed, a person whose house is burgled, and a business at which windows are broken.)*

### Background Information

Violence is a major cause of injury and death in the United States. Since there is no consistent reporting system for injuries that are not fatal, statistics on injury from violence are difficult to obtain. For this reason, public health officials suggest that the rate of violence and victimization among young people may be even greater than statistics indicate.

---

## Lesson 1: Defining Violence

### Objectives

*After reading this lesson, you should be able to*

- explain the short-term and long-term effects of violence
- look at how conflict and violence are alike and different
- identify the most likely victims of violence
- describe the cost of violence to victims and society

**Victim**
*A person who has had a crime committed against him or her*

### What Is Violence?

Violence is actions or words that hurt people or damage things. Violence can take away a person's sense of safety and feelings of self-worth.

Violent behavior may be as simple as pushing, hitting, or name-calling. Violence can involve damaging someone's property. It can involve the deadly use of a weapon. Violence is shown on TV and on the Internet. Many movies include violent scenes and violent special effects. Violence also is a part of many popular video games.

Violence can begin in many different ways. A small difference of opinion can quickly become violent. Violent fights can be about money or drug abuse. People can become the **victims** of family violence. A victim is a person who has had a crime committed against him or her. This can cause great harm to one or more family members. Parents may become violent toward each other. Parents also can become violent toward their children. Often, violent people experienced violence when they were children.

Some violence has no real target. People damage property by spraying paint, throwing rocks, or perhaps setting fires. Sometimes, people are hurt. Other violent acts, such as drive-by shootings, cause the deaths of innocent people.

Gang violence is a growing problem in the United States. Gang members commit many acts of violence. More than 21,000 known gangs are in the United States. Most large cities have gang-related crime. Gangs also are a problem in many small cities, in suburbs, and in the country.

Many people who commit crimes know their victims. Most murders occur between people of the same race. Teen crime is a big concern. In 2002, about 5.7 percent of male teens ages 16 to 19 were victims of violent crimes. The rate for females in that age group was 3.8 percent. Violent crimes are committed more often against teens than against any other age group.

## How Is Conflict Different from Violence?

**Conflict** is a disagreement, or difference of opinion, between two or more people who have different ideas. Conflict can turn into violence. Disagreements among people are normal. Letting conflict turn into violence is not normal.

**Conflict**
A difference of opinion between two or more people who have different ideas

Conflict may be minor, ongoing, internal, or major. You may have a minor conflict with a friend because you hurt the friend's feelings. Your parents may have a minor conflict with each other. These kinds of conflicts can be about doing chores, handling money, or managing family problems.

Many conflicts are ongoing. For example, you may have some constant disagreement with a brother or sister. Sports teams compete against each other. They have an ongoing conflict about which team will win.

Compromising, or each person giving up some things they want, is often needed to settle conflicts peacefully.

You may have a conflict within yourself about making a difficult decision. This is an internal conflict. Maybe someone wants you to go somewhere. You really want to go, but if you go, you will have to give up something else.

Major conflicts can take a long time to resolve. Often, patience and compromise are needed to fix a major conflict. You may face major conflicts with your friends or at home. If you accept others and have respect for everybody's differences, you will have fewer conflicts. If you handle major conflicts well, your conflicts likely will not become violent.

### Link to ➣➣➣
**Language Arts**

The musical *West Side Story* is loosely based on the play *Romeo and Juliet* by William Shakespeare. *West Side Story* is about two enemy gangs. The Sharks are Puerto Rican and the Jets are white. Tony, who once was a Jet, falls in love with Maria, the sister of a Shark. Maria and Tony try to get away from the violence between the two gangs, but fail.

---

 **Warm-Up Activity**

Write the following column headings on the board: TV, Movies, Games; Within a Family; Random—No Real Target; Gangs. For each heading, have students name an act of violence they have known about, heard about, or read about that fits into the category.

### Teaching the Lesson

Help students see that, despite what movies and television show, violence is most often a no-win situation. A person might gain temporary satisfaction from inflicting violence, perhaps for revenge or to "right" a wrong, but that satisfaction is soon replaced by guilt, remorse, and, most likely, legal consequences. Even in cases where violence seems justified, such as defending oneself from an attacker, the unintended result may be injury or death instead of simply the loss of some money or jewelry.

You may want to lead a discussion about whether violence is ever justified, or whether it is ever consequence-free. Encourage students to act as "devil's advocate," searching for arguments against each example that is suggested.

### PRONUNCIATION GUIDE

Use the pronunciation shown here to help students pronounce difficult words in the lesson. Refer to the pronunciation key that appears in the Glossary at the back of the Teacher's Edition for the sounds of the symbols.

compromise (kom′ prə mīz)

### Link to

**Language Arts.** Have a volunteer read aloud the Link to Language Arts on page 419. If time allows, show the movie *West Side Story* over two or three class periods. Discuss with students how hatred and violence became hurtful for all the characters in the story. Ask students what other messages they see in the movie.

# 3 Reinforce and Extend

## ELL/ESL Strategy

**Language Objective:** *To summarize content material*

Have students form pairs composed of a student learning English and a student with strong English skills. Ask the pairs of students to prepare anti-violence graphics—posters, print advertisements, or other visual presentations. Students will present their graphics to the class along with an argument against violence.

## Career Connection

Tell students that statisticians do research, develop surveys, and design experiments to compile data on topics such as "Victims of Violent Crime" (Table 18.1.1 on page 420). Ask students what skills they think might be especially useful for a statistician. *(math skills, attention to detail, computer skills)* Invite students to find out about the educational requirements for a career as a statistician.

## Learning Styles

**Auditory/Verbal**
Provide several short newspaper articles about conflicts involving businesses or government. Challenge students to listen carefully as volunteers read the articles to the class. Then ask questions such as the following about information provided in the articles: What is the cause of the conflict? Who is involved in the conflict? What, if anything, are people doing to resolve the conflict?

## What Is the Trend in Violent Crime?

Violent crime affects everyone. In 1990, almost 2 million people were the victims of violent crimes. In 2003, that number had dropped by 700,000 people. Except for drug-related crimes, the crime rate in the United States has been going down. Table 18.1.1 below shows the percentage of each age group that have been victims of violent crime. This includes violent crimes committed by teens ages 12 to 17. Drug-related crimes among people older than 17, however, keep going up.

| Table 18.1.1 Victims of Violent Crime (approximate percentage of each age group) | | | | | | | |
|---|---|---|---|---|---|---|---|
| Year | Age Group | | | | | | |
|  | 12–15 | 16–19 | 20–24 | 25–34 | 35–49 | 50–64 | 65+ |
| 1990 | 10% | 10% | 9% | 6% | 3% | 1% | 0.4% |
| 1991 | 9 | 12 | 10 | 5 | 4 | 1 | 0.4 |
| 1992 | 11 | 10 | 10 | 6 | 4 | 1 | 0.5 |
| 1993 | 12 | 11 | 9 | 6 | 4 | 2 | 0.6 |
| 1994 | 12 | 12 | 10 | 6 | 4 | 2 | 0.5 |
| 1995 | 11 | 11 | 9 | 6 | 4 | 1 | 0.6 |
| 1996 | 10 | 10 | 7 | 5 | 3 | 2 | 0.5 |
| 1997 | 9 | 10 | 7 | 5 | 3 | 1 | 0.4 |
| 1998 | 8 | 9 | 7 | 4 | 3 | 2 | 0.3 |
| 1999 | 7 | 8 | 7 | 4 | 3 | 1 | 0.4 |
| 2000 | 6 | 6 | 5 | 3 | 2 | 1 | 0.4 |
| 2001 | 6 | 6 | 4 | 3 | 2 | 1 | 0.3 |
| 2002 | 4 | 6 | 5 | 3 | 2 | 1 | 0.3 |
| 2003 | 5 | 5 | 4 | 3 | 2 | 1 | 0.2 |
| 2004 | 5 | 5 | 4 | 2 | 2 | 1 | 0.2 |

*Source: U.S. Department of Justice, Bureau of Justice Statistics,* Criminal Victimization in the United States, *various years.*

**Research and Write**

Does playing violent video games make a person more likely to be violent? Many people think so. Research the link between video games and violent behavior. Work with a partner to create a poster that backs up your opinion.

## What Are the Costs of Violence?

Violence has many costs. One cost is money. Other costs are physical and emotional harm to people.

The victims of violent acts may be hurt permanently. For example, they may have brain damage or become paralyzed. When people are paralyzed, they are unable to move their arms or legs. Victims of violence can also suffer mental health problems. They may always be afraid, angry, or sad. The costs of treating physical and mental health problems can add up to thousands of dollars for each victim.

The friends and relatives of victims often feel sad and angry, too. They may have to take care of the victim. In some cases, they may have to help pay medical costs.

People who commit violent acts may feel guilty about what they have done. They often feel worse about themselves than they did before the crime. Someone who is arrested for a violent act has to go to court and may go to prison. A criminal record or a violent past may limit future job and housing opportunities.

People who see violence may pay a different kind of price. They may lose their peace of mind and may never feel safe again.

### Link to >>>
**Math**

The same information used to create a table also can be shown in pie graphs, bar graphs, or line graphs. Line graphs show trends.

The information from Table 18.1.1 was used to create this line graph. The graph shows that most victims of violent crime were younger than 24 during these years. The highest number of victims were teenagers.

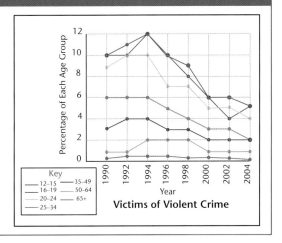

Victims of Violent Crime

## PRONUNCIATION GUIDE

Use the pronunciation shown here to help students pronounce difficult words in the lesson. Refer to the pronunciation key that appears in the Glossary at the back of the Teacher's Edition for the sounds of the symbols.

paralyzed (par´ ə līzd)

## TEACHER ALERT

Tell students that some survivors of violent crimes might have post-traumatic stress disorder, which requires professional treatment. Symptoms of PTSD include terrifying nightmares or flashbacks; extreme emotional numbing; panic attacks, rage, or violence; debilitating worry, compulsions, or obsessions; loss of the ability to feel hope or pleasure; and fragmented thoughts, lack of awareness of surroundings, or amnesia.

### Research and Write

Have students read and conduct the Research and Write feature on page 421. Help students see that video game manufacturers, game players, and groups working against game violence are all poor sources of unbiased data. Urge students to look for sources such as university studies.

### Link to

**Math.** Ask a volunteer to read aloud the Link to Math feature on page 421. Tell students that there are two reasons why the data in the feature is better suited to a line graph than a bar graph. First, the fact that the data points are connected by lines makes it much easier to see how the data change over time. Second, a line graph like this one with seven lines would require seven bars at each of the points. It might be easy to pull one piece of data out of a bar graph like this, but it would be nearly impossible to see how the change in the various age groups compares to each other over time.

## Health in Your Life

Have students read the Health in Your Life feature on page 422. On the Web site of MedlinePlus, a service of the U.S. National Library of Medicine and the National Institutes of Health, there is a wealth of articles about the diagnosis and prevention of domestic violence, as well as protection from domestic violence. The index to these articles can be found at: http://www.nlm.nih.gov/medlineplus/domesticviolence.html.

### Health in Your Life Answers

**1.** Sample answer: The cycle of family violence is a pattern of tension, abuse, and apology. **2.** Sample answer: Understanding this cycle of violence helps victims understand what is happening so that they can best protect themselves. **3.** Sample answer: An important thing to know is that family violence rarely happens only once. This type of violence often happens repeatedly.

---

All people end up paying for crime. The costs of police protection, courts, and prisons go up each year. The government often raises taxes to cover these costs. No matter where, how, or why violence happens, everyone in some way becomes a victim.

### Health in Your Life

**Cycle of Family Violence**

Family violence rarely happens only once. Instead, it often happens in a cycle. Victims of family violence often feel ashamed. Some may feel that they are somehow to blame or deserve the abuse. The cycle of family violence has four phases.

Phase 1: This is a time when tension rises. Minor conflicts between family members explode. The victim feels that he or she has to be very careful about saying or doing anything in the home.

Phase 2: Abuse occurs. The victim is abused physically or verbally. The victim might be slapped, kicked, threatened with weapons, or sexually abused. Usually, no one else sees the violence.

Phase 3: The abuser apologizes and shows sadness for what he or she has done. The abuser promises that it will never happen again. The abuser may give gifts to the victim. The abuser may argue about how violent he or she was or say the victim is making a big deal out of nothing.

Phase 4: The victim again is afraid to say or do something that might set off the abuser. The cycle starts over. The next time is often worse than the last. The time between phases can also shorten, so the abuse happens more often.

Abuse victims must realize that no one has the right to hurt another person. People who experience family violence can take steps to protect themselves and others:

- Talk to a trusted person about the abuse.
- Have a plan in place to leave if someone in the home becomes violent. Have a bag of clothes ready to take when leaving.
- Know where to find community shelters. Know their phone numbers in order to call for help.
- Keep a written record of when and how the abuse happens.
- Remember the cycle of violence when an abuser promises that he or she will never become violent again.
- Call the police or 911 for help when in danger. Acts of violence are crimes.

On a sheet of paper, write your answers to the following questions. Use complete sentences.

1. What is the cycle of family violence?
2. Why is understanding this cycle of violence important?
3. What is an important thing to know about how often family violence occurs?

***Lesson 1 Review*** On a sheet of paper, write the answers to the following questions. Use complete sentences.

1. What is violence?
2. Where are some places violence can occur?
3. Which age group has the highest number of violent crime victims?
4. **Critical Thinking** What is the difference between conflict and violence?
5. **Critical Thinking** How does violence affect a community?

## Health at Work

### Victim or Witness Advocate

People who are victims of crime often feel alone, scared, and sad. They may wonder if anyone can ever be trusted again. Victim or witness advocates work with the county or state justice system. They help victims of crime know their rights under the law. They help victims get restraining orders. A restraining order is a court order that keeps the person who causes harm from coming near the victim.

Victim or witness advocates also help victims with court matters. For example, advocates tells victims when court hearings will take place. Advocates also help victims make victim statements. Victim statements are taken into consideration when a person is sentenced for a crime.

Victim or witness advocates can help survivors of a family member who was murdered get needed services. They can also help victims find resources that will help them return to living a normal life.

*Preventing Violence and Resolving Conflicts* Chapter 18

## Lesson 1 Review Answers

1. Violence is actions or words that hurt people or damage things. 2. Violence occurs in many places, such as at home, in school, in the workplace, and in the community. Violence also is shown on TV, in movies and video games, and on the Internet. 3. Teens have the highest number of crime victims. 4. Sample answer: Conflict is a disagreement or difference of opinion. Violence occurs when someone or something is harmed. 5. Sample answer: Violence results in costs to the community in the form of police, prisons, and health care for victims and those sent to prison. These are financial costs, but the community also is affected by emotional costs because individuals feel less safe.

## Portfolio Assessment

Sample items include:
- Visual presentation from ELL/ESL Strategy
- Posters from Research and Write
- Health in Your Life answers
- Lesson 1 Review answers

### Health at Work

Have students read the Health at Work feature on page 423. Lead a discussion on what types of skills and personality traits would be advantageous to a victim advocate or a witness advocate. Also discuss why victims of crime might find it helpful to have the services of a victim or witness advocate available to them.

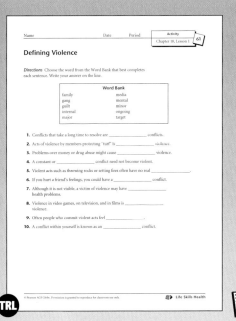

**Activity 61**

**Workbook Activity 61**

# Lesson at a Glance

## Chapter 18 Lesson 2

**Overview** This lesson explores some causes of violence; identifies crimes that happen to teens, including assault and rape; and explains how teens can protect themselves.

## Objectives

- To explain five causes of violence
- To identify teen-related crimes
- To describe three ways to keep from becoming a victim

**Student Pages** 424–428

**Teacher's Resource Library** TRL
  Activity 62
  Alternative Activity 62
  Workbook Activity 62

## Vocabulary

| | |
|---|---|
| date rape | sexual exploitation |
| rape | sexual harassment |
| sexual assault | |

Read aloud dictionary definitions of *rape* and *sexual*. Have students suggest possible meanings for each of the two-word phrases on the list.

## Background Information

Studies have indicated that people who score high for hostility on personality tests are more likely to die of heart disease at a young age. When people feel angry and tense, their body responds by increasing blood pressure, heart rate, and levels of hormones such as adrenaline. These responses are common to all people who have stress, but hostile individuals experience them to a higher degree and for longer periods than others do.

## Health Myth

Ask a volunteer to read aloud the Health Myth feature on page 424. Compare this information with the information in the Health in Your Life feature on page 422. Ask students to suggest ways that a child who has been abused or neglected or who has been exposed to domestic violence might be able to break the cycle of violence.

424  Unit 6  *Injury Prev. & Safety Promo.*

---

# Lesson 2  Causes of Violence

### Objectives

After reading this lesson, you should be able to

- explain five causes of violence
- identify teen-related crimes
- describe three ways to keep from becoming a victim

## How Can Conflict Be Recognized?

You may think that you can simply steer clear of all conflict, but that is not easy. Everyone faces times in which conflict can occur. Once you learn to identify the signs of conflict, you will be better able to avoid it. You will be better able to end the conflict and keep it from getting violent. Here are some warning signs of conflict:

- shouting
- name-calling
- spreading rumors
- giving the "silent treatment"
- using insults
- giving mean looks
- making fun of someone
- getting others to take sides
- making threats or threatening gestures
- pushing or hitting

### Health Myth

**Myth:** All abused and neglected children will become violent adults.

**Fact:** Abused and neglected children do not necessarily become violent adults. As children move into their teen years, the abuse and neglect they experienced may not cause them to become violent.

## What Causes Violence?

In 2003, there were 5.4 million victims of violent crimes in the United States. Some people blame the media for some of these violent crimes. Movies and TV programs often include both verbal and physical violence. For some people, all of the violence in the media makes violence seem like a normal part of life.

Violence within a family is especially harmful for young children. Children who live with violence may grow up thinking violence is the way to solve problems. Many criminals were the victims of family violence. Violent behavior became part of their lives.

**424**  Unit 6  *Injury Prevention and Safety Promotion*

A physical workout, such as running, jogging, walking, or another exercise program can help you deal with anger. The exercise can calm you and help you relax.

Does the media make violence attractive?

Many people buy guns as protection for themselves and their families. They usually do not plan to use them. However, just owning a gun increases the likelihood of being involved in a violent incident. Unfortunately, the more violence occurs, the more people buy guns and the more violence increases.

All schools are gun-free zones. Schools have very strict policies against bringing guns, knives, and other weapons onto school property. Some schools have metal detectors to help enforce these policies.

Alcohol and other drugs make people act in dangerous ways. They also can prevent a person from using good judgment. About one-third of all victims of violent crime believe the person committing the crime used alcohol or other drugs. About one-fourth of all crimes were committed to get money for drugs.

Members of gangs often have troubled lives. A gang may give them a feeling of belonging. Gang members often wear certain colors or clothes, or have tattoos that show they are members of a specific gang. About 6 percent of all violent crime is gang related. Victims of gang crime are often young.

When people have problems, it can cause them great stress and make them feel depressed and angry. People who usually show good judgment may turn to violence during these stressful times.

Sometimes, people take part in violence to show how loyal they are to a group. Some people will go against their own better judgment because of a group. Giving in to such negative peer pressure can get in the way of a person's ability to make good decisions.

Prejudice is a negative opinion that is formed without having enough information or experience. Prejudice means judging people you do not know based on their race, sex, sexual preference, cultural background, or religion. This sometimes leads to hate crimes. Hate crimes are violent acts against people who belong to a particular group. In 2003, more than 9,000 people were the victims of hate crimes. About half of all these hate crimes involved race.

## Warm-Up Activity

Have students identify situations that might lead to conflicts among teenagers. *(situations that involve gangs, drug or alcohol use, weapons, prejudice, discrimination, etc.)* Ask volunteers to describe signs of conflict they have seen, how those conflicts were handled, and how they could have been handled in a nonviolent manner if they weren't in the first place.

## Teaching the Lesson

Some students might be very sensitive about discussions of rape and sexual abuse. It is possible that some students might be victims of sexual abuse. As you work through the lesson, stay alert for students whose reactions to the material seem overly negative or fearful. Discuss these reactions with a school counselor; it is possible that intervention might be necessary.

### PRONUNCIATION GUIDE

Use the pronunciation shown here to help students pronounce difficult words in the lesson. Refer to the pronunciation key that appears in the Glossary at the back of the Teacher's Edition for the sounds of the symbols.

prejudice (prej′ ə dis)

### GLOBAL CONNECTION

Divide students into small groups and assign each group a different country. Have the groups research gun control laws in those countries. Then challenge students to find the rate of violent crime in that country. Have the groups make presentations to the class about the relationship between gun control laws and violent crime in the countries they were assigned.

### 3 Reinforce and Extend

#### ELL/ESL Strategy

**Language Objective:** To summarize subject material

As you work through the lesson with the class, ask students learning English to create an outline. Tell them to include as much information from the lesson as possible. Students' finished outlines will have a similar form to the Chapter 18 Outline in the TRL but will be much more complete. Ask students who are proficient in English to answer any questions ELL students might have as they create their outlines.

### Health Myth

Have students read the Health Myth feature on page 426. Explain to students that while rape is a sexual assault, it is more an act of violence than a sexual act. That is, rape is more about power and control than it is about sexual attraction. Rapists usually rape out of anger and a need for power and control.

#### Teacher Alert

Students might be confused about who is to blame for rape. They might think that rape might sometimes be the fault of the victim. Emphasize to students that rape, incest, and other types of sexual abuse and assault are never the victim's fault. Sex without consent is never acceptable. It does not matter if a perpetrator is drunk or on drugs—he or she is still at fault. Tell students that if a person says "no" to sex, his or her wishes must be respected. Also, it is never acceptable for adults to have sexual contact with children, whether or not they say "no." Rape, incest, and sexual abuse and assault are serious crimes, and perpetrators face lengthy prison terms.

---

**Sexual harassment**
Making unwanted sexual comments or physical advances to another person

**Rape**
The unlawful act of forcing a person to have sex

**Health Myth**

**Myth:** Most people are raped by strangers.

**Fact:** Almost 80 percent of all rapes are committed by people the victims know. The majority of rapes happen in the victims' homes. The same number of rapes happen during the day as at night. Statistics show that one in every six women and one in every 33 men have been victims of rape.

---

A person who has been insulted or hurt may want revenge, or a way to get even. Revenge may become more violent than the act that caused it. This can be especially true in school. Prejudice, rumors, name-calling, and comments about someone's personal life can lead to violent results. For example, revenge for name-calling may be shoving or tripping. Revenge for the shoving or tripping may be a fistfight. Revenge for the fistfight may be a fight with weapons.

**Sexual Harassment**

**Sexual harassment** happens when a person makes unwanted sexual comments or physical advances to another person. Sexual harassment can happen anywhere, including school and the workplace. What is important is the effect the harassment has on the victim, not what the harasser meant to do. It is sexual harassment if someone's comments, suggestions, or physical advances make a person uncomfortable.

About 40 to 70 percent of women and 10 to 20 percent of men believe they have been sexually harassed at work. Students also report high rates of sexual harassment. Yet many victims of sexual harassment choose not to report the harassment.

If you are the victim of sexual harassment, make it clear to the harasser that you find the sexual harassment offensive and uncomfortable. Keep a record of when the sexual harassment happens. If the harassment continues, make a formal complaint to a trusted teacher, counselor, or job supervisor about it.

### What Kinds of Crimes Happen to Teens?

Most teenagers are not violent. However, many teenagers become the victims of different kinds of violence. About one-third of all victims of violent crimes are teens. Murder rates are the highest among people ages 18 to 24. Murder is the second leading cause of death among people ages 15 to 24. Other violent crimes that happen to teens and young adults include robbery, beating, and **rape.** Rape is the unlawful act of forcing a person to have sex.

**Sexual assault**
Forced sexual contact

**Date rape**
Rape committed by someone the victim socializes with

**Sexual exploitation**
Using a person sexually to make money

Several drugs are considered date rape drugs. Date rape drugs are put into victims' drinks by their dates. The drugs cause a victim to become mentally confused and not know what is happening. This makes it hard for the victim to fight against the rapist. In addition to rape, these drugs can cause serious physical problems for the victim.

### Rape

Teens ages 16 to 19 are more than three times as likely as other people to be victims of rape, attempted rape, or some other type of **sexual assault.** Sexual assault is any forced sexual contact. Many rape victims are under the age of 12. Reporting a rape is often very hard for the victim. Experts think that only about half of all rapes are reported. Often, victims of rape are embarrassed or afraid of being blamed for the rape. Sometimes, they know the rapist and are afraid of getting the person in trouble. They may be scared their rapist will come after them again. Many communities have police officers who are trained to help and talk to rape victims.

### Dating Violence and Sexual Exploitation

Sometimes, young people become the victims of dating violence. For example, a person may use threats or aggressive behavior to try to gain control. The person might touch his or her date even if the touching is not wanted. Any touching that is not wanted is abuse.

Some teens are the victims of **date rape,** which is committed by someone with whom the victim socializes. If a person says no or tells his or her date to stop making sexual advances, the person should stop. Warning signs that date rape might happen include threats, insults, controlling behavior, and unwanted touching.

Unwanted sexual contact is always wrong. This type of act does not show the basic respect that every person deserves. Date rape carries serious consequences. People who commit date rape are often expelled from school, charged with a crime, or both.

Another crime that happens to some teens and younger children is **sexual exploitation,** a form of sexual abuse. Sexual exploitation of children happens when an adult forces a young person to do something related to sexual activity. These forced acts include prostitution and taking part in child pornography. The adult involved makes money from this exploitation. Like all sexual abuse, the sexual exploitation of children is illegal. People found guilty of sexual exploitation go to prison.

## TEACHER ALERT

Some students might be confused about the differences among sexual harassment, sexual assault, sexual abuse, and sexual exploitation. If necessary, explain the terms carefully.

Sexual exploitation occurs when money is involved.

Sexual assault involves the use of force.

Sexual harassment consists of unwanted sexual comments or advances.

Finally, any and all of the above examples fall under the category of sexual abuse. Any sort of unwanted sexual activity, whether it involves words, touching, exposure, or force, is sexual abuse.

## PRONUNCIATION GUIDE

Use the pronunciation shown here to help students pronounce difficult words in the lesson. Refer to the pronunciation key that appears in the Glossary at the back of the Teacher's Edition for the sounds of the symbols.

prostitution (pros tə tü´ shən)

pornography (pôr nog´ rə fē)

## AT HOME

Point out to students that being alert and careful can help reduce the chances of being victimized by crime. Have students work with parents or guardians or other family members to create a list of family safety rules. You might want to suggest ideas such as not opening the door to strangers, keeping doors locked, closing the curtains or blinds at night, keeping emergency telephone numbers posted by the phone, and so on. Invite volunteers to share their lists with the class.

## LEARNING STYLES

### Visual/Spatial

Have students illustrate each of the ways, listed on page 428, to keep from becoming a victim of crime. Students may make one drawing or a series of drawings. They may draw cartoons, realistic illustrations, or use any other visual style. If students have access to the appropriate technology, they might want to create their illustrations using computer software programs. Display the students' illustrations in the classroom.

## Lesson 2 Review Answers

**1.** B **2.** D **3.** A **4.** Sample answer: I would feel sexually harassed if someone made a joke about my body. **5.** Sample answer: I can make sure I always walk with a friend in well-lighted areas when walking at night.

## Portfolio Assessment

Sample items include:

- Presentations from Application: Global Connection
- Outline from ELL/ESL Strategy
- Lists from Application: At Home
- Illustrations from Learning Styles: Visual/Spatial
- Lesson 2 Review answers

---

**Figure 18.2.1** *At night, walk quickly and confidently in well-lighted areas with someone you know.*

Here are some ways to keep from becoming a victim of a crime:

- Do not walk alone at night. Walk only in areas or places you know and move quickly and confidently. Stay in well-lighted areas.
- If you are being followed, walk quickly to a place where there are other people.
- Carry a cell phone, if possible.
- Keep out of gangs by becoming involved with school or community activities.
- If dating someone you do not know well, stay out of places where you and your date are alone.
- When driving, park only in well-lighted areas, never pick up strangers, and keep your car doors locked.

**Lesson 2 Review** On a sheet of paper, write the letter of the answer that best completes each sentence.

1. About 6 percent of all violent crime is _____.
   - **A** a hate crime
   - **B** gang related
   - **C** drug related
   - **D** caused by family members

2. Hate crimes always involve _____.
   - **A** date rape
   - **B** nonviolent acts
   - **C** only race
   - **D** prejudice

3. The second leading cause of death among people ages 15 to 24 is _____.
   - **A** murder
   - **B** drug abuse
   - **C** accidents
   - **D** natural causes

On a sheet of paper, write the answers to the following questions. Use complete sentences.

4. **Critical Thinking** What would make you feel sexually harassed?

5. **Critical Thinking** What can you do in your daily life to keep from becoming the victim of a crime?

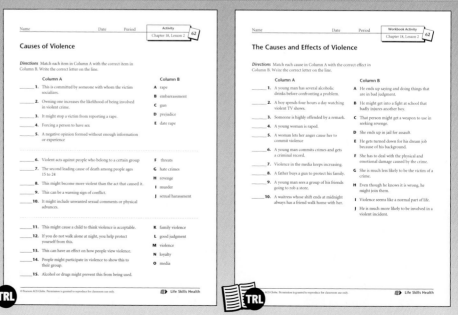

# Lesson 3: Preventing Violence

**Objectives**

*After reading this lesson, you should be able to*

- list three ways to manage anger
- describe how to resolve conflict
- identify seven things a mediator does to help resolve a conflict

You get angry sometimes. Everyone does. A relative, a friend, or someone you hardly know may offend you. You may be angry about something you said or did. When you are angry, your heart beats faster and your blood pressure rises. Your muscles tense. Your face may look strained and red. Anger is a natural feeling. It causes people to feel aggressive and defensive. Anger is important for survival, but it can cause problems if you do not know how to handle it.

## How Can Anger Be Managed?

Managing your anger the right way will help you feel in control. Here are three ways to manage your anger:

- Take a deep breath to calm yourself before you say anything. Say what you need to say in a clear and calm way. Do not curse or use threatening words. Be respectful of the person you are angry with. Also, think about how you hold your body. Shaking your fist or taking a step toward a person shows aggression.

- Control your anger. Breathe deeply. Tell yourself to relax. Picture yourself in control. Think about how you are holding your body. Looking angry might make the other person become defensive and perhaps aggressive. Try to stay calm. Calmly ask the person to explain the problem. Listen quietly. Use humor if you can. Do not be sarcastic.

- Hold back your anger. Think of something positive instead of what is making you angry. Doing this will help calm you. It is important that you later talk with the person about what made you angry. You might also exercise or work on a project, such as mowing the lawn or cleaning your room.

It is important to understand what causes you to get angry. It also is important to control how you react. Acting angrily at the other person often makes things worse.

---

## Lesson at a Glance

### Chapter 18 Lesson 3

**Overview** This lesson provides students with ways to manage anger, resolve conflicts, and help fight crime.

### Objectives

- To list three ways to manage anger
- To describe how to resolve conflict
- To identify seven things a mediator does to help resolve a conflict

**Student Pages** 429–434

**Teacher's Resource Library**

Activity 63
Alternative Activity 63
Workbook Activity 63

### Vocabulary

**mediator**     **negotiation**

Write the two vocabulary words on the board. Under each word, write its Latin derivation:

*negotium*
n. business
v. to confer, bargain, or discuss

*mediare*
to divide in the middle

Discuss what each word might mean based on the Latin sources.

### Background Information

In the past decade, many communities have experimented with radically new policies and programs in an attempt to reduce the crime rate. One of the best-known of these programs is Community Policing, which involves an increase in interaction and cooperation among local police and the people and neighborhoods they serve. Proponents say it causes an increased sense of safety among residents, which in turn results in reduced crime levels.

**Warm-Up Activity**

Bring in newspaper articles about domestic violence situations and have volunteers read them aloud. Ask students to discuss news events in which anger has resulted in family violence. Tell students that people who need help dealing with family violence can call local crisis centers or shelters, which can help protect people from abusive family members.

 **Teaching the Lesson**

As you work through the lesson with the class, bring the topic home by presenting specific scenarios to the students. Ask them, "How can you help your friends stay out of fights? How can you get help with a conflict in this school? What can your community do to help fight crime?" Their responses might lead to an interesting discussion of real local problems, rather than hypothetical situations.

 **Reinforce and Extend**

### LEARNING STYLES

**Logical/Mathematical**

Have half of your students prepare a flowchart that shows the steps to take to avoid a conflict. Have the other half of your students prepare a flowchart that shows the steps to take to peacefully resolve a conflict. Students should take into account all the possible outcomes of each step. Ask students to include illustrations, if appropriate, with the steps on the flowcharts.

### Research and Write

Have students read and conduct the Research and Write feature on page 430. You might want to speak to fellow faculty members or administration officials before students contact them to set up interviews.

What would you do if someone tried to start a fight with you and would not let you walk away?

**Research and Write**

Do you feel your school is safe? What would you do if you saw violence at your school? What kind of school safety plan does your school have? Write a report on safety at your school. Interview teachers and students for your report, and include a visual presentation, such as a poster, brochure, or video.

## How Can Conflict Be Avoided?

Sometimes, the best way to deal with a conflict is to walk away. Ask yourself: Is it really such a big deal? Is the conflict based on a rumor that may not be true? How will I feel if I just forget about it? Think about what is happening. Maybe the other person is having some serious problems at home or at school. Perhaps the person never meant to cause a conflict. Let it go, and realize that the other person will be easier to get along with at another time.

Remember that movements you make and words you use in one culture may mean something very different in another culture. Understanding this will lessen conflict between people from different cultures.

During conflict, people may feel insulted or misjudged. Good listening skills are important. Asking the right questions to clear up any misunderstandings is also important. Do not let your anger get in the way of figuring out what the other person is trying to say. Good communication and listening can keep conflict from getting violent.

## How Can Conflicts Be Resolved?

Some conflicts cannot be ignored. When you have to get to the bottom of a conflict, talking with the person alone is best. Having other people around raises the chance that things will get out of control. Do not talk when the other person has been drinking alcohol or using other drugs.

As you explain your feelings, try to remain calm and keep your feelings under control. Using put-downs will not help you resolve the conflict peacefully. Focus on your feelings rather than what the other person is doing. Give the other person a chance to do the same thing. If you have had a part in hurting the other person, do not be afraid to say, "I'm sorry."

### LEARNING STYLES

 **Body/Kinesthetic**

Have small groups of students write and role-play scenes that involve a potentially violent conflict. Tell students that they must come up with a nonviolent resolution to the conflict. After each group has presented its skit, ask the class for other ideas for preventing violence in the particular situation presented in the skit. Write the ideas on the chalkboard so students can see that there are many ways to peacefully resolve conflicts.

**Mediator**
Someone who helps people find a solution to a conflict

**Negotiation**
Ability to help people reach a compromise

## How Can You Help Others Stay Out of Fights?

If other people get into a fight, do not cheer them on. Often, people in conflict feel they have to follow through when others are watching.

Let people know that you respect them for being able to walk away from a fight. Give respect to someone who is strong enough to apologize.

There may be times when you can put pressure on friends not to fight. Let them know you think they should solve their problems peacefully.

You may be able to help as a **mediator.** A mediator is someone who helps others find a solution to a conflict. To successfully mediate a conflict:

- Let both people know that you are not taking sides.
- Set some rules, such as using calm, reasonable talk.
- Let each person say his or her feelings without being interrupted.
- Allow each person to ask reasonable questions of each other.
- Encourage both people to figure out different ways to solve the problem.
- Ask both people to discuss each way and to agree on one idea.
- Work at finding a compromise. Do not let them give up without really trying.

**Figure 18.3.1** *Peer mediators help others find solutions to conflicts.*

Many schools have peer mediation programs. In these programs, counselors train students to help one another resolve conflicts. **Negotiation,** or the ability to help people reach a compromise, is an important part of peer mediation programs. Find out if your school has such a program.

## ELL/ESL Strategy

**Language Objective:**
*To discuss an assigned topic as part of a small group*

Divide students into small groups and give each group a conflict scenario. (For example, a student finds out that another student is saying nasty things about him behind his back, an employee believes that her boss is treating her unfairly, a teen discovers that her brother has taken something that belongs to her without asking, etc.) Ask the groups to come up with nonviolent solutions to these conflicts. Each group can present its scenario and solution to the class.

## Learning Styles

**Interpersonal/Group Learning**

Have groups of students design a campaign to stop violence in school. The campaign should seek to educate others about the causes of violence and strategies for preventing it. For example, students may design posters for display in the school or present skits to others in the school. Arrange for students to implement their campaigns and help them with details and logistics.

## Health Journal

Have students think of at least two conflicts with which they have been involved recently. Have them describe in their journals how they could have applied the steps in the textbook to resolve each conflict. How would any or all of these steps have helped resolve the conflicts?

## TEACHER ALERT

Help students see that sometimes a conflict may be based entirely on a misunderstanding. You might hear that somebody said something insulting about you and then find out later that he or she did not really say that after all. Misinformation spreads quickly, especially when the people involved in the conflict are from differing cultural backgrounds. Have students suggest scenarios in which conflict arises for no real reason. Ask students to discuss how they can avoid this kind of conflict.

## Decide for Yourself

Have students read the Decide for Yourself feature on page 432. If you wish, assign a point of conflict to two students and have the class direct them how to perform each of the conflict resolution steps.

**Decide for Yourself Answers**

1. Sample answer: To resolve conflict, understanding how a person feels is important. Each person must listen attentively to understand how the other person feels. 2. Sample answer: Compromise is an important aspect of conflict resolution. Cursing and name-calling cause a person to feel defensive and unwilling to compromise. 3. Sample answer: Listening to how the person feels will be the most difficult step for me. I will need to listen without interrupting, and I have a hard time doing that. Thinking of a solution will be the easiest step for me because I do not like conflict.

## How Can a Person Get Help with a Conflict?

When you have a serious conflict, you need to get help. Talk with your parents or a peer mediator, teacher, or counselor. If they cannot help you, they can probably send you to someone who can.

You may hear that a friend has plans for a violent act. You may not want to tell anyone about your friend's plans. Remember, it is always important to stop someone who plans to do harm. Talk with a parent or other trusted adult who may be able to help your friend.

Lots of people have conflicts. Asking for help does not mean that you are weak. It just means that you care enough about yourself and other people to take control of your life.

### Decide for Yourself

Some people think that the only way to resolve a conflict is to fight. Being able to resolve a conflict instead of fighting is a sign of strength and control. Conflict resolution skills give you control in a situation. They lessen the chance of violence among teens and adults.

**Steps to Conflict Resolution**

1. Set the ground rules. Agree to discuss the problem without cursing or name-calling. Avoid accusing the person with whom you are arguing.
2. Allow the person to describe the argument. Listen to how the person feels. Often the conflict is not about what happened but about how a person feels about what happened.
3. Decide what both of you agree on.
4. List the ways you can resolve the conflict. You may find no agreement on what caused it.
5. Think of a solution in which both of you win. Compromise is important at this point. For example, if you disagree on whether you have taken something that belongs to the other person, find a mediator who will make a plan that will be fair to both of you.
6. Come to a specific agreement. Write it down if necessary. Work with a mediator or trusted person to make sure the agreement works.

On a sheet of paper, write your answers to the following questions. Use complete sentences.

1. Why are good listening skills needed during conflict resolution?
2. Why is avoiding name-calling important during conflict resolution?
3. Which of the conflict resolution steps will be the most difficult for you? Which step will be the easiest?

## How Can Communities Help Fight Crime?

Communities are taking steps to fight crime. One way is the Neighborhood Crime Watch. In this program, volunteers are trained to look for unusual activity and crimes that are being committed. Then they report what they see to police. Often, neighbors post signs and stickers that warn would-be criminals that neighbors are watching.

Because many crimes happen at night, community leaders in many places are making important changes. For example, they are planning more nighttime activities for young people. They are putting in bright lights at nighttime hangouts such as playgrounds and parking lots. Some cities have teen curfews that do not allow teenagers on the streets after a certain hour. More people are being added to the police force in some places.

Some other community actions to reduce crime are:

- setting up neighborhood centers where agencies can provide counseling help, recreation, and vocational training
- improving or adding streetlights to help police and neighborhood watch programs
- putting security systems in stores to let police know if a robbery is taking place
- finding and training neighbors to take care of children in danger
- providing crime awareness education to teach people how not to become victims of crime

### Technology and Society

Each year, more than 700 million people fly from one part of the country to another. Each person who wants to get on a plane must go through a series of safety and security checks. This includes checking for sharp metal objects or chemicals that might mean the person has a bomb. Airport security works to prevent violence and terrorist attacks. The Department of Homeland Security controls airport security systems.

### Technology and Society

Have students read the Technology and Society feature on page 433. As the class discusses the subject, tell students what you remember of security procedures at airports before September 11, 2001, and ask them to share their memories as well. Have students search online newspaper and magazine archives for articles that detail how airline security has improved since the tragedy in 2001. Discuss the articles with the class.

### IN THE COMMUNITY

Have a group of students gather information about agencies, counseling services, and self-help groups in their community that can help teenagers and their families cope with problems. The United Way and the agencies it helps are some examples; students might want to begin their research on the Web site of United Way International (www.uwint.org). Have the group present its findings to the class.

### HEALTH JOURNAL

Have students list in their journals people upon whom they can rely for help in resolving a conflict. Students should list several people, as conflicts might arise at home, at school, or elsewhere. Suggest that students begin with family and branch out to friends, teachers, and other trusted adults.

# IN THE ENVIRONMENT

Divide students into two groups for a debate. Present the following topic: A hypothetical town is affected by an environmental problem, such as contaminated waterways or visual pollution, and two groups are arguing about how to solve the problem. Assign opposite positions to each group and allow students time to prepare an argument supporting their point of view. Then ask volunteers to present each group's opinion. Serve as moderator as groups take turns arguing their case.

## Lesson 3 Review Answers

1. mediator 2. mediate 3. compromise
4. Sample answer: I could tell the people involved that I greatly respect anyone who is able to walk away from a fight. I could leave the area so I do not encourage a fight simply by being there.
5. Sample answer: The best way for me to control my anger is to take a deep breath, tell myself to relax, and realize that I do not have to win every argument.

## Portfolio Assessment

Sample items include:

- Flowcharts from Learning Styles: Logical/Mathematical
- Reports and visual presentations from Research and Write
- Skits from Learning Styles: Body/Kinesthetic Conflict solutions from ELL/ESL Strategy
- Campaigns from Learning Styles: Interpersonal/Group Learning
- Entry from Health Journal
- Decide for Yourself answers
- Findings from Application: In the Community
- Entry from Health Journal
- Lesson 3 Review answers

---

Many lawmakers are working for tougher punishments for those found guilty of violent crimes. You may wish to write your local lawmakers with your ideas about how your community can better fight crime.

Schools in many cities now have full-time security guards. The students in these schools must pass through metal detectors. Some schools have video cameras to watch hall activity. Many school districts are hiring more counselors to help students in trouble. Students can work together with teachers and counselors to make sure that their school is safe. Some schools have student-run safety committees. Perhaps the most positive move that many schools have made is to teach courses in accepting and respecting one another. Teens can choose to be peacemakers rather than troublemakers.

**Lesson 3 Review** On a sheet of paper, write the word or phrase in parentheses that best completes each sentence.

1. Using a (mediator, offender, troublemaker) to resolve conflicts is often helpful.
2. One way to stay out of a fight is to (disagree, push, mediate).
3. An important part of conflict resolution is (time pressure, compromise, fault).

On a sheet of paper, answer the following questions. Use complete sentences.

4. **Critical Thinking** What is one way you could help others stay out of a fight?
5. **Critical Thinking** What is the best way for you to control your anger?

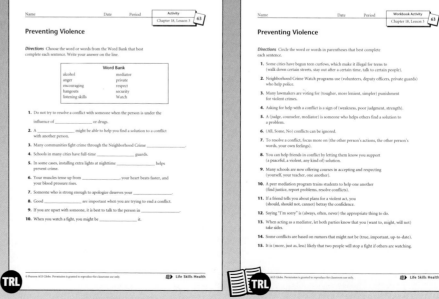

# Chapter 18 SUMMARY

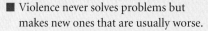

- Violence never solves problems but makes new ones that are usually worse.
- Conflicts are disagreements between two or more people with different ideas. Conflicts may be minor, ongoing, internal, or major. A disagreement is normal conflict, but conflict that becomes violent is not normal.
- Costs of violence are permanent injuries, mental health problems, hurt to loved ones, lost peace of mind, and money that everyone must pay.
- Knowing the signs of conflict helps people avoid or resolve conflicts and reduces the likelihood of violence.
- Some causes of violence are the media, growing up with violence, alcohol and other drugs, and gangs. Other causes are personal problems, group loyalty, prejudice, and revenge.
- Robbery, beating, murder, and rape are crimes that often happen to teens.
- Everyone feels anger at times. People can deal with anger to keep minor conflicts from becoming violent.
- At times, the best way to deal with conflict is to walk away from it. Avoiding conflict shows strength, not weakness.
- Talking with a mediator is one of the best ways to resolve conflict. Both sides in a conflict must remain calm, not insult each other, and focus on feelings rather than actions.
- Two ways to help others not get into fights are not to watch the fight and to respect the person who can walk away from it. You also can pressure others not to fight, and you can act as a mediator.
- Talking with a parent, mediator, teacher, or school counselor can help resolve a conflict. It is always important to stop someone who plans to do harm.
- Neighborhood Crime Watch and other programs help communities fight crime. Local, state, and national lawmakers are making tougher punishments for crime. Schools are taking many actions to fight crime.

| Vocabulary | | |
|---|---|---|
| conflict, 419 | negotiation, 431 | sexual exploitation, 427 |
| date rape, 427 | rape, 426 | sexual harassment, 426 |
| mediator, 431 | sexual assault, 427 | victim, 418 |

## Chapter 18 Summary

Have volunteers read aloud each Summary item on page 435. Ask volunteers to explain the meaning of each item. Direct students' attention to the Vocabulary box on the bottom of page 435. Have them read and review each term and its definition.

### ONLINE CONNECTION

The National Center for Injury Prevention and Control, which is part of the Centers for Disease Control and Prevention (CDC), maintains a Web page devoted to facts about sexual violence. The page also contains a link to a fact sheet about child maltreatment: www.cdc.gov/ncipc/factsheets/svfacts.htm

The Federal Bureau of Investigation devotes a section of its Web site to explain its Investigative Programs: Crimes Against Children. Included are links to national and state sex offender registry Web sites and a link to the National Center for Missing and Exploited Children: www.fbi.gov/hq/cid/cac/crimesmain.htm

## Chapter 18 Review

Use the Chapter Review to prepare students for tests and to reteach content from the chapter.

## Chapter 18 Mastery Test

The Teacher's Resource Library includes two forms of the Chapter 18 Mastery Text. Each test addresses the chapter Goals for Learning. An optional third page of additional critical-thinking items is included for each test. The difficulty level of the two forms is equivalent.

### Review Answers

**Vocabulary Review**

1. date rape 2. conflict 3. victim 4. rape
5. sexual harassment 6. negotiation
7. mediator 8. sexual exploitation
9. sexual assault

**Concept Review**

10. B 11. A 12. D 13. C 14. B 15. A 16. D
17. A

**Critical Thinking**

18. Sample answer: The most serious cause of violence is gangs because innocent people are often hurt by gangs.
19. Sample answer: Violence can leave people with permanent physical injuries. It can leave people with a sense of mistrust, betrayal, and fear. 20. Sample answer: As a mediator, I can let both parties know that I am not taking sides. I can ask the two parties to sit down and talk over the problem.

---

## Chapter 18 REVIEW

**Word Bank**
conflict
date rape
mediator
negotiation
rape
sexual assault
sexual exploitation
sexual harassment
victim

### Vocabulary Review

One a sheet of paper, write the word or phrase from the Word Bank that best completes each sentence.

1. Forced sex with someone the victim socializes with is _____.
2. A disagreement between people is a _____.
3. Injury is one cost for a _____ of violence.
4. A person who forces another person to have sexual intercourse is committing _____.
5. Unwanted sexual comments are a type of _____.
6. Someone who helps people compromise is practicing _____.
7. As a _____, you can help work out conflicts.
8. Using a person sexually to make money is _____.
9. Any forced sexual contact, including rape, is _____.

### Concept Review

On a sheet of paper, write the letter of the answer that best completes each sentence.

10. A negative opinion formed without enough information is _____.
    A a solution          C an educated guess
    B prejudice           D a theory

11. Violence includes actions or words that hurt people or damage things and _____.
    A cause conflict      C result in conflict resolution
    B are not a big deal  D help negotiation

12. Some warning signs of conflict are pushing, name-calling, shouting, and _____.
    A staying calm        C asking for a mediator
    B walking away        D spreading rumors

436 Unit 6 Injury Prevention and Safety Promotion

---

Chapter 18 Mastery Test A, pages 1–3

13. The first step in conflict resolution is _____.
    A asking for a mediator   C setting the ground rules
    B compromising            D listing ideas

14. One way to lessen your chances of being a victim of rape is to _____.
    A walk alone at night
    B avoid being alone with a date
    C date only people you know
    D get in a car with someone you do not know

15. A conflict at home over not doing chores is _____.
    A minor   B internal   C ongoing   D major

16. A conflict that takes a long time to resolve is _____.
    A minor   B internal   C ongoing   D major

17. Many people who commit crimes _____ their victims.
    A know        C have never met
    B date        D are related to

## Critical Thinking

On a sheet of paper, answer the following questions. Use complete sentences.

18. What do you think is the most serious cause of violence? Explain your answer.

19. What effects does violence have on a person?

20. What can you do as a mediator to resolve a conflict?

**Test-Taking Tip** Review your corrected tests. You can learn from mistakes you made earlier.

## Alternative Assessment

Alternative Assessment items correlate to the student Goals for Learning at the beginning of this chapter.

- Have students orally define violence and describe its costs.
- Have students design a poster that describes the warning signs of conflict.
- Have students work in groups to write the script for a public service announcement identifying the causes of violence.
- Have students write and act out a script for an instructional video that explains ways to prevent violence and find answers to conflicts.

# HEALTH IN THE WORLD

Have students read the Health in the World feature on page 438. Review with students the work done by Doctors Without Borders. Ask students what kinds of medical needs the group might address in different kinds of emergencies. *(Floods: bacterial diseases, respiratory illnesses; earthquakes: traumatic injuries, diseases caused by overcrowded conditions in camps; civil war: battle injuries, starvation, diseases caused by overcrowded conditions in camps)*

## Health in the World Answers

1. Sample answer: Doctors Without Borders volunteers are important to world health because they provide medical aid to people worldwide who are affected by natural and human-made emergencies. These volunteers can treat a disease in one area and prevent it from spreading to other areas. 2. Sample answer: Clean water is important for two reasons. The volunteer doctors need pure water to clean wounds and perform surgery in sanitary conditions. Clean drinking water is needed for local residents because impure water can lead to the spread of disease. The volunteers help the local residents by constructing wells for clean drinking water. 3. Sample answer: Getting quick medical help to people affected by emergencies is important to save lives. When medical professionals respond to an emergency, they often learn lessons that can help people prevent future crises.

# HEALTH IN THE WORLD

## Doctors Without Borders

Doctors Without Borders is an international volunteer group. It is made up of doctors, nurses, experts on clean water, and other professionals. Together, they bring medical aid to people affected by emergencies around the world. This includes both natural and human-made emergencies such as floods, earthquakes, and civil war. Doctors Without Borders volunteers can quickly set up mobile operating rooms complete with medical supplies. They also provide tents, cooking kits, and blankets for people affected by emergencies.

Doctors Without Borders began in 1971. Today, the organization responds to more than 3,800 medical missions each year. Along with its emergency programs, Doctors Without Borders sets up worldwide vaccination programs. Volunteers construct wells for clean drinking water. Volunteer doctors perform life-saving operations. They also treat people with infectious diseases such as AIDS and tuberculosis.

Doctors Without Borders has public education programs that help others understand what it is like to live in a refugee camp, be malnourished, or live without medicine. The organization is funded by private donations.

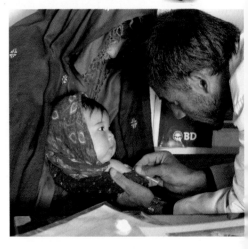

Today, more than 3 million people donate funds to help Doctors Without Borders provide medical aid to people in need.

On a sheet of paper, write the answers to the following questions. Use complete sentences.

1. How are the volunteers of Doctors Without Borders important to world health?
2. Why is clean water important for the work of Doctors Without Borders?
3. Why is getting quick medical help to people affected by emergencies important?

# Unit 6 SUMMARY

- Fires caused by electrical problems, fuel, space heaters, furnaces, and open flames all can be prevented.
- Every home should have one or more smoke detectors and fire extinguishers.
- The Occupational Safety and Health Administration (OSHA) sets standards for safety in the workplace. Employers are responsible for removing dangers in the workplace.
- Learning to swim, survival floating, and following water safety guidelines can prevent drowning.
- Learning safe driving skills and habits, maintaining the speed limit, and adjusting to conditions all help prevent vehicle crashes. Wearing a seat belt can reduce injuries if a crash occurs.
- Wearing helmets, following the rules of the road, and taking other preventive actions can reduce risks of accidents for bicyclists and motorcyclists.
- People can take protective steps in cases of natural disasters such as hurricanes, tornadoes, floods, earthquakes, and lightning strikes.
- In an emergency, stay calm, find out whether the person is conscious and breathing, look for injuries, and call 911 for help.
- Good Samaritan laws protect people who help others in emergencies.
- Following universal precautions prevents rescuers from coming into contact with blood and other body fluids. These precautions also protect wounds from dirt and infections.
- The Heimlich maneuver is used to force an object out of the throat when someone is choking.
- Rescue breathing puts oxygen from the rescuer's lungs into an unconscious person's lungs.
- A person in cardiovascular failure has a hard time breathing and an irregular pulse. A person whose heart has stopped beating and has no pulse is in cardiac arrest. Cardiopulmonary resuscitation (CPR) is needed for both.
- Immediate direct pressure over a wound is the first step in first aid for serious bleeding.
- Knowing the signs of conflict helps people avoid or resolve conflicts and reduces the likelihood of violence.
- At times, the best way to deal with conflict is to walk away from it. Avoiding conflict shows strength, not weakness.
- Talking with a parent, mediator, teacher, or school counselor can help resolve a conflict.
- Neighborhood Crime Watch and other programs help fight crime.

## Unit 6 Review

Use the Unit Review to prepare students for tests and to reteach content from the unit.

## Unit 6 Mastery Test

The Teacher's Resource Library includes two forms of the Unit 6 Mastery Text. An optional third page of additional critical-thinking items is included for each test. The difficulty level of the two forms is equivalent.

## Review Answers

### Vocabulary Review

1. electrical shock 2. cardiopulmonary resuscitation (CPR) 3. Heimlich maneuver 4. hurricane 5. sexual harassment 6. sterile 7. survival floating 8. oral poisoning 9. mediator 10. first aid 11. date rape 12. hazard

### Concept Review

13. B 14. C 15. A 16. D

---

# Unit 6 REVIEW

## Vocabulary Review

On a sheet of paper, write the word or phrase from the Word Bank that best completes each sentence.

**Word Bank**
cardiopulmonary resuscitation (CPR)
date rape
electrical shock
first aid
hazard
Heimlich maneuver
hurricane
mediator
oral poisoning
sexual harassment
sterile
survival floating

1. Using an electric appliance when your clothing is wet can result in _____.

2. Rescue breathing and pressing on a person's chest are both parts of _____.

3. When performing the _____, upward thrusts are made below a person's rib cage to force an object out of the throat.

4. A severe tropical storm with heavy rains that strikes the eastern coast of the United States is a(n) _____.

5. When a person makes unwanted physical advances or says sexual things that make another person uncomfortable, it is an act of _____.

6. A clean bandage that is free from bacteria is _____.

7. One technique that can prevent a person from drowning is _____.

8. When a person swallows a harmful material, he or she could suffer from _____.

9. A(n) _____ is someone who helps others find a solution to a conflict.

10. Immediate emergency care given to an injured person before professional medical care arrives is _____.

11. A person who forces a friend to have sex is committing _____.

12. Leaving books or toys on stairs creates a(n) _____ in the home.

440 Unit 6 Injury Prevention and Safety Promotion

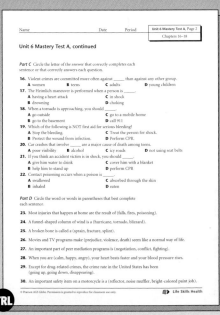

Unit 6 Mastery Test A, pages 1–3

## Concept Review

On a sheet of paper, write the letter of the answer that best completes each sentence.

13. A small grease fire can be put out with _____.
    A water
    B baking soda
    C oxygen
    D milk

14. A person riding a bicycle at night should wear _____ clothing.
    A dark
    B loose
    C reflective
    D lightweight

15. When medical professionals wear protective masks and latex gloves, they are using the _____.
    A universal precautions
    B Good Samaritan laws
    C pressure points
    D universal distress signal

16. People use _____ when they compromise.
    A conflict
    B prejudice
    C anger
    D negotiation

## Critical Thinking

On a sheet of paper, write the answers to the following questions. Use complete sentences.

17. Why should a person never camp alone?

18. Suppose you and a friend are giving first aid to a boy who is seriously bleeding from a cut on his forearm. Your friend suggests that you raise the boy's injured arm above his heart. Do you think this is a good idea? Explain your answer.

19. Why do you think frostbite usually affects the hands, feet, ears, or nose but not other body parts?

20. What is something that could cause violence in your community? How could this issue or situation be handled so that violence does not occur?

## Critical Thinking

17. Sample answer: A person should never camp alone because if the person got lost or became injured, there would be no one around to help. For example, if a tree fell on a person's leg, then that person might become trapped. If another person were there, then he or she could help get the tree off the person or could go for help if he or she could not move it. 18 Sample answer: In this situation, it is a good idea to raise the wound above the level of the injured boy's heart because this will help slow the bleeding. When giving first aid to a person who is suffering from serious bleeding, the main priority is to stop the bleeding. Elevating the wound above the level of the heart can lessen the amount of blood that is lost, and this can keep the person from going into shock. 19. Sample answer: The hands, feet, ears, and nose are parts of the body that have less blood flow. Other parts of the body, such as the chest, have a great deal of blood flow and therefore are not as susceptible to frostbite. Frostbite is a tissue injury that comes from being out in extreme cold without proper clothing. People often get frostbite in these four areas of the body because they are not wearing proper clothing; a lack of gloves for the hands, a lack of adequate footwear for the feet, a lack of earmuffs or a hat for the ears, and a lack of a ski mask or scarf for the nose can lead to frostbite. 20. Sample answer: In my neighborhood, there is a small park where one group of kids likes to play soccer after school, and another group of kids likes to play football. There is not enough room for both groups to play at the same time. This has resulted in some heated arguments and even a couple of fights because both groups wanted to play at the same time. As a mediator, I will suggest that one group use the park on Mondays, Wednesdays, and Fridays, and the other group use the park on Tuesdays, Thursdays, and Saturdays. If both groups follow this schedule, and both groups agree not to use the park on Sundays, then each group will have access to the park three times a week. Following an agreed-upon schedule will prevent situations in which both groups are trying to use the same space at the same time. This could prevent future outbreaks of violence.

Unit 6 Mastery Test B, pages 1–3

# Unit 7

## Planning Guide
## Health and Society

| | Student Pages | Vocabulary | Health in the World | Review | Critical-Thinking Questions | Chapter Summary |
|---|---|---|---|---|---|---|
| **Chapter 19** Consumer Health | 444–465 | ✔ | | ✔ | ✔ | 463 |
| Lesson 1 Being a Wise Consumer | 446–451 | ✔ | | ✔ | ✔ | |
| Lesson 2 Seeking Health Care | 452–456 | ✔ | | ✔ | ✔ | |
| Lesson 3 Paying for Health Care | 457–462 | ✔ | | ✔ | ✔ | |
| **Chapter 20** Public Health | 466–483 | ✔ | | ✔ | ✔ | 481 |
| Lesson 1 Public Health Problems | 468–473 | ✔ | | ✔ | ✔ | |
| Lesson 2 U.S. Public Health Solutions | 474–477 | ✔ | | ✔ | ✔ | |
| Lesson 3 Health Promotion and Volunteer Organizations | 478–480 | ✔ | | ✔ | ✔ | |
| **Chapter 21** Environmental Health | 484–511 | ✔ | | ✔ | ✔ | 509 |
| Lesson 1 Health and the Environment | 486–489 | ✔ | | ✔ | ✔ | |
| Lesson 2 Air Pollution and Health | 490–494 | ✔ | | ✔ | ✔ | |
| Lesson 3 Water Pollution and Health | 495–498 | ✔ | | ✔ | ✔ | |
| Lesson 4 Other Environmental Problems and Health | 499–503 | ✔ | | ✔ | ✔ | |
| Lesson 5 Protecting the Environment | 504–508 | | 512 | ✔ | ✔ | |

### Unit and Chapter Activities

**Student Text**
Unit 7 Summary

**Teacher's Resource Library**
Home Connection 7
Community Connections 19–21

**Teacher's Edition**
Opener Activity, Chapters 19–21

### Assessment Options

**Student Text**
Chapter Reviews, Chapters 19–21
Unit 7 Review

**Teacher's Resource Library**
Self-Assessment Activities 19–21
Chapter Mastery Tests A and B, Chapters 19–21
Unit 7 Mastery Tests A and B

**Teacher's Edition**
Chapter Alternative Assessments, Chapters 19–21

| | Student Text Features | | | | | | | Teaching Strategies | | | | | | Learning Styles | | | | Teacher's Resource Library | | | | |
|---|---|---|---|---|---|---|---|---|---|---|---|---|---|---|---|---|---|---|---|---|---|---|
| Self-Assessment | Health at Work | Health in Your Life | Decide for Yourself | Link to | Research and Write | Health Myth | Technology and Society | Background Information | ELL/ESL Strategy | Health Journal | Applications (Home, Career, Community, Global, Environment) | Online Connection | Teacher Alert | Auditory/Verbal | Body/Kinesthetic | Interpersonal/Group Learning | Logical/Mathematical | Visual/Spatial | Activities | Alternative Activities | Workbook Activities | Self-Study Guide | Chapter Outline |
| 444 | | | | | | | | | | | | 463 | 464 | | | | | | 64–66 | 64–66 | 64–66 | ✓ | ✓ |
| | | | 450 | | 448 | 449 | | 446 | 448 | 448 | 450 | | 450 | 449 | 449 | | 447 | | 64 | 64 | 64 |  |  |
| | | 456 | | | | | | 452 | 454 | 455 | 454, 455 | | 454 | | | | | 458 | 65 | 65 | 65 |  |  |
| | 462 | | | 459, 461 | 458 | 460 | 461 | 457 | 459 | | 459, 461 | | 460 | | | 460 | | | 66 | 66 | 66 |  |  |
| 466 | | | | | | | | | | | | 481 | | | | | | | 67–69 | 67–69 | 67–69 | ✓ | ✓ |
| | | | 472 | 469 | 471, 473 | 470 | 473 | 468 | 469–471 | 470 | 470, 472 | | 469 | | | 471 | | | 67 | 67 | 67 |  |  |
| | 477 | 476 | | 475 | | 474 | | 474 | 475 | | 476 | | | 475 | 476 | | | | 68 | 68 | 68 |  |  |
| | | | | | | | | 478 | 479 | 480 | 479 | | | | | | 480 | 479 | 69 | 69 | 69 |  |  |
| 484 | | | | | | | | | | | | 509 | 510 | | | | | | 70–74 | 70–74 | 70–74 | ✓ | ✓ |
| | | | | 488 | 489 | | | 486 | 487 | | 488 | | 489 | 488 | | 488 | | | 70 | 70 | 70 |  |  |
| | | | | 493 | | 494 | | 490 | 492 | | | | | | | 491, 493 | | 492 | 71 | 71 | 71 |  |  |
| | | 497 | | | | | | 495 | 496 | | 497, 498 | | | | | 497 | | | 72 | 72 | 72 |  |  |
| | | | | | 500 | | 503 | 499 | 501 | 501 | 500, 502 | | | | | | 502 | | 73 | 73 | 73 |  |  |
| | 508 | 507 | | | | | 505 | 504 | 505 | 506 | 505, 506 | | 506 | | | | | 506 | 74 | 74 | 74 |  |  |

## Alternative Activities

The Teacher's Resource Library (TRL) contains a set of lower-level worksheets called Alternative Activities. These worksheets cover the same content as the regular Activities but are written at a second-grade reading level.

## Skill Track

Use Skill Track for *Life Skills Health* to monitor student progress and meet the demands of adequate yearly progress (AYP). Make informed instructional decisions with individual student and class reports of lesson and chapter assessments. With immediate and ongoing feedback, students will also see what they have learned and where they need more reinforcement and practice.

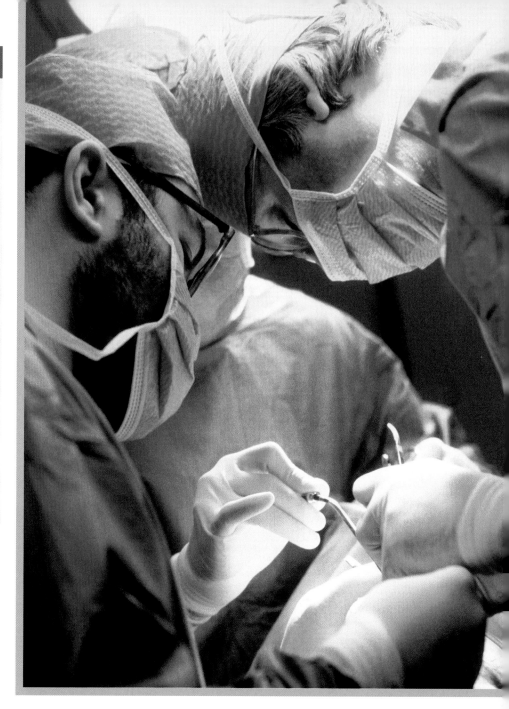

# Unit at a Glance

## Unit 7: Health and Society
pages 442–515

### Chapters

19. **Consumer Health**
    pages 444–465
20. **Public Health**
    pages 466–483
21. **Environmental Health**
    pages 484–511

**Unit 7 Summary** page 513

**Unit 7 Review** pages 514–515

**Audio CD**

**Skill Track for**
**Life Skills Health**

**Teacher's Resource Library** TRL

Home Connection 7

Unit 7 Mastery Tests A and B

(Answer Keys for the Teacher's Resource Library begin on page 559 of this Teacher's Edition.)

## Other Resources

### Books for Teachers

Goldstein, Inge F., and Martin Goldstein. *How Much Risk?: A Guide to Understanding Environmental Health Hazards.* New York: Oxford University Press, 2002. (analysis and assessment of environmental health hazards)

Rees, Alan M., ed. *Consumer Health Information Source Book.* 7th ed. Westport, CT.: Greenwood Press, 2003.

### Books for Students

Osborne, Helen. *Health Literacy from A-Z: Practical Ways to Communicate Your Health.* Boston: Jones and Bartlett Publishers, 2004. (tips for decision-making about health care issues)

Sussman, Art. *Dr. Art's Guide to Planet Earth: For Earthlings Ages 12 to 120.* White River Junction, VT: Chelsea Green Publishing, 2000. (earth systems and their connections)

### CD-ROM/Software

*21st Century Complete Medical Guide to Teen Health Issues.* Progressive Management, 2004. (up-to-date electronic book on teen health issues)

### Videos and DVDs

*Creating a Healthy Home* (82 minutes). Broomfield, CO: Conscious Wave, Inc., 1998. (simple changes to create an environmentally safe and healthy home)

*Critical Condition with Hedrick Smith: How Good Is Your Health Care?* (4-part series, 47-49 minutes each). Chevy Chase, MD: Hedrick Smith Productions, 2000 (1-800-553-7752).

### Web Sites

www.consumer.gov/health.htm (collection of links to federal resources, including health resources)

www.healthpolicy.ucla.edu/ (information on education and public service health programs from the UCLA Center for Health Policy Research)

www.epa.gov/highschool (environmental information for high school students from the U.S. EPA)

# Unit 7: Health and Society

Do you buy products with environmentally safe packaging? Do you volunteer for causes that promote public health? Do you recycle on a regular basis? We all have a responsibility to promote public health and protect our environment. As people who buy and use goods and services, we can choose products that protect the environment. As members of a community, we can help to promote public health.

When buying goods and services, we are faced daily with decisions that affect us physically, socially, and emotionally. These decisions cover a wide range of choices. We might be choosing health care. We might be shopping for food. We might be looking for a way to prevent water pollution. This unit will show how you can make wise health choices. You will learn to promote public health and protect the environment.

## Chapters in Unit 7

Chapter 19: Consumer Health ................................. 444
Chapter 20: Public Health ....................................... 466
Chapter 21: Environmental Health ........................ 484

## Sources of Water Quality Testing Kits

Carolina Biological Supply Company, 2700 York Road, Burlington, NC 27215-3398 (1-800-334-5551)

Science Kit, 777 E. Park Drive, Tonawanda, NY 14150 (1-800-828-7777)

## Introducing the Unit

Have volunteers read aloud the introductory paragraphs on page 443. Ask students to explain how environmental choices are related to health. *(Students might have only vague ideas about this concept before beginning the unit, but they should be able make connections between clear air and clean water and good health.)* Next, ask them to describe some of the decisions they make now and will make in the future about health care. *(Answers will vary but may include decisions about health insurance, health care providers, diet, and exercise, for example.)*

### HOME CONNECTION

The Home Connection unit activity gives students practical experience with concepts taught in the *Life Skills Health* student text. Students complete the Home Connection activity outside the classroom with the help of family members. These worksheets appear on the Life Skills Health Teacher's Resource Library (TRL) CD-ROM.

### CAREER INTEREST INVENTORY

The AGS Publishing Harrington-O'Shea Career Decision-Making System-Revised (CDM) may be used with the chapters in this unit. Students can use the CDM to explore their interests and identify careers. The CDM defines career areas that are indicated by students' responses on the inventory.

Home Connection 7

# Chapter at a Glance

## Chapter 19: Consumer Health
pages 444–465

### Lessons
1. Being a Wise Consumer
   pages 446–451
2. Seeking Health Care
   pages 452–456
3. Paying for Health Care
   pages 457–462

**Chapter 19 Summary** page 463

**Chapter 19 Review** pages 464–465

**Audio CD**

**Skill Track for Life Skills Health**

**Teacher's Resource Library**
   Activities 64–66
   Alternative Activities 64–66
   Workbook Activities 64–66
   Self-Assessment 19
   Community Connection 19
   Chapter 19 Self-Study Guide
   Chapter 19 Outline
   Chapter 19 Mastery Tests A and B

(Answer Keys for the Teacher's Resource Library begin on page 559 of this Teacher's Edition.)

## Opener Activity

Bring to class several issues of *Consumer Reports* magazine. Ask if students are familiar with the magazine. Discuss its purpose—to inform consumers about the quality of the products and services they buy. Choose an article from the health section of the magazine and present it to the class

# Self-Assessment

## Can you answer these questions?

1. Where is the nearest clinic at which people in your neighborhood can get treatment?
2. What kind of health insurance do your parents or guardians have?
3. What are your health care professional's qualifications?
4. What steps should you take if you buy a product that does not work?
5. How do you decide what health products to buy?
6. What are some kinds of health professionals?
7. What is the difference among health care facilities?
8. What types of health insurance are available?
9. What are some situations in which you should see a doctor?
10. How can people lower health care costs?

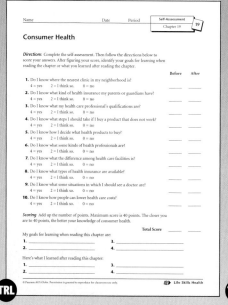

Self-Assessment 19

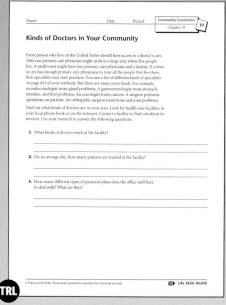

Community Connection 19

444 Unit 7 Health and Society

# Chapter 19: Consumer Health

People buy health products and services every day. Health products and services affect your physical and emotional well-being. From the soap you wash with to the sunscreen you use, you buy health products. You buy services when you see a health care professional such as a doctor, dentist, or nurse.

Making wise choices when buying health products and services takes effort. You need to learn how to judge the products and services you use. You need to find out when and where to seek health care. You need to know how to pay for health care.

In this chapter, you will learn what influences people's health choices and how to choose products and services. You will find out when you need to seek health care from a professional and where to find it. You will learn how health care expenses are paid.

## Goals for Learning

- ◆ To explain the advantages of making wise health choices
- ◆ To describe what influences people's health choices
- ◆ To describe ways to judge products and services and correct problems with them
- ◆ To decide when self-care or professional treatment is best
- ◆ To identify sources of health care
- ◆ To explain ways to pay for health care

## Introducing the Chapter

Have a student volunteer read aloud the Goals for Learning on page 445. Then discuss each goal, asking students what they think they already know about each topic.

### Notes and Questions

Ask volunteers to read the notes and questions that appear in the margins throughout the chapter. Then discuss them with the class.

### Self-Assessment

Have students complete the Self-Assessment worksheet before and after reading the chapter. Before reading the chapter, have students fill in the "Before" column. Ask students to identify their goals for learning. To get ideas for setting goals, students might use the chapter introductory material on page 445, the checklist on page 444, or the questions on the Self-Assessment worksheet. Students can use the back of the worksheet if they need more space to write.

Collect the Self-Assessment worksheets and pass them out again at the end of the chapter. Have students fill in the "After" column. Ask them to identify at least four major points they have learned. Again, suggest they use the back of the worksheet if they need more space to write. You may want to collect and review the worksheets, but return them to students so they have a record of their goals and accomplishments.

Chapter 19 Self-Study Guide, pages 1–2

Chapter 19 Outline, pages 1–2

# Lesson at a Glance

## Chapter 19 Lesson 1

**Overview** This lesson describes what it means to be a wise consumer and what consumers can do on a daily basis to make the best decisions for their health.

## Objectives

- To explain the advantages of being a wise consumer
- To discuss consumers' rights
- To recognize influences on consumers' choices
- To judge health products and services
- To describe how to correct problems with products and services

**Student Pages** 446–451

**Teacher's Resource Library** TRL

  Activity 64

  Alternative Activity 64

  Workbook Activity 64

## Vocabulary

consumer    quackery

Write the vocabulary words on the board. Ask students to provide an example of a consumer and of quackery. Then have students read the definitions of each word. Have them evaluate whether their examples were appropriate and provide new ones if necessary.

## Background Information

The Consumer Product Safety Commission was established in 1973. Since that time, thousands of products have been taken off the market or recalled. When a product is recalled, consumers receive either a refund or a replacement for the product. The company that manufactured the defective product is responsible for disposing of the defective merchandise or correcting the problem.

---

## Lesson 1 — Being a Wise Consumer

### Objectives

*After reading this lesson, you should be able to*

 explain the advantages of being a wise consumer

◆ discuss consumers' rights

◆ recognize influences on consumers' choices

◆ judge health products and services

◆ describe how to correct problems with products and services

**Consumer**
*Someone who buys goods or services*

Part of promoting your own health is being a wise **consumer**. A consumer is someone who buys goods or services. For example, you buy goods such as grooming products or over-the-counter medicines. You buy services such as dental or medical care.

### What Advantages Does a Wise Consumer Have?

Being a wise consumer has several advantages. First, you protect and improve your health. You avoid buying products that are useless or that may be harmful. You recognize symptoms that signal the need for health care and get the help quickly.

Second, you save money. This means getting the best product or service for the least money. Saving money takes some research, but it pays off in the long run. Consumer publications or reviews online help you compare features and prices of different products.

Third, you gain self-confidence. Speaking up for your rights means taking care of yourself. For example, you can make a complaint to the proper source about a product that does not work. You can usually get your money back or have the product replaced.

**Figure 19.1.1** *To make a wise choice, compare prices and ingredients before you buy.*

446  Unit 7  Health and Society

## What Influences Consumers' Choices?

Many factors influence consumers' choices. These include advertising, family and friends, and price. Being a wise consumer means that you judge these factors before you buy. You spot misleading advertisements. You keep in mind that what is right for another person may not be the best product for you. You compare prices.

Businesses use different media sources to advertise their products. You may see or hear ads on television, radio, billboards, newspapers, magazines, and the Internet. Advertisers may use strategies that influence people without their realizing it. For example, you may keep singing a catchy tune that you have heard to advertise a product. Perhaps a famous actor or athlete you admire is promoting the product. Without thinking about the reason, you may buy that product.

The media may influence family and community health, as well as personal health. For example, suppose one parent buys personal care products for the family. If an ad influences that person to buy a certain product, the health of everyone in the family is affected by using the product.

Here are ways to analyze advertising techniques. Ask yourself the following questions before buying a health product. Use the answers to help you decide whether to buy the product.

1. Is the advertisement from a reliable source?
2. Is the health information given in the ad correct?
3. What are the ingredients in the product? Are they effective?
4. What are the health benefits of the product? What are the side effects?
5. Is the advertiser trying to influence my emotions to buy this product?

Another influence on consumers is the advice or opinions of family, friends, and the community. For example, if your friend tells you that a shampoo has solved his dandruff problem, you may decide to try it.

 **Warm-Up Activity**

Have students brainstorm a list of health products and services that teenagers commonly use. Ask students what they would do if a product they bought or a service they used was unsatisfactory. Record their responses on the board.

 **Teaching the Lesson**

As you work through the lesson with students, try to relate the subject matter to consumer purchases they make in their everyday lives. Students already consider certain factors, ranging from cost to trendiness, when deciding whether to make a purchase—or to have their parents make a purchase for them. As they learn what standards a wise consumer should use, help them see that these standards do not apply only to purchasing health care products and services.

**Reinforce and Extend**

### PRONUNCIATION GUIDE

Use the pronunciation shown here to help students pronounce difficult words in the lesson. Refer to the pronunciation key that appears in the Glossary at the back of the Teacher's Edition for the sounds of the symbols.

generic (jə ner´ ik)

### LEARNING STYLES

**Logical/Mathematical**
Write the following ad slogans on the board: "Darnitol has been proven to help 80% of the people who take it." "Thousands of people have been helped by Darnitol." Tell students to imagine they are thinking of buying Darnitol. Then ask what they can tell if they apply deductive reasoning to the first slogan. (*If they take Darnitol, there is an 80 percent chance it will help them.*) Ask what they can tell if they apply inductive reasoning to the second slogan. (*Since Darnitol has helped so many people, it will probably help them.*) Is that a valid conclusion? (*No*)

## ELL/ESL Strategy

**Language Objective:** *To speak in front of a group*

Have students leaf through a local newspaper, looking at the ads. Have them find ads for several different brands of the same product. Ask students to explain to the class which brand they would buy and why. Suggest they consider not just price, but also what they know about the brand, the product, or even the store.

## Research and Write

Have students read and conduct the Research and Write feature on page 448. You may want to help them prepare a chart listing various categories of consumer products. Students could use this chart during their interviews. Or you could arrange with another teacher to set up a time during which your students could interview the students in the other class.

## Health Journal

Ask students to keep track of every purchase they make over the next week—even small items such as a magazine or an after-school snack. Have students describe in their journal what factors made them decide to purchase each item.

---

**Quackery**
*The promotion of medical products or services that are unproved or worthless*

**Research and Write**

Survey several classmates or other students at your school. Ask them which personal care items (such as shampoo and soap) they buy. Go to a drugstore. Compare prices and ingredients on those items. Then prepare a public service announcement informing students how to compare products to get the best buy.

You also might buy a product because your family has always used it. Your community might put messages on the radio and in the newspaper to encourage people to visit a new neighborhood clinic.

Price also influences choices. Brand-name products are familiar but usually cost more than lesser-known brands. The size of a product and the store where a product is purchased also affect price. For example, the cost of a product in a convenience store may be higher than in a drugstore or discount store.

### How Does Quackery Influence Consumers?

**Quackery** is the promotion of medical products or services that are unproved or worthless. Each year, Americans spend billions of dollars on products that claim to cure diseases. People with chronic diseases such as cancer and arthritis often become victims of quackery. People who are overweight or who want to look younger also are targets. At best, people waste their money on these products and services. At worst, people delay getting proper medical treatment.

### How Can Products and Services Be Judged?

You can use a set of standards, or criteria, to help you judge products and services. Use these criteria every time you choose a product or service. They will help you be a wise consumer.

1. Look at the cost benefits. Suppose you want to purchase a pair of running shoes. You need to compare the features with the price on several different brands. Decide which features you really need. Do not pay for features you do not need.

2. Consider recommendations. You can find reviews of exercise equipment and personal grooming products on the Internet and in consumer magazines. A doctor usually can offer information about a product or service. Word-of-mouth recommendations from family and friends can be helpful. Keep in mind, however, that what is right for another person may not work for you.

### Health Myth

**Myth:** Generic drugs are not as good as brand-name drugs.

**Fact:** By federal law, generic drugs contain the same active ingredients as brand-name drugs. Generic drugs are just as safe and effective as brand-name drugs.

3. Watch for false or misleading claims. Avoid any product or service that promises a cure or immediate results. Using library sources or the Internet, do research to make sure the health information given about the product or service is correct.

4. Look at the ingredients. Federal law says that manufacturers must include certain information on the label of every food, drug, and cosmetic or grooming product. Suppose you buy medicine for your allergy symptoms. Your doctor has told you to look for a certain ingredient, so you look for a product that has that ingredient.

5. Compare prices. Compare the price of products with their weight, volume, or count to find the best value. Generic, or nonbrand-name, products are usually less expensive than brand-name products. If two products have the same ingredients, then the generic product is the better buy.

### How Can a Consumer's Problem Be Corrected?

In addition to federal laws, such as for labeling products, states have their own consumer protection laws. These laws protect consumers against false or misleading advertising. Consumers have the right to complain about products that do not work. They can often get their money back. If you buy a product that does not work or have a bad experience with a health care provider, you can take these actions:

1. Write a letter or send an e-mail to the company that sold the product or provided the service. Politely explain what you bought and the problem with it. Ask for your money back or for a fair exchange. If you call, write down the names of people you talked with, the dates, and what they said they would do. Calling does not always protect your rights. A letter or an e-mail is a record that you complained and so it better protects your rights.

## Decide for Yourself

Have students read the Decide for Yourself feature on page 450. Go through the table of symptoms and remedies at the bottom of the page. Have students refer back to the first paragraph of the Decide for Yourself feature to see which symptoms they are supposed to have.

### Decide for Yourself Answers

**1.** Sample answer: I would choose the medicine that is right for my cold, not for all possible cold symptoms. **2.** Sample answer: I would evaluate the outcome of my decision on the basis of the medicine's effectiveness. If the medicine relieved my symptoms, then I made a good decision. I would also check to see whether another store had the same medicine at a lower price. This could influence my shopping habits the next time I need to buy cold medicine.

## AT HOME

 Ask students to bring in warranties from appliances or other products in their home. Have students work in pairs to determine what a warranty for a particular product covers and how easy it would be under the terms of the warranty to get satisfaction if the product is defective. Let pairs share their findings with the class.

---

Keep copies of all sales receipts, letters, and other papers when trying to correct a consumer problem.

**2.** If you do not get a satisfactory response, write a letter to an agency listed on the next page. To get the addresses of government agencies, look in the phone book or on the Internet. In the letter, give the details of your problem. Send a copy of this letter to the company or person the complaint is against. Keep a copy for your records. If you do not get a response within 6 weeks, send a follow-up letter with a copy of the first letter.

### Decide for Yourself

**Define the problem.** You are sitting in class at the end of the day. You have a scratchy throat. You are sneezing and your nose begins to run. You realize you have a cold. You know that cold medicines cannot cure a cold. However, they can help relieve symptoms. You go to the drugstore after school. The over-the-counter medicines you look at contain one or all of the ingredients in the chart below.

**Come up with possible solutions.** (1) Choose the cheapest medicine. (2) Choose the medicine that has everything, because you are getting more for your money. (3) Choose the medicine that has only the ingredients to relieve your symptoms.

**Consider the consequences of each solution.** (1) The cheapest medicine may not contain the right ingredients for your cold. (2) The medicine that has everything has medicine you do not need. (3) The medicine that is right for your cold may be more expensive than some of the other medicines.

**Put your plan into action.** You make your choice, buy the medicine, and use it.

**Evaluate the outcome.** Decide whether you made the best decision for yourself.

On a sheet of paper, write the answers to the following questions. Use complete sentences.

1. Which solution would you choose?
2. How do you think you would evaluate the outcome of your decision?

| Ingredients in Cold Medicines | |
|---|---|
| Analgesics | for headache and body aches |
| Decongestants | for stuffy nose |
| Antihistamines | for sneezing and runny nose |
| Expectorants | for causing coughing to clear bronchial passageways |
| Suppressants | for reducing coughing |

**Sources of Help for the Consumer**

- Complaints about a product or service, false advertising, or irresponsible behavior:
  Better Business Bureau (Check the phone book for the local bureau.)
  Your state attorney general's office (Complaint forms usually are available on the Internet.)

- Complaints about false advertising:
  Federal Trade Commission

- Complaints about foods, drugs, or cosmetics:
  U.S. Food and Drug Administration or your state Department of Agriculture
  U.S. Food Safety and Inspection Service or your state Department of Agriculture

- Complaints about dangerous products:
  U.S. Consumer Product Safety Commission

*Lesson 1 Review* On a sheet of paper, write the answers to the following questions. Use complete sentences.

1. What are three advantages of being a wise consumer?
2. What are two standards to help you judge products and services?
3. What is the first thing you should do to complain about a product or service?

On a sheet of paper, write the answers to the following questions. Use complete sentences.

4. **Critical Thinking** You see an ad for expensive shampoo. Actors with beautiful hair say they use the shampoo. How are the advertisers trying to influence you?
5. **Critical Thinking** Why might people who have chronic diseases be targets of quackery?

Consumer Health  Chapter 19  451

## TEACHER ALERT

Direct students' attention to the first entry under "Sources of Help for the Consumer." Explain that complaining about a company to the Better Business Bureau might help other consumers. The bureau keeps a record of consumer complaints, but it does not take action against a company. Consumers can contact the bureau to see if there have been any complaints about a company they are thinking of doing business with. Depending on the validity and seriousness of the complaints, a state attorney general's office may take action against a company.

## Lesson 1 Review Answers

**1.** Three advantages of being a wise consumer are that these consumers protect and improve their health, save money, and gain self-confidence. **2.** Sample answer: Two standards to help me judge products and services are to look at the ingredients and to consider whether the source of the advertisement is reliable. Another important standard is to consider a product's health benefits and side effects. **3.** The first thing I should do to complain about a product or service is contact the company that provided the product or service. My complaint could be expressed in a letter or an e-mail. **4.** Sample answer: The advertisers are trying to make me believe my hair will look like the hair of one of the actors if I use the shampoo.
**5.** Sample answer: These diseases often cannot be cured, and people are looking for hope and a cure that their doctors don't offer. People who are desperate can be targets of quackery.

## Portfolio Assessment

Sample items include:

- Public service announcement from Research and Write
- Answers from Decide for Yourself
- Lesson 1 Review answers

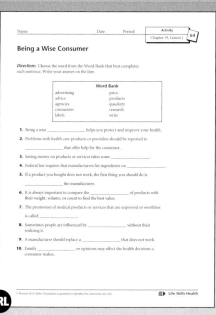

Activity 64

Workbook Activity 64

## Lesson at a Glance

### Chapter 19 Lesson 2

**Overview** This lesson explains when people should seek professional help, identifies where to find health care professionals, and describes some health care facilities.

### Objectives

- To decide when self-care or professional treatment is best
- To describe school health services
- To identify some health care professionals
- To describe some health care facilities

**Student Pages** 452–456

**Teacher's Resource Library**
  Activity 65
  Alternative Activity 65
  Workbook Activity 65

### Vocabulary

| | |
|---|---|
| health care facility | primary care |
| hospice | physician |
| inpatient | rehabilitation |
| nurse practitioner | scoliosis |
| outpatient clinic | specialist |

Write the vocabulary words on the board. Ask students to identify the two words that are neither a person nor a place. *(rehabilitation, scoliosis)* Discuss the meanings of these words with the students. Then have them classify each of the remaining words as a person or a place. Have students describe what they know about the people and places.

### Background Information

As recently as a generation or two ago, medical care differed greatly from current care. Most people simply went to their family doctor, a General Practitioner or GP. The GP would take care of most conditions, and send the patient to a specialist if necessary. Most people did not ask their doctor questions. They explained their problems and then they listened. They did what their doctor told them to do.

---

## Lesson 2  Seeking Health Care

### Objectives

After reading this lesson, you should be able to
- decide when self-care or professional treatment is best
- describe school health services
- identify some health care professionals
- describe some health care facilities

**Scoliosis**
*A condition in which the spine is abnormally curved*

What health services does your school provide?

Wise health consumers know when to use self-care. They also know when it is best not to act as their own doctor.

### When Is Self-Care the Best Action?

Self-care means that you take action to prevent health problems. Good nutrition, regular exercise, and safety practices help to prevent illness or injuries. Not smoking and not breathing in secondhand smoke also are preventive measures. You practice prevention when you avoid tobacco, alcohol, or illegal drug use. Maintaining a healthy body weight helps to prevent heart and joint problems.

You can use self-care for ordinary colds and other minor illnesses. The key word is *ordinary*. For example, good self-care for an ordinary cold is to drink plenty of liquids and get extra rest. Minor illnesses and injuries that do not get better within a few days require professional, or expert, help.

### What Health Services Do Schools Offer?

You might receive certain health services through your school. School health services differ from school to school. Some schools do not have a school nurse. Other schools have health centers with a doctor and nurses. Only 14 states require that a school have a school nurse. Many school districts have school psychologists. Some may have social workers.

School health services differ depending on the medical experts available and the cost of services. School counselors and psychologists help students with emotional or social problems. School nurses may provide care for students with disabilities and emergency care for all students. Nurses may provide information on staying well and preventing disease.

School nurses may also conduct health screenings. These may include screenings of vision, hearing, and blood pressure. School nurses may also do **scoliosis** screening. Scoliosis is a condition in which the spine is abnormally curved.

**Primary care physician**
A doctor who treats people for usual problems

**Specialist**
A doctor who works only in a certain branch of medicine

**Nurse practitioner**
A registered nurse with special training for providing primary health care

---

**Allergist**
Allergies such as hay fever

**Cardiologist**
Heart diseases

**Dermatologist**
Skin problems

**Gynecologist**
Diseases of women

**Obstetrician**
Pregnancy and childbirth

**Ophthalmologist**
Eye problems

**Pediatrician**
Primary care for babies and children

**Psychiatrist**
Mental and emotional problems

---

Taking part in any health screenings your school gives is important. This helps health professionals identify conditions that may need treatment. It also helps you keep track of changes in your health.

## When Should People Seek Professional Help?

Sometimes, not treating your health problem yourself is the best choice. Here are some examples of situations for which you should seek professional help:

- an injury that is more than a minor cut or bruise or a minor blow to the head
- any unusual bleeding, such as blood in the urine or feces
- any severe pain, such as in the abdomen, or a severe headache
- a patch of skin or a mole that has changed shape or color
- a minor problem, such as a cold, flu, cough, sore throat, or body aches, that lasts more than a week
- vomiting that lasts more than a day
- a fever of 103°F for any length of time or a fever of 100.5°F for longer than 3 days
- depression that lasts more than a few days or having thoughts of suicide or harming yourself

## Who Are Some Health Care Professionals?

There are many different kinds of health care professionals. A **primary care physician** treats people for usual problems. For example, a primary care physician may perform checkups or prescribe medicine for a throat infection. A **specialist** is a doctor who works only in a certain branch of medicine. Some types of medical specialists and the types of health problems they treat are listed at left.

A **nurse practitioner** is a registered nurse with special training for providing primary health care. Nurse practitioners perform checkups, detect usual illnesses, and check on chronic conditions.

---

 **Warm-Up Activity**

Discuss with students the importance of having regular health checkups as well as seeking professional help for health problems. Ask students to name tests and procedures they are familiar with that are part of a regular medical checkup. *(taking blood pressure, blood tests, and so on)* Ask why it is important for a doctor to know about a patient's medical history.

 **Teaching the Lesson**

Many people, especially teenagers, are reluctant to seek medical care for something they perceive to be a minor matter. As you work through the lesson with students, reinforce the idea that some situations that seem minor are not, and that even minor problems require treatment if they go on for too long.

**3 Reinforce and Extend**

### PRONUNCIATION GUIDE

Use the pronunciation shown here to help students pronounce difficult words in the lesson. Refer to the pronunciation key that appears in the Glossary at the back of the Teacher's Edition for the sounds of the symbols.

allergist (al′ ər jist)

cardiologist (kär dē ol′ ə jist)

dermatologist (dėr mə tol′ ə jist)

gynecologist (gī nə kol′ ə jist)

obstetrician (ob stə trish′ ən)

ophthalmologist (of thəl mol′ ə jist)

pediatrician (pē dē ə trish′ ən)

psychiatrist (sī kī′ ə trist)

## ELL/ESL Strategy

**Language Objective:** *To teach pronunciations of names of medical specialists*

The names of medical specialists in the list on page 453 can be difficult for students to pronounce, especially those whose primary language is not English. Write the pronunciations provided in the preceding Pronunciation Guide on the board. Have a student proficient in English make use of the pronunciations to help a student learning English correctly pronounce the names of the specialists.

## Teacher Alert

Direct students' attention to the last item in the bulleted list on page 454. Point out this is not as trivial as it may seem. If a doctor does not answer your questions, you probably won't be confident that she or he understands your problem or that she or he is suggesting the right treatment. If you do not feel comfortable with a doctor, you may put off making an appointment, or even avoid going altogether.

## Career Connection

Have students write an essay explaining in which of the health care facilities described in this lesson they would most like to work. Suggest they find out more about the facility before writing their essay. Be sure students include reasons for their choices.

---

**Health care facility**
*A place where people go for medical, dental, and other care*

**Outpatient clinic**
*A place where people receive health care without staying overnight*

Some health conditions are common in members of the same family. Before you visit a doctor the first time, ask your parents or guardian to help you prepare a family history of health problems. Share this information with your doctor.

How can you find out more about a health care professional's qualifications?

---

Doctors sometimes refer people to other kinds of health care providers. For example, a physical therapist helps people regain use of their muscles. A registered dietitian provides nutritional counseling for people with illnesses or special dietary needs.

Regular medical and dental checkups are important. So are vaccinations. Checkups include screening for health problems. Checkups help doctors and dentists prevent problems and find problems early so that treatment is not delayed.

Finding a doctor you are comfortable with may be as simple as asking someone to recommend one. You can get the names and qualifications of physicians in your area from several sources. These include your family's health insurance provider, local medical associations, or directories in the public library or on the Internet. Look for these qualifications:

- a certificate from the appropriate board in the person's field
- an association with a teaching hospital, dental school, or medical center
- experience in treating your particular health problems
- someone who answers your questions and with whom you feel comfortable.

### What Are Some Health Care Facilities?

People see health care professionals in different kinds of places. A **health care facility** is a place where people go for medical, dental, and other care. A dental or medical office is one kind of health care facility. Primary health care is provided in these offices. Primary health care includes checkups and tests to identify diseases and problems. It also includes treatment of common illnesses and injuries.

**Outpatient clinics** provide health care for people who do not need to stay overnight. Outpatient clinics offer primary health care. They also offer same-day surgery, meaning the patient has surgery and goes home the same day. Most clinics treat all kinds of medical problems. Others have a certain focus, such as adolescent health.

*Inpatient*
Referring to someone receiving health care who stays overnight or longer

*Rehabilitation*
Therapy needed to recover from surgery, an illness, or an injury

*Hospice*
A long-term care facility for people who are dying

Other facilities provide **inpatient** care for overnight or longer. Patients can receive nursing care and **rehabilitation**—therapy needed to recover from surgery, illness, or injury.

A hospital is a facility that provides complete health care services. People can receive both outpatient and inpatient care at a hospital. General hospitals provide care for all kinds of illnesses and injuries. Students preparing to be health care professionals receive training in teaching hospitals.

Research hospitals conduct medical studies. Specialty hospitals, such as children's hospitals, deal only with certain illnesses or groups of patients.

Hospital emergency rooms and some urgent care centers offer walk-in emergency care 24 hours a day. These centers are required by law to treat people who do not have health insurance.

Long-term care facilities, such as nursing homes, offer 24-hour nursing care. These facilities are not just for elderly people. Younger people who cannot live independently because of a disability may live there. People recovering from surgery or a serious illness may stay there for a few days or weeks. A **hospice** is another kind of long-term care facility for people who are dying from diseases such as AIDS or cancer.

### What Questions Should You Ask Yourself About a Doctor?

- Does the doctor ask about my health problem and history?
- Does the doctor listen carefully and answer all my questions?
- Do I feel comfortable with the doctor?
- Does the doctor give me enough time and attention?
- Does the doctor clearly explain the purpose of tests and procedures? Does the doctor clearly explain test results or diagnoses?
- Does the doctor encourage me to take preventive action?
- Is the doctor willing to refer me elsewhere when necessary?

## HEALTH JOURNAL

Ask students to imagine they have moved to a new town and are looking for a new doctor. Have students write in their journal questions that they would want to ask a prospective doctor.

## IN THE COMMUNITY

Ask students if they ever have been in a hospital as a patient or a visitor. Invite volunteers to describe the hospital environment and tell about equipment or procedures they observed. Obtain a directory from a local hospital and point out the many departments that make up the hospital. Emphasize that students probably see much more of their school than any particular health care worker sees of the entire hospital.

# Lesson 2 Review Answers

1. a cold 2. specialist 3. nursing home
4. Sample answer: I exercise daily, I eat fruits and vegetables, and I do not smoke. I also avoid illegal drugs. 5. Sample answer: The early detection of health problems usually keeps them from getting worse. For example, if a person had a malignant, cancerous tumor, detecting it early and treating it right away could prevent cancer from spreading to other parts of the body.

## Portfolio Assessment

Sample items include:
- Essay from Career Connection Application
- Answers from Health in Your Life
- Lesson 2 Review answers

## Health in Your Life

Have students read the Health in Your Life feature on page 456. Point out to students that doctors hear all kinds of complaints and see all kinds of situations. A subject may be embarrassing for the patient to talk about, but the doctor has undoubtedly encountered it many, many times. Help students understand that even though a doctor is an adult, it is important they keep asking questions and requesting explanations until they are completely sure of what they are supposed to do.

### Health in Your Life Answers

1. Sample answer: I might feel that my doctor does not care or does not understand, so I may not tell my doctor information that could be important. I might not follow my doctor's orders. Positive, clear communication between a doctor and a patient is very important. Patients usually communicate better when they feel comfortable with their doctors.
2. Sample answer: Can you explain what this medicine is and why I need to take it? Do you have a brochure that will help me understand this medicine's health benefits and side effects?

---

*Lesson 2 Review* On a separate sheet of paper, write the word or phrase in parentheses that best completes each sentence.

1. Self-care is appropriate for (unusual bleeding, vomiting more than a day, a cold).
2. A professional who works only in a certain branch of medicine is a (primary care physician, specialist, nurse).
3. A person who needs short-term physical therapy after surgery might stay for a few days in a(n) (outpatient clinic, hospice, nursing home).

On a sheet of paper, write answers to the following questions. Use complete sentences.

4. **Critical Thinking** What are three actions you take every day to practice self-care?
5. **Critical Thinking** Besides starting treatment early, what is another benefit of finding health problems early?

### Health in Your Life

**Communicating with Your Doctor**

How well you communicate with your doctor is an important part of your health care. Doctors depend on you to talk about your symptoms so they can determine what is wrong. Many teens have questions about sensitive subjects, but they are too embarrassed to ask. Write down your questions before your visit. Then you can read the questions or have the doctor read them.

Make sure you understand your doctor's instructions. Do not be afraid to ask your doctor to repeat an explanation or to tell you again how to take a prescribed medicine. Taking an active role also means that if you feel uncomfortable with a certain doctor, you choose another one.

On a sheet of paper, write answers to the following questions. Use complete sentences.

1. How might your health care be affected if you feel uncomfortable with your doctor?
2. Your doctor prescribes a medicine for you, but you are not sure why. What do you ask your doctor?

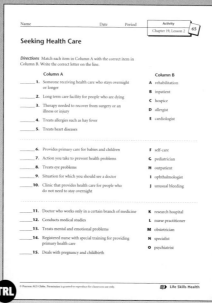

Activity 65

Workbook Activity 65

# Lesson 3: Paying for Health Care

## Objectives

After reading this lesson, you should be able to
- explain how different kinds of health insurance work
- describe ways individuals can lower health care costs
- identify a patient's rights

**Health insurance**
*A plan that pays a major part of medical costs*

**Deductible**
*The first part of each year's medical expenses before the insurance company makes payments*

Health care is expensive. Therefore, understanding the ways in which people pay for health care is important. Some people pay directly for medical expenses. However, most people cannot afford to do this. **Health insurance** is a less expensive way to pay for health care. Health insurance is a plan that pays a major part of medical costs. These costs may include medicines, surgery, tests, hospital stays, and doctors' visits. Health insurance covers major medical expenses that people otherwise could not afford. The main kinds of health insurance are private and government supported.

## How Does Private Health Insurance Work?

With a private health insurance plan, people pay for a policy themselves. Individuals regularly pay a premium, or fee, to an insurance company for a policy. In exchange, the insurance company pays part or all of their health care expenses.

Many employers offer a group plan as a job benefit to workers. Usually, the premiums are lower when people join the plan as a group. Employers usually pay the largest amount. Workers pay a smaller amount each month.

In both plans, each individual usually pays a **deductible**. A deductible is the first part of each year's medical expenses before the insurance company makes payments. If you have a $200 deductible, you must pay the first $200 on your medical expenses each year. For all other expenses that year, the insurance company pays a certain percentage. You pay the rest. For example, suppose you have an additional $1,200 in health care expenses one year. The insurance company pays 80 percent, or $960. You pay the remaining 20 percent, or $240.

Some insurance policies pay only a fixed amount for some procedures or services. They may not cover other procedures or services. For example, some insurance policies may pay 50 percent for eyeglasses. They may pay nothing for a hearing aid. The individual must pay for anything that is not covered.

Consumer Health  Chapter 19  457

---

## Lesson at a Glance

### Chapter 19 Lesson 3

**Overview** This lesson discusses various methods of paying health care expenses, including health insurance, HMOs, PPOs, and government health insurance.

### Objectives

- To explain how different kinds of health insurance work
- To describe ways individuals can lower health care costs
- To identify a patient's rights

**Student Pages** 457–462

**Teacher's Resource Library**
Activity 66
Alternative Activity 66
Workbook Activity 66

### Vocabulary

deductible
health insurance
health maintenance organization
managed care

Medicaid
Medicare
preferred provider organization

Write the vocabulary terms on the board. Ask students which of the terms they have heard and where they have heard them. *(possibly in the news or from their parents)* Have students write their ideas of what each word means on a sheet of paper. Tell them to keep the paper and compare their ideas to the definitions and descriptions of the words in the textbook as they go through the lesson.

### Background Information

Debate about a national health care system in the United States began in the early 20th century. During the Great Depression in the 1930s, President Roosevelt considered national health care. In the end, though, the elements of FDR's New Deal did not address health care. From 1945 through 1952, the Truman administration went a long way toward developing what would become Medicare and Medicaid. President Johnson signed the two programs into law in 1965.

Consumer Health  Chapter 19  457

**Warm-Up Activity**

Ask students to identify different kinds of insurance that people have. *(health, life, automobile, renter's, homeowner's)* Ask how insurance companies have enough money to pay for claims. *(They usually insure large numbers of people. Premiums from people who never collect insurance payments help pay the bills of those who do collect payments.)*

 **Teaching the Lesson**

As you work through the lesson, keep a tree diagram on the board. Start at the top with "Health Care," then add two downward branches marked "Private Health Insurance" and "Government Health Insurance." As you complete each lesson section, have a volunteer add more branches to the diagram.

 **Reinforce and Extend**

### Research and Write

Have students read and conduct the Research and Write feature on page 458. Students can obtain the names and addresses of their senators and representative at www.firstgov.gov/Contact/Elected.shtml.

#### LEARNING STYLES

 **Visual/Spatial**

Review with students the purposes of insurance. Then make copies of a medical insurance claim form and distribute them to the class. Have students fill out the form. Explain sections of the form as necessary so that students fill them out correctly. Ask students what they think might happen if a form is not completed properly. *(The insurance company may return it or not process the claim, causing a delay in payment.)*

---

*Managed care*
A type of private health insurance in which an organization is a go-between for patients and doctors

*Health maintenance organization (HMO)*
A form of managed care in which an organization controls how health care is delivered and its cost

*Preferred provider organization (PPO)*
A form of managed care that has fewer rules and more choices of doctors than an HMO has

 **Research and Write**

Work in small groups to research what Congress is doing on a national level to improve access to health insurance for all Americans. Write a letter to your state's senators expressing your opinion on what can be done.

## How Does Managed Care Work?

**Managed care** is another type of private health insurance. In managed care, an organization is the go-between for the patient and the physician. Managed care tends to cost less than other private insurance. There are two common kinds of managed care.

### Health Maintenance Organizations

A **health maintenance organization (HMO)** is a form of managed care in which the organization controls how health care is delivered and what it costs. Members can be individuals or employers. They pay a fixed amount each month. There are two types of HMOs. Group HMOs require members to go to certain clinics or medical offices. Independent practice associations (IPAs) provide members with a list of doctors and hospitals from which to choose.

HMOs have advantages and drawbacks. HMOs provide preventive health care such as regular physical exams, well-baby care, and routine vaccinations at no extra charge. Private insurance usually charges for preventive care. However, HMO members are limited in their choice of health care providers. Waiting time for appointments can be long. People cannot see a specialist unless their primary care doctor refers them. Members must usually pay all expenses for seeing doctors who do not belong to the HMO.

### Preferred Provider Organizations

A **preferred provider organization (PPO)** is a form of managed care that has fewer rules and more choices of doctors than an HMO has. The doctors are paid a separate fee for each service. Patients pay a fixed amount per visit, and the insurance company pays the rest. Like an HMO, members have a list of doctors who belong to the organization. However, members can go directly to a specialist without a referral. PPOs are generally more expensive than HMOs but less expensive than most private insurance plans.

**Medicare**
*Health insurance for people age 65 or older and some people under age 65 with certain disabilities*

**Medicaid**
*Health insurance for people whose incomes are below a certain level*

## What Is Government Health Insurance?

Government health insurance includes several kinds of insurance for special groups of people. This insurance is paid by federal, state, or local funds.

**Medicare** is health insurance for people age 65 or older. It also covers some people under age 65 with certain disabilities. Medicare pays for some medical expenses. People must pay the rest themselves or with additional private insurance.

**Medicaid** is health insurance for people whose incomes are below a certain level. Medicaid is different from Medicare in that it has no age requirement. It is paid for by state and federal taxes. Individual states operate Medicaid and determine the benefits they will offer. Some people qualify for both Medicare and Medicaid.

### Link to >>>
**Math**

In the United States, the average cost of a hospital stay is $17,300. The average length of a hospital stay is more than 8 days. In 2004, about 16 percent of the total U.S. population did not have any health insurance. How do you think the rising cost of health care affects the cost of health insurance?

The State Children's Health Insurance Program (SCHIP) provides health care for children of low-income parents. The parents make too much money to qualify for Medicaid but generally less than a certain amount. Individual states run this program.

## How Else Does Government Pay for Health Care?

The U.S. Public Health Service (USPHS) oversees the Medicare, Medicaid, and SCHIP programs. The following list describes several more branches of the USPHS:

- The National Institutes of Health (NIH) provides money for research on health and illness. Universities, hospitals, and private institutes conduct the research.

- The Centers for Disease Control and Prevention (CDC) is responsible for the control and treatment of communicable diseases. This includes diseases such as the flu and sexually transmitted diseases.

- The Food and Drug Administration (FDA) enforces laws for the safety, effectiveness, and labeling of food, drugs, and cosmetics.

### ELL/ESL Strategy

**Language Objective:**
*To practice study skills*

As you work through the lesson, have students learning English take thorough notes. Tell English learners that as they finish each section (for example, "How Does Private Health Insurance Work?"), they should look at its heading and see if they can answer the question in their own words. If any students cannot answer a section question, discuss that section with them. Help students understand the content material and find the vocabulary needed to answer the question.

### Link to

**Math.** Have a volunteer read aloud the Link to Math feature on page 459. Ask students to determine the approximate cost of one day in a hospital.
*(17,300 ÷ 8 = 2,162.5)*
*(17,300 ÷ 9 = 1,922.22)*
*One day in a hospital costs, on average, between $1,922.22 and $2,162.50.*

### Global Connection

Challenge students to find out how a national health care system works in another country such as Canada, England, or Sweden. Suggest that students find out how much people spend on medical care, whether they are free to choose their own doctors, how easy it is to obtain care for maintenance treatment or elective surgery, and so on.

## TEACHER ALERT

Some students may have trouble grasping the economics of health insurance. If customers pay the insurance company far less than they would have to pay a doctor or hospital, where does the money to pay a doctor or hospital come from? Explain that many healthy customers pay the insurance company but don't make any claims. The insurance company takes this money, uses some of it to pay the medical expenses of their sick customers, invests some of it, and keeps some of it as profit.

## Health Myth

Have students read the Health Myth feature on page 460. Tell students they can find out more about Medicare, the programs it administers, and the benefits it offers at www.medicare.gov.

## LEARNING STYLES

**Interpersonal/Group Learning**

Have groups of students find newspaper and magazine articles about health care reform. Ask them to choose a position about whether reform is needed, and if so, what kind of reform. Then ask them to prepare and give a presentation about their position. After each group makes a presentation, have group members answer questions from classmates.

---

**Health Myth**

**Myth:** Medicare pays for long-term care for elderly people.

**Fact:** Medicare does not pay for the cost of most long-term care. Medicare pays for the first 20 days if the patient meets certain conditions. After that, patients are expected to pay. Nursing homes cost an average of $66,000 per year in 2003. Many people use up their savings and sell their property to pay for long-term care.

---

- The Substance Abuse and Mental Health Services Administration (SAMHSA) oversees programs to prevent and treat alcoholism, drug addiction, and mental illness.
- The Health Resources and Services Administration (HRSA) supports mother and child health programs, community health education, and training programs for health care workers.

State and local health departments also provide health care services through tax money. These may include vaccinations, blood pressure screening, and vision tests. These services are provided at little or no cost. State and local health departments also operate centers for people with mental and physical disabilities.

### How Can Individuals Lower Health Costs?

The cost of health care in the United States continues to increase every year. Fewer people are able to afford health insurance. Leaders at the state and national level are working to address the problem. However, individuals can do the following to lower health care costs:

- Choose outpatient health care whenever possible. The most costly medical treatment is inpatient care. One low-cost place is a *neighborhood clinic.* These clinics offer many health care services at a low fee. *Alternative birthing centers* (ABCs) usually employ certified nurse-midwives to deliver babies. They offer care for low-risk pregnant women before and after giving birth. The cost is much less than delivery in a hospital. *Hospices* provide dying patients with the medical care they need to live out the rest of their life comfortably. The costs for hospice care are lower than for hospital care.

- Educate yourself and become more active in your own health care. Eat right, exercise, and avoid alcohol, tobacco, and other drugs. Preventing diseases is a major way of keeping down health care costs.

- Ask your doctor to prescribe a generic drug when possible.

### Link to
**Social Studies**

The number of people age 65 and older is expected to almost double in the United States between 2000 and 2030. Because about 80 percent of older adults have at least one chronic condition, this increase may lead to higher health care costs.

### Technology and Society

Fluoride is a substance that helps prevent cavities. In 1945, communities began adding fluoride to public water supplies. For every $1 that is spent on fluoridation, $38 in dental treatment costs are saved. Fluoridation, dental sealants, and toothpaste with fluoride can prevent nearly all cavities.

- Advocate for family and community health. You might volunteer for a charitable health organization, such as the American Lung Association or the American Cancer Society. You also might keep informed about health issues that affect your community. Write a letter or e-mail to your local and state representatives to support available, affordable health care for everyone.

- Take part in health promotion programs that your employer offers. Many employers promote healthy nutrition, physical activity, and avoiding or quitting smoking. Companies can save almost $5 for every $1 spent on these programs. Some companies save millions of dollars in health costs by helping employees prevent chronic diseases.

## What Are Your Rights as a Patient?

The Patients' Bill of Rights was developed by a committee appointed by President Bill Clinton in 1997. Many health plans have adopted these rights and responsibilities. The list below summarizes these rights.

**Patients' Bill of Rights**

1. the right to accurate information about your health plan, health care providers, and health care facilities
2. the right to choose your health care providers
3. the right to receive emergency services without regard to ability to pay
4. the right to know your treatment choices and to take part in decisions about your care
5. the right to considerate and respectful care, without discrimination
6. the right to talk in confidence with health care providers and to have your health care information kept private
7. the right to a fair, fast, and unbiased review of any complaint you have about your health care

### Link to

**Social Studies.** Have a volunteer read aloud the Link to Social Studies feature on page 461. Have students research data concerning the percentage of the U.S. population that was 65 or over from 1900 through the most recent census year. Students may present their findings as a table or graph.

### Technology and Society

Have a volunteer read aloud the Technology and Society feature on page 461. Recently, a movement against the fluoridation of public water supplies has started to grow across the country. The main argument is that, while in 1945, few toothpastes contained fluoride, many consumer products have it today. Have students research both sides of this issue and conduct a class debate.

### IN THE ENVIRONMENT

Tell students that fluorides can occur naturally in water supplies. Water fluoridation involves adjusting the natural levels of fluorides in water to the levels recommended to prevent tooth decay. Have students use research materials and the internet to find out what the recommended level is. *(0.7 to 1.5 milligrams per liter of fluoride in drinking water)* Also, have them find out what the harmful effects of having too high of a level of fluorides in water are. *(causes tooth enamel to become brittle and to chip off)*

## Health at Work

Have students read the Health at Work feature on page 462. Discuss with students the importance of keeping accurate medical records for individuals over the years. Ask what might happen if records or even parts of records are crossed among patients. Have students provide specific examples. (*Crossing records could have grave consequences for patients. For example, a patient may be allergic to penicillin. This part of the patient's medical records may be accidentally crossed with that of a patient who is not allergic to penicillin. The first patient, then, may receive penicillin for an infection and possibly die from the allergic reaction.*)

## Lesson 3 Review Answers

1. Medicaid 2. deductible 3. managed care 4. Sample answer: People without chronic diseases do not need as much health care as people with chronic diseases (who need doctor visits and perhaps medicines and surgeries). Preventive care saves money in the long run by reducing health care costs.
5. Sample answer: The person might go to a nursing home.

## Portfolio Assessment

Sample items include:
- Letter from Research and Write feature
- Medical insurance form from Learning Styles: Visual/Spatial
- Table or graph from Link to Social Studies
- Lesson 3 Review answers

---

**Word Bank**
deductible
managed care
Medicaid

**Lesson 3 Review** On a sheet of paper, write the word or phrase from the Word Bank that best completes each sentence.

1. Health care paid by the government for low-income people is _____.
2. The first part of each year's medical expenses an individual pays to a health insurance company is a _____.
3. HMOs and PPOs are two types of _____.

On a sheet of paper, write the answers to the following questions. Use complete sentences.

4. **Critical Thinking** How does preventing chronic diseases help keep down health care costs?
5. **Critical Thinking** Suppose a person has a stroke and needs rehabilitation. The person is expected to be able to live independently. To which health care facility might the person go right after leaving the hospital?

### Health at Work

#### Medical Records Technician

A career as a medical records technician may be for you if you want to work in a health facility and can pay close attention to details. Medical records technicians organize patient information. Technicians translate disease names and treatments into coding systems. The information is then entered into a computer. Technicians need to be orderly and detailed. They need excellent keyboarding skills. Technicians may work in hospitals, clinics, health agencies, insurance companies, nursing homes, and other health facilities. Medical records technicians usually have a 2-year degree. Most employers require technicians to take an exam that qualifies them as registered health information technicians (RHITs).

462  Unit 7  Health and Society

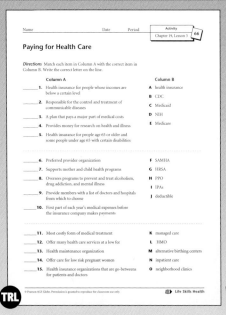

Activity 66

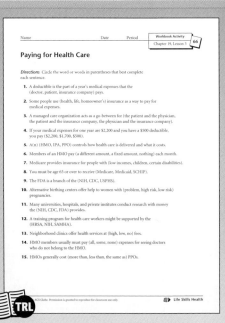

Workbook Activity 66

## Chapter 19 SUMMARY

- Some influences on choices of health consumers are advertising, price, and recommendations of family or friends.

- Criteria to help judge products and services are to look at the cost benefits, consider recommendations, watch for false or misleading claims, look at the ingredients, and compare prices.

- Preventing and taking care of ordinary colds and minor illnesses are wise forms of self-care.

- School nurses may provide care for students with disabilities. They may provide information on staying well and preventing disease.

- Health care professionals include primary care physicians, specialists, nurse practitioners, physical therapists, and registered dieticians.

- Qualifications to look for in a health care professional are board certificates, associations, and experience. The person's comfort with the professional is also important.

- Some health care facilities are dental or medical offices, outpatient clinics, and inpatient facilities for nursing care and rehabilitation. Other health care facilities are hospitals, urgent care centers, nursing homes, and hospices.

- People can pay for health care costs through private individual or group insurance plans. Health maintenance or preferred provider organizations offer less expensive health care.

- Government-supported health insurance includes Medicare for people over age 65, Medicaid for low-income people, and SCHIP for children in low-income families.

- Individuals can help lower health costs. They can choose low-cost outpatient facilities when possible. They can practice prevention, buy generic medicine, and advocate for family and community health. They can also take part in workplace health promotion programs.

### Vocabulary

| | | |
|---|---|---|
| consumer, 446 | inpatient, 455 | preferred provider organization (PPO), 458 |
| deductible, 457 | managed care, 458 | primary care physician, 453 |
| health care facility, 454 | Medicaid, 459 | quackery, 448 |
| health insurance, 457 | Medicare, 459 | rehabilitation, 455 |
| health maintenance organization (HMO), 458 | nurse practitioner, 453 | scoliosis, 452 |
| hospice, 455 | outpatient clinic, 454 | specialist, 453 |

## Chapter 19 Summary

Have volunteers read aloud each Summary item on page 463. Ask volunteers to explain the meaning of each item. Direct students' attention to the Vocabulary box on the bottom of page 463. Have them read and review each term and its definition.

### ONLINE CONNECTION

The Web site of the Department of Health and Human Services has many pages aimed at teens. The index to these pages can be found at:

www.hhs.gov/specificpopulations/index.shtml

The National Institutes of Health also has a number of Web pages aimed at teens. That index can be found at:

www.nlm.nih.gov/medlineplus/teenhealth.html

## Chapter 19 Review

Use the Chapter Review to prepare students for tests and to reteach content from the chapter.

## Chapter 19 Mastery Test

The Teacher's Resource Library includes two forms of the Chapter 19 Mastery Text. Each test addresses the chapter Goals for Learning. An optional third page of additional critical-thinking items is included for each test. The difficulty level of the two forms is equivalent.

## Review Answers
**Vocabulary Review**

1. quackery 2. primary care physician 3. nurse practitioner 4. outpatient clinic 5. inpatient 6. rehabilitation 7. health insurance 8. Medicare 9. consumer 10. scoliosis 11. hospice 12. HMOs 13. PPOs 14. specialist

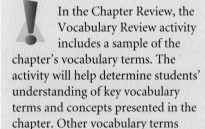

**TEACHER ALERT**

In the Chapter Review, the Vocabulary Review activity includes a sample of the chapter's vocabulary terms. The activity will help determine students' understanding of key vocabulary terms and concepts presented in the chapter. Other vocabulary terms used in the chapter are listed below.

deductible   managed care
health care facility   Medicaid

---

## Chapter 19 REVIEW

**Word Bank**
consumer
health insurance
HMOs
hospice
inpatient
Medicare
nurse practitioner
outpatient clinic
PPOs
primary care physician
quackery
rehabilitation
scoliosis
specialist

### Vocabulary Review
On a sheet of paper, write the word or phrase from the Word Bank that best completes each sentence.

1. People are influenced by _____ when they buy a product that claims to cure a disease such as cancer or arthritis.
2. A(n) _____ is a type of physician who treats people for usual problems.
3. A nurse who provides primary health care is a(n) _____.
4. A facility that provides health care for people who do not stay overnight is a(n) _____.
5. A person who stays overnight or longer in a health care facility is receiving _____ care.
6. Someone recovering from surgery may need _____.
7. Many employers offer _____ as a job benefit.
8. Some people under age 65 with certain disabilities may be covered by _____.
9. When someone buys a product, the person is a(n) _____.
10. School nurses may conduct screenings for _____.
11. A person who is dying may stay in a(n) _____.
12. Members of _____ must visit a primary care physician first before seeing a specialist.
13. Members of _____ may visit a specialist without a referral from primary care physicians.
14. A(n) _____ is a doctor who works only in a certain branch of medicine.

464   Unit 7   Health and Society

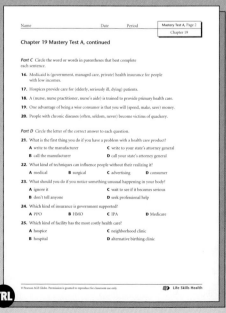

Chapter 19 Mastery Test A, pages 1–3

## Concept Review

On a sheet of paper, write the letter of the answer that best completes each sentence.

15. You are a wise consumer when you _____.
    A buy the least expensive medicine
    B buy generic drugs when available
    C use only the products your best friend uses
    D buy products advertised by your favorite actor

16. You should see a doctor if you _____.
    A have a mild headache
    B bruise your foot on a rock
    C are coughing when you have a cold
    D are depressed for several weeks

17. One form of government health insurance is _____.
    A Medicare   B an HMO   C a PPO   D a hospice

## Critical Thinking

On a sheet of paper, write the answers to the following questions. Use complete sentences.

18. If a person had a disease that could not be cured, why do you think quackery might appeal to the person's emotions instead of reason?

19. What might be one reason that the cost of health care has become so high?

20. Why is it important to recognize how advertisers of health products and services are trying to influence you?

 **Test-Taking Tip** When taking a matching test, match all the items that you know go together for sure. Cross out these items. Then try to match the items that are left.

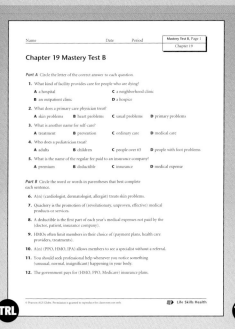
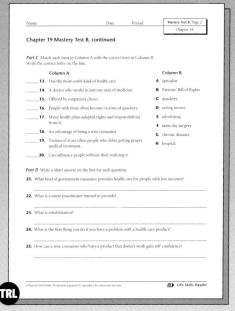

Chapter 19 Mastery Test B, pages 1–3

## Chapter at a Glance

**Chapter 20:**
**Public Health**
pages 466–483

### Lessons

1. **Public Health Problems**
   pages 468–473
2. **U.S. Public Health Solutions**
   pages 474–477
3. **Health Promotion and Volunteer Organizations**
   pages 478–480

**Chapter 20 Summary** page 481

**Chapter 20 Review** pages 482–483

**Audio CD**

**Skill Track for Life Skills Health**

**Teacher's Resource Library** TRL

- Activities 67–69
- Alternative Activities 67–69
- Workbook Activities 67–69
- Self-Assessment 20
- Community Connection 20
- Chapter 20 Self-Study Guide
- Chapter 20 Outline
- Chapter 20 Mastery Tests A and B

(Answer Keys for the Teacher's Resource Library begin on page 559 of this Teacher's Edition.)

## Opener Activity

Ask students what they have learned about people's basic health needs, based on what they have seen and read about real-life disaster situations. Brainstorm a list of who helps people when a disaster strikes. Write students' ideas on the board.

Then shift students' focus to families that have faced adversity. Ask students who helps families when they are facing these kinds of adversity. Encourage students to think about ways they could help as they read the chapter.

466  Unit 7  Health and Society

## Self-Assessment

### Can you answer these questions?

✓ 1. What are some public health goals the U.S. government has set?

✓ 2. What are some public health problems in the United States?

✓ 3. Why is poverty linked to public health problems?

✓ 4. Why is homelessness a community health problem?

✓ 5. What are some groups of people who are at high risk for health problems?

✓ 6. What are some basic health services?

✓ 7. What are some worldwide health concerns?

✓ 8. What organizations help make vaccinations available?

✓ 9. What does a public health department do?

✓ 10. What does the World Health Organization do?

Self-Assessment 20   Community Connection 20

# Chapter 20: Public Health

The health of a community is important because it can affect each individual's personal health. To promote a community's health and well-being, each individual, family, and neighborhood can help. People do this when they become aware of public health problems and cooperate to solve them. Every effort, no matter how small, can affect the health and well-being of individuals and of the whole community.

In this chapter, you will learn about public health problems in the local, national, and international communities. You will learn how the public health system works to solve problems, promote health, and prevent disease. You will learn how you can help promote public health.

## Goals for Learning

- ◆ To identify public health problems nationally and internationally
- ◆ To describe community, state, national, voluntary, and international public health services
- ◆ To explain the role of health promotion and disease prevention in public health
- ◆ To identify individual opportunities to promote public health

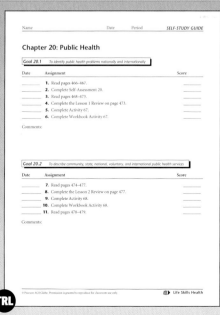

Chapter 20 Self-Study Guide, pages 1–2    Chapter 20 Outline, pages 1–2

## Introducing the Chapter

Have students read the two chapter introductory paragraphs on page 467. Then discuss with students the idea of public health. Challenge them to think of situations, events, or other factors that affect the health of a community. Ask them what public health problems they have heard about on the radio or television. *(hurricanes, earthquakes, terrorist attacks, a potential flu epidemic or pandemic, AIDS, drug abuse, etc.)* Ask them what other public health problems they know about. *(pollution, water and food supply problems, drinking and driving, smoking, domestic violence, etc.)* Keep a list of responses and refer students to it throughout the chapter, letting them make any necessary additions or deletions.

Have students read the Goals for Learning on page 467. Explain that in this chapter students will learn about public health issues and how to promote public health to prevent disease.

## Notes and Questions

Ask volunteers to read the notes and questions that appear in the margins throughout the chapter. Then discuss them with the class.

## Self-Assessment

Have students complete the Self-Assessment worksheet before and after reading the chapter. Before reading the chapter, have students fill in the "Before" column. Ask students to identify their goals for learning. To get ideas for setting goals, students might use the chapter introductory material on page 467, the checklist on page 466, or the questions on the Self-Assessment worksheet. Students can use the back of the worksheet if they need more space to write.

Collect the Self-Assessment worksheets and pass them out again at the end of the chapter. Have students fill in the "After" column. Ask them to identify at least four major points they have learned. Again, suggest they use the back of the worksheet if they need more space to write. You might want to collect and review the worksheets, but return them to students so they have a record of their goals and accomplishments.

# Lesson at a Glance

## Chapter 20 Lesson 1

**Overview** The lesson identifies, describes, and discusses some national and international public health issues.

### Objectives

- To identify some national public health problems
- To describe epidemics
- To discuss poverty, homelessness, and environmental factors as public health concerns
- To describe some worldwide health concerns

**Student Pages** 468–473

**Teacher's Resource Library**

Activity 67

Alternative Activity 67

Workbook Activity 67

### Vocabulary

| | |
|---|---|
| birth defect | public health |
| epidemic | tuberculosis |

Read aloud to students the vocabulary words and their definitions, as presented in the boxes on pages 468, 471, and 472. Invite students to share their prior knowledge of these words.

### Background Information

Population statistics can be complex and misleading. For example, in Table 20.1.1 it looks as if the AIDS epidemic in the United States was about equivalent to the influenza epidemic in 1918. But upon further analysis, it becomes clear that during the flu epidemic, 0.54 percent of the U.S. population died during the course of one year. Compare that with the AIDS epidemic, during which 0.18 percent of the population has died over the course of 20 years. The point is not to minimize or dismiss the AIDS epidemic, but to accurately interpret the data.

---

## Lesson 1 Public Health Problems

### Objectives

After reading this lesson, you should be able to
- identify some national public health problems
- describe epidemics
- discuss poverty, homelessness, and environmental factors as public health concerns
- describe some worldwide health concerns

**Public health**
The practice of protecting and improving the health of a community or a nation

**Epidemic**
The rapid spread of a communicable disease that affects large populations at the same time

Health is more than a personal concern. Health is a concern of whole communities and countries. **Public health** is the practice of protecting and improving the health of a community or a nation. Public health uses such means as preventive medicine, education, and control of communicable diseases to respond to health concerns.

Every 10 years, the U.S. government sets public health goals on which it works over the coming 10-year period. These include the following general goals based on some national public health problems:

- to provide equal health care for all Americans, including children, adolescents, older people, and individuals with disabilities
- to increase health and quality of life for older Americans
- to increase the ability of all Americans to be able to use preventive health services

### What Are Some National Public Health Problems?

You already know that violence and alcohol and other drug abuse are public health problems in the United States. The spread of disease is another problem. Poverty, homelessness, availability of health services, and environmental factors also are public health problems.

### Epidemics

When a communicable disease spreads rapidly, affecting large populations at the same time, it is called an **epidemic.** In the United States, epidemics are not as common as they once were. For example, polio was an epidemic in the 1940s and 1950s. The polio vaccine has nearly eliminated the disease. Nevertheless, epidemics still happen. For example, AIDS has reached epidemic levels in many countries.

468  Unit 7  Health and Society

---

### 1 Warm-Up Activity

Have students examine Table 20.1.1 on page 469. Ask them if they were aware that the United States had suffered through so many epidemics. Tell them that vaccines and antibiotics have removed the threat of these diseases ever reaching epidemic level again, except for influenza and AIDS, both of which are viruses that have defied a cure or a vaccine.

| Link to >>> |
|---|
| **Art**<br>In the movie *Restoration*, a doctor helps to battle the Great Plague of 1665 in England. This bubonic plague epidemic killed 15 percent of London's population. |

Throughout history, organizations such as the Centers for Disease Control and Prevention (CDC) have gathered statistics related to national health. Epidemics have killed thousands of people in the United States. Table 20.1.1 shows the number of people who died in some of the worst epidemics in the United States.

| Table 20.1.1 Deaths from the Worst Epidemics in the United States ||||
|---|---|---|---|
| Date | Disease | Number Who Died | Total U.S. Population at the Time |
| 1832 | cholera | 7,340 | 13,000,000 |
| 1878 | yellow fever | 20,000 | 49,000,000 |
| 1900–1904 | typhoid | 20,520 | 76,000,000 |
| 1900–1904 | tuberculosis | 140,600 | 76,000,000 |
| 1916 | polio | 7,000 | 99,000,000 |
| 1918 | influenza | 550,000 | 102,000,000 |
| 1949–1952 | polio | 6,000 | 151,000,000 |
| 1985–2005 | AIDS | more than 500,000 | 248,000,000–296,000,000 |

Tooth decay is one of the most common infectious diseases among U.S. children. It may cause pain, poor appearance, and weight loss.

### Poverty

Poverty is a public health concern because it puts people at high risk for serious health problems. Many experts believe that poverty can lead to health problems.

When people cannot afford a balanced diet with the right nutrients, malnutrition and other health problems may result. When people cannot afford health care or health insurance, many of their health problems are not treated effectively, if at all.

## Teaching the Lesson

After discussing the concept of public health, ask students to reflect on what they have learned about in previous chapters that affects the public as well as the individual. Challenge them to identify ways in which these topics could be considered public health concerns. (For example, substance abuse affects not only the individual's health but also public health in the form of crime, car accidents, and the mental health toll it takes on families.)

## Reinforce and Extend

### Link to

**Art.** Ask students to read the Link to Art on page 469. Tell them that the movie *Restoration* was released in 1995. Point out that before the bubonic plague epidemic the population of London was about 460,000. Ask students to calculate about how many people in London died of the plague (*15 percent of 460,000 is 69,000*).

### ELL/ESL STRATEGY

**Language Objective:** *To discuss an assigned topic in small groups*

Create a word web on the board with the question head on page 468 in a large central ellipse and the five subheads that answer the question in rectangles connected to the ellipse with lines. Have students work in small groups to add details around each subtopic. Encourage group discussion of the topics.

### TEACHER ALERT

The world is facing a measles epidemic even though this disease is preventable through vaccination. In 2003, between 30 million and 40 million cases occurred worldwide and resulted in 530,000 measles-related deaths.

## Health Myth

Have a volunteer read aloud the Health Myth on page 470. Explain that there is a strong link between health and poverty in the United States. Ask students how they think the problem of poverty could be solved. *(housing reform; better health and dental care, health education programs, and food programs)*

### TEACHER ALERT

Point out that during pregnancy the mother must take special care of her own teeth for two reasons. First, she could pass bacteria from tooth decay to her baby. Second, hormonal changes that take place during pregnancy can affect the gums.

### GLOBAL CONNECTION

Ask students to research information about the effects and causes of homelessness around the world and have them write reports on their findings. Assign students different aspects of homelessness to research (barriers to health care, mental health care, etc.). Have students present their reports to the class.

### HEALTH JOURNAL

Have students write a letter to a hypothetical teenager who is considering leaving home because of a difficult family situation. Students should include suggestions for where the teenager can find help, reasons to avoid homelessness, and the health issues related to homelessness that the teenager might face.

---

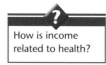
How is income related to health?

### Health Myth

**Myth:** The United States has the best health care in the world.

**Fact:** Americans spend about $2\frac{1}{2}$ times as much for health care as other developed nations. However, the United States has fewer doctors per person, higher infant mortality rates, and lower than average childhood immunization rates than other developed nations.

Poverty is the main reason that many children go without vaccinations and other health care. When pregnant women do not get proper care before a baby is born, both mother and baby can have health problems. Babies may have low birth weight or slowed mental function. To have a healthy pregnancy, mothers should follow these guidelines:

- See a doctor or other health care provider from the start of the pregnancy.
- Do not use tobacco, alcohol, or other drugs.
- Eat healthy foods, including fruits, vegetables, low-fat milk, eggs, cheese, and grains.
- Exercise sensibly.
- Have the baby checked by a doctor or health care provider right after birth and throughout childhood.
- Clean the baby's gums and teeth to prevent tooth decay.

### Homelessness

Homelessness is a community health problem related to poverty. On any given night, about 600,000 Americans are without shelter. Of those, 40 percent are families with children. About 39 percent of homeless people are under age 21.

**Figure 20.1.1** *Volunteer food assistance programs help poor and homeless people.*

*Tuberculosis*
*A communicable lung disease*

People may become homeless because of domestic violence. When homeless shelters are overcrowded, parents and their children sometimes return to an abusive household. Children cannot attend school regularly. They frequently go without any health care.

In some homeless shelters, communicable diseases such as **tuberculosis**—a lung disease—are a growing problem. Poor hygiene also is a problem for homeless people. The other health problems that affect poor people also affect homeless people.

Health disasters such as heat waves do not affect everyone equally. Poor people and elderly people are often the worst hit.

### Getting Health Services

High-risk groups of people often are not able to get basic health services. Poor people and homeless people are two high-risk populations. Teenage parents and their children, people with mental health problems, and elderly people also are high-risk groups. People with disabilities and recent immigrants to the United States are other high-risk groups.

Lack of transportation may limit basic health services. For example, in rural areas, nearby health services are often not available.

**Research and Write**

Using the Internet, research the topic of Americans who do not have health insurance. Prepare an oral report on how this lack of insurance affects dental care, doctor visits, and hospital visits. Explain possible solutions to these problems.

Basic health services include vaccinations and medical care for women during pregnancy. Basic health services also include screening for health problems such as breast cancer. Nutritional counseling and exercise and fitness programs for elderly people are other basic health services. In addition, elderly people often need assistance meeting their daily needs. When the population over age 65 increases, even more health resources will be needed for elderly people.

People who provide health services need to be sensitive to cultural differences. Sometimes, people with health problems cannot speak the same language as the nurse or doctor they contact. In such cases, a trained interpreter, not a family member, should translate. Also, some ethnic groups are more open to problems such as alcoholism or depression. Members of different ethnic groups sometimes metabolize medications differently. This can lead to problems with finding the correct dosage.

## Decide for Yourself

Have students read and complete the Decide for Yourself feature on page 472. Explain that taking care of oneself is important to quality of life as one ages.

### Decide for Yourself Answers

1. Sample answer: Pneumococcal disease is the worst of these diseases. It kills more people than all other vaccine-preventable diseases combined. 2. Sample answer: It is important for elderly people to get flu vaccinations because as a person gets older, his or her immune system may not be as strong as it once was. 3. Sample answer: Some adults may not get vaccinations because of lack of concern.

## IN THE ENVIRONMENT

 Ask students to generate a list of specific environmental crises that could cause a public health problem. For example, an oil spill near a coastal city or industrial waste that was dumped into a lake could cause public health problems. As students make suggestions, write them on the board. Discuss how each crisis affects the environment and public health. Discuss steps that could be taken to prevent similar crises in the future.

## PRONUNCIATION GUIDE

Use the pronunciation shown here to help students pronounce difficult words in the lesson. Refer to the pronunciation key that appears in the Glossary at the back of the Teacher's Edition for the sounds of the symbols.

pneumococcal (nü mə kok´ əl)

---

**Birth defect**
*A deformity or an impaired function of the body that is present from birth*

### Environment

The release of harmful chemicals into the environment is a public health concern. These chemicals harm the food people eat, the water they drink, and the air they breathe. This damage increases the risk of health problems. Some geographical areas are more affected than others. In some areas, for example, cancer rates or the number of **birth defects** are high. Birth defects are deformities or impaired functions of the body that are present from birth.

### Decide for Yourself

Adults can get many diseases. Several of these can be prevented by vaccinations. These diseases include influenza (flu), tetanus, and pneumococcal disease.

Flu is a contagious respiratory disease. In the United States, more than 200,000 people are hospitalized with flu complications every year. About 36,000 people die. To help prevent flu, you can get a vaccination every year.

Tetanus is a disease of the nervous system. It leads to muscle contractions in the face, which is why the disease is also called lockjaw. Another symptom of tetanus is severe muscle spasms throughout the body. It causes death in 30 to 40 percent of cases. After high school, you should get a tetanus shot every 10 years.

Pneumococcal disease can lead to serious infections of the lungs, the blood, and the covering of the brain. Each year pneumococcal disease kills more people than all other vaccine-preventable diseases combined. For many people, one vaccine is all that is needed to make them immune to pneumococcal disease.

On a sheet of paper, write the answers to the following questions. Use complete sentences.

1. Which of the diseases mentioned do you think is the worst? Explain your answer.
2. Why do you think it is important for elderly people to get flu vaccinations?
3. Why do you think many adults do not get vaccinations?

**Research and Write**

Using the Internet, research obesity as an American public health problem. Write two pages on its causes and effects.

Which public health problem do you think should be given the highest priority?

## What Are Some Worldwide Health Concerns?

Developing nations often have little money and low standards of living. Developing nations may not have laws or the resources to protect the environment. When these conditions exist, people cannot meet their basic needs for food, water, and shelter. Problems such as hunger and malnutrition can result. Poverty, AIDS, environmental damage, and diseases such as malaria are public health concerns in many developing nations. Natural disasters and war also are concerns of worldwide public health.

*Lesson 1 Review* On a sheet of paper, write the answers to the following questions. Use complete sentences.

1. What are three goals for public health in the United States?
2. Why is poverty a public health concern?
3. What are two high-risk groups of people who may have trouble getting health services?
4. **Critical Thinking** Why is health a public concern rather than just a personal concern?
5. **Critical Thinking** Why do you think epidemics in the United States are not as common as they used to be?

### Technology and Society

A virus causes poliomyelitis (polio). Polio destroys nerve cells that control muscles. It makes the legs, arms, or body unable to move. It makes swallowing, talking, and breathing difficult. Polio usually occurs in children. In 1943, the best-known person with polio was President Franklin D. Roosevelt. The same year, polio killed 1,151 people in the United States.

In 1954, Jonas Salk developed a polio vaccine. After many laboratory tests, Pittsburgh schoolchildren were vaccinated. In 1955, the medicine was tested in 44 states. Salk said polio would be very rare before long. He was right. By 1969, few people died from polio. The epidemic was over.

*Public Health* Chapter 20 **473**

---

## Research and Write

Have students read and complete the Research and Write feature on page 473. Suggest that they do an advanced search on government sites and type in the phrases "causes of obesity" and "health effects of obesity." Encourage interested students to also research the healthcare costs caused by obesity. Have students work in pairs.

## Technology and Society

Have students read the Technology and Society feature on page 473. Point out that although Franklin D. Roosevelt was crippled by polio in 1921, he went on to become governor of New York and later president of the United States. Roosevelt is the only president to be elected to four terms despite the fact that he used a wheelchair.

## Lesson 1 Review Answers

**1.** Three goals are to provide equal health care for all Americans, to increase health and quality of life for older Americans, and to increase the ability of all Americans to use preventive health services. **2.** Poverty is a public health concern because it puts many people at high risk for serious health problems. **3.** Two high-risk groups of people who might have trouble getting health services are senior citizens and people who are living in poverty. **4.** Sample answer: Health is a public concern because the health of a community can affect each individual's health. **5.** Sample answer: The number of epidemics in the United States has decreased because of the development of vaccines.

## Portfolio Assessment

Sample items include:
- Sentences from Vocabulary
- Word webs from ELL/ESL Strategy
- Reports from Global Connection
- Notes from Research and Write
- Entry from Health Journal
- Decide for Yourself answers
- Reports from Research and Write
- Reports from Technology and Society
- Lesson 1 Review answers

*Public Health* Chapter 20 **473**

# Lesson at a Glance

## Chapter 20 Lesson 2

**Overview** The lesson explains public health solutions in the United States.

### Objectives
- To describe local health service agencies
- To describe national health service agencies

**Student Pages** 474–477

**Teacher's Resource Library**
- Activity 68
- Alternative Activity 68
- Workbook Activity 68

## Vocabulary
**government regulation**
**sanitation**

Read the vocabulary terms and their definitions, as presented in the boxes on pages 474 and 476, aloud to students. Invite students to work in small groups to come up with a list of synonyms for the vocabulary terms. Then ask each group to share one of their synonyms with the rest of the class.

## Background Information
The United States has numerous health service agencies to help Americans. National health services provide some of the funding for state and local programs. National services also establish standards for state and local programs, as well as provide education and research that benefit all programs at all levels.

## Health Myth

Have students read the Health Myth on page 474. Point out that according to the Environmental Protection Agency (EPA), bottled water is not necessarily safer than tap water. Some bottled water is treated more than tap water, while some bottled water is treated less or not at all.

474  Unit 7  Health and Society

---

## Lesson 2  U.S. Public Health Solutions

### Objectives
*After reading this lesson, you should be able to*
- describe local health service agencies
- describe national health service agencies

**Sanitation**
*Measures related to cleanliness that protect public health*

### Health Myth
**Myth:** All bottled water is better than tap water.

**Fact:** City tap water and bottled water are both highly regulated and monitored for quality. Some, but not all, bottled water is purer and safer than some tap water.

Have you ever gotten a vaccination at a school, neighborhood health center, or local drugstore? How was that vaccination made possible? Public health agencies, nonprofit groups, and individuals support organized preventive efforts to ensure the public's health. Most of these health services are provided to individuals at the community level.

### What Are Some Community and State Health Services?

All states have public health departments. Most cities and counties also have health departments. These departments make sure that laws relating to disease and **sanitation** are observed. Sanitation refers to measures related to cleanliness that protect public health. Sanitation tasks include inspecting restaurants, making sure water is clean and safe, and tracking down sources of communicable diseases.

Community and state health departments also work to provide medical care for low-income individuals. The departments provide vaccination programs, routine tests, and physical examinations. They also may provide visiting nurse and nutrition programs, community clinics, and community education programs. For example, community education programs may target young mothers and their babies, focusing on nutrition and education. Senior centers and aging and nutrition programs assist elderly people. School-based and school-linked clinics and services provide students with health care.

Federal and state governments provide local communities with funds to operate prevention and treatment programs. Most local programs, however, are funded through local taxes and private charities.

474  Unit 7  Health and Society

## What Are Some National Health Services?

The Department of Health and Human Services (HHS) is the U.S. government agency with the largest responsibility for public health. HHS gathers and analyzes health information, sets health and safety standards, and sponsors education and research. HHS includes these divisions:

- Centers for Medicare and Medicaid Services (CMS)—controls Medicaid and Medicare
- Centers for Disease Control and Prevention (CDC)—collects and distributes information about public health problems
- Administration for Children and Families (ACF)—oversees programs for low-income families and people with disabilities
- National Institutes of Health (NIH)—conducts research, promotes disease prevention, and carries out national health policies
- Food and Drug Administration (FDA)—regulates food, drugs, medical devices, and blood products

The Environmental Protection Agency (EPA) is another government agency concerned with public health. It sets policy for and monitors the quality of the water and air.

The Occupational Safety and Health Administration (OSHA) and the Consumer Product Safety Commission (CPSC) are government agencies concerned with public health. They set guidelines for safe exposure to toxic substances for workers and consumers.

### Link to >>>
**Social Studies**

Public concern about pollution led to the establishment of the Environmental Protection Agency (EPA) in 1970. The EPA creates and enforces regulations to protect the environment. When a regulation is being created, the EPA requests that members of the public comment on it. The EPA relies on these comments to develop the final wording of the regulation.

---

### Warm-Up Activity

Ask students to identify different health services they know about in your community. List their ideas on the board. Invite students to add to the list as they read and discuss the lesson. The list can be a resource for ideas to use when students complete the "Health in Your Life" feature on page 476.

### Teaching the Lesson

Have students think about things they do to try to keep themselves healthy. Discuss how they might apply those principles toward improving the health of their community. Invite students to compare how they keep their rooms clean with public sanitation that protects the public. What public sanitation programs exist in their community? *(restaurant inspection, ensuring water quality, tracking communicable diseases, keeping streets and alleys free of garbage, keeping sewer lines running, etc.)*

### Reinforce and Extend

#### ELL/ESL Strategy

**Language Objective:** *To create graphic organizers*

Divide students into pairs in which one student is an English language learner. Have each pair create a two-column chart to organize the details from the two question heads in the lesson (What Are Some Community and State Health Services? and What Are Some National Health Services?). Details should be lists and not complete sentences.

### Learning Styles

**Auditory/Verbal**

Invite students to think critically about advertisements and food labels. Students should look for magazine advertisements that focus on food labels. Have students compare ingredients in similar products made by different companies. Let them share their conclusions about what is in the foods they compared.

### Link to

**Social Studies.** Ask a volunteer to read aloud the Link to Social Studies on page 475. Point out that the mission of the EPA is "to protect human health and to safeguard the natural environment—air, water, and land—upon which life depends." Stress that humans cannot exist without a healthy environment.

## Health in Your Life

Have students read Health in Your Life on page 476. Divide the class into pairs of students. Students can use the list they made in the Warm-Up Activity to help them. Let each student pair sign up for the agency they will interview to avoid duplications. Invite student pairs to share their findings with the class.

**Health in Your Life Answers**

1. Answers will vary. Students should identify areas where health services are not available. 2. Answers will vary. Students might find that some agencies provide the same services. 3. Sample answer: One of my neighbors has asthma. She should get information about where flu shots are available because she is in a high-risk group for complications from flu.

### IN THE COMMUNITY

Ask students to work in small groups to research the services offered at nursing homes in their community. Direct them to inquire about the reasons people live in the nursing homes, the health services that are provided, and other benefits that the nursing homes have to offer senior citizens. Ask students to share their findings with the class and use the information as a basis for a discussion on the advantages and disadvantages of nursing home care.

### CAREER CONNECTION

Have students do Internet research about a career in public health. If possible, invite the school career counselor to visit the class to discuss career opportunities in public health. The school career counselor can review with students how to evaluate and make adjustments in their plans about personal and career choices, which students may have begun during the study of Chapter 11.

---

**Government regulation**
Rule that businesses must follow to protect public health

These government agencies fund research and conduct monitoring tasks. They also issue **government regulations**. These are rules that businesses must follow to protect public health. For example, the FDA regulates the ingredients in foods and their labeling. It makes sure that any advertising claims about the effects of a drug are true. Funding for these national health agencies and programs comes from federal income taxes.

### Health in Your Life

**Identifying Health Services in Your Community**

Find out about the different programs for disease prevention, health promotion, and treatment available in your community. List local health agencies and organizations. Check with parents, counselors, librarians, or teachers for ideas. You could also search the phone book, newspaper, or Internet. Work with a partner to gather information about the agencies. Call each agency and ask about the items below. Prepare for the interview in advance. You might record the answers.

- name, address, and phone number of agency
- days and hours of operation
- types of services and health professionals on staff
- who qualifies for services and the fees charged
- whether the services are confidential
- languages in which the services are offered

Use the information to compare the services.

On a sheet of paper, write the answers to the following questions. Use complete sentences.

1. Are there parts of your community in which some services are not available?
2. Do several agencies provide the same services?
3. Do you know anyone who might benefit from knowing about these services? Who are they?

### LEARNING STYLES

**Body/Kinesthetic**

Have students role play how to choose valid health resources from the school and local community. Have students share the information they have gathered about local agencies and act out telephone conversations to make appointments, arriving at the clinics and services for an appointment, and interactions with nurses and doctors. You might wish to include the school nurse in the activity to act as the health care provider in a mock appointment for the class to observe.

**Word Bank**
- Centers for Disease Control and Prevention
- Department of Health and Human Services
- Food and Drug Administration

*Lesson 2 Review* On a sheet of paper, write the phrase from the Word Bank that best completes each sentence.

1. The U.S. government agency with the largest responsibility for public health is the _____.
2. The _____ regulates food, drugs, medical devices, and blood products.
3. The _____ collects and distributes information about public health problems.

On a sheet of paper, write the answers to the following questions. Use complete sentences.

4. **Critical Thinking** Why do you think it is important to have local, as well as national, health services?
5. **Critical Thinking** Why do you think it is necessary for the FDA to monitor drug advertising claims?

### Health at Work

**Home Health Aide**

Living at home is important for most people, especially when they are ill. Home health aides help sick people stay at home by performing routine tasks for them. Aides might change bed linens, prepare meals, clean, do laundry, and run errands. Aides help bathe and clean people. Aides may read aloud or play games with sick people. They may give massages and other treatments.

Usually, health care agencies employ home health aides. Some hospitals run home care programs that employ home health aides. In some cases, individuals hire an aide.

Home health aides should know CPR. Federal law requires home health aides to pass a test covering a wide range of areas. A person can receive training before taking this test. Some states also require home health aides to be licensed.

*Public Health* Chapter 20  477

### Lesson 2 Review Answers

**1.** Department of Health and Human Services **2.** Food and Drug Administration **3.** Centers for Disease Control and Prevention **4.** Sample answer: Local and national health agencies are important because they both have strengths. Local agencies often have effective plans for serving all the neighborhoods in a community. National agencies often have the financial and staffing resources that local agencies lack. **5.** Sample answer: The FDA monitors claims in drug advertising to protect people. Like any other type of business, drug companies have a goal of making money by selling as many products as possible. To make a product appealing, an unethical company might use deceptive or misleading advertising. Drug advertising must be regulated because people could face serious health problems if they took the wrong type of medication.

### Health at Work

Have students read Health at Work on page 477. Discuss who might need a home health aide. *(senior citizens, accident victims, individuals recovering from a major illness, the critically ill)* Ask students to identify some of the public health problems that home health aides could help alleviate. *(less strain on limited hospital and nursing-home facilities, lower costs, people can recover in their own homes)* Find out if any of your students know CPR. Discuss why this is an important skill for a home health aide.

### Portfolio Assessment

Sample items include:
- Charts from ELL/ESL Strategy
- Health in Your Life answers
- Findings from Application: In the Community
- Lesson 2 Review answers

Activity 68

Workbook Activity 68

# Lesson at a Glance

## Chapter 20 Lesson 3

**Overview** The lesson explains how voluntary health organizations and health promotion aid public health.

### Objectives

- To discuss how voluntary organizations aid public health
- To explain how health promotion aids public health

**Student Pages** 478–480

**Teacher's Resource Library**

Activity 69

Alternative Activity 69

Workbook Activity 69

## Vocabulary

**advocacy group**

Read the vocabulary term aloud to students. Ask them to find the term in the text on page 479. Have students use *advocacy group* in a sentence.

## Background Information

Numerous nonprofit organizations provide health services to individuals and businesses around the world. The United Way is a global network of nonprofit organizations funded by donations from individuals, companies, and foundations. The International and American Red Cross are also funded by charitable contributions.

## 1 Warm-Up Activity

Ask students to share their experiences as volunteers, including as many diverse volunteer activities as possible. Ask students for more ideas about how they can help their school and community. Point out that in this lesson, students will learn about voluntary health organizations that help take care of public health needs.

---

## Lesson 3: Health Promotion and Volunteer Organizations

**Objectives**

*After reading this lesson, you should be able to*

- discuss how voluntary organizations aid public health
- explain how health promotion aids public health

Many nonprofit voluntary health organizations respond to public health needs. Voluntary public health agencies are funded by contributions from individuals and businesses.

### What Voluntary Agencies Provide Health Services?

Voluntary organizations that provide health services include the American Heart Association, the American Cancer Society, and the American Lung Association. These groups provide public health support through education programs. Usually, these organizations focus on a single disease. The efforts of these voluntary organizations have helped Americans learn preventive health measures. Anyone can volunteer to assist the efforts of these groups.

In many communities, people with common concerns meet for support and education. Self-help groups provide support. For example, they support people with organ transplants or those who are recovering from an emotional illness.

### What Are Some International Health Services?

Recall that you learned about the WHO in the Health in the World feature on page 92. You read about the International Red Cross in the Health in the World feature on page 246.

Ridding the world of hunger, providing relief after major disasters, and fighting epidemics are shared efforts of the world. One organization that plays a major role in these efforts is the United Nations (UN). Programs of UN agencies such as the World Health Organization (WHO) work to provide public health services.

Other UN agencies provide worldwide health services. The United Nations International Children's Emergency Fund (UNICEF) provides vaccinations, school food programs, and health centers for children. It also helps train nurses. The Food and Agriculture Organization (FAO) works in developing countries to improve food production and to distribute food.

**Advocacy group**
*A prevention organization that helps specific groups get health services*

The United States also sponsors efforts for international public health. Organizations such as the Peace Corps and the U.S. Agency for International Development (USAID) are two examples. The International Red Cross is the world's largest private international public health organization. It provides aid such as food, clothing, medical care, and temporary shelter to victims of natural disasters.

## How Do Health Promotion and Prevention Help?

A healthy population locally, nationally, and internationally has obvious benefits. Healthy people produce more on the job. Health promotion and disease prevention reduce the demands on the health care system. Preventive health care costs less than treatment of illness. Thus, health promotion and disease prevention are goals of many policy makers.

Education for health promotion focuses on sexually transmitted diseases, violence, physical fitness, nutrition, and substance abuse prevention. Those efforts may target specific groups at risk or whole communities. A community-wide nutrition education program, for example, can encourage individuals to lower their fat intake and exercise regularly.

Prevention organizations that work with specific groups are called **advocacy groups.** Advocacy groups may provide health services themselves, or they may help remove health care barriers. They do this by working to change government policies that prevent some people from receiving health care.

Nationally, advocacy groups encourage passage of civil rights laws. For example, laws have been passed that guarantee special education programs in public schools for all children who have disabilities. Advocacy also has produced laws that protect the rights of elderly people.

Many businesses have health promotions for their employees. These businesses educate employees on how to live a healthier life. They encourage people to adopt better health practices concerning diet and exercise. Research has shown that these work-site programs can reduce sick days and health care costs.

How well has your community done in providing barrier-free environments for people in wheelchairs?

## 2 Teaching the Lesson

As students progress through the lesson, have them create a chart to match various health needs and problems with health service agencies, clinics, and other organizations. At the end of this lesson, have students fill in the "After" column of Self-Assessment 20.

## 3 Reinforce and Extend

### ELL/ESL Strategy

**Language Objective:** *To learn vocabulary specific to the lesson*

Have students record the abbreviations and acronyms for the various international agencies included in the lesson. Let them practice saying the abbreviations and acronyms according to how they are commonly used. For example, the United Nations is called the "U-N"; the World Health Organization is called the "W-H-O" and not "*who*"; the United Nations International Children's Emergency Fund (UNICEF) is pronounced "yü nə sef"; the Food and Agriculture Organization of the United Nations is referred to as "F-A-O"; and the U.S. Agency for International Development is called both "*U-S aid*" and "U-S-A-I-D."

### Learning Styles

**Visual/Spatial**
Have students create an educational campaign to promote healthy lifestyle changes to the public. Brainstorm strategies for overcoming barriers when communicating information on health issues, including a needs assessment, community meetings, committee research, a poster campaign, and brochures. Students can choose any cause for which to make a poster, brochure, or informational sheet.

### At Home

Students can create a take-home survey, or needs assessment, to gather data about their parents' or guardians' opinions of public health issues. Have them make a list of the important health issues raised during discussions of the lessons in this chapter. From their list, students can create a series of survey questions to ask their parents or guardians. Then have the class prepare promotional posters to hang around the school or community or speeches to present at a public forum to try to influence public opinion to promote public health.

## HEALTH JOURNAL

Have students write the first draft of a letter as a first step in promoting public health. The topics of their letters may be a letter of introduction to an organization with which they would like to volunteer, a persuasive letter to a lawmaker asking for changes in public health policy, or a letter requesting information from a health care organization.

## LEARNING STYLES

### Logical/Mathematical

Invite students to develop strategies for improving or maintaining health and safety on personal, family, community, and world levels. Divide the class into eight groups, assigning a different level to each of two groups. Encourage students to apply what they have learned in this lesson and in the other lessons in this chapter to develop a strong strategy for providing and promoting health care.

## Lesson 3 Review Answers

1. charitable contributions 2. developing 3. advocacy 4. Sample answer: The dictionary defines the term *advocate* as "one that pleads the cause of another." An advocacy group helps a specific group of people get health services. 5. Sample answer: I could join an advocacy group, encourage people to get vaccinations, and volunteer at a local clinic.

## Portfolio Assessment

Sample items include:
- Charts from Teaching the Lesson
- Surveys and Posters from At Home
- Educational campaigns from Visual/Spatial
- Letters from Health Journal
- Strategies from Logical/Mathematical
- Lesson 3 Review answers

---

### How Can You Promote Public Health?

People get involved in health promotion in their community for many reasons. One reason is that disease prevention and health promotion save society money by preventing illness. Another reason is that communities organized to provide mutual support and promote health have fewer problems, including less crime and drug abuse. A third reason is that individuals increase their self-esteem as they work to improve the health of others.

To get involved, you could investigate existing community resources to identify your community's health needs. You could then choose a place to volunteer. For example, you could volunteer to help with a food drive or get people to sponsor you in a benefit walk for homeless people. You could write letters to lawmakers, asking for changes. You might join a health advocacy group.

Your school also has resources for promoting public health. The school nurse has access to a wide variety of educational materials. Your health teacher also does. You can find a wealth of information on the Internet.

**Lesson 3 Review** On a sheet of paper, write the word or phrase in parentheses that best completes each sentence.

1. Voluntary public health agencies are funded by (only businesses, the government, charitable contributions).
2. Programs of UN agencies mainly target (developing, large, tropical) countries.
3. Prevention organizations that work with specific groups are (international, advocacy, disaster aid) groups.

On a sheet of paper, write the answers to the following questions. Use complete sentences.

4. **Critical Thinking** Look up the word *advocate* in a dictionary. How does its meaning apply to the goals of an advocacy group?
5. **Critical Thinking** What are some examples of how you can promote public health?

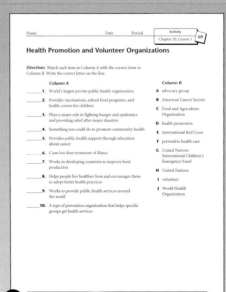

Activity 69

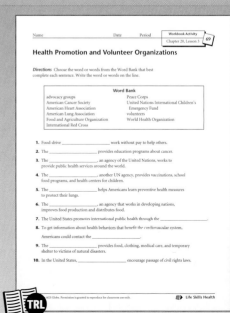

Workbook Activity 69

# Chapter 20 SUMMARY

- Public health means protecting and improving the health of a community or nation.
- Epidemics are public health problems because communicable diseases spread rapidly and affect large populations at the same time.
- Poverty puts people at high risk for health problems. It especially affects children's health.
- Communicable diseases and poor hygiene are major problems for homeless people.
- Poor people, homeless people, and teen parents and their children may have difficulty getting basic health services. Elderly people and people with disabilities or mental health problems also can have difficulty.
- Chemicals in the environment that harm food, water, and air increase the risk of public health problems.
- Poverty, low standards of living, hunger, malnutrition, AIDS, environmental damage, diseases, natural disasters, and war are international public health problems.
- Local and state public health departments make sure that laws relating to disease and sanitation are observed. Public health departments also provide medical care for low-income people.
- The U.S. Department of Health and Human Services gathers and analyzes health information, sets health and safety standards, and sponsors education and research.
- Nonprofit voluntary health organizations provide education programs in most communities.
- The United Nations, some U.S. organizations, and the International Red Cross provide public health services internationally.
- Health promotion and disease prevention through education and advocacy groups reduce costs and relieve demands on the health care system at every level.
- Individuals can promote public health by volunteering, writing letters to lawmakers, and joining public health advocacy groups.

| Vocabulary | | |
|---|---|---|
| advocacy group, 479 | government regulation, 476 | sanitation, 474 |
| birth defect, 472 | | tuberculosis, 471 |
| epidemic, 468 | public health, 468 | |

## Chapter 20 Summary

Have volunteers read aloud each Summary item on page 481. Ask volunteers to explain the meaning of each item. Direct students' attention to the Vocabulary box on the bottom of page 481. Have them read and review each term and its definition.

### ONLINE CONNECTION

 The Focus on Features page of the National Institute of Allergy and Infectious Diseases Web site offers current information about the flu, HIV-AIDS, and tuberculosis: www3.niaid.nih.gov/news/focuson/

The official U.S. government Web site for information on pandemic flu and avian influenza, sponsored by the U.S. Department of Health and Human Services, offers monitoring of outbreaks and information on vaccines, research activities, and how the government plans to respond to a possible pandemic: www.pandemicflu.gov

The United Nations Web site offers many opportunities for learning about the humanitarian activities of the UN, as well as the state of health and living conditions of people around the world: www.un.org/ha

On the official Web site of the World Health Organization, students can learn more about worldwide health issues, including disease outbreaks, natural disasters, and a myriad of WHO projects: www.who.int

## Chapter 20 Review

Use the Chapter Review to prepare students for tests and to reteach content from the chapter.

## Chapter 20 Mastery Test

The Teacher's Resource Library includes two forms of the Chapter 20 Mastery Text. Each test addresses the chapter Goals for Learning. An optional third page of additional critical-thinking items is included for each test. The difficulty level of the two forms is equivalent.

## Review Answers

**Vocabulary Review**

1. public health 2. epidemic
3. tuberculosis 4. sanitation 5. government regulations 6. birth defect
7. advocacy group

**Concept Review**

8. B 9. D 10. C 11. C 12. A 13. C 14. A
15. A 16. B 17. D

# Chapter 20 REVIEW

**Word Bank**
advocacy group
birth defect
epidemic
government regulations
public health
sanitation
tuberculosis

## Vocabulary Review

On a sheet of paper, write the word or phrase from the Word Bank that best completes each sentence.

1. The practice of protecting and improving the health of a community or nation is _____.
2. When an infectious disease spreads rapidly and affects many people, it is a(n) _____.
3. One communicable lung disease is _____.
4. Local and state health departments make sure that laws relating to disease and _____ are observed.
5. The Department of Health and Human Services helps develop _____.
6. A deformity present at birth is a(n) _____.
7. A goal of a(n) _____ is to change government health policies.

## Concept Review

On a sheet of paper, write the letter of the answer that best completes each sentence.

8. A factor that puts people at higher risk for serious health problems is _____.
   A having health insurance  C advocacy
   B poverty                  D clean air

9. High-risk groups include _____ because often they cannot get basic health services.
   A young adults        C people in large cities
   B factory employees   D homeless people

10. On any given night, the number of people who are homeless in the United States is about _____.
    A 6,000     C 600,000
    B 60,000    D 6,000,000

482  Unit 7  Health and Society

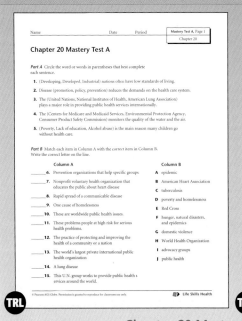
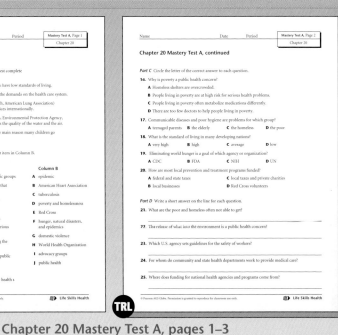

Chapter 20 Mastery Test A, pages 1–3

11. People may become homeless because of _____.
    A lack of transportation  C domestic violence
    B epidemics  D malnutrition

12. Basic health services include _____.
    A vaccinations  B shelter  C surgery  D transportation

13. A type of prevention organization is a(n) _____ group.
    A international  B environmental  C advocacy  D charitable

14. Businesses' work-site programs can reduce _____.
    A sick days  B epidemics  C poverty  D disabilities

15. Poverty can result in _____.
    A malnutrition  C high birth rate
    B multiple sclerosis  D healthy pregnancies

16. A growing problem in homeless shelters is _____.
    A polio  B tuberculosis  C advocacy  D cholera

17. People at high risk for not getting basic health services include _____.
    A insured people  C single people
    B high-income people  D teenage parents

## Critical Thinking

On a sheet of paper, write the answers to the following questions. Use complete sentences.

18. Discuss three international public health problems.

19. How do natural disasters cause public health problems?

20. What are two benefits of health promotion and disease prevention?

**Test-Taking Tip** When you answer multiple-choice questions, read all of the options first. Then eliminate the ones that are not correct.

Public Health  Chapter 20  483

Chapter 20 Mastery Test B, pages 1–3

## Critical Thinking

18. Sample answer: Hunger, natural disasters, and epidemics are three international public health problems. Poverty and AIDS are other problems. International agencies help countries work together to fight these problems. Hunger, AIDS, and poverty are serious problems in parts of Africa. The tsunami of 2004 was a natural disaster that affected several countries, including Thailand. 19. Sample answer: Natural disasters can lead to public health problems because communications networks are disrupted and roads are destroyed. When telephone services are disrupted, emergency workers and doctors have a difficult time locating all the injured and stranded people. When roads are destroyed, ambulances cannot transport people to hospitals quickly. Natural disasters also can cause water and air to become polluted. This pollution can lead to disease. Disposing of hazardous and toxic materials can be difficult in the aftermath of a natural disaster. These materials pose a threat to public health. 20. Sample answer: Health promotion and disease prevention increase levels of public health. Healthier people take fewer sick days. Health promotion and disease prevention reduce the demands on the health care system. Preventive health care costs less than the treatment of illnesses.

## Alternative Assessment

Alternative Assessment items correlate to the student Goals for Learning at the beginning of this chapter.

- Have students orally identify national and international public health problems.
- Have students work in groups to write a one-page report describing community, state, national, voluntary, and international public health services.
- Ask students to draw a graphic organizer to explain the role of health promotion and disease prevention in public health.
- Have students identify individual opportunities to promote public health.

# Chapter at a Glance

## Chapter 21: Environmental Health
pages 484–511

## Lessons

1. Health and the Environment
   pages 486–489
2. Air Pollution and Health
   pages 490–494
3. Water Pollution and Health
   pages 495–498
4. Other Environmental Problems and Health
   pages 499–503
5. Protecting the Environment
   pages 504–508

**Chapter 21 Summary** page 509

**Chapter 21 Review** pages 510–511

**Audio CD**

**Skill Track for Life Skills Health**

**Teacher's Resource Library** TRL

Activities 70–74

Alternative Activities 70–74

Workbook Activities 70–74

Self-Assessment 21

Community Connection 21

Chapter 21 Self-Study Guide

Chapter 21 Outline

Chapter 21 Mastery Tests A and B

(Answer Keys for the Teacher's Resource Library begin on page 559 of this Teacher's Edition.)

## Opener Activity

Ask students to describe their environment. Tell them to write a paragraph or list about things that they associate with their environment. Remind them to use all their senses when writing their paragraph or list; tell them to include things they can see, hear, smell, taste, and touch. Ask them to write another paragraph or list about ways that their environment has been damaged, and then to write a third paragraph or list about things that further threaten their environment.

484  Unit 7  Health and Society

# Self-Assessment

## Can you answer these questions?

1. What is the balance of nature?
2. How does the balance of nature happen?
3. How can human activity change the balance of nature?
4. How does air pollution affect health?
5. How does air pollution affect Earth?
6. What effect does acid rain have on animal and plant life?
7. What are five causes of water pollution?
8. How does deforestation affect health?
9. What is noise pollution?
10. How can the environment be protected?

Self-Assessment 21   Community Connection 21

# Chapter 21
## Environmental Health

The environment is made up of natural resources. Natural resources include Earth's air, water, land, and living things. Earth's natural resources work together. What happens to any one resource can directly affect other resources. For example, an oil spill in the ocean near Alaska can affect the fish and animals that live in and along the Alaskan coastline. Earth's environment is protected when its natural resources are protected.

In this chapter, you will learn about people's effect on the environment. You will learn how the environment affects your health. You will learn about ways people can protect the environment. You also will learn about ways the government protects the environment. Protecting the environment is one way to protect your health.

### Goals for Learning

- ◆ To explain how the environment affects health
- ◆ To identify causes of air and water pollution
- ◆ To describe other environmental problems
- ◆ To identify actions that protect the environment

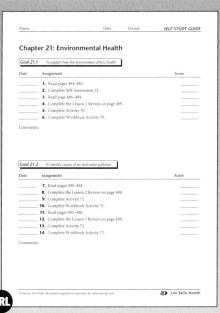

Chapter 21 Self-Study Guide, pages 1–2     Chapter 21 Outline, pages 1–2

## Introducing the Chapter

Have students read the introductory material on page 485. Discuss the meaning of *environment*. Ask students to describe the ideal environment. Remind them that such an environment would meet all their needs. (Lead students to include among their descriptions not only clean air, water, and land but also such elements as safety, adequate space, pleasant surroundings, and so on.) Then ask students how close their real environment comes to matching the ideal environment. Invite students to suggest ideas for improving their environment.

## Notes and Questions

Ask volunteers to read the notes and questions that appear in the margins throughout the chapter. Then discuss them with the class.

## Self-Assessment

Have students complete the Self-Assessment worksheet before and after reading the chapter. Before reading the chapter, have students fill in the "Before" column. Ask students to identify their goals for learning. To get ideas for setting goals, students might use the chapter introductory material on page 485, the checklist on page 484, or the questions on the Self-Assessment worksheet. Students can use the back of the worksheet if they need more space to write.

Collect the Self-Assessment worksheets and pass them out again at the end of the chapter. Have students fill in the "After" column. Ask them to identify at least four major points they have learned. Again, suggest they use the back of the worksheet if they need more space to write. You might want to collect and review the worksheets, but return them to students so they have a record of their goals and accomplishments.

*Environmental Health* Chapter 21

# Lesson at a Glance

## Chapter 21 Lesson 1

**Overview** This lesson explains how living things depend on the air, land, and water in the environment and what happens to the quality of life when wastes pollute the environment.

## Objectives

- To explain the term *balance of nature*
- To describe three ways to disturb the balance of nature
- To explain how damage to the environment affects health
- To explain why understanding the effects of pollution on the environment is important

**Student Pages** 486–489

**Teacher's Resource Library**

Activity 70

Alternative Activity 70

Workbook Activity 70

## Vocabulary

ecology    pollution
fragile    soil erosion
pesticide

Ask volunteers to read aloud the vocabulary words and their definitions. Challenge students to use all five vocabulary words to write a paragraph that describes the content of this lesson. Have volunteers read their paragraphs to the class.

## Background Information

Problems with air quality have been linked to asthma occurrences in the United States. Death rates from asthma have risen steadily in the last 20 years. Rates have risen most dramatically in urban areas. Asthma is now the most common chronic childhood illness in the United States. In the United States, nearly 5 million children under age 18 have asthma. Most children who have asthma develop it by age 5, although anyone of any age can develop the disease.

---

## Lesson 1 — Health and the Environment

**Objectives**

*After reading this lesson, you should be able to*

- explain the term *balance of nature*
- describe three ways to disturb the balance of nature
- explain how damage to the environment affects health
- explain why understanding the effects of pollution on the environment is important

All living things depend on their environment for their needs. The environment is a system of natural resources, including air, land, water, plants, and animals. Living things depend on air, water, and land to live and grow.

### What Is the Balance of Nature?

Humans contribute to the environment, just as the environment contributes to humans. For example, humans breathe out carbon dioxide. Plants need carbon dioxide to live and grow. Plants in turn give off oxygen, which humans need to live and grow. When all parts of the environment contribute to one another in this way, a healthy balance of nature is maintained. When any part of the environment is hurt or destroyed, this balance is changed.

**Figure 21.1.1** *Natural and human-made changes, such as forest fires, affect the balance of nature.*

**Soil erosion**
Soil moved by wind, water, or ice

**Pollution**
Harmful materials in the soil, water, or air

## What Disturbs the Balance of Nature?

Natural events are always disturbing the balance of nature. Earthquakes, volcanoes, hurricanes, fires, and floods sometimes cause serious damage. However, nature usually recovers. In fact, the changes that natural events cause can be good for the environment over time.

Humans also can disturb the balance of nature. For years, humans have changed the environment by building houses, farming, and using Earth's resources. As the number of people on Earth grows, more demands are made on its resources. For example, metals and minerals are taken from the ground and used for many things. Coal and oil are used to heat homes and fuel cars and trucks. Trees are cut to make houses and paper.

Changes in the environment that humans cause can upset the balance of nature for many years. Mining can make the land unusable. Stripping the land of trees can cause **soil erosion**. Soil erosion happens when the soil is moved by wind, water, or ice. Using harmful chemicals also can make the land unsafe for people and animals.

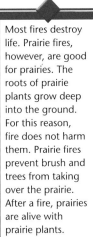

Most fires destroy life. Prairie fires, however, are good for prairies. The roots of prairie plants grow deep into the ground. For this reason, fire does not harm them. Prairie fires prevent brush and trees from taking over the prairie. After a fire, prairies are alive with prairie plants.

## How Does Environmental Damage Affect Health?

The wastes that people make have an immediate effect on health. Human waste usually does not cause problems because it is broken down in nature. Factories, however, can give off dangerous chemicals and other harmful wastes. When these harmful wastes get into the air or water or are buried under the ground, they cause **pollution**. Pollution happens when harmful materials get into the soil, water, or air. For example, an oil spill on the water can kill fish and birds that live there. If that water is used in people's homes, they are also affected.

### 1 Warm-Up Activity

Invite students to name as many examples of balance as they can. Have them explain how they think balance relates to the environment. Have students read about the balance of nature on pages 486-487. Then return to the discussion on balance and let students refine some of their ideas.

### 2 Teaching the Lesson

Tell students to read the lesson. When they are finished, have them go back and review the headings in the lesson. Each heading asks a question. Have the students write a one-sentence summary answer to each question. Divide the class into small groups and have students share their sentences with the members of their group. Visit the different groups and listen as students share their sentences. Correct any misconceptions that you hear.

### 3 Reinforce and Extend

### ELL/ESL Strategy

**Language Objective:** *To reinforce vocabulary through visuals*

Bring in as many old magazines as you can find. (Get old magazines at recycling centers, yard sales, or used bookstores.) Divide the class into groups that contain both students learning English and students who have a strong grasp of English. Ask the groups to find and cut out photos or graphics in the magazines or to create graphics that illustrate each vocabulary term. Students can draw graphics or create collage illustrations; if they have access to the appropriate technology, they can create graphics using computer software programs. Collect the visuals from all the groups and display them in the classroom along with the terms they represent.

## LEARNING STYLES

**Auditory/Verbal**

Ask students to interview people of different age groups about which environmental concerns are most important to them. Age groups should include the student's age group, that of their parents, and that of their grandparents. Have students listen carefully to interviewees' responses, take notes as needed, and report the results of the interviews to the class. Then lead a class discussion about whether, or how, the results varied among age groups.

## Link to

**Math.** Have students read the Link to Math on page 488. Ask students to calculate how much trash would be saved from landfills every year if a person recycled just 10 percent of his or her trash. *(Multiply 4 pounds by 365 days. That equals 1,460 pounds. Multiply 1,460 by .10. That equals 146 pounds, which is the amount of trash that would be saved from landfills per year if a person recycled just 10 percent of his or her trash.)*

## IN THE COMMUNITY

Brainstorm with students a list of environmental concerns in their community. Then point out that they as individuals and as a group can make a difference for the environment. Have students develop a plan aimed at protecting or cleaning up the local environment. They might plan a recycling or antilittering campaign, clean up a particular area, or help educate consumers about how their choices affect the environment. Encourage students to carry out their plan. They might need to seek cooperation and help from local government or nonprofit organizations. Students can perform this activity in conjunction with the Decide for Yourself activity on page 497.

---

**Pesticide**
A chemical used to kill bugs

**Fragile**
Easily broken, destroyed, or damaged

Why is the environment so fragile?

Pollution can be so harmful that it puts human life in danger. For example, chemicals are used in factories that make products for people to use every day. These same chemicals can make the factory workers sick. Wastes from factories, power plants, and cars can harm the air and water that people need to live. Pollution can be especially harmful to a growing fetus. For example, **pesticides** are chemicals that are used to kill bugs. Studies have shown that pesticides can pass from a mother's blood to the fetus. Some chemicals are known to cause cancer. Others cause brain and nerve damage or have an effect on how a fetus develops.

### How Can Understanding the Effects of Pollution Help?

Damage to the environment can be turned around. The first step is understanding the problem. There is only so much of Earth's resources. Someday the resources will run out. Many people understand that the environment that gives us life—rivers, forests, soil, and air—is **fragile,** or easily damaged. The conditions that are needed to keep humans alive—air, water, animals, and soil—are all linked together. Keeping this fragile existence going is an important health issue.

> **Link to >>>**
>
> **Math**
>
> On average, every person in the United States throws out about 4 pounds of trash each day. About how much trash does one person throw out in a year?
>
> Most trash ends up in special places set aside for garbage. These places have low-lying ground where waste is buried between layers of dirt. Some trash decomposes much faster than other trash. Trash that takes many years to decompose can pollute the land over time.

## LEARNING STYLES

**Interpersonal/Group Learning**

Direct groups of students to research an endangered animal species and find out what people have been doing to help protect it. Groups should create presentations, including visuals, to share with the class. If students have access to the appropriate technology, they can create multimedia reports.

**Ecology**
*The study of how living things and the environment work together*

People have begun to make changes to protect the environment. For example, lakes that were once believed to be dead have been brought back to life through careful management. The fish and plants that once lived there are coming back. The lives of bald eagles were once in danger. After the use of certain pesticides was stopped, the numbers of bald eagles grew.

Whole new branches of science that deal with protecting the environment have come about. **Ecology,** for example, is the study of how living things and the environment work together and depend on each other. Clinical ecology is a new science that deals with the harmful effects of human-made and natural materials on humans.

**Word Bank**
balance
environment
flood

*Lesson 1 Review* On a sheet of paper, write the word or phrase from the Word Bank that best completes each sentence.

1. All living things depend on their _____ to live.
2. All parts of the environment add to the _____ of nature.
3. One natural event that disturbs the balance of nature is a(n) _____.

On a sheet of paper, write the answers to the following questions. Use complete sentences.

4. **Critical Thinking** How can humans cause pollution?
5. **Critical Thinking** Why is it important to maintain the balance of nature?

### Research and Write
Humans have had a critical effect on the environment. Visit the library and research a plant or animal that no longer exists because of what humans have done. How did humans affect the plant or animal? Make a poster about what you learn.

*Environmental Health* Chapter 21  **489**

### Research and Write

Ask a volunteer to read aloud the Research and Write feature on page 489. Have students use the library or the Internet to research a plant or animal that has become extinct. Make sure that students include on their posters information on what changes occurred in the environment to cause the extinctions.

### TEACHER ALERT

Ask students to name places where they have seen indoor plants. Then discuss how they think such plants change the appearance of an indoor environment. Do students think an inviting setting affects the overall health—physical and emotional—of people indoors? Tell students that visual pollution is a negative visual environment including outdoor things like trash, weeds, graffiti, and run-down buildings and indoor things like clutter and dirt. Students can improve their indoor visual environment—as well as their physical and emotional health—by keeping things neat and clean and by including attractive plants.

### Lesson 1 Review Answers

1. environment 2. balance 3. flood
4. Sample answer: Pollution can be caused by several human activities, such as mining, burning coal and oil, and using harmful chemicals. 5. Sample answer: A healthy environment depends on a healthy balance of nature. If people do not maintain the balance of nature, then Earth may become an uninhabitable planet.

### Portfolio Assessment

• Paragraphs from Vocabulary
• Sentences from Teaching the Lesson
• Graphics from ELL/ESL Strategy
• Interview results from Learning Styles: Auditory/Verbal
• Plan from Application: In the Community
• Presentations from Learning Styles: Interpersonal and Group Learning
• Posters from Research and Write
• Lesson 1 Review answers

*Environmental Health* Chapter 21  **489**

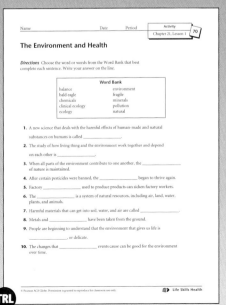

Activity 70 | Workbook Activity 70

# Lesson at a Glance

## Chapter 21 Lesson 2

**Overview** This lesson describes causes of air pollution and explains how air pollution affects the human body and the entire Earth.

### Objectives

- To list three causes of air pollution
- To describe how air pollution affects human health
- To describe how air pollution affects Earth
- To explain the harmful effects of acid rain

**Student Pages** 490–494

**Teacher's Resource Library**

Activity 71
Alternative Activity 71
Workbook Activity 71

## Vocabulary

| | |
|---|---|
| acid rain | ozone layer |
| asbestos | particulate |
| bronchitis | photochemical |
| fossil fuel | smog |
| greenhouse effect | radon |

Read and discuss the definitions of the vocabulary words. Then ask each student to write a question for which one of the vocabulary words is the answer. Collect the questions and read a selection of them to the class, one at a time. Ask volunteers for answers.

## Background Information

The first major air pollution disaster in the United States occurred in 1948 in Donora, Pennsylvania. Pollution from steel mills, coke plants, and a zinc smelter settled over the town. Seven thousand of Donora's 14,000 inhabitants became ill, and 20 of them died. This incident brought the health problems of air pollution to public attention.

---

## Lesson 2: Air Pollution and Health

**Objectives**

*After reading this lesson, you should be able to*

- list three causes of air pollution
- describe how air pollution affects human health
- describe how air pollution affects Earth
- explain the harmful effects of acid rain

**Photochemical smog**
*A form of air pollution*

**Bronchitis**
*A lung disease that makes breathing hard and causes coughing*

Every day, you breathe in about 16,000 quarts of air. If the air you breathe is polluted, toxins get into your body. If you breathe in polluted air over months or years, your lungs may be damaged. Air pollution can affect your health, whether you are outdoors or indoors.

### What Causes Outdoor Air Pollution?

The main cause of outdoor air pollution is burning. Whenever a material is burned, harmful gases get into the air. For example, factories that burn oil and coal make smoke. Cars and trucks that burn gas make a kind of smoke called exhaust. Other sources of air pollution are tobacco smoke, furnaces that burn waste, and leaf burning.

### How Does Outdoor Air Pollution Affect Health?

Hydrocarbons are materials made of hydrogen and carbon. They are found in the exhaust that comes out of cars and trucks. They are also found in natural gas and other liquid fuels that get into the air. Hydrocarbons in the air can sting your nose and throat and make your eyes water. Some hydrocarbons can cause cancer.

**Photochemical smog** occurs when sunlight mixes hydrocarbons and nitrogen oxides together. Photochemical smog makes ozone. Ozone is a gas. Ozone found in the air you breathe keeps your body from getting oxygen from the lungs to your bloodstream. It can hurt your respiratory system and give you lung diseases such as asthma. It can also cause emphysema and **bronchitis,** a lung disease that makes breathing hard. Ozone can irritate your eyes.

Carbon monoxide is a poisonous gas that has no color or smell. It is found in the exhaust of cars and trucks. It is also found in tobacco smoke. Breathing in carbon monoxide affects the red cells in your blood. It keeps you from getting enough oxygen to your other body cells. This can result in headaches, dizziness, blurred vision, and tiredness. Over many years, carbon monoxide can make heart and lung disease worse.

---

## PRONUNCIATION GUIDE

Use the pronunciation shown here to help students pronounce difficult words in the lesson. Refer to the pronunciation key that appears in the Glossary at the back of the Teacher's Edition for the sounds of the symbols.

emphysema (em fə sē´ mə)

hydrocarbon (hī drō kär´ bən)

nitrogen oxide (nī´ trə jən ok´ sīd)

sulfur dioxide (sul´ fər dī ok´ sīd)

**Fossil fuel**
*A material that is made from plant or animal remains that can be burned*

**Particulate**
*Tiny piece of solid material*

Sulfur dioxide is a poisonous gas that gets into the air from factories and power plants that burn **fossil fuels.** Fossil fuels are burnable materials such as coal, oil, and natural gas that form in the earth from plant or animal remains. Sulfur dioxide can irritate your eyes, throat, and lungs, making breathing hard. It makes asthma, emphysema, and bronchitis worse.

**Particulates** are tiny pieces of solid materials such as dust, ash, dirt, and soot that float in the air. Some particulates carry gases and pesticides that can harm your body when you breathe them in. Particulates may bother your eyes, nose, throat, and lungs.

Lead is an example of a particulate. Leaded gasoline raises the amount of lead in the air. Only unleaded gasoline can be used in the United States. Leaded gasoline is still used in other countries. Lead gets into your bloodstream through your lungs or gets absorbed into your skin. Too much lead can cause high blood pressure, loss of appetite, and damage to your brain and nervous system.

Outdoor air pollution affects everyone. In some areas, a smog alert is given when there is a lot of pollution in the air. People with serious health problems should stay indoors on days when the air is poor. Even healthy people should not do heavy exercise on those days.

Many large cities post ozone codes. Some cities show the ozone codes in the chart at the right on electric message boards. Some cities put daily ozone levels on the Internet.

| Ozone Codes | |
|---|---|
| Air Quality Rating | Air Quality Index (AQI) |
| Green—good | 0–50 |
| Yellow—moderate | 51–100 |
| Orange—unhealthy for sensitive groups | 101–150 |
| Red—unhealthy | 151–200 |
| Purple—very unhealthy | 201–300 |

### Warm-Up Activity

Ask students to identify causes of outdoor air pollution in their community. *(pollution from automobiles and other vehicles, pollution from factories, tobacco smoke, etc.)* List their suggestions on the chalkboard. Have students read about the causes of outdoor air pollution and its effects on pages 490-491. Have them adjust the list to reflect information from the text.

### Teaching the Lesson

As students read the lesson, have them create a table with these three headings: Problem, Cause(s), and Effects. Have them fill out the table with information about each type of air pollution about which they read. Tell students that the chart will be helpful to them when they are studying for quizzes and tests.

### Reinforce and Extend

**LEARNING STYLES**

**Interpersonal/Group Learning**

Have students work in groups to investigate the syndrome in which people have adverse reactions to chemicals used in furniture, carpeting, and various materials used in buildings. Have students research the symptoms of the affected people and how common the syndrome is. Then have each group form and state its position about who should be responsible for the medical expenses of affected people—the individuals themselves; the owners of the building; the manufacturers of the furniture, carpeting, and similar items; or the manufacturers of the chemicals. Each group should defend its position.

## LEARNING STYLES

**Visual/Spatial**

Have students create a concept map for causes and effects of indoor air pollution. Make sure they include such causes as building materials and household chemicals, including asbestos; tobacco smoke; broken heaters and appliances; and radon. Also make sure they include such effects as difficulty breathing; coughing; irritation of the nose and throat, lung and respiratory diseases; headaches; and other effects.

## ELL/ESL STRATEGY

**Language Objective:** *To use hands-on artifacts to increase comprehension*

Gather the students near a window on a sunny day and have them look at the air that is illuminated by the sunlight. They should be able to see lots of tiny dust particles swirling around in the air. If the windowsills or other surfaces in your rooms have not been dusted for a while, students also should be able to see dust resting on the windowsills or other surfaces. Tell them that air pollution and particulates are in the air all around us and that they can even collect visibly on surfaces.

---

**Asbestos**
*A material used for building materials*

**Radon**
*A gas with no smell or color that comes from the radioactive decay of radium underground*

## How Does Indoor Air Pollution Affect Health?

Air inside homes and buildings also can affect health. This can happen when homes and buildings are closed up too tightly. Window and door seals are made to save energy and keep the outside air from getting in. Sometimes what happens, though, is that no new air gets in. All of the pollution is trapped inside the home or building.

Some building materials and household chemicals are toxic. For example, some glues used in wood furniture and paneling give off harmful gases. Household chemicals such as paints, drain cleaners, and pesticides give off gases that can cause respiratory problems. A building material that gives off particulates when crumbled is **asbestos**. Asbestos is a material that was used in older homes and buildings. Moving asbestos around puts tiny bits of it into the air. Asbestos can cause lung cancer. Asbestos is rarely used today. State laws say that workers must follow rules when removing asbestos from public places, such as schools.

There are other indoor air pollutants. Tobacco smoke causes respiratory and other problems for both smokers and nonsmokers. Broken heaters and appliances such as ovens and refrigerators can allow carbon monoxide and nitrogen dioxide to get into the air. **Radon** is another gas with no smell or color that comes from the radioactive decay of radium underground. Radon can be a serious indoor air pollutant when it gets into houses through walls and basement floors. Radon may cause lung cancer.

Indoor air pollution can make it hard for you to breathe, give you a cough that will not go away, and bother your nose and throat. Some indoor air pollutants can cause lung and respiratory diseases. Some indoor air pollutants can cause certain cancers. Indoor air pollution also can bother your eyes and give you an upset stomach and headaches. Some people cannot sleep. They might also get depressed, dizzy, cranky, and tired when they breathe in certain pollutants.

**Greenhouse effect**
*The warming of Earth's atmosphere*

**Ozone layer**
*An area above Earth that guards and protects Earth from the sun's harmful rays*

### Link to >>>
**Earth Science**

Global warming refers to a worldwide climate change. Some scientists say that there is less Arctic ice than there once was. They conclude that less Arctic ice means that temperatures around the world are rising. Temperatures have been rising over the past 100 years. Some scientists believe global warming will upset the balance of nature and make serious changes in the weather.

## How Does Air Pollution Affect Earth?

Air pollution affects not only people's health but also adds to long-term changes in Earth's atmosphere. Air pollution causes harm in several ways.

### The Greenhouse Effect

Many scientists believe that all the carbon dioxide in the air is causing Earth to act like a greenhouse. A greenhouse is warm and damp inside. It has a glass ceiling that traps the sun's heat. A carbon dioxide cloud caused by air pollution surrounds Earth. This cloud acts like the glass ceiling in a greenhouse, trapping heat and dampness. The result is that Earth's atmosphere is warm, which allows life to exist on Earth. This is called the **greenhouse effect.**

Even small rises in the temperatures around the world affect rainfall everywhere. Less rainfall in some areas could cause crops to fail and make more deserts. More rainfall in other areas could cause serious floods and soil erosion. Sea levels could rise, flooding areas along the coast and mixing saltwater with freshwater.

### The Ozone Layer

Scientists think that air pollution has made the **ozone layer** thinner. The ozone layer is an area 6 to 30 miles above Earth. It is made of naturally occurring gases. It guards and protects Earth from the sun's harmful rays. In 1985, scientists discovered a hole the size of the United States in the ozone layer over Antarctica. Later studies found an ozone loss over the Arctic polar region.

Scientists know that natural events such as volcanoes can affect the ozone layer. They also think that chlorofluorocarbons (CFCs) in the atmosphere have an effect. CFCs are chemicals that once were used in refrigerators and air conditioners. CFCs stay in the atmosphere for years. They turn ozone molecules into oxygen molecules. Oxygen molecules cannot protect Earth from the sun's harmful rays. Because of this, most products with CFCs are no longer used in the United States.

A thinner ozone layer allows more of the sun's harmful rays to reach Earth. These rays may damage crops, forest growth, animal life, and human health.

### Link to
**Earth Science.** Have students read the Link to Earth Science on page 493. Tell them that some scientists and policy makers dispute that global warming is occurring because of air pollution. Have the students break into groups and do research to find opposing viewpoints on global warming. Groups can share and discuss the opposing viewpoints with the rest of the class.

### LEARNING STYLES

**Interpersonal/Group Learning**

Acid rain damages nonliving things, including bridges, statues, tombstones, and buildings. Invite interested students to find out about structures that were damaged by acid rain and give reports to the class. Possibilities include famous monuments such as Cleopatra's Needle in New York City, the Parthenon in Greece, and pyramids and sphinxes near Cairo, Egypt.

### Health Myth

Have students read the Health Myth feature on page 494. Tell students that it is not always other countries that are sending air pollution our way; often, air pollution created in the United States drifts over other countries. Air pollution from industrial states in the northeastern United States often drifts into Canada and creates acid rain.

## Lesson 2 Review Answers

**1.** B **2.** D **3.** C **4.** Sample answer: I think car exhaust causes the most pollution because so many cars and trucks are being driven all the time. **5.** Sample answer: I think we can reduce our dependence on cars by carpooling, biking, or walking to school and to work. We also can promote the idea of using public transportation. Reducing car exhaust would lessen air pollution.

### Portfolio Assessment

Sample items include:
- Questions from Vocabulary
- Tables from Teaching the Lesson
- Concept maps from Learning Styles: Visual/Spatial
- Reports from Learning Styles: Interpersonal/Group Learning
- Lesson 2 Review answers

---

**Acid rain**
Rain, snow, sleet, or hail that contains large amounts of sulfuric or nitric acids

**Health Myth**

**Myth:** The United States is not affected if a faraway country uses products with CFCs.

**Fact:** Winds lift and mix CFCs. CFCs released far away can eventually make their way to the United States.

What might happen if some countries in the world do not work together with other countries to protect the environment?

### Acid Rain

Acid rain is precipitation—rain, snow, sleet, or hail—that contains large amounts of sulfuric or nitric acids. Sulfur and nitrogen from cars, trucks, factories, and home heating mix with the dampness in the air to form sulfuric and nitric acids. These acids spread through the atmosphere and come back to Earth in the form of acid rain. Acid rain may fall far from where the acids were made. For example, factories in the Midwest make large amounts of these acids. The acid rain falls as far away as the Southeast, New England, and eastern Canada.

Acid rain harms rivers, lakes, and water and plant life. As the water gets full of acid, fish do not lay eggs and may die. Water with a lot of acid pulls mercury and lead out of the lake beds and into the water. When the mercury and lead get into the fish, they become unsafe for humans to eat. In Germany, major forests have been destroyed by acid rain.

*Lesson 2 Review* On a sheet of paper, write the letter of the answer that best completes each sentence.

1. The main cause of air pollution is _____.
   A acid rain          C photochemical fog
   B burning            D ozone

2. A building material that gives off cancer-causing particulates is _____.
   A ozone              C coal
   B radon              D asbestos

3. A rise in the amount of carbon dioxide causes _____.
   A acid rain          C the greenhouse effect
   B the ozone layer    D photochemical smog

On a sheet of paper, write the answers to the following questions. Use complete sentences.

4. **Critical Thinking** What do you think causes the most air pollution?

5. **Critical Thinking** What can be done to lessen air pollution?

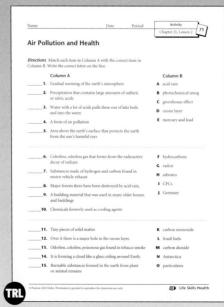

Activity 71

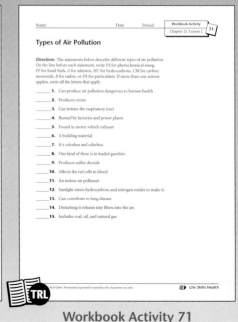

Workbook Activity 71

## Lesson 3  Water Pollution and Health

**Objectives**

*After reading this lesson, you should be able to*

- list five causes of water pollution
- describe the effect of chemical fertilizers on lakes and streams
- explain three ways freshwater can be conserved

**Sewage**
*Liquid and solid waste from drains and toilets*

While three-quarters of Earth's surface is covered with water, only a small fraction of that water is freshwater. Most of Earth's water is in oceans, which contain saltwater. Saltwater cannot be used for drinking or growing crops. Freshwater is as important as air is in supporting human life.

Groundwater is water beneath Earth's surface. Groundwater is the largest source of freshwater. It supplies springs and wells. People depend on groundwater for drinking, cooking, and bathing. Groundwater is an important natural resource. It must be kept clean and free of pollution.

### What Causes Water Pollution?

For many years, countries have dumped their wastes into oceans, lakes, and rivers. Many of these wastes cannot be broken down. They stay there for a long time and do permanent harm. Waste from factories, drains and toilets, oil leaks and spills, farm and household chemicals, and acid rain are the most common ways water gets polluted.

#### Factory Wastes

Toxic materials from factories cause water pollution when they get into the water supply. Most factory wastes such as mercury, lead, and acids cannot be broken down. Often the wastes were dumped into bodies of water or into groundwater. Mining has also caused a great deal of water pollution in the United States. The federal and state governments have passed laws to stop factories from polluting freshwater.

#### Sewage

**Sewage** is waste from drains and toilets. Most sewage comes from human body wastes. Sewage contains bacteria and life-forms that can cause disease. If sewage gets into groundwater and bodies of water such as ponds, lakes, and rivers, it can kill animals and plants. Humans can get diseases such as typhoid fever and cholera from water that has been polluted with sewage.

---

## Lesson at a Glance

### Chapter 21 Lesson 3

**Overview** This lesson describes causes of water pollution, including industrial wastes, sewage, oil leaks, agricultural runoff, and household chemicals. It also explains how freshwater can be conserved.

### Objectives

- To list five causes of water pollution
- To describe the effect of chemical fertilizers on lakes and streams
- To explain three ways freshwater can be conserved

**Student Pages** 495–498

**Teacher's Resource Library**

Activity 72

Alternative Activity 72

Workbook Activity 72

### Vocabulary

algae          sewage

Write the words *sewage* and *algae* on the board. Call on students to describe these items, using as many senses as possible. For example: Where have they seen these items? What colors were they? Did they smell? Have they seen algae in aquariums or in ponds? Did they ever touch algae? After students describe these words, read the definitions aloud and point out how the descriptions supplement the definitions.

### Background Information

Polluted water can cause many health problems. Hepatitis A and E, viruses that inflame the liver, can be transmitted by water which is contaminated by fecal matter. People can get lead poisoning from water that has run through lead pipes, which are common in old buildings. Lead poisoning can damage the nervous and reproductive systems and the kidneys, it can harm the brains of developing fetuses and young children, and it can cause death. Cholera, a disease caused by bacteria that can cause severe dehydration and death, is transmitted by water which is contaminated by the fecal matter of people who have cholera.

**Warm-Up Activity**

Show students a tank, jar, or glass of clean water. Place a few drops of red or blue food coloring in the water and have them watch as the food coloring spreads through the water. Tell them that this illustrates how even small amounts of pollutants can spread quickly throughout a water system.

 **Teaching the Lesson**

Draw a circle on the board. Tell students that the circle represents all the water on Earth. Model how such a circle is used as a pie graph to divide the parts of a whole. Ask several volunteers to draw a pie piece showing how much of the water on Earth they think is usable freshwater. Show students a globe or wall map of the world. Have a volunteer point out the oceans. Emphasize that the oceans are made of saltwater, which cannot be used for drinking, growing crops, washing, or even in industrial machines. Have another student point out sources of freshwater—lakes and rivers. Note how comparatively little freshwater there is, even taking into account groundwater. Let students revise the pie graph if they wish. Then color in a thin section of the pie representing 3 percent of the whole. This is the amount of water on Earth that is usable freshwater.

 **Reinforce and Extend**

### ELL/ESL Strategy

 **Language Objective:** *To speak in front of a group*

Have students form small groups. Each group should include students learning English and students who have a firm grasp of English. Each person in each group should pick one form of water pollution and give a very brief presentation on it to the rest of his or her group. Students should describe the form of pollution, tell where it comes from, and describe the effects of the pollution.

---

**Algae**
*A large group of plants that live in water*

In 1977, the U.S. government passed a law that said sewage in the United States had to be treated before it was put into the environment. As a treatment, tiny helpful life-forms were added to the sewage to break down bacteria and harmful life-forms. Since then, the U.S. government has passed other clean water laws. In 1987, for example, it made stronger laws about sewage treatment and how clean the water had to be. More changes to the law were made in the 1990s and more recently. It is the job of the Environmental Protection Agency (EPA) to make sure factories do not break any clean water laws.

### Oil Leaks and Spills

The worldwide need for oil and petroleum, which is a mixture of hydrocarbons, is great. To meet the need, large oil tankers move a large amount of oil and petroleum across oceans. Sometimes, these tankers leak, causing oil spills on the water. The spills kill animals and leave behind carcinogens. These carcinogens can affect humans.

### Agricultural Runoff

Farmers use chemicals to make plants grow better. They use pesticides and herbicides to control bugs and weeds. When it rains, these chemicals run off into streams, rivers, lakes, and groundwater. The fertilizers speed up the growth of **algae,** a large group of plants that live in water. These plants quickly use up the oxygen supply in the water so other plants and fish cannot live. Rain runoff can also wash the chemicals in the fertilizers into the local water supply. When these chemicals get into water, they can kill fish in nearby rivers and lakes. They can also ruin the groundwater that people drink.

**Figure 21.3.1** *Oil spills kill animals and leave behind carcinogens.*

Wastes from animals also make agricultural runoff that is harmful to the environment and humans. U.S. and state governments watch the effects of agricultural runoff. They give researchers money to find ways of preventing water pollution that agricultural runoff causes.

### Household Chemicals

Some household cleaners contain chemicals that are harmful to plants and humans if they get into the water supply. Many cleaning products now must be sold without these harmful chemicals. For example, laundry soaps must now be sold without phosphates. They must be able to break down in the environment and not pollute water sources. Even so, reading labels on household products to learn how to throw them away safely is important.

### Decide for Yourself

Some communities have strong laws about the environment. For example, some communities do not allow gasoline to be sold on a lakefront. Other communities may have rules for motorboats that are on a lake. Still other communities may give out fines for polluting the water. It may be against the law in some places to throw trash into a lake or allow farm runoff to reach a river or stream.

Polluting a community can mean that there is trash in a park or on a street. It could also mean there is air or water pollution from factories. Every community, small and large, has some type of pollution problem. What can you do as a person or as a class to make your community cleaner?

**Steps to Environmental Health**

1. First, find the problems. Think about how you can make your community cleaner. List the problems you find.
2. Decide on a problem you feel you can correct.
3. Create a plan to correct the problem. You may need the help of other students. You may need to work with community officials. Decide how long it will take and who must be involved for the project to be a success.
4. Carry out your plan.
5. Decide how successful you were.

On a sheet of paper, write your answers to the following questions. Use complete sentences.

1. How hard was it to decide on a community project? Explain your answer.
2. Whom did you need to involve for your project to be successful?
3. Do you think the problem would not have existed if there was already a law in place? Explain your answer.

## Decide for Yourself

Have students read the Decide for Yourself feature on page 497. Encourage them to contact the local Environmental Protection Agency or talk to community officials for information on problems in their community.

**Decide for Yourself Answers**

1. Sample answer: It was not hard to decide on a community project because we felt that students have a responsibility to keep the school grounds clean. 2. Sample answer: We needed to involve students and school administrators. We also asked parents to help. 3. Sample answer: We might not be able to get a law passed regarding the polluting of school grounds. Therefore, another option would be for the school district to establish a rule that would require each class to take turns cleaning the school grounds.

## LEARNING STYLES

### Body/Kinesthetic

Bring to class water samples from different sources in the community, or supervise students in taking such samples. Sources include ponds, streams, rainwater, and faucets at school and at home. The sources of the samples should be labeled. Use simple testing kits to determine the quality of the water. See Other Resources on page 368 for places to obtain testing kits.

## CAREER CONNECTION

Inform students that many individuals are needed to operate a sewage treatment plant. Operators might repair equipment, perform chemical tests on water samples, or manage the operation of the entire plant. Have students find out more about the job responsibilities and educational requirements of a sewage treatment plant operator. Then have students write a one-page report on the topic.

## GLOBAL CONNECTION

Ask students if they consider the availability of their water to be a luxury. Have them compare their use of water to that of other places around the world. Point out, for example, that in many developing nations in Asia, Africa, and South America, people must carry water in buckets from a well or river. Discuss how the convenience of freshwater affects one's attitude toward it and actions regarding it. *(People often take water for granted when it is so convenient. People are much more likely to conserve water when it becomes scarce or is more difficult to obtain.)*

## Lesson 3 Review Answers

1. groundwater 2. cannot 3. herbicides 4. Sample answer: Agricultural runoff causes a change in the balance of nature in lakes and streams. It also causes the growth of algae, which can be harmful to plants and fish. Agricultural runoff can put harmful chemicals in freshwater lakes. These chemicals can kill fish and ruin groundwater so that people cannot drink it. 5. Sample answer: Three things I can do to help conserve water are turning the water off when I brush my teeth, turning the water off when I'm washing dishes and just rinsing them all at the same time, and filling the bathtub only halfway. The hardest change will be to remember to turn off the water when I'm brushing my teeth.

## Portfolio Assessment

Sample items include:
- Decide for Yourself answers
- Reports from Application: Career Connection
- Lesson 3 Review answers

## How Can Freshwater Be Conserved?

Many places do not have a lot of freshwater. Therefore, protecting it is important. One way to protect freshwater is to conserve it, or use it carefully.

Here are some ways you can conserve water:

- Take showers no longer than 5 minutes.
- Fill the tub only halfway when you take a bath.
- Wet the toothbrush when you brush your teeth and turn off the water until you rinse.
- Wait to run the clothes washer or dishwasher until there is a full load.
- Fix leaking faucets.

**Lesson 3 Review** On a sheet of paper, write the word or phrase in parentheses that best completes each sentence.

1. The largest supply of freshwater is (clean water, groundwater, saltwater).
2. Factory wastes such as mercury and acid (can, cannot, might) be broken down.
3. Agricultural runoff comes from pesticides, (herbicides, germicides, petroleum), and animals.

On a sheet of paper, write the answers to the following questions. Use complete sentences.

4. **Critical Thinking** Why is agricultural runoff harmful to places where freshwater is found?
5. **Critical Thinking** What are three habits you can change to help conserve water? Which habit will be the hardest to change?

498  Unit 7  Health and Society

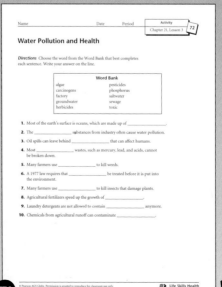

Activity 72

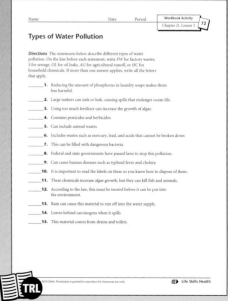

Workbook Activity 72

# Lesson 4: Other Environmental Problems and Health

**Objectives**

*After reading this lesson, you should be able to*

- explain how world population affects health
- explain how pollution affects health
- describe what is meant by hazardous waste
- identify three symptoms of radiation sickness

**Deforestation**
*Cutting down and clearing trees*

**Famine**
*A time when there is not enough food for people*

The world's growing population is a problem for the environment. So is **deforestation**, which is the cutting down and clearing of trees. Solid waste, noise pollution, and radiation can also affect the environment. Because these problems affect the environment, they also can affect people's health.

## How Does Population Size Affect Health?

The world's population is more than 6 billion people. Growing enough crops to feed so many people is a concern. See Table C.7 in Appendix C. Many countries have already gone through long periods of **famine**, a time when there is not enough food for people. Famines in Africa, for example, have come after long periods without enough rainfall to grow crops.

Many concerns seem to come from the growing world population. The more people there are on Earth, the more cars, trucks, and factories there will be. With more cars, trucks, and factories, there will be more air pollution. The more air pollution there is, the more global warming there will be, and the ozone layer may get thinner. If the ozone layer becomes thinner, more harmful rays from the sun will get into the atmosphere. With more harmful rays, more people will get skin cancers. You can see that everything in the environment is connected.

## How Does Deforestation Affect Health?

Some people believe that long periods with no rain and the famine in Africa came from deforestation. Over history, two-thirds of the world's forests have been lost to deforestation. Some African countries have lost all of their forests to make room for people, crops, and cattle.

*Environmental Health* Chapter 21

---

# Lesson at a Glance

## Chapter 21 Lesson 4

**Overview** This lesson discusses the effects on health of population size, deforestation, solid wastes, noise pollution, light pollution, and radiation.

### Objectives

- To explain how world population affects health
- To explain how pollution affects health
- To describe what is meant by hazardous waste
- To identify three symptoms of radiation sickness

**Student Pages** 499–503

**Teacher's Resource Library**

Activity 73

Alternative Activity 73

Workbook Activity 73

### Vocabulary

| | |
|---|---|
| deforestation | radiation sickness |
| famine | sensorineural |
| hazardous waste | deafness |
| landfill | solid waste |

Ask students to write sentences in which they use a vocabulary word. Have them underline the word. Instruct students to exchange papers and rewrite the sentences, substituting the definitions for the underlined words. Students should write the definitions in their own words.

### Background Information

Repeated exposure to excessive noise can cause high blood pressure, headaches, ulcers, and hearing loss. The most dangerous noises—such as those made by sirens, jet engines, rivet guns, gunshots, and some musical concerts—are those that are 130 decibels or above. They can cause immediate and permanent hearing damage. Some people whose hearing has been damaged develop tinnitus, which is a constant buzzing or ringing sound in the ears. Many teenagers put themselves at risk of hearing loss by having the volume too high when they are wearing headphones or ear buds to listen to music.

**Warm-Up Activity**

Make a list on the board of environmental problems that will be discussed in this lesson. When the list is complete, challenge students to use what they have learned in class so far to identify the effects each problem might have on health. After teaching the lesson, return to the list to make any necessary additions or omissions.

 **Teaching the Lesson**

Have students make an outline of the lesson using the headings as the framework. Underneath each heading, have students list the important ideas under the section and any subheadings as well as notes. For example, under the heading How Does Solid Waste Affect Health?, have them create the subheadings Hazardous Waste and Solid Waste.

 **Reinforce and Extend**

### AT HOME

Challenge students to find out how much trash their family produces each week. They might bag the trash, weigh it each day, and add up the totals. Compare and discuss these totals. Do some people create significantly more trash than others? What are some ways families might reduce the amount of trash they produce?

### Research and Write

Have students read and complete the Research and Write feature on page 500. You might want to have students work with partners or in small groups to complete this activity. Encourage the students to include visuals such as maps and photos in their reports.

---

**Solid waste**
*Garbage, trash*

Earth needs trees to keep its fragile balance. When people cut down forests to help support a growing world population, rains wash away the topsoil. The soil underneath does not hold water well. Soon the land cannot support the growth of crops. This causes famine and human suffering.

Tropical rain forests cover only about 6 percent of Earth's surface. However, they include half of all Earth's plants. Because of deforestation, only half of Earth's original rain forests are left. Each year, people burn millions of acres of tropical rain forest. Rain forest trees are burned or cleared to plant crops, to raise cattle, or to sell the wood.

Destroying tropical rain forests affects human and environmental health. Plants in the rain forests give us more than one-quarter of the medicines we use. Also, rain forest plants may hold the key to new medicines. Rain forest plants act as Earth's lungs by taking in carbon dioxide. Too much carbon dioxide adds to global warming.

### How Does Solid Waste Affect Health?

Developing nations get rid of few materials because there are not many resources. In more developed nations, people tend to throw things away after a short time. Then they buy more. In the United States, for example, people dump billions of tons of **solid waste**—garbage or trash—each year. Getting rid of solid waste has an effect on the environment and therefore, on human health.

### Research and Write

Wetlands such as swamps and marshes make a good home for plants and animals. Wetlands are important in flood control. More than half of all U.S. wetlands have been lost. Today, scientists are making new wetlands. Using the Internet, research wetlands in the United States. Make a graph showing what you learned.

**Landfill**
*Low-lying ground where waste is buried between layers of dirt*

**Hazardous waste**
*A material that can be harmful to the environment, human health, and other living things*

**Sensorineural deafness**
*Hearing loss in which the tiny cells in the inner ear are destroyed*

Many communities have dumped solid waste in **landfills.** Landfills are low-lying ground where waste is buried between layers of dirt. One problem, however, is that landfill space is running out in the United States. Another problem is that toxic materials called **hazardous waste** are sometimes dumped. Hazardous waste can be harmful to the environment, human health, and other living things. Factories are the largest makers of hazardous waste. Toxic materials that come from hazardous waste sometimes leak into nearby ground and pollute the groundwater.

Sometimes, people notice the effect of polluted groundwater right away. Other times, the effect is not noticed for years. For example, between the late 1930s and early 1940s, a factory in Love Canal, New York, dumped toxic chemicals into the ground. Later, people built homes there. In 1978, heavy rains flooded basements in those homes. People noticed oily liquids in the floodwater. Hazardous wastes had leaked into the soil and groundwater. The homes were polluted, and all the people had to move.

Every person can help to reduce hazardous waste. Paint, car batteries, and car oil can be thrown away separately from the rest of the garbage. Check with your community to learn how to dispose of hazardous waste.

## How Does Noise Pollution Affect Health?

As the population and number of factories have risen, so has noise. Noise pollution happens when sounds in the environment become too great. Listening to too much loud noise over a long time can cause **sensorineural deafness.** This is a hearing loss in which the tiny cells in the inner ear are destroyed. Many Americans have sensorineural deafness that comes from noise pollution. Too much noise also may cause stress and problems such as high blood pressure, tension, anger, and tiredness.

## ELL/ESL Strategy

**Language Objective:** *To speak in front of a group*

Have students form pairs composed of a student learning English and a student with strong English skills. Ask the pairs of students to prepare anti-noise pollution graphics—posters, print advertisements, or other visual presentations. If students have access to the appropriate technology, they can create multimedia presentations. Students will present their graphics to the class along with an argument against noise pollution. During their presentations, students should discuss how teenagers might expose themselves to noise pollution and the consequences of noise pollution.

## Health Journal

Have students write a story about a forest or natural environment that is disturbed and damaged by some outside force. Students should include information on how the plants and animals are affected by the disturbance and how balance could be restored to the natural environment.

## LEARNING STYLES

### Logical/Mathematical

Have groups of students debate the use of nuclear power to generate electricity. They should investigate whether alternative energy sources such as solar, geothermal, wind, or water power can fill the needs of most people in our society. Assign each group a position and have students conduct their debates for the class. Remind students to be respectful of each other during the debates.

## IN THE ENVIRONMENT

 Have students investigate how the traditional attitudes of American Indians toward the environment influence how they treat it. Students will discover that American Indians generally have had great respect for the environment. To American Indians, the environment is something they are a part of and live with rather than something to conquer. Have students give oral reports on their findings, including graphics if possible.

**Radiation sickness**
*Sickness caused by being around high levels of radiation*

### How Does Light Pollution Affect Health?

Light pollution is the glow you see above a city at night. As communities have grown, so has their use of light in parks, streets, and factory areas. Although many people think lights are needed to make places safe, other people do not like this light. Environmental studies have shown that this light is harmful for some animals. When an area is light all the time, animals that are most active at night get confused. Migrating birds have been known to fly into lighted towers. Some communities are taking steps to lessen the glow in the night sky. Others are using special lights that shine downward but guard against light that goes up.

### How Does Radiation Affect Health?

Radiation sends energy in the form of waves. It happens naturally in the environment. Usually, people are exposed only to low levels of radiation, such as X-rays. Some factories use high levels of radiation. Nuclear power plants, for example, use radiation to make electricity for homes and businesses. People who work in nuclear power plants are exposed to high levels of radiation.

Too much radiation may cause cancer in humans. People who are exposed to very high levels of radiation may get **radiation sickness.** Some symptoms of radiation sickness are upset stomach, vomiting, headache, diarrhea, hair loss, and tiredness. Radiation sickness can lead to death. The U.S. government sets exposure safety rules for people who work with or near radiation.

Nuclear power plants are carefully controlled. Still, accidents have happened at these plants. For example, an accident happened in 1986 in the nuclear plant at Chernobyl in the former Soviet Union. Many people died immediately. Others died of radiation sickness weeks later. Wind carried radiation across much of Europe. Some scientists believe that many Europeans will die of cancer because of what happened. Today, the villages and forests around Chernobyl are deserted and unsafe. It was the worst nuclear accident ever.

**Lesson 4 Review** On a sheet of paper, write the answers to the following questions. Use complete sentences.

1. What are two effects of a growing world population?
2. What is deforestation?
3. How can groundwater become polluted?
4. **Critical Thinking** Why are nuclear power plants both a resource and a danger?
5. **Critical Thinking** Which type of pollution do you believe is the greatest danger to health? Explain your answer.

### Technology and Society

Agriculture biotechnology is science's way of making better plants and animals. Agriculture biotechnology has been used to make new medicines. It also has helped farmers use less pesticides and herbicides. The Environmental Protection Agency (EPA) and the Food and Drug Administration (FDA) are in charge of biotechnology. Food that is created through biotechnology must be tested before it can be sold.

Environmental Health   Chapter 21   503

## Lesson 4 Review Answers

1. Sample answer: Two effects of a growing world population are famine and air pollution. 2. Deforestation is the cutting down and clearing of trees. 3. Groundwater can become polluted when toxic materials that come from hazardous waste leak into the ground. 4. Sample answer: Nuclear power plants produce electricity for homes and businesses, but these plants are also sources of radiation. 5. Sample answer: I believe the greatest danger to health is hazardous waste, because it can pollute groundwater used by many people.

### Portfolio Assessment

Sample items include:
- Sentences from Vocabulary
- Outlines from Teaching the Lesson
- Figures from Application: At Home
- Graphs from Research and Write
- Presentations from ELL/ESL Strategy
- Stories from Health Journal
- Oral reports from Application: In the Environment
- Lesson 4 Review answers

### Technology and Society

Have students read the Technology and Society feature on page 503. Tell students that all medicines and food, even if they are not created through agriculture biotechnology, must be tested to uncover any harm they might cause people who consume them. Foods, food additives, and medicines that show any health risks in testing are examined thoroughly and labeled with their risks if they are approved for use.

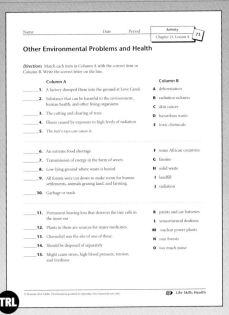

Activity 73

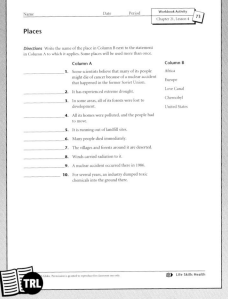

Workbook Activity 73

Environmental Health   Chapter 21   503

# Lesson at a Glance

## Chapter 21 Lesson 5

**Overview** This lesson identifies laws and agencies that protect the environment, describes future environmental plans, and explains how individuals can protect the environment.

**Objectives**

- To explain how U.S. laws protect the environment
- To explain how world governments work together to protect the environment
- To describe how you can protect the environment

**Student Pages** 504–508

**Teacher's Resource Library** TRL

Activity 74

Alternative Activity 74

Workbook Activity 74

## Background Information

One way to protect the environment is to throw away less trash that will end up in landfills. One way to throw away less trash is to buy products that do not have a lot of packaging. Buy in bulk when you can, because products packaged in bulk produce less waste. Choose products packaged in reusable or recyclable packages, such as glass or aluminum.

 **Warm-Up Activity**

Ask students to name government agencies that protect the environment. Have students write a few paragraphs describing how they believe the government helps provide protection.

 **Teaching the Lesson**

Create a chart on the board with three columns. Title the columns Government, Industry, and Individuals. Ask students to brainstorm ways that governments, industries, and individuals can work to protect the environment.

---

## Lesson 5 — Protecting the Environment

**Objectives**

*After reading this lesson, you should be able to*

- explain how U.S. laws protect the environment
- explain how world governments work together to protect the environment
- describe how you can protect the environment

Protecting the environment is everyone's job—the world community, governments, businesses and factories, local communities, and individuals.

### What Laws Protect the Environment?

The U.S. government has passed several laws to protect the environment. All of these laws are also meant to protect human health.

The Clean Air Act was first passed in 1970 and has been changed several times since then. It says how much air pollution there can be. It also makes each state have a plan to meet its standards.

The National Environmental Policy Act requires federal agencies to think about how their plans and activities will affect the environment. Before an agency can begin a new project, it must file an environmental impact statement. This statement explains how the project will affect the environment. The agency also must hold public meetings about the effect of the project on the environment. Many states have passed similar laws.

Congress has passed other environmental laws. These laws set standards for water safety and ways to get rid of solid and hazardous wastes. They require polluters to manage the cleanup of chemical spills and toxic waste.

### How Do Governments Protect the Environment?

In the United States, the Environmental Protection Agency (EPA) writes and carries out the environmental laws Congress passes. Its job is to protect the environment.

Because so many environmental problems affect the entire world, nations have joined together to try to solve the problems. For example, an international meeting in Senegal in 2005 looked at ways to keep from thinning the ozone layer. Many nations have agreed to stop using chlorofluorocarbons (CFCs) and other materials that may thin the ozone layer.

In 1992, the United Nations (UN) General Assembly started the Conference on Environment and Development. It brought together leaders from 178 countries. They talked about ways to live well and protect the environment at the same time. Today, the UN keeps looking for ways to protect the environment.

## What Are Future Environmental Plans?

Scientists are working on ways to protect the environment from further harm. For example, they are searching for ways to use solar energy, or energy from the sun. Solar energy is a clean and environmentally safe energy source. Wind energy is another energy source. Most of our oil, gas, and electricity now comes from fossil fuels. If we used solar and wind energy, we would lessen our use of fossil fuels. Some scientists say we could lessen it as much as 70 percent.

Scientists have also found other fuels to lessen exhaust from cars and trucks. One of these fuels is ethanol, or alcohol made from corn—a renewable resource. Some states already use ethanol mixed with regular gasoline. Hydrogen and natural gas are two other kinds of fuels. Hybrid cars use a mix of gasoline and electricity. These cars save gasoline and lessen harmful emissions, or exhaust. Clean energy, such as solar energy, may one day be a common way to power cars.

California has passed a law that says new cars have to use less energy. Other states have high-occupancy vehicles (HOV) lanes. HOV lanes encourage people to ride to work together. Having fewer cars on the road is one way to lessen car exhaust. Some states have laws about how much exhaust cars can make. In these states, cars must pass an emissions test every year.

> **Health Myth**
>
> **Myth:** Biodegradable items do not harm the environment.
>
> **Fact:** When they are thrown away, biodegradable items release methane into the atmosphere as they break down. Methane is one of several greenhouse gases. A good option is to use biodegradable products that can be turned into compost. A better option is to use recyclable products or products that come in refillable containers.

## 3 Reinforce and Extend

### ELL/ESL Strategy

**Language Objective:**
*To use language for functional purposes*

Have students learning English work with students proficient in English to make lists of environmental slogans they have heard and to create new slogans of their own. They should write the slogans in both languages. Invite students to use their slogans to make a bilingual bulletin board display for the classroom or other area of the school. Encourage students to find or create illustrations to accompany their slogans.

### In the Environment

Suggest that students start an environmental awareness campaign in school. First students should identify activities they can do. Activities might include not littering, using less paper and fewer disposable items, recycling, and so on. Then encourage students to make posters, flyers, or newsletters for the school; to plan a school assembly or environmental awareness week; to make sure recycling bins are easily accessible; and so on. Have students make a plan, carry it out, and then evaluate its success.

### Health Myth

Ask a volunteer to read aloud the Health Myth on page 505. Tell students that most biodegradable items are an improvement over items made of plastic and other materials that do not break down. However, when materials break down, they can be introduced to the soil or water we use. Some organic items that break down can have the same effect as releasing a chemical fertilizer.

## LEARNING STYLES

**Visual/Spatial**

Have interested students create a timeline that shows key developments in the struggle in the United States to protect the environment. The timeline might begin with the establishment of national parks and forest preserves in the nineteenth century. Other key events might include pollution disasters such as the ones at Donora, Pennsylvania, in 1948; Love Canal in 1978; and the oil spill from the tanker *Exxon Valdez* in 1989. The timeline also should include laws that state legislatures and the federal government passed to protect the environment. Display the timeline in the classroom or in a hallway.

## HEALTH JOURNAL

Have students find out how recycling is handled at their school and in their community. Students should write a brief entry in their journals that includes their recommendations on how recycling efforts could be increased.

## TEACHER ALERT

Ask students what they know about wetlands in the United States. Point out that wetlands include swamps, marshes, and bogs. Ask students if they think preserving wetlands is important to the environment. Point out that wetlands provide many benefits. They provide habitat for wildlife, control flooding by absorbing runoff, act as storm breakers along the coasts, filter pollution from the waters that flow through the wetland, and replenish groundwater while slowly releasing water during dry times. Discuss with students the wetlands in your community, if there are any, and talk about how they benefit your community.

### How Can You Help Protect the Environment?

Everything you do makes a difference. Every individual can work to protect the environment. Here are some actions you and others can take to protect the environment:

- Walk, ride a bike, take the bus, or carpool to school or work.
- If you drive, keep your car tuned up, have the exhaust equipment checked regularly, and keep tires properly inflated.
- Buy a car that gets good gas mileage.
- Turn off lights when you leave a room.
- Keep air conditioning at a steady temperature on the automatic setting, and keep windows and doors closed.
- Set the furnace at a steady low temperature—perhaps 68 degrees Fahrenheit—and wear layers of clothing.
- Take shorter showers.
- Do not run hot water if you do not need to.
- Use fewer things that need to be thrown away. For example, use a glass cup instead of a paper or plastic cup, or use cloth towels instead of paper towels.
- Recycle newspapers, cans, bottles, plastic, and paper. Find out what recycling services are in your community. Reuse items such as paper clips, too.
- Buy only what you need.
- In general, buy products that have the least amount of packaging. Buy products that have refillable containers, such as beverages in glass bottles.
- Look for products with the recycled symbol. This symbol shows that recycled materials were used in making them.
- Find other products for your home that do not contain toxins. For example, instead of using oven cleaners with lye, use baking soda and water. Wash windows with white vinegar and water.

- If you must buy household products with toxins in them, throw them away, using the instructions on the package.
- Write to lawmakers about your concerns for the environment.
- Plant some trees to take in more carbon dioxide and reduce the greenhouse effect.

### Health in Your Life

**Recycling**

Recycling means reusing materials instead of buying new ones. Recycling saves natural resources and makes less solid waste. Newspapers, cardboard, aluminum cans, glass, and some paper can be recycled. Many communities pick up recyclable products at the curb. Other communities have recycling centers. In many communities, people can get cash for aluminum and other metals.

Scientists have found ways to make materials like plastic from biodegradable materials such as corn. Plastic made from hydrocarbons takes hundreds of years to biodegrade. Plastic made from corn, however, takes about 47 days to biodegrade. Plastic cups made from corn can be crushed into compost after use. The compost becomes dirt. Combining recycling with biodegradable plastics helps to lengthen the life of landfills. It also protects the environment.

Each community needs to have a plan to recycle and use packaging that can biodegrade into compost when possible.

What recycling plan does your community have? Can the plan be improved? How can you help your community save natural resources?

On a sheet of paper, write your answers to the following questions. Use complete sentences.

1. What is recycling?
2. What products can be recycled?
3. How can mixing recyclable products and products that can biodegrade into compost help the environment?

### Health in Your Life

Have students read the Health in Your Life feature on page 507 and answer the questions. Discuss the recycling programs that are available in your area. Ask students to identify which materials they recycle at home.

**Health in Your Life Answers**

1. Recycling is reusing materials instead of buying new ones. 2. Products that can be recycled include newspapers, cardboard, aluminum cans, and glass. 3. Sample answer: Mixing recyclable products with biodegradable products will help lengthen the life of landfills. Landfills will become full we if we keep discarding plastic containers that take years to biodegrade. Extending the life of a landfill helps the environment.

### AT HOME

Discuss how families working together can play an important role in improving the environment. Ask students to plan a family meeting to discuss what they can do together to help protect the earth. Suggest that students get each member of the family to come to the meeting with at least two suggestions of ways to improve the environment. Tell students that as a family, they should devise a plan of action for incorporating some of their suggestions into their lives. Ask for volunteers to share their family's plan

# Lesson 5 Review Answers

1. EPA  2. corn  3. recycle  4. Sample answer: State laws about car exhaust help to protect the environment by reducing the amount of car exhaust in the environment. Carbon monoxide is found in the exhaust of cars and trucks. Breathing in carbon monoxide can be harmful to people's health, and it can make heart and lung disease worse.  5. Sample answer: Scientists keep trying to find new ways to protect the environment because people continue to harm the environment, and our good health depends on having a healthy environment.

## Portfolio Assessment

Sample items include:
- Paragraphs from Warm-Up Activity
- Slogans from ELL/ESL Strategy
- Campaign from Application: In the Environment
- Timeline from Learning Styles: Visual/Spatial
- Entry from Health Journal
- Health in Your Life answers
- Family plans from Application: At Home
- Lesson 5 Review answers

## Health at Work

Have students read the Health at Work feature on page 508. Ask students how important they think it is for someone to try to figure out how to best use the land without harming the environment. Then ask what they think a conservation scientist actually does. After reading the feature, ask students if the number and kinds of responsibilities associated with the work is surprising to them.

508  Unit 7  Health and Society

---

**Lesson 5 Review** On a sheet of paper, write the word or phrase in parentheses that best completes each sentence.

1. The (UN, EPA, CDC) makes sure factories follow clean air and water laws.
2. An example of a renewable resource is (corn, aluminum, gasoline).
3. One way to help protect the environment is to (recycle, throw away, buy more) items.

On a sheet of paper, write the answers to the following questions. Use complete sentences.

4. **Critical Thinking** How do state laws about car exhaust help protect the environment and keep people from getting sick?
5. **Critical Thinking** Why do scientists keep trying to find new ways to protect the environment?

### Health at Work

**Conservation Scientist**

Conservation scientists find ways to make the best use of the land without harming the environment. There are different types of conservation scientists. Some look for ways to keep water supplies clean. Others find the best use of farmland or manage cattle ranges. Still others work directly with animal management.

A conservation scientist needs to have a 4-year college degree in ecology, environmental science, forestry, or some related field. Along with working in the field, a conservation scientist must know how to write reports and be able to work with others to solve problems.

508  Unit 7  Health and Society

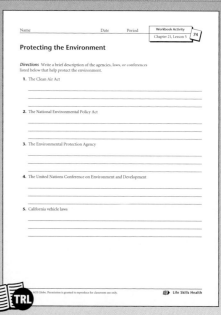

Activity 74                     Workbook Activity 74

# Chapter 21 SUMMARY

- When all parts of the environment work together, a healthy balance of nature is maintained.
- Human activity has upset the balance of nature by overusing Earth's resources and producing pollution.
- Because pollution can put human health and life in danger, people are finding some solutions.
- Burning is the main cause of outdoor air pollution. Cars, trucks, and factories make the most air pollution.
- Hydrocarbons, nitrogen oxides, ozone, carbon monoxide, and particulates are some outdoor air pollutants.
- Some indoor air pollutants are gases and particulates from building materials and household chemicals, tobacco smoke, broken appliances, and radon.
- Respiratory and lung problems are common health effects of air pollution.
- Results of air pollution on Earth are the greenhouse effect, thinning of the ozone layer, and acid rain.
- Since the supply of necessary freshwater is limited, saving and protecting it is important.
- Some causes of water pollution are factory wastes, sewage, oil leaks and spills, agricultural runoff, and toxic chemicals.
- The growing world population has caused many environmental problems, including deforestation, solid waste, noise pollution, and radiation.
- The federal and state governments have passed several laws that protect the environment.
- Nations and individuals can take many actions that help protect the environment, including using energy and water carefully.

### Vocabulary

| | | |
|---|---|---|
| acid rain, 494 | fragile, 488 | pollution, 487 |
| algae, 496 | greenhouse effect, 493 | radiation sickness, 502 |
| asbestos, 492 | hazardous waste, 501 | radon, 492 |
| bronchitis, 490 | landfill, 501 | sensorineural deafness, 501 |
| deforestation, 499 | ozone layer, 493 | sewage, 495 |
| ecology, 489 | particulate, 491 | soil erosion, 487 |
| famine, 499 | pesticide, 488 | solid waste, 500 |
| fossil fuel, 491 | photochemical smog, 490 | |

## Chapter 21 Summary

Have volunteers read aloud each Summary item on page 509. Ask volunteers to explain the meaning of each item. Direct students' attention to the Vocabulary box on the bottom of page 509. Have them read and review each term and its definition.

### ONLINE CONNECTION

The U.S. Environmental Protection Agency maintains a Web site that contains news on environmental issues, information about environmental programs, explanations of environments laws and regulations, and more: www.epa.gov

The Smithsonian Environmental Research Center's Web site provides information about scientific research on environmental protection: www.serc.si.edu/

## Chapter 21 Review

Use the Chapter Review to prepare students for tests and to reteach content from the chapter.

## Chapter 21 Mastery Test TRL

The Teacher's Resource Library includes two forms of the Chapter 21 Mastery Test. Each test addresses the chapter Goals for Learning. An optional third page of additional critical-thinking items is included for each test. The difficulty level of the two forms is equivalent.

## Review Answers

**Vocabulary Review**

1. radon 2. pollution 3. ecology
4. asbestos 5. ozone layer 6. pesticides
7. sewage 8. deforestation 9. particulates
10. hazardous waste 11. radiation sickness 12. fossil fuels 13. acid rain

### TEACHER ALERT

In the Chapter Review, the Vocabulary Review activity includes a sample of the chapter's vocabulary terms. The activity will help determine students' understanding of key vocabulary terms and concepts presented in the chapter. Other vocabulary terms used in the chapter are listed below.

algae
bronchitis
famine
fragile
greenhouse effect
landfill
photochemical smog
sensorineural deafness
soil erosion
solid waste

## Chapter 21 REVIEW

**Word Bank**
acid rain
asbestos
deforestation
ecology
fossil fuels
hazardous waste
ozone layer
particulates
pesticides
pollution
radiation sickness
radon
sewage

### Vocabulary Review

On a sheet of paper, write the word or phrase from the Word Bank that best completes each sentence.

1. A gas produced by radioactive radium is _____.
2. Wastes that cannot be broken down in nature make _____.
3. The study of how living things and the environment work together is _____.
4. Schools and other buildings no longer use _____ as a building material.
5. The _____ keeps the sun's rays from harming Earth.
6. Farmers use _____ to get rid of bugs.
7. Typhoid fever and cholera are diseases people get from water polluted with _____.
8. Cutting down and clearing away all the trees in an area is _____.
9. When you breathe in _____, they can harm your body.
10. Toxic materials dumped in landfills are _____.
11. Many people who lived near Chernobyl died from _____.
12. The remains of dead animals are used for _____.
13. Snow and sleet that contain harmful chemicals as they fall are examples of _____.

Chapter 21 Mastery Test A, pages 1–3

## Concept Review

On a sheet of paper, write the letter of the answer that best completes each sentence.

**14.** Earthquakes, volcanoes, and floods can upset the _____.

A ozone layer
B creation of oxygen
C chlorofluorocarbons
D balance of nature

**15.** A slow warming of Earth's atmosphere is called _____.

A acid rain
B the greenhouse effect
C radiation sickness
D chlorofluorocarbons

**16.** One cause of water pollution is _____.

A sewage
B famine
C wetlands
D groundwater

**17.** The destruction of tropical rain forests affects both human health and _____.

A landfills
B deforestation
C environmental health
D solid waste

## Critical Thinking

On a sheet of paper, write the answers to the following questions. Use complete sentences.

**18.** What can people do to lessen noise pollution?

**19.** What do you think is the most important action people can take to protect air quality?

**20.** What role do you think government should play in protecting the environment?

 **Test-Taking Tip** When taking a short-answer test, first answer the questions you know. Then go back to spend time on the questions about which you are less sure.

Environmental Health  Chapter 21  511

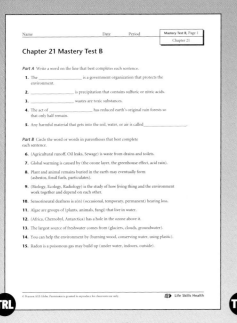

Chapter 21 Mastery Test B, pages 1–3

## Concept Review

**14.** D  **15.** B  **16.** A  **17.** C

## Critical Thinking

**18.** Sample answer: People can turn down the music in their cars and keep their cars in good repair so that they are less noisy. **19.** Sample answer: I think that the most important action people can take to protect air quality is to limit the use of cars by riding bikes or carpooling with friends. **20.** Sample answer: I believe government should pass and enforce laws that balance the need to use natural resources with the need to protect them for the future. I also believe that government officials should pass laws that encourage businesses to use solar energy and wind power.

## Alternative Assessment

Alternative Assessment items correlate to the student Goals for Learning at the beginning of this chapter.

- Have students orally explain how the environment can affect a person's health.
- Have students create a graphic organizer that identifies the causes of air and water pollution.
- Have students make a collage of images and words that describes other environmental problems.
- Have students work in groups to identify actions that protect the environment.

Environmental Health  Chapter 21  511

# HEALTH IN THE WORLD

Have students read the Health in the World feature on page 512. Invite them to describe experiences they might have had with aging relatives and acquaintances. How do the health challenges of aging people affect them and their families? *(Answers will vary but may include increasing costs of health care, increased reliance on family members for help, and increased stress for all.)* What kinds of resources are available to aging people and their families? *(Local communities might have visiting nurses, adult daycare options, and social workers.)* Ask students to explain why an aging population is a global public health issue. *(Communities need additional resources to cope with the needs of the aging; those resources might otherwise have been used to meet other needs.)*

### Health in the World Answers

1. Sample answer: The population of older people is increasing because birthrates are decreasing and people are living longer because of improvements in public health and medical technology.
2. Sample answer: People with disabilities might need someone to take care of them, or they might have to live in a long-term care facility. People with disabilities might need special medicine or equipment. All of these things can be expensive.
3. Sample answer: Young people who lower their risk of chronic diseases will stay healthier as they age and need fewer health care services. Young people who exercise regularly and avoid tobacco might not get some of the chronic diseases that affect people who use tobacco and who do not exercise. Medical treatment for chronic diseases related to cardiovascular problems, for example, can be very expensive.

# HEALTH IN THE WORLD

## Health Care and Aging

About 60 percent of people over age 65 live in developing countries such as Chile, China, and India. Worldwide, older people as a percentage of the total population is increasing. One reason for this increase is lower birthrates. Another reason is improvements in public health and medical technology, such as vaccinations. These improvements help people live longer. Between 2000 and 2025, the number of people age 60 and over will double to 1.2 billion.

Health care costs for older people are higher than for younger people. About 80 percent of Americans over age 65 have at least one chronic condition such as diabetes, heart disease, or cancer. People with chronic diseases make more doctor visits, take more prescription medicines, and get other medical treatments. In addition, people with chronic diseases sometimes have disabilities their disease causes.

Leaders of many countries are working to address the problem of increased health care costs as their populations age. The World Health Organization (WHO), for example, holds conferences on aging. Policies are being developed to help older people live a healthier lifestyle and improve access to health services.

In the United States, the Centers for Disease Control and Prevention (CDC) promotes healthy behaviors in people of all ages, including older people. These include getting regular exercise, keeping a healthy weight, staying involved with family and community, and not smoking. The WHO calls these behaviors lifestyle choices for active aging. People who practice healthy behaviors throughout their lives tend to have a lower risk of chronic diseases as they age.

On a sheet of paper, write the answers to the following questions. Use complete sentences.

1. What do you think is the main reason that the population of older people is increasing?
2. Why do you think disabilities in older people add to health care costs?
3. How do young people who reduce their risk of chronic disease help to reduce health care costs in the future?

# Unit 7 SUMMARY

- Criteria to help judge products and services are to look at the cost benefits, consider recommendations, watch for false or misleading claims, look at the ingredients, and compare prices.

- Health care professionals include primary care physicians, specialists, nurse practitioners, physical therapists, and registered dieticians.

- Qualifications to look for in a health care professional are board certificates, professional associations, and experience. The patient's comfort with the professional is also important.

- People can pay for health care costs through private individual or group insurance plans. Health maintenance organizations (HMOs) and preferred provider organizations (PPOs) offer less expensive health care.

- Government-supported health insurance includes Medicare for people over age 65, and Medicaid for people whose incomes are below a certain level. Another government program is the State Children's Health Insurance Program (SCHIP) for children of low-income parents.

- Individuals can help lower health care costs. They can choose low-cost outpatient facilities when possible. They can practice prevention, buy generic medicine, and advocate for family and community health. They can also take part in workplace health promotion programs.

- Groups of people who may have difficulty getting basic health services include poor people, elderly people, and homeless people. Teen parents and their children, and people with mental health problems are also included.

- Local and state public health departments make sure that laws relating to disease and sanitation are observed.

- Health promotion and disease prevention through education and advocacy groups reduce costs and relieve demands on the health care system.

- Burning is the main cause of outdoor air pollution. Cars, trucks, and factories make the most air pollution.

- Some results of air pollution are the greenhouse effect, acid rain, and thinning of the ozone layer.

- Some causes of water pollution are factory wastes, sewage, oil leaks and spills, and agricultural runoff.

- The growing world population has caused many environmental problems, including deforestation, solid waste, noise pollution, and light pollution.

- Nations and individuals can take many actions that help protect the environment. Recycling materials, conserving energy, and using water carefully can help.

# Unit 7 Review

Use the Unit Review to prepare students for tests and to reteach content from the unit.

## Unit 7 Mastery Test

The Teacher's Resource Library includes two forms of the Unit 7 Mastery Text. An optional third page of additional critical-thinking items is included for each test. The difficulty level of the two forms is equivalent.

## Review Answers

**Vocabulary Review**

1. Medicaid 2. consumer 3. famine
4. sanitation 5. acid rain 6. rehabilitation
7. public health 8. epidemic 9. landfill
10. greenhouse effect 11. advocacy group
12. primary care physician

**Concept Review**

13. C 14. D 15. B 16. B

---

# Unit 7 REVIEW

**Word Bank**
- acid rain
- advocacy group
- consumer
- epidemic
- famine
- greenhouse effect
- landfill
- Medicaid
- primary care physician
- public health
- rehabilitation
- sanitation

## Vocabulary Review

On a sheet of paper, write the word or phrase from the Word Bank that best completes each sentence.

1. Health insurance for people whose incomes are below a certain level is _____.

2. A person who pays for a service or buys a product is a(n) _____.

3. A period of _____ is a time when there is not enough food for people.

4. Restaurant owners who make sure that their kitchen equipment is cleaned properly are following laws related to _____.

5. Precipitation that contains large amounts of sulfuric or nitric acids is _____.

6. Therapy needed to recover from surgery, an illness, or an injury is _____.

7. Education programs, the use of preventive medicine, and the control of diseases are all parts of _____.

8. Before a vaccine for polio was developed, there was a(n) _____ in the United States because thousands of people had this disease.

9. A(n) _____ is an area where waste is buried between layers of dirt.

10. The slow warming of Earth's atmosphere is the _____.

11. A prevention organization that helps specific groups of people get health services is a(n) _____.

12. A doctor that a patient visits for a regular checkup is a(n) _____.

514   Unit 7   Health and Society

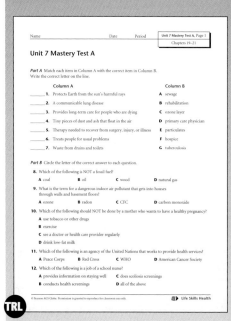
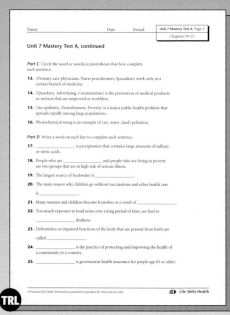

Unit 7 Mastery Test A, pages 1–3

## Concept Review

On a sheet of paper, write the letter of the answer that best completes each sentence.

13. If your health insurance plan has a $300 _____, you must pay the first $300 of your medical costs.
    A PPO   B hospice   C deductible   D premium

14. To have a healthy pregnancy, a mother should avoid _____.
    A fruits   B exercise   C cheese   D tobacco

15. One voluntary public health organization that has education programs is the _____.
    A Food and Drug Administration
    B American Heart Association
    C National Institutes of Health
    D Consumer Product Safety Commission

16. One of the problems with deforestation is that it can lead to _____.
    A oil leaks            C light pollution
    B soil erosion         D radiation sickness

## Critical Thinking

On a sheet of paper, write the answers to the following questions. Use complete sentences.

17. Why would a primary care physician tell a patient to visit a specialist? Describe a situation in which this might happen.

18. Why do you think the spread of tuberculosis might be a problem in a homeless shelter? Explain your answer.

19. How has the balance of nature been disturbed in your community? What were the results?

20. Do you think the problem of solid waste will be better or worse in the year 2050? Explain your answer.

Unit 7 Review   515

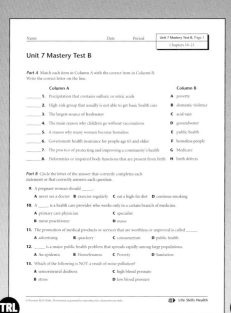

Unit 7 Mastery Test B, pages 1–3

## Critical Thinking

17. Sample answer: A primary care physician might tell a patient to visit a specialist because that specialist knows a great deal about a certain branch of medicine. For example, I might visit my primary care physician if I had soreness in my right eye. After examining me, my primary care physician might recommend that I visit an ophthalmologist because an eye doctor has expertise in treating specific medical conditions related to eyes and vision. 18. Sample answer: Tuberculosis is a communicable disease. It is spread through the air when someone who has TB coughs or sneezes. People in a homeless shelter might come into close contact with people who have this disease. Because some homeless people have no access to health care, they might not even realize that they have a communicable disease and are spreading it to other people. 19. Sample answer: Two years ago, there was a patch of trees near my friend's house. Birds, squirrels, and other animals lived in these trees. These trees produced oxygen for people to breathe. The balance of nature was disturbed when these trees were cut down to make room for a new apartment building. Two negative results of this disturbance in the balance of nature were that some animals died and some trees were destroyed. On the other hand, one of the positive results is that three families now have a new place to live. The owner of the apartment building has planted some new trees, and this could help restore the balance of nature. 20. Sample answer: If history is a guide, then the problem of solid waste will only get worse in the coming decades. As the world population increases, the amount of garbage produced also will increase. By the year 2050, many landfills will be completely full. In the future, people will have to find new ways of disposing of garbage. Perhaps scientists will invent inexpensive rockets that will carry huge amounts of garbage into outer space.

Health and Society   Unit 7   515

# Appendix A: Body Systems

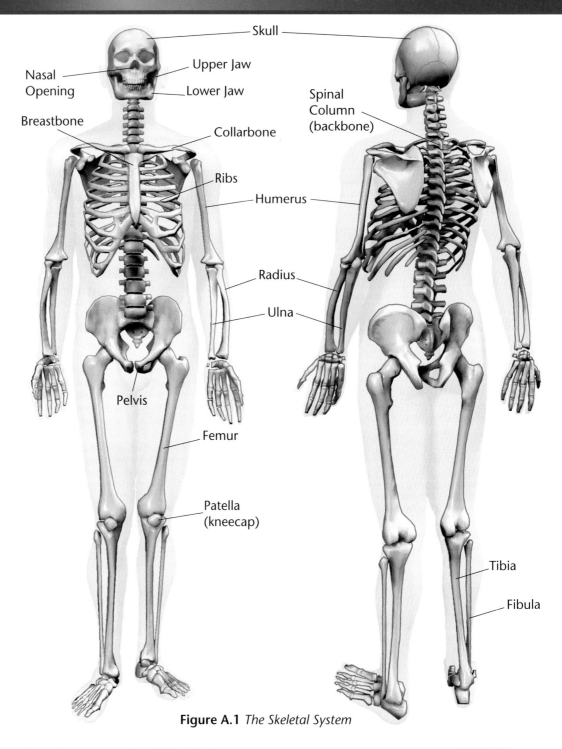

**Figure A.1** *The Skeletal System*

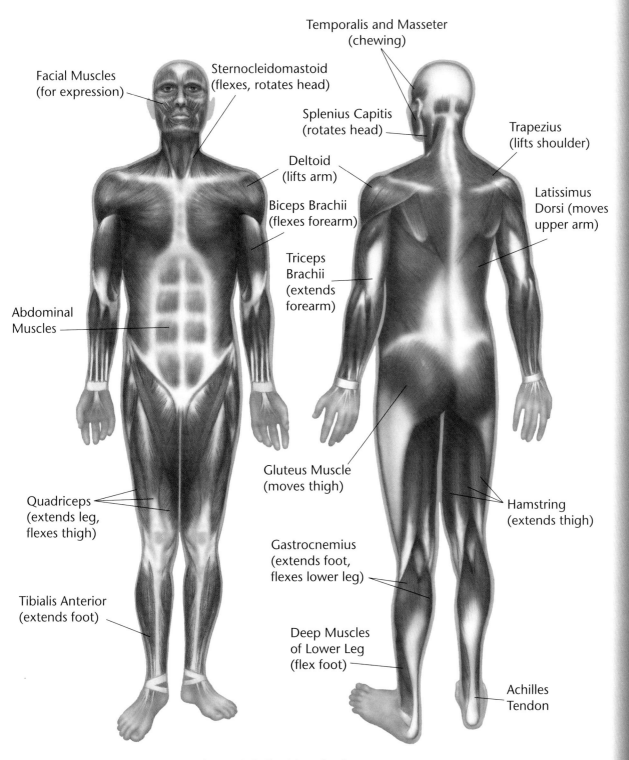

**Figure A.2** *The Muscular System*

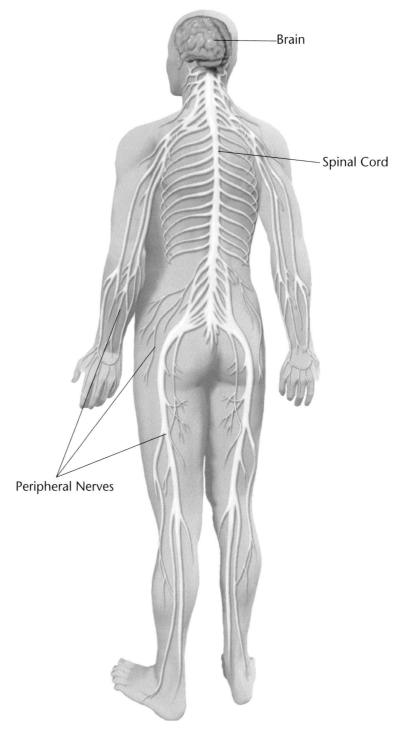

**Figure A.3** *The Nervous System*

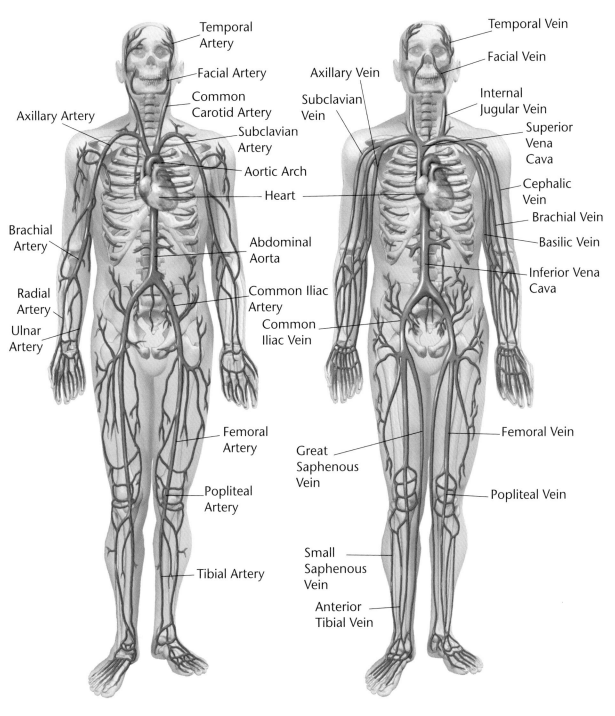

**Figure A.4** *The Circulatory System*

# Appendix B: Nutrition Tables

## Table B.1 The Six Essential Nutrient Classes

| Nutrient | Best Food Sources | Why They Are Needed |
|---|---|---|
| **Proteins** | | |
| | Cheese, eggs, fish, meat, milk, poultry, soybeans, nuts, dry beans, and lentils | To promote body growth; to repair and maintain tissue |
| **Carbohydrates** | | |
| | Bread, cereal, flour, potatoes, rice, sugar, dry beans, fruit | To supply energy; to furnish heat; to save proteins to build and regulate cells |
| **Fats** | | |
| | Butter, margarine, cream, oils, meat, whole milk, nuts, avocado | To supply energy; to furnish heat; to save proteins to build and regulate cells; to supply necessary fat-soluble vitamins and other nutrients |
| **Minerals** | | |
| Calcium | Milk, cheese, leafy green vegetables, oysters, almonds | To give rigidity and hardness to bones and teeth; for blood clotting, osmosis, action of heart and other muscles, and nerve response |
| Iron | Meats (especially liver), oysters, leafy green vegetables, legumes, dried apricots or peaches, prunes, raisins | To carry oxygen in the blood |
| Iodine | Seafood, iodized salt | To help the thyroid gland regulate cell activities for physical and mental health |
| **Vitamins** | | |
| Vitamin A | Whole milk, cream, butter, liver, egg yolk, leafy green vegetables, dark yellow fruits and vegetables | To promote health of epithelial tissues and eyes; to promote development of teeth |
| Thiamin | Present in many foods, abundant in few: pork, some animal organs, some nuts, whole grains, yeast, dry beans, and peas | To promote healthy nerves, appetite, digestion, and growth; for metabolism of carbohydrates |
| Riboflavin | Milk, glandular organs, lean meats, cheese, eggs, leafy green vegetables, whole grains | To promote better development, greater vitality, freedom from disease; to aid metabolism of carbohydrates, fats, and proteins |
| Niacin | Lean meats, liver, poultry, peanuts, legumes, yeasts | To promote good digestion, healthy skin, and a well-functioning nervous system |
| Vitamin C | Citrus fruits, strawberries, tomatoes, broccoli, cabbage, green peppers | To enhance iron absorption; for deposit of intercellular cement in tissues and bone |
| Vitamin D | Milk, salmon, tuna | To help absorb and use calcium and phosphorus |
| **Water** | | |
| | Drinking water, foods | To supply body fluids; to regulate body temperature |

## Table B.2 Fiber Content of Selected Foods

| Food | Serving Size | Dietary Fiber (g) |
|---|---|---|
| **Grains** | | |
| Bread, white | 1 slice | 0.6 |
| Bread, whole wheat | 1 slice | 1.5 |
| Oat bran, dry | ⅓ cup | 4.5 |
| Oatmeal, dry | ⅓ cup | 2.7 |
| Rice, brown, cooked | ½ cup | 2.4 |
| Rice, white, cooked | ½ cup | 0.8 |
| **Fruits** | | |
| Apple, with skin | 1 small | 2.8 |
| Apricots, with skin | 4 | 3.5 |
| Banana | 1 small | 2.2 |
| Blueberries | ¾ cup | 2.9 |
| Figs, dried | 3 | 5.3 |
| Grapefruit | ½ | 1.6 |
| Pear, with skin | 1 large | 5.8 |
| Prunes, dried | 3 medium | 1.7 |
| **Vegetables** | | |
| Asparagus, cooked | ½ cup | 1.8 |
| Broccoli, cooked | ½ cup | 2.4 |
| Carrots, cooked, sliced | ½ cup | 2.1 |
| Peas, green, frozen, cooked | ½ cup | 4.3 |
| Potato, with skin, raw | ½ cup | 1.5 |
| Tomato, raw | 1 medium | 1.0 |
| **Legumes** | | |
| Beans, white, cooked | ½ cup | 5.8 |
| Kidney beans, cooked | ½ cup | 6.9 |
| Lentils, cooked | ½ cup | 7.8 |
| Lima beans, canned | ½ cup | 6.6 |
| Peas, blackeye, canned | ½ cup | 5.6 |
| Pinto beans, cooked | ½ cup | 7.4 |

## Table B.3 Calcium and Fat Content of Dairy Products*

| Product | Calories | Fat (grams) | Calcium (milligrams) |
|---|---|---|---|
| **Skim milk (also called nonfat or fat free)** | | | |
| Plain | 86 | 0 | 301 |
| Chocolate | 144 | 1 | 292 |
| **1% milk (also called lowfat or light)** | | | |
| Plain | 102 | $2\frac{1}{2}$ | 300 |
| Chocolate | 158 | $2\frac{2}{3}$ | 288 |
| **2% milk (now called reduced fat, not lowfat)** | | | |
| Plain | 121 | 5 | 298 |
| Chocolate | 179 | 5 | 285 |
| **Whole milk** | | | |
| Plain | 150 | 8 | 290 |
| Chocolate | 209 | 8 | 280 |
| **Buttermilk (lowfat)** | 100 | 2 | 300 |
| **Buttermilk (whole)** | 150 | 8 | 300 |
| **Sweetened condensed milk** | 246 | 7 | 217 |

*All nutrition information is based on a 1-cup serving, except sweetened condensed milk, which is based on a quarter-cup serving.

# Appendix C: Fact Bank

## Frame Size

To determine your frame size:

1. Extend your arm in front of your body bending your elbow at a 90° angle to your body. (Your arm will be parallel to your body.)

2. Keep your fingers straight and turn the inside of your wrist to your body.

3. Place your thumb and index finger on the two prominent bones on either side of your elbow. Measure the distance between the bones with a tape measure or calipers.

4. Compare to the medium frame chart below. Select your height based on what you are in 1-inch heels. If you are below the listed inches, your frame is small. If you are above, your frame is large.

| Elbow Measurements for Medium Frame | | |
|---|---|---|
| **Sex** | **Height in 1-Inch Heels** | **Elbow Breadth** |
| **Male** | 5'2"–5'3" | $2\frac{1}{2}"$–$2\frac{7}{8}"$ |
| | 5'4"–5'7" | $2\frac{5}{8}"$–$2\frac{7}{8}"$ |
| | 5'8"–5'11" | $2\frac{3}{4}"$–$3"$ |
| | 6'0"–6'3" | $2\frac{3}{4}"$–$3\frac{1}{8}"$ |
| | 6'4" | $2\frac{7}{8}"$–$3\frac{1}{4}"$ |
| **Female** | 4'10"–4'11" | $2\frac{1}{4}"$–$2\frac{1}{2}"$ |
| | 5'0"–5'3" | $2\frac{1}{4}"$–$2\frac{1}{2}"$ |
| | 5'4"–5'7" | $2\frac{3}{8}"$–$2\frac{5}{8}"$ |
| | 5'8"–5'11" | $2\frac{3}{8}"$–$2\frac{5}{8}"$ |
| | 6'0" | $2\frac{1}{2}"$–$2\frac{3}{4}"$ |

### Table C.1 Top 10 Fat-Blasting Exercises

| Activity | Calories Burned (per 30 minutes) | Activity | Calories Burned (per 30 minutes) |
|---|---|---|---|
| Bicycling, vigorous (15 MPH) | 340 | Spinning class (indoor cycling) | 312 |
| Jogging (10- to 12-minute miles) | 340 to 272 | Jumping rope, slowly | 340 |
| Swimming, vigorous | 340 | Tennis, singles | 240 |
| Cross-country ski machine | 323 | Hiking, uphill | 238 |
| Inline skating | 238 | Walking, uphill (3.5 MPH) | 204 |

### Table C.2 Recommended Height and Weight for Women

| Height (feet/inches) | Small Frame | Medium Frame | Large Frame |
|---|---|---|---|
| 4'10" | 102–111 | 109–121 | 118–131 |
| 4'11" | 103–113 | 111–123 | 120–134 |
| 5'0" | 104–115 | 113–126 | 122–137 |
| 5'1" | 106–118 | 115–129 | 125–140 |
| 5'2" | 108–121 | 118–132 | 128–143 |
| 5'3" | 111–124 | 121–135 | 131–147 |
| 5'4" | 114–127 | 124–138 | 134–151 |
| 5'5" | 117–130 | 127–141 | 137–155 |
| 5'6" | 120–133 | 130–144 | 140–159 |
| 5'7" | 123–136 | 133–147 | 143–163 |
| 5'8" | 126–139 | 136–150 | 146–167 |
| 5'9" | 129–142 | 139–153 | 149–170 |
| 5'10" | 132–145 | 142–156 | 152–173 |
| 5'11" | 135–148 | 145–159 | 155–176 |
| 6'0" | 138–151 | 148–162 | 158–179 |

Weights at ages 25–59 based on lowest mortality. Weight in pounds according to frame (in indoor clothing weighing 3 lb; shoes with 1-inch heels)

### Table C.3 Recommended Height and Weight for Men

| Height (feet/inches) | Small Frame | Medium Frame | Large Frame |
|---|---|---|---|
| 5'2" | 128–134 | 131–141 | 138–150 |
| 5'3" | 130–136 | 133–143 | 140–153 |
| 5'4" | 132–138 | 135–145 | 142–156 |
| 5'5" | 134–140 | 137–148 | 144–160 |
| 5'6" | 136–142 | 139–151 | 146–164 |
| 5'7" | 138–145 | 142–154 | 149–168 |
| 5'8" | 140–148 | 145–157 | 152–172 |
| 5'9" | 142–151 | 148–160 | 155–176 |
| 5'10" | 144–154 | 151–163 | 158–180 |
| 5'11" | 146–157 | 154–166 | 161–184 |
| 6'0" | 149–160 | 157–170 | 164–188 |
| 6'1" | 152–164 | 160–174 | 168–192 |
| 6'2" | 155–168 | 162–178 | 172–197 |
| 6'3" | 158–172 | 167–182 | 176–202 |
| 6'4" | 162–176 | 171–187 | 181–207 |

Weights at ages 25–59 based on lowest mortality. Weight in pounds according to frame (in indoor clothing weighing 5 lb; shoes with 1-inch heels)

## The Five Leading Causes of Death in the United States by Age Group

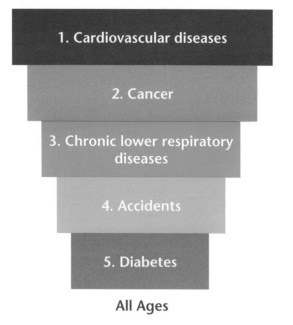

All Ages:
1. Cardiovascular diseases
2. Cancer
3. Chronic lower respiratory diseases
4. Accidents
5. Diabetes

Ages 15–24:
1. Accidents
2. Suicide
3. Cancer
4. Cardiovascular diseases
5. Chronic lower respiratory diseases

Source: American Heart Association, 2002.

| Table C.4 Estimated U.S. Cancer Cases for 2004 ||||
|---|---|---|---|
| Men | Percentage of 699,560 Total Cases | Women | Percentage of 668,470 Total Cases |
| Prostate | 33% | Breast | 32% |
| Lung and bronchus | 13% | Lung and bronchus | 12% |
| Colon and rectum | 11% | Colon and rectum | 11% |
| Urinary bladder | 6% | Uterus | 6% |
| Melanoma of skin | 4% | Ovary | 4% |
| Non-Hodgkin's lymphoma | 4% | Non-Hodgkin's lymphoma | 4% |
| Kidney | 3% | Melanoma of skin | 4% |
| Oral cavity | 3% | Thyroid | 3% |
| Leukemia | 3% | Pancreas | 2% |
| Pancreas | 2% | Urinary bladder | 2% |
| All other sites | 18% | All other sites | 20% |

Source: American Cancer Society, 2004.

# The ABCDs of Melanoma

Melanoma is usually curable if you find it early. Follow this ABCD self-examination guide adapted from the American Academy of Dermatology.

- **A** *is for asymmetry*—Symmetrical round or oval growths are usually benign. Look for irregular shapes where one half is a different shape than the other half.

- **B** *is for border*—Irregular, notched, scalloped, or vaguely defined borders need to be checked out.

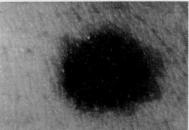

- **C** *is for color*—Look for growths that have many colors or an uneven distribution of color. Generally, growths that are the same color all over are usually benign.

- **D** *is for diameter*—Have your doctor check out any growths that are larger than 6 millimeters, about the diameter of a pencil eraser. The mole shown at the right measures about 1 millimeter.

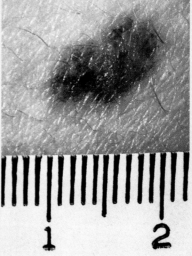

### Table C.5 Barriers to Helmet Use

| Barrier | Parents | Children | Physician |
|---|---|---|---|
| Lack of perceived need | X | X | X |
| Lack of physician's recommendation | X | | X |
| Do not think child will wear helmet | X | | |
| Cost of helmet | X | | |
| Peer pressure | | X | |
| Lack of "coolness" | | X | |
| Discomfort | | X | |

### Table C.6 State-of-the-Art Care for Five Common Conditions

| Condition | Old Recommendation | New Recommendation |
|---|---|---|
| Lower back pain | Bed rest, stretching exercises | Continuation of normal activities; ibuprofen or acetaminophen |
| Early stage breast cancer | Mastectomy | Lumpectomy plus radiation therapy |
| Ulcers | Drugs to suppress stomach acid secretion | Antibiotics plus drugs to suppress stomach acid secretion |
| Simple ovarian cysts | Removal by surgery | Depending on age and family history, watchful waiting with periodic pelvic and ultrasound exams |
| Type II diabetes | Insulin or sulfonylureas | Diet and exercise; if not effective, insulin often combined with new oral drugs that prompt insulin secretion |

## Table C.7 World Population and Food Supply Production

| Year | World Population (billions) | Wheat (millions of metric tons) | Rice (millions of metric tons) | Vegetables and melons (millions of metric tons) | Fruits (millions of metric tons) | Beef and pork (millions of metric tons) |
|---|---|---|---|---|---|---|
| 1990 | 5.28 | 588 | 352 | 462 | 353 | 117 |
| 1995 | 5.69 | 538 | 371 | 565 | 411 | 118 |
| 2000 | 6.08 | 582 | 398 | 746 | 472 | 132 |
| 2003 | 6.30 | 549 | 391 | 842 | 480 | 139 |

Sources: U.S. Census Bureau, *Total Midyear Population for the World, 1950–2050*; and *Statistical Abstract of the United States: 2004–2005*, page 861, Table No. 1355, World Food Production by Commodity: 1990 to 2003.

## Table C.8 Population Growth in Select Urban Areas

| Year | Tokyo, Japan (millions) | São Paolo, Brazil (millions) | Mexico City, Mexico (millions) | New York City, United States (millions) | Shanghai, China (millions) |
|---|---|---|---|---|---|
| 1975 | 19.8 | 10.3 | 10.7 | 15.9 | 11.4 |
| 2001 | 26.5 | 18.3 | 18.3 | 16.8 | 12.8 |
| 2015 (estimate) | 27.2 | 21.2 | 20.4 | 17.9 | 13.6 |

Source: United Nations Population Division, *World Urbanization Prospects: The 2001 Revision.*

# Glossary

## A

**Abnormal** (ab nôr´ məl) Unusual or different from normal (p. 76)

**Abnormality** (ab nôr mal´ ə tē) A condition that is not usual (p. 286)

**Abstain** (ab stān´) Choose not to use or do something (p. 15)

**Abstinence** (ab´ stə nəns) The choice not to have a sexual relationship (p. 178)

**Abuse** (ə byüs´) Physical or emotional mistreatment; actions that harm someone (p. 189)

**Acid rain** (as´ id rān) Rain, snow, sleet, or hail that contains large amounts of sulfuric or nitric acids (p. 494)

**Acne** (ak´ nē) Clogged skin pores that cause pimples and blackheads (p. 130)

**Acquire** (ə kwīr´) To get after being born (p. 254)

**Acquired immunodeficiency syndrome (AIDS)** (ə kwīrd´ i myü nō di fish´ ən sē sin´ drōm) A disorder of the immune system (p. 270)

**Added sugar** (ad´ id shug´ ər) Sugar that is added to a food (p. 211)

**Addicted** (ə dik´ tid) Unable to stop using alcohol or another drug (p. 80)

**Additive** (ad´ ə tiv) A substance added in small amounts to foods that changes them in some way (p. 241)

**Adolescence** (ad l es´ ns) The years between ages 13 and 19 (p. 151)

**Adoptive family** (ə dop´ tiv fam´ ə lē) Family that includes parents and an adopted child or children (p. 185)

**Adrenal gland** (ə drē´ nl gland) Part of the endocrine system that releases several hormones (p. 111)

**Adrenaline** (ə dren´ l ən) Hormone released into the bloodstream that causes the heart to beat faster and blood pressure to rise in an emergency (p. 111)

**Advertise** (ad´ vər tīz) Communicate to draw attention to a product (p. 233)

**Advocacy group** (ad´ və kə sē grüp) A prevention organization that helps specific groups get health services (p. 479)

**Aerobic exercise** (âr ō´ bik ek´ sər sīz) Activity that raises your heart rate (p. 135)

**Affective disorder** (ə fek´ tiv dis ôr´ dər) A mental problem in which a person has disturbed or uncontrolled emotions (p. 81)

**Aggression** (ə gresh´ ən) Any action that intends harm to someone (p. 39)

**Alcohol** (al´ kə hȯl) A drug that slows down the central nervous system (p. 8)

**Alcoholism** (al´ kə hȯ liz əm) A dependence on alcohol and lack of control in drinking (p. 330)

**Algae** (al´ jē) A large group of plants that live in water (p. 496)

**Allergy** (al´ ər jē) A bad reaction of the body to a food or to something in the air (p. 129)

**Alveolus** (al vē´ ə ləs) Air sac with thin walls (plural is *alveoli*) (p. 115)

**Amino acid** (ə mē´ nō as´ id) A small chemical unit that makes up protein (p. 214)

**Amphetamine** (am fet´ ə mēn) A central nervous system stimulant (p. 333)

**Anabolic steroid** (an ə bol´ ik stâr´ oid) A human-made drug that is like the natural male sex hormone, testosterone (p. 340)

**Anaerobic exercise** (an ə rō´ bik ek´ sər sīz) Activity that quickly uses up oxygen in your body (p. 135)

**Analgesic** (an l jē´ zik) A medicine that eases or gets rid of pain (p. 319)

**Anger** (ang´ gər) Strong feeling of displeasure (p. 39)

**Angina pectoris** (an jī´ nə pek´ tər əs) Severe pain that happens when the heart does not receive enough blood (p. 289)

**Antibiotic** (an ti bī ot´ ik) Substance that can destroy the growth of certain germs (p. 256)

**Antibody** (an´ ti bod ē) A protein that kills a particular pathogen (p. 260)

**Antiperspirant** (an ti pėr´ spər ənt) A product that helps control perspiration by closing the pores (p. 130)

**Anus** (ā´ nəs) Opening through which solid waste leaves the body (p. 118)

**Anxiety** (ang zī´ ə tē) A feeling of uneasiness or fearful concern (p. 38)

**Anxiety disorder** (ang zī′ ə tē dis ôr′ dər) A mental problem that makes normal life difficult because of intense anxiety (p. 80)

**Arterial** (är tir′ ē əl) Relating to the arteries (p. 286)

**Arteriosclerosis** (är tir ē ō sklə rō′ sis) Hardening of the arteries because of a buildup of fat (p. 287)

**Artery** (är′ tər ē) Vessel that carries blood away from the heart (p. 113)

**Arthritis** (är thrī′ tis) A group of diseases that causes swelling of the joints (p. 302)

**Asbestos** (as bes′ təs) A material used for building materials (p. 492)

**Association** (ə sō sē ā′ shən) Something that reminds a person of something else (p. 231)

**Asthma** (az′ mə) A respiratory disease that makes breathing difficult (p. 304)

**Atherosclerosis** (ath ər ō sklər ō′ sis) A form of arteriosclerosis in which the large and medium arteries become narrow from a fat buildup along their walls (p. 287)

**Athlete's foot** (ath′ lēts fút) Itching or cracked skin between the toes (p. 131)

**Attachment** (ə tach′ mənt) Emotional bond or tie to others (p. 30)

**Auditory nerve** (ȯ də tôr ē nėrv) Part of the ear that sends sound information to the brain (p. 108)

## B

**Bacteria** (bak tir′ ē ə) Germs that cause disease (p. 121)

**Barbiturate** (bär bich′ ər it) A category of sedative-hypnotic drugs (p. 335)

**Barrier** (bar′ ē ər) Something that blocks passage (p. 259)

**Behavior modification** (bi hā′ vyər mod ə fə kā′ shən) Form of psychotherapy that teaches a person to replace a less effective behavior pattern with a more effective one (p. 85)

**Benign** (bi nīn′) Not harmful to health (p. 292)

**Bile** (bīl) Liquid in the liver that breaks down fats (p. 117)

**Binge drink** (binj dringk) To drink alcohol heavily, then stop for a time, then drink heavily again (p. 330)

**Bipolar disorder** (bī pō′ lər dis ôr′ dər) An affective disorder involving an uncontrolled shift from feeling too much energetic emotion to feeling very depressed (p. 81)

**Birth defect** (bėrth di fekt′) A deformity or an impaired function of the body that is present from birth (p. 472)

**Blended family** (blen′ dəd fam′ ə lē) Family made when parents live with children from an earlier marriage or marriages (p. 184)

**Blizzard** (bliz′ ərd) A violent snowstorm with winds above 35 miles per hour that lasts for at least 3 hours (p. 388)

**Blood cholesterol** (blud kə les′ tə rol) Cholesterol in the bloodstream (p. 213)

**Blood pressure** (blud presh′ ər) Force of blood on the walls of blood vessels (p. 113)

**Body language** (bod′ ē lang′ gwij) The way people move and hold their heads, arms, and legs that gives clues about what they think and feel (p. 28)

**Brain stem** (brān stem) The part of the brain that connects the cerebrum to the spinal cord (p. 105)

**Bronchiole** (brong′ kē ōl) A tube that branches off the bronchus (p. 114)

**Bronchitis** (brong kī′ tis) A lung disease that makes breathing hard and causes coughing (p. 490)

**Bronchus** (brong′ kəs) Breathing tube (plural is *bronchi*) (p. 114)

## C

**Calcium** (kal′ sē əm) Mineral important for keeping bones and teeth strong (p. 217)

**Calorie** (kal′ ər ē) A unit that measures the amount of energy in food (p. 135)

**Cancer** (kan′ sər) A harmful growth that destroys healthy cells (p. 169)

**Capillary** (kap′ ə ler ē) Tiny blood vessel that connects arteries and veins (p. 113)

**Carbohydrate** (kär bō hī′ drāt) Chemical in foods that provides starches and sugars (p. 210)

| a | hat | e | let | ī | ice | ȯ | order | u̇ | put | sh | she | ə { | a in about |
|---|---|---|---|---|---|---|---|---|---|---|---|---|---|
| ā | age | ē | equal | o | hot | oi | oil | ü | rule | th | thin | | e in taken |
| ä | far | ėr | term | ō | open | ou | out | ch | child | ᴛʜ | then | | i in pencil |
| â | care | i | it | ȯ | saw | u | cup | ng | long | zh | measure | | o in lemon |
| | | | | | | | | | | | | | u in circus |

**Carbon dioxide** (kär′ bən dī ok′ sīd) A gas that is breathed out of the lungs (p. 112)

**Carcinogen** (kär sin′ ə jən) Material that causes cancer (p. 295)

**Cardiac arrest** (kär′ dē ak ə rest′) A condition in which the heart has stopped beating and there is no pulse (p. 404)

**Cardiopulmonary resuscitation (CPR)** (kär dē ō pul′ mə ner ē ri sus ə tā′ shən) A procedure for cardiovascular failure (p. 404)

**Cardiovascular** (kär dē ō vas′ kyə lər) Relating to the heart and blood vessels (p. 286)

**Cartilage** (kär′ tl ij) A cushion, or padding, in the joints (p. 302)

**Cerebellum** (ser ə bel′ əm) The part of the brain that controls balance and helps coordinate muscular activities (p. 105)

**Cerebrum** (sə rē′ brəm) The part of the brain that lets a person read, think, and remember (p. 104)

**Cervix** (sėr′ viks) Narrow outer end of the uterus (p. 165)

**Cesarean childbirth** (si zâr′ ē ən chīld′ bėrth) The delivery of a baby through a cut in the abdomen and uterus (p. 165)

**Chancre** (shang′ kər) A hard sore that is the first sign of syphilis (p. 276)

**Chemotherapy** (kē mō ther′ ə pē) A treatment that involves taking drugs to kill cancer cells (p. 295)

**Chest cavity** (chest kav′ ə tē) Part of the body that includes ribs and muscles that surround the heart and lungs (p. 115)

**Chest compression** (chest kəm presh′ ən) When doing CPR, the act of pressing against the chest over and over (p. 404)

**Chlamydia** (klə mid′ ē ə) A sexually transmitted disease (p. 276)

**Cholesterol** (kə les′ tə rol) A waxy, fatlike material in your body cells that helps with certain body functions (p. 213)

**Chromosome** (krō′ mə sōm) Tiny structure in cells that contains hereditary information (p. 167)

**Chronic** (kron′ ik) Lasting for a long time (p. 277)

**Clinical depression** (klin′ ə kəl di presh′ ən) An affective disorder involving long-lasting, intense sadness (p. 81)

**Clotting** (klot′ ing) Sticking together to form a plug and stop a wound from bleeding (p. 112)

**Cocaine** (kō kān′) A highly addictive illegal stimulant (p. 333)

**Cognitive therapy** (kog′ nə tiv ther′ ə pē) Form of psychotherapy that teaches a person to replace destructive thoughts with positive ones (p. 85)

**Collaborative** (kə lab′ ə rə tiv) Working together as a team (p. 180)

**Coma** (kō′ mə) A condition in which a person loses consciousness (p. 298)

**Commitment** (kə mit′ mənt) A promise (p. 178)

**Communicable** (kə myü′ nə kə bəl) Infectious (p. 254)

**Complex carbohydrate** (kom′ pleks kär bō hī′ drāt) Carbohydrate that contains fiber; the best source of dietary carbohydrates (p. 210)

**Compromise** (kom′ prə mīz) Agree by giving in a little (p. 57)

**Conceive** (kən sēv′) To become pregnant (p. 157)

**Conflict** (kon′ flikt) A difference of opinion between two or more people who have different ideas (p. 419)

**Consequence** (kon′ sə kwens) Something that happens as a result of something else (p. 183)

**Consumer** (kən sü′ mər) Someone who buys goods or services (p. 446)

**Contact poisoning** (kon′ takt poi′ zn ing) Poisoning that happens when harmful materials are absorbed through the skin (p. 408)

**Contract** (kən trakt′) Shorten (p. 101)

**Cope** (kōp) To deal with or overcome problems and difficulties (p. 41)

**Cornea** (kôr′ nē ə) Part of the eye that light passes through (p. 106)

**Crack** (krak) A form of cocaine that is smoked (p. 334)

**Culture** (kul′ chər) Beliefs, way of living, and language of a group of people (p. 13)

**Custodial parent** (ku stō′ dē əl pâr′ ənt) The parent who is responsible for a child after a divorce (p. 185)

**Custody** (kus′ tə dē) The legal right and responsibility to care for a child (p. 185)

## D

**Daily Values (DV)** (dā′ lē val′ yüz) Section of a nutrition label that provides information about the percentage of nutrients in a product (p. 236)

**Dandruff** (dan′ drəf) Flaking from the scalp (p. 132)

**Date rape** (dāt rāp) Rape committed by someone the victim socializes with (p. 427)

**Deductible** (di duk′ tə bəl) The first part of each year's medical expenses before the insurance company makes payments (p. 457)

**Defense mechanism** (di fens′ mek′ ə niz əm) A mental device one uses to protect oneself (p. 77)

**Deficiency** (di fish′ ən sē) A lack of something (p. 223)

**Deforestation** (dē fôr əs tā′ shən) Cutting down and clearing trees (p. 499)

**Deformity** (di fôr mə tē) Abnormally formed body structure (p. 257)

**Delusion** (di lü′ zhən) A false belief (p. 82)

**Denial** (di nī′ əl) A refusal to believe or accept something negative or threatening (p. 77)

**Deodorant** (dē ō′ dər ənt) A product that covers body odor (p. 130)

**Dependent** (di pen′ dənt) Having developed a need for a medicine or drug (p. 321)

**Depressant** (di pres′ nt) A drug that slows down the central nervous system (p. 328)

**Depression** (di presh′ ən) A state of deep sadness (p. 40)

**Dermis** (dėr′ mis) Inner layer of skin made up of living cells (p. 120)

**Designer drug** (di zī′ nər drug) A drug with a slightly different chemical makeup from a legal drug (p. 339)

**Deteriorate** (di tir′ ē ə rāt) Wear away a little bit at a time (p. 15)

**Detoxification** (di tok sə fə kā′ shən) Removing addictive drugs from the body (p. 352)

**Diabetes** (dī ə bē′ tis) A disease caused by lack of insulin in the body or when the body cannot use insulin effectively (p. 298)

**Diaphragm** (dī′ ə fram) Band of muscle tissue beneath the respiratory organs (p. 115)

**Dietary cholesterol** (dī′ ə ter ē kə les′ tə rol) Cholesterol found in some foods (p. 213)

**Digestion** (dī jes′ chən) Breaking down food into smaller parts and changing it to a form that cells can use (p. 116)

**Disadvantage** (dis əd van′ tij) A condition or situation that makes reaching a goal harder (p. 58)

**Discipline** (dis′ ə plin) A fair and safe way to teach a child how to behave correctly (p. 183)

**Discriminate** (dis krim′ ə nāt) Treat differently on the basis of something other than individual worth (p. 67)

**Disease** (də zēz′) A condition that harms or stops the normal working of the body (p. 7)

**Displacement** (dis plās′ mənt) Shifting an emotion from its real object to a safer or more immediate one (p. 77)

**Divorce** (də vôrs′) The legal end of a marriage (p. 180)

**Domestic violence** (də mes′ tik vī′ ə ləns) Violence that happens mostly at home (p. 190)

**Dominant** (dom′ ə nənt) Having the most control (p. 168)

**Dressing** (dres′ ing) Layers of cloth and other materials that protect a wound (p. 398)

**Drug** (drug) A chemical substance other than food that changes the way the mind and body work (p. 8)

**Dysfunctional** (dis fungk′ shən əl) Harmed or unhealthy (p. 77)

### E

**Eardrum** (ir′ drum) Thin piece of tissue stretched across the ear canal (p. 108)

**Earthquake** (ėrth′ kwāk) The shaking caused when the major slabs of rock in the earth's crust shift and move (p. 387)

**Eating disorder** (ēt′ ing dis ôr′ dər) An attempt to cope with psychological problems through eating habits (p. 83)

**Ecology** (ē kol′ ə jē) The study of how living things and the environment work together (p. 489)

**Ecstasy** (ek′ stə sē) An illegal, psychoactive drug (p. 336)

**Electrical shock** (i lek′ trə kəl shok) A flow of electric current through the body (p. 373)

**Embryo** (em′ brē ō) Fertilized egg after implantation (p. 163)

**Emotion** (i mō′ shən) Feeling (p. 28)

**Empathy** (em′ pə thē) Recognition and understanding of other people's feelings (p. 68)

**Emphysema** (em fə sē′ mə) A serious respiratory disease that makes breathing difficult (p. 324)

| a | hat | e | let | ī | ice | ô | order | u̇ | put | sh | she | | ə { | a in about |
|---|---|---|---|---|---|---|---|---|---|---|---|---|---|---|
| ā | age | ē | equal | o | hot | oi | oil | ü | rule | th | thin | | | e in taken |
| ä | far | ėr | term | ō | open | ou | out | ch | child | ᴛʜ | then | | | i in pencil |
| â | care | i | it | ȯ | saw | u | cup | ng | long | zh | measure | | | o in lemon |
| | | | | | | | | | | | | | | u in circus |

**Enabling** (en ā′ bling) Actions of a friend or family member that help a person continue to use drugs (p. 349)

**Endurance** (en dùr′ əns) The ability to stay with an activity for a long time (p. 134)

**Enzyme** (en′ zīm) Special chemical that breaks down food (p. 116)

**Epidemic** (ep ə dem′ ik) The rapid spread of a communicable disease that affects large populations at the same time (p. 468)

**Epidermis** (ep ə dėr′ mis) Outer layer of skin (p. 120)

**Epilepsy** (ep′ ə lep sē) A chronic disease that is caused by disordered brain activity and that causes seizures (p. 303)

**Erection** (i rek′ shən) Condition in which the penis becomes hard and larger (p. 159)

**Esophagus** (i sof′ ə gəs) Long tube that connects the mouth and the stomach (p. 117)

**Essential amino acid** (ə sen′ shəl ə mē′ nō as′ id) An amino acid the body cannot make by itself but that can be obtained from food (p. 214)

**Essential nutrient** (ə sen′ shəl nü′ trē ənt) Chemical in foods that the body cannot make (p. 206)

**Esteem** (e stēm′) Value or worth; how one sees oneself or others (p. 29)

**Estrogen** (es′ trə jən) Female sex hormone (p. 155)

**Excretory system** (ek′ skrə tôr ē sis′ təm) System that allows the body to eliminate liquid and solid waste (p. 118)

**Exhale** (els hāl′) Breathe out (p. 115)

**Extended family** (ek sten′ dəd fam′ ə lē) Family that includes parents, children, and other relatives (p. 185)

## F

**Fainting** (fān′ ting) Short-term loss of consciousness caused by too little blood to the brain (p. 408)

**Fallopian tube** (fə lō′ pē ən tüb) Passage through which mature egg cells pass from the ovaries to the uterus (p. 157)

**Family love** (fam′ ə lē luv) Love based on attachment and support (p. 30)

**Famine** (fam′ ən) A time when there is not enough food for people (p. 499)

**Fat** (fat) Stored energy (p. 210)

**Fatty acid** (fat′ ē as′ id) Acid that includes saturated and unsaturated fats (p. 212)

**Fear** (fir) A strong feeling of fright; awareness of danger (p. 38)

**Feces** (fē′ sēz) Solid waste material remaining in the large intestine after digestion (p. 118)

**Fertilize** (fėr′ tl īz) An egg cell and a sperm join together (p. 159)

**Fetal alcohol syndrome** (fē′ tl al′ kə hòl sin′ drōm) A condition caused by drinking alcohol during pregnancy (p. 161)

**Fetus** (fē′ təs) Unborn baby from 8 weeks after fertilization until birth (p. 163)

**Fiber** (fi′ bər) The coarse material in fruits, vegetables, and grains that helps you digest your food (p. 211)

**Fire extinguisher** (fir ek sting′ gwish ər) A portable device containing chemicals that put out small fires (p. 371)

**First aid** (fėrst ād) Immediate emergency care given to a sick or injured person before professional medical care arrives (p. 396)

**Flashback** (flash′ bak) Experiencing the effects of a drug without using it again (p. 336)

**Flash flood** (flash flud) A dangerous flood that occurs suddenly without warning because of heavy rains (p. 387)

**Flexibility** (flek sə bil′ ə tē) The ability to twist, turn, bend, and stretch easily (p. 134)

**Flood** (flud) A natural disaster that occurs when a body of water overflows and covers normally dry land (p. 387)

**Food Guide Pyramid** (füd gīd pir′ ə mid) A chart that helps you decide how much and what kinds of foods to eat (p. 204)

**Fossil fuel** (fos′ əl fyü′ əl) A material that is made from plant or animal remains that can be burned (p. 491)

**Foster family** (fo′ stėr fam′ ə lē) Family that cares for children who are in need of short-term parenting from people other than their birth parents (p. 185)

**Fracture** (frak′ chər) A broken bone (p. 409)

**Fragile** (fraj′ əl) Easily broken, destroyed, or damaged (p. 488)

**Friendship** (frend′ ship) Love based on choices (p. 30)

**Frostbite** (fròst′ bīt) A tissue injury caused by being in extreme cold (p. 411)

**Fructose** (fruk′ tōs) The natural sugar found in fruit (p. 211)

**Frustrate** (frus′ trāt) To block from something wanted (p. 13)

**Fume** (fyüm) Gas (p. 407)

**Fungus** (fung´ gəs) An organism that grows in damp places (p. 131)

## G

**Gallbladder** (gȯl blad´ ər) Digestive organ attached to the liver that stores bile (p. 117)

**Gender** (jen´ dər) The condition of being male or female; sex (p. 167)

**Gene** (jēn) A tiny structure in chromosomes that controls the transfer of traits from parents to children (p. 167)

**Genetic counselor** (jə net´ ik koun´ sə lər) Health care professional who helps people determine the chance of passing inherited disorders to their children (p. 168)

**Genetic disorder** (jə net´ ik dis ȯr´ dər) A disease that is caused by abnormal genes or chromosomes (p. 168)

**Genetics** (jə net´ iks) The science that deals with heredity and inherited characteristics (p. 167)

**Genital herpes** (jen´ ə təl hėr´ pēz) A sexually transmitted disease (p. 277)

**Genitals** (jen´ ə təlz) Reproductive, or sex, organs (p. 154)

**Genital warts** (jen´ ə təl wȯrts) Bumps caused by a virus spread through sexual contact (p. 278)

**Gestation** (je stā´ shən) The period of development in the uterus from the time the egg is fertilized until birth; pregnancy (p. 160)

**Gingivitis** (jin jə vī´ tis) First stage of gum disease (p. 132)

**Gland** (gland) A group of cells or an organ that produces a substance and sends it into the body (p. 110)

**Glucose** (glü´ kōs) The main sugar in blood; the major energy source for your body (p. 212)

**Glycogen** (glī´ kə jən) Glucose that is stored in your body (p. 212)

**Gonorrhea** (gon ə rē´ ə) A sexually transmitted disease (p. 275)

**Good Samaritan laws** (gu̇d sə mâr´ ə tən lȯz) Laws that protect people who help others in emergencies (p. 398)

**Government regulation** (guv´ ərn mənt reg yə lā´ shən) Rule that businesses must follow to protect public health (p. 476)

**Grand mal seizure** (grän mal´ sē´ zhər) A type of epileptic seizure in which a person passes out (p. 303)

**Greenhouse effect** (grēn´ hous ə fekt´) The warming of Earth's atmosphere (p. 493)

**Grief** (grēf) A feeling of sorrow; feelings after a loss (p. 40)

**Guilt** (gilt) A feeling of having done something wrong (p. 33)

## H

**Halfway house** (haf´ wā hous) A place for recovering individuals to get help easing back into society (p. 354)

**Hallucination** (hə lü sn ā´ shən) A twisted idea about an unreal object or event (p. 82)

**Hallucinogen** (hə lü´ sn ə jən) A psychoactive drug that confuses the central nervous system and makes a person see, hear, feel, and experience things that are not real (p. 336)

**Hazard** (haz´ ərd) Something that is dangerous (p. 374)

**Hazardous waste** (haz´ ər dəs wāst) A material that can be harmful to the environment, human health, and other living things (p. 501)

**Health** (helth) A state of being well that has three parts physical, social, and emotional (p. 4)

**Health care facility** (helth kâr fə sil´ ə tē) A place where people go for medical, dental, and other care (p. 454)

**Health insurance** (helth in shu̇r´ əns) A plan that pays a major part of medical costs (p. 457)

**Health literacy** (helth lit´ ər ə sē) Gathering, understanding, and using information to improve health results (p. 18)

**Health maintenance organization (HMO)** (helth mān´ tə nəns ȯr gə nə zā´ shən) A form of managed care in which an organization controls how health care is delivered and its cost (p. 458)

| a | hat | e | let | ī | ice | ȯ | order | u̇ | put | sh | she | | a | in about |
|---|---|---|---|---|---|---|---|---|---|---|---|---|---|---|
| ā | age | ē | equal | o | hot | oi | oil | ü | rule | th | thin | ə | e | in taken |
| ä | far | ėr | term | ō | open | ou | out | ch | child | ᴛʜ | then | | i | in pencil |
| â | care | i | it | ȯ | saw | u | cup | ng | long | zh | measure | | o | in lemon |
| | | | | | | | | | | | | | u | in circus |

**Health risk** (helth risk) An action or a condition that contributes to injury or disease (p. 11)

**Heart attack** (härt ə tak´) A condition in which the blood supply and the nutrients to the heart are greatly reduced (p. 287)

**Heat exhaustion** (hēt´ eg zȯs´ chən) A condition that comes from hard physical work in a very hot place (p. 411)

**Heatstroke** (hēt´ strōk) A condition that comes from being in high heat too long (p. 411)

**Heimlich maneuver** (hīm´ lik mə nü´ vər) A way of giving firm upward thrusts below a person's rib cage to force out an object from the throat (p. 401)

**Heredity** (hə red´ ə tē) Passage of physical characteristics from parents to children (p. 11)

**Heroin** (her´ ō ən) An illegal narcotic made from morphine (p. 336)

**Hierarchy** (hī´ ə rär kē) Order from most to least important (p. 29)

**Hodgkin's disease** (hoj´ kinz də zēz´) A lymph gland cancer that often affects young adults (p. 295)

**Hormone** (hôr´ mōn) A chemical messenger that helps control how body parts do their jobs (p. 110)

**Hospice** (hos´ pis) A long-term care facility for people who are dying (p. 455)

**Human immunodeficiency virus (HIV)** (hyü´ mən i myü nō di fish´ ən sē vī´ rəs) A virus that infects body cells and severely damages the human immune system, causing AIDS (p. 270)

**Hurricane** (hėr´ ə kān) A severe tropical storm with heavy rains and winds above 74 miles per hour (p. 385)

**Hygiene** (hī´ jēn) Practices that promote cleanliness and good health (p. 128)

**Hypertension** (hī pər ten´ shən) High blood pressure (p. 286)

**Hypothermia** (hī pō thėr´ mē ə) A serious loss of body heat that comes from being in the cold too long (p. 411)

## I

**Immune system** (i myün´ sis´ təm) Combination of body defenses that fight pathogens (p. 255)

**Immunization** (im yə nī zā´ shən) The process of making the body immune to a disease (p. 261)

**Impair** (im pâr´) Have a negative effect on (p. 15)

**Implantation** (im plan tā´ shən) Process during which the fertilized egg plants itself in the lining of the uterus (p. 162)

**Impulse** (im´ puls) The urge to react before taking time to think (p. 57)

**Incubation period** (ing kyə bā´ shən pir´ ē əd) Time it takes after a person is infected for symptoms of a disease to appear (p. 225)

**Infection** (in fek´ shən) A sickness caused by a germ in the body (p. 129)

**Infectious** (in fek´ shəs) Passed from a germ carrier to another living thing (p. 254)

**Inferior** (in fir´ ē ər) Not as good as others (p. 151)

**Inflammation** (in flə mā´ shən) Redness, soreness, and swelling after tissue damage (p. 260)

**Influenza** (in flü en´ zə) A common illness known as the flu (p. 254)

**Inhalant** (in hā´ lənt) A chemical that is breathed in (p. 339)

**Inhalation poisoning** (in hə lā´ shən poi´ zn ing) Poisoning caused by breathing in harmful fumes (p. 407)

**Inhale** (in hāl´) Breathe in (p. 115)

**Inherit** (in her´ it) To receive something passed down from parents (p. 254)

**Inpatient** (in´ pā shənt) Referring to someone receiving health care who stays overnight or longer (p. 455)

**Insulin** (in´ sə lən) Hormone produced in the pancreas that helps cells use sugar (p. 117)

**Intimate** (in´ tə mit) Very personal or private (p. 61)

**Intoxication** (in tok sə kā´ shən) Excitement or stimulation caused by use of a chemical material (p. 330)

**Iris** (ī´ ris) Colored part of the eye around the pupil (p. 107)

**Isokinetic exercise** (ī sə ki net´ ik ek´ sər sīz) Activity that builds muscle strength by using resistance (p. 135)

**Isometric exercise** (ī sə met´ rik ek´ sər sīz) Activity that builds muscle strength by using tension (p. 135)

**Isotonic exercise** (ī sə ton´ ik ek´ sər sīz) Activity that builds muscle strength by using weights and repeated movements (p. 135)

## K

**Kaposi's sarcoma** (kä´ pə sēz sär kō´ mə) A rare type of cancer (p. 272)

**Keratin** (ker´ ə tin) A type of protein that makes nails hard (p. 121)

**Kidney** (kid´ nē) Organ in the excretory system where urine forms (p. 118)

## L

**Lactose** (lak´ tōs) The natural sugar found in milk (p. 211)

**Landfill** (land´ fil) Low-lying ground where waste is buried between layers of dirt (p. 501)

**Large intestine** (lärj  in tes´ tən) Tube connected to the small intestine (p. 117)

**Latex** (lā teks) A kind of rubber used to make thin gloves for health care professionals (p. 397)

**Lens** (lenz) The soft, clear tissue behind the pupil (p. 107)

**Ligament** (lig´ ə mənt) Tough band of stretchy tissue that holds joints together or keeps organs in place (p. 101)

**Lightning** (līt´ ning) An electrical discharge that commonly occurs during thunderstorms (p. 389)

**Liver** (liv´ ər) A large organ that produces bile (p. 117)

**Look-alike drug** (lùk´ ə līk  drug) A drug made from legal chemicals to look like a common illegal drug (p. 339)

**LSD** (el es dē) A human-made hallucinogen (p. 337)

**Lymph** (limf) A colorless liquid that feeds tissue (p. 293)

## M

**Malignant** (mə lig´ nənt) Harmful to health (p. 292)

**Malnutrition** (mal nü trish´ ən) Condition you get from a diet that lacks some or all nutrients (p. 222)

**Managed care** (man´ ijd  kâr) A type of private health insurance in which an organization is a go-between for patients and doctors (p. 458)

**Marijuana** (mâr ə wä´ nə) A drug from the hemp plant, *Cannabis sativa* (p. 338)

**Marriage counselor** (mar´ ij  koun´ sə lər) A person trained in helping couples work on their relationship with each other (p. 190)

**Marrow** (mar´ ō) The matter inside bones, which forms blood (p. 101)

**Maximum heart rate** (mak´ sə məm  härt  rāt) The number of heartbeats in a minute when one exercises as hard, fast, and long as possible (p. 136)

**Media** (mē´ dē ə) Ways people can communicate information; TV, Web sites, newspapers, radio, magazines, and so on (p. 233)

**Mediator** (mē´ dē ā tər) Someone who helps people find a solution to a conflict (p. 431)

**Medicaid** (med´ ə kād) Health insurance for people whose incomes are below a certain level (p. 459)

**Medicare** (med´ ə kâr) Health insurance for people age 65 or older and some people under age 65 with certain disabilities (p. 459)

**Medicine** (med´ ə sən) A drug that is used to relieve, cure, or keep a person from getting a disease or problem (p. 318)

**Medulla** (mi dul´ ə) The part of the brain that controls the body's automatic activities (p. 105)

**Melanin** (mel´ ə nən) Substance that gives skin its color (p. 121)

**Menstrual cycle** (men´ strü əl  sī´ kəl) The monthly cycle of changes in a female's reproductive system (p. 158)

**Menstruation** (men strü ā´ shən) The process in females during which blood tissue leaves the lining of the uterus (p. 158)

**Metabolism** (mə tab´ ə liz əm) Rate at which cells produce energy (p. 110)

**Methamphetamine** (meth´ am fet ə mēn) A highly addictive, illegal amphetamine (p. 333)

**Mineral** (min´ ər əl) A natural material needed for fluid balance, digestion, and other bodily functions (p. 216)

**Mouth-to-mouth resuscitation** (mouth  tü  mouth  ri sus ə tā´ shən) Rescue breathing (p. 403)

**Mucous membrane** (myü´ kəs  mem´ brān) The moist lining of the mouth and other body passages (p. 259)

**Mucus** (myü´ kəs) The fluid that mucous membranes produce (p. 259)

**Mutation** (myü tā´ shən) Change (p. 296)

## N

**Narcotic** (när kot´ ik) A pain-killing drug made from the opium poppy and in laboratories (p. 335)

**Natural sugar** (nach´ ər əl  shùg´ ər) Sugar that exists in a food in nature (p. 211)

| a | hat | e | let | ī | ice | ò | order | ù | put | sh | she | ə { | a | in about |
|---|---|---|---|---|---|---|---|---|---|---|---|---|---|---|
| ā | age | ē | equal | o | hot | oi | oil | ü | rule | th | thin | | e | in taken |
| ä | far | ėr | term | ō | open | ou | out | ch | child | ᴛʜ | then | | i | in pencil |
| â | care | i | it | ȯ | saw | u | cup | ng | long | zh | measure | | o | in lemon |
| | | | | | | | | | | | | | u | in circus |

**Neglect** (ni glekt´) Failure to take care of a person's basic needs; regular lack of care (p. 189)

**Negotiation** (ni gō shē ā´ shən) Ability to help people reach a compromise (p. 431)

**Nicotine** (nik´ ə tēn) A chemical in tobacco to which a person can become addicted (p. 324)

**Nonverbal** (non vėr´ bəl) Unspoken (p. 13)

**Nonverbal communication** (non vėr´ bəl kə myü nə kā´ shən) Using one's body to send messages (p. 63)

**Nuclear family** (nü´ klē ər fam´ ə lē) Family made up of a mother, a father, and their children (p. 184)

**Nurse practitioner** (nėrs prak tish´ ə nər) A registered nurse with special training for providing primary health care (p. 453)

**Nurture** (nėr´ chər) Help someone or something grow or develop (p. 61)

**Nutrient** (nü´ trē ənt) A part of food that helps the body function and grow (p. 100)

**Nutrient-dense** (nü´ trē ənt dens) Having a good source of nutrients but being low in calories (p. 207)

**Nutrition Facts** (nü trish´ ən faks) Section of a food label that provides information about the product (p. 240)

## O

**Obesity** (ō bē´ sə tē) Condition of being extremely overweight (p. 299)

**Opportunistic pathogen** (op ər tü nis´ tik path´ ə jən) Pathogen that the body would normally fight off but that takes advantage of a weakened immune system (p. 270)

**Optic nerve** (op´ tik nėrv) Part of the eye that sends information to the brain (p. 107)

**Optimism** (op´ tə miz əm) Tending to expect the best possible outcome (p. 56)

**Oral poisoning** (ôr´ əl poi´ zn ing) Poisoning that happens after a harmful material is swallowed (p. 407)

**Organic food** (ôr gan´ ik füd) Food raised without pesticides, unnatural fertilizers, antibiotics, or growth hormones (p. 234)

**Osteoarthritis** (os tē ō är thrī´ tis) A condition in which the cartilage in the joints wears away (p. 302)

**Outpatient clinic** (out´ pā shənt klin´ ik) A place where people receive health care without staying overnight (p. 454)

**Outpatient treatment center** (out´ pā shənt trēt´ mənt sen´ tər) A place where individuals go for drug-dependence treatment while living at home (p. 354)

**Ovaries** (ō´ vər ēz) Two organs in females that produce egg cells (p. 157)

**Overdose** (ō´ vər dōs) Too much of a drug (p. 334)

**Over-the-counter (OTC) medicine** (ō´ vər thə koun´ tər med´ ə sən) Medicine that can be bought without a prescription (p. 318)

**Ovulation** (ov yə lā´ shən) The release of a mature egg cell (p. 157)

**Ozone layer** (ō´ zōn lā´ ər) An area above Earth that guards and protects Earth from the sun's harmful rays (p. 493)

## P

**Pancreas** (pan´ krē əs) A gland that produces insulin and enzymes (p. 117)

**Panic attack** (pan´ ik ə tak´) A feeling of terror that comes without warning and includes symptoms of chest pain, rapid heartbeat, sweating, shaking, or shortness of breath (p. 80)

**Particulate** (pär tik´ yə lit) Tiny piece of solid material (p. 491)

**Pathogen** (path´ ə jən) Germ that causes infectious diseases (p. 255)

**PCP** (pē sē pē) A human-made hallucinogen (p. 337)

**Penis** (pē´ nis) Male reproductive organ (p. 159)

**Perinatal care** (per ə nā´ tl kâr) Care of the baby around the time of birth and through the first month (p. 166)

**Peripheral nervous system** (pə rif´ ər əl nėr´ vəs sis´ təm) Network of peripheral nerves carrying messages between the brain and spinal cord and the rest of the body (p. 105)

**Personality** (pėr sə nal´ ə tē) All of one's behavioral, mental, and emotional characteristics (p. 50)

**Perspiration** (pėr spə rā´ shən) Sweat (p. 121)

**Pessimism** (pes´ ə miz əm) Tending to suspect the worst possible outcome (p. 56)

**Pesticide** (pes´ tə sīd) A chemical used to kill bugs (p. 488)

**Petit mal seizure** (pə tē´ mal´ sē´ zhər) A type of epileptic seizure in which a person does not pass out (p. 303)

**Pharmacist** (fär´ mə sist) Druggist (p. 318)

**Phobia** (fō´ bē ə) Excessive fear of an object or a specific situation (p. 81)

**Phosphorus** (fos´ fər əs) Mineral that works with calcium to keep bones and teeth strong (p. 217)

**Photochemical smog** (fō tō kem´ ə kəl smog) A form of air pollution (p. 490)

**Physical environment** (fiz´ ə kəl en vī´ rən mənt) The area around a person (p. 11)

**Pimple** (pim´ pəl) An inflamed swelling of the skin (p. 130)

**Placenta** (plə sen´ tə) An organ lining the uterus that surrounds the embryo or fetus (p. 163)

**Platelet** (plāt´ lit) Small part in the blood that helps with clotting (p. 112)

**Pneumonia** (nü mō´ nyə) A lung infection (p. 272)

**Pollution** (pə lü´ shən) Harmful materials in the soil, water, or air (p. 487)

**Pore** (pôr) Tiny opening in the epidermis (p. 121)

**Postpartum** (pōst pär´ təm) Following birth (p. 166)

**Preferred provider organization (PPO)** (pri fėrd´ prə vī´ dėr ôr gə nə zā´ shən) A form of managed care that has fewer rules and more choices of doctors than an HMO has (p. 458)

**Pregnant** (preg´ nənt) Carrying a developing baby in the female body (p. 159)

**Prejudice** (prej´ ə dis) Negative, or unfavorable, opinions formed about something or someone without enough experience or knowledge (p. 67)

**Prenatal care** (prē nā´ tl kâr) Health care during a pregnancy (p. 161)

**Prescription** (pri skrip´ shən) A written order from a doctor for a medicine (p. 318)

**Preservative** (pri zėr´ və tiv) A substance added to foods to prevent them from spoiling (p. 241)

**Pressure point** (presh´ ər point) A place on the body where an artery can be pressed against a bone to stop bleeding in an arm or a leg (p. 406)

**Primary care physician** (prī´ mer ē kâr fə zish´ ən) A doctor who treats people for usual problems (p. 453)

**Product** (prod´ əkt) Item for sale made by someone or something (p. 233)

**Progesterone** (prō jes´ tə rōn) Female sex hormone (p. 155)

**Projection** (prə jek´ shən) Assuming that another person has one's own attitudes, feelings, or purposes (p. 77)

**Protein** (prō´ tēn) A material in food needed for growth and repair of body tissues (p. 210)

**Psychoactive** (sī kō ak´ tiv) Affecting the mind or mental processes (p. 328)

**Psychologist** (sī kol´ ə jist) A person who studies mental and behavioral characteristics (p. 76)

**Psychotherapy** (sī kō ther´ ə pē) Psychological treatment for mental or emotional disorders (p. 85)

**Puberty** (pyü´ bər tē) Period at the beginning of adolescence when children begin to develop into adults and are able to have children (p. 152)

**Public health** (pub´ lik helth) The practice of protecting and improving the health of a community or a nation (p. 468)

**Pulse** (puls) Regular beat of the heart (p. 112)

**Pupil** (pyü´ pəl) Dark center part of the eye that adjusts to let in the correct amount of light (p. 106)

## Q

**Quackery** (kwak´ ər ē) The promotion of medical products or services that are unproved or worthless (p. 448)

## R

**Rabies** (rā´ bēz) A serious disease that affects the nervous system (p. 408)

**Radiation** (rā dē ā´ shən) A cancer treatment that involves sending energy in the form of waves (p. 295)

**Radiation sickness** (rā dē ā´ shən sik´ nis) Sickness caused by being around high levels of radiation (p. 502)

**Radon** (rā´ don) A gas with no smell or color that comes from the radioactive decay of radium underground (p. 492)

**Rape** (rāp) The unlawful act of forcing a person to have sex (p. 426)

**Rational** (rash´ ə nəl) Realistic or reasonable (p. 57)

**Reaction** (rē ak´ shən) Response (p. 28)

**Realistic** (rē ə lis´ tik) Practical or reasonable (p. 56)

| a | hat | e | let | ī | ice | ô | order | ú | put | sh | she | ə | a in about |
|---|---|---|---|---|---|---|---|---|---|---|---|---|---|
| ā | age | ē | equal | o | hot | oi | oil | ü | rule | th | thin | | e in taken |
| ä | far | ėr | term | ō | open | ou | out | ch | child | ᵺ | then | | i in pencil |
| â | care | i | it | ò | saw | u | cup | ng | long | zh | measure | | o in lemon |
| | | | | | | | | | | | | | u in circus |

*Life Skills Health* Glossary  537

**Receptor cell** (ri sep′ tər sel) Cell that receives information (p. 109)

**Recessive** (ri ses′ iv) Hidden (p. 168)

**Rectum** (rek′ təm) Lower part of the large intestine (p. 118)

**Reflex** (rē′ fleks) Automatic response (p. 106)

**Rehabilitation** (rē hə bil ə tā′ shən) Therapy needed to recover from surgery, illness, or an injury (p. 455)

**Relief** (ri lēf′) A light, pleasant feeling after something painful or distressing is gone (p. 40)

**Repression** (ri presh′ ən) A refusal to think about something that upsets you (p. 77)

**Reproduction** (rē prə duk′ shən) The process through which two human beings produce a child (p. 156)

**Reproductive organ** (rē prə duk′ tiv ôr′ gən) Organ in the body that allows humans to mature sexually and to have children (p. 154)

**Rescue breathing** (res′ kyü brē ŦHing) Putting oxygen from a rescuer's lungs into an unconscious person's lungs to help the person breathe (p. 403)

**Residential treatment center** (rez ə den′ shəl trēt′ mənt sen′ tər) A place where individuals live while being treated for drug dependence (p. 354)

**Resilient** (ri zil′ yənt) Able to bounce back from misfortune or hardship (p. 59)

**Resistance** (ri zis′ təns) The pushing back of something (p. 134)

**Resistance skill** (ri zis′ təns skil) A way to resist negative peer pressure (p. 356)

**Respiration** (res pə rā′ shən) Breathing in oxygen and breathing out carbon dioxide (p. 114)

**Respiratory failure** (res′ pər ə tôr ē fā′ lyər) A condition in which breathing has stopped (p. 402)

**Retina** (ret′ n ə) Part of the eye that sends information about light (p. 107)

**Rheumatoid arthritis** (rü′ mə toid är thrī′ tis) A condition in which the joints swell badly and become stiff and weak (p. 302)

**Risk factor** (risk fak′ tər) A habit or trait known to increase the chances of getting a disease (p. 290)

**Role confusion** (rōl kən fyü′ zhən) Being unsure about who one is and one's goals as an adult (p. 151)

**Romantic love** (rō man′ tik luv) Strong physical and emotional attraction between two people (p. 30)

## S

**Saliva** (sə lī′ və) A liquid in the mouth (p. 116)

**Sanitation** (san ə tā shən) Measures related to cleanliness that protect public health (p. 474)

**Saturated fat** (sach′ ə rā təd fat) A material in food that comes from animal products and that can lead to health problems (p. 212)

**Scoliosis** (skō lē ō′ sis) A condition in which the spine is abnormally curved (p. 452)

**Secondary sex characteristic** (sek′ ən der ē seks kar ik tə ris′ tik) Feature that signals the beginning of adulthood (p. 154)

**Secondhand smoke** (sek′ ənd hand smōk) Smoke from a cigarette or cigar someone else is smoking (p. 15)

**Sedative-hypnotic drug** (sed′ ə tiv hip not′ ik drug) A prescribed depressant that lessens anxiety or helps a person sleep (p. 335)

**Sedentary** (sed′ n ter ē) Including little or no exercise (p. 14)

**Seizure** (sē′ zhər) A physical reaction to an instance of mixed-up brain activity (p. 303)

**Self-actualization** (self ak chü ə lī zā′ shən) Achieving one's possibilities (p. 32)

**Self-awareness** (self ə wâr′ nəs) Understanding oneself as an individual (p. 66)

**Self-concept** (self kon sept′) Ideas one has about oneself (p. 31)

**Self-defeating behavior** (self di fēt′ ing bi hā′ vyer) An action that blocks a person's efforts to reach goals (p. 76)

**Self-esteem** (self e stēm′) Self-respect; how one feels about oneself (p. 31)

**Sensorineural deafness** (sen sər ē nùr′ əl def′ nis) Hearing loss in which the tiny cells in the inner ear are destroyed (p. 501)

**Separation** (sep ə rā′ shən) A couple's agreement, or a court decision made for the couple, to stop living together (p. 187)

**Sewage** (sü′ ij) Liquid and solid waste from drains and toilets (p. 495)

**Sex-linked** (seks lingkd) Carried by a sex chromosome (p. 168)

**Sexual abuse** (sek′ shü əl ə byüs) Any sexual contact that is forced on a person, including unwanted talk and touching (p. 189)

**Sexual assault** (sek′ shü əl ə sȯlt′) Forced sexual contact (p. 427)

**Sexual exploitation** (sek′ shü əl ek sploi tā′ shən) Using a person sexually to make money (p. 427)

**Sexual harassment** (sek´ shü əl har´ əs mənt) Making unwanted sexual comments or physical advances to another person (p. 426)

**Sexual intercourse** (sek´ shü əl in´ tər kôrs) Insertion of the penis into the vagina (p. 159)

**Sexually transmitted disease** (sek´ shü ə lē tranz mit´ əd də zēz) Any disease that is spread through sexual activity (p. 275)

**Shock** (shok) Failure of the circulatory system to provide enough blood to the body (p. 405)

**Side effect** (sīd ə fekt´) Reaction to a medicine other than what it is intended for (p. 318)

**Simple carbohydrate** (sim´ pəl kär bō hī´ drāt) A sugar the body uses for quick energy (p. 210)

**Single-parent family** (sing´ gəl pâr´ ənt fam´ ə lē) Family that includes a child or children and one adult (p. 184)

**Small intestine** (smól in tes´ tən) A curled-up tube just below the stomach where most of the breakdown of food takes place (p. 117)

**Smoke detector** (smōk di tek´ tər) A device that sounds loudly when it senses smoke or high amounts of heat (p. 371)

**Social comparison** (sō´ shəl kəm par´ ə sən) Observing other people to determine how to behave (p. 66)

**Social environment** (sō shəl en vī´ rən mənt) Community resources such as access to doctors, hospitals, family counseling, and after-school programs (p. 12)

**Social esteem** (sō´ shəl e stēm´) How others value a person (p. 51)

**Socializing** (sō´ shə lī zing) Getting together with others during free time (p. 176)

**Social support** (sō´ shəl sə pôrt´) Help for those you care about (p. 69)

**Sodium** (sō´ dē əm) Mineral that controls fluids in the body (p. 218)

**Soil erosion** (soil i rō´ zhən) Soil moved by wind, water, or ice (p. 487)

**Solid waste** (sol´ id wāst) Garbage, trash (p. 500)

**Specialist** (spesh´ ə list) A doctor who works only in a certain branch of medicine (p. 453)

**Sperm** (spėrm) Male sex cells produced by the testes (p. 156)

**Spinal column** (spī´ nl kol´ əm) Series of small bones that surround and protect the spinal cord (p. 105)

**Spinal cord** (spī´ nl kôrd) Major pathway the brain uses to send messages to the body (p. 105)

**Splint** (splint) An unbendable object that keeps a broken arm, leg, or finger in place (p. 409)

**Sprain** (sprān) The tearing or stretching of tendons or ligaments that connect joints (p. 410)

**Sterile** (ster´ əl) Free from bacteria (p. 406)

**Sterility** (stə ril´ ə tē) Inability to produce a child (p. 276)

**Stimulant** (stim´ yə lənt) A drug that speeds up the central nervous system (p. 324)

**Stress** (stres) A state of physical or emotional pressure (p. 34)

**Stress response** (stres ri spons´) Automatic physical reactions to stress (p. 34)

**Stroke** (strōk) A cardiovascular disease that happens when the blood supply to the brain is suddenly stopped (p. 289)

**Subcutaneous layer** (sub kyü tā´ nē us lā´ ər) Deepest layer of skin (p. 121)

**Substance abuse disorder** (sub´ stəns ə byüs´ dis ôr´ dər) An unhealthy dependence on alcohol or other drugs (p. 80)

**Sucrose** (sü´ krōs) Table sugar (p. 211)

**Suicide** (sü´ ə sīd) The act of killing oneself (p. 81)

**Surgery** (sėr´ jər ē) Opening up the patient's body (p. 288)

**Survival floating** (sər vī´ vəl flō´ ting) A technique that can help prevent drowning (p. 378)

**Symptom** (simp´ təm) A bodily reaction that signals an illness or a physical problem (p. 59)

**Syphilis** (sif´ ə lis) A sexually transmitted disease (p. 276)

## T

**Taste bud** (tāst bud) Receptor on the tongue that recognizes tastes (p. 109)

**Temperament** (tem´ pər ə mənt) A person's emotional makeup (p. 50)

**Tendon** (ten´ dən) A strong set of fibers joining muscle to bone or muscle to muscle (p. 102)

| a | hat | e | let | ī | ice | ȯ | order | u̇ | put | sh | she | ə | a in about |
|---|---|---|---|---|---|---|---|---|---|---|---|---|---|
| ā | age | ē | equal | o | hot | oi | oil | ü | rule | th | thin | | e in taken |
| ä | far | ėr | term | ō | open | ou | out | ch | child | ᴛʜ | then | | i in pencil |
| â | care | | i | it | ȯ | saw | u | cup | ng | long | zh | measure | o in lemon |
| | | | | | | | | | | | | | u in circus |

**Testes** (tes′ tēz) Two glands in males that produce sperm cells (p. 156)

**Testosterone** (te stos′ tə rōn) Male sex hormone (p. 155)

**THC** (tē āch sē) The psychoactive chemical in marijuana (p. 338)

**Thought disorder** (thȯt dis ȯr′ dər) A mental problem in which a person has twisted or false ideas and beliefs (p. 82)

**Tissue** (tish′ ü) Material in the body that is made up of many cells that are somewhat alike in what they are and what they do (p. 287)

**Tobacco** (tə bak′ ō) Dried leaves from certain plants that are used for smoking or chewing (p. 14)

**Tolerance** (tol′ ər əns) The need for more of a drug to get the effect that once occurred with less of it (p. 334)

**Tornado** (tȯr nā′ dō) A funnel-shaped column of wind with wind speeds up to 300 miles per hour (p. 385)

**Toxic** (tok′ sik) Harmful to human health or to the environment (p. 295)

**Trachea** (trā′ kē ə) A long tube running from the nose to the chest; windpipe (p. 114)

**Trait** (trāt) Quality (p. 167)

**Tranquilizer** (trang′ kwə lī zər) A category of sedative-hypnotic drugs (p. 335)

**Trans fat** (tranz fat) Type of fat from processed vegetable oil (p. 236)

**Transfusion** (tran sfyü′ zhən) Transfer of blood from one person to another (p. 272)

**Trimester** (trī mes′ tər) A period of 3 months (p. 163)

**Tuberculosis** (tü bėr kyə lō′ sis) A communicable lung disease (p. 471)

**Tumor** (tü′ mər) A mass of tissue made from the abnormal growth of cells (p. 292)

**Type I diabetes** (tīp wun dī ə bē′ tis) A condition in which insulin is needed to stay alive (p. 298)

**Type II diabetes** (tīp tü dī ə bē′ tis) A condition in which the pancreas makes insulin that the body cells cannot use correctly (p. 299)

## U

**Ultrasound** (ul′ trə sound) The use of sound waves to show pictures of what is inside the body (p. 160)

**Umbilical cord** (um bil′ ə kəl kȯrd) Structure that joins the embryo or fetus with the placenta (p. 163)

**Universal distress signal** (yü nə vėr′ səl dis tres′ sig′ nəl) A sign made by grabbing the throat with the hand or hands to show a person is choking (p. 401)

**Universal precautions** (yü nə vėr′ səl pri kȯ′ shəns) Methods of self-protection that prevent contact with a person's blood or body fluids (p. 398)

**Unsaturated fat** (un sach′ ə rā tid fat) Material in food from vegetable and fish oils that helps lower cholesterol (p. 213)

**Ureter** (yu̇ rē′ tər) Tube through which urine passes from a kidney to the urinary bladder (p. 119)

**Urethra** (yu̇ rē′ thrə) Tube through which urine passes out of the body (p. 119)

**Urinary bladder** (yu̇r′ ə nȧr ē blad′ ər) Bag that stores urine (p. 119)

**Urine** (yu̇r′ ən) Liquid waste product formed in the kidneys (p. 119)

**Uterus** (yü′ tər əs) Female organ that holds a growing baby (p. 157)

## V

**Vaccination** (vak sə nā′ shən) A means of getting dead or weakened pathogens into the body (p. 261)

**Vagina** (və jī′ nə) Canal connecting the uterus to the outside of a female's body (p. 157)

**Vein** (vān) Vessel that carries blood back to the heart (p. 113)

**Victim** (vik′ təm) A person who has had a crime committed against him or her (p. 418)

**Villus** (vil′ əs) Fingerlike bulge that absorbs food after it has been broken down into chemicals (plural is *villi*) (p. 117)

**Violence** (vī′ ə ləns) Actions or words that hurt people (p. 189)

**Virus** (vī′ rəs) Small germ that reproduces only inside the cells of living things (p. 255)

**Vitamin** (vī′ tə mən) A material needed in small amounts for growth and activity (p. 216)

## W

**Well-being** (wel′ bē ing) State of being physically and emotionally healthy (p. 38)

**Wellness** (wel′ nis) An active state of health (p. 9)

**Withdrawal** (wiŧH drȯ′ əl) A physical reaction to not having a drug in the body (p. 326)

# Index

## A

Abnormal behavior, 76
Abnormality, 286
Abstinence, 15, 178
Abuse, effect of, on family, 189–190
Abused children, 424
Acceptance, grief and, 40
Acid rain, 494
Acne, 130–131
Added sugar, 211–212
Addiction, 80, 324, 346,347
Administration for Children and Families (ACF), 475
Adolescence, 151. *See also* Teenagers
    good health habits in, 153
Adolescent counselor, 192
Adoptive family, 185
Adrenal glands, 111
Adrenaline, 111
Advertising
    for depression relief, 86
    effect on food choices, 233–235
    influence on consumers, 447
Advocacy groups, 479, 480
Aerobic exercise, 135
Affective disorders, 81, 86
Afterbirth, 165
Agency for International Development (USAID), 479
Aggression, 39
Aging, effect of, on family, 187–188
Agricultural runoff, 496–497
Agriculture, U.S. Department of (USDA), 293
Agriculture biotechnology, 503
AIDS (acquired immunodeficiency syndrome), 92, 256, 270–274
    acquisition of, 271–272, 273
    causes of, 270
    drug dependence and, 271, 350
    future for people with, 274
    symptoms of, 272
    worldwide effects of, 271
Air ambulances, 405
Air bags, 383
Air pollution, 490–494
Airport security, 433
Al-Anon, 353
Alateen, 353
Alatot, 353

Alcohol, 8, 347
    addiction to, 80
    drinking, 153
    effects of, 14, 328–331
    fetal alcohol syndrome and, 161
    use of, with depressants, 334
    violence and, 425
Alcoholics Anonymous (AA), 353
Alcoholism, 330, 346
Algae, 496
Allergies, 129, 304
Allergist, 453
Alternative birthing centers (ABCs), 460
Alveoli, 115
Ambulances, air, 405
American Cancer Society, 461, 478
American Diabetes Association, 300
American Heart Association, 478
American Lung Association, 461, 478
American Social Health Association (ASHA), 278
Amino acids, 214
Amphetamines, 333, 347
Anabolic steroids, 340
Anaerobic exercise, 135
Analgesics, 319
Anger, 39
    grief and, 40
    management of, 429
Angina pectoris, 289
Animal bites, 408
Anorexia, 83
Antibacterial soap, 258
Antibiotics, 256, 259, 275, 319
Antibodies, 260, 261, 262
Antiperspirants, 130
Anus, 118
Anvil, 108
Anxiety, 38
Anxiety disorders, 80–81, 85
Arterial diseases, 286, 287, 289
Arteries, 113
Arteriosclerosis, 287
Arthritis, 302–303
    rheumatoid, 302
Arts, Link to, 108, 139, 190, 271, 382, 469
Asbestos, 297, 492
Association, food choices and, 231
Asthma, 161, 304–305, 490

Astigmatism, 128
Atherosclerosis, 287
Athletes, special dietary needs of, 222
Athlete's foot, 131
Attachment, 30
Attention deficit disorder (ADD), stopping, 326
Auditory nerve, 108

## B

Baby, decision to have, 182
Bacteria, 121, 255, 256
Balance of nature, 486–487, 488–489, 500
Ball-and-socket joint, 100
Barbiturates, 335, 347
Bargaining, grief and, 40
Barriers, 259
Beans, in Food Guide Pyramid, 206
Behavioral risks, causes of, 14–16
Behavior modification, 85
Behaviors
    changing, 18–20
    self-defeating, 76–79
Belonging needs, 29, 30
Benign tumor, 292
Bicycle safety, increasing, 383
Bile, 117
Binge drink, 330
Binge eating, 83
Biodegradable materials, 505, 507
Biofeedback, 35
Biology, Link to, 20, 37, 51, 326, 407
Bipolar disorder, 81, 86
Birth defects, 472
    fetal alcohol syndrome as cause of, 329
Blackheads, 130, 131
Bleeding, first aid for, 405–406, 409
Blended family, 184
Blizzard, preparing for, 388
Blood, 112–113
    movement through body, 112–113
    safety of supply of, 272–273
Blood alcohol concentration (BAC), 329, 330
Blood cholesterol, 213, 291
Blood pressure, 113, 286
Blushing, 37
Body, role of, in fighting infection, 259
Body language, 28, 64
Body mass index (BMI), 210
Bones, 100–101
    injuries to, 409
Brain, 104–105

Brain stem, 105
Breast cancer, 294
Breast-feeding, 166
Breast milk, 262
Breast self-exam, 294
Bronchi, 114
Bronchioles, 114
Bronchitis, 490
Bulimia, 83
Bullying, 32
Burns, 410
    preventing, 374
B vitamins, 217

## C

Caffeine, 232
Calcium, 217, 221
Calories, 135
    burning, 140
Cancer, 169, 292–296, 297
    causes of, 292
    location of, 292–293
    risk factors for, 295–296
    symptoms of, 293
    treatment of, 295
    warning signs of, 293–294
Capillaries, 113
Carbohydrates, 206, 210–212
    complex, 210–211, 230
    need for, 9
Carbon dioxide, 112, 493
Carbon monoxide, 325, 490, 492
    detection of, 16
Carcinogens, 295, 297
Cardiac arrest, 404
Cardiac muscles, 102
Cardiologist, 453
Cardiopulmonary resuscitation (CPR), 404
Cardiovascular disease, 286
    prevention of, 286
    risk factors for, 290–291
Cardiovascular failure, first aid for, 403–404
Cartilage, 302
Cells, 100
Centers for Disease Control and Prevention (CDC), 459, 469, 475
Centers for Medicare and Medicaid Services (CMS), 475
Central nervous system, 104
Cerebellum, 105
Cerebrum, 104

Cervix, 165
    dilation of, 165
Cesarean childbirth, 165
Chancre, 276
Chemotherapy, 295
Chernobyl, 502
Chest cavity, 115
Chest compressions, 404
Childbirth, 165, 166
Children
    abused and neglected, 424
    responsibilities of parents for, 183–184
    safety when caring for, 374
Chlamydia, 276
Chlorofluorocarbons (CFCs), 493, 494, 504
Choking, first aid for, 401–402
Cholera, 196
Cholesterol, 213, 232, 234, 288
    blood, 213
    cardiovascular problems and, 291
    dietary, 213
Chromosomes, 167, 257
Chronic diseases, 277
Circulatory system, 112–115
Citizenship skills, 53
Claustrophobia, 81
Clean Air Act (1970), 504
Clinical depression, 81, 86
Clotting, 112
Cocaine, 333–334, 347, 348, 350
    pregnancy and, 162
Cocaine Anonymous, 353
Codeine, 335
Cognitive therapy, 85
Cold, exposure to extreme, 411
Collaborative approach to life, 180
Color blindness, 128
Coma, 298
Commitment, 178
Communicable diseases, 254, 269
Communication
    in dating, 177
    nonverbal, 64
    skills in, 53
    successful, 62–64
Community
    health services in, 474, 476
    preparing for disasters, 389
    role of, in fighting crime, 433–434
Complex carbohydrates, 210–211, 230
Compromise, 57
Compulsive overeating, 83

Conceive, 157
Conference on Environment and Development, 505
Conflict
    avoiding, 430
    comparison to violence, 419
    getting help with, 432
    recognizing, 424
    solving, 57
Conflict resolution, 57, 430, 432
Consequences, 183
Conservation scientist, 508
Consumer health, 446–462
Consumer Product Safety Commission (CPSC), 475
Consumers, 446, 447
    being wise, 446–451
    correcting problems of, 449–451
Contact poisoning, 408
Contract, 101
Cooldown, 102, 138
Coping, 58–60
    with stress, 41–42
Cornea, 106
Counseling, drug dependence and, 353
Couples, decision to have a baby, 182
Crack, 334
Crime
    drug dependence and, 350
    role of communities in fighting, 433–434
    violence and, 420
Cultural background, food choices and, 231
Cultural risks, decreasing, 13
Culture, 13, 65
Custodial parent, 185
Custody, 185

## D

Daily Values (DV), 236, 240
Dandruff, 132
Date rape, 427
Date rape drugs, 427
Dating, 176–177
    difficult parts of, 177–178
    violence and, 427
Day care centers, 184
Death, effect of, on family, 188
Decide for Yourself, 6, 32, 56, 88, 119, 142, 154, 184, 214, 237, 257, 279, 300, 331, 384, 412, 432, 450, 472, 497
Decisions, making, 18–20
Deductible, 457
Defense mechanisms, 77–78
Defibrillator, 288

Deficiency, 223
Deforestation, 499–500
Deformities, 257
Delusion, 82
Denial, 77
    grief and, 40
Dental care, 8, 454
Deodorant, 130
Dependence, 321
    physical, 332, 346–347
    psychological, 332, 347
Depressants, 328
    effects of, 334–335
Depression
    clinical, 81
    grief and, 40
    postpartum, 166
Depression relief, advertising for, 86
Dermatologist, 453
Dermis, 120
Designer drug, 339
Deterioration, 15
Detoxification, recovering from, 352
Diabetes, 298–300
    risk factors for, 300
    treatment of, 299
Diaphragm, 115
Diary, 79
Diet
    disease and, 223–224
    eating a balanced, 356
    effect of fads on, 237
    effect of fast-food industry on, 235–236
    effect of poor, on health, 222–223
    health and, 204–224
    healthy, 204–208
    high blood pressure and, 287
    in weight control, 222
Dietary aide, 238
Dietary cholesterol, 213
Dietary guidelines, 204
    food groups in, 205, 207
Digestion, 116
Digestive system, 116–117
Diphtheria, 260
Disadvantages, 58
Disasters, community preparations for, 389
Discipline, 183
Discrimination, 67
Diseases
    acquired, 254
    avoiding germs, 219
    causes of, 254–257
    diet and, 223–224
    infectious, 255–256
    inherited, 256–257
    role of body in protecting itself from, 259–261
Displacement, 77
Divorce, 180
    effect on family, 187
Doctor, 438
    communicating with, 456
    questions to ask about, 455
Doctors Without Borders, 438
Domestic violence, 190, 471
Down syndrome, 257
Dressings, 398
Drinking, avoiding, 331
Drowning, preventing, 378
Drug dependence, 346–358, 352
    AIDS and, 271, 350
    costs to society, 349–350
    defined, 346
    effect on family, 348–349
    psychological, 347
    signs of, 348
    sources of help, 353–354
Drugs, 8
    alternatives to, 357, 358
    avoiding negative effects of, 358
    effects of, 15
    generic, 449
    refusing, 356
    testing for, 354
    violence and, 425
Dysentery, 196
Dysfunctional relationship, 77, 78–79

## E

Eardrum, 108
Early adulthood, 152
Earplugs, 129
Ears, taking care of, 129
Earthquakes, preparing for, 387–388
Earth Science, Link to, 493
Eating disorders, 83, 87
Eating patterns
    effect of stress on, 232
    media influences on, 233–238
Ecology, 489
Ecstasy, 336
Electrical shock, reducing risk of, 373
Electrocardiograph machines, 405
Electrocardiograph (EKG) treadmill, 115

Electronic defibrillators, 405
Embryo, 163
Emotional health, 4, 7, 55–60
    achieving, 66–69
    assessing, 21
    coping and, 58–60
Emotional well-being, 52, 53–54
Emotions
    defined, 28
    effects of, 28
    handling stressful, 41
    managing, 57
    needs and, 33
    stress and, 38–40
Empathy, 68
Emphysema, 324, 490
Employers, drug testing by, 354
Enabling, 349
Endocrine glands, 110
Endocrine system, 110–111, 154
Endurance, 134
Entertainment, healthy forms of, 53
Environment, 472
    effect of, on health, 5
    protecting, 504–507
    social, 12
Environmental carcinogens, 297
Environmental damage, effect on health, 487–488
Environmental health, 485–508
Environmental plans, future, 505
Environmental Protection Agency (EPA), 12, 475, 496, 503, 504
Environmental protection specialist, 12
Environmental science, Link to, 161, 256
Enzyme, 116
Epidemics, 468–469
Epidermis, 120
Epilepsy, 303–304
Erection, 159
Erikson, Erik, 150
Escape routes, planning, 375
Esophagus, 117
Essential amino acids, 214
Essential nutrients, 206
Esteem, 29
Esteem needs, 29, 31
Estrogen, 155
Ethanol, 505
Excretory system, 118–119, 120
Exercise(s), 5, 17, 20
    benefits of, 134–135
    cooldown, 102, 138

    effect on eating patterns, 231
    in improving flexibility, 138
    in improving muscular fitness, 138
    need for regular, 287
    program for, 136–138
    purpose of, 135
    stress and, 35
    types of, 135
    warm up, 102, 136
Exhale, 115
Extended family, 185
Eyes
    objects in, 410
    taking care of, 128–129

## F

Factory wastes, 495
Fads, effect of, on diet, 237
Fainting, 408
Fallopian tubes, 157
Falls, reducing risk of, 370
Family, 175–192
    adoptive, 185
    belief systems for, 191
    blended, 184
    changes in, 185
    cycle of violence in, 422
    effect of abuse on, 189–190
    effect of aging on, 187–188
    effect of drug dependence on, 348–349
    effect of illness and death on, 188
    effect of separation and divorce on, 187
    extended, 185
    finding help for troubled, 190–192
    foster, 185
    nuclear, 184
    protecting, 399
    signs of healthy, 186
    single-parent, 184, 185
    violence within, 422, 424
Family life cycle, 176–181
Family love, 30
Famine, 499
Farsightedness, 128
Fast-food industry, effect on diet, 235–236
Fats, 205, 210, 212–213, 235, 236
    saturated, 212–213, 236
    unsaturated, 213
Fatty acids, 212
Fatty foods, 236
Fear, 38
Feces, 118

Federal Emergency Management Agency (FEMA), 389
Females, changes during puberty, 157–158
Fertility drugs, 182
Fertilization, 159
Fertilized egg, growth of, 162–163
Fetal alcohol syndrome (FAS), 161, 329
Fetus, 163
Fiber, 211, 212
Fights, staying out of, 431
Fire
    planning escape route for, 375
    reducing risk of, 370–371
Fire drills, 372, 377
Fire extinguishers, 16, 371–372
Fire safety devices, 371–372
First aid, 396, 408–412
    basic guidelines for, 396–397
    for cardiovascular failure, 403–404
    for choking, 401–402
    for poisoning, 407–408
    for respiratory failure, 402–403
    for serious bleeding, 405–406
    for shock, 405
First aid kit, putting together, 399
Fitness, importance of rest and sleep for, 140, 143
Fitness instructor, 143
Flashbacks, 336
Flash floods, 387
Flat bones, 100
Flexibility, 134
    exercises to improve, 138
Flood, preparing for, 387
Flu, 472
Fluoride, 461
Folic acid, need for, in pregnancy, 161
Food additives, 241
Food and Agriculture Organization (FAO), 478
Food and Drug Administration (FDA), 236, 239, 241
    322, 459, 475, 476, 503
Food and Nutrition Board of the National Research
    Council, 239
Food choices
    effect of advertising on, 233–235
    effects on, 231–235
Food Guide Pyramid, 204–206, 207, 208, 221
Food label, information on, 239–241
Food poisoning, 373
Food safety, role of government, 239, 241
Fossil fuels, 491, 505
Foster family, 185
Fracture, 409
Fragile, 488

Freckles, 121
Freshwater, conservation of, 498
Friendships, 30, 53, 61
Frostbite, 411
Fructose, 211–212
Fruits, 212
    in Food Guide Pyramid, 206
Frustration, 13
FSH, 158
Fungus, 131, 255

## G

Gallbladder, 117
Gang violence, 418, 425
Gender, determination of, 167
Generic drugs, 449, 460
Genes, 167, 257
Gene therapy, 352
Genetic disorders, 168, 256–257
Genetics, 167
Genital herpes, 277
Genitals, 154
Genital warts, 278
Geography, 82
Germs, avoiding, 219
Gestation, 160
Gingivitis, 132
Glands, 110
Global warming, 493, 499
Glucose, 212, 298
Glycogen, 212
Goals
    importance of, 56
    setting, 17–18
Gonads, 110
Gonorrhea, 275–276
Good Samaritan laws, 398
Government
    development of Food Guide Pyramid, 204
    role of, in food safety, 239, 241
Government health insurance, 459
Government regulations, 476
Grains, in Food Guide Pyramid, 206
Grand mal seizure, 303
Greenhouse effect, 493
Grief, 40
Groundwater, 495
Growth hormone, 110
Growth spurts, 152–153
Guilt, 33
Gun, keeping in the home, 374
Gynecologist, 453

## H

Hair
- structure of, 121
- taking care of, 132

Halfway houses, 354
Hallucinations, 82
Hallucinogens, effects of, 336–337
Hammer, 108
Hate crimes, 425
Hazardous waste, 501
HDL cholesterol, 288
Health
- achieving good, 7
- defined, 4
- diet and, 204–224
- effect of environment on, 5, 486–489
- emotional, 4, 7, 55–60
- hygiene for good, 128–132
- overpopulation and, 196
- physical, 4, 7, 53
- setting goals for, 6
- social, 4, 7
- steps in maintaining, 8
- taking charge of, 17–20
- worldwide concerns, 473

Health assessment chart, 21
Health at Work
- adolescent counselor, 192
- conservation scientists, 508
- dietary aide, 238
- environmental protection specialist, 12
- fitness instructor, 143
- health unit coordinator, 264
- high school counselor, 69
- home health aide, 477
- industrial safety specialist, 376
- laboratory assistant, 274
- medical records technician, 462
- medical technologist, 155
- mental health specialist, 43
- 911 operator, 400
- nutrition technician, 223
- pharmacy technician, 322
- resident assistant, 78
- substance abuse counselor, 355
- ultrasound technician, 305
- victim or witness advocate, 423

Health care, 8
- costs of, 470
- paying for, 457–461

Health care facilities, 454–455
Health care professionals, 453–454
- prevention of AIDS in, 273

Health costs, lowering, 460–461
Health habits, good, in adolescence, 153
Health insurance, 457
Health in the World, 92, 196, 246, 310, 362, 438, 512
Health in Your Life, 5, 44, 52, 86, 122, 133, 153, 191, 219, 235, 261, 280, 297, 327, 351, 375, 399, 422, 456, 476, 507
Health literacy, 18
Health maintenance organization (HMO), 458
Health Myths, 9, 17, 35, 36, 55, 86, 87, 101, 107, 131, 143, 159, 178, 181, 218, 222, 234, 236, 258, 260, 271, 273, 289, 294, 304, 326, 330, 348, 353, 374, 383, 410, 411, 424, 426, 449, 460, 470, 474, 494, 505
Health promotion, 461, 479
Health Resources and Services Administration (HRSA), 460
Health risk, 11
Health screenings, 452–453
Health services, 476
- getting, 471
- school, 452–453

Health unit coordinator, 264
Healthy eating patterns, 230–232
Healthy relationships, 61–65
Hearing
- noise and, 133
- sense of, 108
- taking care of, 129

Heart, 112, 115
Heart attacks, 232, 287–288
Heart disease, risk of, 135
Heart valves, 288
Heat, exposure to extreme, 411
Heat exhaustion, 411
Heatstroke, 411
Heimlich maneuver, 401–402
Helmets, 383, 384
Help, giving and accepting, 69
Hemophilia, 256
Hepatitis B, 256
- vaccine for, 261

Hepatitis C, 256
Heredity, 11, 167
- disease and, 11
- drug dependence and, 346
- as risk factor for cancer, 296

Heroin, 336, 347
Hierarchy, 29
Hierarchy of needs, Maslow's, 29, 31
High blood pressure, 286–287, 340
High school counselor, 69

Hinge joint, 101
HIV/AIDS Act (2003), 273
Hodgkin's disease, 295
Home, reducing risks at, 16, 370–375
Home health aide, 477
Homeland Security, U.S. Department of, 389, 433
Homelessness, 470–471
Hormones, 110, 130, 154–155, 156, 158
Hospices, 455, 460
Hospital, 455
Household chemicals, 497
Household dangers, identifying, 374
Human immunodeficiency virus (HIV), 270, 271, 272
Human papilloma virus (HPV), 278
Hurricane, preparing for, 385
Hygiene, 128
Hypertension, 286
Hypothalamus, 110
Hypothermia, 411

Illness, effect of, on family, 188
"I" messages, 39, 63, 79
Immune system, 255, 256, 260–262
Immunization, 261
Impairment, 15
Implantation, 162
Impulses, 57
Incubation period, 255
Independent practice associations (IPAs), 458
Individual differences, accepting, 67–68
Indoor air pollution, 492
Industrial safety specialist, 376
Infant mortality rate (IMR), 166
Infection, 129
    role of body in fighting, 259–261
Infectious diseases, 254, 255–256, 258
Inferior, 151
Infertility, 182
Inflammation, 260
Influenza, 254
Inhalants, 339, 348
Inhalation poisoning, 407–408
Inhale, 115
Inherited diseases, causes of, 256–257
Injury reduction, 369–389
Inpatient care, 455
Insect stings, 409
Instinctive thinking, 104
Insulin, 117, 298, 299
International health services, 478–479

International Red Cross, 246, 479
Intimate, 61
Intoxication, 330
Iris, 107
Iron, 218, 221
Isokinetic exercise, 135
Isometric exercise, 135
Isotonic exercise, 135

## J

Joints, 100–101
    injuries to, 409–410
Journaling, 78, 79

## K

Kaposi's sarcoma, 272
Keratin, 121
Kidney disease, 286, 340
Kidneys, 118
Kidney stones, 131
Knuckles, cracking, 101

## L

Laboratory assistant, 274
Lacto-ovo vegetarian diet, 230
Lactose, 211
Landfills, 501
Language Arts, Link to, 22, 67, 79, 183, 347, 419
Large intestine, 117–118
Larrey, Dominique-Jean, 397
Lasers, 131
Latex gloves, 397
LDL cholesterol, 288
Lead, 161, 491, 495
Lens, 107
LH, 158
Life cycle, family, 176
Life events, as cause of stress, 36
Life-threatening emergencies, first aid for, 401–406
Ligaments, 101
Lightning strikes, 389
Light pollution, 502
Listening, effective, 63–64
Liver, 117
Liver cancer, 340
Lockjaw, 472
Long bones, 100
Long-term care facilities, 455
Long-term goal, 17
Look-alike drugs, 339

Love
    family, 30
    romantic, 30
LSD, 337
Lungs, 112–113
Lymph, 293

## M

Males, changes during puberty, 156
Malignant tumor, 292
Malnutrition, 196, 222
Managed care, 458
Marijuana, 338, 347
    pregnancy and, 162
Marriage
    reasons for, 178–179
    signs of healthy, 180–181
    timing of, 179–180
Marriage counselor, 190
Marrow, 101
Maslow's hierarchy of needs, 29, 31
Math, Link to, 207, 240, 291, 329, 350, 421, 459, 488
Maximum heart rate, 136–138
Measles, 260, 310
Measles-Mumps-Rubella (MMR) vaccine, 261
Meat, in Food Guide Pyramid, 206
Media
    influences of, 51, 233–235, 447
    messages in, 88
Mediator, 431
Medicaid, 459
Medical records technician, 462
Medical technologist, 155
Medicare, 459, 460
Medicines, 318–323
    effect on body, 320–321
    purposes of, 319–320
    role of government in keeping safe, 322
    taking, 320, 321–322, 323
    types of, 318
Medulla, 105
Melanin, 121
Menstrual cycle, 158
Menstruation, 158–159
Mental disorders, 80–88
Mental health,
    characteristics of poor, 76
    cocaine and, 334
    influences on, 50–54
Mental health counselor, 70
Mental health specialist, 43
Mental illness, society and, 87
Mercury, 495
Metabolism, 110–111
Methamphetamine, 333, 348
Middle adulthood, 152
Milk, in Food Guide Pyramid, 206
Minerals, 205, 216
    importance of, 217–218
Mining, 495
Morphine, 335
Motorcycle safety, increasing, 383
Mouth-to-mouth resuscitation, 403
Mucous membranes, 259
Mucus, 259
Muscles
    cardiac, 102
    contraction of, 101
    skeletal, 102
    smooth, 102
    working, 103
Muscle tone, 103
Muscular fitness, 134
    exercises to improve, 138
Muscular system, 100–103
Mutation, 296

## N

Nails
    structure of, 121
    taking care of, 132
Narcotics, 335, 347
    effects of, 335–336
Narcotics Anonymous, 353
National Drug and Alcohol Treatment Hotline, 357
National Environmental Policy Act, 504
National health services, 475–476
National Institutes of Health (NIH), 459, 475
National Weather Service, 385, 389
Natural disasters, 246
    safety during, 385–389
Natural sugar, 211–212
Nearsightedness, 128
Needs, basic, 29–32
Negative self-concept, 51
Neglect, 189
Neglected children, 424
Negotiation, 431
Neighborhood clinic, 460
Nervous system, 104
    central, 104
    peripheral, 104, 105–106

Neurotransmitters, 326
Niacin, 217
Nicotine, 324, 332
    breaking an addiction to, 327
Niger, World Health Organization programs in, 92
911 operator, 400
Nitrogen dioxide, 492
Noise, hearing and, 129, 133
Noise pollution, 501
Nonverbal communication, 13, 63, 64, 65
Nosebleeds, 411
Nuclear family, 184
Nuclear power plants, 502
Nurse practitioner, 453
Nursing homes, 455
Nurture, 61
Nutrient-dense foods, 207
Nutrients, 100
    minerals as, 217–218
    vitamins as, 216
Nutrition facts, 240
Nutrition labels, 207

## O

Obesity, 299
Obstetrician, 453
Occupational Safety and Health Administration (OSHA), 376–377, 475
Oil glands, 154
Oil spills, 496
Older adulthood, 152
Olfactory nerve, 109
Ophthalmologist, 453
Opportunistic pathogens, 270
Optician, 107
Optic nerve, 107
Optimism, 56, 58
Oral poisoning, 407
Organic food, 234
Osteoarthritis, 302
Outdoor air pollution, 490–491
Outpatient clinics, 454
Outpatient treatment centers, 354
Ovaries, 157
Overdose, 334
Overeating, 5, 34, 83
Overpopulation, health and, 196
Over-the-counter (OTC) medicines, 15, 318, 319, 321
Ovulation, 157
Ozone, 490, 491, 504
Ozone layer, 493, 499, 504

## P

Pacemaker, 288
Pancreas, 110, 117
Panic attack, 80
Parents
    custodial, 185
    responsibilities of, for children, 183–184
Particulates, 491
Pathogens, 255, 259, 260
Patients, rights of, 280, 461
PCP, 337
Peace Corps, 479
Pediatrician, 453
Penis, 159
People, getting to know a variety of, 68
Perinatal care, 166
Peripheral nervous system, 104, 105–106
Personal fitness plan, 140–143
Personality, 50
Perspiration, 121, 130
Pertussis, 310
Pessimism, 56
Pesticides, 488
Petit mal seizure, 303
Pharmacist, 318
Pharmacy technician, 322
Phobia, 81
Phosphorus, 217
Photochemical smog, 490
Physical dependence, 332, 346–347
Physical environment, risks in, 11
Physical health, 4, 7, 53
    assessing, 21
Physical inactivity, cardiovascular problems and, 291
Physical needs, 29, 30
Physical reactions to stress, 36
Physical therapist, 454
Physical well-being, 52, 53–54
Pimples, 130
Pineal gland, 110
Pituitary gland, 110, 154
Pivot joints, 101
Placenta, 163, 262
Platelets, 112
Pneumococcal disease, 472
Pneumonia, 259, 272
Poisoning
    first aid for, 407–408
    reducing risk of, 372–373
Polio, 260, 310, 473
    vaccine for, 473

Pollution, 487–488
    noise, 501
    understanding effects of, 488–489
Population size, effect on health, 499
Pores, 121
Positive self-concept, 50–51
Positive thinking, 60
Postpartum depression, 166
Poverty, 469–470
Preferred provider organization (PPO), 458
Pregnancy, 159
    changes during, 164
    drug dependence and, 350
    drug use and, 162
    home test for, 161
    signs of, 160
    smoking and, 326
    tests for, 160
Prejudice, 67, 425–426
Prenatal care, 161
Prescription medicines, 318, 319, 321
Preservatives, 241
Pressure point, 406
Preventive health care, 479
Primary care physician, 453
Problems, solving, 18–20
Products, 233
    judging, 448–449
Progesterone, 155
Projection, 77
Proteins, 206, 210, 213–214
Psychiatrists, 453
Psychoactive drugs, 328, 332, 347
Psychological dependence, 332, 347
Psychological development, 150
Psychologist, 76
Psychotherapy, 85
Puberty, changes during, 152–154, 156–158
Public health, 468–480
Public Health Service, 459
Public service announcement (PSA), 153
Pulse, 112
    taking your, 291
Pupil, 106

## Q

Quackery, 448
Questions, asking, 82

## R

Rabies, 408
Radiation, 295
    effect on health, 502
Radiation sickness, 502
Radon, 492
Rape, 426, 427
Rational thinking, 57
Reactions, 28
Reading, in dim light, 107
Realistic optimism, 58
Receptor cells, 109
Recipes, baked apples, 214
Recreational injuries, reducing, 377–379
Rectum, 118
Recycling, 507
Red blood cells, 112
Red Cross, 389
Reflexes, 106
Registered dietitian, 454
Rehabilitation, 455
Relationships
    conflicts in, 55, 57
    dysfunctional, 77, 78–79
    maintaining, 55
    self-acceptance and, 66–67
    self-esteem and, 65
Relief, 40
Repression, 77
Reproduction, 156–159
Reproductive organs, 154
Rescue breathing, 403
Research and write, 12, 31, 39, 65, 81, 82, 113, 132, 166, 186, 213, 218, 230, 235, 258, 262, 273, 274, 290, 296, 325, 340, 350, 357, 380, 384, 396, 412, 421, 430, 448, 458, 471, 473, 489, 500
Resident assistant, 78
Residential treatment centers, 354
Resilient people, 59
Resistance, 134
Resistance skills, 356
Respiration, 114
Respiratory failure, first aid for, 402–403
Respiratory system, 112–115
Rest, importance of, to fitness, 143
Retina, 107
Revenge, 426
Rheumatoid arthritis, 302
Riboflavin, 217
Risk factors, 130
    cardiovascular, 290

Role confusion, 151
Romantic love, 30

## S

Safety
    during natural disasters, 385–389
    on school bus, 382
    when caring for children, 374
Safety hazards, 374
Safety needs, 29, 30
Safety rules, failure to observe, in the home, 16
Saliva, 116
Salk, Jonas, 473
Sanitation, 474
Saturated fat, 212–213, 236
School
    health services at, 452–453
    reducing injuries at, 377
School bus, staying safe on, 382
School nurses, 452
Scoliosis, 452
Scurvy, 220
Seasonal affective disorder (SAD), 82
Seat belts, need for, 381, 383
Secondary sex characteristics, 154
Secondhand smoke, 15, 297, 325
Sedative-hypnotic drugs, 335
Sedentary lifestyle, 14
Seizures, 303
Self-acceptance, 62, 66–67
Self-actualization, 29, 32
Self-awareness, 66
Self-care, 452
Self-concept, 31, 50–51
Self-defeating behaviors, 76–79
Self-esteem, 31, 65
Self-help groups, 478
Self-talk, 39
Sense receptors, 109
Sensorineural deafness, 501
Separation, effect on family, 187
Services, judging, 448–449
Serving size, 207
Sewage, 495–496
Sexual abstinence, 15, 178
Sexual abuse, 189
Sexual assault, 427
Sexual exploitation, 427
Sexual harassment, 426
Sexual intercourse, 159
Sexual relationship, deciding on, in dating, 177–178

Sexually transmitted diseases, 275–279
    drug dependence and, 271, 350
Shame, 33
Shock, first aid for, 405
Short bones, 100
Short-term goal, 17
Side effects, 318, 320
Sight
    sense of, 106–107
    taking care of, 128–129
Simple carbohydrates, 210, 211
Single-parent families, 184, 185
Skeletal muscles, 102
Skeletal system, 100–103
Skin
    protecting, 121, 122, 130
    structure of, 120–121
Skin problems, taking care of, 130–131
Sleep, 111
    importance of, to fitness, 143
Small intestine, 117
Smallpox, 310
Smell, sense of, 109
Smog alert, 491
Smoke detector, 371
Smokeless tobacco, 325
Smoking, 113, 161
    cardiovascular problems and, 290
Smooth muscles, 102
Snacks, 214, 230, 237
Snakebites, 409
Social anxiety disorder, 80
Social comparison, 66
Social development, 150
Social environment, 12
Social esteem, 51
Social health, 4, 7
    assessing, 21
Socializing, 176–177
Social messages, 50
Social Studies, Link to, 41, 101, 163, 220, 232, 259, 273, 289, 371, 397, 461, 475
Social support, 69
Social well-being, 52, 53–54
Society
    costs of drug dependence to, 349–350
    mental illness and, 87
Sodium, 218
Soil erosion, 487
Solar energy, 505
Solid waste, 500–501

Special dietary needs
    of athletes, 222
    of teenagers, 221
Specialist, 453
Speed, 333
Sperm, 156
Sphygmomanometers, 289
Spinal column, 105
Spinal cord, 105
Splint, 409
Sports safety, 379
Sprain, 410
Stages of development
    adolescence, 151
    ages 2 to 3, 151
    ages 4 to 6, 151
    ages 7 to 12, 151
    birth to 12 months, 150–151
    early adulthood, 152
    middle adulthood, 152
    older adulthood, 152
State Children's Health Insurance Program (SCHIP), 459
State health services, 474
Sterility, 276
Stimulants, 324
    effects of, 332–334
Stirrup, 108
Strabismus, 128
Stress
    defined, 34
    effect of, on eating patterns, 232
    emotions and, 38–40
    exercise and, 35
    life events causing, 36
    management of, 38–40, 44
    reactions to, 34–36
    relieving, 41–43
Stress response, 34
Stroke, 289
Subcutaneous layer, 121
Substance Abuse and Mental Health Services Administration (SAMHSA), 460
Substance abuse counselor, 355
Substance abuse disorders, 80, 85, 346
Sucrose, 211
Sudden infant death syndrome (SIDS), 166
Suicide, 81
Sulfure dioxide, 491
Sunlight, as risk factor for cancer, 296
Sun protection factor (SPF), 122, 130
Sunscreen, 122

Support groups, 191–192
    drug dependence and, 353
Support system, creating, 52
Surgery, 288
Survival floating, 378
Sweat, 131
Sweat glands, 154
Symptoms, 59
Syphilis, 276–277

### T

Taste, sense of, 109
Taste buds, 109
Technology and Society, 10, 35, 64, 84, 115, 131, 163, 182, 210, 242, 263, 276, 288, 322, 352, 386, 405, 433, 461, 473, 503
Teenagers. *See also* Adolescence
    crime and, 418, 426–428
    special dietary needs of, 221
Teeth, taking care of, 132
Temperament, 50, 51
Tendon, 102, 103
Testes, 156
    self-exam of, 294
Testicular cancers, 294
Testosterone, 155
Tetanus, 310
Tetanus-Diphtheria vaccine, 261, 472
THC, 338
Thiamin, 217
Thinking
    healthy ways of, 58–60
    instinctive, 104
    rational, 57
Thought disorders, 82, 87
Thymus gland, 110
Thyroid gland, 110
Tissues, 287
Tobacco, 14, 15, 324–327
    physical effects of, 324–325
    reasons for use of, 324
    smokeless, 325
Tolerance, 334, 346
Tongue, 109
Tooth decay, 469
Tornado, preparing for, 385–386
Touch, sense of, 109
Toxic chemicals, 295
Toxins, 490
Trachea, 114
Traits, 167
Tranquilizers, 335

Trans fat, 236
Transfusions, 272–273
Trimesters, 163
Tuberculosis, 471
Tumor, 292
Type I diabetes, 298
Type II diabetes, 299
Typhoid fever, 196

## U

Ulcers, stress and, 36
Ultrasound, 160, 163, 305
Ultrasound technician, 305
Umbilical cord, 163
United Nations (UN), 478, 505
United Nations International Children's Emergency Fund (UNICEF), 478
United Way, 389
Universal distress signal, 401
Universal precautions, 398
Unsaturated fats, 213
Ureters, 119
Urethra, 119
Urgent care centers, 455
Urinary bladder, 119
Urine, 119
Uterus, 157, 163

## V

Vaccines, 260, 261, 263, 310, 319, 320, 454, 472
Vagina, 157, 159
Values, meaningful, 58
Varicella (chickenpox) vaccine, 261
Vegan diet, 230
Vegetables, in Food Guide Pyramid, 206
Vegetarians, 230
Vehicle crashes, reducing, 381
Veins, 113
Victim or witness advocate, 423
Victims, 418
Villi, 117
Violence, 189
    causes of, 424–426
    comparison to conflict, 419
    costs of, 421–422
    crime and, 420
    dating, 427
    defined, 418
    domestic, 190, 422
    gang, 418, 425
    guns and, 425
    preventing, 429–434
    victims of, 418, 423
Viruses, 255
    as risk factor for cancer, 296
Vitamin A, 217, 230
Vitamin C, 217, 220, 221, 230
Vitamin D, 217, 221
Vitamin E, 217
Vitamin K, 217
Vitamins, 206, 216, 217
Voice, tone of, 64
Voluntary organizations, health, 478
Volunteering, 31

## W

Warm-up, 102, 136
Water, 206, 216, 218
    bottled, 474
    importance of, 218
    need for, 8
    safety in, 378–379
Water pollution, 495–498
Weight, proper, 204, 291
Weight control, importance of diet in, 222
Weight loss, 234
Well-being, 38
    actions promoting, 53–54
    defined, 51–52
    emotional, 52, 53–54
    physical, 52 53–54
    relationships between forms of, 53
    social, 52, 53–54
Wellness, 9
West Nile virus, 256
Wetlands, 500
Wheezing, 304
White blood cells, 112, 260
Whiteheads, 130
Whooping cough, 310
Willpower, 87
Wind energy, 505
Withdrawal, 326, 335, 346
Women, after childbirth, 166
Work, reducing injuries at, 376–377
World Health Organization (WHO), 92, 478

## Y

"You" messages, 63

# Photo and Illustration Credits

Cover photos: background, © Photodisc/Getty Images, Inc.; left inset, © SuperStock, Inc.; middle inset, right inset, © Digital Vision/Getty Images, Inc.; pp. x, 1, © Photonica/Getty Images, Inc.; pp. 7, 8, © David Young-Wolff/PhotoEdit; pp. 12, 31, © A. Ramey/PhotoEdit; p. 42, © David Young-Wolff/PhotoEdit; p. 43, © Tom Stewart/Corbis; p. 51, © David Young-Wolff/PhotoEdit; p. 52, Pearson Learning for AGS Publishing; p. 59, © John Kolesidis/Reuters/Corbis; p. 62, © Chuck Savage/Corbis; p. 68, © Tony Freeman/PhotoEdit; p. 69, © David Young-Wolff/PhotoEdit; p. 78, © Michael Pole/Corbis; p. 79, Pearson Learning for AGS Publishing; p. 86, © David Young-Wolff/PhotoEdit; p. 92, © Patrick Robert/Corbis; p. 96, Pearson Learning for AGS Publishing; p. 102, © Nancy Ney/Digital Vision/Getty Images, Inc.; p. 107, © Michael A. Keller/Zefa/Corbis; p. 122, © Pixland/Index Stock Imagery, Inc.; p. 133, © Charly Franklin/Taxi/Getty Images, Inc.; p. 138, © Michael Newman/PhotoEdit; p. 140, © David Young-Wolff/PhotoEdit; p. 142, © ThinkStock LLC/Index Stock Imagery, Inc.; p. 143, © Inti St Clair/Photodisc Red/Getty Images, Inc.; p. 153, © giantstepinc/Photonica/Getty Images, Inc.; p. 155, © Jacob Halaska/Index Stock Imagery, Inc.; p. 160, © Eurelios/Phototake; p. 177, Pearson Learning for AGS Publishing; p. 191, © Bruce Ayers/Stone/Getty Images, Inc.; p. 192, © Michelle D. Bridwell/PhotoEdit; p. 196, © Dinodia/The Image Works Inc.; p. 200, © Jenny Mills/PhotoLibrary.com; p. 219, © SW Productions/PhotoDisc, Inc.; p. 223, © Corbis Digital Stock; p. 231, © David Young-Wolff/PhotoEdit; p. 238, © Michael Heron/Prentice Hall; p. 246, © AP/Wide World Photo; p. 250, © David Arky/Corbis; p. 257, © Bill Losh/Taxi/Getty Images, Inc.; p. 261, © Stockbyte/Getty Images, Inc.; p. 262, © Blair M. Seltz/Creative Eye/Mira; p. 264, © Kim Steele/PhotoDisc, Inc.; p. 272, © PNC/Brand X Pictures/Getty Images, Inc.; p. 274, © Jay Freis/Image Bank/Getty Images, Inc.; p. 277, © Keith Brofsky/PhotoDisc, Inc.; p. 280, © Michael Newman/PhotoEdit; p. 297, © Phil Savoie/Index Stock Imagery, Inc.; p. 299, © Mark Harmel/Photographer's Choice/Getty Images, Inc.; p. 300, Pearson Learning for AGS Publishing; p. 304, © Michael P. Gadomski/Photo Researchers, Inc; p. 305, © Jeff Kaufman/Taxi/Getty Images, Inc.; p. 310, © AFP/Getty Images; p. 314, © John Wilkes/Taxi/Getty Images, Inc.; p. 322, © Jose Luis Pelaez, Inc./Corbis; p. 327, © Profimedia/SuperStock, Inc.; p. 332, © Royalty-Free/Corbis; p. 349, © SW Productions/Getty Images, Inc.; p. 351, © Bill Aron/PhotoEdit; p. 357, © Mary Kate Denny/PhotoEdit.; p. 358, Pearson Learning for AGS Publishing; p. 362, © Sandy Huffaker/Getty Images; p. 366, © Dennis O'Clair/Stone/Getty Images, Inc; p. 375, Pearson Learning for AGS Publishing; p. 376, © Jeff Greenberg/PhotoEdit; p. 382, © EyeWire Collection/Getty Images, Inc.; p. 386, © PhotoDisc, Inc.; p. 399, Pearson Learning for AGS Publishing; p. 400, © Lester Lefkowitz/Corbis; p. 401, © Keith/Custom Medical Stock Photo, Inc.; p. 410, © Dennis MacDonald/PhotoEdit; p. 423, © Creasource/Corbis; pp. 428, 431, © Pearson Learning for AGS Publishing; p. 438, © Jean-Marc Giboux/Getty Images; p. 442, © Color Day Production/Image Bank/Getty Images, Inc.; p. 446, © David Young-Wolff/PhotoEdit; p. 456, © Will & Deni McIntyre/Photo Researchers, Inc.; p. 462, © Myrleen Ferguson Cate/PhotoEdit; p. 470, © Jeff Greenberg/PhotoEdit; p. 472, © Larry Downing/Reuters/Corbis; p. 476, © David Young-Wolff/PhotoEdit; p. 486, © George Fray-Pool/Getty Images, Inc.; p. 496, © Andy Levin/Photo Researchers, Inc.; p. 507, © Bob Daemmrich/PhotoEdit; p. 508, © Annie Griffiths Belt/Corbis; p. 512, © V.C.L/Taxi/Getty Images, Inc.; Illustrations: John Edwards Illustration

# Midterm Mastery Test

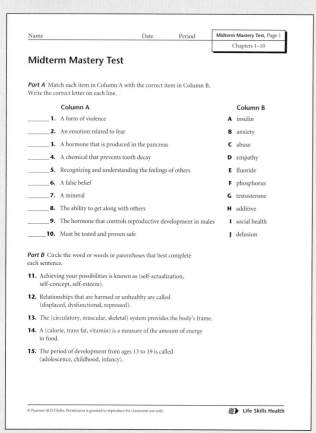

Midterm Mastery Test Page 1

## Midterm Mastery Test

**Part A** Match each item in Column A with the correct item in Column B. Write the correct letter on each line.

| Column A | Column B |
|---|---|
| _____ 1. A form of violence | **A** insulin |
| _____ 2. An emotion related to fear | **B** anxiety |
| _____ 3. A hormone that is produced in the pancreas | **C** abuse |
| _____ 4. A chemical that prevents tooth decay | **D** empathy |
| _____ 5. Recognizing and understanding the feelings of others | **E** fluoride |
| _____ 6. A false belief | **F** phosphorus |
| _____ 7. A mineral | **G** testosterone |
| _____ 8. The ability to get along with others | **H** additive |
| _____ 9. The hormone that controls reproductive development in males | **I** social health |
| _____ 10. Must be tested and proven safe | **J** delusion |

**Part B** Circle the word or words in parentheses that best complete each sentence.

11. Achieving your possibilities is known as (self-actualization, self-concept, self-esteem).
12. Relationships that are harmed or unhealthy are called (displaced, dysfunctional, repressed).
13. The (circulatory, muscular, skeletal) system provides the body's frame.
14. A (calorie, trans fat, vitamin) is a measure of the amount of energy in food.
15. The period of development from ages 13 to 19 is called (adolescence, childhood, infancy).

Midterm Mastery Test Page 2

## Midterm Mastery Test, continued

**Part C** Choose the word or words from the Word Bank that best complete each sentence. Write your answer on the line.

**Word Bank**
heredity    rational    temperament
preservatives    single-parent

16. In a _____ family, one adult raises and cares for the children.
17. Sugar and salt are both _____.
18. Your _____ is your emotional makeup.
19. When you are reasonable and realistic, you are being _____.
20. The passage of physical characteristics from parents to children is _____.

**Part D** Write the letter of the answer that correctly completes each sentence on the line.

21. Emotions are _____ to events and experiences.
    **A** attachments  **B** needs  **C** reactions  **D** signals
22. The most serious kind of mental disorder is a(n) _____ disorder.
    **A** affective  **B** anxiety  **C** bipolar  **D** thought
23. The tongue is a _____ organ.
    **A** circulatory  **B** respiratory  **C** sense  **D** skeletal
24. Slow stretching exercises can improve _____.
    **A** endurance  **B** flexibility  **C** stamina  **D** strength
25. In the female reproductive system, a released egg travels into one of two _____ tubes to the uterus.
    **A** bladder  **B** fallopian  **C** ovulation  **D** vaginal

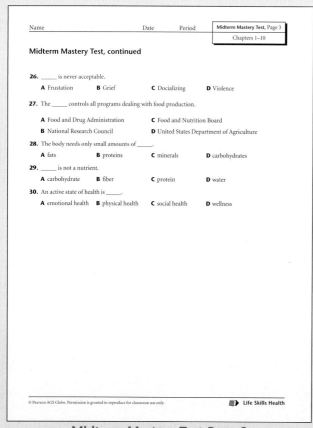

Midterm Mastery Test Page 3

## Midterm Mastery Test, continued

26. _____ is never acceptable.
    **A** Frustration  **B** Grief  **C** Socializing  **D** Violence
27. The _____ controls all programs dealing with food production.
    **A** Food and Drug Administration  **C** Food and Nutrition Board
    **B** National Research Council  **D** United States Department of Agriculture
28. The body needs only small amounts of _____.
    **A** fats  **B** proteins  **C** minerals  **D** carbohydrates
29. _____ is not a nutrient.
    **A** carbohydrate  **B** fiber  **C** protein  **D** water
30. An active state of health is _____.
    **A** emotional health  **B** physical health  **C** social health  **D** wellness

Midterm Mastery Test Page 4

## Midterm Mastery Test, continued

**Part E** Write your answer to each question. Use complete sentences. Support each answer with facts and examples from the textbook.

31. Explain how stress can affect physical health. (2 points)

32. Compare and contrast anorexia and bulimia. (2 points)

**Part F** Write a paragraph for each topic. Include a topic sentence, body, and conclusion in the paragraph. Support your answers with facts and examples from the textbook.

33. How do the circulatory system and the respiratory system work together? (3 points)

34. Describe three of the six essential nutrients and give examples of foods in which each can be found. (3 points)

# Final Mastery Test

Name _____ Date _____ Period _____  Final Mastery Test, Page 1 — Chapters 1–21

## Final Mastery Test

**Part A** Match each item in Column A with the correct item in Column B. Write the correct letter on the line.

| Column A | Column B |
|---|---|
| ___ 1. Helps others find a solution to a conflict | A syphilis |
| ___ 2. Rapid spread of a communicable disease | B malignant |
| ___ 3. Progresses in three stages if left untreated | C mediator |
| ___ 4. Burnable substances formed in the earth from plant or animal remains | D epidemic |
| ___ 5. A tumor that is harmful to health | E fossil fuels |
| ___ 6. Increases the range of motion in joints | F shock |
| ___ 7. Removing addictive drugs from the body | G detoxification |
| ___ 8. Time between infection and the appearance of symptoms | H premium |
| ___ 9. Physical reaction to severe injury that affects the circulatory system | I incubation period |
| ___ 10. Regular fee paid for an insurance policy | J flexibility |
| ___ 11. Includes air, water, land, plants, animals, and buildings | K intoxication |
| ___ 12. Provide information about the percentage of nutrients in food | L earthquake |
| ___ 13. Excitement caused by the use of a chemical substance | M physical environment |
| ___ 14. To stay safe during this, duck under a strong table or desk. | N panic attack |
| ___ 15. An extreme symptom of anxiety disorder | O Daily Values |

---

Name _____ Date _____ Period _____  Final Mastery Test, Page 2 — Chapters 1–21

## Final Mastery Test, continued

**Part B** Circle the word or words in parentheses that best complete each sentence.

16. A disagreement between two or more people who have different ideas is (conflict, negotiation, violence).

17. Physical reactions to stress normally help people to (increase, resist, ignore) stress.

18. Owning a gun (doesn't affect, decreases, increases) the likelihood of being involved in a violent incident.

19. Disease (prevention, promotion, preference) reduces demands on the health care system.

20. The promotion of unproved or worthless medical products or services is (advertising, hierarchy, quackery).

21. The (dominant, recessive, bipolar) gene determines which trait a child will inherit.

22. A(n) (vaccine, analgesic, antibiotic) reduces or destroys pathogens.

23. A person in cardiac arrest has no (temperature, pulse, pressure point).

24. (Pathogens, Antibodies, Viruses) form in the bloodstream when foreign substances enter it.

25. Places that are equipped to provide complete health care services are (outpatient clinics, hospices, hospitals).

26. Chemicals in foods that the body cannot make are (nutrient-dense, essential nutrients, deficiencies).

27. HIV is known to be transmitted through (saliva, semen, sweat).

28. A (heart attack, petit mal seizure, stroke) happens when the blood supply to the brain is suddenly stopped.

29. A place that individuals recovering from drug dependence can get help easing back into society is a (halfway house, outpatient treatment center, residential treatment center).

30. During a (flash flood, thunderstorm, tornado), move to higher ground for safety.

---

Name _____ Date _____ Period _____  Final Mastery Test, Page 3 — Chapters 1–21

## Final Mastery Test, continued

**Part C** Choose the word or words from the Word Bank that best complete each sentence. Write your answer on the line.

| Word Bank | |
|---|---|
| bacteria | Nutrition Facts |
| gonorrhea | pathogens |
| harm | physical health |
| insulin | prejudice |
| medulla | self-defeating |

31. Negative opinions formed about something or someone without enough experience or knowledge is known as _____.

32. Germs that cause infectious diseases are _____.

33. Discipline should be fair and never cause _____.

34. Skin prevents _____ from entering the body.

35. Antibiotics can be used to cure _____.

36. The _____ is the part of the brain stem that controls the body's automatic activities.

37. Type 1 diabetes is treated with _____.

38. The body's ability to meet the demands of daily living is _____.

39. Behaviors that block a person from reaching his or her own goals are _____.

40. The size of one serving and how many servings are in a package of food is found under _____.

---

Name _____ Date _____ Period _____  Final Mastery Test, Page 4 — Chapters 1–21

## Final Mastery Test, continued

**Part D** Write the letter of the answer that correctly completes each sentence on the line.

41. People have ___ basic needs.
    A two   B five   C ten   D twelve

42. The need for more of a drug to get the effect that used to occur with less of it is called ___.
    A flashback   B heredity   C tolerance   D withdrawal

43. Bicyclists should wear ___ clothing at night in order to be seen.
    A deflective   B effective   C reflective   D selective

44. The main reason adolescents use drugs is ___ pressure.
    A enabling   B family   C peer   D school

45. A person whose heart has stopped beating is in ___.
    A cardiac arrest   B heatstroke   C hypothermia   D respiratory failure

46. Because of ___ over two-thirds of the world's forests have been lost.
    A deforestation   B drought   C famine   D pollution

47. The ozone layer is an area 6 to 30 ___ above Earth's surface.
    A feet   B kilometers   C meters   D miles

48. The main reason children go without health care is ___.
    A lack of education   B homelessness   C poverty   D violence

49. A(n) ___ develops from an embryo.
    A cervix   B fetus   C ovary   D placenta

50. Drugs that slow down the central nervous system are ___.
    A amphetamines   B depressants   C hallucinogens   D stimulants

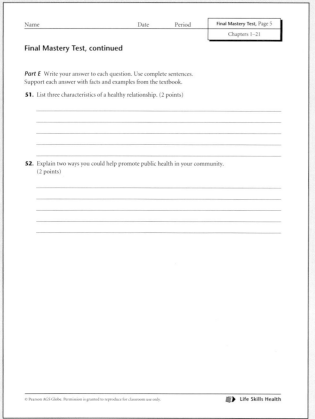

Final Mastery Test Page 5

Final Mastery Test Page 6

The lists below show how items from the Midterm and Final correlate to the chapters in the student edition.

**Midterm Mastery Test**
Chapter 1: 8, 20, 30
Chapter 2: 2, 11, 21
Chapter 3: 5, 18, 19
Chapter 4: 6, 12, 22
Chapter 5: 3, 13, 23
Chapter 6: 4, 14, 24
Chapter 7: 9, 15, 25
Chapter 8: 1, 16, 26
Chapter 9: 7, 28, 29
Chapter 10: 10, 17, 27

**Critical Thinking Items:**
31. Chapter 2 (how stress affects physical health)
32. Chapter 4 (eating disorders)
33. Chapter 5 (respiratory and circulatory systems)
34. Chapter 9 (essential nutrients)

**Final Mastery Test**
Chapter 1: 11, 38
Chapter 2: 17, 41
Chapter 3: 31
Chapter 4: 15, 39
Chapter 5: 34, 36
Chapter 6: 6
Chapter 7: 21, 49
Chapter 8: 33
Chapter 9: 26
Chapter 10: 12, 40
Chapter 11: 8, 22, 24
Chapter 12: 3, 27, 35
Chapter 13: 5, 28, 37
Chapter 14: 13, 42, 50
Chapter 15: 7, 29, 44
Chapter 16: 14, 30, 43
Chapter 17: 9, 23, 45
Chapter 18: 1, 16, 18

Chapter 19: 10, 20, 25
Chapter 20: 2, 19, 48
Chapter 21: 4, 46, 47

**Critical Thinking Items:**
51. Chapter 3 (healthy relationships)
52. Chapter 20 (promote public health)
53. Chapter 11 (vaccinations)
54. Chapter 18 (anger management) and function of the digestive system)

# Teacher's Resource Library Answer Key

## Activities

**Activity 1—The Three Parts of Good Health**
1. health 2. social health 3. lifestyle 4. surroundings 5. tasks
6. smoking 7. physical 8. exercise 9. B 10. E 11. G 12. A 13. F
14. C 15. D

**Activity 2—Good Health and Wellness**
1. B 2. A 3. F 4. C 5. D 6. E 7. H 8. G 9. health 10. eight
11. breakfast 12. wellness 13. exercise 14. self-control 15. emotional

**Activity 3—Health Risks**
1. risk 2. long-term 3. smoke 4. characteristics 5. beliefs
6. impair 7. environment 8. cultures 9. heredity 10. behavior
11. social environment 12. physical environment 13. physical environment 14. culture 15. behavior

**Activity 4—Take Charge of Your Health**
1. E 2. D 3. B 4. C 5. A 6. B 7. D 8. C 9. D 10. A

**Activity 5—Emotions and Needs**
1. D 2. A 3. E 4. C 5. B 6. F 7. H 8. J 9. G 10. I 11. L 12. N
13. K 14. O 15. M

**Activity 6—The Body and Stress**
1. emotions 2. needs 3. stress 4. good 5. bad 6. respond
7. automatic 8. nervousness 9. flight 10. threatening 11. brain
12. chemicals 13. heart 14. sweat 15. muscles

**Activity 7—Managing Stress**
1. B 2. H 3. A 4. D 5. M 6. I 7. C 8. J 9. I 10. E 11. G 12. L
13. F 14. K 15. G

**Activity 8—Mental Health**
1. D 2. E 3. A 4. C 5. B 6. H 7. F 8. G 9. J 10. I 11. O 12. K
13. M 14. L 15. N

**Activity 9—Emotional Health**
1. communicate 2. balance 3. pessimism 4. compromise
5. frustration 6. optimism 7. rational 8. consequences 9. stress
10. coping 11. mistakes 12. symptoms 13. values 14. impulse
15. positive

**Activity 10—Healthy Relationships**
1. E 2. B 3. D 4. A 5. C 6. F 7. J 8. G 9. I 10. H 11. M 12. O
13. L 14. K 15. N

**Activity 11—Becoming More Emotionally Healthy**
1. self-awareness 2. goals 3. others 4. social comparison 5. self-accepting 6. differences 7. prejudice 8. discriminate 9. good will
10. empathy 11. cooperation 12. cultural 13. advice 14. friends
15. support

**Activity 12—Characteristics of Poor Mental Health**
1. healthy 2. normal 3. thinking 4. changes 5. self-defeating
6. help 7. you 8. yourself 9. defense mechanism 10. repression
11. denial 12. projection 13. displacement 14. dysfunctional
15. "I" messages

**Activity 13—Mental Disorders**
1. D 2. E 3. A 4. B 5. C 6. G 7. F 8. H 9. J 10. I 11. O 12. M
13. K 14. N 15. L

**Activity 14—Treating Mental Disorders**
1. substance abuse 2. physical 3. psychotherapy 4. cognitive
5. physical activity 6. medicines 7. bipolar 8. thought 9. symptoms
10. resist 11. feelings 12. body 13. changed 14. fear 15. prejudices

**Activity 15—The Skeletal and Muscular Systems**
1. B 2. D 3. E 4. A 5. L 6. J 7. F 8. H 9. G 10. I 11. L 12. O
13. K 14. M 15. N

**Activity 16—The Nervous System and Sense Organs**
1. optic nerve 2. cerebellum 3. cornea 4. cerebrum 5. reflex
6. spinal cord 7. pupil 8. receptor cells 9. auditory nerve
10. peripheral nervous system 11. brain stem 12. medulla
13. eardrum 14. iris 15. retina

**Activity 17—The Endocrine System**
1. A 2. E 3. P 4. T 5. A 6. P 7. A 8. E 9. T 10. A 11. T 12. E
13. T 14. P 15. E

**Activity 18—The Circulatory and Respiratory Systems**
1. B 2. C 3. D 4. A 5. E 6. H 7. J 8. G 9. F 10. I 11. N 12. O
13. L 14. M 15. K

**Activity 19—The Digestive and Excretory Systems**
1. E 2. C 3. D 4. A 5. B 6. J 7. I 8. F 9. G 10. H 11. M 12. N
13. O 14. K 15. L

**Activity 20—The Body's Protective Covering**
1. dermis 2. sweat glands 3. keratin 4. melanin 5. roots
6. epidermis 7. subcutaneous layer 8. bacteria 9. skin 10. pores

**Activity 21—Hygiene for Good Health**
1. B 2. C 3. E 4. D 5. A 6. G 7. H 8. F 9. I 10. J 11. M 12. O
13. L 14. N 15. K

**Activity 22—Exercise and Physical Fitness**
1. endurance 2. aerobic 3. maximum 4. isometric 5. calorie
6. strengthens 7. warm up 8. resistance 9. flexibility 10. isotonic

**Activity 23—Personal Fitness Plan**
1. exercising 2. balance 3. doctor 4. makeup 5. comfortable
6. friend 7. water 8. rest 9. eight 10. relax

**Activity 24—The Life Cycle and Adolescence**
1. testosterone 2. genitals 3. puberty 4. reproductive organs
5. pituitary 6. secondary sex characteristics 7. progesterone 8. self-absorbed 9. adolescence 10. role confusion

**Activity 25—Reproduction**
1. E 2. A 3. C 4. B 5. D 6. H 7. I 8. G 9. F 10. J 11. K 12. N
13. M 14. L 15. O

**Activity 26—Pregnancy and Childbirth**
1. trimester 2. cervix 3. fetus 4. gestation 5. liquid 6. umbilical cord 7. ultrasound 8. postpartum 9. embryo 10. labor 11. placenta
12. pregnancy 13. drinking 14. implantation 15. dilated

**Activity 27—Heredity and Genetics**
1. B 2. E 3. C 4. A 5. D 6. G 7. H 8. F 9. I 10. J 11. O 12. N
13. K 14. L 15. M

**Activity 28—The Family Life Cycle, Dating, and Marriage**
1. B 2. D 3. A 4. E 5. C 6. I 7. H 8. J 9. G 10. F 11. L 12. N
13. M 14. O 15. K

**Activity 29—Parenting and Family Systems**
1. foster family 2. teachers 3. adoptive family 4. custody 5. joint
6. commitment 7. blended family 8. coping 9. single-parent family
10. life 11. discipline 12. nuclear family 13. extended family
14. custodial parent 15. beliefs

**Activity 30—Problems in Families**
1. E 2. C 3. A 4. B 5. D 6. J 7. I 8. H 9. F 10. G 11. L 12. O
13. K 14. M 15. N

**Activity 31—A Healthy Diet**
1. U 2. H 3. U 4. H 5. H 6. U 7. U 8. H 9. H 10. U 11. U
12. H 13. U 14. H 15. U

### Activity 32—Carbohydrates, Fats, and Protein
1. C  2. F  3. A  4. C  5. F  6. P  7. C  8. C  9. P  10. A  11. F  12. C  13. A  14. C  15. P

### Activity 33—Vitamins, Minerals, and Water
1. A  2. M  3. V  4. A  5. V, M  6. W  7. A  8. V  9. M, W  10. A  11. A  12. W  13. M  14. M  15. W

### Activity 34—Special Dietary Needs
1. malnutrition  2. nutrition  3. balanced  4. high blood pressure  5. weight  6. deficiency  7. diseases  8. calcium  9. vitamin C  10. iron

### Activity 35—Healthy Eating Patterns and Food Choices
1. stress  2. breakfast  3. association  4. snacks  5. cultural background  6. cost  7. vitamins  8. pattern  9. fat  10. sugar

### Activity 36—How the Media Influences Eating Patterns
1. C  2. A  3. D  4. E  5. B  6. I  7. H  8. J  9. G  10. F  11. O  12. N  13. K  14. C  15. M

### Activity 37—Food Labels and Food Additives
1. E  2. C  3. B  4. A  5. D  6. H  7. G  8. I  9. F  10. J  11. O  12. K  13. M  14. L  15. N

### Activity 38—Causes of Disease
1. D  2. B  3. E  4. C  5. A  6. H  7. F  8. J  9. I  10. G  11. L  12. O  13. N  14. K  15. M

### Activity 39—How the Body Protects Itself from Disease
1. mucous membranes  2. immunization  3. antibody  4. virus  5. mucus  6. defense  7. bloodstream  8. vaccination  9. placenta  10. inflammation

### Activity 40—AIDS
1. B  2. D  3. C  4. A  5. E  6. I  7. J  8. H  9. G  10. F  11. M  12. K  13. O  14. L  15. N

### Activity 41—Sexually Transmitted Diseases
1. E  2. A  3. B  4. D  5. C  6. G  7. F  8. J  9. I  10. H  11. N  12. K  13. L  14. M  15. O

### Activity 42—Cardiovascular Diseases and Problems
1. atherosclerosis  2. chronic  3. risk factor  4. oxygen  5. blood pressure  6. stroke  7. kidney disease  8. arteriosclerosis  9. smoke  10. surgery  11. hypertension  12. family medical history  13. ageing  14. angina pectoris  15. heart attack

### Activity 43—Cancer
1. E  2. A  3. C  4. B  5. D  6. I  7. F  8. G  9. H  10. J  11. M  12. K  13. L  14. N  15. O

### Activity 44—Diabetes
1. diabetes  2. energy  3. noninsulin-dependent  4. eyesight  5. coma  6. pancreas  7. urination  8. blood  9. insulin-dependent  10. glucose

### Activity 45—Arthritis, Epilepsy, and Asthma
1. D  2. C  3. A  4. B  5. E  6. G  7. F  8. J  9. I  10. H  11. O  12. K  13. L  14. N  15. M

### Activity 46—Medicines
1. A  2. B  3. E  4. D  5. C  6. I  7. G  8. J  9. F  10. H  11. M  12. L  13. O  14. K  15. N

### Activity 47—Tobacco
1. nicotine  2. stimulant  3. emphysema  4. chemicals  5. tars  6. smokeless  7. gums  8. secondhand  9. effects  10. public  11. television  12. poison  13. withdrawal  14. gums  15. support

### Activity 48—Alcohol
1. C  2. A  3. B  4. A  5. D  6. C  7. D  8. A  9. C  10. B

### Activity 49—Stimulants, Depressants, Narcotics, and Hallucinogens
1. D  2. C  3. A  4. B  5. E  6. F  7. G  8. I  9. J  10. H  11. N  12. M  13. O  14. K  15. L

### Activity 50—Other Dangerous Drugs
1. anabolic steroid  2. look alike drug  3. THC  4. lungs  5. designer drug  6. inhalant  7. china white  8. fatty tissues  9. liver cancer  10. marijuana

### Activity 51—Drug Dependence—The Problems
1. genetic  2. physical  3. withdrawal  4. psychoactive  5. pattern  6. drug dependence  7. tolerance  8. pregnant  9. psychological  10. enabling

### Activity 52—Drug Dependence—The Solutions
1. C  2. A  3. D  4. E  5. B  6. G  7. F  8. H  9. J  10. I  11. L  12. M  13. K  14. O  15. N

### Activity 53—Avoiding Drug Use
1. B  2. C  3. C  4. D  5. A  6. B  7. C  8. E  9. A  10. D

### Activity 54—Reducing Risks at Home
1. stepladder  2. sumac  3. shock  4. electrician  5. poisonous  6. injuries  7. nonskid  8. detector  9. extinguisher  10. drills

### Activity 55—Reducing Risks Away from Home
1. floating  2. weather  3. protective  4. regulations  5. preserver  6. OSHA  7. sports  8. productivity  9. drowning  10. rescue

### Activity 56—Reducing Risks on the Road
1. requirements  2. training course  3. influence  4. posted speed  5. head  6. judgment  7. maintaining  8. single file  9. reflectors  10. wheels

### Activity 57—Safety During Natural Disasters
1. E  2. B  3. C  4. A  5. D  6. I  7. J  8. G  9. F  10. H  11. L  12. N  13. M  14. O  15. K

### Activity 58—First Aid Basics
1. identification  2. warm  3. care  4. universal  5. latex gloves  6. airway  7. Good Samaritan Laws  8. 911  9. fluids  10. hands

### Activity 59—First Aid for Life-Threatening Emergencies
1. resuscitation  2. choking  3. rescue  4. electrical  5. Heimlich  6. stroke  7. cardiac arrest  8. bleeding  9. pressure point  10. shock

### Activity 60—First Aid for Poisoning and Other Problems
1. E  2. D  3. A  4. C  5. B  6. I  7. F  8. G  9. J  10. H  11. N  12. M  13. O  14. L  15. K

### Activity 61—Defining Violence
1. major  2. gang  3. family  4. ongoing  5. target  6. minor  7. mental  8. media  9. guilt  10. internal

### Activity 62—Causes of Violence
1. E  2. C  3. B  4. A  5. D  6. G  7. I  8. H  9. F  10. J  11. K  12. M  13. O  14. N  15. L

### Activity 63—Preventing Violence
1. alcohol  2. mediator  3. Watch  4. security  5. hangouts  6. anger  7. respect  8. listening skills  9. private  10. encouraging

### Activity 64—Being a Wise Consumer
1. consumers  2. agencies  3. research  4. labels  5. write  6. price  7. quackery  8. advertising  9. products  10. advice

### Activity 65—Seeking Health Care
1. B  2. C  3. A  4. D  5. E  6. G  7. F  8. I  9. J  10. H  11. N  12. K  13. O  14. L  15. M

### Activity 66—Paying for Health Care
1. C  2. B  3. A  4. D  5. E  6. H  7. G  8. F  9. I  10. J  11. N  12. O  13. L  14. M  15. K

### Activity 67—Public Health Problems
1. birth defects  2. poverty  3. public health  4. high-risk  5. tuberculosis  6. standards  7. polio  8. preventive  9. epidemic  10. breast cancer

### Activity 68—U.S. Public Health Solutions
1. C  2. B  3. D  4. B  5. D

### Activity 69—Health Promotion and Volunteer Organizations
1. E 2. G 3. H 4. I 5. B 6. F 7. C 8. D 9. J 10. A

### Activity 70—Health and the Environment
1. clinical ecology 2. ecology 3. balance 4. bald eagle 5. chemicals 6. environment 7. pollution 8. minerals 9. fragile 10. natural

### Activity 71—Air Pollution and Health
1. C 2. A 3. E 4. B 5. D 6. G 7. F 8. J 9. H 10. I 11. O 12. N 13. K 14. M 15. L

### Activity 72—Water Pollution and Health
1. saltwater 2. toxic 3. carcinogens 4. factory 5. herbicides 6. sewage 7. pesticides 8. algae 9. phosphorus 10. groundwater

### Activity 73—Other Environmental Problems and Health
1. E 2. D 3. A 4. B 5. C 6. G 7. J 8. I 9. F 10. H 11. L 12. N 13. M 14. K 15. O

### Activity 74—Protecting the Environment
1. Environmental Protection Agency 2. Clean Air Act 3. solar 4. Conference on Environment and Development 5. CFCs 6. alternative 7. National Environmental Policy Act 8. public meetings 9. impact statement 10. hydrogen

## Alternative Activities

### Alternative Activity 1—Good Health
1. lifestyle 2. surroundings 3. tasks 4. social 5. health 6. physical 7. A 8. C 9. F 10. E 11. B 12. D

### Alternative Activity 2—Good Health and Wellness
1. C 2. D 3. E 4. B 5. F 6. A 7. alcohol 8. eight 9. breakfast 10. scale 11. exercise 12. emotional

### Alternative Activity 3—Health Risks
1. characteristics 2. environment 3. risk 4. smoke 5. alcohol 6. beliefs 7. heredity 8. social environment 9. physical environment 10. physical environment 11. culture 12. behavior

### Alternative Activity 4—Take Charge of Your Health
1. E 2. B 3. G 4. F 5. C 6. D 7. A 8. B 9. D 10. D 11. C 12. C

### Alternative Activity 5—Emotions and Needs
1. B 2. C 3. A 4. D 5. E 6. F 7. H 8. G 9. J 10. I 11. L 12. K

### Alternative Activity 6—The Body and Stress
1. normal 2. stress 3. good 4. stress response 5. flight 6. brain 7. chemicals 8. heart 9. sweat 10. react

### Alternative Activity 7—Managing Stress
1. A 2. G 3. H 4. D 5. L 6. I 7. C 8. J 9. B 10. E 11. F 12. K

### Alternative Activity 8—Mental Health
1. B 2. C 3. D 4. A 5. F 6. E 7. G 8. H 9. I 10. J 11. L 12. K

### Alternative Activity 9—Emotional Health
1. experience 2. impulse 3. pessimism 4. compromise 5. anxiety 6. optimism 7. rational 8. consequences 9. coping 10. failure 11. values 12. symptoms

### Alternative Activity 10—Healthy Relationships
1. D 2. A 3. B 4. C 5. G 6. F 7. H 8. E 9. I 10. J 11. K 12. L

### Alternative Activity 11—Becoming More Emotionally Healthy
1. richness 2. comfortable 3. social comparison 4. differences 5. prejudice 6. discriminate 7. good will 8. empathy 9. cooperation 10. cultural 11. advice 12. friend

### Alternative Activity 12—Characteristics of Poor Mental Health
1. problems 2. psychologist 3. thinking 4. changes 5. self-defeating 6. you 7. defense mechanisms 8. repression 9. denial 10. projection 11. displacement 12. dysfunctional

### Alternative Activity 13—Mental Disorders
1. D 2. B 3. C 4. A 5. F 6. G 7. H 8. E 9. J 10. L 11. K 12. I

### Alternative Activity 14—Treating Mental Disorders
1. professionals 2. relieved 3. physical 4. cognitive 5. physical activity 6. thought 7. symptoms 8. resist 9. feelings 10. changed 11. fear 12. prejudices

### Activity 15—The Skeletal and Muscular Systems
1. D 2. C 3. A 4. B 5. G 6. E 7. H 8. F 9. K 10. L 11. J 12. I

### Alternative Activity 16—The Nervous System and Sense Organs
1. eardrum 2. iris 3. retina 4. optic nerve 5. medulla 6. cerebrum 7. cerebellum 8. pupil 9. brain stem 10. receptor cells 11. reflex 12. auditory nerve

### Alternative Activity 17—The Endocrine System
1. T 2. P 3. A 4. P 5. E 6. E 7. E 8. A 9. T 10. A 11. T 12. A

### Alternative Activity 18—The Circulatory and Respiratory Systems
1. C 2. D 3. B 4. A 5. F 6. E 7. H 8. G 9. I 10. K 11. J 12. L

### Alternative Activity 19—The Digestive and Excretory Systems
1. L 2. E 3. F 4. J 5. G 6. H 7. D 8. B 9. A 10. I 11. K 12. C

### Alternative Activity 20—The Body's Protective Covering
1. skin 2. nails 3. keratin 4. melanin 5. perspiration 6. epidermis 7. subcutaneous layer 8. dermis

### Alternative Activity 21—Hygiene for Good Health
1. B 2. C 3. D 4. A 5. E 6. G 7. F 8. H 9. K 10. J 11. L 12. I

### Alternative Activity 22—Exercise and Physical Fitness
1. better 2. cooldown 3. endurance 4. aerobic 5. maximum 6. calorie 7. strengthens 8. warm up

### Alternative Activity 23—Personal Fitness Plan
1. goals 2. rest 3. balance 4. doctor 5. makeup 6. shoes 7. friend 8. water

### Alternative Activity 24—The Life Cycle and Adolescence
1. life stages 2. changes 3. testosterone 4. genitals 5. pituitary 6. progesterone 7. adolescence 8. role confusion

### Alternative Activity 25—Reproduction
1. B 2. D 3. A 4. C 5. E 6. F 7. G 8. H 9. J 10. I 11. L 12. K

### Alternative Activity 26—Pregnancy and Childbirth
1. health 2. childbirth 3. trimester 4. cervix 5. fetus 6. nine 7. umbilical cord 8. ultrasound 9. labor 10. placenta 11. drink 12. dilates

### Alternative Activity 27—Heredity and Genetics
1. B 2. D 3. C 4. A 5. F 6. E 7. G 8. H 9. L 10. I 11. J 12. K

### Alternative Activity 28—The Family Life Cycle, Dating, and Marriage
1. D 2. A 3. B 4. C 5. F 6. E 7. H 8. G 9. L 10. I 11. J 12. K

### Alternative Activity 29—Parenting and Family Systems
1. single-parent family 2. life 3. foster family 4. adoptive family 5. custody 6. joint 7. commitment 8. blended family 9. discipline 10. nuclear family 11. extended family 12. custodial parent

### Alternative Activity 30—Problems in Families
1. C 2. B 3. D 4. A 5. G 6. H 7. F 8. E 9. I 10. J 11. L 12. K

### Alternative Activity 31—A Healthy Diet
1. U 2. U 3. U 4. H 5. U 6. H 7. H 8. U 9. H 10. H 11. U 12. U

### Alternative Activity 32—Carbohydrates, Fats, and Protein
1. C 2. F 3. A 4. C 5. P 6. C 7. C 8. A 9. C 10. A 11. C 12. P

### Alternative Activity 33—Vitamins, Minerals, and Water
1. M  2. V  3. A  4. A  5. W  6. A  7. V  8. M, W  9. A  10. W  11. M  12. M

### Alternative Activity 34—Special Dietary Needs
1. calcium  2. vitamin C  3. balanced  4. high blood pressure  5. weight  6. deficiency  7. diseases  8. malnutrition

### Alternative Activity 35—Healthy Eating Patterns and Food Choices
1. snacks  2. cost  3. vitamins  4. fat  5. stress  6. sugar  7. breakfast  8. pattern

### Alternative Activity 36—How the Media Influences Eating Patterns
1. B  2. A  3. C  4. D  5. G  6. H  7. E  8. F  9. J  10. L  11. I  12. K

### Alternative Activity 37—Food Labels and Food Additives
1. D  2. B  3. A  4. C  5. F  6. E  7. H  8. G  9. K  10. L  11. J  12. I

### Alternative Activity 38—Causes of Disease
1. C  2. A  3. D  4. B  5. H  6. E  7. G  8. F  9. L  10. D  11. I  12. J

### Alternative Activity 39—How the Body Protects Itself from Disease
1. barrier  2. saliva  3. antibody  4. mucus  5. bloodstream  6. vaccination  7. placenta  8. inflammation

### Alternative Activity 40—AIDS
1. D  2. A  3. B  4. C  5. H  6. E  7. F  8. G  9. I  10. K  11. L  12. J

### Alternative Activity 41—Sexually Transmitted Diseases
1. B  2. A  3. D  4. C  5. E  6. H  7. F  8. G  9. L  10. J  11. K  12. I

### Alternative Activity 42 Cardiovascular Diseases and Problems
1. blood pressure  2. cardiovascular  3. chronic  4. stroke  5. ageing  6. smoking  7. surgery  8. arteriosclerosis  9. family medical history  10. hypertension  11. angina pectoris  12. heart attack

### Alternative Activity 43—Cancer
1. B  2. A  3. D  4. C  5. G  6. H  7. E  8. F  9. J  10. I  11. K  12. L

### Alternative Activity 44—Diabetes
1. insulin  2. vessels  3. energy  4. eyesight  5. coma  6. pancreas  7. blood  8. glucose

### Alternative Activity 45—Arthritis, Epilepsy, and Asthma
1. B  2. A  3. D  4. C  5. E  6. F  7. H  8. G  9. L  10. I  11. J  12. K

### Alternative Activity 46—Medicines
1. B  2. A  3. D  4. C  5. F  6. H  7. E  8. G  9. J  10. L  11. I  12. K

### Alternative Activity 47—Tobacco
1. cigarettes  2. chew  3. nicotine  4. emphysema  5. chemicals  6. tars  7. gums  8. secondhand  9. public  10. television  11. poison  12. support

### Alternative Activity 48—Alcohol
1. C  2. C  3. A  4. A  5. C  6. B  7. A  8. B

### Alternative Activity 49—Stimulants, Depressants, Narcotics, and Hallucinogens
1. B  2. A  3. D  4. C  5. E  6. G  7. H  8. F  9. K  10. L  11. I  12. J

### Alternative Activity 50—Other Dangerous Drugs
1. mind  2. illegal  3. THC  4. lungs  5. inhalant  6. china white  7. fatty tissues  8. liver cancer

### Alternative Activity 51—Drug Dependence—The Problems
1. psychoactive  2. pattern  3. withdrawal  4. genetic  5. physical  6. drug dependence  7. tolerance  8. pregnant

### Alternative Activity 52—Drug Dependence—The Solutions
1. B  2. D  3. C  4. A  5. H  6. G  7. E  8. F  9. K  10. L  11. I  12. J

### Alternative Activity 53—Avoiding Drug Use
1. B  2. A  3. D  4. B  5. D  6. A  7. B  8. C

### Alternative Activity 54—Reducing Risks at Home
1. drills  2. spills  3. stepladder  4. electrician  5. injuries  6. nonskid  7. detector  8. extinguisher

### Alternative Activity 55—Reducing Risks Away from Home
1. injuries  2. exits  3. floating  4. weather  5. protective  6. preserver  7. sports  8. rescue

### Alternative Activity 56—Reducing Risks on the Road
1. seat belt  2. helmet  3. requirements  4. training course  5. posted speed  6. head  7. single file  8. reflectors

### Alternative Activity 57—Safety in Natural Disasters
1. A  2. C  3. D  4. B  5. H  6. E  7. F  8. G  9. K  10. J  11. I  12. L

### Alternative Activity 58—First Aid Basics
1. awake  2. dressings  3. warm  4. universal  5. airway  6. 911  7. diabetes  8. hands

### Alternative Activity 59—First Aid for Life-Threatening Emergencies
1. failure  2. sterile  3. rescue  4. chest  5. Heimlich  6. stroke  7. bleeding  8. choking

### Alternative Activity 60—First Aid for Poisoning and Other Problems
1. B  2. D  3. A  4. C  5. H  6. F  7. G  8. E  9. K  10. J  11. I  12. L

### Alternative Activity 61—Defining Violence
1. violence  2. conflict  3. ongoing  4. target  5. minor  6. mental  7. media  8. internal

### Alternative Activity 62—Causes of Violence
1. C  2. D  3. B  4. A  5. E  6. G  7. F  8. H  9. L  10. J  11. K  12. I

### Alternative Activity 63—Preventing Violence
1. angry  2. problems  3. mediator  4. security  5. respect  6. listening skills  7. private  8. encouraging

### Alternative Activity 64—Being a Wise Consumer
1. consumer  2. standards  3. research  4. labels  5. write  6. product  7. quackery  8. advice

### Alternative Activity 65—Seeking Health Care
1. C  2. D  3. A  4. B  5. H  6. F  7. G  8. E  9. K  10. L  11. I  12. J

### Alternative Activity 66—Paying for Health Care
1. B  2. D  3. A  4. C  5. E  6. F  7. G  8. H  9. K  10. L  11. J  12. I

### Alternative Activity 67—Public Health Problems
1. cancer  2. pregnant women  3. poverty  4. public health  5. high-risk  6. standards  7. polio  8. epidemic

### Alternative Activity 68—U.S. Public Health Solutions
1. C  2. D  3. A  4. B  5. A

### Alternative Activity 69—Health Promotion and Volunteer Organizations
1. A  2. E  3. F  4. G  5. D  6. B  7. C  8. H

### Alternative Activity 70—Health and the Environment
1. pesticide  2. bald eagles  3. chemicals  4. environment  5. pollution  6. minerals  7. fragile  8. natural

### Alternative Activity 71—Air Pollution and Health
1. C  2. D  3. A  4. B  5. F  6. E  7. G  8. H  9. K  10. I  11. L  12. J

### Alternative Activity 72—Water Pollution and Health
1. toxic  2. oil spill  3. carcinogens  4. herbicides  5. pesticides  6. algae  7. phosphorus  8. groundwater

### Alternative Activity 73—Other Environmental Problems and Health
1. D  2. C  3. A  4. B  5. F  6. E  7. H  8. G  9. I  10. L  11. J  12. K

### Alternative Activity 74—Protecting the Environment
1. Clean Air Act  2. Environmental Protection Agency  3. solar  4. CFCs  5. ethanol  6. National Environmental Policy Act  7. public meetings  8. hydrogen

# Workbook Activities

### Workbook Activity 1—The Three Parts of Good Health
1. daily life  2. problems and pressures  3. cooperate  4. environment  5. emotional health  6. A  7. C  8. B  9. A  10. B

### Workbook Activity 2—Health and Wellness Word Scramble
1. principles  2. chemical  3. exercising  4. behaviors  5. snacks  6. sleep  7. wellness  8. lifestyle  9. different  10. dental  11. water  12. weight  13. To achieve wellness a person must work toward balancing physical, social, and emotional health.  14. Any three of the following: exercise three to five times per week; eat three balanced, healthy meals each day, including breakfast; choose to eat healthy snacks; drink enough water every day; maintain a normal weight for height and age; sleep at least eight hours a night; do not smoke, drink alcohol, or use other drugs a health care professional has not recommended; seek medical and dental advice from health care professionals  15. The Wellness Scale shows how to chart a person's level of wellness. It includes physical, social, and emotional health. Good health is on the right side of the scale, poor health is on the left side of the scale.

### Workbook Activity 3—Types of Health Risks
1. social environment  2. heredity  3. culture  4. physical environment  5. behavior  6. social environment  7. behavior  8. physical environment  9. culture  10. behavior  11. lifestyle  12. community  13. pregnancy  14. resources  15. legal

### Workbook Activity 4—Changing Behaviors
1. S  2. L  3. S  4. S  5. S  6. L  7. S  8. S  9. L  10. A  11. K  12. A  13. A  14. K  15. K

### Workbook Activity 5—What Is Maslow's Hierarchy of Needs?
Answers will vary. Students' stories should reflect a clear grasp of the five levels of Maslow's hierarchy of basic needs.

### Workbook Activity 6—Identify the Statements
1. SR  2. SR  3. S  4. PR  5. PR  6. S  7. SR  8. S  9. PR  10. PR  11. S  12. SR  13. S  14. S  15. PR

### Workbook Activity 7—Emotions and Stress
1. grief  2. relief  3. coping  4. anger  5. fear  6. relief  7. anger  8. anxiety  9. coping  10. grief  11. fear  12. coping  13. frustration  14. anger  15. grief

### Workbook Activity 8—Mental Health: Personality, Well-Being, and Self Concept
1. B  2. A  3. C  4. D  5. B  6. D  7. A  8. D  9. A  10. A

### Workbook Activity 9—Emotional Statements
1. P  2. O  3. O  4. P  5. P  6. O  7. O  8. C  9. MV  10. C  11. RO  12. C  13. MV  14. HE  15. HE

### Workbook Activity 10—Relationships
1. UR  2. HR  3. UR  4. UR  5. HR  6. HR  7. UR  8. HR  9. UR  10. HR  11. HR  12. UR  13. UR  14. UR  15. HR

### Workbook Activity 11—Cause and Effect
1. A  2. D  3. C  4. B  5. E  6. J  7. I  8. G  9. F  10. H  11. N  12. M  13. O  14. L  15. K

### Workbook Activity 12—Defense Mechanisms
1. projection  2. repression  3. projection  4. denial  5. displacement  6. displacement  7. repression  8. denial  9. repression  10. denial  11. projection  12. displacement  13. displacement  14. projection  15. denial

### Workbook Activity 13—Mental Disorders
1. ANX  2. SA  3. AFF  4. E  5. TH  6. TH  7. E  8. AFF  9. ANX  10. TH  11. E  12. AFF  13. TH  14. E  15. AFF

### Workbook Activity 14—Treating Mental Disorders
1. ANX  2. SA  3. TH  4. E  5. ANX  6. AFF  7. SA  8. E  9. SA  10. TH  11. SA  12. E  13. AFF  14. ANX  15. SA, ANX, AFF, TH, E

### Workbook Activity 15—The Skeletal and Muscular Systems
1. hinge  2. 600  3. food you eat  4. hold joints together  5. three  6. Bone marrow  7. Muscles  8. heart  9. Skeletal  10. tendon  11. muscle tone  12. either muscle to bone or muscle to muscle  13. elbows  14. Muscles  15. Smooth

### Workbook Activity 16—The Nervous System and Sense Organs
**Brain:** cerebrum, cerebellum, medulla, spinal cord  **Nervous System:** peripheral nerves, reflexes, stress response  **Eyes:** cornea, iris, optic nerve, pupil, retina  **Ears:** anvil, auditory, hammer, stirrup, vibrations  **Smell and Taste:** buds, olfactory, receptor cells

### Workbook Activity 17—The Endocrine Web

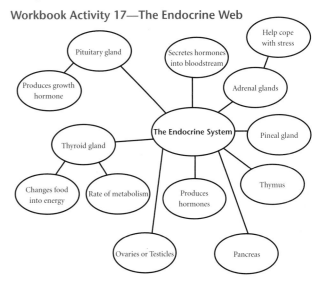

### Workbook Activity 18—The Circulatory and Respiratory Systems
1. four  2. Red blood cells  3. three  4. arteries to veins  5. Veins  6. White  7. oxygen  8. trachea  9. bronchioles  10. below  11. contracts  12. twenty  13. lungs  14. 300 million  15. force

### Workbook Activity 19—The Digestive and Excretory Systems
1. esophagus  2. saliva  3. liver  4. villi  5. pancreas  6. insulin  7. kidneys  8. urethra  9. ureters  10. feces  11. enzyme  12. gallbladder  13. bile  14. anus  15. rectum

### Workbook Activity 20—The Skin, Hair, and Nails
1. excretory  2. organ  3. three  4. hair  5. subcutaneous layer  6. bacteria  7. epidermis  8. The hair shaft  9. more  10. with  11. melanin  12. dermis  13. epidermis  14. glands  15. Keratin

### Workbook Activity 21—Positive Health Practices
1. H  2. P  3. H  4. P  5. P  6. P  7. H  8. P  9. H  10. P  11. P  12. H  13. P  14. H  15. H

### Workbook Activity 22—Writing About Fitness
Answers will vary. Students should incorporate the given words into sentences that show a clear grasp of the material.

### Workbook Activity 23—Parts of a Fitness Plan
1. 1a  2. 2b  3. 3b  4. 4a  5. 5a  6. 6b  7. stressed  8. disease  9. healthy  10. rest

### Workbook Activity 24—Life Stages
1. If babies learn that their basic needs are being met, they learn to trust.  2. Children begin to do some things on their own.  3. Children begin to explore the world, use their imagination, and initiate activities.  4. Children begin to create things and respond to positive and negative comments about their efforts.  5. Teenagers work to discover their own identity.  6. People work on developing a close relationship with another person.  7. People turn their

attention to their families. **8.** People reflect on their life and accomplishments.

## Workbook Activity 25—Reproduction
**1.** C **2.** A **3.** E **4.** B **5.** D **6.** F **7.** I **8.** G **9.** H **10.** J **11.** L **12.** O **13.** M **14.** K **15.** N

## Workbook Activity 26—Order of Pregnancy and Childbirth
**A.** 7 **B.** 12 **C.** 9 **D.** 1 **E.** 4 **F.** 15 **G.** 2 **H.** 14 **I.** 11 **J.** 3 **K.** 5 **L.** 8 **M.** 10 **N.** 6 **O.** 13

## Workbook Activity 27—Heredity and Genetics
**1.** chromosomes **2.** an X **3.** genes **4.** a pair of genes **5.** an X or a Y **6.** the dominant **7.** sex **8.** 23 **9.** recessive **10.** Two **11.** either parent **12.** XX **13.** Chemical codes **14.** sex-linked **15.** thousands

## Workbook Activity 28—The Family Life Cycle, Dating, and Marriage
**1.** dating **2.** communication **3.** marry **4.** relationship **5.** children **6.** socializing **7.** pregnancy **8.** common **9.** life cycle **10.** unhealthy **11.** collaborative **12.** commitment **13.** education **14.** abstinence **15.** divorce

## Workbook Activity 29—Parenting and Family Systems
Answers will vary. Students should show a clear understanding of Chapter 8 Lesson 2 content.

## Workbook Activity 30—Problems in Families
Answers will vary. Students should include the words listed in parentheses and show a grasp of Chapter 8 Lesson 3 content.

## Workbook Activity 31—A Plan for a Healthy Diet
Answers will vary. Students should show a firm grasp of the dietary guidelines as well as the contents of the Food Guide Pyramid.

## Workbook Activity 32—Comparing Foods to Determine Nutritional Value
**1.** The meal would be more nutritious if Jack replaced the corn chips with whole grain crackers, replaced the candy bar with a piece of fruit, and added a vegetable. **2.** The pizza has complex carbohydrates. The cola has simple carbohydrates. Margaret could make this meal more nutritious if she made sure the pizza crust was whole grain and if she replaced the cola with milk or fruit juice. **3.** The saturated fat in this meal is very high, with saturated fat in the cheeseburger and milkshake and most likely in the fries. **4.** This meal is healthy because it includes protein and unsaturated fat (fish), vegetables with a low-fat dressing (salad), and fruit (apple). The meal is quite low in carbohydrates. LeeJay could add some carbohydrates by eating some bread. **5.** The meal would be more nutritious with some added source of protein, perhaps from eggs, beans, nuts, meat or fish; some dairy, perhaps from a glass of milk; and some carbohydrates, perhaps from bread.

## Workbook Activity 33—Vitamins, Minerals, and Water
**1.** F **2.** B **3.** I **4.** D **5.** G **6.** A **7.** J **8.** C **9.** H **10.** E

## Workbook Activity 34—Dietary Needs
**1.** John might need to eat more calories and more foods that contain calcium and iron than his father does. Teenagers need large amount of calories, calcium, and iron because they are still growing. **2.** People who play sports need more calories than do people who do not play sports. People who play sports also need to replace the water they lost while playing. **3.** Katrina should eat more calories in order to gain weight, and she should eat more protein in order to build her muscles. **4.** Ralph's doctor asked him to stop eating potato chips because they are high in sodium. Too much sodium in your diet can lead to high blood pressure. **5.** Emilio should calculate how many calories he is consuming and evaluate how healthy his diet is. He also should see a doctor before he begins to alter his diet. He then can work on eating fewer calories each week and eat healthier foods in order to lose weight.

## Workbook Activity 35—Food Choices
**1.** D **2.** D **3.** B **4.** A **5.** C **6.** energy **7.** nutritious **8.** sleep **9.** caffeine **10.** cholesterol

## Workbook Activity 36—Advertising Food Products
Students' advertisements will vary. Sample advertisements: **1.** New and improved cookies are better than ever. You will really taste the difference. **2.** Remember when Grandma made sweet potato pie? The pie in this package will bring back all those wonderful memories. **3.** You won't gain weight with Perky Peanut Butter, a delicious, fat-free snack food! **4.** Yes, Breakfast Crunch cereal costs more, but it's the best. Don't you deserve the best? **5.** Why put chemicals into your body when you can eat Crunchies, an organic cereal? Crunchies is all-natural, healthful, and tastes great too.

## Workbook Activity 37—Analyzing a Food Label
**1.** DV stands for Daily Value. It tells you the amount of a nutrient in the food compared with the total amount you need each day. **2.** Answers may vary, but the cereal is probably a good choice. Students should consider other food choices they will make during the day to determine how good the cereal choice is with respect to sodium. **3.** Corn meal is the major ingredient in this food. It is listed first in the ingredients list. **4.** The cereal is called "natural" cereal. This may be misleading if the food is highly processed. **5.** The cereal has good nutritional value. It is low in fat and contains no cholesterol. It is a little high in sodium, but it has a reasonable amount of carbohydrates.

## Workbook Activity 38—Causes of Disease
**1.** E **2.** P **3.** P **4.** B **5.** I **6.** B **7.** I **8.** E **9.** P **10.** P **11.** I **12.** I **13.** P **14.** E **15.** I

## Workbook Activity 39—How the Body Protects Itself from Disease
**1.** A **2.** A **3.** C **4.** C **5.** A **6.** B **7.** C **8.** C **9.** A **10.** B **11.** A **12.** C **13.** C **14.** A **15.** C

## Workbook Activity 40—What Is AIDS?
**1.** Students should respond that the symptoms of AIDS are nothing like those of the flu. **2.** Most people know that trips abroad require vaccinations. If Latisha had AIDS, a vaccination would do no good. Vaccinations are meant to prevent disease, not cure it. There is no known cure for AIDS. **3.** All blood donated for transfusion is tested for HIV. Health care workers who draw the blood use precautions to protect themselves and the donors. **4.** AIDS cannot be passed by sneezing, coughing, shaking hands, hugging, or any other casual social contact. **5.** The HIV virus cannot exist outside of the body. Surfaces such as sinks and toilets cannot transmit the virus.

## Workbook Activity 41—Facts About Sexually Transmitted Diseases
**1.** Answers will vary. Students should reflect the lesson content regarding genital herpes, gonorrhea, chlamydia, and syphilis. **2.** Answers will vary. Students should include serious mention of responsible sexual behavior.

## Workbook Activity 42—What Is Cardiovascular Disease?
**1.** Due to the patient's race and family history, he may likely have high blood pressure. His smoking, high-fat diet, and age may also have contributed. **2.** The heart pains might have been a warning. Also, regular visits to the doctor prior to the heart attack may have detected some other warning signs. **3.** Response should include eliminating cigarette smoking entirely and seriously cutting down on high-fat foods. Regular exercise and visits to the doctor are essential, as well as ongoing medication in some cases.

## Workbook Activity 43—What Is Cancer?
**1.** cells **2.** tumor **3.** young adults **4.** malignant **5.** drugs **6.** kills **7.** sunlight **8.** heredity **9.** AIDS **10.** carcinogens **11.** warning sign **12.** month **13.** risk factors **14.** skin cancer **15.** bloodstream

### Workbook Activity 44—Diabetes
1. insulin  2. type I diabetes  3. immune system  4. pancreas
5. extreme thirst  6. adulthood  7. blood vessels  8. obesity
9. carbohydrates  10. glucose

### Workbook Activity 45—Arthritis, Epilepsy, and Asthma
1. Ar  2. E  3. As  4. E  5. Ar  6. E  7. Ar  8. E  9. Ar  10. As  11. E
12. As  13. E  14. As  15. E

### Workbook Activity 46—Medicines
1. The Food and Drug Administration divides medicines into prescription or over the counter types.  2. Prescription medicines are generally stronger than OTC medicines and require a doctor's written order.  3. Accept any five of the following: Five purposes for medicines are to get rid of pain, control or destroy bacteria, make the heart and blood vessels healthy, prevent a disease, relieve the symptoms of a disease, replace body chemicals, or lessen anxiety.  4. Accept any five of the following: Medicines may be taken by swallowing, shots, breathing in, or in creams, lotions, or patches put on the skin.  5. Medicine labels must give the purpose of the medicine, the suggested amount, possible unwanted side effects, and cautions.

### Workbook Activity 47—Tobacco
1. A  2. A  3. A  4. SL  5. SL  6. S  7. SL  8. S  9. S  10. A  11. A  12. A
13. S  14. S  15. S

### Workbook Activity 48—Alcohol
1. abuse  2. does not  3. does not  4. does not  5. negative  6. liver
7. 1/2 ounce  8. 0.6 ounces  9. hallucinations  10. stop

### Workbook Activity 49—Classifying Drugs
1. S  2. D  3. D  4. A  5. N  6. H  7. S  8. N  9. S  10. N  11. S  12. H
13. A  14. D  15. S

### Workbook Activity 50—Other Dangerous Drugs
1. grow or use  2. 400  3. 30  4. raise  5. the first  6. to reduce stress  7. a different purpose than it is meant for  8. stronger than  9. anything  10. abuse  11. both medically and illegally  12. birth defects

### Workbook Activity 51—Drug Dependence Problems
1. B  2. PH  3. PH  4. B  5. PS  6. PS  7. PH  8. B  9. B  10. PS  11. B
12. PH  13. PH  14. B  15. PS

### Workbook Activity 52—Drug Dependence Solutions
1. The first step in recovery is for a person to recognize that he or she has a problem with drugs. The second step is going through detoxification. The third step is learning to live day to day without drugs.  2. The goal of recovery from drug dependence is abstinence. This is the goal because drug dependence has no cure.  3. At a residential treatment center, doctors use a medical approach to treatment by taking people off drugs slowly and by sometimes using medicines to help people through withdrawal. People who are treated at outpatient treatment centers do not live there.  4. Alcoholics Anonymous offers support to people who are addicted to alcohol. Three other groups offer support to the family members of people who are addicted to alcohol: Al-Anon is for adults, Alateen is for 12- to 18-year-olds, and Alatot is for 6- to 12-year-olds.  5. Counselors at mental health centers direct people to treatment centers. They also provide help and encouragement after treatment and for family members.

### Workbook Activity 53—Alternatives to Drug Use
1. Worthwhile alternatives to drug use include physical activity; getting involved in a cause that helps others and adds purpose to your life; joining an interest or activity group; working at a part-time job; choosing entertainment that promotes mental and physical health; and supporting efforts by your school and community to prevent drug use.  2. Peer pressure is the main reason that adolescents start using drugs.  3. Changing the subject and repeating that you do not want to use drugs are two refusal techniques for avoiding drug use.  4. Drug hotlines link people with treatment centers in their own community.  5. A healthful way a person can cope with stress and other problems is by eating a balanced diet every day. A balanced diet increases energy levels and maintains normal brain chemistry.

### Workbook Activity 54—Risks at Home
1. R  2. I  3. I  4. I  5. R  6. R  7. I  8. R  9. I  10. R  11. R  12. I  13. I
14. I  15. R

### Workbook Activity 55—Risks Away from Home
1. R  2. R  3. I  4. I  5. I  6. R  7. I  8. R  9. I  10. R  11. I  12. R  13. I
14. I  15. R

### Workbook Activity 56—Risks on the Road
1. R  2. I  3. I  4. I  5. R  6. I  7. R  8. R  9. I  10. R  11. I  12. I  13. R
14. I  15. R

### Workbook Activity 57—Safety in Natural Disasters
1. F  2. E  3. H  4. F; H; T; B  5. E  6. B  7. H; F  8. F  9. T  10. E
11. B  12. H  13. T  14. B  15. T

### Workbook Activity 58—First Aid
1. No  2. Yes  3. No  4. Yes  5. No  6. No  7. Yes  8. No  9. Yes  10. No

### Workbook Activity 59—Saving Lives
1. RB  2. HM  3. RB  4. CPR  5. HM  6. CPR  7. FS  8. HM  9. CPR
10. HM

### Workbook Activity 60—First Aid Choices
1. CP  2. IS  3. IP  4. HE  5. IS  6. FB  7. HE  8. CP  9. SB  10. CP
11. FB  12. OP  13. OP  14. IP  15. SB

### Workbook Activity 61—What Is Violence?
1. Gangs  2. Random  3. are not  4. ongoing  5. 1.3 million  6. teens
7. mental and/or physical  8. might  9. 5.7  10. never  11. children
12. guilty  13. can  14. normal  15. Drug-related

### Workbook Activity 62—The Causes and Effects of Violence
1. A  2. B  3. C  4. F  5. D  6. J  7. I  8. E  9. H  10. G

### Workbook Activity 63—Preventing Violence
1. stay out after a certain time  2. volunteers  3. tougher  4. strength
5. mediator  6. Some  7. your own feelings  8. a peaceful  9. one another  10. resolve conflicts  11. should  12. often  13. will not
14. true  15. less

### Workbook Activity 64—Wise or Unwise Consumer
1. W  2. B  3. U  4. B  5. U  6. U  7. W  8. W  9. B  10. U  11. U  12. B
13. U  14. W  15. W

### Workbook Activity 65—Medical Specialists
1. cardiologist  2. dermatologist  3. psychiatrist  4. pediatrician
5. gynecologist  6. psychiatrist  7. ophthalmologist  8. allergist
9. obstetrician  10. pediatrician

### Workbook Activity 66—Paying for Health Care
1. patient  2. health  3. the patient and the physician  4. $500  5. HMO
6. a fixed amount  7. certain disabilities  8. Medicare  9. USPHS
10. low risk  11. NIH  12. HRSA  13. low  14. all  15. less than

### Workbook Activity 67—Public Health Problems
1. HO  2. EP  3. PO, HO  4. HO  5. PO  6. EN  7. EP  8. EP  9. HO
10. EP  11. PO  12. PO  13. PO, HO, GHS  14. EP, EN  15. PO, HO, GHS, EN, EP

### Workbook Activity 68—U.S. Public Health Solutions
1. B  2. C  3. E  4. D  5. A  6. C  7. B  8. C  9. B  10. C

### Workbook Activity 69—Health Promotion and Volunteer Organizations
1. volunteers  2. American Cancer Society  3. World Health Organization  4. United Nations International Children's Emergency Fund  5. American Lung Association  6. Food and Agriculture Organization  7. Peace Corps  8. American Heart Association  9. International Red Cross  10. advocacy groups

### Workbook Activity 70—Health and the Environment
1. D  2. D  3. M  4. M  5. D  6. D  7. M  8. D  9. D  10. M  11. M
12. D  13. D  14. M  15. M

### Workbook Activity 71—Types of Air Pollution
1. PS; FF; A; HC; CM; R; PA  2. PS  3. FF; PS; A; HC; PA  4. FF
5. HC; CM  6. A  7. CM; R  8. PA  9. FF  10. CM  11. A; R; PA  12. PS
13. PS; FF; A; CM; R  14. A  15. FF

### Workbook Activity 72—Types of Water Pollution
1. HC  2. OL  3. AG  4. AG  5. AG  6. FW  7. S  8. FW, S, AG  9. S
10. HC  11. AG  12. S  13. AG  14. OL  15. S

### Workbook Activity 73—Places
1. Europe  2. Africa  3. Africa  4. Love Canal  5. United States
6. Chernobyl  7. Chernobyl  8. Europe  9. Chernobyl  10. Love Canal

### Workbook Activity 74—Protecting the Environment
1. Sample answer: The Clean Air Act, first passed in 1970, says how much air pollution there can be. It makes each state have a plan to meet its standards.  2. Sample answer: The National Environmental Policy Act requires federal agencies to think about how their plans and activities will affect the environment.  3. Sample answer: The Environmental Protection Agency writes and carries out the environmental laws that Congress passes. Its job is to protect the environment.  4. Sample answer: The United Nations (UN) General Assembly started the Conference on Environment and Development in 1992. It brought together leaders from 178 countries to talk about how to live well and protect the environment at the same time.  5. Sample answer: California has passed a law that says that new cars have to use less energy.

## Community Connections
Completed activities will vary for each student. Community Connection activities are real-life activities that students complete outside the classroom.

## Home Connections
Answers for Home Connections 1–3 and 7 will vary.

### Home Connection 4—Family AIDS Awareness Assessment
1. False  2. True  3. False  4. False  5. Ways AIDS is acquired include through exchange of body fluids, sexual activity with an infected partner, sharing needles among drug users, and passage from mother to child.  6. Ways AIDS is not acquired are casual social contact such as shaking hands, hugging, crying, coughing, sneezing, sharing eating utensils, bathing in the same water, and bites from sucking insects.  7. Sexually transmitted diseases are prevented by avoiding contact with the pathogens that cause them or abstinence from sexual activity

### Home Connection 5—Family Quiz on Substances
1–8. True

### Home Connection 6—Family Quiz on Accidents and Injuries
1. True  2. True  3. True  4. True  5. False  6. True  7. True  8. False
9. False

## Self-Assessments
Self-Assessments provide an opportunity for students to gauge their knowledge of a chapter's content before and after reading the chapter. Answers will vary.

## Self-Study Guides
Self Study Guides outline suggested sections from the text. These assignment guides provide flexibility for individualized instruction or independent study.

# Chapter Outlines

## Chapter 1 Outline
I.  A. 1. healthy; three
    B. 1. tasks; tired or ill
    C. 1. friends; friend
       2. behaviors; trust; respect; fairness
    D. 1. problems; pressures
       2. good
II. A. 1. principles
    B. 1. three; five
       2. three; meals
       3. smoke; drugs
    C. 1. active
       2. physical; social; emotional
III. A. 1. action; condition
    B. 1. characteristics
       2. diseases
    C. 1. risks
       2. physical; person
       3. air; land; animals
       4. community
       5. poor
    D. 1. beliefs; language
       2. culture; understand; health
    E. 1. health; others
       2. risks; lifestyle
       3. system; body
       4. secondhand
IV. A. 1. short-term
       2. goals; accomplish
    B. 1. aware; signals
    C. 1. awareness; knowledge

## Chapter 2 Outline
I.  A. 1. reactions; events; experiences
    B. 1. five
    C. 1. necessary; alive
    D. 1. protection
    E. 1. friendships
       2. attachment; support
       3. physical; emotional
    F. 1. self-respect
    G. 1. achieving one's possibilities
II. A. 1. normal; good; bad
    B. 1. control
    C. 1. chemical; brain
       2. resist
III. A. 1. fear; anxiety, anger; grief; relief
    B. 1. well-being; threatened
       2. anxiety
    C. 1. aggressive
       2. why; want to happen
    D. 1. denial; anger; bargaining; depression; acceptance
    E. 1. gone
    F. 1. overcome
       2. identifying; solve; carrying out

## Chapter 3 Outline
I.  A. 1. thoughts; feelings; behavior
       2. experiences
    B. 1. ideas; we are
       2. social esteem
    C. 1. physical; emotional; social; personal
    D. 1. work together
    E. 1. healthy
    F. 1. stress
    G. 1. communication; friendship; citizenship
II. A. 1. experience
       2. mistakes
    B. 1. Change
    C. 1. pessimism; optimism
    D. 1. compromise
    E. 1. impulses; appropriately
    F. 1. adjust; reach
III. A. 1. attachment; mutual; needs
    B. 1. communication
       2. accept
IV. A. 1. individual
       2. others
    B. 1. behave
    C. 1. surface; visible
       2. prejudice
    D. 1. recognizing; understand
       2. basis

## Chapter 4 Outline
I.  A. 1. psychologists
       2. normal; abnormal
    B. 1. goals
       2. defense mechanisms; dysfunctional relationships
    C. 1. hide from their problems
       2. repression; denial; projection; displacement
    D. 1. destructive
II. A. 1. dependence
    B. 1. uneasiness; fearful
       2. panic attack
       3. phobia
    C. 1. disturbed; uncontrolled
       2. clinical depression
       3. suicide
    D. 1. serious
       2. hallucinations
       3. delusions
       4. frightening; dangerous
    E. 1. psychological problems; eating habits
       2. anorexia; bulimia; compulsive overeating
II. A. 1. individual counseling; behavior change programs; support groups
    B. 1. different
       2. physical; psychological
       3. medicines; psychotherapy
    C. 1. chemistry
    D. 1. cured
    E. 1. professional; resists
    F. 1. fear

## Chapter 5 Outline
I.  A. 1. structure; protect; store; blood cells
       2. ligaments
       3. marrow; red blood cells; white blood cells
    B. 1. contract
       2. smooth; skeletal; cardiac
       3. tendons
       4. involuntary; voluntary
II. A. 1. central nervous system; brain; spinal cord; peripheral nervous system
    B. 1. three; cerebrum; cerebellum; brain stem
       2. medulla
    C. 1. pathway; messages
    D. 1. messages; rest of the body
       2. reflexes
    E. 1. outside
       2. eyes; ears; tongue; nose; skin
III. A. 1. nervous
       2. glands; hormones; bloodstream
IV. A. 1. heart; blood; arteries; capillaries; veins
       2. blood
    B. 1. tubes; organs
       2. oxygen; carbon dioxide
V.  A. 1. food; cells
       2. enzymes
    B. 1. eliminate; waste
       2. blood; kidneys
VI. A. 1. three; epidermis; dermis; subcutaneous layer
       2. bacteria
       3. pores

## Chapter 6 Outline
I.  A. 1. cleanliness
       2. eyes; ears; skin; hair; nails; teeth
    B. 1. dirt; sweat; direct light
       2. eye exams
    C. 1. loud noise; infection; heredity; allergies; too much wax; wound
       2. earplugs
    D. 1. clean; protected
       2. sunscreen
    E. 1. skin problem; teenagers
       2. clogged
    F. 1. shampooing; brushing
       2. dandruff
    G. 1. clean
    H. 1. lifetime
       2. brushed; flossed
II. A. 1. emotional; social; physical
       2. three; endurance; muscular; flexibility
    B. 1. increased; strength; flexibility; energy
       2. better
       3. calories
    C. 1. purposes
       2. warm-up; aerobic; fitness; flexibility; cooldown
       3. aerobic; anaerobic; isometric; isotonic; isokinetic
III. A. 1. eating well; exercising
       2. activities; goals
    B. 1. paying attention; stressed; disease; injury

## Chapter 7 Outline
I.  A. 1. eight; stages of life
    B. 1. birth; 12 months
       2. 12
    C. 1. discover things about themselves
    D. 1. 20; 40
    E. 1. middle; older
    F. 1. adolescence
II. A. 1. testosterone
       2. testes
    B. 1. Hormones; ovaries; egg
       2. menstrual cycle
    C. 1. male sperm; female egg cell
III. A. 1. egg cell; join; fertilized
       2. nine
    B. 1. sick; very tired; breasts; menstrual cycle
    C. 1. health care

2. Smoking; drinking alcohol; using other drugs; substances in the environment
 D. 1. trimesters
 E. 1. three; dilation; birth; uterus; placenta
IV. A. 1. science; heredity
  2. chemical codes; traits
  3. X and Y chromosomes; gender
  4. gene pairs; dominant; recessive
 B. 1. diseases; disorders

## Chapter 8 Outline
I. A. 1. life cycle; marry
 B. 1. others; enjoy; activities
 C. 1. communication; physical; breaking up
 D. 1. pregnancy; disease
 E. 1. understand; each other; happy; satisfied
 F. 1. longer; education
  2. longer; older
 G. 1. money; children; jobs; parents
II. A. 1. health; safety; well-being
  2. pay; costs
  3. parenting skills
 B. 1. common
  2. nuclear; children
  3. blended family; single-parent family
 C. 1. healthy; unhealthy
III. A. 1. agreement; court
 B. 1. grief; denial; anger; bargaining; depression; acceptance
 C. 1. actions; words; hurt; things
 D. 1. form; violence
  2. physical; emotional
  3. neglect
 E. 1. suggestions; problems

## Chapter 9 Outline
I. A. 1. foods; eats; drinks
 B. 1. different kinds; activity
 C. 1. chart; how much; what kinds
  2. number of servings; each day
 D. 1. nutrients
  2. essential nutrients
  3. six
  4. nutrient-dense
  5. calorie
II. A. 1. complex; simple
  2. starches; sugars; plant
  3. energy
  4. glucose; blood
 B. 1. stored
  2. body cells
  3. fatty acids; saturated; unsaturated
  4. cholesterol
 C. 1. nutrient; muscle; repair; tissues
  2. amino acids
  3. essential amino acids
III. A. 1. minerals; water; food; energy
  2. grow; healthy
 B. 1. earth
  2. Cannot
 C. 1. mineral
  2. bones; teeth
 D. 1. calcium; bones
 E. 1. nutrients; hormones; waste products
  2. 60 percent

IV. A. 1. calories; calcium; iron; growing; changing
 B. 1. calories; energy; water
 C. 1. malnutrition; one; essential nutrients
  2. deficiency
 D. 1. lower; risk

## Chapter 10 Outline
I. A. 1. set time; regular times
  2. Breakfast; active
  3. Snacks; regular meals
 B. 1. where you live; the time of year; your religion; your cultural background
  2. cost
  3. Association
 C. 1. stress
  2. enough, much
II. A. 1. new product; old-fashioned; expensive and the best; inexpensive and a good deal; quick and easy to prepare; contains no cholesterol; fat-free or low carb; organic foods
 B. 1. weight problems; fat
  2. Saturated
  3. Unsaturated
  4. Trans fat; lengthen; spoiling
  5. Daily Values (DV)
 C. 1. one food; unhealthy
III. A. 1. labeled, safe
  2. USDA, FDA
 B. 1. Nutrition Facts
  2. 2,000
  3. percent
  4. weight; most; least
  5. serving
 C. 1. additive; color; flavor; texture
  2. Preservatives
  3. small

## Chapter 11 Outline
I. A. 1. infections; human behaviors; environment
  2. infectious (or communicable); infection
 B. 1. germs; pathogens
  2. stages; incubation period
  3. immune system
 C. 1. genes; parents; children
  2. three; abnormal gene; abnormal genes; environment; abnormal chromosomes
II. A. 1. disease; infection
  2. physical; chemical
  3. immune
 B. 1. skin; mucous membranes
  2. pathogens; surface
 C. 1. saliva; tears; sweat
  2. acids; urine
 D. 1. immune system
  2. pathogens
 E. 1. blood; proteins
  2. recovered
  3. infection; immunization (or vaccination)

## Chapter 12 Outline
I. A. 1. acquired immunodeficiency syndrome

  2. immune system
  3. human immunodeficiency virus; HIV
  4. virus; communicable
  5. incubation; 6; 10
  6. vaccine; cure
 B. 1. infects; weakens; immune
  2. opportunistic; body
  3. exchange; body fluids; sexual; needles; drug users; mother; child
 C. 1. social; HIV; outside
 D. 1. pneumonia; lung; Kaposi's sarcoma
 E. 1. transfer
  2. blood; HIV
II. A. 1. sexual activity
  2. serious; diseases; treated
  3. AIDS; gonorrhea; chlamydia; syphilis; genital herpes
 B. 1. antibiotics
  2. sterility
 C. 1. symptoms; gonorrhea
  2. antibiotic
 D. 1. three
 E. 1. chronic; lasting
  2. cure; controlled
 F. 1. cured; early
  2. abstinence

## Chapter 13 Outline
I. A. 1. group; heart; blood vessels
  2. blood; artery
  3. hypertension
  4. medicine; diet; exercise
 B. 1. artery; harden
  2. arteries; fat
  3. heart attack; oxygen; nutrients
  4. blood, heart
  5. stroke
 C. 1. family medical history; being male; aging; race
  2. smoking; cholesterol; being overweight; physical inactivity
II. A. 1. harmful, cells
  2. malignant
  3. benign
 B. 1. bladder; sore; bleeding; lump; swallowing; mole; hoarseness
 C. 1. chemotherapy; surgery; radiation
  2. smoke; sunlight; viruses
III. A. 1. insulin
  2. insulin-dependent; noninsulin-dependent
  3. insulin; diet
IV. A. 1. rheumatoid arthritis; swelling
  2. cartilage
 B. 1. brain; seizures
  a. grand mal
  b. petit mal
 C. 1. breathing
  2. bronchioles

## Chapter 14 Outline
I. A. 1. prescription; over the counter (OTC)
  2. analgesics
  3. bacteria; antibiotics; sulfa
  4. vaccines; immunity
  5. cold; over the counter
 B. 1. side effects; need; mixing; misusing

C. 1. purpose; suggested; unwanted; cautions
II. A. 1. change; harmful
 B. 1. chemical; tobacco; addiction
III. A. 1. depressant; slows down; central nervous; psychoactive
 2. negative; major system
 B. 1. defects; learning; vision; hearing
 C. 1. dependence; control
IV. A. 1. synthetic
 B. 1. addictive
 C. 1. sedate
 D. 1. morphine; codeine
 E. 1. Confuse
V. A. 1. 400; THC
 B. 1. chemicals; breathe
 C. 1. legal
 D. 1. legal; illegal
 E. 1. synthetic; natural; testosterone

## Chapter 15 Outline
I. A. 1. need
 2. tolerance
 3. withdrawal
 4. genetic
 5. narcotics; barbiturates; alcohol
 B. 1. Psychological dependence
 2. psychoactive
 C. 1. drug
 2. drug dependence
 D. 1. enabling
 2. stress; abused
 3. alcohol; cocaine
 4. job; college
II. A. 1. Recovery; recognizing; detoxification; drugs
 2. abstinence
 B. 1. Counselors
 2. Alcoholics Anonymous; Al-Anon; Alatot; Alateen
 3. residential
 4. outpatient
 5. halfway houses
III. A. 1. Resistance skills
 2. refusal technique
 B. 1. responsible
 2. Hotlines
 C. 1. self-esteem

## Chapter 16 Outline
I. A. 1. falls; fires; poisoning; electrical shock
 B. 1. tubs; showers
 C. 1. electrical; starters; fireplaces; space; furnaces
 2. detectors; extinguisher
 3. easily
 D. 1. injury; illness; death
 2. babies; young; mouth
 E 1. flow; current; body
 2. wet; properly; too many
II. A. 1. OSHA; protect; workers
 B. 1. inform; regulations
 C. 1. follow; procedures; learn; equipment; clothing; devices
 D. 1. third; accidental
 2. prevented; swim
 3. survival floating
 E. 1. properly; gear; learning; skills
III. A 1. highest; deaths
 2. safe; skills; habits; observing; maintaining; conditions
 B. 1. helmets; obey
IV. A. 1. emergencies; nature
 2. may not; predicted
 3. hurricanes; tornadoes; floods; earthquakes; blizzards; lightning

## Chapter 17 Outline
I. A. 1. emergency; sick; injured; professional
 B. 1. calm
 2. conscious; breathing; pulse; open
 3. bleeding; 911
 C. 1. protect
 D. 1. medical workers; lessen; contact; blood; fluids
II. A. 1. universal distress signal; throat; hands
 2. Heimlich maneuver
 B. 1. Rescue breathing; mouth-to-mouth resuscitation
 C. 1. CPR; rescue breathing; compressions
 D. 1. Shock; physical; injury
 E. 1. stop; infection; shock
III. A. 1. Oral; swallowed
 2. Inhalation
 3. Contact
 B. 1. brain
 C. 1. Animal; diseases
 2. calm
 3. allergic; bees
 D. 1. splint
 2. sprain; tendons; ligaments; torn
 E. 1. types; degrees
 F. 1. physical work
 2. heatstroke
 H. 1. serious; heat
 2. death

## Chapter 18 Outline
I. A. 1. actions; words; hurt; damage
 B. 1. disagreement; different
 2. minor; ongoing; internal; major
 C. 1. everyone
 D. 1. money; physical; emotional
II. A. 1. media; families
 2. alcohol; peer
 B. 1. sexual; physical
 C. 1. robbery; assault; rape; murder
 D. 1. unlawful; forcing; sex
 E. 1. Date rape; socializes
 2. money; sexual exploitation
III. A. 1. deep breath; waiting; avoiding
 B. 1. communication; listening
 2. resolved; alone
 C. 1. encourage; watching
 2. respect; walk away
 D. 1. parents; mediator; teacher; counselor
 E. 1. Neighborhood Crime Watch
 2. teachers; counselors

## Chapter 19 Outline
I. A. 1. goods; services
 2. advertising; family; friends; price
 B. 1. promotion; unproved; worthless
 C. 1. complaining; products
II. A. 1. action; prevent
 B. 1. primary care; specialists; practitioners
 2. checkups
 C. 1. places; medical; dental
III. A. 1. expensive
 2. private; government
 B. 1. fee; part; all
 C. 1. Health Maintenance Organizations (HMOs); Preferred Provider Organizations (PPOs)
 D. 1. Medicare; 65; Medicaid
 E. 1. outpatient
 2. educating
 F. 1. information; choose; decisions; private

## Chapter 20 Outline
I. A. 1. preventive; education; control
 B. 1. communicable; rapidly
 2. common
 C. 1. high risk; serious
 D. 1. community; poverty
 E. 1. not; basic
 F. 1. harmful; concern
 2. birth
 G. 1. nations; little; low
II. A. 1. laws; disease; sanitation
 B. 1. analyzes; information; standards; education; research
 2. Environmental Protection Agency; Occupational Safety and Health Administration; Consumer Product Safety Commission
III. A. 1. contributions; individuals; businesses
 2. education
 B. 1. role; hunger; relief; epidemics
 C. 1. Peace Corps; U.S. Agency for International Development
 D. 1. private
 2. aid; natural
 E. 1. prevention; groups
 F. 1. money; illness
 2. support; promote; problems

## Chapter 21 Outline
I. A. 1. system; natural resources
 2. living things; needs
 B. 1. environment; human
 C. 1. soil; water; air; harmful
II. A. 1. burned
 2. hydrocarbons; cancers
 B. 1. closed up; air
 C. 1. carbon dioxide
 2. warming; atmosphere
 D. 1. 6; 30; surface; guards; protects; rays
 E. 1. precipitation; sulfuric; nitric
 2. rivers; lakes; plant
III. A. 1. freshwater
 B. 1. toxic; factories
 C. 1. waste; drains; toilets
 D. 1. oil; animals; carcinogens
 2. Agricultural runoff; algae; groundwater
IV. A. 1. food; 6; countries; famines
 B. 1. cutting; clearing
 C. 1. garbage; trash
 2. landfill
 D. 1. Sensorineural
 E. 1. Light
 2. Radiation; energy; waves
V. A. 1. writes; carries out

Answer Key 569

# Mastery Tests

See page 576 for the Scoring Rubric for Short-Response Items in Part E and Essay items in Part F of the Mastery Tests.

## Chapter Mastery Tests

**Chapter 1 Mastery Test A**
Part A  1. B  2. D  3. B  4. A  5. C
Part B  6. emotional  7. heredity  8. culture  9. behavioral  10. goals  11. aware
Part C  12. F  13. C  14. B  15. E  16. D  17. A
Part D  18. Nonsmokers can be injured through secondhand smoke, smoke from a cigarette or cigar someone else is smoking.  19. The social environment is made up of community resources.  20. A sedentary lifestyle includes little or no exercise.
Part E  21. Answers will vary. Students should define and give examples of physical and social health, and show how they affect each other.  22. Answers will vary but should include two of the following: exercising; eating three balanced, healthy meals; eating healthy snacks; drinking enough water; sleeping at least eight hours; and not smoking, drinking alcohol, or using drugs.
Part F  23. Answers will vary. Students should name heredity, environment, and behavior as causes of health risks and give examples of each. Students should explain which risks can be controlled.  24. The first thing she should do is define the problem. She can read books, use the Internet, and talk with experts like a family health care provider. Then she should come up with possible solutions. Before making any changes she should consider the consequences of each action. She should choose a diet and exercise that she will enjoy so she will stick to her plan. Now she is ready to put her plan into action. After a few weeks she should evaluate the outcome of what she has done. Then she can adjust and change whatever she must so she can reach her goal.

**Chapter 1 Mastery Test B**
Part A  1. C  2. D  3. B  4. E  5. F  6. A
Part B  7. Heredity is the passage of physical characteristics from parents to children.  8. Health literacy is gathering, understanding, and using information to improve health results.  9. Wellness is an active state of health.
Part C  10. physical  11. fruit  12. behavioral  13. abstain  14. risk  15. social
Part D  16. D  17. A  18. B  19. C  20. A
Part E  21. You must gain awareness and knowledge. You gain awareness by paying attention to the signals your mind and body send you. You gain knowledge by reading, watching DVDs, and talking to knowledgeable people.  22. Sample answer: People often fail to abstain from dangerous activity. I can decide to abstain from using alcohol and other drugs not prescribed by a health care professional. People also fail to observe safety rules in the home. I can make sure outlets are not overloaded.
Part F  23. Answers will vary. Students should explain three basic ways to stay healthy and give examples of each.  24. Answers will vary. Students should name and define physical, social, and emotional health and give examples of how they can be applied to daily life.

**Chapter 2 Mastery Test A**
Part A  1. D  2. E  3. A  4. B  5. C
Part B  6. emotions  7. Abraham Maslow  8. depression  9. self-concept  10. physical needs  11. friendship  12. bargaining
Part C  13. C  14. B  15. C  16. B  17. B
Part D  18. harm  19. Denial  20. relieve  21. attachment  22. high-level  23. grief  24. blocked  25. stress response
Part E  26. Emotions are a person's individual reactions to an experience. Examples of emotions include anger, grief, and fear. Emotions start with thoughts. Thinking about how you feel about something triggers an emotion.  27. Emotions may cause physical changes or reactions. For example, when you are feeling fear, your muscles may tighten or your hands may shake. When you feel anxiety you may have a stomachache.
Part F  28. The five human needs are physical needs, safety needs, belonging needs, esteem needs, and self-actualization. During a normal day a person would focus on meeting physical and safety needs but also belonging needs by spending time with friends and self-actualization needs by volunteering. During a disaster a person may focus more closely on meeting physical and safety needs.  29. Student answers will vary but may include feeling anxious about not fitting in, not being able to make friends, keeping up in class, keeping up with homework, or not knowing his way around the school.

**Chapter 2 Mastery Test B**
Part A  1. C  2. D  3. A  4. C  5. D
Part B  6. self-actualization  7. coping  8. depression  9. friendship  10. emotions  11. human needs  12. cannot
Part C  13. self-actualization  14. romantic love  15. hierarchy  16. stress response  17. denial
Part D  18. D  19. H  20. F  21. G  22. C  23. E  24. B  25. A
Part E  26. Emotions help you realize what you need. When one of your needs is not being met, you may feel emotions. If a person's belonging needs are not being met he or she may feel frustrated or angry or embarrassed.  27. Emotions can cause physical changes or reactions in the body. The brain triggers the release of chemicals into the bloodstream that can cause a faster heartbeat, sweating, and a rush of blood to your arms, legs, and head. When you are feeling fear, your muscles may tighten or your hands may shake.
Part F  28. The five human needs are physical needs, safety needs, belonging needs, esteem needs, and self-actualization. To meet physical needs a person eats food and drinks water. To meet safety needs a person may look before crossing the street or avoid walking after dark. To meet belonging needs a person will make friends and join groups. To meet esteem needs a person will develop a good self-concept and do things that make him or her feel happy. To meet self-actualization needs a person will help others, such as volunteering or donating money to a charity.  29. She may feel stress because she has a lot of things to do and not enough time. She can make a stress management plan and carry it out, including walking to relieve stress or cutting back on her hours at work. She may feel anxiety at having to play in front of a large group of people for band. She could practice to build confidence.

**Chapter 3 Mastery Test A**
Part A  1. B  2. C  3. A  4. C  5. B
Part B  6. personality  7. help  8. twisted sending  9. "I"  10. compromise
Part C  11. I  12. H  13. E  14. A  15. D  16. J  17. C  18. G  19. B  20. F
Part D  21. the messages you get from others  22. how others value you  23. communication  24. empathy  25. nurturing
Part E  26. Emotional well-being is your ability to handle stress and problems. If you cannot handle stress, you do not have emotional well-being. One way to handle stress is to do some physical activity like walking.  27. Having meaningful values helps you face and overcome disadvantages like not being good in math. A value of doing your best in school would make you try harder and not give up. Meaningful values also help you find meaning in your work and actions.
Part F  28. Answers will vary. Students should give examples of effective speaking and listening skills.  29. Answers will vary. Students may list meeting new people, working together, and working with people of different cultures.

**Chapter 3 Mastery Test B**
Part A  1. C  2. B  3. D  4. A  5. A
Part B  6. Well-being  7. impulse  8. Coping  9. social comparison  10. nonverbal
Part C  11. D  12. C  13. H  14. J  15. I  16. F  17. E  18. G  19. A  20. B
Part D  21. citizenship skills  22. realistic  23. symptom  24. friendship  25. "I" messages
Part E  26. Citizenship skills include respecting others, doing your part, and volunteering to help. All of these skills promote social well-being, which is your ability to get along well with others.  27. As an effective listener, I would not interrupt the speaker. I would give the speaker my full attention. I would think about what the speaker is saying and not what I'm going to say next. I would repeat what the

speaker said in my own words to make sure I understood it correctly.
**Part F** **28.** Answers will vary. Students should list a pessimistic goal, and optimistic goal, and a realistic goal. **29.** Answers will vary. Students may mention meeting new people, accepting differences, learning about people who are different, and accepting and giving help.

### Chapter 4 Mastery Test A
**Part A** **1.** phobia **2.** fear **3.** death **4.** thought **5.** "I" **6.** panic attack **7.** Bulimia **8.** clinical depression **9.** third **10.** thought
**Part B** **11.** yes **12.** anxiety **13.** nothing **14.** projection **15.** problems
**Part C** **16.** D **17.** B **18.** A **19.** C **20.** D
**Part D** **21.** Defense mechanisms and dysfunctional relationships **22.** Suicide **23.** alcohol or other drugs **24.** unreal **25.** goal-directed activity
**Part E** **26.** Possible answers include two of the following: Repression—refusing to think about something that upsets you; Denial—Refusing to believe something that is negative or threatening; Projection—accusing someone of your attitudes, feelings, or purposes; Displacement—shifting an emotion from its real object to a safer one **27.** Characteristics of poor mental health include difficulty getting along with others, depending too much on others, and being easily defeated by failure.
**Part F** **28.** It seems likely that your friend suffers from an eating disorder. You should encourage your friend to get professional help. You should let the person know you care about his or her well-being. You should encourage the person to talk about his or her feelings. **29.** Answers will vary. Students should explain what Alice can expect depends upon what kind of counseling she is getting.

### Chapter 4 Mastery Test B
**Part A** **1.** D **2.** C **3.** A **4.** B **5.** B
**Part B** **6.** a physical **7.** substance abuse **8.** Thought **9.** psychological **10.** behavior modification
**Part C** **11.** C **12.** B **13.** G **14.** I **15.** H **16.** A **17.** D **18.** F **19.** J **20.** E
**Part D** **21.** People fear what they do not understand. **22.** thought disorders **23.** dysfunctional **24.** mental and behavioral characteristics **25.** suicide
**Part E** **26.** A person with poor mental health may have difficulty getting along with others. The person may depend too much on others. **27.** You should encourage your friend to get professional help, let the person know you care about his or her well-being, and encourage the person to talk about his or her feelings.
**Part F** **28.** Because anxiety disorders may be physical or psychological, you would need to hire two different professionals. You would hire a medical doctor to treat physical problems with medicines. You would hire a psychologist to treat psychological problems with psychotherapy. **29.** Answers will vary. Students should name a defense mechanism and explain a specific use of it.

### Chapter 5 Mastery Test A
**Part A** **1.** B **2.** C **3.** B **4.** C **5.** B
**Part B** **6.** three **7.** sense **8.** cornea **9.** cerebrum **10.** Ligaments
**Part C** **11.** I **12.** H **13.** E **14.** A **15.** D **16.** J **17.** C **18.** G **19.** B **20.** F
**Part D** **21.** A band of muscle tissue; beneath the lungs **22.** Cardiac muscles **23.** Insulin **24.** Arteries **25.** The skin
**Part E** **26.** The arteries carry blood away from the heart. The capillaries connect arteries to veins and bring blood to every cell in the body. The veins carry blood back to the heart. The heart pumps the blood through the system. Blood gives oxygen and nutrients to cells and carries waste products away from cells. **27.** Light from an object passes through the cornea and the pupil to the lens, which directs light energy onto the retina. Receptor cells on the retina send light information to the brain through the optic nerve. The brain changes light information into understandable pictures.
**Part F** **28.** Answers will vary. Students should name and explain what these glands do: pituitary, thyroid, adrenal. **29.** Answers will vary. Students should explain how a torn ligament affects movement of the shoulder.

### Chapter 5 Mastery Test B
**Part A** **1.** C **2.** B **3.** A **4.** A **5.** D
**Part B** **6.** marrow **7.** endocrine **8.** fingertips **9.** liver **10.** platelets
**Part C** **11.** D **12.** C **13.** H **14.** J **15.** I **16.** A **17.** E **18.** G **19.** F **20.** B
**Part D** **21.** Smooth, skeletal, and cardiac; skeletal **22.** Platelets **23.** Metabolism **24.** Three **25.** Your cerebrum
**Part E** **26.** Ligaments hold joints together. Tendons connect muscles to bones or muscles to muscles. Muscles move bones by pulling on the attached tendons. **27.** The pituitary gland produces a growth hormone and other hormones. The thyroid gland regulates the body's metabolism. The adrenal glands regulate the body's water balance and produce adrenaline and other hormones.
**Part F** **28.** Answers will vary. Students should explain possible reasons someone may have trouble seeing by talking about parts of the eye and their functions. **29.** Answers will vary. Students should explain how blood flows through the circulatory system, using arteries, capillaries, blood cells, and veins.

### Chapter 6 Mastery Test A
**Part A** **1.** A **2.** C **3.** D **4.** B **5.** C
**Part B** **6.** warm up **7.** maximum heart rate **8.** cooldown **9.** personal fitness plan **10.** rest
**Part C** **11.** wax **12.** antiperspirant **13.** dandruff **14.** resistance **15.** calories
**Part D** **16.** C **17.** E **18.** D **19.** A **20.** B
**Part E** **21.** Good hygiene can have a positive effect on personal health and personal and work-related relationships. Good hygiene habits can help prevent infection and injury to eyes, ears, skin, hair, nails, and teeth. **22.** Sleep and rest are basic to the body's well-being. Without enough sleep and rest, your body is at greater risk of disease and injury. When you are tired, you have a harder time paying attention and you often feel more stressed.
**Part F** **23.** They wore the earplugs and safety glasses to protect their ears and eyes from injury. The noise of a jet engine is loud enough to cause hearing loss if ears are unprotected. Flying objects could injure the eyes. **24.** Answers will vary. Students should name and give examples of these five parts of a workout: warm-up, aerobic exercise, exercise for muscular fitness, flexibility exercise, and cooldown.

### Chapter 6 Mastery Test B
**Part A** **1.** C **2.** E **3.** B **4.** A **5.** D
**Part B** **6.** aerobic **7.** anaerobic **8.** isotonic **9.** isometric **10.** isokinetic
**Part C** **11.** loud noise **12.** acne **13.** endurance **14.** dandruff **15.** deodorant
**Part D** **16.** D **17.** A **18.** C **19.** D **20.** A
**Part E** **21.** Ways to protect eyes from injury include keeping sharp objects away from eyes; wearing safety glasses, goggles, or a helmet during sports, and avoiding being in bright light for long periods of time. Ways to protect ears from injury include avoiding loud noise, cleaning ears with a damp washcloth on your fingertip, and wearing a guard or helmet during contact sports. **22.** Both are important parts of an exercise workout because they help prevent injuries. A warm-up is done at the beginning of a workout to allow the heart rate to increase gradually and get more blood flowing to muscles. A cooldown is done at the end of a workout to allow the body to slow down gradually.
**Part F** **23.** Answers will vary. Students should define hygiene, give examples good hygiene practices, and explain of how they can affect health. **24.** Answers will vary. Students should explain how tiredness affects health and give specific examples.

### Chapter 7 Mastery Test A
**Part A** **1.** ultrasound **2.** fetus **3.** first **4.** dominant **5.** Hemophilia **6.** Fallopian **7.** normal **8.** Testosterone **9.** testes **10.** chromosome
**Part B** **11.** D **12.** H **13.** G **14.** J **15.** I **16.** A **17.** E **18.** C **19.** B **20.** F
**Part C** **21.** B **22.** C **23.** D **24.** D **25.** C
**Part D** **26.** estrogen, progesterone **27.** adolescence **28.** a sex-linked trait **29.** perinatal care **30.** one
**Part E** **31.** As adults get older, their immune systems do not work as well, so they are more likely to get sick. Also, parts of the ears and the eyes weaken or break down, so older adults often have hearing and vision problems. **32.** She should see a doctor regularly. She should eat healthful foods and take folic acid. She should exercise regularly, but only the type and length of time that a doctor recommends.

Part F 33. Answers will vary. Students should explain reasons why adolescence can be a difficult time for people. Students should use specific examples. 34. Answers will vary. Students should explain possible complications of childbirth, using specific examples and providing solutions.

## Chapter 7 Mastery Test B
Part A 1. D 2. C 3. B 4. A 5. B
Part B 6. placenta 7. birth defects 8. puberty 9. Y 10. sperm 11. months 12. fertilized egg 13. ovary 14. first 15. Perinatal
Part C 16. B 17. A 18. G 19. I 20. H 21. J 22. C 23. F 24. D 25. E
Part D 26. embryo, fetus 27. dominant 28. cervix 29. Fallopian tube 30. testes
Part E 31. Adolescents might be embarrassed if they don't reach puberty at the same time as their friends. They might also be embarrassed because growth spurts have made them feel clumsy or awkward. Adolescents should accept their differences and focus on strengths and abilities instead. 32. Labor might be lasting too long. The baby's head might be too big to fit through the birth canal.
Part F 33. Answers will vary. Students should name and explain what health problems older adults can expect, giving specific examples. 34. Answers will vary. Students should list and explain ways to protect the health of an unborn child, using specific examples.

## Chapter 8 Mastery Test A
Part A 1. dating 2. single-parent 3. abstinence 4. communication 5. encourage
Part B 6. C 7. C 8. A 9. B 10. A
Part C 11. B 12. D 13. E 14. C 15. A
Part D 16. marriage 17. older 18. adoptive 19. single-parent 20. harmful
Part E 21. It is an extended family; Answers will vary but could include: Jenna helping her brother with homework, Jenna's brother helps with chores around the house, Jenna's grandparents watch her and her brother while their parents are at work. 22. Tim and Rhoda's actions are signs of an unhealthy marriage, they are arguing about two important things that in a healthy marriage couples should agree on—money and children.
Part F 23. Answers will vary. Students should define separation and provide possible solutions. 24. Answers will vary. Students should explain why socializing is important, using specific examples.

## Chapter 8 Mastery Test B
Part A 1. C 2. E 3. A 4. D 5. B
Part B 6. dating 7. single-parent 8. abstinence 9. communication 10. encourage
Part C 11. marriage 12. older 13. adoptive 14. single-parent 15. harmful
Part D 16. D 17. B 18. C 19. B 20. C
Part E 21. Families that are having problems can visit a family counselor or a counseling center for help. Children can talk to teachers or a school counselor. Other family members can join support groups. 22. They can become closer and learn more about each other. Dating some other people can give Jim and Trisha the chance to get to know about them.
Part F 23. A custodial parent is the parent who is responsible for children after a divorce. They will have a blended family. 24. Answers may vary but should include four of the following: Make sure they agree on money matters, have shared interests, have common goals, accept and support each other, agree about having children and how to raise them, and agree about sharing jobs around the house.

## Chapter 9 Mastery Test A
Part A 1. food 2. carbohydrates 3. glucose 4. protein 5. calcium and phosphorus 6. sixty percent 7. sodium 8. malnutrition
Part B 9. C 10. D 11. B 12. A 13. E
Part C 14. C 15. B 16. A 17. D 18. B 19. D 20. C 21. D 22. B 23. B
Part D 24. calories 25. meat 26. unsaturated 27. carbohydrates 28. vitamins 29. sodium 30. more calories than
Part E 31. Answers will vary. Students should list carbohydrates, fats, proteins, vitamins, minerals, and water as the six nutrients. They should include an example of a food with each nutrient. 32. Teenagers and athletes need to consume more calories than adults or teenagers who do not play sports. Teenagers need more calcium and iron than do adults.
Part F 33. Food provides calories, which your body uses for heat, movement, growth, and repair. Food also provides essential nutrients, which are chemicals in food that the body cannot make. Nutrients provide energy for the body and help the body change food into energy. 34. Answers will vary. Students should compare and contrast the health of people who eat healthy diets and those who do not, using specific examples.

## Chapter 9 Mastery Test B
Part A 1. nutrient-dense 2. fiber 3. glycogen 4. meat 5. complex 6. phosphorus 7. unsaturated 8. fruit
Part B 9. D 10. B 11. A 12. E 13. C
Part C 14. C 15. B 16. B 17. B 18. A 19. C 20. D 21. B 22. B 23. D
Part D 24. calcium and phosphorus 25. glucose 26. C 27. minerals 28. gain weight 29. Protein 30. higher risk of disease
Part E 31. Answers will vary. Students should explain how food choice affects caloric and nutrient intake, using specific examples. 32. Answers will vary. Students should explain why Sarah's food choices are unhealthy and provide suggestions for how she could make healthier choices.
Part F 33. Answers will vary. Students should give possible doctor's recommendations for healthy eating for an athlete, using specific examples. 34. Answers will vary. Students should include in their lists foods that contain carbohydrates, fats, proteins, vitamins, minerals, and water in the amounts recommended by the Food Guide Pyramid.

## Chapter 10 Mastery Test A
Part A 1. A 2. D 3. B 4. C 5. B
Part B 6. C 7. E 8. A 9. H 10. I 11. D 12. J 13. B 14. F 15. G
Part C 16. unsaturated fat 17. trans fat 18. media 19. organic 20. 2,000
Part D 21. serving 22. ingredients 23. additive 24. fad diet 25. pattern
Part E 26. This means that nothing has been added to the food to change its color, flavor, or texture. This is not likely because cheese and pasta usually have additives. The ad probably means that nothing was added to the finished product. 27. You will eat 330 calories, 9 grams of fat, 15 percent of the DV for fat, and 0 milligrams of cholesterol.
Part F 28. Answers will vary. Students should list these factors, and provide examples of each: location, time of year, religion, and cultural background. 29. Answers will vary. Students should name and explain what the following entities do: USDA, FDA, Food and Nutrition Board, National Research Council.

## Chapter 10 Mastery Test B
Part A 1. C 2. C 3. A 4. D 5. A
Part B 6. D 7. E 8. A 9. H 10. I 11. C 12. F 13. B 14. J 15. G
Part C 16. meats, dairy products 17. ingredients 18. nuts, seafood, olive oil 19. 2,000 20. ingredients list
Part D 21. calories 22. largest 23. additive 24. fad diet 25. healthy
Part E 26. This means that no substances were added to the salsa for the purpose of preventing growth of bacteria that would spoil the food. The ad is probably honest because the salsa will not spoil if it is refrigerated after opening. 27. You will be eating 110 calories, 3 grams of fat, 5 percent of DV for fat, and 5 mg of cholesterol.
Part F 28. Answers will vary. Students should explain comparison shopping to find a food's cost and nutrition facts. 29. Answers will vary. Student should explain specific things the FDA does.

## Chapter 11 Mastery Test A
Part A 1. A 2. D 3. D 4. B 5. B
Part B 6. E 7. C 8. D 9. B 10. A
Part C 11. chemical barriers 12. disease 13. immune system 14. AIDS 15. inherited 16. incubation or incubation period 17. antibodies 18. through the placenta 19. recovery 20. death
Part D 21. barriers 22. incubation 23. illness 24. recovery 25. death
Part E 26. Answers will vary. Students should name physical and

chemical barriers such as skin and mucus, and explain how they protect the body. **27.** Answers will vary. Students should explain how acquired and inherited diseases are alike and different, using specific examples.
**Part F 28.** Answers will vary. Students should explain how the immune system works, giving specific examples. **29.** Answers will vary. Students' letters should include specific reasons that vaccinations are important.

### Chapter 11 Mastery Test B
**Part A 1.** B **2.** A **3.** D **4.** A **5.** C
**Part B 6.** deformity **7.** vaccination **8.** Down syndrome
**9.** pathogens **10.** saliva
**Part C 11.** F **12.** B **13.** D **14.** G **15.** J **16.** C **17.** A **18.** H **19.** I **20.** E
**Part D 21.** barriers **22.** incubation period **23.** illness **24.** recovery **25.** death
**Part E 26.** Answers will vary. Students should name physical and chemical barriers, such as skin and mucus, and explain how they protect the body. **27.** Answers will vary. Students should explain what vaccinations do and why they are important.
**Part F 28.** Answers will vary. Students' articles should explain how the immune system works, giving specific examples. **29.** Answers will vary. Students should define both acquired and inherited diseases, and give specific examples of how they are alike and different.

### Chapter 12 Mastery Test A
**Part A 1.** D **2.** E **3.** H **4.** G **5.** B **6.** A **7.** I **8.** C **9.** J **10.** F
**Part B 11.** Chlamydia **12.** Gonorrhea **13.** Syphilis **14.** semen **15.** cannot **16.** weakens **17.** detect **18.** chronic **19.** incubation **20.** genital herpes
**Part C 21.** C **22.** B **23.** A **24.** C **25.** D
**Part D 26.** G **27.** B **28.** A **29.** A **30.** G **31.** B **32.** G **33.** B **34.** A **35.** A
**Part E 36.** HIV is the virus that causes AIDS. Some ways that AIDS can be spread is through sexual activity, sharing needles, or some transfer of blood with the virus in it. **37.** A sexually transmitted disease is a disease that is spread through sexual contact. You can get help for sexually transmitted diseases from doctors, local health departments, and family planning clinics.
**Part F 38.** Answers will vary. Students should mention drugs and vaccines being developed to fight AIDS. **39.** Answers will vary. Students should explain how STDs are treated and/or cured, using specific examples.

### Chapter 12 Mastery Test B
**Part A 1.** C **2.** B **3.** D **4.** A **5.** D
**Part B 6.** A **7.** B **8.** A **9.** A **10.** G **11.** B **12.** G **13.** B **14.** A **15.** G
**Part C 16.** cannot **17.** incubation **18.** genital herpes **19.** first **20.** AIDS **21.** weakens **22.** detect **23.** chronic **24.** Gonorrhea **25.** Syphilis
**Part D 26.** I **27.** C **28.** D **29.** J **30.** B **31.** E **32.** H **33.** A **34.** D **35.** F
**Part E 36.** Answers will vary. Students should explain how STDs are treated and/or cured, using specific examples. **37.** Answers will vary. Students should explain how HIV and AIDS are related to each other.
**Part F 38.** Answers will vary. Students should mention drugs and vaccines being developed to fight AIDS. **39.** Answers will vary. Students should define STDs and list places to go for help or treatment.

### Chapter 13 Mastery Test A
**Part A 1.** B **2.** D **3.** C **4.** A **5.** B
**Part B 6.** E **7.** F **8.** H **9.** J **10.** C **11.** I **12.** B **13.** D **14.** A **15.** G
**Part C 16.** heart attack **17.** benign **18.** osteoarthritis **19.** grand mal **20.** cardiovascular **21.** arteriosclerosis **22.** tissue **23.** cancer **24.** abnormality **25.** asthma
**Part D 26.** asthma **27.** rheumatoid arthritis **28.** diabetes **29.** a stroke **30.** cancer
**Part E 31.** Angina is caused by atherosclerosis, the build-up of fat deposits on the walls of the arteries. The arteries become narrowed, and not enough blood reaches the heart. The pain of angina is caused when not enough oxygen-rich blood reaches the heart. **32.** The signs depend on what part of the brain was affected and to what degree. Some people may have a decreased ability to speak, write, or walk. Some people may lose the use of an arm or have a loss of memory.
**Part F 33.** Students should refer to information about dealing with stress, exercising, and eating a low-fat, low-sodium diet. Other ways to decrease the risk of hypertension include not smoking. **34.** Students may suggest using a sunblock with a high UV rating, wearing hats and sunglasses, and wearing clothing with long sleeves. They may also suggest avoiding the sun during the part of the day when the sun is strongest.

### Chapter 13 Mastery Test B
**Part A 1.** malignant **2.** petit mal **3.** angina pectoris **4.** tumorl **5.** epilepsy **6.** type I diabetes **7.** lymph gland **8.** joints **9.** glucose **10.** arteriosclerosis
**Part B 11.** B **12.** C **13.** B **14.** A **15.** D
**Part C 16.** F **17.** C **18.** I **19.** G **20.** E **21.** A **22.** J **23.** B **24.** D **25.** H
**Part D 26.** asthma **27.** cancer **28.** cancer **29.** hypertension **30.** atherosclerosis
**Part E 31.** Yes, even if you have none of the risk factors that you cannot control, you are still increasing your chances of cardiovascular disease if you ignore the risk factors that you can control. **32.** Both diseases result from the death of part of a vital organ caused by the blockage of blood to that organ. Stroke involves death of cells in the brain; heart attack involves death of cells in the heart.
**Part F 33.** The artery walls become thicker. If the pressure in the arteries is higher than normal, the heart must pump harder than it normally does. This can cause other kinds of cardiovascular disease. **34.** Cells from a malignant tumor can break away, get into the bloodstream, and reach other parts of the body. There, they can grow new tumors. Cells of a benign tumor can not break off and spread, so they are not harmful to your health.

### Chapter 14 Mastery Test A
**Part A 1.** D **2.** C **3.** G **4.** F **5.** A **6.** E **7.** H **8.** B
**Part B 9.** doctor **10.** sulfa drug **11.** fastest **12.** misuse **13.** withdrawal
**Part C 14.** B **15.** D **16.** A **17.** B **18.** D **19.** C **20.** A
**Part D 21.** a dependence on alcohol and a lack of control in drinking **22.** synthetic (human made) stimulants **23.** the opium poppy **24.** all of them **25.** alcohol and tobacco
**Part E 26.** Answers will vary. Students should name four specific ways drug or alcohol use can affect an unborn baby. **27.** Answers will vary. Students should name reasons why people might take medicine, giving specific examples of how specific types of medicines work.
**Part F 28.** Answers will vary. Students should state that she should stop drinking alcohol immediately. Students should explain specific effects alcohol use could have on the baby. **29.** Answers will vary. Students may mention side effects, dependence, or overdose. Students should also give examples of how to avoid adverse effects of medicines.

### Chapter 14 Mastery Test B
**Part A 1.** B **2.** A **3.** D **4.** D **5.** C
**Part B 6.** PCP **7.** steroids **8.** no **9.** personality **10.** emphysema
**Part C 11.** D **12.** C **13.** H **14.** J **15.** I **16.** A **17.** E **18.** G **19.** F **20.** B
**Part D 21.** in the way it will start working the fastest **22.** a physical reaction to not having a drug in the body **23.** to get rid of pain **24.** chemicals the body cannot produce **25.** unintended reactions to a medicine
**Part E 26.** Answers will vary. Students should name six specific effects FAS can have on a baby. **27.** Answers will vary. Students may mention side effects, dependence, or overdose.
**Part F 28.** Answers will vary. Students should name reasons why people might take medicine, giving specific examples of how specific types of medicines work. **29.** Answers will vary. Students should state that she should stop smoking immediately. Students should explain specific effects smoking could have on the baby.

### Chapter 15 Mastery Test A
**Part A 1.** C **2.** A **3.** D **4.** C **5.** D
**Part B 6.** D **7.** B **8.** A **9.** E **10.** C
**Part C 11.** Psychoactive **12.** withdrawal **13.** detoxification **14.** alcohol **15.** refusal technique **16.** enabling **17.** vote; hold office **18.** hotline **19.** self-esteem **20.** Resistance skills
**Part D 21.** tolerance **22.** unpredictable **23.** residential treatment

centers  24. risky  25. peer
**Part E**  26. Recovery begins when a person recognizes that he or she has a problem with drugs. The second step, detoxification, removes the drug from the person's body. The third step is learning to live day-to-day without drugs.  27. The risk of alcoholism tends to run in families. If a person has an alcoholic parent, he or she may have genetically inherited the tendency to become addicted to alcohol. This person should avoid drinking alcohol to be safe.
**Part F**  28. Answers will vary. Students should explain how physical drug dependence affects the body, using specific examples.  29. Answers will vary. Students may mention that different people react differently to drugs and treatment.

### Chapter 15 Mastery Test B
**Part A**  1. C  2. B  3. D  4. B  5. D
**Part B**  6. I  7. J  8. G  9. F  10. C  11. A  12. D  13. B  14. H  15. E
**Part C**  16. enabling  17. hotline  18. resistance skill  19. detoxification  20. refusal technique  21. psychoactive  22. convicted  23. withdrawal  24. self-esteem  25. alcohol
**Part D**  26. higher insurance costs  27. interest in new hobbies or activities  28. abstinence  29. developing tolerance for a drug  30. Al-Anon
**Part E**  31. The person might use a drug and notice that its effects deaden the emotional pain that a problem gives them. When the drug wears off, emotional pain comes back, and the person might use the drug to deaden the pain again.  32. Most students will think there is no one drug treatment program that is effective for everyone. Students might discuss how different individuals have different needs. For example, some people might respond better to individual treatment, while others might respond better to group support.
**Part F**  33. Answers will vary. Students will likely suggest that Steven avoid alcohol because of his family history.  34. Answers will vary. Students should explain why learning to live everyday life without drugs can be a challenge during recovery.

### Chapter 16 Mastery Test A
**Part A**  1. E  2. G  3. J  4. A  5. B  6. I  7. D  8. C  9. F  10. H
**Part B**  11. poisonings  12. shock  13. hurricane  14. metal  15. swim  16. water  17. smoke detector  18. screen  19. depth  20. reflective
**Part C**  21. B  22. D  23. B  24. C  25. D
**Part D**  26. heat or smoke  27. loose  28. separated  29. catch fire  30. grease  31. life jacket  32. dusk to sunrise  33. tornado  34. flash  35. plates
**Part E**  36. Answers can include staying in a warm and safe shelter, dressing warmly, keeping a supply of batteries on hand, and maintaining your vehicle. You can prepare an emergency gear pack in your car containing blankets, a flashlight, snacks, and warm clothing.  37. Answers can include staying away from traffic while waiting for the bus, staying quietly in your seat, keeping your hands and head inside the window, keeping the aisles clear, and waiting for the bus to come to a complete stop before getting off.
**Part F**  38. Answers will vary. Students should explain how to extinguish the fire depending on what type it is, or how to evacuate the house.  39. Answers will vary. Students should explain what OSHA does, giving specific examples.

### Chapter 16 Mastery Test B
**Part A**  1. G  2. D  3. A  4. H  5. B  6. E  7. C  8. F  9. J  10. I
**Part B**  11. detector  12. separate  13. headlights  14. earthquake  15. tornado  16. grease  17. loose  18. drop  19. life jacket  20. flash floods
**Part C**  21. floating  22. flood  23. OSHA  24. shock  25. baking soda  26. FEMA  27. blizzard  28. lightning  29. tornado  30. extinguisher
**Part D**  31. A  32. B  33. C  34. D  35. D
**Part E**  36. Answers include: You should always wear a safety helmet and ride with traffic or in bicycle lanes. You should signal all turns and wear bright clothing in the day and reflective clothing at night. You should maintain your bicycle.  37. Answers include: Storing emergency supplies such as food, water, flashlights, extra batteries and a first aid kit. Learning where the gas valve is in the home and shutting it off. Securing pictures and mirrors, and finding safe places to go in the house.
**Part F**  38. Answers will vary. Students should explain what the National Weather Service does, giving specific examples.  39. Answers will vary. Students should explain what the National Weather Service does, giving specific examples.

### Chapter 17 Mastery Test A
**Part A**  1. emergency  2. Good Samaritan Laws  3. calm  4. CPR  5. universal precautions
**Part B**  6. D  7. B  8. C  9. B  10. A
**Part C**  11. C  12. A  13. E  14. D  15. B
**Part D**  16. Heimlich maneuver  17. frostbite  18. cardiac arrest  19. mouth-to-mouth resuscitation  20. sprain
**Part E**  21. Students should mention the following: look around for danger; check for consciousness, breathing, and pulse; clear airway; check for other injuries; stop any bleeding; call 911.  22. A choking person cannot speak, cough, or breathe. The person might use the universal distress signal for choking.
**Part F**  23. The person should receive CPR. Call 911 immediately. The person should be given rescue breathes along with chest compressions. Do not perform CPR unless you are trained to do so.  24. Your friend could be suffering from heat exhaustion. Remove him from the heat. Loosen his clothes, cool him using wet clothes, and give him cool water to sip. Call 911 if he loses consciousness.

### Chapter 17 Mastery Test B
**Part A**  1. heatstroke  2. hypothermia  3. shock  4. rabies  5. pressure point
**Part B**  6. B  7. D  8. A  9. D  10. D
**Part C**  11. D  12. E  13. C  14. A  15. B
**Part D**  16. oral poisoning  17. fainting  18. frostbite  19. fracture  20. first-degree
**Part E**  21. CPR is cardiopulmonary respiration. It involves giving rescue breathes along with chest compressions. It should be used as first aid when a person goes into cardiac arrest.  22. Heat exhaustion comes from hard physical work in a very hot place. The signs include weakness, heavy sweating, muscle cramps, headache, and dizziness.
**Part F**  23. Students should mention the following: look around for danger; check for consciousness, breathing, and pulse; clear airway; check for other injuries; stop any bleeding; call 911.  24. Students should mention asking if the person is choking and performing the Heimlich maneuver, giving specific examples.

### Chapter 18 Mastery Test A
**Part A**  1. never  2. strength  3. will not  4. victim  5. all around us  6. sometimes  7. gun  8. big  9. Alcohol and other drugs  10. victims
**Part B**  11. rape  12. in private  13. peacemakers  14. when you know someone plans violence  15. about one-third
**Part C**  16. D  17. C  18. A  19. B  20. E
**Part D**  21. B  22. D  23. A  24. B  25. C
**Part E**  26. The warning signs of date rape include threats, insults, controlling behavior, and unwanted touching.  27. Some schools have full-time security guards. Some schools have people pass through metal detectors on their way in. Some have video cameras to watch hall activity.
**Part F**  28. Answers will vary. Students should name ways to control anger, giving specific examples.  29. Answers will vary. Students should define sexual harassment, explain when and how is can happen, and provide examples of what victims of it can do.

### Chapter 18 Mastery Test B
**Part A**  1. always  2. physical or mental  3. Insults  4. resolve  5. conflict  6. no  7. police protection, courts, and prisons
**Part B**  8. It increases them.  9. They prevent you from using good judgment.  10. never  11. a conflict within yourself about making a difficult decision  12. It might limit educational and job possibilities.
**Part C**  13. G  14. H  15. B  16. F  17. C  18. E  19. D  20. A
**Part D**  21. A  22. D  23. C  24. C  25. B
**Part E**  26. Make it clear to the harasser that you find the behavior offensive or uncomfortable. Keep a record of the incidents. If the harassment continues, make a complaint to a teacher, counselor, or supervisor.  27. Speak in a clear and calm way. Do not curse or

threaten. Be respectful of the other person. Calmly ask the person to explain the problem. Use humor, but not sarcastic humor. **Part F 28.** Answers will vary. Students should define date rape, give possible warning signs, and list its consequences for the victim and the perpetrator. **29.** Answers will vary. Students should list steps schools can take to prevent violence, and provide specific examples.

## Chapter 19 Mastery Test A
**Part A 1.** F **2.** D **3.** G **4.** H **5.** A **6.** I **7.** C **8.** B **9.** J **10.** E
**Part B 11.** Patients' Bill of Rights **12.** quackery **13.** specialist **14.** skin problems **15.** outpatient clinic
**Part C 16.** government **17.** dying **18.** nurse practitioner **19.** save **20.** often
**Part D 21.** A **22.** C **23.** D **24.** D **25.** B
**Part E 26.** Students should list any five of these: more than minor cuts, bruises, or blows to the head; unusual bleeding; severe pain; change in a mole or skin; vomiting lasting more than a day; high fever for more than three days; thoughts of suicide; severe depression. **27.** Answers will vary. Students should list questions that relate to the ad's source, facts, reliability, and effectiveness.
**Part F 28.** Answers will vary. Students should list patient's rights and explain that all rights come down to being treated fairly and with respect. **29.** Answers will vary. Students should give specific ways they can reduce the cost of their own health care.

## Chapter 19 Mastery Test B
**Part A 1.** D **2.** C **3.** B **4.** B **5.** A
**Part B 6.** dermatologist **7.** unproven **8.** insurance company **9.** health care providers **10.** PPO **11.** unusual **12.** Medicare
**Part C 13.** H **14.** A **15.** F **16.** G **17.** B **18.** D **19.** C **20.** E
**Part D 21.** Medicaid **22.** primary health care **23.** therapy needed to recover from surgery, injury, or illness **24.** write to the manufacturer **25.** by complaining to the manufacturer
**Part E 26.** Answers will vary. Students should give specific ways they can reduce the cost of their own health care. **27.** Answers will vary. Students should list five patient's rights.
**Part F 28.** Answers will vary. Students may mention more than minor cuts, bruises, or blows to the head; unusual bleeding; severe pain; change in a mole or skin; vomiting lasting more than a day; high fever for more than three days; thoughts of suicide; severe depression. **29.** Answers will vary. Students should list questions that relate to the ad's source, facts, reliability, and effectiveness.

## Chapter 20 Mastery Test A
**Part A 1.** Developing **2.** prevention **3.** United Nations **4.** Environmental Protection Agency **5.** Poverty
**Part B 6.** I **7.** B **8.** A **9.** G **10.** F **11.** D **12.** J **13.** E **14.** C **15.** H
**Part C 16.** B **17.** C **18.** D **19.** D **20.** C **Part D 21.** basic health services. **22.** harmful chemicals **23.** Occupational Safety and Health Administration or OSHA **24.** low-income individuals **25.** federal income taxes
**Part E 26.** Answers will vary. Students should support their opinions what facts about health and health promotion. **27.** Answers will vary. Students should list state and community, national, and volunteer health organizations.
**Part F 28.** Answers will vary. Students' letters should explain the benefits of blood donation. **29.** Answers will vary. Students should give specific reasons why people cannot get the healthcare they need.

## Chapter 20 Mastery Test B
**Part A 1.** an epidemic **2.** American Cancer Society **3.** Tuberculosis **4.** Poverty **5.** Red Cross
**Part B 6.** I **7.** C **8.** J **9.** A **10.** H **11.** B **12.** E **13.** G **14.** F **15.** D
**Part C 16.** B **17.** C **18.** D **19.** C **20.** D
**Part D 21.** People living in poverty are at high risk for serious health problems. **22.** the homeless **23.** low **24.** United Nations **25.** local taxes and private charities
**Part E 26.** Answers will vary. Students should explain specific ways people can promote public health. **27.** Answers will vary. Students should list national and international health problems.
**Part F 28.** Answers will vary. Students' letters should explain why health promotion is important, using specific examples. **29.** Answers will vary. Students should list state and community, national, and volunteer health organizations. Students should provide examples of what these organizations do.

## Chapter 21 Mastery Test A
**Part A 1.** balance **2.** clinical **3.** ozone **4.** indoors **5.** lung **6.** freshwater **7.** algae **8.** population **9.** waste **10.** waves
**Part B 11.** EPA **12.** Hazardous wastes **13.** deforestation **14.** precipitation **15.** Lead
**Part C 16.** E **17.** I **18.** G **19.** F **20.** J **21.** H **22.** D **23.** C **24.** B **25.** A
**Part D 26.** B **27.** D **28.** B **29.** D **30.** B
**Part E 31.** Asbestos and radon are common indoor pollutants. Exposure to these pollutants can lead to lung cancer. **32.** Answers will vary. Sample answer: The U.S. government created the EPA to write and carry out environmental laws.
**Part F 33.** Answers will vary. Students should explain why maintaining the balance of nature is important, using specific examples. **34.** Answers will vary. Students should mention the limited water supply.

## Chapter 21 Mastery Test B
**Part A 1.** EPA **2.** Acid rain **3.** Hazardous **4.** deforestation **5.** pollution
**Part B 6.** Sewage **7.** the greenhouse effect **8.** fossil fuels **9.** Ecology **10.** permanent **11.** plants **12.** Antarctica **13.** groundwater **14.** conserving water **15.** indoors
**Part C 16.** B **17.** G **18.** H **19.** J **20.** C **21.** A **22.** F **23.** I **24.** D **25.** E
**Part D 26.** A **27.** C **28.** B **29.** D **30.** D
**Part E 31.** Global warming changes climates and environments all over the world. It might decrease rainfall in dry areas and increase it in others, creating droughts or floods. **32.** Answers will vary. Students should explain specific ways pollution affects the balance of nature.
**Part F 33.** Answers will vary. Students may mention new technologies that are better for the environment or renewable energy sources. **34.** Answers will vary. Students should explain that controlling pollution is important to the balance of nature and give a specific example of air pollution and its effects.

## Scoring Rubric for Short-Response Items in Part E of Mastery Tests

**2 points**—The student demonstrates a solid understanding of the content by providing:
- a complete set of accurate facts that support the answer
- a clearly-stated answer to the question

**1 point**—The student demonstrates a partial understanding of the content by providing one of the following:
- a complete set of accurate facts that support the answer
- a clearly-stated answer to the question

**0 points**—The student fails to demonstrate understanding of the content by doing one of the following:
- includes no facts or incorrect facts, and fails to provide a clearly-stated answers or
- provides no answer at all

## Scoring Rubric for Essay Items in Part F of Mastery Test

**3 points**—The student demonstrates a solid understanding of the content by providing:
- a complete set of accurate facts that support the answer
- a clearly-stated answer to the essay's primary question
- a standard essay response (topic sentence, introduction, body, conclusion)

**2 points**—The student demonstrates a good understanding of the content by providing two of the following:
- a complete set of accurate facts that support the answer
- a clearly-stated answer to the essay's primary question
- a standard essay response (topic sentence, introduction, body, conclusion)

**1 point**—The student demonstrates a partial understanding of the content by providing one of the following:
- a complete set of accurate facts that support the answer
- a clearly-stated answer to the essay's primary question
- a standard essay response (topic sentence, introduction, body, conclusion)

**0 points**—The student fails to demonstrate understanding of the content by doing one of the following:
- includes no facts or incorrect facts, fails to answer the essay's primary question, and fails to include a standard essay response (topic sentence, introduction, body, conclusion) or
- provides no answer at all

## Unit Mastery Tests

### Unit 1 Mastery Test A
Part A 1. stress response 2. self-actualization 3. projection 4. wellness 5. relief
Part B 6. D 7. C 8. A 9. E 10. B
Part C 11. physical health 12. place 13. shame 14. chemicals 15. denial
Part D 16. A 17. B 18. C 19. B 20. C
Part E 21. Answers will vary. Students should define self-defeating behaviors and give two specific examples. 22. Answers will vary. Students should list three specific ways to improve social well-being.
Part F 23. Answers will vary. Students should explain that Josie is improving her physical health and list three more things she could do. 24. Answers will vary. Student should explain that Tony is meeting his belonging needs and explain how forming friendships will help him.

### Unit 1 Mastery Test B
Part A 1. physical health 2. belonging needs 3. shame 4. chemicals 5. denial
Part B 6. E 7. A 8. C 9. D 10. B
Part C 11. stress response 12. self-actualization 13. projection 14. wellness 15. relief
Part D 16. C 17. B 18. C 19. C 20. B
Part E 21. Answers will vary. Students should define emotional health and give specific examples. 22. Answers will vary. Students should define physical needs and give specific example of how to meet them.
Part F 23. Answers will vary. Students should define self-defeating behaviors and give specific examples. 24. Answers will vary. Students should list physical, emotional, social, and personal well-being. Students should list three ways to promote well-being.

### Unit 2 Mastery Test A
Part A 1. C 2. E 3. B 4. D 5. A
Part B 6. ultrasound 7. fetus 8. dominant 9. neglect 10. adolescence
Part C 11. B 12. B 13. C 14. D 15. C
Part D 16. allergy 17. platelets 18. commitment 19. marriage counselors 20. domestic violence
Part E 21. Bones provide structure for the body. The protect organs, store minerals, and produce certain blood cells. 22. Answers will vary. Sample answer: My hearing could be harmed by loud noises. I could wear earplugs if I am going to be near loud noise.
Part F 23. Prenatal care is important because if women stay healthy during pregnancy, there is a better chance that their babies will be healthy too. Good prenatal care includes taking a multivitamin that has folic acid every day; eating healthy foods; exercising regularly; and getting regular checkups. 24. Dating is how you get to know someone. Getting married is a step people might take after dating. A family life cycle begins when two people get married. The birth of a child is also part of the family life cycle.

### Unit 2 Mastery Test B
Part A 1. A 2. C 3. E 4. D 5. B
Part B 6. Fallopian 7. hemophilia 8. testes 9. secondary sex characteristic 10. nuclear family
Part C 11. B 12. B 13. A 14. C 15. A
Part D 16. esophagus 17. medulla 18. fluoride 19. thyroid 20. flexibility
Part E 21. Prenatal care is health care during pregnancy. All of the following are examples of good prenatal care behaviors: taking a multivitamin that has folic acid every day; eating healthy foods; exercising regularly; and getting regular checkups. 22. The family life cycle includes events such as marriage and the birth of children. A new family life cycle begins when two people decide to marry.
Part F 23. Joints allow the skeletal frame to move, but all body movements depend on muscles. Tendons connect skeletal muscles to the skeletal system. When a muscle contracts, it pulls on a tendon, which acts on a bone to produce a movement. 24. The football player should wear a helmet while playing. The swimmer should swim only in clean water and dry her ears after swimming.

### Unit 3 Mastery Test A
Part A 1. C 2. E 3. D 4. A 5. B
Part B 6. serving 7. malnutrition 8. balanced 9. fiber 10. organic food
Part C 11. Food Guide Pyramid 12. Daily Value 13. additive 14. Nutrition Facts 15. lactose
Part D 16. B 17. A 18. B 19. A 20. D
Part E 21. Answers will vary. Students may discuss any two of these: carbohydrates, fats, proteins, vitamins, minerals, or water. 22. Signs of malnutrition include tiredness, frequent headaches, stomachaches, or depression. Malnutrition can be avoided by eating a balanced diet.
Part F 23. An association is something that reminds a person of something else. For instance, you might dislike hot dogs because you once got sick after eating them. Other things that can affect food choices include cost, where you live, the time of year, your religion, your cultural background, and your family and friends. 24. An additive is something that is added to food that changes it in some way. Additives may be added to prevent food from spoiling or to increase its nutritional value. Read the Nutrition Facts section of food labels to check for additives.

## Unit 3 Mastery Test B
**Part A** 1. D  2. E  3. C  4. B  5. A
**Part B** 6. glucose  7. fat  8. phosphorus  9. amino acids  10. Food and Drug Administration
**Part C** 11. D  12. C  13. A  14. B  15. D
**Part D** 16. diet  17. preservative  18. advertise  19. protein  20. water
**Part E** 21. The friend has formed an association—something that reminds a person of something else—with peanut butter. The peanut butter reminds her of a bad experience. Another example of an association is to like a certain food because it reminds you of a time you enjoyed.  22. Sugars and salt are sometimes added to food to keep it from spoiling. Nutrients such as vitamin C and calcium sometimes are added to fruit juices to increase the amount of nutrients in the food.
**Part F** 23. Carbohydrates are starches and sugars that come mainly from plants. Simple carbohydrates are sugars that the body uses for quick energy. Complex carbohydrates contain fiber; they are the best source of dietary carbohydrates. Simple carbohydrates are found in sugar, fruits, jelly, and syrup. Complex carbohydrates are found in bread, pasta, beans, and potatoes.  24. Deficiencies can lead to serious health problems, including damage to the nervous system or heart. People who do not get enough iron in their diet might have a lack of energy and, possibly, heart problems. A calcium deficiency can lead to weak bones and teeth.

## Unit 4 Mastery Test A
**Part A** 1. C  2. D  3. E  4. B  5. A
**Part B** 6. skin  7. gonorrhea  8. respiratory  9. blood  10. hypertension
**Part C** 11. antibodies  12. syphilis  13. stroke  14. diabetes  15. malignant
**Part D** 16. B  17. D  18. B  19. B  20. B
**Part E** 21. Cigarette smoke is a known carcinogen. Smokers have a much higher risk of cancer than nonsmokers do. Sunlight is a natural source of radiation that can cause skin cancer. People who sunbathe have an increased risk of developing malignant melanoma.  22. To avoid getting a sexually transmitted disease, a person can practice abstinence or have sexual activity only with one partner who is known to be uninfected with any sexually transmitted diseases and who is not having sex with anyone else.
**Part F** 23. Both are diseases that harm or stop the normal workings of the body. Acquired diseases are diseases that a person gets after they are born. An inherited disease is a disease that is passed physically from parents to children.  24. Antibodies are proteins that kill particular pathogens. Some antibodies remain in the body after a person has recovered from a disease. These antibodies give a person immunity, so he or she will not get the disease again.

## Unit 4 Mastery Test B
**Part A** 1. D  2. E  3. B  4. A  5. C
**Part B** 6. heart attack  7. vaccination  8. chlamydia  9. arteriosclerosis  10. insulin
**Part C** 11. pathogens  12. rheumatoid arthritis  13. HIV  14. epilepsy  15. incubation period
**Part D** 16. C  17. A  18. B  19. A  20. C
**Part E** 21. Acquired diseases are caused by pathogens, health behaviors, or environmental conditions. Inherited diseases are caused by abnormal genes or chromosomes or a combination of abnormal genes and environment. An example of an acquired disease is a cold or the measles. An example of an inherited disease is Down syndrome or hemophilia.  22. When pathogens enter the bloodstream, antibodies are produced. Antibodies are proteins that kill particular pathogens. Some antibodies remain in the body after a person has recovered from a disease. These antibodies give a person immunity, so he or she will not get the disease again.
**Part F** 23. Answers will vary. Students should list three risk factors for cancer and explain how to avoid them.  24. There are no vaccines to prevent sexually transmitted diseases, and the body does not build up immunity to sexually transmitted diseases. Practice abstinence to avoid getting a sexually transmitted disease.

## Unit 5 Mastery Test A
**Part A** 1. D  2. E  3. A  4. H  5. C  6. B  7. J  8. I  9. F  10. G
**Part B** 11. negative  12. Antibiotics  13. withdrawal  14. adolescents  15. oxidation
**Part C** 16. A  17. D  18. B  19. C  20. D
**Part D** 21. Crack  22. Stimulants  23. nicotine  24. alcohol  25. detoxification
**Part E** 26. By reading a label on a medicine package, you can learn the purpose of the medicine, the suggested amount you should take, possible unwanted side effects, and any cautions you must observe.  27. Answers will vary. Students should explain how a person recovering from drug dependence can get help, using specific examples.
**Part F** 28. Student answers will vary. Most students will agree that drug dependence cannot be cured. A person who has recovered from drug dependence can begin to use drugs again. This is why the process of recovery never stops.  29. Answers will vary. Students should list specific ways to misuse a medicine and explain the possible effects of misuse.

## Unit 5 Mastery Test B
**Part A** 1. A  2. E  3. D  4. H  5. C  6. B  7. J  8. I  9. F  10. G
**Part B** 11. cocaine  12. slow down  13. nicotine  14. alcohol  15. detoxification
**Part C** 16. B  17. D  18. A  19. C  20. D
**Part D** 21. stomach  22. bacteria  23. withdrawal  24. peer pressure  25. oxygen
**Part E** 26. A person who is recovering from drug dependence might begin to use drugs again. Abstinence is the only way a person can be sure he or she will never again become drug dependent. This is why the process of recovery never stops.  27. Marla's strep throat infection will most likely come back. Because she stopped taking the antibiotic too soon, not all the bacteria were killed. They will start to grow again, and the infection will return.
**Part F** 28. The FDA requires that every medicine label must give the purpose of the medicine, the suggested amount people should take, possible unwanted side effects, and any cautions people must observe.  29. Answers will vary. Students should explain how a person recovering from drug dependence can get help, using specific examples.

## Unit 6 Mastery Test A
**Part A** 1. F  2. E  3. D  4. H  5. G  6. A  7. C  8. B
**Part B** 9. extinguisher  10. prejudice  11. guilty  12. gun  13. Good Samaritan  14. pressure  15. Shock
**Part C** 16. B  17. D  18. B  19. D  20. B  21. C  22. C
**Part D** 23. falls  24. tornado  25. fracture  26. violence  27. negotiation  28. angry  29. going down  30. reflector
**Part E** 31. Answers will vary. Students should name places where falling is likely and give specific ways to make these places safer.  32. Rescue breathing is mouth-to-mouth resuscitation. It is needed when a person is not breathing but has blood circulation and a pulse. It puts oxygen from a rescuer's lungs into an unconscious person's lungs to help the person breathe.
**Part F** 33. Answers will vary. Students may list the following: gloves, proper disposal of waste, washing hands.  34. Answers will vary. Students should define conflict and violence and give specific examples of each, explaining how the two are related.

## Unit 6 Mastery Test B
**Part A** 1. B  2. A  3. D  4. G  5. E  6. F  7. C  8. H
**Part B** 9. chimney  10. Violence  11. conflict  12. rescue breathing  13. occupational  14. universal precautions  15. guilty
**Part C** 16. A  17. B  18. A  19. D  20. B  21. C  22. C
**Part D** 23. teens  24. choking  25. go to a basement  26. alcohol  27. cold and damp skin  28. on a pressure point  29. absorbed through the skin  30. sprain
**Part E** 31. Answers will vary. Students should list universal precautions such as gloves, proper disposal of waste, and washing hands.  32. Answers will vary. Answers will vary. Students should define conflict and violence and give specific examples of each, explaining how the two are related.
**Part F** 33. Answers will vary. Students should give specific examples

of fire hazards in the home and list ways to prevent such fires. **34.** Rescue breathing transfers oxygen from a rescuer's lungs to an unconscious person's lungs through mouth-to-mouth resuscitation. It is done when a person stops breathing but has a pulse and circulation. It is used when a person is choking, has had an electrical shock, or is drowning.

## Unit 7 Mastery Test A
**Part A** 1. C  2. G  3. F  4. E  5. B  6. D  7. A
**Part B** 8. C  9. B  10. A  11. C  12. D
**Part C** 13. Specialists  14. Quackery  15. An epidemic  16. air
**Part D** 17. Acid rain  18. homeless  19. groundwater  20. poverty  21. domestic violence  22. sensorineural  23. birth defects  24. Public health  25. Medicare
**Part E** 26. Mr. and Mrs. Chang have to pay the $100 deductible plus 20 percent of the remaining $1100, or $100 plus $220 for a total of $320. Their insurance company pays 80 percent of $1100 (after the deductible), or $880.  27. Answers will vary. Students should explain two specific ways humans can disrupt the balance of nature.
**Part F** 28. Answers will vary. Students' letters should state the problem clearly and politely and include a suggested solution.  29. Answers will vary. Students should explain what kind of care hospitals, nursing homes, and hospices provide. Students should give specific examples.

## Unit 7 Mastery Test B
**Part A** 1. C  2. F  3. D  4. A  5. B  6. G  7. E  8. H
**Part B** 9. B  10. C  11. B  12. A  13. D
**Part C** 14. Wood  15. radon  16. WHO  17. school nurse  18. photochemical smog
**Part D** 19. ozone layer  20. sewage  21. primary care physician  22. Rehabilitation  23. Particulates  24. hospice  25. Tuberculosis
**Part E** 26. Answers will vary. Students' letters should state the problem clearly and politely and include a suggested solution.  27. Answers will vary. Students may suggest asking about qualifications, certifications, experience, listening skills, and level of comfort with the doctor.
**Part F** 28. Mr. and Mrs. Wallace have to pay the $200 deductible plus 20 percent of the remaining $2,500, or $200 plus $500 for a total of $700. Their insurance company pays 80 percent of $2,500 (after the deductible), or $2,000.  29. Sewage, or human waste, can disturb the balance of nature because it contains bacteria that can cause disease. If sewage gets into groundwater or other bodies of water, it can kill animals and plants and give diseases such as typhoid fever and cholera to humans who drink the contaminated water.

## Chapters 1–10 Midterm Mastery Test
**Part A** 1. C  2. B  3. A  4. E  5. D  6. J  7. F  8. I  9. G  10. H
**Part B** 11. self-actualization  12. dysfunctional  13. skeletal  14. calorie  15. adolescence
**Part C** 16. single-parent  17. preservatives  18. temperament  19. rational  20. heredity
**Part D** 21. C  22. D  23. C  24. B  25. B  26. D  27. D  28. C  29. B  30. D
**Part E** 31. Stress can cause physical reactions such as increased heart rate, increased breathing rate, and increased sweating. A stress response causes blood to rush to the arms, legs, and head. Other chemical messages strengthen the muscles for endurance. Stress also can cause headaches, stomach pain, and sleeping problems.  32. Both are eating disorders in which people attempt to cope with psychological problems through eating. Anorexia is characterized by extreme dieting, compulsive exercising, and severe weight loss. Bulimia is characterized by eating large amounts of food in a short amount of time (bingeing) followed by ridding one's body of the food (purging).
**Part F** 33. When a person inhales, the respiratory system brings oxygen into the lungs. The oxygen is transferred to the blood. The circulatory system pumps the oxygenated blood throughout the body. Carbon dioxide, a waste material, is brought to the lungs by the circulatory system. The carbon dioxide exits the body when a person exhales.  34. Students may describe any three of the following: Carbohydrates are chemicals in food that provide starches and sugars. They can be found in potatoes, bread, and pasta. Proteins are made of amino acids and are used to build and repair body tissues. They can be found in fish, meat, eggs, and poultry. Fats are made of fatty acids. They are stored energy for the body. Fats can be found in meat, ice cream, fish, and vegetable oils. The body needs vitamins in small amounts for growth and activity. Vitamin C can be found in citrus fruits. Minerals are needed for fluid balance and digestion. Calcium can be found in milk and green, leafy vegetables. Water makes up about 60 percent of body weight. It carries nutrients, hormones, and waste products to and from cells in the body. Water is found in soups and juices.

## Final Mastery Test
**Part A** 1. C  2. D  3. A  4. E  5. B  6. J  7. G  8. I  9. F  10. H  11. M  12. O  13. K  14. L  15. N
**Part B** 16. conflict  17. resist  18. increases  19. prevention  20. quackery  21. dominant  22. antibiotic  23. pulse  24. antibody  25. hospitals  26. essential nutrients  27. semen  28. stroke  29. halfway house  30. flash flood
**Part C** 31. prejudice  32. pathogens  33. harm  34. bacteria  35. gonorrhea  36. medulla  37. insulin  38. physical health  39. self-defeating  40. nutrition facts
**Part D** 41. B  42. C  43. C  44. C  45. A  46. A  47. D  48. C  49. B  50. B
**Part E** 51. A healthy relationship has three characteristics: emotional attachment, mutual dependence between partners, and the satisfaction of both partner's needs.  52. Answers will vary but may include trying to identify the community's health needs, being a volunteer at a food drive, finding sponsors for a benefit walk, writing letters to lawmakers, or joining a health advocacy group.
**Part F** 53. Vaccinations are a means of getting dead or weakened pathogens into the body for the purpose of making a person immune to a disease. An example of a vaccine is the measles vaccine. After the pathogens from the vaccine enter the body, the body produces antibodies and develops immunity.  54. Anger management is an important part of preventing violence. If a person can keep control of his or her anger, conflict and possibly violence can be avoided.